# Handbook of Experimental Pharmacology

Continuation of Handbuch der experimentellen Pharmakologie

# Vol. 66/I

# Glucagon I


Contributors

T. T. Aoki · T. L. Blundell · C. Bordi · J. Brange · W. W. Bromer
G. F. Cahill, Jr. · A. D. Cherrington · J.-L. Chiasson · T. H. Claus
J. W. Ensinck · S. Falkmer · D. W. Foster · L. G. Heding
C. Hellerström · B. de Hemptinne · J. J. Holst · K. S. Koch
P. J. Lad · P. J. Lefèbvre · H. L. Leffert · J. D. McGarry
R. B. Merrifield · A. J. Moody · S. Mojsov · R. Nosadini
L. Orci · C. R. Park · A. Perrelet · S. J. Pilkis · M. Pingel
M. Rodbell · E. Samols · B. Skelbaek-Pedersen · H. Skelly
R. J. Smith · W. Stalmans · L. Thim · A. Tiengo · R. H. Unger
I. Valverde · S. Van Noorden

Editor

P. J. Lefèbvre



Springer-Verlag
Berlin Heidelberg New York Tokyo 1983

Professor PIERRE J. LEFEBVRE, M.D.
Professor of Medicine, University of Liège
Guest Professor, University of Brussels (V.U.B.)
Corresponding Member, Royal Academy of Medicine of Belgium
Chairman, Medical Policlinics, University of Liège
Head, Division of Diabetes, University of Liège
Institut de Médecine, Hôpital de Bavière
Boulevard de la Constitution, 66
4020 Liège, Belgium

With 136 Figures

ISBN-13: 978-3-642-68868-3 e-ISBN-13: 978-3-642-68866-9
DOI: 10.1007/978-3-642-68866-9

Library of Congress Cataloging in Publication Data. Main entry under title: Glucagon. (Handbook of experimental pharmacology; v. 66) Bibliography: p. Includes index. 1. Glucagon – Addresses, essays, lectures. I. Lefèbvre, Pierre J. II. Series. [DNLM: 1. Glucagon. W1 HA51L vol. 66 pt. 1–2/WK 801 G5656] QP905.H3 vol. 66 [QP572.G5], 615′.1s [612′.34] 83-583



Softcover reprint of the hardcover 1st edition 1983



2122/3130-543210

# List of Contributors

T. T. Aoki, Joslin Diabetes Center, One Joslin Place, Boston, MA 02215, USA

T. L. Blundell, Laboratory of Molecular Biology, Department of Crystallography, Birkbeck College, University of London, Malet Street, London WC1E 7HX, Great Britain

C. Bordi, Istituto di Anatomia ed Istologia Patologica, Università di Parma, Parma, Italy

J. Brange, Novo Research Institute, Novo Allée, 2880 Bagsvaerd, Denmark

W. W. Bromer, Research Advisor, Lilly Research Laboratories, Indianapolis, IN 46285, USA

G. F. Cahill, Jr., Director of Research, Howard Hughes Medical Institute, 398 Brookline Avenue, Suite 8, Boston, MA 02215, USA

A. D. Cherrington, Department of Physiology, Howard Hughes Medical Institute, Vanderbilt University, 702 Light Hall, Nashville, TN 37232, USA

J.-L. Chiasson, Director, Research Laboratory in Diabetes and Carbohydrates Metabolism, Clinical Research Institute of Montreal, 110 Avenue des Pins Ouest, Montreal, Qué. H2W 1R7, Canada

T. H. Claus, American Cyanamid Company, Medical Research Division, Lederle Laboratories, Pearl River, NY 10965, USA

J. W. Ensinck, Department of Medicine, Program Director, Clinical Research Center, University Hospital, RC-14, University of Washington, Seattle, WA 98195, USA

S. Falkmer, Department of Pathology, Malmö General Hospital, 21401 Malmö, Sweden

D. W. Foster, Department of Internal Medicine, University of Texas Health Science Center at Dallas, Southwestern Medical School, 5323 Harry Hines Boulevard, Dallas, TX 75235, USA

L. G. Heding, Medical Department, Novo Research Institute, Novo Allée, 2880 Bagsvaerd, Denmark

C. Hellerström, University of Uppsala, Department of Medical Cell Biology, Biomedicum, Box 571, 75123 Uppsala, Sweden

B. de Hemptinne, Laboratory of experimental Surgery, Catholic University of Louvain, 1200 Brussels, Belgium

J. J. Holst, Institute of Medical Physiology C, University of Copenhagen, The Panum Institute, Blegdamsvej 3, 2200 Copenhagen N, Denmark

K. S. Koch, Department of Medicine, M – 013 H, Division of Pharmacology, School of Medicine, University of California at San Diego, La Jolla, CA 92093, USA

P. J. Lad, Department of Medicine, M – 013 H, Division of Pharmacology, School of Medicine, University of California at San Diego, La Jolla, CA 92093, USA

P. J. Lefebvre, Head of the Division of Diabetes, Université de Liège, Hôpital Universitaire de Bavière, Institut de Médecine, Diabétologie, Boulevard de la Constitution, 66, 4020 Liège, Belgium

H. L. Leffert, Department of Medicine, M–013 H, Division of Pharmacology, School of Medicine, University of California at San Diego, La Jolla, CA 92093, USA

J. D. McGarry, Departments of Internal Medicine and Biochemistry, University of Texas Health Science Center at Dallas, Southwestern Medical School, 5323 Harry Hines Boulevard, Dallas, TX 75235, USA

R. B. Merrifield, The Rockefeller University, 1230 York Avenue, New York, NY 10021, USA

A. J. Moody, Novo Research Institute, Novo Allée, 2880 Bagsvaerd, Denmark

S. Mojsov, The Rockefeller University, 1230 York Avenue, New York, NY 10021, USA

R. Nosadini, Istituto di Medicina Clinica dell'Università di Padova, Via Giustiniani, 2, 35100 Padova, Italy

L. Orci, Institute of Histology and Embryology, University of Geneva Medical School, 1211 Geneva 4, Switzerland

C. R. Park, Chairman, Department of Physiology, Vanderbilt University, School of Medicine, Nashville, TN 37232, USA

A. Perrelet, Institute of Histology and Embryology, University of Geneva, Medical School, 1211 Geneva 4, Switzerland

S. J. Pilkis, Department of Physiology, Vanderbilt University, School of Medicine, Nashville, TN 37232, USA

M. Pingel, Novo Research Institute, Novo Allée, 2880 Bagsvaerd, Denmark

M. Rodbell, Section on Membrane Regulation, National Institute of Arthritis, Diabetes, Digestive, and Kidney Diseases, Bethesda, MD 20205, USA

E. Samols, Division of Endocrinology, Metabolism and Radionuclide Studies, Department of Medicine, VA Medical Center, 800 Zorn Avenue, Louisville, KY 40202, USA

B. Skelbaek-Pedersen, Novo Research Institute, Novo Allée, 2880 Bagsvaerd, Denmark

H. Skelly, Department of Medicine, M – 013 H, Division of Pharmacology, School of Medicine, University of California at San Diego, La Jolla, CA 92093, USA

R. J. Smith, Howard Hughes Medical Institute, Laboratories at the Joslin Diabetes Center, One Joslin Place, Boston, MA 02215, USA

W. Stalmans, Afdeling Biochemie, Faculteit Geneeskunde, Katholieke Universiteit Leuven, Herestraat 49, 3000 Leuven, Belgium

L. Thim, Novo Research Institute, Novo Allée, 2880 Bagsvaerd, Denmark

A. Tiengo, Istituto di Medicina Clinica dell'Università di Padova, Cattedra di Malattie del Ricambio, Via Giustiniani, 2, 35100 Padova, Italy

R. H. Unger, Senior Medical Investigator, Dallas VA Medical Center, 5323 Harry Hines Boulevard, Dallas, TX 75235, USA

I. Valverde, Fundación Jiménez Diaz, Universidad Autónoma de Madrid, Avda. Reyes Católicos 2, Ciudad Universitaria, Madrid 3, Spain

S. Van Noorden, Department of Histopathology, Royal Postgraduate Medical School, Hammersmith Hospital, London, Great Britain

# Preface

The Editorial Board of the Handbook of Experimental Pharmacology apparently did not hurry in suggesting production of a volume on glucagon since the present opus is number sixty-six in the series. This fact is even more striking if we consider that 34 volumes published over about eight years will separate the books on glucagon from those on insulin on library shelves, whereas only a few microns separate the cells manufacturing these two polypeptides within the islets of Langerhans in the pancreas!

Numerous factors have probably caused this dicrimination; four of them are:

*First*, insulin deficiency or resistance is the cause of one of the most serious and distressing diseases, diabetes mellitus, which affects millions of people, whereas glucagon deficiency is apparently an extremely rare disorder, for which detailed reports are published of individual cases whenever they occur.

*Second*, since its discovery in 1921 by BANTING and BEST, insulin has been irreplaceable for the treatment of the most severe forms of diabetes, whereas, in contrast, glucagon was until recently considered a relatively minor therapeutic agent.

*Third*, whereas insulin is a compound which has been well characterized since the pioneering work of SANGER and its biosynthesis clearly identified by STEINER and his co-workers, glucagon, also well characterized chemically, has suffered from its parenthood with the so-called "glucagon-like immunoreactive substances", an incompletely defined series of immunologically related polypeptides present in the gut, the pancreas and some other parts of the body.

*Fourth*, while the insulin-producing cell, the B-cell of the islets of Langerhans, has received a great amount of attention from morphologists for many years, the glucagon-producing cell or A-cell has long been considered as the "parent pauvre."

The credentials of glucagon are numerous however. It was among the very first polypeptide hormones to be isolated, purified, sequenced, and synthetized. Thanks to the efforts of UNGER, it was the second polypeptidic hormone to become measurable by radioimmunoassay, a few months after insulin. It has served as a very valuable tool which permitted SUTHERLAND and his co-workers to discover cyclic AMP and RODBELL and his associates to investigate cell membrane receptors.

Our knowledge of physiology of glucagon is based on the work of STAHL, FOÀ, SUTHERLAND, and DE DUVE, to mention only few of the pioneers, and on work performed since the late fifties in Dallas by UNGER and his co-workers. In the seventies UNGER and ORCI, combining their physiological and morphological expertises, renewed the study of the pathopysiology of diabetes; they considered the microanatomy of the islets of Langerhans and demonstrated both the interaction and the partition of the various cell types involved.

When asked to serve as editor for a volume on glucagon in the present series, I immediately decided to accept since there was a great need for a comprehensive book on this subject. As a matter of fact, my task of selecting and contacting contributors has been easy: all are personal friends, all are undisputed experts in their field, and all accepted my invitation.

The only major problem arose when all the contributions had been collected, namely the total amount of pages exceeded by far that which could be reasonably gathered in a single volume. Analysis of the content showed that shortening the individual contributions was definitely unacceptable. Another possibility was to renounce the comprehensive character of the book by leaving out several topics, a position which was acceptable neither to the authors, who had already written the chapters, nor to the editor, who had indicated to the contributors that comprehensiveness was a major characteristic of the whole project. The help came from the publisher, who accepted a third possible solution, that of producing *Glucagon* in two volumes.

The first volume gathers the more "basic" contributions on the chemistry and physicochemistry of glucagon, the morphology of the A-cells, the biosynthesis of the hormone, its production, its immunogenicity, and its assay procedures. It also contains ten chapers describing the actions of the hormone at the molecular level. The second volume deals with the various factors controlling glucagon secretion, the question of extrapancreatic glucagon, the place of glucagon in physiology and in pathology, the catabolism of the hormone, the pharmacological effects of glucagon, and its use in diagnosis and in therapy.

Two critiques often made on a work like this are redundancy and not being up to date. We have done our best to eliminate both.

About *redundancy*, great efforts have been made to delineate the topics to be treated by the various contributors, and direct contacts have been encouraged between contributors having to deal with closely related topics. At the editorial level, redundant paragraphs or figures have been deleted. Yet, it has been our deliberate policy to preserve homogeneity of each chapter and, as a consequence, some topics have been considered in two chapters on a different scope. For instance, glicentin, the newly recognized precursor of glucagon, has been considered both in the chapter on the biosynthesis of glucagon and in the one dealing with the various glucagon-related peptides. Similarly, the chapter on "glucagon and liver output in vivo" is a physiological review of mechanisms dealt with in a more biochemical manner in the contributions on "liver glycogen metabolism" and on "gluconeogenesis." With such an approach, we think that the reader will find both detailed analyses of the mechanisms involved and comprehensive and integregated views on the phenomena as they occur in the organism as a whole.

About *up-dating*, producing a multiauthored book like this takes time; this is not, however, an excuse for it to be out of date when in print. Every effort has been made by the contributors and by the editor to cover the literature completely up to the very last possible moment. Newly published data have been introduced, when necessary, at the various steps of scientific editing, copy editing, and galley-proof correcting. We hope that this effort will be appreciated by the readers.

I wish to thank all those who participated in this enterprise: the authors for their comprehensive contributions and their willingness to abide by the rules of the game, my secretary Mrs. VAESSEN-PETIT in Liège who helped me in the editorial process, Mr. EMERSON, in Great Britain, whose copy editing was outstanding and the publisher, in Heidelberg, from whom I have received help from Mr. BERGSTEDT, Mrs. WALKER, and Mr. BISCHOFF.

Our readers will be the judge of our joint efforts; let us hope that they will not be disappointed.

PIERRE J. LEFEBVRE

# Contents

## Chemistry and Physicochemistry of Glucagon

CHAPTER 1

**Chemical Characteristics of Glucagon.** W. W. BROMER. With 2 Figures



CHAPTER 2

**The Chemical Synthesis of Glucagon.** R. B. MERRIFIELD and S. MOJSOV
With 8 Figures

CHAPTER 6

**The Biosynthesis of Glucagon.** C. HELLERSTRÖM. With 6 Figures



CHAPTER 7

**Glucagon, Glicentin, and Related Peptides**
A. J. MOODY and L. THIM. With 4 Figures

## Production and Assay of Glucagon

CHAPTER 8

**Glucagon Preparations.** M. PINGEL, B. SKELBAEK-PEDERSEN, and J. BRANGE
With 3 Figures

CHAPTER 9

**The Immunogenicity of Glucagon.** L. G. HEDING. With 4 Figures



CHAPTER 10

**Immunoassays for Glucagon.** J. W. ENSINCK. With 3 Figures

CHAPTER 11

**Heterogeneity of Circulating Glucagon and Glucagon-Like Immunoreactivity**
I. VALVERDE. With 10 Figures



CHAPTER 12

**Radioreceptorassays for Glucagon.** J. J. HOLST. With 4 Figures

## Actions of Glucagon

CHAPTER 13

**The Actions of Glucagon at Its Receptor: Regulation of Adenylate Cyclase**
M. Rodbell. With 3 Figures



CHAPTER 14

**Glucagon and Liver Glycogen Metabolism.** W. Stalmans. With 7 Figures

CHAPTER 22

**Glucagon and Insulin Secretion.** E. SAMOLS

# Contents
# Part II: Glucagon II

# Chemistry and Physicochemistry of Glucagon

CHAPTER 1

# Chemical Characteristics of Glucagon

W. W. BROMER

## A. Introduction

The object of this chapter is to describe the covalent chemistry of pancreatic glucagon to provide a background for subsequent chapters. Modifications of the primary structure are considered with respect to biologic activity, but not immunologic activity which is discussed in Chap. 9. Conformation and function are considered in Chap. 3.

## B. Isolation and Purification

Glucagon is effectively extracted from animal pancreas by any of the acidic ethanol methods commonly employed to extract insulin (e.g., SUTHERLAND and DE DUVE 1948; KENNY 1955). Such methods have been successfully used with mammalian, avian, and piscine pancreas (for details see Chap. 8). The glucagon content of freshly collected bovine or porcine pancreas is about 5–10 µg/g wet tissue. RHOTEN (1976) found as much as 5 mg/g glucagon in splenic pancreas from lizards. In a typical insulin purification scheme, glucagon and insulin are purified together to the point of insulin crystallization, where most of the glucagon is found in the mother liquor (SUNDBY and MARKUSSEN 1971). Even at this stage the glucagon fraction is less than 2% pure. The first successful purification of glucagon by STAUB et al. (1955) employed repeated precipitations culminating in crystallization, which was in itself an excellent purification step. In more recent studies, purification was often achieved by separations based more directly on molecular size and on charge. For example, in the purification of chicken glucagon, POLLOCK and KIMMEL (1975) used ion exchange chromatography on columns of CM-Sephadex and QAE-Sephadex. A particulary elegant procedure applicable to small-scale purification was developed by SUNDBY and MARKUSSEN (1971) for rat glucagon and was applied subsequently to glucagon from several other species: (a) the mother liquor fraction from the citrate crystallization of insulin was brought to about half-saturation with NaCl; (b) the resulting precipitate was dissolved in acidic urea solution and gel-filtered in dilute acetic acid; (c) glucagon fractions (4% pure) were lyophilized and fractionated by preparative isoelectric focusing in polyacrylamide gels; (d) the glucagon fraction was eluted and crystallized. If desirable, the isoelectric focusing step may most likely be replaced by ion exchange chromatography using both anion and cation exchange media.

The first glucagon crystals contained about 10% of a minor acidic component which was separable by electrophoresis (STAUB et al. 1955); reelectrophoresis of the major glucagon fraction showed that the it reappeared to the extent of about 10%, indicating that the minor component was generated from glucagon by the conditions of the experiments. COLE (1960), using cation exchange chromatography in buffers containing urea, also observed more negatively charged minor components in crystalline glucagon. Variations of COLE's approach have been found to be particularly beneficial for the purification of glucagon; use of buffers containing urea not only permits chromatographic fractionation at pH values wherein glucagon normally has poor solubility, but also helps reduce protein self-association that makes the separation of closely related molecules difficult. Using anion exchange chromatography in 7 *M* urea buffers, BROMER et al. (1972) isolated and characterized the acidic impurity previously observed by STAUB et al. (1955), COLE (1960), and LOCHNER et al. (1964). This impurity, with one more negative charge than glucagon, was found to contain nearly equal proportions of three isomeric monodeamidated glucagons, each having a different glutamine hydrolyzed to glutamic acid. Essentially all of the other minor impurities in crystalline glucagon were found to resemble the parent hormone closely. For example, two minor impurities were identified as [desHis$^{1}$]glucagon and [desMet$^{27}$,Asn$^{28}$,Thr$^{29}$]glucagon (BROMER 1976). Glucagon of very high purity can be obtained by a variety of methods. These include ion exchange chromatography in buffers containing urea and preparative isoelectric focusing, both already discussed. Partition chromatography (HRUBY and GROGINSKY 1971) and reverse-phase high pressure liquid chromatography (BIEMOND et al. 1979; HANCOCK et al. 1978; BURGUS and RIVIER 1976) are two other powerful techniques that have been applied successfully to glucagon.

## C. Properties

In concentrations of about 1–10 mg/ml, mammalian glucagon crystallizes in the pH range between 6 (SUNDBY and MARKUSSEN 1971) and 8.6 (STAUB et al. 1955). The crystals contain no metals or carbohydrate, and are sharply defined rhombic dodecahedra that dissolve poorly in aqueous solutions from pH 4 to 9 (STAUB et al. 1955). Some selected properties of mammalian glucagon are given in Table 1. Crystalline glucagon retains 5%–15% water under normal vacuum-drying conditions. For this reason, concentrations of the highly purified hormone may best be determined by measuring optical density at 278 nm and using the extinction coefficient (see Table 1).

Although glucagon crystallizes readily in the absence of metals, complexes such as zinc glucagon are easily prepared. Zinc glucagon is somewhat less soluble than glucagon, and exhibits prolonged biologic effects (WEINGES 1959; SOKAL 1960; BROMER and CHANCE 1969; TARDING et al. 1969; ASSAN and DELAUNAY 1972). If protamine is added to the zinc glucagon suspension, a greater prolongation of action is observed, even though the protamine is not bound to the zinc glucagon (TARDING et al. 1969).

The molecular weight range of 3,500–4,000 daltons for porcine glucagon was established by quantitative end group analysis and ultracentrifugation in 2 *M* guanidine solutions (BROMER et al. 1957). These data, along with the amino acid

**Table 1.** Selected properties of mammalian [a] glucagon

| Property or condition | Value | Reference |
|---|---|---|
| $E^{1\%}_{1\,cm}$ (pH 2, 278 nm) | 23.8 | GRATZER et al. (1967) |
| Molecular weight [b] | 3,485 daltons | BROMER et al. (1957) |
| Isoelectric point | pH ~7 | ZIEGLER and LIPPMANN (1968), SUNDBY and MARKUSSEN (1971) |
| Solubility (aqueous) | | |
| pH 3.5–8.5 | ≦50 μg/ml | STAUB et al. (1955) |
| pH 2–3 and 9.5–10.5 | >10 mg/ml | |
| 7 *M* urea, pH 8.5 | | BROMER et al. (1972) |
| 60% propylene glycol, pH 8 | | CONTAXIS and EPAND (1974) |
| Fibril formation | | |
| acid pH; increased | | STAUB et al. (1955) |
| ionic strength, glucagon concentration, agitation, and temperature to 30 °C | Enhancement | BEAVEN et al. (1969) |
| Stability [c] (0.2–1 mg/ml) | | STAUB et al. (1955) |
| pH 2.5–3 | 7 months, 4 °C | |
| pH 10 | 50% after about 4 months, 4 °C | |

[a] Except guinea pig glucagon
[b] Calculated from amino acid composition of porcine glucagon
[c] As measured by hyperglycemia in cats; desamidation occurs slowly

composition, permitted the calculation of 3,485 daltons as the molecular weight for porcine glucagon.

As will be discussed in later chapters, glucagon in solution readily shifts into a $\beta$ conformation which aggregates into visible fibrils and gels (BEAVEN et al. 1969). The fibrillar form of glucagon is very insoluble except at pH > 10, a condition which permits recovery of most of the glucagon in soluble form. The formation of fibrils may be inhibited not only by avoiding the conditions in Table 1 but also by adding small amounts of dioxane or guanidine to the solution (BEAVEN et al. 1969).

The amino acid compositions of glucagon from various species is listed in Table 2. Few, if any, hormones exhibit identical amino acid compositions in the variety of mammalian species represented in the first column. Guinea pig glucagon is a dramatic exception, differing from the known mammalian glucagons even more than do fish or fowl glucagon. Cysteine is not present in glucagon from any of the species studied; however, anglerfish glucagon contains proline and guinea pig glucagon has isoleucine, two amino acids not present in any other species.

## D. Amino Acid Sequence

The amino acid sequence (Fig. 1) is identical for glucagon isolated from the pancreas of pigs (BROMER et al. 1956), cattle (BROMER et al. 1971), and humans (THOMSEN et al. 1972). The amino acid composition, crystalline forms, and physiochemical properties of rat (SUNDBY and MARKUSSEN 1971), rabbit (SUNDBY and MARKUSSEN 1972), camel (SUNDBY et al. 1974), and sheep (R.L. JACKSON 1981, per-

**Table 2.** The amino acid composition of glucagon from various species

| Amino acid | Residues[a]/mol | | | | | |
|---|---|---|---|---|---|---|
| | Mammals[b] | Guinea pig[c] | Angler fish[d] | Shark[c] | Turkey[e] or chicken[f] | Duck[g] |
| Aspartic acid | 4 | 4 | **3** | 4 | **3** | **3** |
| Threonine | 3 | **2** | **2** | 3 | 3 | **4** |
| Serine | 4 | 4 | **3** | 4 | **5** | 4 |
| Glutamic acid | 3 | **4** | 3 | **2** | 3 | 3 |
| Proline | 0 | 0 | **1** | 0 | 0 | 0 |
| Glycine | 1 | **2** | **2** | 1 | 1 | 1 |
| Alanine | 1 | **2** | **2** | 1 | 1 | 1 |
| Valine | 1 | **2** | 1 | 1 | 1 | 1 |
| Methionine | 1 | **0** | **0** | **2** | 1 | 1 |
| Isoleucine | 0 | **1** | 0 | 0 | 0 | 0 |
| Leucine | 2 | **5** | 2 | **1** | 2 | 2 |
| Tyrosine | 2 | **3** | 2 | 2 | 2 | 2 |
| Phenylalanine | 2 | **3** | 2 | 2 | 2 | 2 |
| Histidine | 1 | 1 | 1 | 1 | 1 | 1 |
| Lysine | 1 | **3** | **2** | **2** | 1 | 1 |
| Arginine | 2 | **3** | 2 | 2 | 2 | 2 |
| Tryptophan | 1 | +[h] | 1 | n.d.[j] | 1 | 1 |
| Total | 29 | **40** | 29 | 28 (29) | 29 | 29 |

[a] Bold numbers indicate differences from mammalian glucagon
[b] Except for guinea pig. Pig (BROMER et al. 1957), cattle (BROMER et al. 1971), rat (SUNDBY and MARKUSSEN 1971), rabbit (SUNDBY and MARKUSSEN 1972), human (THOMSEN et al. 1972), camel (SUNDBY et al. 1974), sheep (R. L. JACKSON 1981, personal communication)
[c] SUNDBY (1976)
[d] TRAKATELLIS et al. (1975)
[e] MARKUSSEN et al. (1972)
[f] POLLOCK and KIMMEL (1975)
[g] SUNDBY et al. (1972)
[h] Present; probably one residue
[j] Not determined but probably one residue

$H_2N$-His-Ser-Gln-Gly-Thr-Phe-Thr-Ser-Asp-Tyr-Ser-Lys-Tyr-Leu-Asp-
1 2 3 4 5 6 7 8 9 10 11 12 13 14 15

Ser-Arg-Arg-Ala-Gln-Asp-Phe-Val-Gln-Trp-Leu-Met-Asn-Thr-COOH
16 17 18 19 20 21 22 23 24 25 26 27 28 29

**Fig. 1.** The amino acid sequence of porcine glucagon

sonal communication) glucagon are all the same as porcine glucagon, strongly suggesting that all these mammalian glucagons have identical amino acid sequences. As noted in Table 2, guinea pig glucagon is quite different from all other glucagons that have been isolated. Avian glucagons differ from the predominant mammalian glucagon only by a few conservative replacements: duck glucagon (SUNDBY et al.

1972) contains threonine in position 16 and serine in position 28; turkey glucagon (MARKUSSEN et al. 1972) also has serine in position 28, and partial sequences of chicken glucagon (POLLOCK and KIMMEL 1975) strongly indicate identity with turkey glucagon. This extreme conservation of primary structure exhibited by glucagon can at least be termed unusual and at most, extraordinary. No other hormones that have been studied exhibit such stringent conservation of primary structure. One might assume from this that the physiologic function of glucagon is more important than it actually appears to be. In any case, these sequence data are in general agreement with Sect. E which shows that few chemical and no enzymatic modifications can be made without appreciable loss of function.

# E. Covalent Chemical Modification and Biologic Function

## I. Limitations of the Approach

The basic assumption underlying this topic is that covalent modification of the glucagon structure will help identify those specific chemical groups or side chains responsible for biologic function. This assumption is probably valid, but the approach has several limitations that need clarification before the data are considered.

First of all, the recognition (RODBELL et al. 1971 a) of specific membrane receptors requires consideration that modification of a functional group of glucagon may affect either or both the receptor-binding process (i.e., affinity) and the process of transmission of the biologic signal (i.e., potency). Stated in the most elementary fashion (see Chap. 13 for a description of this complex process), glucagon binds to a specific receptor subunit and initiates a signal that is transmitted to a catalytic subunit, adenylate cyclase, causing activation of the enzyme and leading to a measureable biologic response. Those portions of the glucagon molecule responsible for binding may or may not be responsible for transmission of the biologic signal, and vice versa. Obviously, if a derivative has no binding affinity for the receptor, it cannot initiate a biologic response. Instead of a simple cause–effect relationship, several different events may occur.

The native hormone is a full agonist, binding the receptor with high affinity and eliciting a maximal activation of hepatic adenylate cyclase at hormone concentrations of about 100 n*M*; the concentrations of glucagon required for half-maximal activation and for half-maximal displacement of glucagon [$^{125}$I] from the receptor are the same, i.e., about 5 n*M* (RODBELL et al. 1971 a). However, BIRNBAUMER and POHL (1973) reported that glucagon produces a maximal response in the liver membrane system when only about 10%–20% of the receptors are occupied. ROSSELIN et al. (1974) and SONNE et al. (1978) obtained similar results with intact hepatocytes. These data are in agreement with current concepts of partial receptor occupancy for strong agonists (TALLARIDA and JACOB 1979), but further studies of RODBELL et al. (1974) confirmed their original data (RODBELL et al. 1971 a) that full receptor occupation is required for maximal activity. The process of binding and signal transmission is complex; because receptor occupancy is not reported for any of the glucagon derivatives it will not be considered further in this chapter.

Some derivatives of glucagon are less effective full agonists, binding the receptor with less affinity, but eliciting the same maximal response as glucagon, albeit at a higher concentration than that required for the native hormone. In these cases, the loss of potency (as measured by concentration of the derivative required to give 50% of the maximal response of native glucagon) is most likely a direct consequence of loss of affinity (as measured by the concentration of the derivative required to displace 50% of glucagon $[^{125}I]$ from the receptor). With less effective full agonists, affinity and potency decrease proportionally.

Other derivatives of glucagon bind to the receptor, but are incapable of initiating a full biologic response. These modifications are partial agonists/antagonists and, if a suitable ratio of binding to activation exists, could give potent competitive inhibition of the glucagon effect. Such inhibitors would be of interest in probing the purported role of hyperglucagonemia in diabetes (see Chap. 44). In some cases, partial binding affinity is observed, but is accompanied by no measurable potency.

Still another observation involves relatively more enhancement of potency than affinity, which can be explained if the derivative is not degraded as rapidly as glucagon. The degradation (POHL et al. 1972) is a significant factor, and the biologic response at any given time is dependent on the hormonal concentration (BIRNBAUMER and POHL 1973). Clearly, the high rate of destruction of glucagon complicates interpretation of the data.

In addition, on a molecular basis, little is known about the receptor other than that it functions in the hydrophobic environment of the cell membrane. Even less is known about the molecular events responsible for transmitting the biologic signal. Interpretations are thus limited to one component of the system, namely glucagon.

Aside from these interpretive problems imposed by the biologic mechanism, two other limitations need consideration. One obvious limitation pertains to the purity of the modified glucagon. Any covalent modification of glucagon, no matter how carefully it is purified, almost certainly retains some finite contamination with native glucagon or glucagon-like materials. It is incumbent on the investigator to prove that such contamination is not a factor in the biologic or binding activity that is observed. Because of the inherent limitations of quantitative chemical characterization, a level of contamination at or below about 1% can be difficult to detect with certainty. These considerations therefore become crucial when binding or biologic activity are about 1% or less. An additional complication flows from use of assay systems which have either very high, or perhaps very different, sensitivities or where it is impossible (e.g., because of solubility) to test higher concentrations of hormone; thus, a derivative may be "inactive" in some systems, but have finite activity in others, causing ambiguities in determining either the maximal response or the minimum active structure.

Another limitation relates to possible perturbation of the conformation by covalent modification. If biologic activity is dependent on native conformation, any change in biologic activity observed with a covalent modification could be related to altered conformation rather than altered covalent structure. The argument can be made that glucagon at physiologic concentrations is almost surely a random coil, but an equally persuasive argument can be made (see Chap. 2) that one or

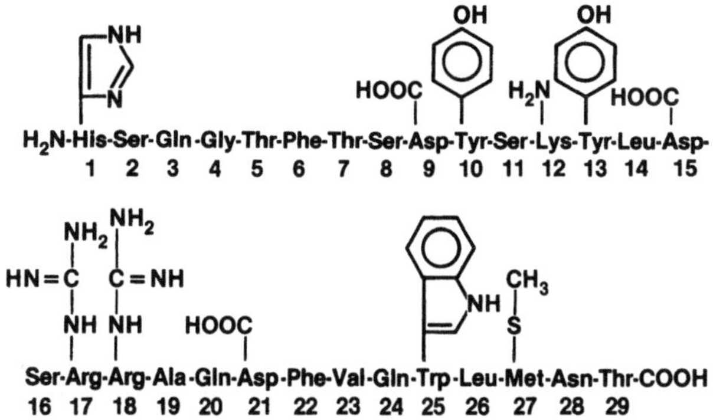


**Fig. 2.** The amino acid sequence of porcine glucagon with reactive functional groups indicated

more conformers of glucagon may be stabilized by hydrophobic interactions with the receptor. Thus, the issue is clouded, but the safest course would be to determine whether or not a covalent modification appreciably alters the native conformation of the hormone; if it does, interpretations may be qualified appropriately. In general, conformational aspects will not be considered because they will be discussed in Chap. 3.

## II. General Considerations

To facilitate discussion, the covalent structure of glucagon is given in Fig. 2 and the activities of known derivatives are listed in Table 3. Those functional groups that are readily modified are emphasized in Fig. 2. Glucagon contains only one $\alpha$-amino group and one $\varepsilon$-amino group; therefore, modifications of these groups are described in Table 3 as $N^{\alpha}$- and $N^{\varepsilon}$- respectively. Table 3 is arranged in a general manner, beginning at the amino terminus of glucagon and proceeding towards the carboxyl terminus. An effort was made to use modern terminology and concepts in discussing binding affinity, potency, agonism, and antagonism (see TALLARIDA and JACOB 1979).

## III. The Question of Several Active Sites

Glucagon elicits a wide variety of hormonal effects both in vivo and in vitro (e.g., see Chaps. 13–22), many of which have been used to measure the activity of glucagon derivatives (e.g., ASSAN and SLUSHER 1972). The question may logically be raised whether those structures within the glucagon molecule responsible for biologic activity are identical for all activities. The available data suggest that the active site is identical, with one possible exception. Numerous studies have shown (see Table 3) that removal of a few COOH terminal residues markedly diminishes hepatic adenylate cyclase or glycogenolytic activity; however FELTS et al. (1970) using glucagon 1–21 and several laboratories (ASSAN and SLUSHER 1972; MITZNEGG et al. 1976), using synthetic glucagon 1–23, observed quite significant lipolytic activity in rat or human tissue or in human subjects. On the other hand, DESBUQUOIS

**Table 3.** The activity of glucagon derivatives

| Glucagon derivative | % Activity | | | Reference |
|---|---|---|---|---|
| | Relative binding affinity[a] | Maximal response | Relative potency | |
| Glucagon | 100 | 100 | 100 | |
| [desHis[1]]$N^{\varepsilon}$-phenyl-thiocarbamoyl- | | | 0[b] | FELTS et al. (1970) |
| | ~ 16 | 0 | 0[c] | RODBELL et al. (1971 b) |
| | ~ 6 | 0 | 0[c] | BREGMAN and HRUBY (1979) |
| | | 100 | ~ 0.3[d] | KAHN et al. (1980) |
| | ~ 1[e] | 32[e] | ~ 1[e] | SONNE et al. (1978) |
| [desHis[1]]- | | | 0[c] | LANDE et al. (1972) |
| | < 2 | | < 2[d] | BROMER (1976) |
| | 7 | 70 | 2[c] | LIN et al. (1975) |
| | | 56[c] | | BREGMAN and HRUBY (1979) |
| [desHis[1], Ser[2]]- | | | 0[d] | MCDONALD et al. (1969) |
| [desHis[1], homoArg[12]]- | | 22 | 0.1[c] | BREGMAN et al. (1980) |
| $N^{\alpha}$-carbamoyl- | | 27 | 5.5[c] | BREGMAN et al. (1980) |
| | | | < 5[b] | GRANDE et al. (1972) |
| | 33 | | 6[c] | EPAND et al. (1981) |
| $N^{\alpha}$,$N^{\varepsilon}$-biscarbamoyl- | | 17 | 0.4[c] | BREGMAN et al. (1980) |
| $N^{\alpha}$-acetyl- | 12 | 100 | 11[c] | DESBUQUOIS (1975a) |
| | | | 17[d] | W.W. BROMER (1981, unpublished work) |
| $N^{\alpha}$,$N^{\varepsilon}$-bisacetyl- | 1 | 100 | 1[c] | DESBUQUOIS (1975a) |
| | 3.5 | | 0.15[c] | EPAND et al. (1981) |
| | | | 4[d] | W.W. BROMER (1981, unpublished work) |
| $N^{\alpha}$-trifluoroacetyl- | | | 30–40[c] | LANDE et al. (1972) |
| $N^{\alpha}$-trinitrophenyl- | 7 | | 0.1[c] | EPAND et al. (1981) |
| $N^{\alpha}$,$N^{\varepsilon}$-bisacetamidino- | 5–10 | 80 | 5–10[c] | WRIGHT and RODBELL (1980a) |
| $N^{\alpha}$,$N^{\varepsilon}$-bis(iodoacetyl)- | 0 | | 0[c] | HRUBY et al. (1976) |
| $N^{\alpha}$,$N^{\varepsilon}$-bis(4-mercapto-butylamidino)- | 5–10 | 90 | 5–10[c] | WRIGHT and RODBELL (1980a) |
| $N^{\alpha}$,$N^{\varepsilon}$-bis(trinitrophenyl)- | 0.8 | | 0.3[c] | EPAND et al. (1981) |
| $N^{\alpha}$,$N^{\varepsilon}$-bisacetyl-$N^{\text{imid}}$-ethoxyformyl- | 0.4 | | 0.15[c] | EPAND et al. (1981) |
| Monodesamido- | | 100 | ~ 70[d] | BROMER et al. (1972) |
| [HomoArg[12]]- | | | 52[d] | W.W. BROMER (1981, unpublished work) |
| | | 100 | 20[c] | BREGMAN et al. (1980) |
| | 38 | | 17[c] | EPAND et al. (1981) |
| | | 100 | 14[b] | ROSS et al. (1979) |

**Table 3** (continued)

| Glucagon derivative | % Activity | | | Reference |
|---|---|---|---|---|
| | Relative binding affinity[a] | Maximal response | Relative potency | |
| $N^{\varepsilon}$-butoxycarbonyl- | | | 10–15[c] | LANDE et al. (1972) |
| | | | 14[d] | W.W. BROMER (1981, unpublished work) |
| $N^{\varepsilon}$-acetyl- | | | 20[d] | W.W. BROMER (1981, unpublished work) |
| $N^{\varepsilon}$-acetamidino- | 100 | 100 | 100[c] | WRIGHT and RODBELL (1980a) |
| $N^{\varepsilon}$-4-hydroxphenyl amidino- | ~ 50 | 100 | ~ 50[c] | WRIGHT and RODBELL (1980a) |
| $N^{\varepsilon}$-4-azido-2-nitrophenyl-$^{125}I$ | 100 | 0 | 0[c] | BREGMAN and LEVY (1977) |
| | | pH dependent | | |
| Mono-iodo-[f] | >100 | 100 | > 100[c] | LIN et al. (1975, 1976) |
| | 142 | 100 | 250[c] | DESBUQUOIS (1975b) |
| | | 100 | 480[c] | BROMER et al. (1973) |
| | | | 183[d] | |
| Purified monoiodo- | 172 | 100 | 330[c] | DESBUQUOIS (1975b) |
| diiodo-[f] | 154 | 100 | 350[c] | DESBUQUOIS (1975b) |
| | | 100 | 570[c] | BROMER et al. (1973) |
| | | | > 100[d] | |
| Purified diiodo- | 195 | 100 | 370[c] | DESBUQUOIS (1975b) |
| Triiodo-[f] | 167 | 100 | | DESBUQUOIS (1975b) |
| | | 100 | 1,000[c] | BROMER et al. (1973) |
| | | | > 100[d] | |
| Tetraiodo-[f] | 95 | 100 | 500[c] | DESBUQUOIS (1975b) |
| | | 100 | 1,000[c] | BROMER et al. (1973) |
| | | | > 100[d] | |
| Pentaiodo-[f] | 42 | 100 | 140[c] | DESBUQUOIS (1975b) |
| | | 100 | 1,000[c] | BROMER et al. (1973) |
| | | | > 100[d] | |
| [nitroTyr$^{13}$]- | | 100 | 77[c] | PATTERSON and BROMER (1973) |
| | | | 231[d] | |
| [nitroTyr$^{10}$,nitroTyr$^{13}$]- | | 100 | 22[c] | PATTERSON and BROMER (1973) |
| | | | 145[d] | |
| [aminoTyr$^{13}$]- | | 100 | 40[c] | PATTERSON and BROMER (1973) |
| | | | 56[d] | |
| [aminoTyr$^{10}$,aminoTyr$^{13}$]- | | 100 | 14[c] | PATTERSON and BROMER (1973) |
| | | | 28[d] | |
| [CHDArg$^{17}$ or $^{18}$]- | | 100 | 24[c] | W.W. BROMER (1981, unpublished work) |
| | | | 78[d] | |

**Table 3** (continued)

| Glucagon derivative | % Activity | | | Reference |
|---|---|---|---|---|
| | Relative binding affinity [a] | Maximal response | Relative potency | |
| [CHDArg$^{17}$,CHDArg$^{18}$]- | | 100 | 5 [c]<br>34 [d] | W.W. BROMER (1981, unpublished work) |
| [5-nitro-2-pyrimidylOrn$^{17,18}$]- | 20 | | > 100 [c]<br>> 100 [d] | EPAND et al. (1976)<br>EPAND and WHEELER (1975) |
| [2,4-dinitrophenyl-sulfenylTrp$^{25}$]- | 100 | 100 | 100 [c] | WRIGHT and RODBELL (1980b) |
| [2-thiolTrp$^{25}$]- | 100 | 100 | 100 [c] | WRIGHT and RODBELL (1980b) |
| [disulfideTrp$^{25}$(dimer)]- | 25 | 100 | 100 [c] | WRIGHT and RODBELL (1980b) |
| [2-nitro-4-azidophenyl-sulfenylTrp$^{25}$]- | | 100 | 100 [c] | DEMOLIOU and EPAND (1980) |
| [2-nitrophenyl-sulfenylTrp$^{25}$]- | | 100 | > 100 [c] | EPAND and COTE (1976) |
| [2,4-dinitrophenyl-1,5-disulfenylTrp$^{25}$(dimer)]- | | 100 | 100 [c] | EPAND and COTE (1976) |
| [S-methylMet$^{27}$]- | | 100<br>100 | ~ 0.2 [c]<br>~ 0.2 [d] | ROTHGEB et al. (1977)<br>NOOIJEN and KEMPEN (1979) |
| [Met$^{27}$sulfoxide]- | | 100<br><br>100<br> | ~ 1 [d]<br>100 [b]<br>30 [c]<br>100 [d] | NOOIJEN and KEMPEN (1979)<br>FELTS et al. (1970)<br>W.W. BROMER (1981, unpublished work) |
| [desAsn$^{28}$,Thr$^{29}$]-[homoSer$^{27}$]- | <br>3 [g]<br><br>2–2.5<br><br>≪100<br>2.5<br><br> | <br><br><br>100<br>100<br><br><br><br> | 100 [b]<br><br><br>2–2.5 [c]<br>~ 1 [c]<br><br>5 [c]<br>2 [c]<br>4 [d] | FELTS et al. (1980)<br>DESBUQUOIS and LAUDAT (1974)<br>LIN et al. (1975)<br>ENGLAND and GURD (1980)<br>JONES and GURD (1981)<br>EPAND et al. (1981)<br>W.W. BROMER (1981, unpublished work) |
| [desAsn$^{28}$,Thr$^{29}$]-[S-methylMet$^{27}$]- | | 100 | ~ 0.05 [c] | ENGLAND and GURD (1980) |
| [desAsn$^{28}$,Thr$^{29}$]-(1-27) | | 100 | ~ 1 [c] | ENGLAND and GURD (1980) |
| [desMet$^{27}$,Asn$^{28}$,Thr$^{29}$]-(1–26) | 0.5 | | 1 [d] | BROMER (1976) |
| [desThr$^{29}$][homoSer$^{27}$,Gly$^{28}$]- | ~ 2.5 | 100 | 3–4 [c] | WRIGHT et al. (1980) |

**Table 3** (continued)

| Glucagon derivative | % Activity | | | Reference |
|---|---|---|---|---|
| | Relative binding affinity[a] | Maximal response | Relative potency | |
| [desAsn$^{28}$,Thr$^{29}$]-[homoSer$^{27}$hydrazide]- | ~ 2.5 | 100 | ~ 2[c] | WRIGHT et al. (1980) |
| [desAsn$^{28}$,Thr$^{29}$]-[homoSer$^{27}$*n*-butylamide]- | ~ 2.5 | 100 | ~ 2[c] | WRIGHT et al. (1980) |
| [desAsn$^{28}$,Thr$^{29}$]-[homoSer$^{27}$biotinyl-amide]- | ~ 2.5 | 100 | < 1[c] | WRIGHT et al. (1980) |
| [taurine$^{30}$]- | 2.5 | | 2.3[c] | EPAND et al. (1981) |
| Tetrataurine- | | | 0.1[c] | WHEELER et al. (1974) |
| Tetramethylester- | 0.02 | | 0.01[c] | EPAND et al. (1981) |
| Tetraglycineamide- | 0.05 | | 0.004[c] | EPAND et al. (1981) |
| 1–23 | < 0.01[g] | | | DESBUQUOIS and LAUDAUT (1974) |
| | | | 50–100[b] | ASSAN and SLUSHER (1972) |
| | | | 0[d] | ASSAN and SLUSHER (1972) |
| | | | ~ 100[b] | GUY-GRAND and ASSAN (1973) |
| | | < 100 | ~ 50[b] | MITZNEGG et al. (1976) |
| | 0 | 0 | 0[c] | RODBELL et al. (1971 b) |
| 1–21 | | | 100[b] | FELTS et al. (1970) |
| | 0.1 | 100 | 0.1[c] | WRIGHT et al. (1978) |
| | | | 2[d] | W.W. BROMER (1981, unpublished work) |
| 1–18 | | | 0[b] | FELTS et al. (1970) |
| | | | < 1[d] | W.W. BROMER (1981, unpublished work) |
| 1–12 | | | < 1[d] | W.W. BROMER (1981, unpublished work) |
| 1–6 | 0.001 | 75 | 0.001[b] | WRIGHT and RODBELL (1979) |
| 19–29 | | | 0[b] | FELTS et al. (1970) |
| 20–29 | 0 | 0 | 0[c] | RODBELL et al. (1971 b) |
| 22–29 | 0 | 0 | 0[c] | RODBELL et al. (1971 b) |

[a] Binding to liver cell membranes unless otherwise noted
[b] Lipolysis
[c] Activation of hepatic adenylate cyclase
[d] Glycogenolysis
[e] Intact rat hepatocytes
[f] A mixture containing, on the average, the indicated number of iodine atoms
[g] Binding to fat cell membranes

and Laudat (1974) found synthetic glucagon 1–23 to be over four orders of magnitude less potent than glucagon in binding to fat cell membranes. The question is clearly not resolved, but the preponderance of evidence suggests that the six carboxyl terminal residues of glucagon are more important for glycogenolysis than for lipolysis. In all other cases that have been studied (see Table 3), the primary structural requirements of glucagon appear to be reasonably similar for all biologic activities. Another supporting line of evidence too cumbersome to include in Table 3 is the observation (W. W. Bromer 1981, unpublished work) that glucagon 1–18 retains < 1% potency in activation of adenylate cyclase, glycogenolysis, lowering of blood amino acids, lipolysis, cardiac contractility, liver enzyme induction, and inhibition of gastric secretion.

## IV. Amino Terminal and Diamino Modifications

It is apparent from Table 3 that removal of amino terminal histidine drastically lowers relative potency; binding affinity is also depressed, but not to the same extent as potency. Purified [desHis$^1$] glucagon is a very weak partial agonist/antagonist, giving about 60%–70% of the maximal response (Lin et al. 1975; Bregman and Hruby 1979) with low (7% or less) binding affinity. However, when lysine is also blocked, as in [desHis$^1$]$N^\varepsilon$-phenylthiocarbamoylglucagon, all detectable activation of adenylate cyclase is lost with retention of 6%–16% receptor-binding affinity (Rodbell et al. 1971 b; Bregman and Hruby 1979). This "pure" antagonist in the liver membrane system proved inexplicably to be a very weak full agonist when tested in the perfused rat liver (Khan et al. 1980). It is interesting that Sonne et al. (1978) found this derivative to be a partial agonist with little affinity or potency in rat hepatocytes. At the present time, it is impossible to rationalize these three different results in hepatic membranes, cells, and tissue; however the lack of antagonism in glycogenolysis will be discussed later.

Another analog with histidine removed and lysine modified, [desHis$^1$, homoArg$^{12}$] glucagon, retains a positive charge on lysine that apparently permits retention of very weak potency (Bregman et al. 1980). This analog is probably similar in action to [desHis$^1$] glucagon, but shows even less partial agonism. With derivatization of only the $\alpha$-amino group ($N^\alpha$-carbamoyl-, $N^\alpha$-acetyl-, $N^\alpha$-trifluoroacetyl-, and $N^\alpha$-trinitrophenylglucagon) potency is sharply diminished, usually more than loss of receptor binding affinity. In cases where maximal response was studied, $N^\alpha$-carbamoylglucagon gave only 27% of the maximum (Bregman et al. 1980), but $N^\alpha$-acetylglucagon gave 100% (Desbuquois 1975 a). However, in the latter study, the derivative contained a small amount of unreacted glucagon that would account for the maximal response. The relatively high potency (17%) of $N^\alpha$-acetylglucagon in glycogenolysis (W. W. Bromer 1981, unpublished work) may have resulted in part from deacetylation in vivo. In general, the more hydrophilic blocking groups (carbamoyl, acetyl, and trifluoroacetyl) allowed higher retention of potency than the hydrophobic trinitrophenyl group (Epand et al. 1981). Cote and Epand (1979) studied this latter derivative as an antagonist of glucagon and these studies will be discussed in Sect. E.V.

Several analogs have been studied wherein both amino groups of glucagon are derivatized. In cases where direct comparisons can be made between $N^\alpha$- and

$N^{\alpha}$,$N^{\varepsilon}$-derivatives (carbamoyl and acetyl), the derivatization of the ε-amino group lowers both potency and binding affinity approximately one order of magnitude. Introduction of the bulky hydrophobic iodine atom ($N^{\alpha}$,$N^{\varepsilon}$-bisiodoacetyl) eliminates all binding and potency (HRUBY et al. 1976). In all cases except the amidino derivatives reported by WRIGHT and RODBELL (1980a), both potency and affinity are very low, usually <1% that of glucagon. The amidino derivatives ($N^{\alpha}$,$N^{\varepsilon}$-diacetamidino- and $N^{\alpha}$,$N^{\varepsilon}$-4-mercaptobutylamidinoglucagon), unlike all the others examined, preserve a positive charge; this apparently accounts for retention of 5%–10% potency and affinity as well as near maximal (80%–90%) ability to stimulate adenylate cyclase. Another interesting analog in this series has both amino groups acetylated and the imidazole side chain formylated (EPAND et al. 1981); derivatization of the imidazole did not appreciably affect the already low potency, but appeared to lower affinity for the receptor even further.

## V. Inhibitors of Glucagon

Modification of both amino groups of glucagon often results in derivatives that retain some binding affinity, but have little potency. Therefore it is appropriate at this point to discuss attempts to prepare competitive inhibitors of glucagon. As already noted, such compounds would be of value in elucidating the role of relative hyperglucagonemia in diabetes. The most suitable assay system to search for inhibitors is the purified rat liver membrane preparation wherein relative potency (generation of cAMP), maximal stimulation, and receptor binding affinity (competition with labeled glucagon) can be measured. Already noted is the antagonistic function of [desHis$^{1}$]$N^{\varepsilon}$phenylthiocarbamoylglucagon in this system and the perplexing lack of antagonism (weak full agonist) when tested for glycogenolysis in the perfused rat liver (KHAN et al. 1980). A similar result was observed by COTE and EPAND (1979) with $N^{\alpha}$-trinitrophenylglucagon. In the rat liver membrane system, this derivative had very little potency with about 7% relative binding affinity, and it acted as a competitive inhibitor of glucagon; but in isolated hepatocytes it gave a maximal glycogenolytic response with a relative potency of 1%–2%, despite an apparent depression in cAMP levels and despite its ability to inhibit the effect of glucagon on activation of adenylate cyclase. COTE and EPAND (1979) consider the possibility that glucagon promotes glycogenolysis via a cAMP-independent process. An alternative proposal is raised by BREGMAN et al. (1980) that an undetectable rise in cAMP may provide a full biologic response. Thus, to provide an inhibition of glycogenolysis, an antagonist with nearly the same receptor affinity as glucagon might be required. To this end, BREGMAN et al. (1980) noted that hydrophobic amino-blocking groups lowered potency much more than affinity, and they prepared several potential inhibitors given in Table 4. The $N^{\alpha}$-trinitrophenylglucagon derivative of COTE and EPAND (1979) probably has an inhibitory strength similar to [desHis$^{1}$]$N^{\varepsilon}$-phenylthiocarbamoylglucagon, and both inhibit about one-tenth as well as $N^{\alpha}$-trinitrophenyl-[homoArg$^{12}$]glucagon. As this chapter was being completed, HRUBY et al. (1981) reported that the latter derivative substantially lowers blood glucose levels when infused into diabetic rats (see also HRUBY 1982). Their data suggest further that this inhibitor binds the receptor in a fashion different from glucagon. These interesting results open up the possibility of obtaining even better inhibitors.

**Table 4.** Some inhibitors of glucagon

| Analog | Relative inhibitory strength[a] |
|---|---|
| $N^\alpha$-trinitrophenyl[homoArg$^{12}$]glucagon | 0.54 |
| [desHis$^1$][$N^\alpha$-trinitrophenylhomoArg$^{12}$]glucagon | 0.1 |
| [desHis$^1$]$N^\varepsilon$-phenythiocarbamoylglucagon | 0.05 |
| $N^\alpha$-carbamoyl-$N^\varepsilon$-trinitrophenylglucagon | 0.01 |

[a] Determined by measuring the inhibition of glucagon-stimulated adenylate cyclase in rat liver membrane preparations

## VI. Modifications of Glutamyl, Lysyl, Arginyl, and Tryptophyl Residues

As previously mentioned, purified monodesamidoglucagon comprises a nearly equal mixture of isomers, involving desamidation of glutamine residues 3, 20, and 24 (BROMER et al. 1972). The mixture gives full agonism, but with 30% reduced potency; desamidation of one site could account for the loss of potency or each desamidated molecule may be about 30% less active.

Conversion of lysine to homoarginine retains the positive charge, giving a full agonist with up to 50% retention of potency in glycogenolysis in vivo. Two other lysine derivatives which retain a positive charge ($N^\varepsilon$-acetamidino- and $N^\varepsilon$-4-hydroxyphenylamidinoglucagon) gave unusually high affinity and potency, with the simpler derivative being indistinguishable from glucagon in the liver membrane system (WRIGHT and RODBELL 1980a). Neutralization of the positive charge of the ε-amino group ($N^\varepsilon$-acetyl- and $N^\varepsilon$-butoxycarbonylglucagon) lowered potency 5–10-fold (LANDE et al. 1972; W. W. BROMER 1981, unpublished work). The photoaffinity probe, $N^\varepsilon$-4-azido-2-nitrophenylglucagon $^{125}$I, exhibited remarkable properties of full binding affinity with no ability to activate adenylate cyclase (BREGMAN and LEVY 1977; BREGMAN et al. 1978), suggesting that this derivative or variations of it may be excellent competitive inhibitors. This interesting modification differs dramatically from all other derivatives considered in this section, and clearly warrants additional exploration.

EPAND and WHEELER (1975) reacted both arginine residues of glucagon with nitromalondialdehyde, giving [5-nitro-2-pyrimidylOrn$^{17,18}$] glucagon; this analog exhibits enhanced activity with about one-fifth the affinity of glucagon (EPAND and WHEELER 1975; EPAND et al. 1976). Perhaps these modifications of the guanidino groups impart greater resistance to proteolytic degradation. When glucagon is reacted with 1,2-cyclohexanedione (CHD) in dilute alkali at O °C for a few hours, mono- and disubstituted CHD derivatives of the arginines are obtained (W. W. BROMER 1981, unpublished work). Despite elimination of the positively charged guanidinium group (or groups), the analogs behaved as full agonists with reduced potency. Retention of significant potency for both of these modifications suggests that the guanidinium groups are not as important as the amino groups for binding or for expression of activity.

Several sulfenyl derivatives of the indole moiety of tryptophan in position 25 have been prepared (WRIGHT and RODBELL 1980b; DEMOLIOU and EPAND 1980;

EPAND and COTE 1976) with complete retention of binding affinity and potency. In fact, ([2-nitrophenylsulfenylTrp$^{25}$] glucagon) appears to have higher potency than glucagon (EPAND and COTE 1976). Reaction of glucagon with a disulfenyl reagent provided ([2,4-dinitrophenyl-1,5-disulfenylTrp$^{25}$] glucagon)$_2$ dimer that was equipotent with glucagon on a molar basis (EPAND and COTE 1976). Conversion of [2,4-dinitrophenylsulfenylTrp$^{25}$]glucagon to [2-thiolTrp$^{25}$]glucagon did not affect potency, and the oxidation of the latter analog to the disulfide dimer of glucagon (via position 2 of the indole) gives an analog with the same potency as glucagon on a molar basis, but with one-fourth the binding affinity (WRIGHT and RODBELL 1980b). These interesting analogs show clearly that bulky substituents (even glucagon itself) attached to tryptophan have little or no effect on biologic function. Conversely, it appears very unlikely that the indole side chain participates in a specific fashion, either in binding or in signal transmission.

## VII. Modifications of Tyrosyl Residues

The tyrosine (positions 10 and 13) derivatives of glucagon are a very interesting series wherein iodination or nitration of the phenolic ring results in greater potency than glucagon itself (BROMER et al. 1973; DESBUQUOIS 1975b; PATTERSON and BROMER 1973). In the iodo series, a quantitative measure of glycogenolytic potency could not be obtained because the dose–response curves were not parallel to that of glucagon (BROMER et al. 1973); however, each increment of iodine gave corresponding increases in extent and duration of glycogenolysis. DESBUQUOIS (1975b) found the enhanced potency of iodoglucagon to be a consequence both of higher receptor affinity (see Table 3) and of resistance to proteolytic degradation. Receptor affinity for iodoglucagon is difficult to study in an unambiguous manner since affinity is usually related to competition with glucagon $^{125}$I for receptor sites. Use of glucagon $^{3}$H would be ideal, but this label does not appear to be very stable (LIN et al. 1977). Introduction of iodine atoms imparts a more hydrophobic character to glucagon, and the enhanced receptor affinity may be a consequence of greater hydrophobic interactions with the receptor. In this respect, LIN et al. (1976) made the interesting observation that monoiodoglucagon at pH 7 shows threefold greater affinity and potency than glucagon, but gives the same result as glucagon at pH 8.5. Since monoiodination lowers the pK of the tyrosine hydroxyl from 10 to 8.2, LIN et al. (1976) reasoned that ionization of the phenoxy group at pH 8.5 lowered the enhanced binding affinity of iodoglucagon.

Although this hypothesis helps explain earlier reports (LIN et al. 1975) of the equivalent activities of glucagon and monoiodoglucagon, other factors must also be involved in the enhanced glycogenolytic activity of tetra- and pentaiodoglucagon and of [nitroTyr$^{13}$]glucagon. With these heavily iodinated glucagons, most of the tyrosines are diiodinated, lowering the pK of the phenolic hydroxyl to 6.5; nitration depresses the pK to 6.7. Thus, these tyrosine derivations are mostly ionized at pH 7–7.4, but both kinds of analogs are more potent than glucagon (BROMER et al. 1973; DESBUQUOIS 1975b; PATTERSON and BROMER 1973). Clearly, protonation of the phenoxy group at physiologic pH is not a major requirement for enhanced potency. In fact, one could interpret the bulk of the data (except for that of LIN et al. 1976) to suggest that, with each increment of iodine the concentration of ionized tyrosyl hydroxyl at pH 7 increases, with a subsequent increase of potency. To help resolve this question, the entire iodoglucagon series could be studied for pH dependency in a manner similar to the monoiodoglucagon study of LIN et al. (1976). Increased hydrophobicity of these derivatives is not an entirely adequate explanation either because the nitro group is considerably more hydrophilic than iodine, and [nitroTyr$^{13}$]glucagon is a more potent glycogenolytic agent than glucagon itself. In this case, we must consider that resistance to proteolysis is playing an important role, even though [nitroTyr$^{13}$]glucagon is hydrolyzed by chymotrypsin under one set of conditions at the same rate as glucagon (PATTERSON and BROMER 1973).

At the highest levels of iodination, DESBUQUOIS (1975 b) observed that tryptophan in position 25 is oxidized, a modification that is detrimental to potency. This finding probably explains the lower affinity and potency he found with pentaiodoglucagon compared with that found by BROMER et al. (1973). The latter workers iodinated at lower temperatures for shorter times, with nearly stoichiometric amounts of ICl, preserving the integrity of the tryptophan. Since the histidine residue is also iodinated in pentaiodoglucagon (DESBUQUOIS 1975 b), this analog may also be considered a modification of the imidazole side chain. Of particular interest is the high activity of this analog, suggesting again that the imidazole group of histidine is not vital for biologic function.

Reduction of the nitrophenol function to aminophenols (see Table 3) raises the pK of the tyrosylphenoxy group back to 10 and introduces an aromatic amino group with a pK of about 5. Both [aminoTyr]glucagon analogs are full agonists with less potency than glucagon or the corresponding [nitroTyr]analogs. In view of the instability of o-aminophenols, perhaps the most reasonable explanation for the drop in potency would be simply that reduction of the nitro groups provides analogs more similar to glucagon, but chemically less stable.

## VIII. Methionyl Residue and Carboxyl Terminal Modifications

Methylation of methionine residue 27 (ROTHGEB et al. 1977; NOOIJEN and KEMPEN 1979) converts the hydrophobic thioether side chain to a positively charged sulfonium group and gives a derivative that is quite soluble in water at pH 7. This [*S*-methylMet$^{27}$]glucagon is a full agonist with very low relative potency. On the other hand, peroxide oxidation of methionine to the sulfoxide form has little effect on potency or on other properties of glucagon (FELTS et al. 1970; W. W. BROMER 1981, unpublished work). NOOIJEN and KEMPEN (1979) oxidized the methionine with chloramine-T and the low potency of their sulfoxide derivative may be best explained by the probable concomitant oxidation of tryptophan residue 25 (DESBUQUOIS 1975 b). From [*S*-methylMet$^{27}$]glucagon ENGLAND and GURD (1980) prepared [desAsn$^{28}$,Thr$^{29}$][*S*-methylMet$^{27}$]glucagon and [desAsn$^{28}$,Thr$^{29}$]glucagon (glucagon 1–27), both of which are full agonists with very little potency. BROMER (1976) isolated a degradation product, [desMet$^{27}$,Asn$^{28}$,Thr$^{29}$]glucagon (glucagon 1–26), that also retained little potency or affinity. From these data it is clear that an intact thioether side chain is not vital for maximal response, although the positively charged sulfonium derivative is a very weak agonist. The similar potencies of glucagon 1–26 and glucagon 1–27, along with the low receptor-binding of glucagon 1–26, suggest that the two carboxyl terminal residues are very important for binding to the receptor. Perhaps the location of the α-COOH group is a key factor.

From Table 3 it is apparent that seven different laboratories have reacted glucagon with BrCN, converting methionine residue 27 to homoserine (and/or homoserine lactone) and shortening the glucagon chain by two residues. The high lipolytic potency (FELTS et al. 1970) of this derivative was discussed previously. However, activation of adenylate cyclase and glycogenolysis were both reduced about 50-fold, almost certainly a result of a 50-fold drop in receptor affinity. JONES and GURD (1981) reported that the lactone form had 20-fold less affinity for the receptor than the homoserine moiety. In general, these data agree with those discussed previously, suggesting that methionine is not required for maximal response and that such truncated modifications are full agonists with low potency because receptor affinity is markedly diminished.

Starting with [desAsn$^{28}$,Thr$^{29}$][homoSer$^{27}$]glucagon, WRIGHT et al. (1980) addèd glycine, hydrazine, butylamide, and biotinylamide to the homoserine 27 residue; all these derivatives were substantially like the starting modification, namely fully agonists with weak potency and affinity. Since glucagon 1–27 also behaved similarly, a homoserine analog with asparagine 28 and threonine 29 replaced by this approach would provide more direct evidence for the function of the thioether side chain of methoinine.

Addition of taurine to the COOH terminus of glucagon (EPAND et al. 1981) retains a negative charge from the sulfonic acid group of taurine but reduces affinity and potency about 40-fold. Addition of taurine, methyl esters, or glycineamide to all four carboxyl groups of glucagon (see Fig. 2) markedly reduces potency with a somewhat reduced affinity. Perhaps some of the β-carboxyl groups of the aspartic acid residues are more important for function than for binding. The low affinity and relative potency of [taurine$^{30}$]glucagon emphasizes the importance of the carboxyl terminal residues for binding to the receptor.

Other COOH terminal truncated forms of glucagon (1–6, 1–12, 1–18, 1–21, 1–23) provide an interesting series. As already discussed, lipolysis is observed for glucagon 1–21 and glucagon 1–23, but very little adenylate cyclase and glycogenolytic potency is retained. Glucagon 1–18 has no lipolytic potency, indicating that residues 19–21 are important for this function. Although RODBELL et al. (1971b) in early studies reported no cyclase activity for glucagon 1–23, extensive liver membrane studies from the same laboratory (WRIGHT et al. 1978; WRIGHT and RODBELL 1979) showed that glucagon 1–21 is a full agonist in the cyclase system with 1/1,000 the potency and affinity, and synthetic glucagon 1–6 is almost a full agonist with markedly, but equally reduced affinity and potency. Although glucagon 1–6 is not quite a full agonist, the equal reduction, both in affinity and potency, suggest that it is not a classical partial agonist; an alternative interpretation is that glucagon 1–6 retains the ability to transmit the biologic signal and its very weak potency is due primarily to loss of affinity for the receptor. It would be very interesting to know if glucagon 1–6 retains glycogenolytic potency. The COOH terminal fragments of glucagon (19–29, 20–29, and 22–29), showed no activity in lipolysis or in activation of adenylate cyclase (FELTS et al. 1970; RODBELL et al. 1971b).

## IX. Summary of Covalent Modifications and Function

Despite the incompleteness of the data in Table 3 and the previously stated limitations inherent in this approach, a few tentative generalizations are presented as a basis for future work. The side chains of lysine residue 12, tyrosines 10 and 13, tryptophan 25, and methionine 27 all can be modified with complete retention of affinity and potency, indicating that these side chains are probably not involved in *highly specific* interactions, either with the receptor or with signal transmission. A partial exception is lysine residue 12 wherein full retention of affinity and potency is observed only when the positive charge is retained and when the blocking group itself is small and hydrophilic, i.e., the acetamidino group. If the blocking group is large and hydrophobic or if the ε-amino function is converted to homoarginine, relatively more potency is lost than affinity, suggesting that the lysyl side chain may participate particularly in signal transmission. Another partial exception is the enhancement of affinity and potency with resistance to degradation observed in the iodotyrosyl series. The tyrosines appear to be involved in some manner with receptor interaction, although the bulkiness of the iodine atoms would argue that this interaction is probably not highly specific. Increased ionization of the tyrosyl hydroxyls at pH 7 may also be involved in function, but more data are needed to determine if ionization enhances or repress activity. On the basis of limited data, arginine residues 17 and 18 appear to fall in the same category as these other interior residues, namely that a surprising level of potency is retained despite drastic modification of the guanidinium group with loss of a positive charge.

The amino terminal portion of glucagon appears to be concerned both with potency (i.e., biologic signal transmission) and with affinity (i.e., receptor-binding), but the carboxyl terminal segment is probably responsible mainly for affinity. For example, removal of amino terminal histidine gives a partial agonist with relatively more receptor affinity than potency. Blocking the α-amino group of histidine with a hydrophobic reagent (trinitrophenyl, TNP) has the same effect. The most potent inhibitor of glucagon yet developed (highest ratio of affinity:potency) bears this $N^{\alpha}$-TNP group with lysine residue 12 converted to homoarginine. Considering all the data in Tables 3 and 4, the only partial agonists reported involved modifications of either or both histidine residue 1 and lysine 12. With respect to truncated derivatives, glucagon 1–21 is a weak full agonist and glucagon 1–6 has almost full agonist activity; both modifications have markedly reduced relative affinities that correspond closely to and probably are responsible for their reduced relative potencies. By implication, residues 22–29 (and probably 7–29) contain structures much more important for receptor-binding affinity than for signal transmission. Taking all the data available for glucagon 1–26, glucagon 1–27, and other BrCN-truncated forms, one must conclude that loss of two or more carboxyl terminal residues leaves full agonist activity, but markedly lowers affinity with a corresponding drop in potency. The importance of these seemingly innocuous terminal residues to receptor-binding is emphasized even more when one considers that the indole side chain of nearby tryptophan residue 25 can be modified with a variety of bulky substitutents, including glucagon itself, with complete retention of affinity and potency. These data, along with the large loss of relative affinity and potency for [taurine$^{30}$]glucagon, suggest that the α-carboxyl of glucagon may interact in a highly specific manner with the receptor.

Having emphasized the glucagon derivatives that *retain* binding and potency, one must not lose sight of the fact that the vast majority of modifications in nearly any portion of glucagon actually *reduce* both binding and potency. Clearly, the entire molecule is important for full hormonal function. This conclusion is in agreement with the strict conservation of primary structure already noted.

EPAND (1980; see also EPAND et al. 1981) has proposed an interesting model in which some of the hydrophobic side chains of glucagon ($Phe^{6,22}$,$Tyr^{10,13}$, and $Trp^{25}$) interact directly, but not very specifically with membrane phospholipids. These side chains are considered to participate primarily in receptor-binding and the polar residues (especially $His^{1}$,$Lys^{12}$) are thought to contribute to biologic signal transmission as well as to binding. In broad terms, EPAND's model is consistent with the data and provides a useful hypothesis for further studies. The following chapter emphasizes conformational aspects of the biologic interactions of glucagon.

## References


Assan R, Delaunay J (1972) Activités biologiques comparées de différents glucagons-retard et de glucagon ordinaire. Pathol Biol (Paris) 20:979–984

Assan R, Slusher N (1972) Structure/function and structure/immunoreactivity relationships of the glucagon molecule and related synthetic peptides. Diabetes 16:843–855

Beaven GH, Gratzer WB, Davies HG (1969) Formation and structure of gels and fibrils from glucagon. Eur J Biochem 11:37–42

Biemond MEF, Sipman WA, Olivie J (1979) Quantitative determination of polypeptides by gradient elution high pressure liquid chromatography. J Liq Chromatogr 2:1407–1435

Birnbaumer L, Pohl SL (1973) Relation of glucagon-specific binding sites to glucagon-dependent stimulation of adenylyl cyclase activity in plasma membranes of rat liver. J Biol Chem 248:2056–2061

Bregman MD, Hruby VJ (1979) Synthesis and isolation of a glucagon antagonist. FEBS Letters 101:191–194

Bregman MD, Levy D (1977) Labeling of glucagon binding components in hepatocyte plasma membranes. Biochem Biophys Res Commun 78:584–590

Bregman MD, Cheng S, Levy D (1978) Synthesis and properties of a photoaffinity probe for the glucagon receptor in hepatocyte plasma membranes. Biochim Biophys Acta 539:489–495

Bregman MD, Trivedi D, Hruby VJ (1980) Glucagon amino groups: evaluation of modifications leading to antagonism and agonism. J Biol Chem 255:11725–11731

Bromer WW (1976) Studies with glucagon analogs. Metabolism [Suppl 1] 25:1315–1316

Bromer WW, Chance RE (1969) Zinc glucagon depression of blood amino acids in rabbits. Diabetes 18:748–754

Bromer WW, Sinn LG, Staub A, Behrens OK (1956) The amino acid sequence of glucagon. J Am Chem Soc 78:3858–3859

Bromer WW, Staub A, Diller ER, Bird HL, Sinn LG, Behrens OK (1957) The amino acid sequence of glucagon. I. the amino acid composition and terminal amino acid analyses. J Am Chem Soc 79:2794–2798

Bromer WW, Boucher ME, Koffenberger JE (1971) Amino acid sequence of bovine glucagon. J Biol Chem 246:2822–2827

Bromer WW, Boucher ME, Patterson JM, Pekar AH, Frank BH (1972) Glucagon structure and function. I. Purification and properties of bovine glucagon and monodesamidoglucagon. J Biol Chem 247:2581–2585

Bromer WW, Boucher ME, Patterson JM (1973) Glucagon structure and function. II. Increased activity of iodoglucagon. Biochem Biophys Res Commun 53:134–139

Burgus R, Rivier J (1976) Use of high pressure liquid chromatography in the purification of peptides. In: Loffett A (ed) Proceedings 14th European peptide symposium. Editions de l'Université de Bruxelles, Brussels, p 85
Cole RD (1960) Ion exchange chromatography of glucagon in urea-containing buffers. J Biol Chem 235:2300–2301
Contaxis CC, Epand RM (1974) A study of the conformational properties of glucagon in the presence of glycols. Can J Biochem 52:456–468
Cote TE, Epand RM (1979) $N^{\alpha}$-trinitrophenylglucagon. An inhibition of glucagon-stimulated cyclic AMP production and its effects on glycogenolysis. Biochim Biophys Acta 582:295–306
Demoliou CD, Epand RM (1980) Synthesis and characterization of a heterobifunctional photoaffinity reagent for modification of tryptophan residues and its application to the preparation of a photoreactive glucagon derivative. Biochemistry 19:4539–4546
Desbuquois B (1975a) Acetylglucagon: preparation and characterization. Eur J Biochem 60:335–347
Desbuquois B (1975b) Iodoglucagon: preparation and characterization. Eur J Biochem 53:569–580
Desbuquois B, Laudat M-H (1974) Glucagon-receptor interactions in fat cell membranes. Mol Cell Endocrinol 1:355–370
England RD, Gurd RS (1980) Structure-function relationships in glucagon: a comparative study of four COOH-terminal derivatives (Abstr). Clin Res 28:798
Epand R (1980) The activation of adenylate cyclase by chemically modified forms of glucagon and its relationship to receptor binding. In: Brandenburg D, Wollmer A (eds) Insulin: Chemistry, structure, and function of insulin and related hormones. de Gruyter, Berlin New York, p 363
Epand RM, Cote TE (1976) Conformational and biological properties of a covalently-linked dimer of glucagon: reaction of mono- and bi-functional sulfenyl halides. Biochim Biophys Acta 453:365–373
Epand RM, Wheeler GE (1975) Conformational and biological properties of a glucagon derivative with enhanced biological activity (Abstr) Fed Proc 34:660
Epand RM, Cote TE, Hoa DHB, Rosselin G, Schreier S (1976) Biologic activity and conformational properties of glucagon and glucagon analogs. Metabolism [Suppl 1] 25:1317–1318
Epand RM, Rosselin G, Hoa DHB, Cote TE, Laburthe M (1981) Structural requirements for glucagon receptor binding and activation of adenylate cyclase in liver. J Biol Chem 256:1128–1132
Felts DW, Ferguson MEC, Hagey KA, Stitt ES, Mitchell WM (1970) Studies on the structure-function relationships of glucagon (Abstr). Diabetologia 6:44–45
Grande F, Grisola S, Diederich D (1972) On the biological and chemical reactivity of carbamylated glucagon. Proc Soc Exp Biol Med 139:855–860
Gratzer WB, Bailey E, Beaven GH (1967) Conformational states of glucagon. Biochem Biophys Res Commun 28:914–919
Guy-Grand B, Assan R (1973) Lipolytic responses of human adipose tissue in vitro to natural glucagon and to synthetic [1–23] glucagon peptide (Abstr). Diabetologia 9:70
Hancock WS, Bishop CA, Prestidge RL, Harding DRK, Hearn MTW (1978) Reversed-phase, high-pressure liquid chromatography of peptides and proteins with ion-pairing reagents. Science 200:1168–1170
Hruby VJ (1982) Structure-conformation-activity studies of glucagon and semi-synthetic glucagon analogs. Mol Cell Biochem 44:49–64
Hruby VJ, Groginsky CM (1971) Partition chromatography of glucagon and secretin on sephadex. J Chromatogr 63:423–428
Hruby VJ, Wright DE, Lin MC, Rodbell M (1976) Semisynthetic glucagon derivatives for structure-function studies. Metabolism [Suppl 1] 25:1323–1325
Hruby VJ, Bregman MD, Johnson DG, Ulichny C, Trivedi D (1981) In vivo and in vitro studies with glucagon inhibitors. In: Rich DH, Aross E (eds) Peptides: synthesis-structure-function, proceedings 7th American peptide symposium. Pierce Chemical Co, Rockford, pp 813–816

Jones BN, Gurd RS (1981) Ultrapure cyanogen bromide-cleaved glucagon: isolation in high yield by ion exchange chromatography. Arch Biochem Biophys 206:227–331

Kahn BA, Bregman MD, Nugent CA, Hruby VJ, Brendel K (1980) (des-His$^1$) ($N^\varepsilon$-phenylthiocarbamoylysine$^{12}$)-glucagon: effects on glycogenolysis in perfused rat liver. Biochem Biophys Res Comm 93:729–736

Kenny AJ (1955) Extractable glucagon of the human pancreas. J Clin Endocrinol Metab 15:1089–1105

Lande S, Gorman R, Bitensky M (1972) Selectively blocked and des-histidine-glucagons: preparation and effects on hepatic adenylate cyclase activity. Endocrinology 90:597–604

Lin MC, Wright DE, Hruby VJ, Rodbell M (1975) Structure-function relationships in glucagon: properties of highly purified des His$^1$,-monoiodo-, and [desAsn$^{28}$,Thr$^{29}$] (homoserine lactone$^{27}$)-glucagon. Biochemistry 14:1559–1563

Lin MC, Nicosia S, Rodbell M (1976) Effects of iodination of tyrosyl residues on the binding and action of glucagon at its receptor. Biochemistry 15:4537–4540

Lin MC, Nicosia S, Lad PM, Rodbell M (1977) Effects of GTP on binding of [$^3$H]glucagon to receptors in rat hepatic plasma membranes. J Biol Chem 252:2790–2792

Lochner JdeV, Esterhuizen AC, Unger RH (1964) Separation of human insulin, glucagon, and other pancreatic proteins. Diabetes 13:387–394

Markussen J, Frandsen E, Heding LG, Sundby F (1972) Turkey glucagon: crystallization, amino acid composition and immunology. Horm Metab Res 4:360–363

McDonald JK, Callahan PX, Zeitman BB, Ellis S (1969) Inactivation and degradation of glucagon by dipeptidyl aminopeptidase I (cathepsin C) of rat liver. J Biol Chem 244:6199–6208

Mitznegg P, Domschke W, Domschke S, Sprugel W, Estler C-J, Wunsch E, Jaeger E, Demling L (1976) Effect of glucagon and its 1–23 peptide fragment on lipolysis in isolated rat and human fat cells. Biochem Pharmacol 25:210–211

Nooijen WJ, Kempen HJ (1979) Immunogenicity and bioactivity of glucagon, modified at methionine-27. Horm Metab Res 11:459–463

Patterson JM, Bromer WW (1973) Glucagon structure and function: preparation and characterization of nitroglucagon and aminoglucagon. J Biol Chem 248:8337–8342

Pohl SL, Krans HMJ, Birnbaumer L, Rodbell M (1972) Inactivation of glucagon by plasma membranes of rat liver. J Biol Chem 247:2295–2301

Pollock HG, Kimmel JR (1975) Chicken glucagon. Isolation and amino acid sequence studies. J Biol Chem 250:9377–9380

Rhoten WB (1976) Glucagon levels in pancreatic extracts and plasma of the lizard. Am J Anat 147:131–137

Rodbell M, Krans HMJ, Pohl SL, Birnbaumer L (1971 a) The glucagon-sensitive adenyl cyclase system in plasma membranes of rat liver. J Biol Chem 246:1861–1871

Rodbell M, Birnbaumer L, Pohl SL, Sundby F (1971 b) The reaction of glucagon with its receptor: evidence for discrete regions of activity and binding in the glucagon molecule. Proc Natl Acad Sci USA 68:909–913

Rodbell M, Lin MC, Salomon Y (1974) Evidence for interdependent action of glucagon and nucleotides on the hepatic adenylate cyclase systems. J Biol Chem 249:59–65

Ross JBA, Rousslang KW, de Haen C, Lavis VR, Deranleau DA (1979) 12-Homoarginine glucagon: synthesis and observations on conformation, biological activity, and copper-mediated peptide cleavage. Biochim Biophys Acta 576:372–384

Rosselin G, Freychet P, Fouchereau M, Rancon F, Broer Y (1974) Interactions of insulin and glucagon with isolated rat liver cells II – Dynamic changes in the cyclic AMP induced by hormones. Horm Metab Res 5:78–86

Rothgeb MT, Jones BN, Hayes DF, Gurd RS (1977) Methylation of glucagon, characterization of the sulfonium derivative, and regeneration of the native covalent structure. Biochemistry 16:5813–5818

Sokal JE (1960) The duration of glucagon effect. Endocrinology 67:774–783

Sonne O, Berg T, Christoffersen T (1978) Binding of $^{125}$I-labeled glucagon and glucagon-stimulated accumulation of adenosine 3′:5′-monophosphate in isolated intact rat hepatocytes. Evidence for receptor heterogeneity. J Biol Chem 253:3203–3210

Staub F, Sinn L, Behrens OK (1955) Purification and crystallization of glucagon. J Biol Chem 214:619–632
Sundby F (1976) Species variation in the primary structure of glucagon. Metabolism [Suppl 1] 25:1319–1321
Sundby F, Markussen J (1971) Isolation crystallization and amino acid composition of rat glucagon. Horm Metab Res 3:184–187
Sundby F, Markussen J (1972) Rabbit glucagon: isolation, crystallization and amino acid composition. Horm Metab Res 4:56
Sundby F, Frandsen EK, Thomsen J, Kristiansen K, Brunfeldt K (1972) Crystallization and amino acid sequence of duck glucagon. FEBS Lett 26:289–293
Sundby F, Markussen J, Danho W (1974) Camel glucagon: isolation, crystallization and amino acid composition. Horm Metab Res 6:425
Sutherland EW, de Duve C (1948) Origin and distribution of the hyperglycemic-glycogenolytic factor of the pancreas. J Biol Chem 175:663–674
Tallarida RJ, Jacob LS (1979) The dose-response relation in pharmacology, 1st edn. Springer, Berlin Heidelberg New York
Tarding F, Nielsen P, Pingel M, Volund Aa (1969) Biological and chemical properties of two glucagon preparations with prolonged action. Eur J Pharmacol 7:206–210
Thomsen J, Kristiansen K, Brunfeldt K (1972) The amino acid sequence of human glucagon. FEBS Lett 21:315–319
Trakatellis AC, Tada K, Yamaji K, Gardiki-Kouidou P (1975) Isolation and partial characterization of anglerfish proglucagon. Biochemistry 14:1508–1512
Weinges KF (1959) Der Einfluß eines protrahiert wirkenden Glucagons auf den Blutzucker, das anorganische Serumphosphat und die Gesamtaminosäuren im Serum. Naunyn Schmiedebergs Arch Pharmacol 237:22–26
Wheeler GE, Epand RM, Barrett D (1974) Non-equivelance of the carboxyl groups of glucagon in the carbodiimide-promoted reaction with nucleophiles and the role of carboxyl groups in the ability of glucagon to stimulate adenyl cyclase of rat liver. Biochim Biophys Acta 372:440–449
Wright DE, Rodbell M (1979) Glucagon$_{1-6}$ binds to the glucagon receptor and activates hepatic adenylate cyclase. J Biol Chem 254:268–269
Wright DE, Rodbell M (1980a) Properties of amidinated glucagons. Eur J Biochem 111:11–16
Wright DE, Rodbell M (1980b) Preparation of 2-thioltryptophan-glucagon and (tryptophan-S-Glucagon)$_2$. J Biol Chem 255:10884–10887
Wright DE, Hruby VJ, Rodbell M (1978) A reassessment of structure-function relationships in glucagon: glucagon$_{1-21}$ is a full agonist. J Biol Chem 253:6338–6340
Wright DE, Hruby VJ, Rodbell M (1980) Preparation and properties of glucagon analogs prepared by semi-synthesis from CNBr-glucagon. Biochim Biophys Acta 631:49–58
Ziegler M, Lippmann HG (1968) Quantitative electrophoretic separation of insulin and glucagon. Naturwissenschaften 55:181

CHAPTER 2

# The Chemical Synthesis of Glucagon

R.B. MERRIFIELD and S. MOJSOV

## A. Introduction

With the determination of the complete amino acid sequence of porcine glucagon by BROMER et al. (1957), the way was open for the chemical synthesis of this important peptide hormone. A synthesis would serve several purposes: it would, first of all, provide a classical proof of the proposed structure; secondly, it could lead to a large scale source of the hormone to replace the natural material, and, especially, it would provide a route to structural analogs designed to be inhibitors or superagonists that would be useful, both in defining the mechanism of action of glucagon and as potential clinical agents.

Although peptide synthesis was in a state of rapid development in 1957, and complex molecules were being prepared, glucagon was a severe challenge to the peptide chemist, and it was more than 10 years before a total synthesis was accomplished. Of the 29 amino acid residues, 22 are trifunctional and therefore present special synthetic difficulties. Notable among these are the single residues of tryptophan, methionine, and histidine, the three aspartic acids, three glutamines, nine hydroxy amino acids, and the problem sequences arginyl-arginine, aspartyl-serine, and asparaginyl-threonine.

The approaches to the synthesis of a peptide such as glucagon may be divided into two broad strategies: (1) classical solution synthesis; and (2) solid phase synthesis. These may be further divided into two substrategies: (a) fragment assembly; and (b) stepwise assembly. The principal advantage of a solution synthesis is that the intermediates can be isolated, purified, and characterized at every stage, while the main difficulties are usually associated with problems of insolubility, slow and incomplete reactions, and excessive amounts of labor. The advantages of solid phase synthesis, in which the peptide chain is covalently anchored to an insoluble solid support, are rapid, high yield reactions and simplicity of operation, while the main disadvantage is the lack of an opportunity to isolate and purify the product at any intermediate stage of the synthesis and the resulting requirement for near quantitative reactions with minimal side reactions. Initially, the fragment approach was used for syntheses in solution and the stepwise approach was used for solid phase syntheses, but subsequently both procedures have been adapted to each strategy. The total synthesis of glucagon has been achieved in four different laboratories using three of the four general strategies just described.

An entirely different approach to the preparation of analogs or variants of proteins or peptides such as glucagon is also possible. It involves semisynthesis, in which the naturally occurring parent molecule is directly derivatized or modified, or is first partly degraded and then reconstructed by the addition of synthetic peptides or other components. Although a number of useful glucagon derivatives have been prepared by these methods, e.g., BREGMAN and HRUBY (1979) and WRIGHT et al. (1980) they cannot be described here (see Chaps. 1 and 13).

## B. Early Synthetic Efforts

Beginning in 1962, the syntheses of several small fragments of glucagon were reported (BEYERMAN and BONTEKOE 1962, 1964; WÜNSCH and ZWICK 1964, 1966; SCHRÖDER 1964a–c; WÜNSCH and DREES 1966; HYLTON et al. 1966; NITECKI et al. 1971). In 1967, the results of a major effort by the Schering group in Berlin (SCHRÖDER 1967) leading to the preparation of several large protected fragments covering the entire sequence of glucagon were published. However, because of severe solubility problems, the final coupling reactions between fragments 1–19 and 20–29 or fragments 1–11 and 12–29 could not be achieved. By further modification of the fragment scheme, a crude product was obtained in low yield from the condensation of fragments 1–14 and 15–29 that contained a component migrating in electrophoresis with the same mobility as natural glucagon. Similar difficulties were encountered by WÜNSCH and his collaborators in Münich. They adopted the general strategy that had been so successfully applied to $\beta$-corticotropin by SCHWYZER and SIEBER (1963), but after 3 years of effort they were forced to revise their synthetic plan completely.

## C. The First Total Synthesis

In 1968, WÜNSCH reported the first successful total synthesis of glucagon (WÜNSCH and WENDLBERGER 1968; WÜNSCH et al. 1968a, b). The synthetic strategy involved the preparation of four protected peptide fragments, 1–6, 7–15, 16–21, and 22–29, and their sequential condensation, beginning with the COOH terminal 22–29 fragment as shown in Fig. 1. The four intermediates were each prepared primarily by stepwise elongation procedures. Thus, this overall synthesis applied both the fragment and stepwise substrategies to the main strategy of classical solution synthesis. The success of the synthesis depended on the development and application of several new methods and of an astute combination of new and old methods.

The major advance was the use of the newly developed dicyclohexylcarbodiimide/hydroxysuccinimide (DCC/HOSu) procedure (WEYGAND et al. 1966), which allows the coupling of peptide fragments in high yield without racemization, even when an optically active COOH-terminal residue is activated. This avoided the requirement for coupling by the azide method that had not been satisfactory in the earlier work on glucagon.

Of equal importance was the reliance on the acid-labile *tert*-butyl-based protecting groups for the trifunctional amino acid residues; *tert*-butyl ethers for Ser, Thr, and *tert*-butyl esters for Asp and Glu, and *tert*-butyloxycarbonyl urethane for Lys. The o-nitrophenylsulfenyl (Nps) group (ZERVAS et al. 1963) was chosen for $N^{\alpha}$ protection since it can be selectively removed by very mild acid in the presence of *tert*-butyl derivatives. The thioindole derivative that it gives with tryptophan

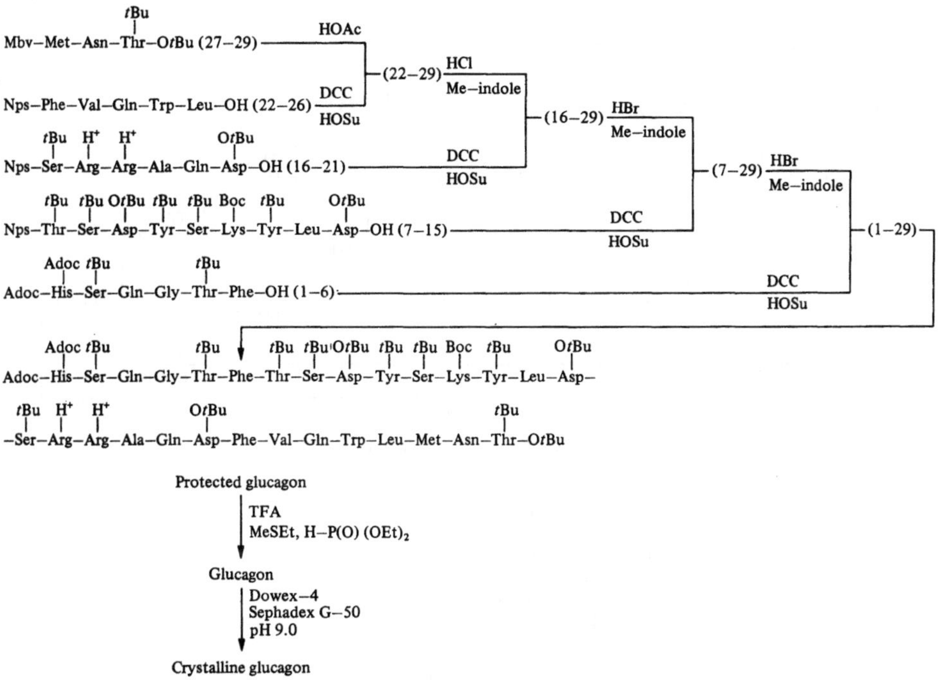


**Fig. 1.** Synthetic scheme for the first total synthesis of glucagon. *Mbv*, 1-methyl-2-benzoylvinyl. WÜNSCH et al. (1968 b)

during acidolytic removal was largely avoided by the addition of an excess of methylindole scavenger. To avoid solubility problems with the products, the arginine residues were incorporated as the mono- and tribenzyloxycarbonyl derivatives and were carried through the later stages of the synthesis with protection of the guanidine side chain only by protonation.

The final condensation of fragments 1–6 and 7–29 gave an 84% yield of the fully protected 1–29, nonacosapeptide $C_{232}H_{367}N_{43}O_{55}SBr_2 \cdot 6\,H_2O$, molecular weight 4,938.8 daltons. A total of 2.06 g (0.42 mmol peptide, 12.2 mmol amino acid residues) was obtained from approximately 141 mmol protected amino acid starting material, giving a remarkably high overall synthetic efficiency of 8.6%. The final deprotection was with trifluoroacetic acid in the presence of methylethyl sulfide and diethyl phosphite as scavengers of *tert*-butyl carbonium ions that otherwise would cause extensive side reactions with methionine and tryptophan. After partial purification on an ion exchange resin, the crude product showed, by thin layer chromatography, a main band corresponding to natural glucagon and several less well-defined components in smaller amounts. The main component was resolved into two fractions by gel filtration on Sephadex G-50. The principal fraction 320 mg, 4.26% overall yield, was homogeneous by several criteria. Crystallization at pH 9.3 gave 140 mg glucagon in the form of rhombic dodecahedra, which were indistinguishable from crystalline natural glucagon. Identity with the natural hormone was further established by amino acid analysis of acid and enzymatic hydrolysates, ultraviolet absorption spectra, and optical rotatory dispersion curves.

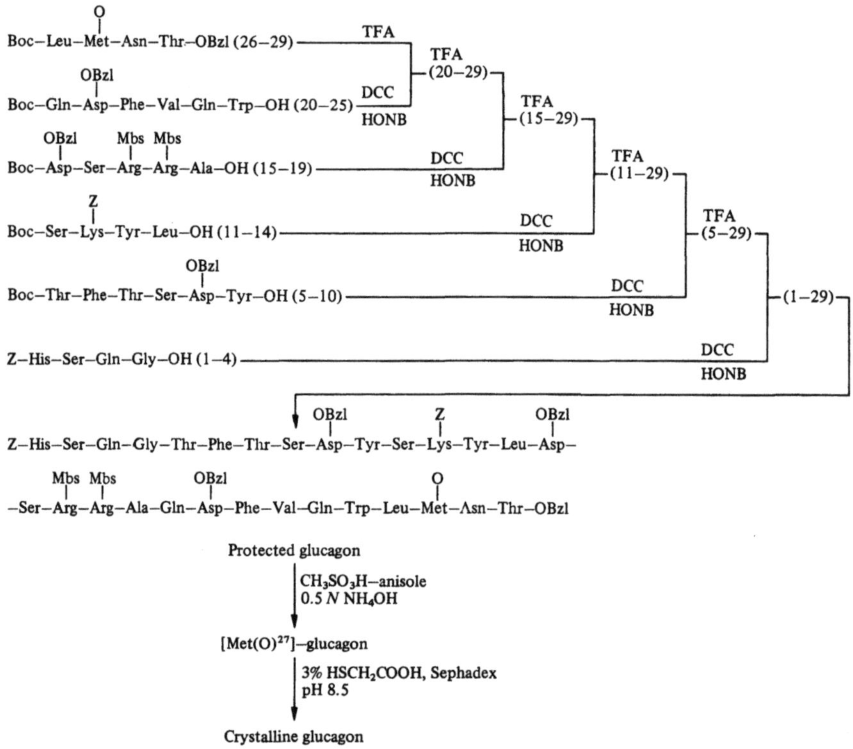

**Fig. 2.** Synthetic scheme for the fragment synthesis of mammalian glucagon. *Mbs*, 4-methoxybenzenesulfonyl. FUJINO et al. (1978)

The biologic activity and immunologic specificity of the synthetic glucagon were also shown to be indistinguishable from those of natural glucagon. The tests included glycogenolysis in rat liver slices, lipolysis in isolated fat cells, changes in blood glucose and insulin levels in humans, and radioimmunoassays. Thus, by all the physical, chemical, and biologic criteria examined, the final purified synthetic glucagon was identical with the natural hormone. This synthesis is considered to be a milestone in the field and established standards with which other syntheses of peptides must be compared.

## D. Further Syntheses by Fragment Condensation in Solution

The second synthesis of mammalian glucagon by the fragment strategy in solution was reported some 10 years after the Wünsch synthesis by a combined group from Takeda Chemical Industries and Kyoto University (FUJINO et al. 1978). The design of the synthesis is shown in Fig. 2. It involved the preparation of six protected fragments, 1–4, 5–10, 11–14, 15–19, 20–25, and 26–29, and their sequential coupling, beginning with fragment 26–29. The individual fragments were made by stepwise assembly, starting with the COOH terminal amino acid or its ester and coupling with *N*-hydroxy-5-norbornene-2,3-dicarboximide (HONb) esters of the succeeding *tert*-butyloxycarbonyl- (Boc) or benzyloxycarbonyl- (Z) amino acids. The latter were either preformed or prepared in situ with DCC. In some instances, the intermediates were purified by crystallization and in others by precipitation and washing. The fragments were also coupled by the DCC/HONb method in dimethylformamide (DMF). In this synthesis, the Ser, Thr, and Tyr hydroxyls, the His imidazole nitrogen, and the Trp indole nitrogen were all unprotected. The β-benzyl ester was used for Asp, Z for Lys, the sulfoxide for Met, and p-methoxybenzenesulfonyl for Arg.

To produce 0.93 g partially protected glucagon (0.21 mmol peptide, 6.1 mmol amino acid residues) required approximately 110 mmol protected amino acid derivatives as starting material; yield 5.5%. The protected 29-residue peptide was deprotected by treatment with methanesulfonic acid (MSA) containing 7.5% anisole, at room temperature for 1 h. An N→O acyl migration at serine was reversed by treatment with 0.5 *M* $NH_4OH$ and the product was purified by passing through columns of Amberlite IRA-410, Sephadex LH-20, and CM-cellulose. The deprotection and workup caused rather large losses, with a recovery of only 6.4%, or an overall yield of 0.35%. After reductive removal of the sulfoxide of methionine by thioglycolic acid, the yield was 0.23%. The lyophilized product was dissolved in dilute aqueous NaCl at pH 10.5, adjusted to pH 8.5, and allowed to stand for 2 days. Crystalline glucagon, 8.3 mg (0.13% overall yield), was obtained (see Fig. 6b). By deprotecting in HF, the yields were increased to 0.45%, 0.31%, and 0.16% respectively. The physical, chemical, and biologic properties of the synthetic glucagon were all identical with those of natural glucagon.

YAJIMA and co-workers (OGAWA and YAJIMA 1978; OGAWA et al. 1978) have also completed the total synthesis of duck glucagon by the same general approach, but using seven fragments. These were selected in order to permit the entire assembly to be formed by Rudinger's azide procedure (HONZL and RUDINGER 1961; Fig. 3). The side chain protecting groups were similar to those used previously. The Boc group was removed in the presence of tryptophan by 4 *N* ethanesulfonic acid in trifluoroethanol or by trifluoroacetic acid in the presence of skatole and anisole containing 0.1% ethanedithiol. For the tetrapeptide fragment containing Lys(Z), the $NH_2$ terminal hydrazide was protected with the β,β,β-trichloroethyloxycarbonyl group, which could be selectively removed by treatment with zinc in acetic acid.

The deprotection, workup, and purification was similar to that for the mammalian preparation and gave 49 mg product. Crystals (4.3 mg) could be obtained from this synthetic duck glucagon. The lipolytic activity on adipocytes was equivalent to that of natural porcine glucagon. The immunologic specificity of duck glucagon is known to be different from that of porcine glucagon (SUNDBY et al. 1972), even though they differ only at positions 16 and 28. A radioimmunoassay using anti-porcine glucagon rabbit serum showed the synthetic duck glucagon to be 46% active.

This and other syntheses from the laboratory of YAJIMA continue to support the view that the most satisfactory approach to the assembly of large peptides by the fragment strategy is by the stepwise addition of many small fragments to the growing peptide chain. A fully stepwise synthesis from single amino acids by classical solution methods, analogous for example to the one for secretin (BODANSZKY et al. 1967), has not been reported for glucagon.

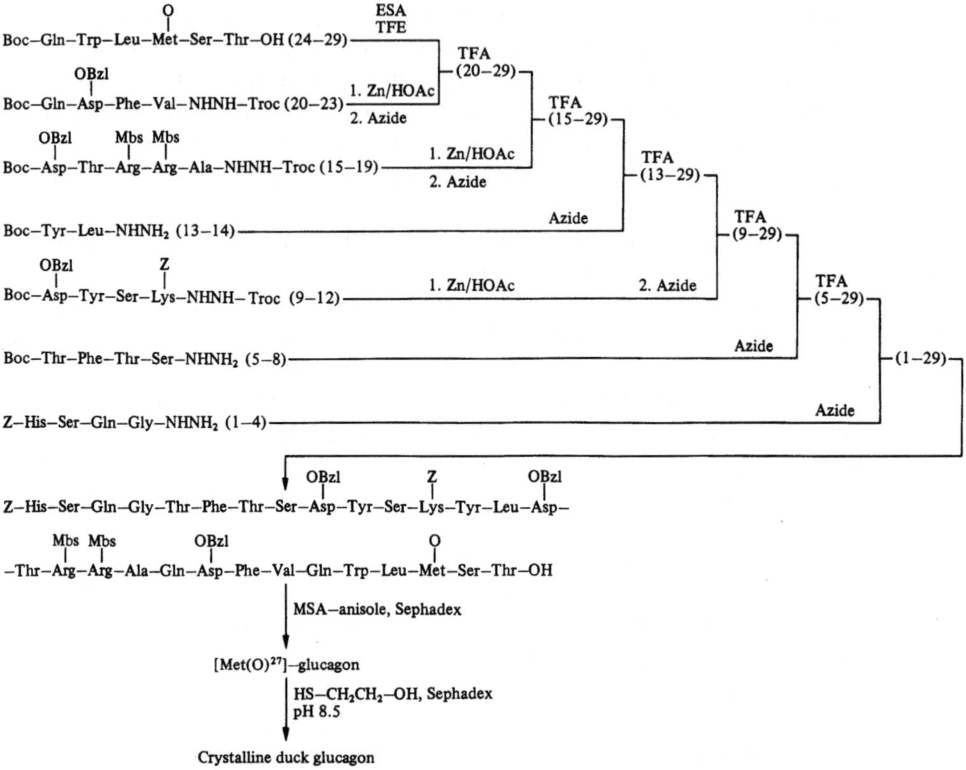


**Fig. 3.** Synthetic scheme for the fragment synthesis of duck glucagon. *Troc*, *β,β,β*-trichloroethyloxycarbonyl. OGAWA et al. (1978)

# E. Solid Phase Fragment Synthesis

A synthesis of glucagon by the PROTEIN SYNTHESIS GROUP, SHANGHAI INSTITUTE OF BIOCHEMISTRY (1975) combined the solid phase approach to peptide synthesis (MERRIFIELD 1963) with the fragment strategy (WEYGAND 1968). They prepared four protected peptides by solution methods with only small variations from the scheme used by WÜNSCH. They divided the molecule in a similar way, giving fragments 1–6, 7–15, 16–21, and 22–26 and they also relied heavily on *tert*-butyl groups and on the DCC/hydroxybenzotriazole (DCC/HOBt) coupling agent (KÖNIG and GEIGER 1970), which is closely related to the DCC/HOSu method. The very acid-labile biphenylisopropyloxycarbonyl (Bpoc) group replaced NpS; and both arginines as well as histidine were protected with adamantyloxycarbonyl (Adoc) groups (HAAS et al. 1966). The fourth fragment was three residues shorter than the one used by WÜNSCH to permit the fragment synthesis to be conducted on a solid support.

Thus, Boc-Thr(Bzl)-OH (where Bzl indicates a benzyl radical) was esterified as a benzyl ester to a pellicular resin (HORVATH et al. 1967), comprised of a highly cross-linked polystyrene core, surrounded by a thin shell of slightly cross-linked polystyrene. Residues of Asn and Met were then added to give H-Met-Asn-

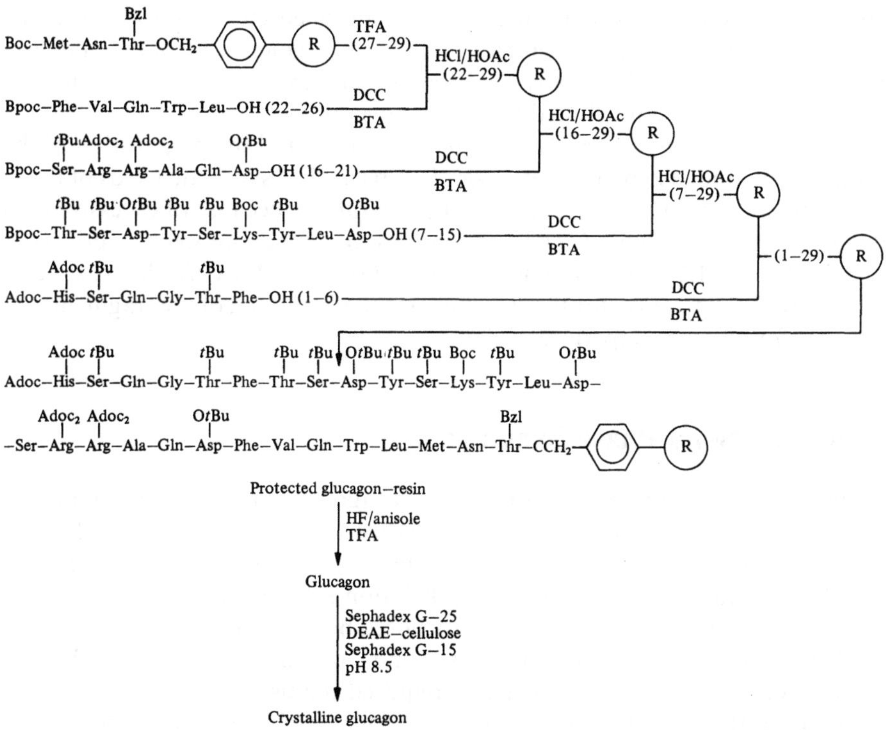


**Fig. 4.** Synthetic scheme for the solid phase fragment synthesis of glucagon. Protein Synthesis Group, Shanghai Institute of Biochemistry (1975)

Thr(Bzl)-$OCH_2C_6H_4$-resin. To this short peptide handle the four fragments were coupled one at a time (Fig. 4). Each of the couplings was reported to be essentially quantitative, which is quite remarkable because the assembly of the peptide was restricted to the outer shell of the pellicular support, comprising about 30% of the mass of the polystyrene beads, and the effective loading of peptide chains was therefore about 1.7 mmol/g. This is far above the usual loading and might have been expected to result in extreme crowding and slow or incomplete coupling. The final protected peptide–resin consisted of 94 mg dense, highly cross-linked polystyrene core and 40 mg pellicular, slightly cross-linked shell to which was attached 350 mg protected peptide. The peptide was cleaved from the resin and fully deprotected by treatment with HF–anisole (9:1). Two peptide components were obtained by gel filtration, and the glucagon-containing fraction was further purified by diethylaminoethyl-(DEAE)-cellulose chromatography, giving 3.4 mg homogeneous glucagon from which typical crystals could be obtained (see Fig. 6c). The synthetic product was indistinguishable from natural glucagon by gel electrophoresis and was essentially fully active in the rabbit blood glucose assay.

The yield of the protected peptide resin, after coupling the four fragments, was 85%, based on the first threonine residue, and the yield of the pure product after cleavage and chromatography was 17%. However, this number decreases to approximately 0.34% if the overall yield includes the total amount of protected amino acids required to produce the fragment intermediates. Even with high synthetic yields at all stages of the preparation of the fragments, the large number of steps reduced the yield drastically, and the 3–4-fold excess of the fragments used in the couplings also contributed to the low overall yield.

DU et al. (1980) have recently described in more detail the preparation of their pellicular resin and also of a radiation-grafted resin. The synthetic yields of the latter were not as good as with the pellicular preparations because of excessive cross-linking and poor swelling. A comparison was also made between *N*-ethoxycarbonyl-2-ethyloxy-1,2-dihydroquinoline (EEDQ) and DCC/$N^1$-hydroxybenzotriazole (HOBt) activation for the rate and extent of coupling of the fragments. The latter was found to be superior.

## F. Stepwise Solid Phase Synthesis

The first stepwise solid phase synthesis of glucagon was reported by MERRIFIELD et al. (1977) and was described in detail by MOJSOV and MERRIFIELD (1981). It was based on a mild acidolytic scheme of deprotection to avoid $\alpha$–$\beta$ rearrangement at the Asp-Ser sequence and to minimize alkylation side reactions. The COOH terminal Bpoc-Thr(*t*-Bu)-OH residue was anchored by an ester bond to p-alkoxybenzyl alcohol resin (WANG 1973). This bond, which was cleavable by 50% trifluoroacetic acid in dichloromethane, required the use of a temporary $N^\alpha$-protecting group that could be removed selectively at each cycle of the synthesis by very mild acid. The Bpoc group (SIEBER and ISELIN 1968) was chosen for that purpose. Side chain protection was largely based on the *tert*-butyl group. The two arginines were protected with nitro groups, which allowed the option of their removal by catalytic hydrogenation or by HF, or of their retention in the final product. Racemization during coupling of histidine (WINDRIDGE and JORGENSEN 1971) was prevented by protecting the imidazole ring with a 2,4-dinitrophenyl (Dnp) group. Leucine $^{14}C$ was introduced at position 26 to aid in quantitation of the synthesis and isolation of the product. Special precautions were taken against nitrile and amidine formation by coupling asparagine and glutamine with DCC/HOBt (MOJSOV et al. 1980). All other residues were coupled with DCC alone. The synthetic scheme is shown in Fig. 5.

The synthesis began with 5.1 g (0.20 mmol/g) Bpoc-Thr(*t*Bu)-$OCH_2$-$C_6H_4$-$OCH_2$-resin and was carried out in an automated peptide synthesizer, using a double coupling protocol (5.5-h cycle). The extent of each coupling was monitored by the fluorescamine test (FELIX and JIMENEZ 1973) and the losses of peptide chains from the resin were quantitated by counting $^{14}C$ in the filtrates. The average loss of peptide chains was 0.2% per cycle. Extensive chain termination in the early stages of the synthesis by irreversible acylation lowered the growing chains to 48% of the starting value. The workup involved the removal of the Dnp from His by thiolysis, and cleavage of the chain from the resin by 50% $CF_3COOH/CH_2Cl_2$ in the presence of indole (35% yield). 47% of the crude product was recovered in a single peak from Sephadex G-25. The nitro groups were quantitatively removed from Arg by treatment with HF at O °C for 30 min. Further purification on a DEAE-cellulose cation exchange column gave a principal peak in 57% yield, which

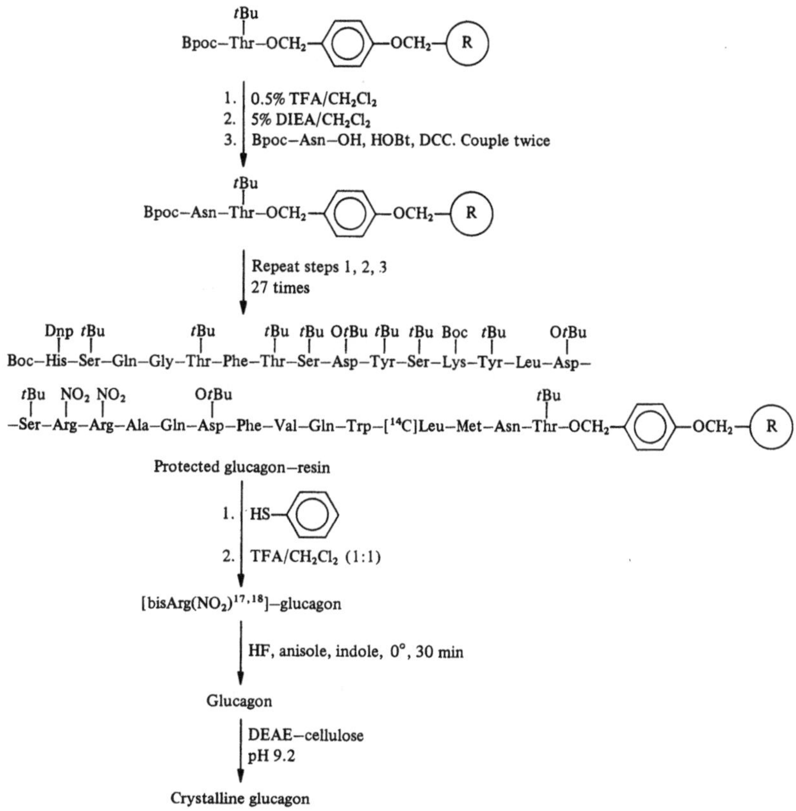


**Fig. 5.** Synthetic scheme for the stepwise solid phase synthesis of glucagon. MOSJOV and MERRIFIELD (1981)

corresponded with the main component of natural glucagon run on the same column. The yield from the first amino acid (calculated on limiting reactants) was 4.3%, but if corrected for the excess of protected amino acid reagents used in the synthesis, it was reduced to 0.72%.

Rechromatography of the synthetic product gave a single, symmetric, homogeneous peak. Cochromatography with natural glucagon showed them to behave as a single compound. Homogeneity and identity were further confirmed by polyacrylamide gel electrophoresis and reverse-phase high pressure liquid chromatography on a C-18 column. Amino acid analysis of acid hydrolysates gave excellent agreement with the natural hormone, and for the 29 residues the mean deviation of the individual values from equimolar ratios was only $\pm 3\%$. The molar ratio Tyr/Trp was 1.99: 1 (theoretical ratio 2: 1). The ultraviolet and fluorescence spectra of natural and synthetic glucagon were identical.

The synthetic glucagon was crystallized twice, using two separate conditions; (a) at pH 8.8 in dilute NaCl; (b) at pH 9.2 in 0.2 $N$ potassium phosphate. The crystals had the typical hexagonal appearance (Fig. 6d) corresponding to the rhombic dodecahedron structure of glucagon (STAUB et al. 1955) and were birefringent.

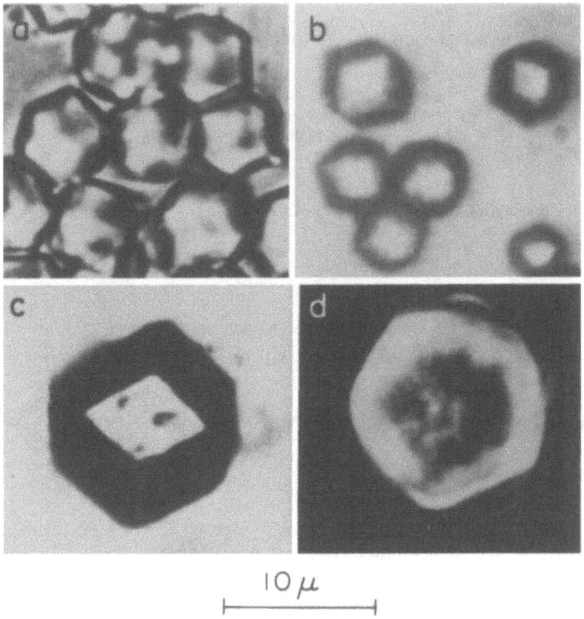


**Fig. 6a–d.** Crystals of synthetic glucagon prepared by **a** WÜNSCH et al. (1968b); **b** FUJINO et al. (1978); **c** PROTEIN SYNTHESIS GROUP, SHANGHAI INSTITUTE OF BIOCHEMISTRY (1975); **d** MOJSOV and MERRIFIELD (1981). The scale applies to **a, c,** and **d**

The hyperglycemic response of fasted rabbits to intravenous glucagon is shown in Figs. 7 and 8. In dose–response experiments, both the time required to reach a maximum blood glucose level and the average rise in glucose were found to be equivalent for the natural and synthetic hormones.

A new stepwise solid phase synthesis of glucagon has recently been completed in the authors' laboratory (MOJSOV et al. 1982) in which the same stepwise solid phase strategy was applied, but entirely different tactics were used. A relatively acid-stable benzyl ester linkage to the polystyrene resin was selected, namely the aminoacyloxymethylphenylacetamidomethyl–resin, I (MITCHELL et al. 1976):

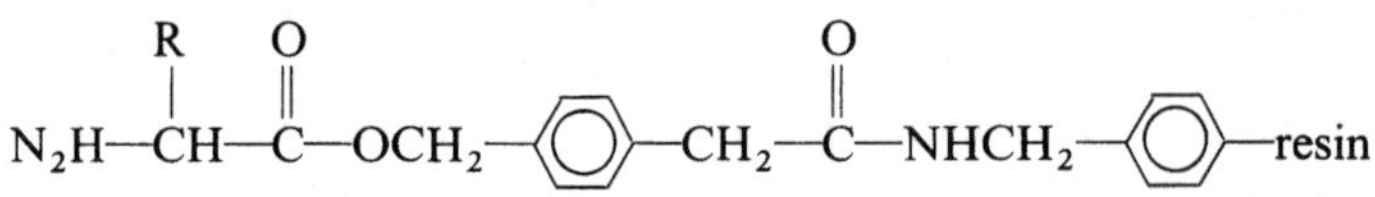

Aminoacyloxymethyl-Pam-resin I

This allowed the use of Boc groups for temporary $N^{\alpha}$ protection, and benzyl-based derivatives for most of the side chains. Couplings were with either symmetric anhydrides or HOBt esters, and cleavage and deprotection was with a recently modified HF technique (TAM et al. 1982). The stepwise assembly of the peptide chain started with 2 g Boc-Thr(Bzl)-$OCH_2$-Pam–resin (0.35 mmol) and proceeded in 79% yield to the fully protected peptide–resin. Cleavage and deprotection went in 80% yield and the chromatographic purification gave a 45% recovery. This means

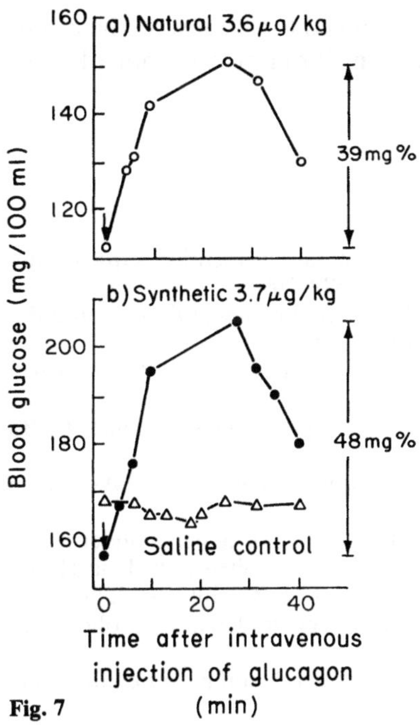


Fig. 7

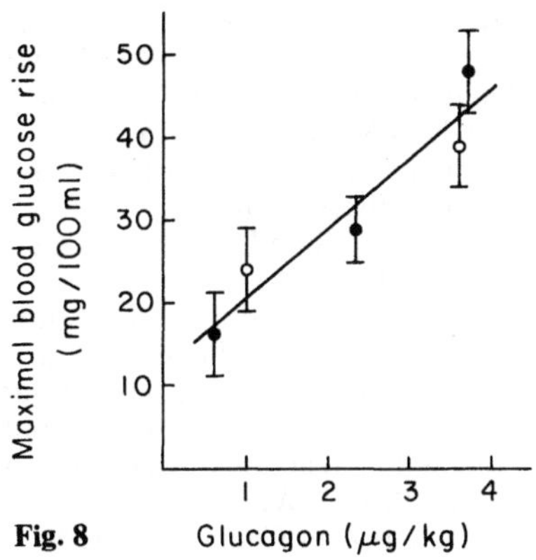


Fig. 8

**Fig. 7 a, b.** Blood glucose response to intravenous injection of **a** natural and **b** synthetic glucagon into rabbits

**Fig. 8.** Dose–response curve for glucagon in the blood glucose assay. *Open circles*, natural glucagon; *Full circles*, synthetic glucagon. Averages of three separate determinations

that a small laboratory scale run with 2 g resin can give 500 mg highly purified glucagon that is indistinguishable from natural glucagon by a host of criteria. The overall yield of pure glucagon calculated from the COOH terminal threonine residue was 28%, and it is unlikely that this can be improved by more than a factor of 2–3 by any existing method. When based on the total excess amino acid derivatives used in the coupling reactions the synthetic efficiency was 4.8%. Although it is not usually done in a small scale synthesis, it has been demonstrated in the laboratory of WIELAND (FLOR et al. 1973) that 80%–90% of the unused amino acid reagents can be recovered in a stepwise solid phase synthesis of this design.

## G. Conclusions

It has been clearly demonstrated that glucagon can be synthesized by a variety of methods. Both solution and solid phase approaches in their fragment or stepwise modes have given highly purified material that is homogeneous and indistinguishable from natural glucagon by a range of sensitive analytic methods. All reported syntheses have given crystalline peptide, and all have led to products with the full specific activity of the natural hormone. The synthetic work has provided a classical proof of the structure of glucagon. Most importantly, the synthetic methods are now adequate for the preparation of analogs designed for studies on the mechanism of action of the hormone and as potential drugs for improved control of diabetes mellitus (see Chaps. 1, 44, and 54).

## References


Beyerman HC, Bontekoe JS (1962) Synthesis of the C-terminal tetrapeptide sequence derived from glucagon. Rec Trav Chim Pays-Bas Bel 81:699–709

Beyerman HC, Bontekoe JS (1964) Synthesis of the N-terminal octapeptide sequence derived from glucagon. Rec Trav Chim Pays-Bas Bel 83:255–275

Bodansky M, Ondetti MA, Levine SD, Williams NJ (1967) Synthesis of secretin. II. J Am Chem Soc 89:6753–6757

Bregman MD, Hruby VJ (1979) Synthesis and isolation of a glucagon antagonist. FEBS Lett 101:191–194

Bromer WW, Sinn L, Behrens OK (1957) The amino acid sequence of glucagon. J Am Chem Soc 79:2807–2810

Du YC, Shen JH, Shi JP (1980) Studies on the method of solid phase fragment condensation – preparation and comparison of two kinds of thin layer resin. Sheng Wu Hua Hsueh Yu Sheng Wu Wu Li Hsueh Pao (in Chinese). Acta Biochim Biophys Sin 12:93–99

Felix AM, Jimenez MH (1973) Rapid fluorometric detection for completeness in solid phase coupling reactions. Anal Biochem 52:377–381

Flor F, Birr C, Wieland T (1973) Verwendung symmetrischer Boc-aminosäureanhydride. Liebigs Ann Chem 1601–1604

Fujino M, Wakimasu M, Shinagawa S, Kitada C, Yajima H (1978) Synthesis of the nonacosapeptide corresponding to mammalian glucagon. Chem Pharm Bull (Tokyo) 26:539–548

Haas WL, Krumkalns EV, Gerzon K (1966) Adamantyloxycarbonyl, a new blocking group. Preparation of l-adamantyl chloroformate. J Am Chem Soc 88:1988–1992

Honzl J, Rudinger J (1961) Nitrosyl chloride and butyl nitrite as reagents in peptide synthesis by the azide method; suppression of amide formation. Collect Czech Chem Commun 26:2333–2344

Horvath CG, Preiss BA, Lipsky SR (1967) Fast liquid chromatography: an investigation of operating parameters and the separation of nucleotides on pellicular ion exchangers. Anal Chem 39:1422–1428

Hylton T, Preston J, Weinstein B (1966) Synthesis of a heptapeptide sequence ($A_{20}$–$A_{26}$) of glucagon. J Org Chem 31:3400–3402

König W, Geiger R (1970) Eine neue Methode zur Synthese von Peptiden: Aktivierung der Carboxylgruppe mit Dicyclohexylcarbodiimid unter Zusatz von l-Hydroxybenzotriazolen. Chem Ber 103:788–798

Merrifield RB (1963) Solid phase peptide synthesis. J Am Chem Soc 85:2149–2154

Merrifield RB, Barany G, Cosand WL, Engelhard M, Mojsov S (1977) Some recent developments in solid phase peptide synthesis. In: Goodman M, Meienhofer J (eds) Peptides. John Wiley, New York Chichester, pp 489–502

Mitchell AR, Erickson BW, Ryabtsev MN, Hodges RS, Merrifield RB (1976) tert-Butoxycarbonylaminoacyl-4-(oxymethyl)phenylacetamidomethyl-resin, a more acid-resistant support for solid phase peptide synthesis. J Am Chem Soc 98:7357–7362

Mojsov S, Merrifield RB (1981) Solid phase synthesis of crystalline glucagon. Biochemistry 20:2950–2956

Mojsov S, Mitchell AR, Merrifield RB (1980) A quantitative evaluation of methods for coupling asparagine. J Org Chem 45:555–560

Mojsov S, Lu GS, Merrifield RB (1982) (Manuscript in preparation)

Nitecki DE, Senyk G, Williams EB, Goodman JW (1971) Immunologically active peptides of glucagon. Intrasci Chem Rep 5:295–303

Ogawa H, Yajima H (1978) Synthesis of the protected heptadecapeptide corresponding to positions 13–29 of avian glucagon (duck). Chem Pharm Bull (Tokyo) 26:1540–1548

Ogawa H, Sugiura M, Yajima H, Sakurai H, Tsuda K (1978) Synthesis of the nonacosapeptide corresponding to the entire amino acid sequence of avain glucagon (duck). Chem Pharm Bull (Tokyo) 26:1549–1557

Protein Synthesis Group, Shanghai Institute of Biochemistry (1975) Total synthesis of crystalline glucagon by the method of solid phase condensation of fragments. Sci Sin 18:745–768

Schröder E (1965a) Synthese der N-terminalen Glukagon-Teilsequenz 1–11. Liebigs Ann Chem 681:231–240

Schröder E (1965b) Synthese der Glukagon-Teilsequenzen 12–19 und 1–19. Liebigs Ann Chem 681:241–249

Schröder E (1965c) Synthese C-terminaler Glukagon-Teilsequenzen. Liebigs Ann Chem 688:250–264

Schröder E (1967) Attempts for a synthesis of glucagon. In: Beyerman HC, van de Linde A, Maassen van den Brink W (eds) Peptides. North Holland, Amsterdam, pp 245–275

Schwyzer R, Sieber P (1963) Total synthesis of Adrenocorticotrophic hormone. Nature 199:172–174

Sieber P, Iselin B (1968) Peptidsynthesen unter Verwendung der 2-(p-Diphenyl)-isopropyloxycarbonyl (Dpoc)-Aminoschutzgruppe. Helv Chim Acta 51:622–632

Staub A, Sinn L, Behrens OK (1955) Purification and crystallization of glucagon. J Biol Chem 214:619–632

Sundby F, Frandsen EK, Thomsen J, Kristiansen K, Brunfeldt K (1972) Crystallization and amino acid sequence of duck glucagon. FEBS Lett 26:289–292

Tam JP, Heath W, Merrifield RB (1982) Improved deprotection in solid phase peptide synthesis: Removal of protecting groups from synthetic peptides by an $S_N2$ mechanism with low concentrations of HF in dimethylsulfide. Tetrahedron Lett. 23:4435–4438

Wang SS (1973) p-Alkoxybenzyl alcohol resin and p-alkoxy-benzyloxycarbonyl hydrazide resin for solid phase synthesis of protected peptide fragments. J Am Chem Soc 95:1328–1333

Weygand F (1968) Bromacetyliertes Polystyrol als Ausgangsmaterial für Peptidsynthesen am festen Träger. In: Bricas E (ed) Peptides. North Holland, Amsterdam Oxford New York, pp 183–184

Weygand F, Hoffmann D, Wünsch E (1966) Peptidsynthesen mit Dicyclohexylcarbodiimid unter Zusatz von N-Hydroxysuccinimide. Z Naturforsch 21b:426–428

Windridge GC, Jorgensen EC (1971) Racemization in the solid phase synthesis of histidine-containing peptides. Intrascience Chem Rep 5:375–380

Wright DE, Hruby VJ, Rodbell M (1980) Preparation and properties of glucagon analogs prepared by semi-synthesis from CNBr-glucagon. Biochim Biophys Acta 631:49–58

Wünsch E, Drees F (1966) Zur Synthese des Glucagons. X. Darstellung der Sequenz 22–29. Chem Ber 99:110–120

Wünsch E, Wendlberger G (1968) Zur Synthese des Glucagons. XVIII. Darstellung der Gesamtsequenz. Chem Ber 101:3659–3663

Wünsch E, Zwick A (1964) Zur Synthese des Glucagons. V. Darstellung der Sequenz 16–19. Chem Ber 97:3312–3316

Wünsch E, Zwick A (1966) Zur Synthese des Glucagons. IX. Darstellung der Sequenz 9–15. Chem Ber 99:105–109

Wünsch E, Jaeger E, Scharf R (1968a) Zur Synthese des Glucagons. XIX. Chem Ber 101:3664–3670

Wünsch E, Wendlberger G, Jaeger E, Scharf R (1968b) Zur Total-Synthese des biologisch-aktiven Pankreas-Hormons Glucagon. In: Bricas E (ed) Peptides. North Holland, Amsterdam Oxford New York, pp 229–236

Zervas L, Borovas D, Gazis E (1963) New methods in peptide synthesis. I. Tritylsulfenyl and o-nitrophenylsulfenyl groups as N-protecting groups. J Am Chem Soc 85:3660–3666

CHAPTER 3

# The Conformation of Glucagon

T.L. Blundell

## A. Introduction

The biological activity of glucagon is mediated through binding, with high affinity and specificity, to a membrane receptor, implying extensive and well-defined intermolecular interactions (see Chap. 13). However, in dilute aqueous solutions glucagon has little defined secondary structure and almost certainly exists as a population of conformers in equilibrium. The formation of the receptor–hormone complex must involve either selection of one conformer from the population or induction of a conformer as the interaction takes place. Whatever the mechanism, the definition of the receptor-bound conformer, as well as the nature of the hormone–receptor interactions must be a primary objective in the understanding of the biology of glucagon and the design of glucagon agonists and inhibitors (Blundell 1979; Blundell and Humbel 1980). Unfortunately, the receptor has yet to be defined biochemically and so direct study of the receptor–hormone complex is not possible at present. Instead we must examine the conformation of glucagon in aqueous solution, in crystals, in lipid micelles and other environments in order to establish the nature of the conformational dependence on intermolecular interactions. Here, I first describe recent developments, especially in the use of X-ray diffraction and proton nuclear magnetic resonance (NMR) spectroscopy which have allowed description of the conformation in great detail under varied conditions. I then discuss the relevance of these conformations to the molecular biology of glucagon.

## B. The Crystal Structure

### I. Crystals

It is convenient to begin a description of the conformation of glucagon by reviewing the crystal structure analysis, for in the crystals the glucagon molecule has less flexibility, as it is stabilised by numerous intermolecular interactions, some of which may be maintained in other environments. The relatively ordered state allows a medium resolution analysis using the techniques of protein X-ray crystallography.

King (1959) showed that rhombic dodecahedral crystals of glucagon contain 12 molecules packed with cubic symmetry (space group $P2_13$). Preliminary results (King 1965; Blanchard and King 1966) suggested a packing of $\alpha$-helical rods and

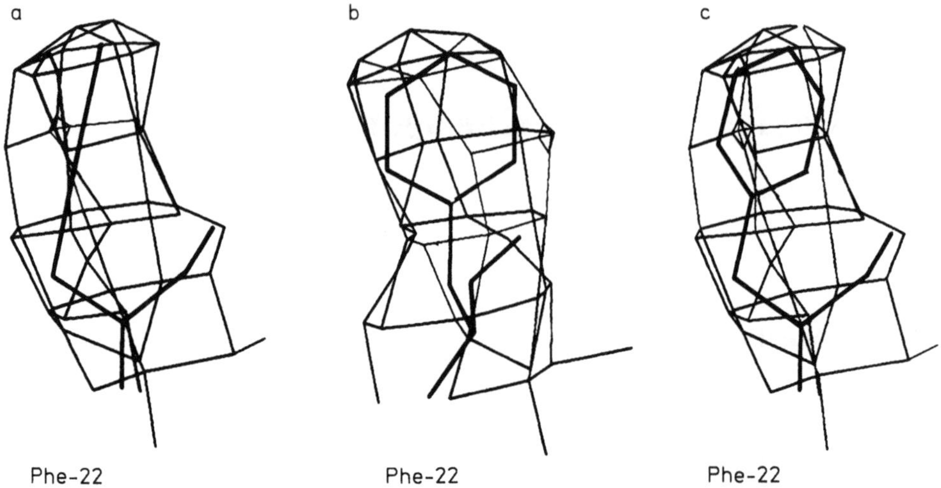


**Fig. 1 a–c.** The electron density at ~3 Å resolution for Phe-22. Orthogonal views (**a, b**) of the refined fit of the side chain to the density; rotation of 50° (**c**) around the $\beta$–$\gamma$ bond (viewed in the same direction as **a**) leaves the side chain in the density, indicating rotational disorder in the crystals

this was detailed by SASAKI et al. (1975) who carried out a medium resolution X-ray analysis using the methods of isomorphous replacement and anomalous scattering.

Crystals of glucagon are most easily obtained at high pH (~9.2) where the glucagon is more soluble than at physiological pH. However, SASAKI et al. (1975) showed that these crystals undergo a phase change on lowering the pH to around 6.5, involving a retention of the cubic symmetry, but a shortening of the cubic cell parameter from 47.9 to 47.1 Å. Similar crystals are obtained by careful crystallisation at pH 6.5. A further crystal form of the same symmetry, but of a different cell dimension (48.7 Å) is found at lower pH (~3) (DOCKERILL 1978). These observations are a reflection of the flexibility of the glucagon conformation and especially its dependence on the pH which is also observed in solution. Various metal ion complexes cause similar changes in the crystals, giving rise to complications in the use of the method of isomorphous replacement.

Even the best crystals are disordered relative to those of most globular proteins. The nominal resolution is ~3 Å, which implies that the general conformation can be defined although the details are often unclear. For instance, the precise orientation of the carbonyls of the peptide groups and the side chains are not defined. Figures 1 and 2 show some typical side chain electron densities. Figure 1 shows that the phenyl ring of Phe-22 may be rotated by ±50° around the $\beta$–$\gamma$ bond from the optimal position and still remain in the electron density. As the resolution of the crystals is only ~3 Å, this implies that even in the better ordered parts of the crystal structure there is considerable disorder. Other groups such as Leu-26 have a poorly defined conformation and the crystal probably contains a population of each of the three conformers with a staggered arrangement around the $\beta$–$\gamma$ bond. Other parts of the molecule are even more disordered and these include residues 1–4 at the $NH_2$-terminus and 28 and 29 at the COOH-terminus as well as the side chains of Asp-9, Lys-12, and Arg-18.

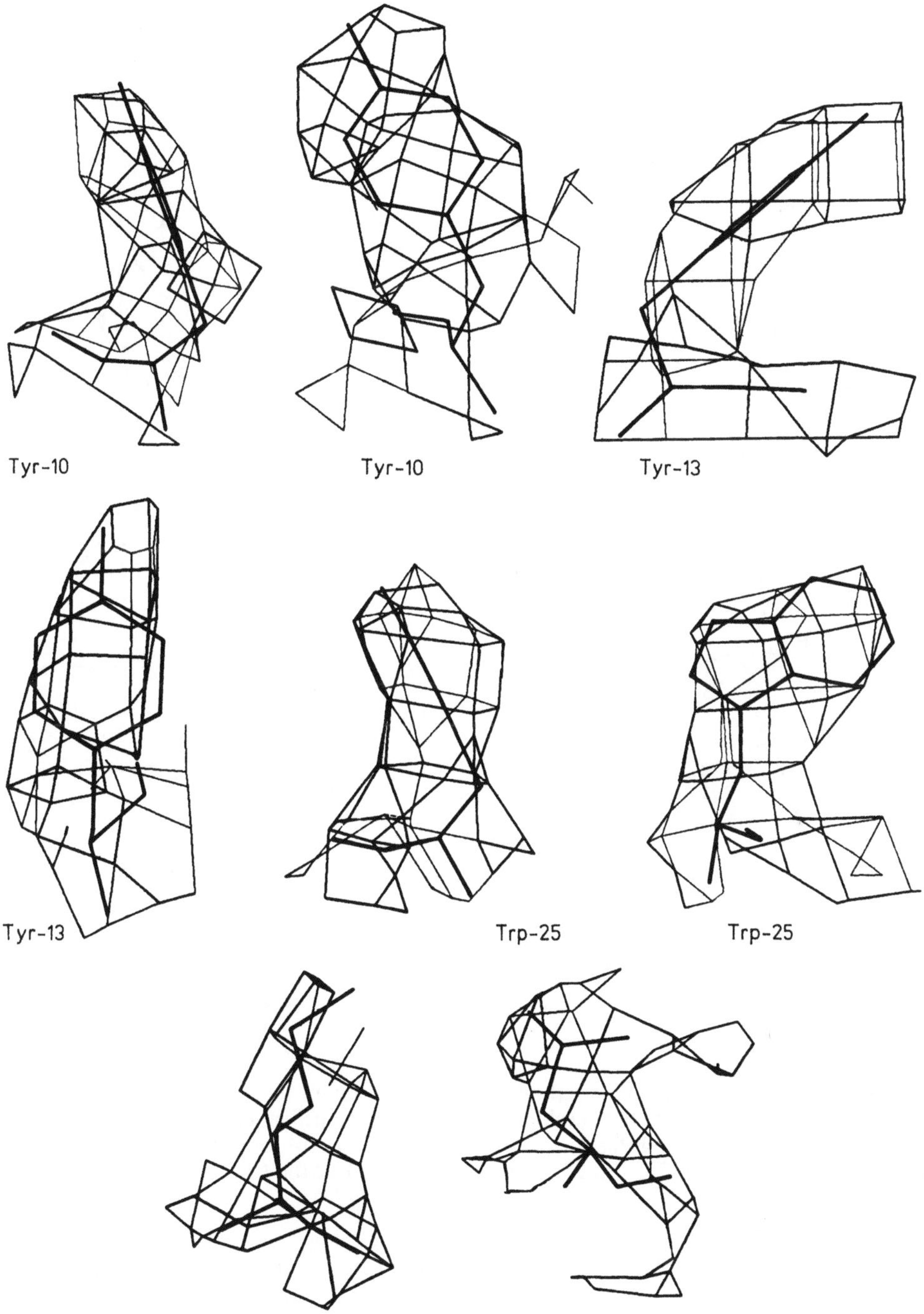


**Fig. 2.** The electron densities for a selection of side chains at ~3 Å resolution with the side chain positions from the least-squares refinement. Note that the ends of some side chains, e.g. Met-27 $C^{\varepsilon}$ have no density, indicating complete disorder, while others such as Leu-26 are not unambiguously defined, indicating several possible conformations

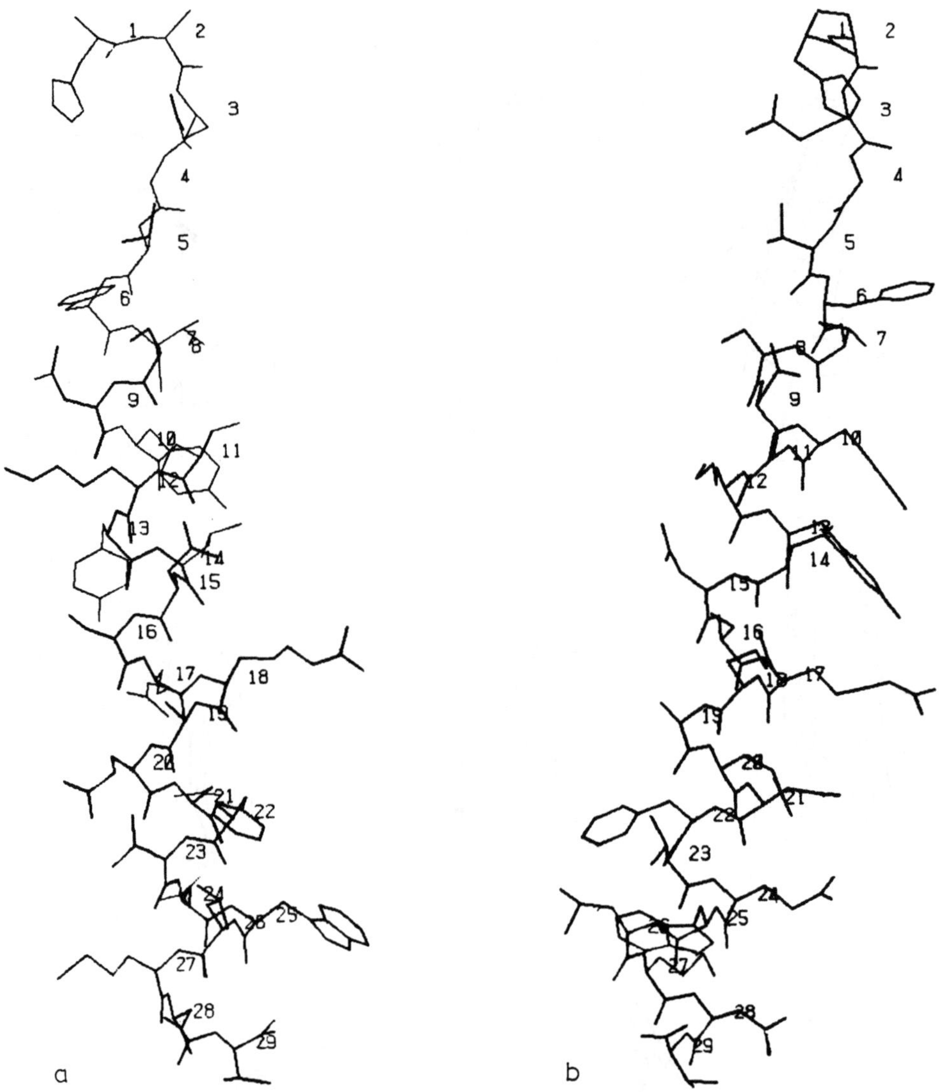


**Fig. 3 a, b.** Two orthogonal views (**a, b**) of the conformation of the glucagon molecule (protomer) in the crystals. Note the amphipathic nature of the helical conformer with two hydrophobic regions involving Phe-6, Tyr-10, Tyr-13, and Leu-14 towards the $NH_2$ terminus and Ala-19, Phe-22, Val-23, Trp-25, Leu-26, and Met-27 at the COOH terminus

The model defined by X-ray analysis and described in this chapter is an average of the conformers present. In fact, the poor resolution of the crystals poses problems in the refinement of the molecular structure using least-squares techniques. The model presented by SASAKI et al. (1975) was a best fit to the electron density at pH 6.5, obtained by the method of isomorphous replacement with anomalous scattering using an optical comparator (see BLUNDELL and JOHNSON 1976 for a review). The first set of coordinates deposited with the Brookhaven Protein Structure

Data Bank were refined using the methods considered optimal in 1975: cycles of real space refinement and calculation of electron density. More recently, J. MOREIRA, I.J. TICKLE, and T.L. BLUNDELL (1981) unpublished work) have shown that restrained least-squares refinement gives improved coordinates and these are used in this discussion.

## II. Protomer Conformation

Figure 3 illustrates the conformation of the glucagon molecule (a protomer) in the cubic crystals. Table 1 gives the main chain torsion angles ($\phi$, $\psi$, and $\omega$) which show that residues 6–28 are in an approximately $\alpha$-helical conformation. Table 2 gives the lengths and angles subtended at the peptide oxygen and hydrogen atoms of possible hydrogen bonds. The variations in lengths and angles are partly a reflection of the conformational disorder although the existence of a $3_{10}$ helix in residues 5–11 is probably real. At Gly-4 the chain becomes poorly defined, indicating

**Table 1.** The torsion angles defining the main chain conformation of the glucagon molecule from the medium resolution X-ray analysis of crystals at pH ~ 6.5

| Residue | $\Phi$ | $\Psi$ | $\omega$ |
|---|---|---|---|
| His-1 | – | 117 | –173 |
| Ser-2 | – 90 | 6 | –177 |
| Gln-3 | –109 | – 69 | –177 |
| Gly-4 | 114 | 137 | –177 |
| Thr-5 | –162 | 8 | 178 |
| Phe-6 | – 53 | – 29 | 178 |
| Thr-7 | – 31 | – 49 | –179 |
| Ser-8 | – 39 | – 61 | –179 |
| Asp-9 | – 63 | – 14 | 174 |
| Tyr-10 | – 71 | – 56 | 177 |
| Ser-11 | – 52 | – 28 | 178 |
| Lys-12 | – 73 | – 53 | 178 |
| Tyr-13 | – 46 | – 39 | –179 |
| Leu-14 | – 68 | – 22 | 180 |
| Asp-15 | – 76 | – 42 | –179 |
| Ser-16 | – 62 | – 30 | 180 |
| Arg-17 | – 85 | – 46 | 179 |
| Arg-18 | – 49 | – 42 | –176 |
| Ala-19 | – 79 | – 44 | –180 |
| Gln-20 | – 66 | – 31 | 179 |
| Asp-21 | – 80 | – 26 | 179 |
| Phe-22 | – 77 | – 22 | 177 |
| Val-23 | – 72 | – 16 | 178 |
| Gln-24 | – 76 | – 48 | 178 |
| Trp-25 | – 67 | – 55 | –178 |
| Leu-26 | – 51 | – 54 | –176 |
| Met-27 | – 76 | – 62 | –179 |
| Asn-28 | – 35 | – 66 | –180 |
| Thr-29 | 174 | –169 | |

**Table 2.** Possible hydrogen bonds between peptide N-H and C=O groups

| Donor | Acceptor | N...O | N...O=O | N...H...O | H...O |
|---|---|---|---|---|---|
| 8 | 5 | 2.6 | 129 | 140 | 1.8 |
| 9 | 6 | 2.6 | 142 | 122 | 2.0 |
| 10 | 6 | 3.4 | 150 | 146 | 2.5 |
| 10 | 7 | 2.6 | 123 | 122 | 1.9 |
| 11 | 7 | 2.9 | 147 | 138 | 2.1 |
| 12 | 8 | 2.9 | 148 | 138 | 2.2 |
| 13 | 9 | 2.6 | 136 | 143 | 1.7 |
| 14 | 10 | 2.7 | 165 | 155 | 1.8 |
| 15 | 12 | 2.5 | 123 | 147 | 1.6 |
| 16 | 12 | 3.3 | 167 | 155 | 2.4 |
| 17 | 13 | 3.3 | 152 | 138 | 2.5 |
| 18 | 14 | 3.2 | 141 | 162 | 2.2 |
| 19 | 15 | 3.0 | 154 | 149 | 2.1 |
| 20 | 16 | 3.2 | 134 | 150 | 2.3 |
| 21 | 17 | 3.0 | 156 | 153 | 2.0 |
| 22 | 18 | 3.5 | 128 | 127 | 2.3 |
| 23 | 19 | 3.0 | 147 | 133 | 2.3 |
| 24 | 20 | 3.4 | 131 | 115 | 2.8 |
| 25 | 21 | 3.2 | 144 | 143 | 2.3 |
| 26 | 22 | 2.6 | 152 | 174 | 1.6 |
| 27 | 23 | 2.8 | 129 | 151 | 1.9 |
| 28 | 24 | 3.2 | 132 | 133 | 2.4 |
| 29 | 25 | 3.5 | 148 | 148 | 2.6 |

that the glycine endows the chain with flexibility; consequently the placing of the $NH_2$-terminus is somewhat arbitrary.

The α-helical conformation has few stabilising interactions other than the hydrogen bonds of the main chain. However, certain hydroxyl groups, in particular those of Thr-29 and Thr-5, hydrogen bond to carbonyl groups, and the accessibility of other residues – Tyr-10, Tyr-13, Leu-14, Arg-17, Asp-21, Phe-22, and Gln-24 – to solvent is significantly decreased through van der Waals' contacts in the helical conformer (R. WEST and B. GELLATLY 1980, unpublished work). Table 3 gives the accessibility of side chains to a water molecule calculated by the methods of LEE and RICHARDS (1971) and FINNEY (1978). The helical conformer brings together the hydrophobic groups in two "patches" comprising Phe-6, Tyr-10, Tyr-13, and Leu-14 towards the $NH_2$-terminus, and Ala-19, Phe-22, Val-23, Trp-25, Leu-26, and Met-27 towards the COOH-terminus. These hydrophobic patches are responsible for the intermolecular interactions which stabilise the crystal form.

## III. Trimer Conformation

The cubic arrangement of mainly α-helical glucagon molecules involves three-fold rotation and two-fold screw axes of symmetry leading to a continuous oligomeric structure. This may be considered in terms of a series of trimers which are not mutually exclusive and each of which contains a perfect three-fold axis. One such trimer (trimer 1) is shown in Fig. 4. The three molecules are related by applying

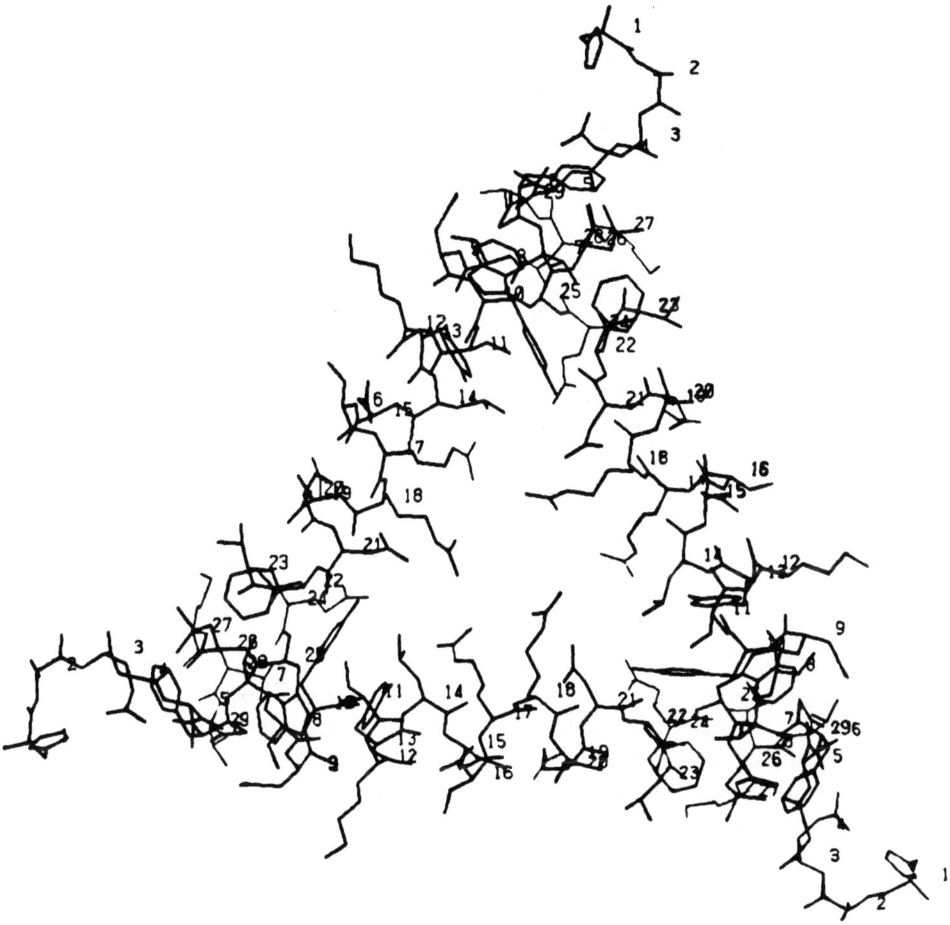

**Fig. 4.** A trimeric arrangement (trimer 1) found in glucagon crystals formed from heterologous interactions between the hydrophobic patches at the $NH_2$ and COOH termini. The central hydrophilic region contains charged groups such as Asp-21, Arg-17, and Arg-18

twice the symmetry operation: $x,y,z \rightarrow 1-y,\ ½+z,\ ½-x$, where $x$, $y$, and $z$ are fractional coordinates. In the trimer, parts of the hydrophobic region towards the $NH_2$-terminus involving Phe-6, Tyr-10, Tyr-13, and Leu-14 are brought into contact with the hydrophobic region at the COOH-terminus involving Phe-22, Trp-25, and Leu-26 (Fig. 5). The centre of the "triangular" trimer is hydrophilic and includes charged groups Asp-9, Arg-17, Arg-18, and Asp-21, most of which appear to be relatively disordered. There is evidence for a hydrogen bond interaction between the Tyr-10 side chain hydroxyl and the Arg-18 main chain. Solvent accessibilities are listed in Table 3 and side chains in van der Waals' contact in Table 4. The angle between the helices is $\sim 103°$.

A second trimer (trimer 2) is shown in Fig. 6, in which molecules are related by applying twice the symmetry operation $x,y,z \rightarrow z,x,y$. The contact region involving only the COOH-terminal hydrophobic region – residues Ala-19, Phe-22, Val-23,

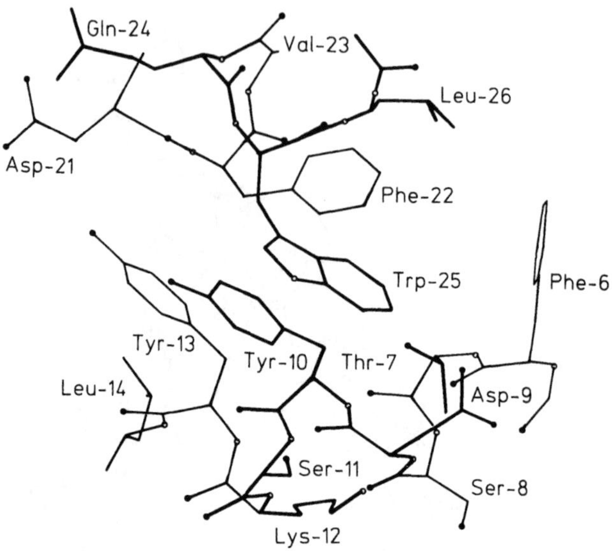


**Fig. 5.** The intermolecular, mainly hydrophobic interactions which stabilise trimer 1, shown in Fig. 4

Leu-26, and Met-27 – is shown in Fig. 7. There is a hydrogen bond between Gln-20 side chain amide and Ser-16 side chain hydroxyl groups. Table 3 gives the change in accessibility of these residues to solvent compared with the isolated helical molecule. The angle between the helices in the trimer is 76°.

In view of the possibility of either of these trimers existing in solution (see Sect. C.II) it is relevant to consider factors which may affect their stability. Clearly the major interactions are nonpolar and hydrophobic. Trimer 1 involves a decrease of accessibility to solvent of 18.4% of the groups compared with an isolated helical conformer. The decrease of accessibility for trimer 2 is 14%. If the free energy gain by hydrophobic interactions is dependent only on the surface area made inaccessible to solvent, then trimer 1 would be most stabilised. However, trimer 1 involves the necessary formation of helix between residues 6 and 26 involving unfavourable entropy terms, whereas the interactions between molecules in trimer 2 require helix formation only between residues 16 and 27 (see Sect. C.II for further discussion).

Trimer 1 would appear to be much more pH dependent in its formation. In particular, protonation of Asp-9 or Asp-21 might disturb the balance of charge and may be expected to destabilise this trimer in acidic conditions. Perhaps of greater importance would be the ionisation in alkaline solutions (above pH 10) of tyrosyl hydroxyl groups, especially that of Tyr-10 which is buried in trimer 1. These observations may be the cause of difficulty in forming crystals outside the range, say pH 3 to pH 11, when there would be few species of the correct charge.

The dependence of the crystal cell dimensions on the pH in the range 7.5–8.5 probably arises from rather different interactions within the crystal. The $NH_2$-terminus occupies a complicated polar region including two COOH termini, in addition to Asp-9 and Asp-15 of adjacent molecules. A change of pH from ~9.0 to ~6.0 will lead to protonation of the $N^{\alpha}$ and the $N^{\varepsilon}$ of His-1, leading to small rearrangements. A similar transition occurs in the crystals of adenylate kinase, in which

**Table 3.** The accessibility to water (assumed to be a sphere of 1.4 Å radius) of amino acid residues in different conformations. The values for random coil have been calculated from those of a residue (X) in a tripeptide Ala-X-Ala in LEE and RICHARDS (1971). The helical conformer and the two trimers (1 and 2) are those defined by X-ray analysis of crystals at pH 6.5 before restrained refinement

| Residue | Accessibility[a] | | | | | | |
|---|---|---|---|---|---|---|---|
| | LR Random coil | LR Helical conformer | F Helical conformer | LR Trimer 1 | F Trimer 1 | LR Trimer 2 | F Trimer 2 |
| His-1 | 186 | 198 | 123 | 198 | 123 | 198 | 123 |
| Ser-2 | 108 | 121 | 73 | 121 | 73 | 121 | 73 |
| Gln-3 | 162 | 160 | 102 | 160 | 102 | 160 | 102 |
| Gly-4 | 72 | 37 | 43 | 37 | 43 | 37 | 43 |
| Thr-5 | 123 | 117 | 87 | 117 | 87 | 117 | 87 |
| Phe-6 | 165 | 152 | 119 | 99[c] | 102 | 152 | 119 |
| Thr-7 | 123 | 107 | 84 | 104 | 84 | 107 | 84 |
| Ser-8 | 108 | 78 | 67 | 78 | 67 | 78 | 67 |
| Asp-9 | 133 | 103 | 77 | 91 | 74 | 103 | 77 |
| Tyr-10 | 183 | 97[b] | 92 | 4[c] | 36 | 97 | 92 |
| Ser-11 | 108 | 81 | 70 | 62 | 67 | 81 | 70 |
| Lys-12 | 173 | 160 | 107 | 160 | 107 | 160 | 107 |
| Tyr-13 | 183 | 122[b] | 108 | 80[c] | 92 | 122 | 108 |
| Leu-14 | 135 | 61[b] | 79 | 12[c] | 68 | 61 | 79 |
| Asp-15 | 133 | 115 | 85 | 115 | 85 | 115 | 85 |
| Ser-16 | 108 | 74 | 55 | 74 | 55 | 44[c] | 55 |
| Arg-17 | 205 | 148[b] | 108 | 112[c] | 108 | 148 | 108 |
| Arg-18 | 205 | 196 | 128 | 123[c] | 114 | 151[c] | 117 |
| Ala-19 | 93 | 52 | 45 | 52 | 45 | 13[c] | 30 |
| Gln-20 | 162 | 130 | 92 | 130 | 92 | 96[c] | 90 |
| Asp-21 | 133 | 76[b] | 71 | 40[c] | 71 | 76 | 77 |
| Phe-22 | 165 | 100[b] | 93 | 50[c] | 64 | 34[c] | 64 |
| Val-23 | 125 | 93 | 75 | 93 | 75 | 14[c] | 45 |
| Gln-24 | 162 | 96[b] | 71 | 76 | 71 | 96 | 70 |
| Trp-25 | 202 | 209 | 146 | 98[c] | 105 | 209 | 146 |
| Leu-26 | 135 | 140 | 98 | 108[c] | 84 | 76[c] | 70 |
| Met-27 | 162 | 124 | 99 | 124 | 99 | 30[c] | 58 |
| Asn-28 | 133 | 96 | 72 | 96 | 72 | 93 | 72 |
| Thr-29 | 153 | 163 | 104 | 163 | 104 | 140 | 101 |
| Totals | 4,268 | 3,406 | 2,573 | 2,778 | 2,369 | 2,929 | 2,419 |

[a] Accessibilities are calculated by A. WEST and B. GELLATLY (1979, unpublished work) using the methods of LEE and RICHARDS (1971) indicated by LR and by FINNEY (1978) indicated by F. The methods give very different results, indicating different assumptions in the computational approach

[b] Significant accessibility decrease for side chain on formation of the helical conformer

[c] Significant accessibility decrease on trimer formation

an imidazole and a carboxylate group are also close together. In both glucagon and adenylate kinase, a similar transition is effected by metal ions. For example, binding of $Ag^{+}$ and $Hg^{2+}$ between the imidazole of His-1 and the sulphur of Met-27 stabilises the lower pH form; the metal cations appear to be playing the role of a proton.

**Table 4.** Residues and atoms in van der Waal's contact in trimers 1 and 2 (Fig. 4–7)

| | Residue | Atom(s) | Residue | Atom(s) |
|---|---|---|---|---|
| Trimer 1 | Leu-26 | $C^{\delta_1}$ | Phe-6 | $C^{\varepsilon_2}$, $C^{\zeta}$ |
| | Trp-25 | $C^{\beta}$, $C^{\gamma}$, $C^{\zeta_2}$ | Tyr-10 | $C^{\gamma_2}$ |
| | | $C^{\gamma}$, $C^{\delta_1}$, $N^{\varepsilon_1}$ | Tyr-13 | $C^{\beta}$, $C^{\gamma}$, $C^{\delta_2}$ |
| | | $C^{\zeta_2}$, $C^{\varepsilon}H2$ | Asp-9 | $O2^{\delta_2}$ |
| | | $C^{\varepsilon}H2$ | Phe-6 | O |
| | Phe-22 | $C^{\delta_2}$ | Tyr-10 | $C^{\delta_2}$ |
| | Arg-18 | O | Tyr-10 | OH |
| Trimer 2 | Met-27 | $C^{\beta}$ | Phe-22 | $C^{\varepsilon_1}$, $C^{\delta_1}$ |
| | Leu-26 | $C^{\delta_2}$ | Leu-26 | $C^{\delta_2}$ |
| | Val-23 | $C^{\gamma_1}$ | Val-23 | $C^{\gamma_2}$ |
| | | $C^{\gamma_1}$ | Ala-19 | $C^{\beta}$ |
| | | $C^{\gamma_1}$ | Phe-22 | $C^{\delta_1}$ |
| | Gln-20 | $O^{\varepsilon_1}/N^{\varepsilon_2}$ | Ser-16 | $O^{\gamma}$ |

## C. The Solution Structure

### I. Monomer

Although some earlier experiments favoured a globular structure for the monomer, most recent experimental data indicate a predominantly flexible structure with little defined secondary structure. Circular dichroism (PANIJPAN and GRATZER 1974; SRERE and BROOKS 1969) and fluorescence (EDELHOCH and LIPPOLDT 1969) are consistent with 10%–15% of helix.

Recent proton NMR studies on dilute aqueous solutions (BOESCH et al. 1978; WAGMAN et al. 1980; WAGMAN 1981) are consistent with a largely unstructured and flexible chain. This is indicated by the narrow line-widths and a single set of chemical shifts which, with the exception of the aromatic tyrosines, are independent of temperature and phosphate concentration. Those chemical shifts which are pH dependent are attributable to protonation (WAGMAN 1981). A mainly flexible and nonspherical structure is also indicated by viscosity measurements (WAGMAN 1981), although not by the earlier results of EPAND (1971); this intrinsic viscosity is found to be insensitive to denaturating agents such as urea (WAGMAN 1981). Most chemical shifts are not affected by 8 *M* urea and are similar to those expected for random coil, the exceptions being those of Val-23 methyl resonances, Val-23 $H^{\beta}$ and Leu-26 $H^{\gamma}$. These have been attributed by BOESCH et al. (1978) to a proximity between the side chain of Val-23 and the aromatic ring of Trp-25 in about 20% of the population of conformers, and a similar interaction occurs in a peptide fragment corresponding to residues 22–26; these residues cannot have a helical conformation. There may also be some structure in a proportion of the conformers involving Tyr-10 and Tyr-13 which unfolds with increasing temperature (WAGMAN 1981). In summary, it appears that in dilute aqueous solutions, the glucagon monomer exists either as a mainly flexible structure with a small section of ordered

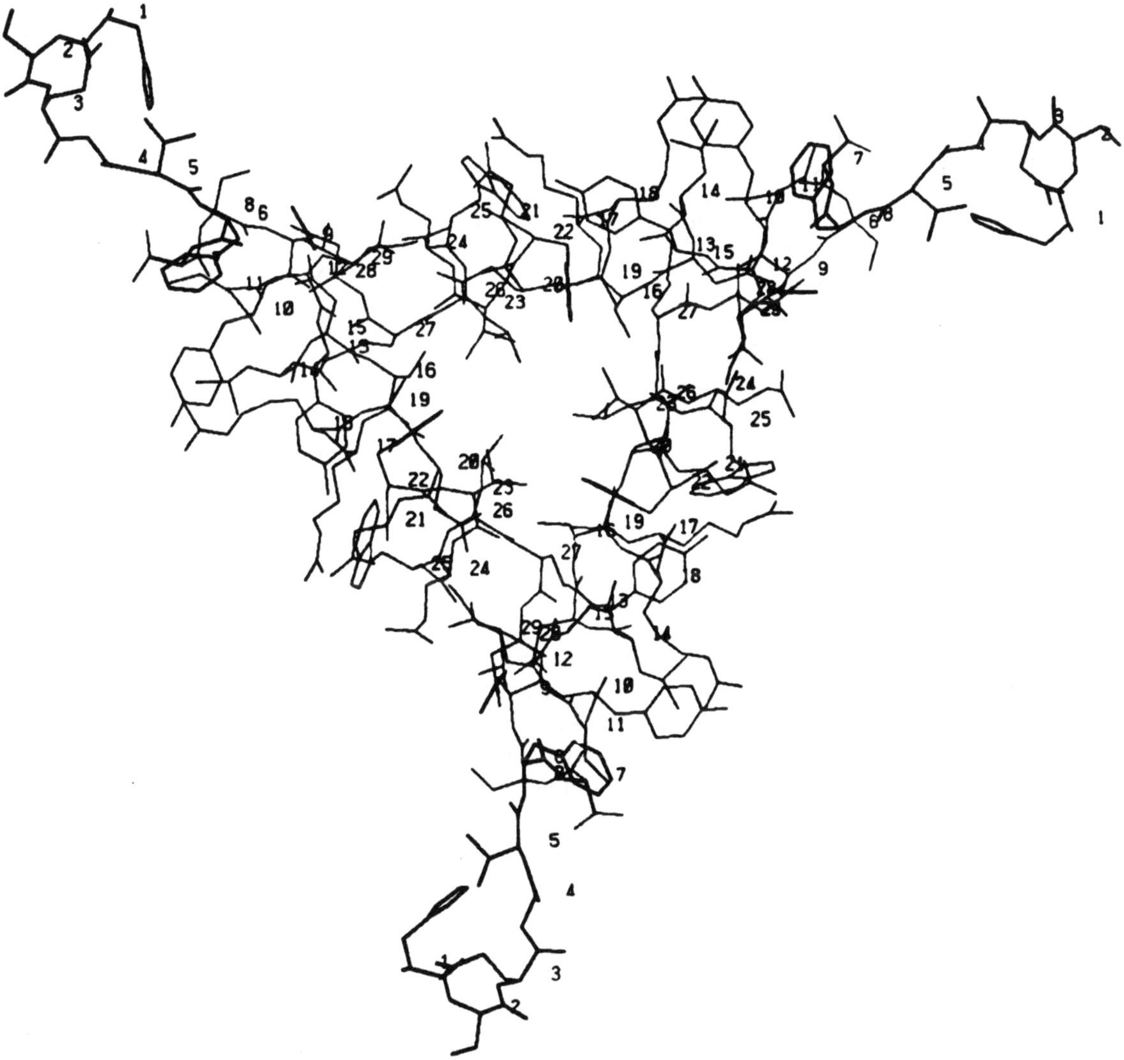

**Fig. 6.** A trimeric arrangement (trimer 2) found in glucagon crystals involving hydrophobic contacts between equivalent regions towards the COOH terminus (Phe-22, Val-23, Leu-26, and Met-27). These interactions may be maintained in the absence of helix in residues 6–18, and such a partly helical trimer probably exists in concentrated aqueous solutions

secondary structure, or as a series of conformers in equilibrium, none of which dominates and some of which may involve structure in residues Tyr-10 to Tyr-13 and Phe-22 to Leu-26.

## II. Trimers

The concentration dependence of most spectroscopic probes – circular dichroism, optical rotary dispersion, optical detection of magnetic resonance and NMR – in-

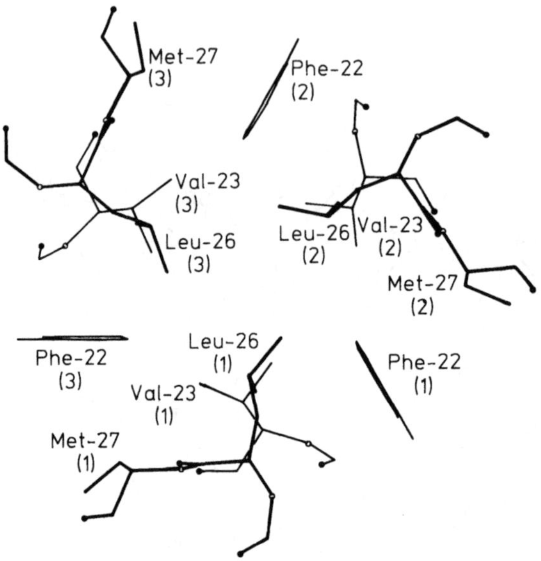


**Fig. 7.** The hydrophobic interactions which stabilise trimer 2 shown in Fig. 6

dicate that glucagon self-associates with an accompanying change in secondary structure. Although the nature and structure of the oligomers has been the subject of much debate, it now appears that at least in the pH ranges 2–4.5 and 9–11, a well-defined trimer is formed which is partly helical and involves intermolecular interactions similar to those of the crystalline trimer 2 (Wagman et al. 1980; Wagman 1981).

Circular dichroism and optical rotary dispersion (Srere and Brooks 1969; Gratzer and Beaven 1969) indicate that conformers of 35% α-helical content are induced in concentrated solutions. Fluorescence studies of rhodamine 6G dye bound to glucagon at pH 10.6 (Formisano et al. 1978a) and optical detection of magnetic resonance (Ross et al. 1976, 1977) favour preferential formation of secondary structure in the COOH terminal region close to the tryptophan (Trp-25). Proton NMR studies in concentrated solutions at high pH also suggest that aggregation involves COOH terminal residues and a detailed assignment of resonance (Wagman 1981; Wagman et al. 1980) has shown that the largest chemical shifts and line-width broadening occur in COOH terminal residues (22–29).

Blanchard and King (1966) postulated trimer formation in solution on the basis of their early X-ray studies of crystals, and this was confirmed by Gratzer et al. (1972) who cross-linked glucagon solutions using dimethyl suberimidate. Although Swann and Hammes (1969) had favoured a monomer–dimer–hexamer equilibrium on the basis of sedimentation studies, the concentration dependence of the chemical shifts in the proton NMR can be fully accounted for by a two-state model involving a monomer–trimer equilibrium (Wagman et al. 1980).

The constants for trimerisation ($K_\alpha$) at pH 10.6 for glucagon in $D_2O$ with 0.2 $M$ sodium phosphate are $1.08(\pm 0.42)\times 10^6 M^{-2}$ at 30 °C, and $5.86(\pm 1.07)\times 10^4 M^{-2}$ at 50 °C (Wagman 1981). Similar values are obtained from the concentration dependence of the circular dichroism at different temperatures (FORMISANO et al. 1978b). Analysis of the temperature dependence of the association constants (WAGMAN 1981; FORMISANO et al. 1978b) and calorimetry (JOHNSON et al. 1979) show that glucagon association is characterised by large negative values of both $\Delta H^0$ and $\Delta S^0$ which are temperature dependent, decreasing with increasing temperature. $\Delta C_p{}^0$ has a large negative value, indicating hydrophobic interactions (FORMISANO et al. 1978b). These data are consistent with a dominant negative entropy of a coil–helix transition on association which is greater than the positive entropy change resulting from hydrophobic interactions (WAGMAN 1981). The negative enthalpy change makes the overall free energy change favourable. The formation of trimers is favoured by phosphate (FORMISANO et al. 1978b; WAGMAN 1981), is inhibited by urea (WAGMAN 1981) and is strongly pH dependent in the pH range 9.5–10.5 where $K_a$ decreases by a factor of 25. However, the association is similar at pH 2 and 10.6 (WAGMAN 1981).

These data are consistent with a trimeric structure similar to trimer 2 involving hydrophobic interactions but with a well-defined structure only in the COOH terminal region. This is supported by nuclear Overhauser enhancement evidence, indicating no interactions between residues widely separated in the glucagon sequence of the kind found in trimer 1. To define further the interactions in the trimer, WAGMAN (1981) has calculated the secondary shifts mainly resulting from ring current effects and the transverse relaxation times which might result from interactions observed in the crystal trimer 2, and in closely related structures obtained by changing the conformation to minimise the energy. In general there is a reasonable correlation of the observed shifts with those calculated for trimer 2, especially for Ala-19, Val-23, Met-27, and Thr-29. Although some observed shifts such as those for the Leu-26 methyl carbons are not in good agreement with those calculated on the basis of trimer 2, an improved agreement is obtained by small rotations around bonds such as the $\beta$–$\gamma$ bond of Phe-22 (WAGMAN 1981). However, these rotations may not represent a real difference between solution and crystal structure as the crystals are clearly poorly ordered (resolution $\sim$3 Å) and rotations of up to $\pm 50°$ may be made consistent with the electron density (see Fig. 1).

Although the evidence is clear that trimer 1 does not exist in acidic or alkaline solutions, a small fraction of such conformers may exist in neutral conditions. The increase in oligomer formation on decreasing the pH corresponds to the p$K$ of the tyrosines and the same tyrosine involvement in self-association is observed at lower pH. However, WAGMAN et al. (1980) find no evidence for an abnormal p$K$ for the tyrosines which would be expected of trimer 1.

In summary, in concentrated solutions there is an equilibrium between a largely unstructured polypeptide and a partly helical trimer. The helical structure is stabilised by hydrophobic interactions and possibly some hydrogen bonds, such as that between Ser-16 and Gln-20. On increasing concentrations – and most easily at neutral solutions – these trimers become involved in further intermolecular interactions leading to an increased $\alpha$-helical content and crystallisation.

## III. Fibrils

On standing in acid solution, the viscosity of glucagon increases and a birefringent gel is formed (GRATZER et al. 1968; BEAVEN et al. 1969). Sedimentation studies indicate the formation of large aggregates. On further standing or warming, a precipitate appears which has the appearance of long fibrils in the electron microscope. Infrared spectra of the gel, of solid films and of the precipitated material show that in all these states the glucagon is in the form of antiparallel $\beta$-pleated sheets. The ability of glucagon to form both $\alpha$-helical and $\beta$-pleated sheet conformers is reflected in the prediction of secondary structure by CHOU and FASMAN (1975), who find that the sequence favours both $\alpha$-helix and $\beta$-pleated sheet formation, and suggest that the conformation is delicately balanced between these two conformations.

## D. Conformation of Micelle-Bound Glucagon

GRATZER et al. (1968) showed that glucagon assumes a more ordered structure – possibly 90% $\alpha$-helical – in chloroethanol. As glucagon is amphipathic in a helical conformation (BLANCHARD and KING 1966; SCHIFFER and EDMUNDSEN 1967), hydrophobic surfaces favour helix formation. Indeed, fluorescence and circular dichroism studies (SCHNEIDER and EDELHOCH 1972; EPAND et al. 1977) show that detergents and surfactant micelles bind glucagon with a concomitant formation of ordered secondary structure. WU and YANG (1980) have estimated that in 25 m$M$ sodium dodecylsulphate below pH 4, glucagon may be 50% $\alpha$-helical. They have shown that sodium decylsulphate is equally good at inducing secondary structure, but that dodecylammonium chloride and dodecylheptaoxyethylene are less effective. They have also shown that $NH_2$ terminal fragments of glucagon are not affected, whereas structure is induced in COOH terminal fragments such as Ala-19 to Thr-29, the part of the sequence that has the highest propensity for helical structure as shown by CHOU and FASMAN (1975). It is possible that this region may also be $\alpha$-helical when the complete molecule binds to a surfactant.

More recently, WÜTHRICH and co-workers (BOESCH et al. 1982) have undertaken a complete conformational analysis of glucagon bound to micelles using high resolution NMR techniques. They show that one glucagon molecule binds to 40 detergent molecules (perdeuterated dodecylphosphocholine) with a well-defined and extended conformation. Electron paramagnetic resonance (EPR) and NMR studies indicate that the molecule is parallel to the micelle surface (BROWN et al. 1981). In a beautifully designed study, they have combined the use of proton–proton Overhauser enhancements and distance geometry algorithms using upper limits for selected proton–proton distances to define the conformation of the glucagon molecules bound to the micelle (BRAUN et al. 1981). They present three structures (see Fig. 8) which meet the distance criteria for the segment Ala-19 to Met-27 given in Table 5. Although certain distances are inconsistent with the existence of the helical conformer found in the crystals, it is possible to build a helical conformer with these interatomic distances by moving the side chain of Met-27, which is restricted by intermolecular interactions in trimer 2, and by rotating some side chains through small angles which are not precluded at 3.0 Å resolution by the X-ray analysis. The details of the conformer bound to the micelle will be further defined as the NMR analysis proceeds.

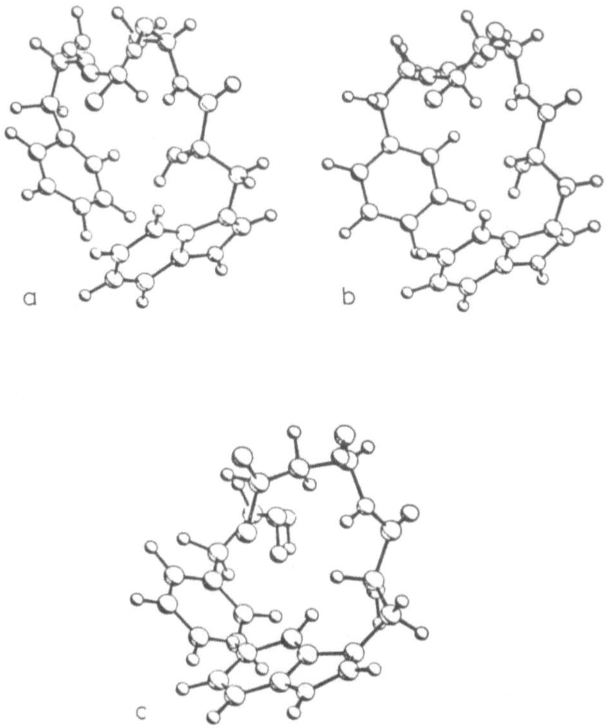


**Fig. 8 a–c.** Computer drawings of the residues Phe-22, Trp-25, and Leu-26 in three structures (**a, b, c**) predicted from NMR and geometry algorithms. BRAUN et al. (1981)

## E. Conformation and Storage Granules

Glucagon is stored in granules which are usually amorphous in character. However, the glucagon granules of teleosts are crystalline rhombic dodecahedra (LANGE and KLEIN 1974; LANGE 1976, 1979; LANGE and KOBAYASHI 1980). Figure 9 is a reproduction of the results of LANGE and KLEIN using tilting stage microscopy while Fig. 10 ist a drawing of a rhombic dodecahedral crystal of glucagon in a general view and along [100], [111], and [110]. The central view of the electron micrograph appears to be along [100] while the other views [111] or equivalent axes (the angle of tilt should be 54.7°) and [110] or equivalent axes (the angle of tilt should be 45°). The angles are compressed in the thin sections. These electron micrographs show that the glucagon granule has 12 faces and that it is a rhombic dodecahedron (LANGE and KLEIN 1974). LANGE (1976, 1979) has shown by optical transforms that the cell dimension varies between 41 and 48 Å. Values smaller than that observed for porcine glucagon crystals bathed in their solvent of crystallisation at pH 6.5 are expected as a result of dehydration necessary for electron microscopy. These crystalline granules almost certainly have a structure similar to that of the crystals studied by X-ray analysis and contain glucagon trimers.

Xanthydrol, a reagent specific for tryptophan residues, produces an unusual blue–grey colour in A-cell granules (BUSSOLATI et al. 1971) as distinct from the nor-

**Table 5.** Upper bounds for the distances used in the distance geometry algorithm as obtained from truncated driven Overhauser effect measurements. All observed Overhauser effects were attributed to proton–proton distances of 5.0 Å or less. Where individual protons could not be resolved in the $^1$H NMR spectrum, e.g. the ring protons of Ph-22 or $\delta_1$ $CH_3$ and $\delta_2$ $CH_3$ of Leu-26, a proton–proton distance sufficiently large to include all indistinguishable protons was used (BRAUN et al. 1981)

| Observed Overhauser effects | | | | Maximum distance (Å) |
|---|---|---|---|---|
| Between ... | | ...and... | | |
| Ala-19 | $\beta$ $CH_3$ | Phe-22 | ring | 10.0 |
| Phe-22 | ring | Val-23 | $\alpha$ CH | 10.0 |
| | | Val-23 | $\gamma_1$ $CH_3$ | 10.0 |
| | | Trp-25 | C5H | 10.0 |
| | | Trp-25 | C7H | 10.0 |
| | | Leu-26 | $\delta_1, \delta_2$ $CH_3$ | 12.5 |
| Val-23 | $\alpha$ CH | Leu-26 | $\gamma$ CH | 5.0 |
| | | Leu-26 | $\delta_1, \delta_2$ $CH_3$ | 7.5 |
| Val-23 | $\gamma_1$ $CH_3$ | Leu-26 | $\delta_1, \delta_2$ $CH_3$ | 7.5 |
| | | Met-27 | $\varepsilon$ $CH_3$ | 5.0 |
| Val-23 | $\gamma_2$ $CH_3$ | Leu-26 | $\gamma$ CH | 5.0 |
| | | Leu-26 | $\delta_1, \delta_2$ $CH_3$ | 7.5 |
| | | Met-27 | $\varepsilon$ $CH_3$ | 5.0 |
| Val-23 | $\alpha$ CH [a] | Val-23 | $\beta$ CH | 2.87 [a] |
| | | Val-23 | $\gamma_1, \gamma_2$ C | 2.71 [a] |
| Val-23 | $\beta$ CH [a] | Val-23 | N | 2.65 [a] |
| | | Val-23 | C′ | 2.74 [a] |
| Trp-25 | C2H | Leu-26 | $\delta_1, \delta_2$ $CH_3$ | 7.5 |
| Trp-25 | C4H | Leu-26 | $\gamma$ CH | 5.0 |
| | | Leu-26 | $\delta_1, \delta_2$ CH | 7.5 |
| Trp-25 | C5H | Leu-26 | $\delta_1, \delta_2$ $CH_3$ | 7.5 |
| Trp-25 | C6H | Leu-26 | $\delta_1, \delta_2$ $CH_3$ | 7.5 |
| Leu-26 | $\delta_1, \delta_2$ $CH_3$ | Met-27 | $\varepsilon$ $CH_3$ | 7.5 |

[a] For Val-23, the $^3J_{\alpha\beta}$ coupling constant was 10.5 Hz, which indicates that $\alpha$ CH and $\beta$ CH are *trans*. This information has been included in the input data as fixed values of the distances from the $\alpha$ proton to the $\beta$ proton and the $\gamma$ carbons, and from the $\beta$ proton to the backbone amide nitrogen and carbonyl carbon of Val-23

mal violet colour of the other structures containing tryptophan. Glucagon crystals also stain this unusual colour, suggesting that the environment of the tryptophan is the same in both. Thus it seems likely that the crystal trimer also exists in the amorphous granules. At neutral pH, glucagon is very insoluble, and the high rate of precipitation makes crystals very difficult to obtain (see Sect. B.I). Thus it is not surprising that the granules are often amorphous in mammalian A-cells, even though conditions (and possibly the chemical sequence) may be conducive to crystallisation in the teleost.

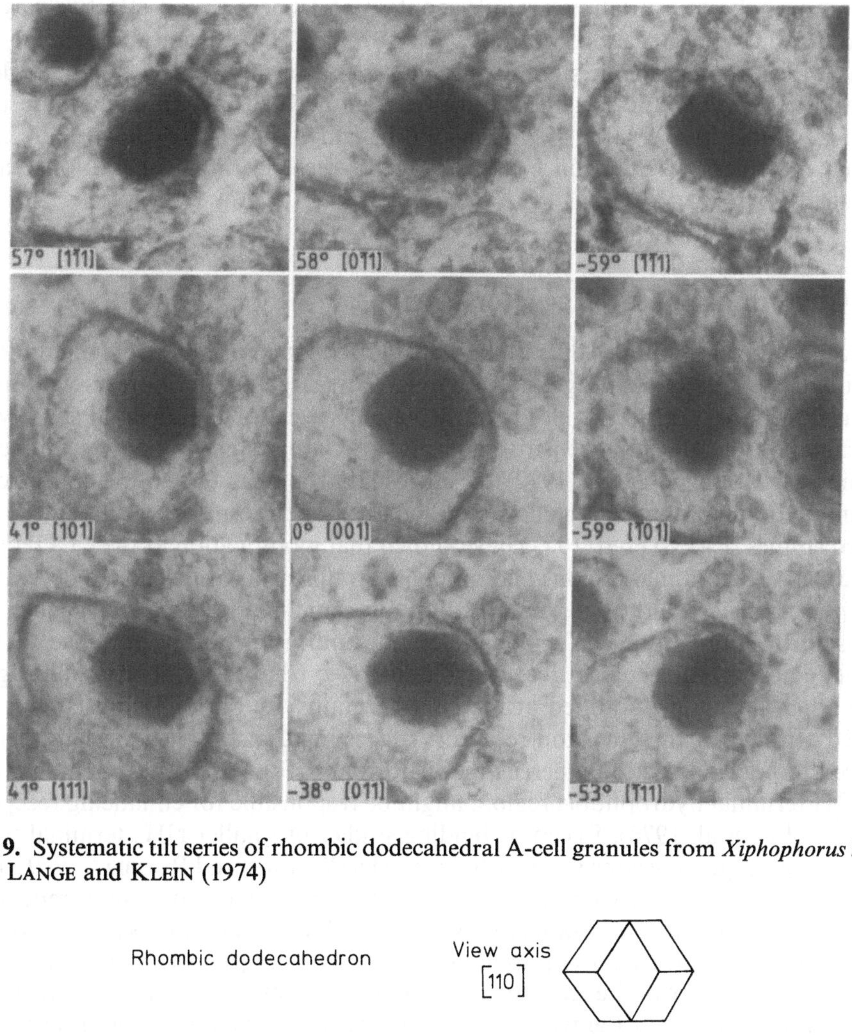


**Fig. 9.** Systematic tilt series of rhombic dodecahedral A-cell granules from *Xiphophorus helleri*. LANGE and KLEIN (1974)

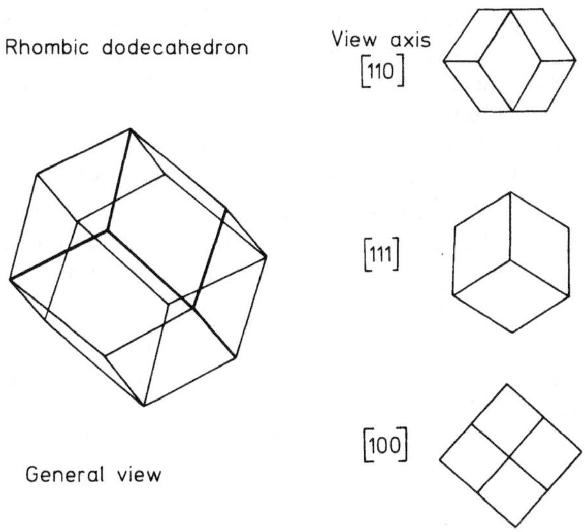


**Fig. 10.** Schematic representation of the rhombic dodecahedron characteristic of cubic glucagon crystals

Amorphous or crystalline granules containing trimers are an effective way of concentrating the glucagon molecules for storage. The existence of trimers increases the thermodynamic stability and makes the hormone less available to degradation by proteolytic enzymes (BLUNDELL et al. 1976, 1978). Thus there is a strong parallel in the roles of zinc insulin hexamers of insulin and trimers and higher oligomers of glucagon in the storage of these hormones (BLUNDELL and HUMBEL 1980).

## F. Conformation and Receptor Binding

Glucagon storage granules, like those of insulin, become unstable at high dilutions in circulation and circulating glucagon must exist largely as monomers with little defined secondary structure. Study of the data on receptor-binding and activity of glucagon on the glucagon-sensitive adenylate cyclase system of hepatocyte membranes led RODBELL et al. (1972) to suggest that almost the entire molecule is required for full biological potency (see Chap. 13). Their results and subsequent detailed studies (EPAND et al. 1977) have shown that modification at the COOH terminal or $NH_2$ terminal regions and on polar groups invariably leads to decreased receptor potency, although modification of some nonpolar groups (such as tyrosine by iodination) can enhance activity, indicating the importance of the whole glucagon molecule to receptor binding (see Chap. 13). [desHis$^1$]Glucagon and other glucagons modified at the $NH_2$ terminus are partial agonists with reduced affinity for the glucagon receptor, suggesting that this region might be involved in hormone action whereas the major part of the molecule, including the COOH terminal hydrophobic region, might be responsible for enhancing receptor affinity (LIN et al. 1976). Receptor-binding studies of smaller $NH_2$ terminal fragments of glucagon are consistent with this model (WRIGHT and RODBELL 1979). Evidence that interaction with the receptor is entropy driven suggests a model in which a helical conformer, stabilised by hydrophobic interactions, is induced or selected at the receptor from a population of conformers (SASAKI et al. 1975).

The exact nature of the receptor-bound conformer must await further biochemical characterisation of the receptor. In the long term it is hoped that a receptor–glucagon complex might be crystallised in a two- or three-dimensional lattice and studied by electron or X-ray diffraction. The results using NMR on the glucagon–micelle complex are also encouraging; it is possible that the glucagon may be studied bound to a deuterated receptor molecule. Finally, the use of low angle neutron diffraction with receptor or glucagon selectively deuterated may also be useful in defining the nature of the hormone–receptor complex.

*Acknowledgments.* I am grateful to Professor M. KARPLUS, Dr. C. DOBSON, Dr. M. WAGMAN, Professor K. WÜTHRICH, Dr. I.J. TICKLE, Dr. K. SASAKI, Dr. B. GELLATLY, Dr. S. DOCKERILL, G. MOREIRA, A. WEST, Dr. W. BRAUN, Dr. L. BROWN, Professor R. LANGE, and Professor R. EPAND for making available unpublished data and for useful discussions.

## References


Beaven GH, Gratzer WB, Davies HG (1969) Formation and structure of gels and fibrils from glucagon. Eur J Biochem 11:37–42

Blanchard MH, King MV (1966) Evidence of association of glucagon from optical rotary dispersion and concentration difference spectra. Biochem Biophys Res Commun 25:298–303

Blundell TL (1979) Conformation and molecular biology of pancreatic hormones. II. Glucagon. Trends Biochem Sci 4:80–83

Blundell TL, Humbel RE (1980) Hormone families: pancreatic hormones and homologous growth factors. Nature 287:781–787

Blundell TL, Johnson LN (1976) Protein crystallography. Academic Press, London New York

Blundell TL, Dockerill S, Sasaki K, Tickle IJ, Wood SP (1976) The relation of structure to storage and receptor binding of glucagon. Metabolism 25:1331–1336

Blundell TL, Dockerill S, Pitts JE, Wood SP, Tickle IJ (1978) Glucagon and pancreatic hormone. III. X-ray analysis, conformation and receptor binding. In: Nichols P (ed) Membrane proteins. Pergamon, Oxford New York, pp 249–257

Boesch C, Bundi A, Oppliger M, Wüthrich K (1978) $^1$H NMR studies of the molecular conformation of monomeric glucagon in aqueous solution. Eur J Biochem 91:209–214

Boesch C, Brown LR, Wüthrich K (1982) Physicochemical characterisation of glucagon-containing lipid micelles, unpublished results

Braun W, Boesch C, Brown LR, Gō N, Wüthrich K (1981) Combined use of protein-proton Overhauser enhancements and a distance geometry algorithm for determination of polypeptide conformations: application to micelle-bound glucagon. Biochim Biophys Acta 667:377–396

Brown LR, Boesch C, Wüthrich K (1981) Location and orientation relative to the micelle surface for glucagon in mixed micelles with dodecylphosphocholine EPR and NMR studies, unpublished results

Bussolati G, Capella C, Vassallo G, Solcia E (1971) Xanthydrol staining of pancreatic A-cell granules. Diabetologia 7:181–188

Chou PY, Fasman GD (1975) The conformation of glucagon: predictions and consequences. Biochemistry 14:2536–2541

Dockerill S (1978) Structure function studies with insulin and glucagon. DPhil thesis, Sussex University, Brighton

Edelhoch H, Lippoldt RE (1969) The conformation of glucagon in detergents. Biol Chem 244:3876–3883

Epand RM (1971) Studies of the conformation of glucagon. Can J Biochem 49:166–169

Epand RM, Jones AJS, Sayer B (1977) Molecular interactions in the model lipoprotein complex formed between glucagon and dimyristoylglycerophosphocholine. Biochemistry 16:4360–4368

Finney JL (1978) Volume occupation, environment and accessibility in proteins. J Mol Biol 119:415–441

Formisano S, Johnson ML, Edelhoch H (1978 a) The effect of Hofmeister salts on the self-association of glucagon. Biochemistry 17:1468–1473

Formisano S, Johnson ML, Edelhoch H (1978 b) Thermodynamics of the self-association of glucagon. Proc Natl Acad Sci USA 74:3340–3344

Gratzer WB, Beaven GH (1969) Relation between conformation and association state. J Biol Chem 244:6675–6679

Gratzer WB, Bailey E, Beaven GH (1967) Conformational states of glucagon. Biochem Biophys Res Commun 28:914–919

Gratzer WB, Beaven GH, Rattle HWE, Bradbury EM (1968) A conformational study of glucagon. Eur J Biochem 3:276–283

Gratzer WB, 2Creeth JM, Beaven GH (1972) Presence of trimers in glucagon solution. Eur J Biochem 31:505–509

Johnson RE, Hruby VJ, Rupley JA (1979) A calorimetry study of glucagon association. Biochemistry 18:1176–1179

King MV (1959) The unit cell and space group of cubic glucagon. J Mol Biol 1:375–378

King MV (1965) A low resolution structural model for cubic glucagon based on packing of cylinders. J Mol Biol 11:549–561
Lange RH (1976) Crystallography of islet secretory granules. In: Fujita T (ed) Endocrine gut and pancreas. Elsevier, Amsterdam Oxford New York, pp 167–178
Lange RH (1979) Distribution of molecule numbers per secretion granule. A study of crystals in glucagon secreting cells. Eur J Cell Biol 20:71–75
Lange RH, Klein C (1974) Rhombic dodecahedral secretory granules in glucagon producing islet cells. Cell Tiss Res 148:561–563
Lange RH, Kobayashi K (1980) Cubic crystals in endocrine pancreatic A-cells of a teleost, *Fugu rubripes*. J Ultrastruct Res 72:20–26
Lee B, Richards FM (1971) Interpretation of protein structures: estimation of static accessibility. J Mol Biol 55:379–400
Lin MC, Nicosia S, Rodbell M (1976) Effects of iodination of tyrosyl residues on the binding and action of glucagon at its receptor. Biochemistry 15:4537–4540
Panijpan B, Gratzer WB (1974) Circular dichroism studies of glucagon. Eur J Biochem 45:547–553
Patel DJ, (1970) Proton NMR of glucagon. Macromolecules 3:448–449
Rodbell M, Birnbaumer L, Pohl SL, Sundby F (1971) The reaction of glucagon with its receptor. Proc Natl Acad Sci USA 68:909–913
Ross JBA, Rousslang KW, Deranleau DA, Kwiram AL (1976) Optical detection of magnetic resonance of glucagon. A conformation change related to critical length of the peptide chain. J Am Chem Soc 98:6781–6782
Ross JBA, Rousslang KW, Deranleau DA, Kwiram AL (1977) Glucagon conformation: use of optically detected magnetic resonance and phosphorescence of tryptophan. Biochemistry 16:5398–5402
Sasaki K, Dockerill S, Adamiak DA, Tickle IJ, Blundell TL (1975) X-ray analysis of glucagon and its relationships to receptor binding. Nature 257:751–757
Schiffer N, Edmundsen AB (1967) A helical wheel for glucagon. Biophys J 7:121–134
Schneider AB, Edelhoch H (1972) Conformational changes of glucagon bound to lysolecithin. J Biol Chem 247:4992–4995
Srere PA, Brooks GC (1969) The circular dichroism of glucagon solutions. Arch Biochem Biophys 129:708–710
Swann JC, Hammes GC (1969) Self-association of glucagon. Equilibrium studies. Biochemistry 8:1–7
Wagman ME (1981) Proton NMR studies of glucagon association in solution. PhD thesis, Harvard University, Cambridge, Massachusetts
Wagman ME, Dobson CM, Karplus M (1980) Proton NMR studies of the association and folding of glucagon in solution. FEBS Lett 119:265–270
Wright DE, Rodbell M (1979) $Glucagon_{16}$ binds to the glucagon receptor and activates hepatic adenylate cyclase. J Biol Chem 254:268–269
Wu CSC, Yang JT (1980) Helical conformation of glucagon in surfactant solutions. Biochemistry 19:2117–2122

# Morphology of the A-cell of Islets of Langerhans, Biosynthesis of Glucagon and Related Peptides

CHAPTER 4

# Glucagon- and Glicentin-Producing Cells

L. ORCI, C. BORDI, R. H. UNGER and A. PERRELET

*Qu'on ne dise pas que je n'ai rien dit de nouveau,*
*la disposition des matières est nouvelle.*
*Blaise Pascal (1623–1662)*

## A. Introduction

Although the A-cell of pancreatic islets was identified around the beginning of this century (for a review of the landmarks in its identification, see MUNGER 1972; WELLMANN and VOLK 1977), this islet cell type has not received as much attention as the insulin-producing B-cell. However, in the last decade, a mounting interest in A-cell structure and function has partially bridged the gap. Interest has been built around the application of different new investigative techniques of cell biology to the A-cell, mainly freeze-fracture electron microscopy, immunocytochemistry, autoradiography, cell culture, and intracellular microinjection, coupled to precise sampling of animal and human pancreas. Since a review on the same subject appeared very recently (ORCI and PERRELET 1981), this chapter will consist primarily of an updated survey of the work, published and unpublished, carried out by our laboratory.

## B. Pancreas

### I. Morphological Features

This section has been deliberately kept to a summary format and the data will be mentioned only to cover the relevant literature. Before the advent of immunocytochemistry (see Sect. B.II), the A-cell was distinguished from the other islet cell types mainly by its affinity for acid dyes and silver-containing stains (for review see MUNGER 1972; LANGE 1973). Ultrastructurally, the A-cell can be unmistakably distinguished from the other islet cell types by the characteristic appearance of its secretory granules showing a somewhat eccentric, dense roundish core surrounded by a paler mantle, tightly applied to the limiting membrane (Fig. 1). The mantle region also shows distinct silver-binding properties (BUSSOLATI et al. 1971; BAETENS et al. 1976). As a polypeptide-secreting cell, the A-cell is well endowed with organelles concerned with protein synthesis and protein processing for export, i.e., the rough endoplasmic reticulum and the Golgi apparatus, respectively (for review see PALADE 1975) and a characteristic timetable for the intracellular processing of tryptophan $^3$H-labeled polypeptides has been established (HOWELL et al. 1974). In this study, the production of glucagon from its precursors by the A-cell was noted to be slower than that determined for insulin in the B-cell (ORCI et al. 1973; ORCI

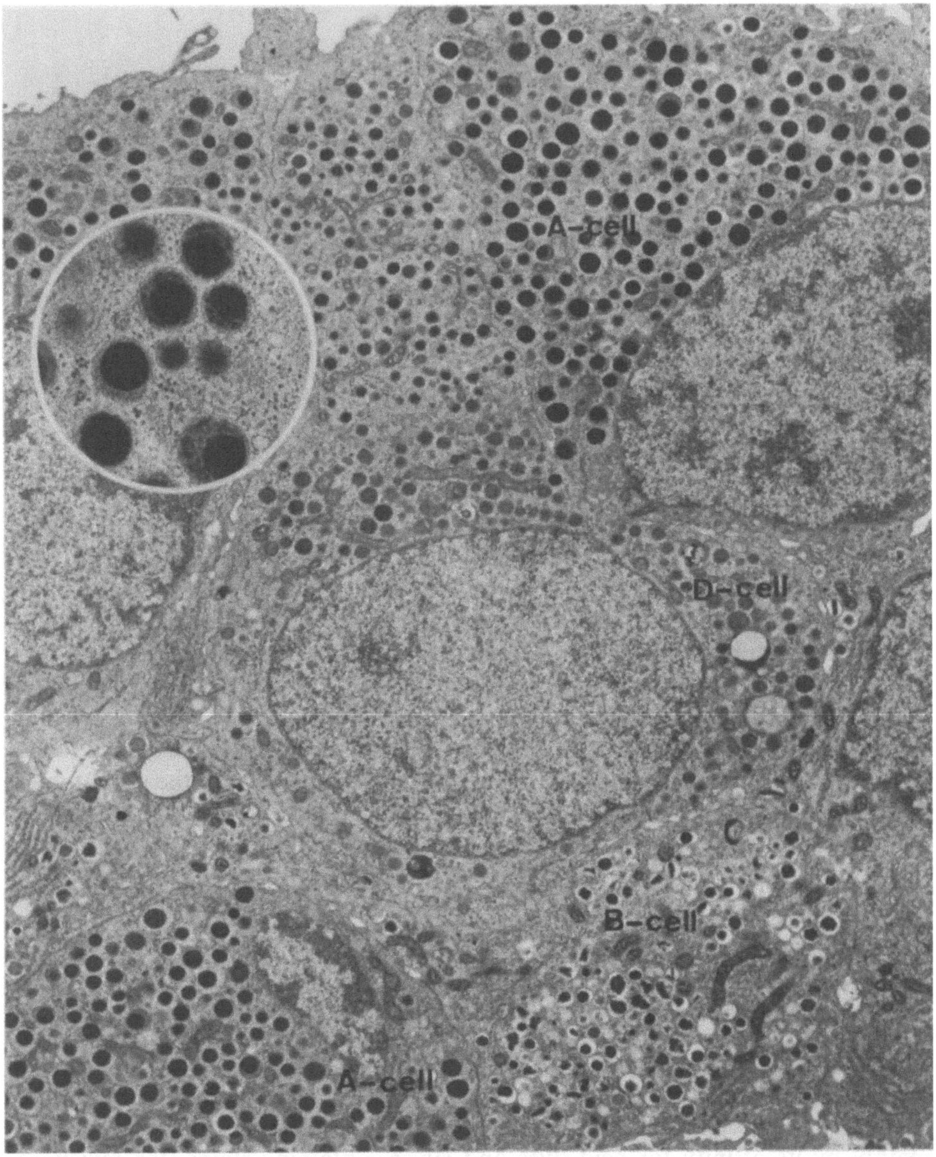


**Fig. 1.** Thin section of the periphery of an islet of Langerhans from the human pancreas. This image shows the characteristic appearance of A-, B-, and D-cells which can be distinguished by the morphology of their respective specific secretory granules. Those of an A-cell are shown at high magnification in the *inset*: α-granules are round and have an eccentric, round dense core surrounded by a mantle of variable electron density. ×5,600; *inset* ×24,000

1974). Other functional differences between A- and B-cells pertain to the specific secretory granules: unlike α-granules, β-granules contain acid phosphatase (ORCI et al. 1971) and may store exogenous horseradish peroxidase added to the medium for incubating isolated islets (ORCI et al. 1978a).

Although not studied to the same extent as that of the B-cell (for review see ORCI and PERRELET 1977), the plasma membrane of the A-cell has been subjected to the freeze-fracture technique which allows the determination of the concentration

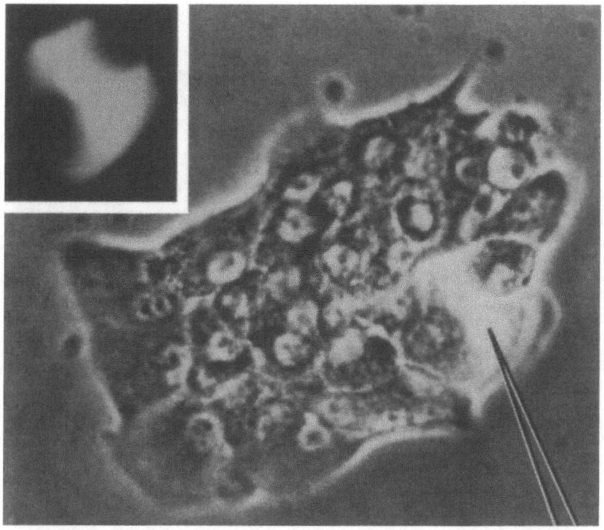

**Fig. 2.** Light micrograph of a cluster of endocrine pancreatic cells in monolayer culture after injection of 6-carboxyfluorescein (molecular weight 376 daltons). From the injection site situated at the lower right edge of the cluster (marked by the drawing of the electrode tip), the fluorescent dye spread in a limited area of the cluster: electron microscopic identification of the involved cells revealed an injection site in an A-cell and dye transfer in at least one neighboring B-cell (see Fig. A, D, and E in MEDA et al. 1981). The picture of the cluster was taken in phase contrast and fluorescent light microscopy. The exact delimitation of the communicating territory is outlined in the inset taken in fluorescence microscopy alone. ×540

and distribution of membrane integral proteins (for review see DAEMER 1977). This approach was used to establish the plasma membrane protein content of A-cells of nondiabetic and diabetic Chinese hamsters (ORCI et al. 1974) and to determine in the rat the contribution of exocytosis to the process of glucagon release by the A-cell under various stimulatory conditions (CARPENTIER et al. 1977). In addition, this technique confirmed the presence in these cells of specific sites of the plasma membrane, the nexus or gap junctions, which represent the structural basis for cell-to-cell communication (ORCI et al. 1975a, b; for a review on gap junctions see HERTZBERG et al. 1981). A recent advance in this field has been the demonstration of the effective transfer of exogenous dyes or endogenous metabolites between A- and B-cells (Fig. 2; MICHAELS and SHERIDAN 1981; MEDA et al. 1981, 1982; and for review see ORCI 1982), raising interesting possibilities of cellular interactions in the framework of the hypothesis that the topography and distribution of islet cells is not random and serves the integrated islet function (ORCI and UNGER 1975; UNGER and ORCI 1981a, b).

## II. Distribution of A-Cells

Immunocytochemistry at both the light and electron microscopic levels have become powerful tools for the characterization of polypeptide-containing cells (Fig. 3). Since its use in demonstrating glucagon at the A-cell level (BAUM et al.

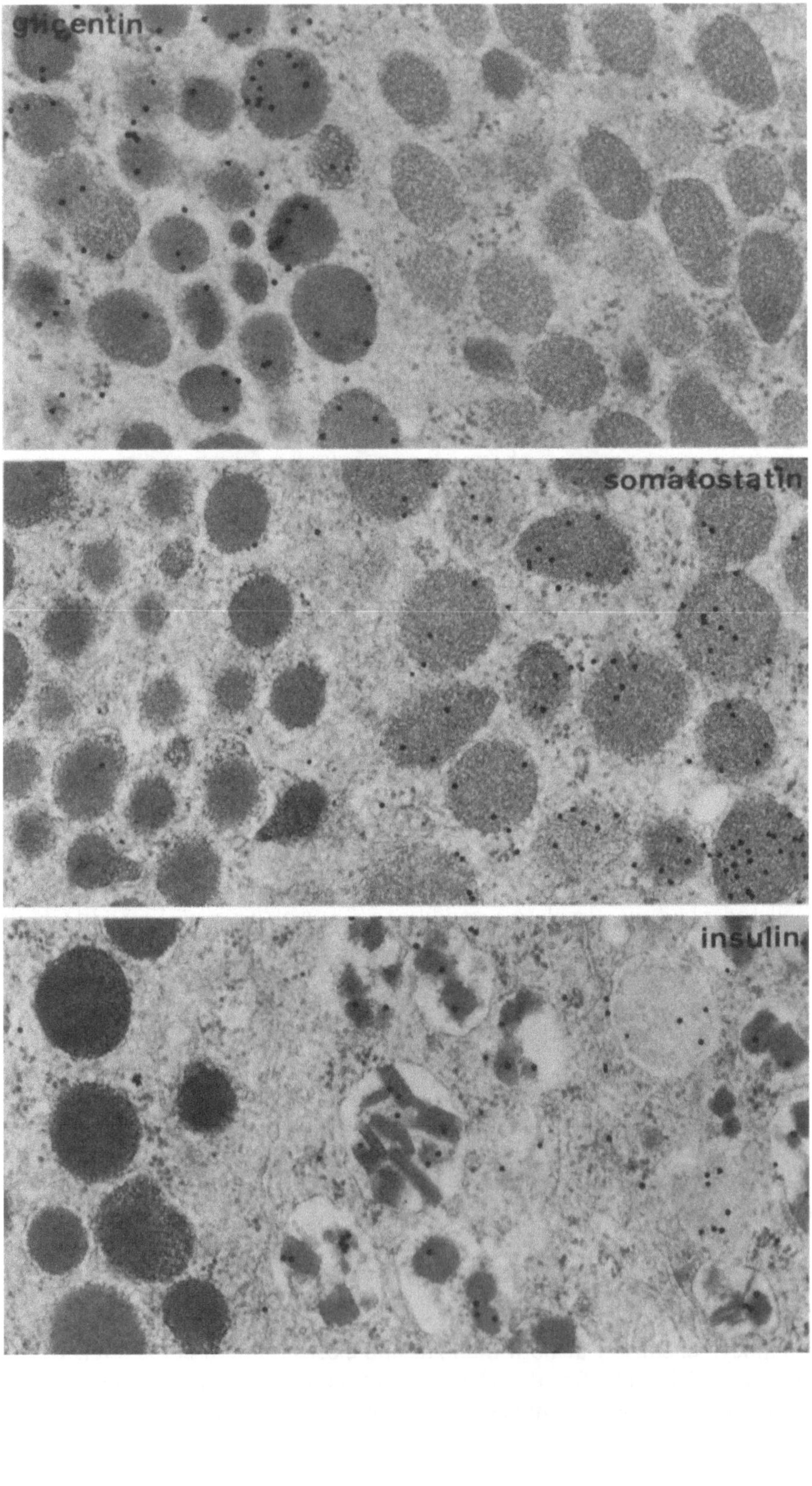
glicentin
somatostatin
insulin

1962), light microscope immunofluorescence with anti-insulin, anti-glucagon, anti-somatostatin, and anti-pancreatic polypeptide (anti-PP) antisera led to the finding that not all pancreatic islets have the same endocrine cell content and that within each islet, endocrine cells assume specific relationships one with another (Fig. 4; ORCI and UNGER 1975; ORCI et al. 1976a; BAETENS et al. 1979; ORCI and PERRELET 1979). From these studies, a model has emerged of an adult pancreas resulting from the fusion of a large dorsal primordium which is rich in glucagon-containing cells and poor in pancreatic polypeptide-containing cells, with a smaller ventral primordium which is poor in glucagon-containing cells and rich in pancreatic polypeptide-containing cells and forms a distinct lobe at the posterior part of the pancreatic head (MALAISSE-LAGAE et al. 1979a, b).

The identification of such a specific pattern, together with the possible role of the topography of the respective endocrine cell types in islet functioning, prompted the search for changes of this pattern in states of altered glucose homeostasis. The search has been conducted in animal models of diabetes (ORCI et al. 1976b; BAETENS et al. 1978; STEFAN et al. 1978), and was extended to the human pancreas. The recent collection of a number of pancreas from control or "normal" subjects (i.e., patients who died without known pancreatic pathology or abnormality of glucose homeostasis), coupled to a quantitative mapping of insulin, glucagon, somatostatin, and pancreatic polypeptide immunofluorescent cells in precisely sampled pancreatic regions, allowed the establishment of the first endocrine cell profile of whole pancreas. Such a profile has also been established in a limited number of diabetic pancreas (STEFAN et al. 1982a). This approach represents an important advance made over the last few years in the effort to understand structure–function correlates in the human pancreas. Figure 5 and Table 1 represent typical data obtained, emphasis in the discussion of the results being limited to the A-cell.

In nondiabetic subjects, the glucagon cell volume ranges from 16.3 to 154.3 µl (i.e., 5.0%–18.0% of the total pancreatic endocrine volume). In two diabetic subjects of the maturity onset type (or noninsulin-dependent diabetes mellitus, NIDDM), this proportion was within the range of control values; in two diabetics of the juvenile onset type (or insulin-dependent diabetes mellitus, IDDM), however, owing to the marked decrease of insulin cells, the glucagon cells (98.2 and 156.4 µl, respectively – not different from control values) become in proportion the most abundant cell type (67.5% and 49.3% of the total endocrine volume, respectively), largely above any of the control values of this parameter (STEFAN et al. 1982a).[1]

1 Contrary to previously published data (GEPTS et al. 1977), PP cells were not found elevated in juvenile-onset diabetics; the former result may have been linked to a sampling bias, since in the PP-rich lobe (posterior part of the head) PP cells will always appear more numerous than any other cell type (see Table 1)

◄ **Fig. 3.** Thin sections of endocrine cell pairs in human pancreas with a field of their respective cytoplasms showing secretory granules following immunostaining with the protein A gold technique. The antiserum used for immunostaining is indicated on each illustration. The specificity of the staining, as well as the good spatial resolution given by the black gold particles revealing the antigenic sites, are manifest. Glicentin × 41,000; somatostatin and insulin × 39,000

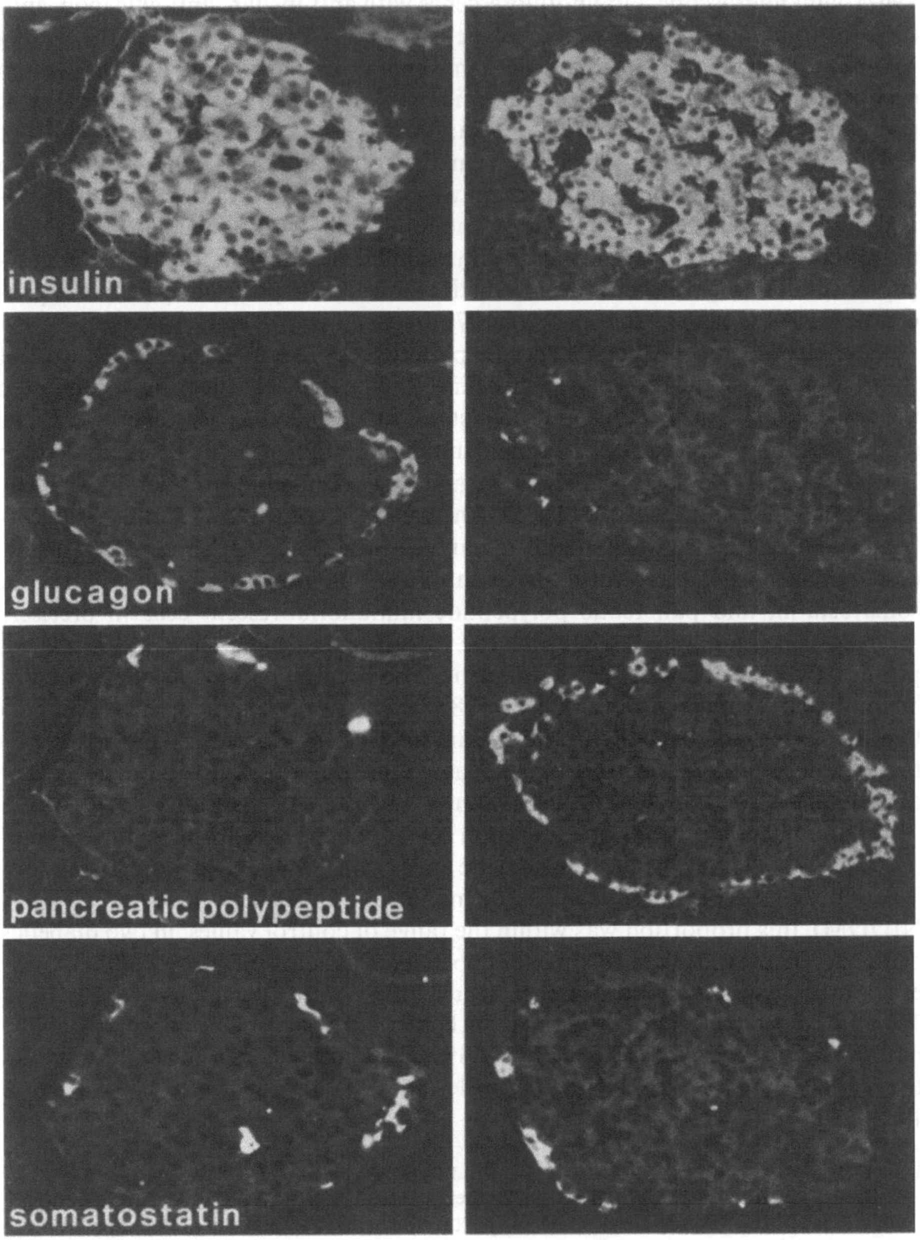


**Fig. 4.** Light micrographs of consecutive serial sections of two rat islets immunostained with anti-insulin, anti-glucagon, anti-pancreatic polypeptide, and anti-somatostatin antisera, respectively. The islet on the left is from the dorsal pancreatic region and is characterized by a high glucagon and low pancreatic polypeptide-immunofluorescent cell content; the islet on the right is from the ventral pancreatic region (in the rat, it corresponds to the inferior part of the pancreatic head) and appears poor in glucagon-containing cells, but rich in pancreatic polypeptide-containing cells. Dorsal islet ×270; ventral islet ×220

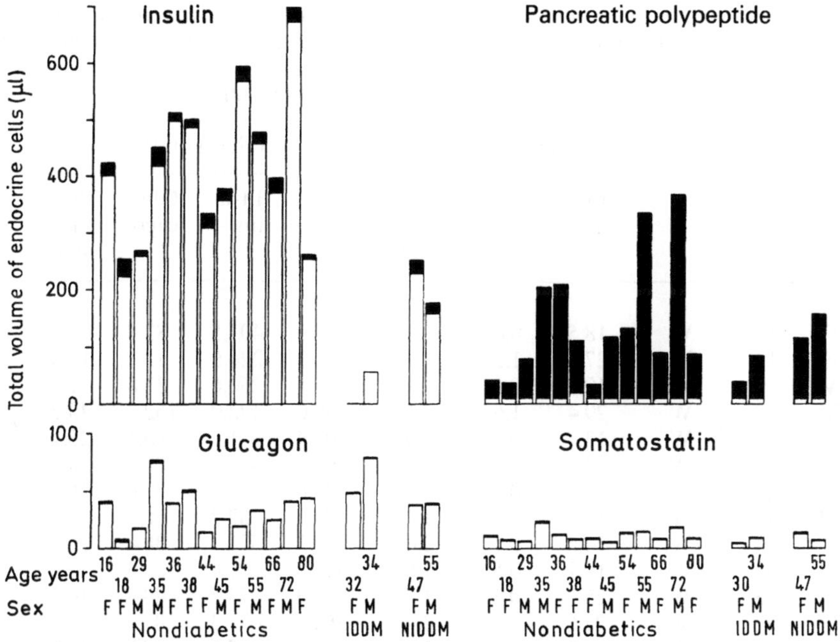


**Fig. 5.** Cumulative histogram of the volume of immunofluorescent cell populations in nondiabetic and diabetic subjects. Each column corresponds to one subject identified by sex and age at the bottom of the figure. In each column, the white part represents the volume of immunofluorescent cells in the glucagon-rich region of the pancreas, the black part, the volume of the same cell type in the pancreatic polypeptide-rich region. For pancreatic polypeptide, except for F 38, the contribution of the glucagon-rich region to each column was drawn arbitrarily at the limit of visibility permitted by the photographic reduction. For actual values, see Table 1. This applies also to some *black bars* at the top of columns for glucagon and somatostatin, which, drawn to scale, would not have been visible. *IDDM*, insulin-dependent diabetes mellitus, *NIDDM*, noninsulin-dependent diabetes mellitus

The establishment of the immunofluorescent hormonal profile in the developing human pancreas yielded interesting data pertaining to the relative content of glucagon and glicentin immunoreactivity. The study of the developing pancreas confirmed the qualitative observations made previously on A-cells (LIKE and ORCI 1972), but in addition, it revealed two different populations of glucagon/glicentin-immunoreactive cells (Table 2). In the early fetal stages (3–4 cm crown-rump), a majority of cells reacted with the anti-glicentin antiserum, but only part of these cells were stained with the anti-glucagon antiserum (Fig. 6). Cells reacting only with anti-glicentin antiserum (remember that such cells are not detectable in the adult pancreas) remained equally unreactive to COOH or $NH_2$ terminally directed anti-glucagon antiserum following trypsin and carboxypeptidase B digestion, a procedure known to convert glicentin-containing L-cells of the digestive mucosa (see Sect. C) to glucagon-containing cells (RAVAZZOLA and ORCI 1980a). At present, it is not known whether the secretory product of cells reacting exclusively with the anti-glicentin antiserum is a glucagon-free derivative, or whether it does contain a masked glucagon sequence. Moreover, the fact that the cells reacting exclu-

**Table 1.** Total volume of insulin-, glucagon-, somatostatin-, and pancreatic polypeptide-containing cells in pancreatic polypeptide-rich and glucagon-rich regions of the pancreas

| Sex | Age (years) | Race | Volume (μl) | | | | | | | |
|---|---|---|---|---|---|---|---|---|---|---|
| | | | PP-rich lobules | | | | Glucagon-rich lobules | | | |
| | | | Insulin | Glucagon | Somatostatin | PP | Insulin | Glucagon | Somatostatin | PP |
| *Nondiabetics* | | | | | | | | | | |
| 1. F | 16 | W | 18.5 | 0.8 | 1.2 | 37.9 | 407.5 | 82.8 | 19.9 | 3.8 |
| 2. F | 18 | W | 27.2 | 1.2 | 1.4 | 35.3 | 226.6 | 15.1 | 12.8 | 3.6 |
| 3. M | 29 | W | 10.7 | 0.5 | 1.2 | 77.7 | 260.6 | 33.8 | 11.3 | 1.8 |
| 4. M | 35 | W | 29.4 | 1.6 | 3.1 | 198.6 | 422.9 | 152.7 | 43.2 | 5.6 |
| 5. F | 36 | W | 13.7 | 0.3 | 2.0 | 202.4 | 501.7 | 80.1 | 23.1 | 7.9 |
| 6. F | 38 | W | 13.5 | 0.8 | 1.5 | 90.0 | 490.6 | 100.5 | 14.4 | 21.5 |
| 7. F | 44 | W | 25.3 | 0.3 | 1.3 | 33.6 | 311.1 | 27.6 | 16.7 | 1.0 |
| 8. M | 45 | W | 16.6 | 1.4 | 1.4 | 109.6 | 362.2 | 50.7 | 9.2 | 8.6 |
| 9. F | 54 | W | 24.6 | 0.1 | 2.4 | 130.3 | 571.7 | 38.8 | 25.0 | 1.7 |
| 10. M | 55 | W | 19.7 | 0.5 | 1.6 | 325.8 | 461.4 | 66.4 | 27.8 | 8.2 |
| 11. F | 66 | W | 25.3 | 0.3 | 1.7 | 88.3 | 373.5 | 49.8 | 14.2 | 1.2 |
| 12. M | 72 | W | 21.6 | 0.8 | 3.2 | 364.0 | 677.3 | 81.7 | 32.4 | 4.1 |
| 13. F | 80 | W | 8.0 | 0.3 | 1.6 | 82.3 | 255.5 | 87.7 | 14.6 | 5.2 |
| *Juvenile-onset-diabetes* (IDDM) | | | | | | | | | | |
| 14. F | 30 | W | 0.0 | 0.1 | 0.8 | 38.0 | <0.2 | 98.1 | 7.0 | 1.3 |
| 15. M | 34 | W | 0.0 | 0.7 | 4.6 | 80.9 | 58.6 | 155.7 | 12.3 | 4.5 |
| *Maturity-onset-diabetes* (NIDDM) | | | | | | | | | | |
| 16. F | 47 | W | 22.5 | 4.1 | 3.8 | 111.4 | 229.5 | 72.1 | 20.3 | 4.9 |
| 17. M | 55 | P | 19.3 | 1.8 | 1.0 | 146.9 | 159.5 | 79.4 | 9.8 | 4.2 |

Subjects are numbered according to age in the three categories studied. *W*, white; *P*, Pima Indian. Fig. 5 gives a graphic representation of this data. For the data collection methods, see Appendix. STEFAN et al. 1982a

**Table 2.** Volume density $V_v$ of endocrine tissue in the pancreas of fetus of various ages and relative contribution of each immunofluorescent cell type to the total volume density

| | Endocrine tissue $V_v$ | Glicentin cells (%) | Glicentin/glucagon cells (%) | Insulin cells (%) | Somatostatin cells (%) | PP cells (%) |
|---|---|---|---|---|---|---|
| Fetus I 3.0 cm | 0.00297 | 46.5 | 0.7 | 40.1 | 11.8 | 1.0 |
| Fetus II 4.3 cm | 0.00572 | 33.0 | 11.7 | 39.3 | 12.9 | 3.0 |
| Fetus III 4.3 cm | 0.01078 | 14.3 | 15.9 | 58.4 | 8.5 | 2.9 |
| Fetus IV 6.0 cm | 0.01228 | 12.4 | 8.4 | 66.1 | 9.7 | 3.4 |
| Fetus V 9.6 cm | 0.03704 | 3.1 | 23.6 | 49.4 | 14.3 | 9.6 |

This Table shows the decrease of cells immunofluorescent exclusively to glicentin antiserum in the older fetus and the parallel increase of cells immunofluorescent to glicentin and glucagon antisera. Glicentin cell values were obtained by the subtraction of glucagon cells from total glicentin-immunoreactive cells (the latter comprise glicentin/glucagon cells). See Appendix for the significance of $V_v$. STEFAN et al. 1982b

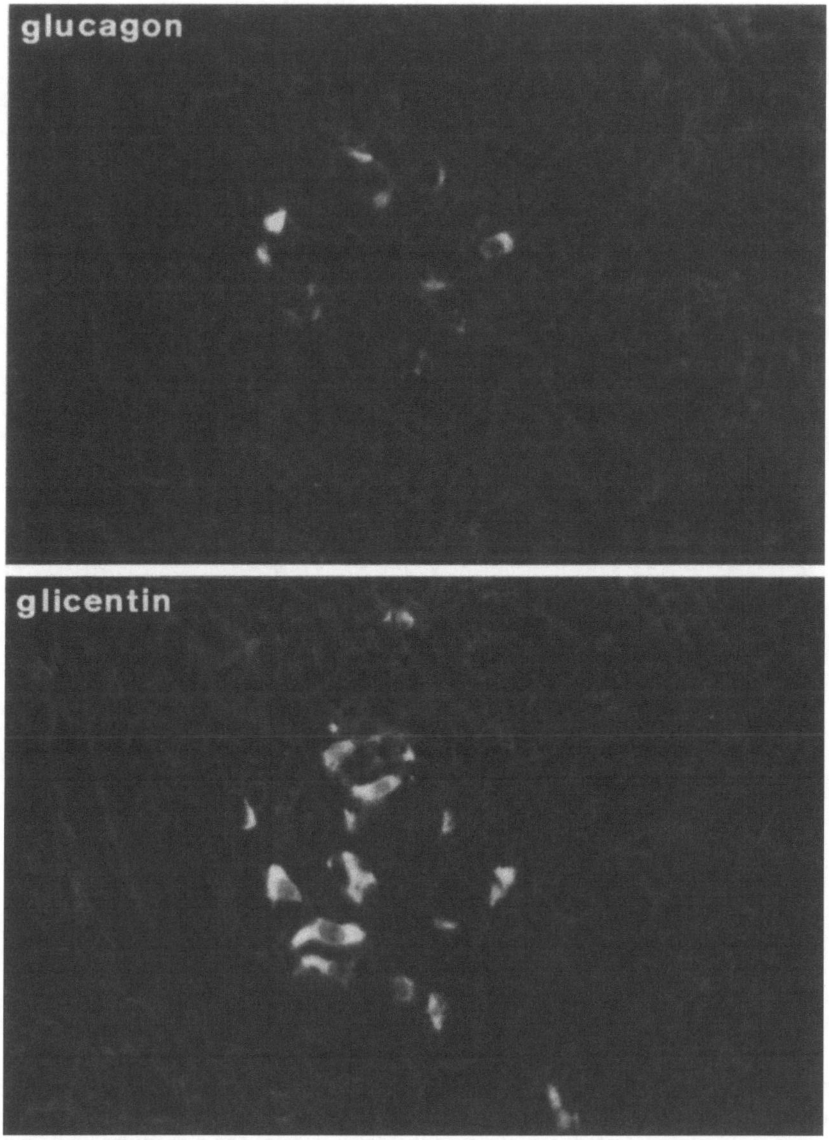


**Fig. 6.** Light micrographs of consecutive serial sections of the human pancreas from a 4.3-cm crown-rump fetus. The two sections, immunostained with anti-glucagon and anti-glicentin antisera, show a marked difference in the immunofluorescent reaction elicited by each antiserum (for the quantitative evaluation of this material, see Table 2, fetus III). ×460

sively with the anti-glicentin antiserum outnumber the glucagon-immunoreactive cells in the early stages, but that the former cells are no longer found in later stages and in the adult pancreas, leaves open the question as to whether they represent a specific transient cell population or a differentiation stage of the A-cell (STEFAN et al. 1982b).

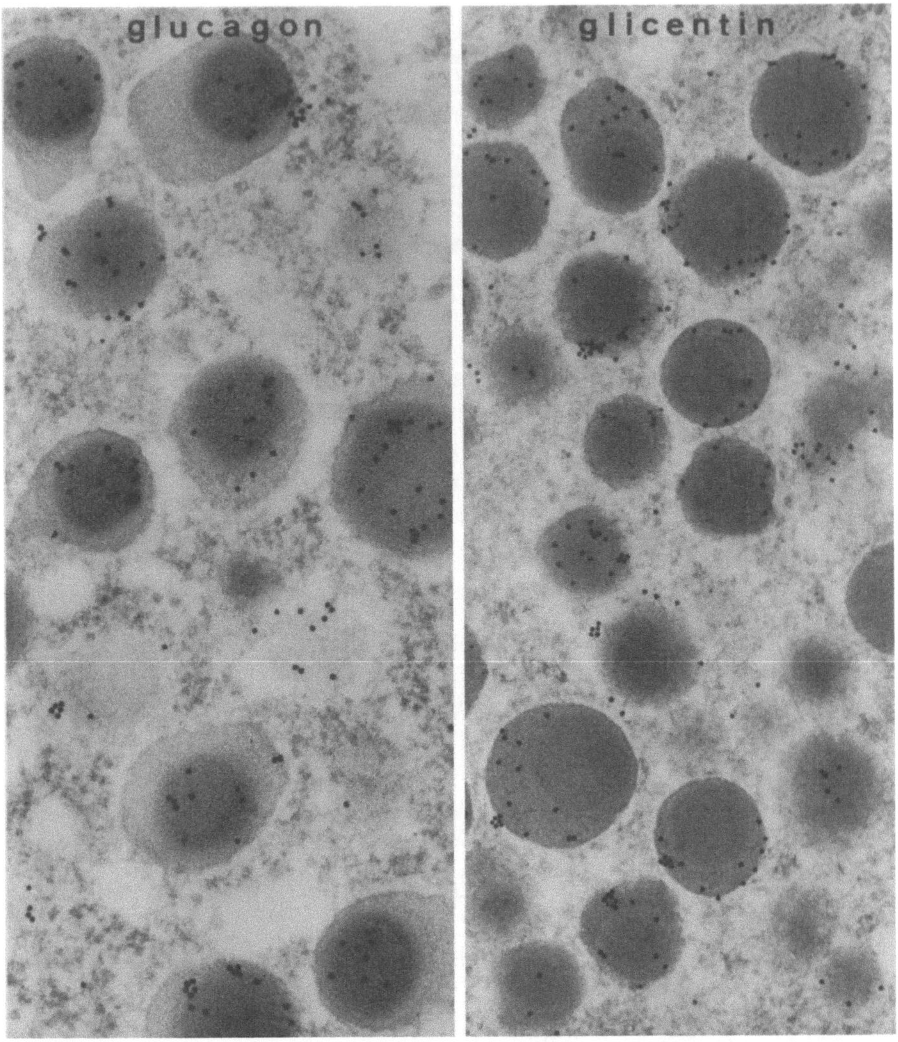


**Fig. 7.** Thin sections of human pancreatic A-cells stained respectively with anti-glucagon and anti-glicentin antisera revealed by the protein A gold technique. With the anti-glucagon antiserum, gold particles predominate over the dense core of the α-granules while with the anti-glicentin antiserum, antigenic sites detected by the gold particles are more abundant over the granule's mantle. RAVAZZOLA and ORCI (1979) Glucagon ×55,000; glicentin ×53,000

## III. Intracellular Distribution of Secretory Polypeptides

Following the identification of glucagon-like immunoreactivity (GLI) in extracts of the digestive mucosa (for review see CONLON 1980, 1981), an antiserum was made available against one of them, glicentin, a 69 amino acid polypeptide, purified from the pig intestine (THIM and MOODY 1981; see also Chap. 7). When applied to the pancreas, this antiserum elicited glicentin immunofluorescence in the same

**Table 3.** Quantitative evaluation of the immunoreactivity of different A-cell compartments to anti-glucagon and anti-glicentin antisera

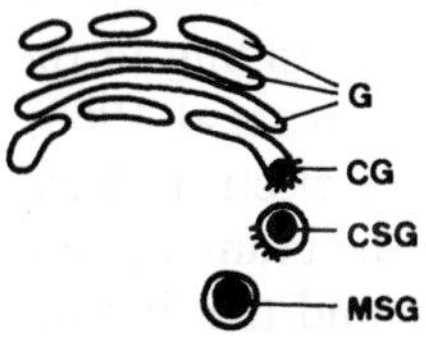


| | Gold particles/$\mu m^2$ | | |
|---|---|---|---|
| | COOH terminal anti-glucagon | $NH_2$ terminal anti-glucagon | R64 anti-glicentin |
| G | $2.19 \pm 0.35$ | $1.79 \pm 0.19$ | $3.58 \pm 0.36$ |
| CG | N.S. ⌐ $6.51 \pm 1.66$ | N.S. ⌐ $42.26 \pm 7.61$ | $70.60 \pm 7.21$ |
| CSG | ⌙ $7.02 \pm 1.53$ | ⌙ $57.35 \pm 7.14$ | $91.62 \pm 7.19$ |
| MSG | $66.76 \pm 4.99$ | $76.82 \pm 9.19$ | $45.17 \pm 5.32$ |

*G*, Golgi cisternae; *CG*, condensing granules in Golgi cisternae; *CSG*, coated immature secretory granules; *MSG*, mature secretory granules.
From the condensing material in Golgi cisternae to mature secretory granules, the Table shows a decrease in glicentin-immunoreactive sites and an increase in glucagon immunoreactivity. Data are mean ± standard error. *N.S.*, difference not significant. The inset schematizes the different compartments evaluated. RAVAZZOLA et al. 1981b

cells showing glucagon immunoreactivity with both specific, COOH terminally directed, and nonspecific, $NH_2$ terminally directed antisera, i.e., the A-cells (RAVAZZOLA et al. 1979a, b). Using a high resolution immunocytochemical method, the protein A gold technique [2] (ROTH et al. 1978), it was possible subsequently to show that glucagon and glicentin immunoreactivities were segregated in specific regions of the α-granule, glucagon being predominant in the dense core, glicentin in the mantle (Fig. 7; RAVAZZOLA and ORCI 1979, 1980b). Moreover, this technique allowed the quantitation of antigenic binding sites (represented by gold particles) as a function of the surface area of other A-cell compartments containing these polypeptides, namely the Golgi apparatus and the immature secretory granules (the latter are characterized by a coated segment on their limiting membrane and assumed to give rise to the mature α-granules; ORCI 1977).

The result of this quantitative evaluation (Table 3) shows that, from the condensing material in Golgi cisternae to the mature α-granule, there is a marked increase in the number of both COOH terminal and $NH_2$ terminal anti-glucagon antigenic sites, and a reverse, although less pronounced, decrease in anti-glicentin immunoreactivity. This result is in favor of a partial conversion of glicentin to glucagon during the maturation of α-secretory granules (RAVAZZOLA et al. 1981b). This agrees with the previous demonstration (PATZELT et al. 1979; for review see TAGER 1981; Chap. 6), that glucagon arises from the conversion of a larger precursor (proglucagon) in the A-cell, the latter probably containing the glicentin molecule and other immunodeterminants common to gastric inhibitory peptide (SMITH et al. 1977; ALUMETS et al. 1978; LARSSON and MOODY 1980), cholecystokinin

2 This technique is based on the affinity of the protein A of Staphylococcus aureus to bind to the Fc portion of G-class immunoglobulins. The immunocytochemical detection of antigens at the ultrastructural level is carried out on thin sections of fixed and embedded tissue incubated with the desired antiserum (anti-glucagon for example), the bound antiserum being then visualized with gold particles coated with protein A

(GRUBE et al. 1978 b), or $\beta$-endorphin (GRUBE et al. 1978 a) claimed to be present in the A-cell on the basis of positive immunofluorescent staining.

# C. Digestive Tract

The demonstration, due to SUTHERLAND and DE DUVE (1948), of a source of glucagon outside the pancreas preceded by 20 years the ultrastructural finding in the gastrointestinal mucosa of a cell resembling the pancreatic A-cell (ORCI et al. 1968). Subsequently, endocrine cells of the digestive tract with round electron-dense granules were classified as A-like cells, irrespective of the size of the granules and/or the presence or absence of a mantle between the dense core and the limiting membrane. Further, another cell type, characterized by large dense and round granules, was described in the intestinal mucosa and named the L-cell (for large granules; for review see SOLCIA et al. 1975). A detailed cytochemical and immunocytochemical study performed at the light and electron microscope levels helped to clarify the relationships between gastrointestinal A-cells, A-like cells, and L-cells (BAETENS et al. 1976). The oxyntic mucosa of animals (BAETENS et al. 1976; GRIMELIUS et al. 1976; HELMSTAEDTER et al. 1977) and fetal humans (ITO et al. 1981; RAVAZZOLA et al. 1981 a) – fresh, undigested mucosa of an adult subject has not yet been available for study – contains A-cells indistinguishable from pancreatic A-cells on cytologic, cytochemical, and immunocytochemical criteria (see also Chap. 33).

The L-cells, on the other hand, detectable in the postduodenal mucosa, stain with anti-glicentin antiserum (as do the pancreatic and gastric A-cells), but not with the specific COOH terminally directed anti-glucagon antiserum. However, also in common with the pancreatic and gastric A-cell, the L-cell stains with nonspecific $NH_2$ terminally directed anti-glucagon antiserum (RAVAZZOLA et al. 1979 a, b). The hypothesis mentioned earlier (that glicentin could be converted to glucagon) has received support from the fact that glicentin-containing L-cells of the dog ileum can be made reactive to COOH terminally directed anti-glucagon antiserum following a proteolytic treatment with trypsin and carboxypeptidase B (Fig. 8; RAVAZZOLA and ORCI 1980 a). The recent demonstration of a spontaneous conversion of infused glucagon-like immunoreactivity into pancreatic-type glucagon in anesthetized animals suggests that the enzymatic breakdown is made extracellularly in vivo (KORANYI et al. 1981).

# D. Pathology of A- and L-cells

The existence of tumors originating from the pancreatic A-cells has been known in humans for more than 20 years. The presence of glucagon in these tumors has

**Fig. 8.** Light micrographs of three consecutive serial sections of the rat ileal mucosa immunostained with anti-glucagon and anti-glicentin antisera, before and after digestion of the sections with proteolytic enzymes. Without proteolytic digestion, fluorescence is elicited in cells of the mucosa by the anti-glicentin antiserum, but not with the anti-glucagon antiserum. After digestion with trypsin (*T*) and carboxypeptidase B (*CB*), specific glucagon immunofluorescence appears in the same cells which were fluorescent with the anti-glicentin antiserum before digestion. RAVAZZOLA and ORCI (1980 a) × 700 ▶

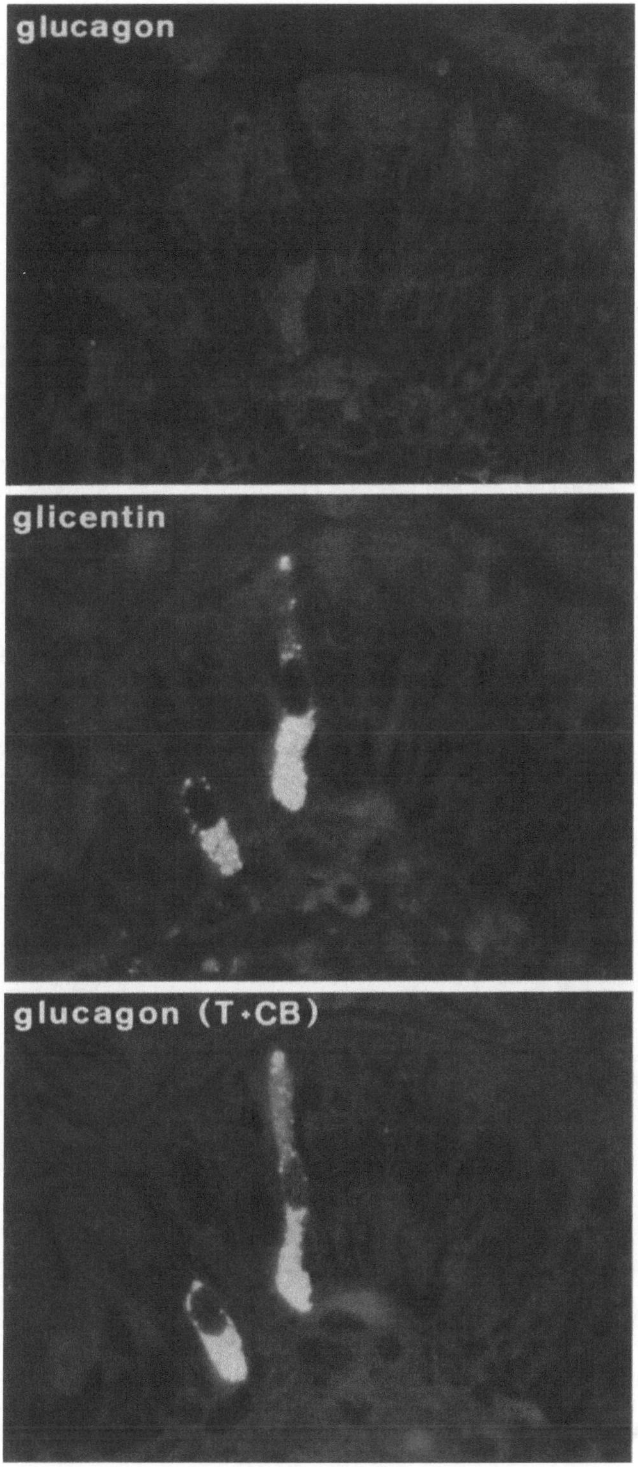
glucagon
glicentin
glucagon (T+CB)

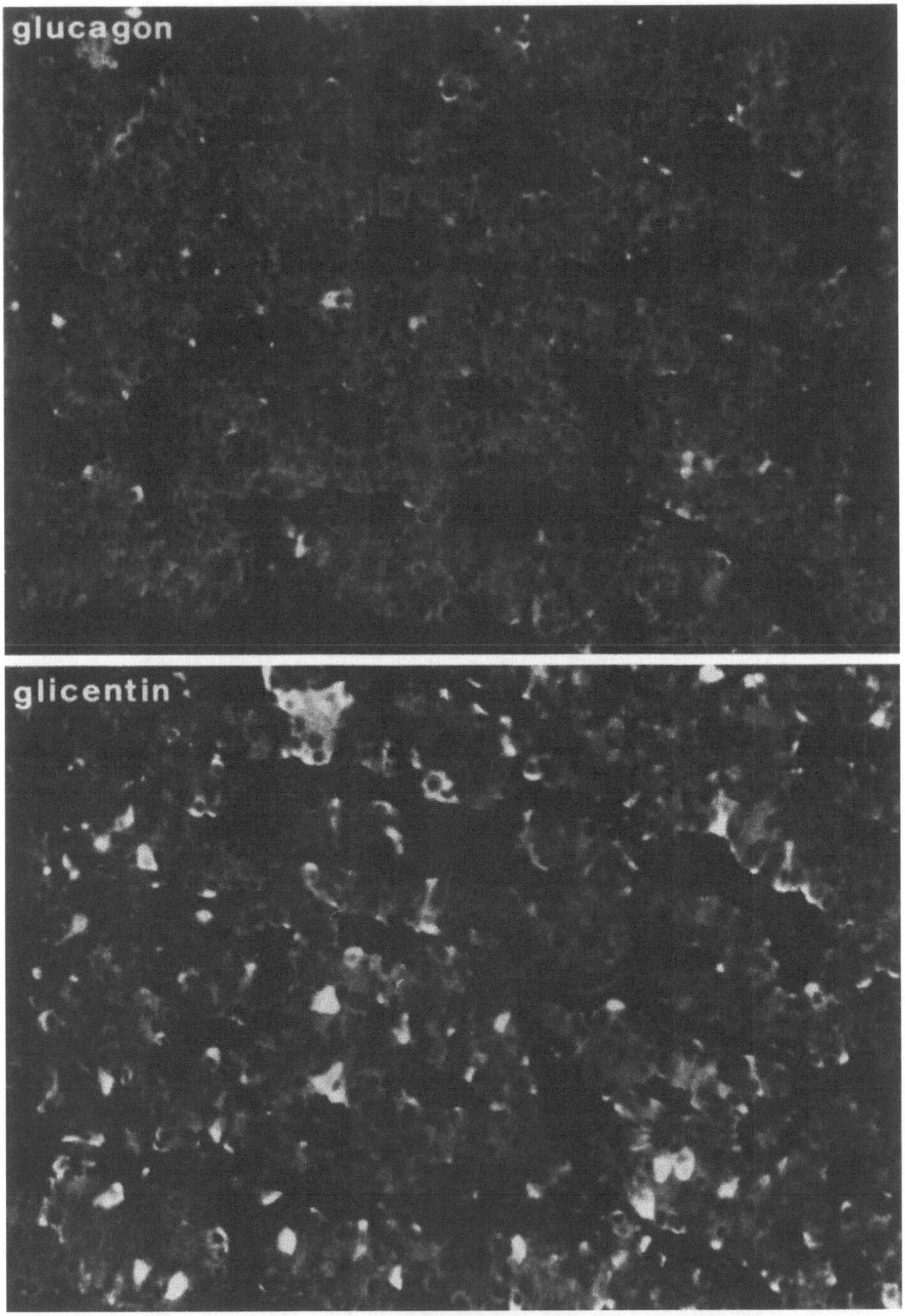


**Fig. 9.** Light micrographs of consecutive sections taken through a pancreatic tumor accompanied by the glucagonoma syndrome. When immunostained with anti-glucagon and anti-glicentin antisera respectively, sections of the tumor show a scarcity of glucagon-immunoreactive cells but a relative abundance of glicentin-immunoreactive cells. BORDI et al. (1979) ×250

also been known for a long time and, more recently, they were divided into two classes on the basis of their clinical outcome: (1) tumors associated with the so-called glucagonoma syndrome (see Chap. 43); and (2) tumors not manifested by clinical signs of glucagon hypersecretion (for reviews of the literature see BORDI et al. 1979, 1981; LOKICH et al. 1980; RUTTMAN et al. 1980). In light microscopy, tumors associated with the glucagonoma syndrome (these are usually large, single, and malignant) appear most commonly formed by irregular association of trabecular and diffuse ("solid" or "medullary") arrays of cells with a wide variation of cytoplasmic and nuclear shapes. When assessed by immunocytochemistry with anti-glucagon and anti-glicentin antiserum (Fig. 9), glicentin-immunofluorescent cells frequently outnumber glucagon-reactive cells, a situation much alike that already described in the developing normal pancreas. Moreover, as in the latter case, glucagon-immunoreactive cells in the tumor always show superposable glicentin immunofluorescence. Ultrastructurally, the dense core granules observed in the tumor cells usually do not resemble the typical mature α-granule.

A-cell tumors not associated with the glucagonoma syndrome (usually small, benign, and multiple), may be found in patients with polyendocrine adenomatosis or with other pancreatic endocrine tumors (for reviews see BORDI et al. 1979, 1981; RUTTMAN et al. 1980). Microscopically, they also differ sharply from the malignant form. They appear composed of thin ribbons of cells, virtually all of them showing a superposable glucagon and glicentin immunofluorescence as in the normal adult pancreatic A-cells (Fig. 10). Ultrastructurally, the secretory granules of these cells usually resemble typical mature α-granules. Whether these small, not functionally manifested tumors may evolve in the larger malignant type is not known (BORDI et al. 1981).

As amply shown at the pancreatic level, the assessment of abnormal changes in an endocrine cell population (tumoral growth aside) is dependent on an adequate sampling procedure which ensures that the values obtained reflect properly the inherent regional differences (or stated differently, that these values are not biased by inherent regional differences). While the completing of such a procedure took several years for the pancreas, the size and the complexity of the compartment harboring extrapancreatic A-cells and related L-cells, the digestive tract, makes it a formidable task at this level. For this reason, the few data available at present are at best semiquantitative and far from clear-cut (POLAK et al. 1975). Qualitatively, L-cells have been described in intestinal metaplasia of the stomach, a level where they are not normally present (BORDI et al. 1978; BORDI and RAVAZZOLA 1979). In tumoral growth of the intestinal endocrine system, GLI-producing cells appear a frequent component of the carcinoids of the ileocecal and rectal regions (WILANDER et al. 1977, 1979; ALUMETS et al. 1980; FIOCCA et al. 1980). Against this extremely scanty background, one can only acknowledge a total ignorance of structure–function relationships in the cases described.

## E. Conclusions

One may summarize the main points of this chapter: (1) ultrastructurally, cytochemically, and immunocytochemically, the normal pancreatic A-cell and its secretory polypeptide content may now be considered as reasonably well characterized;

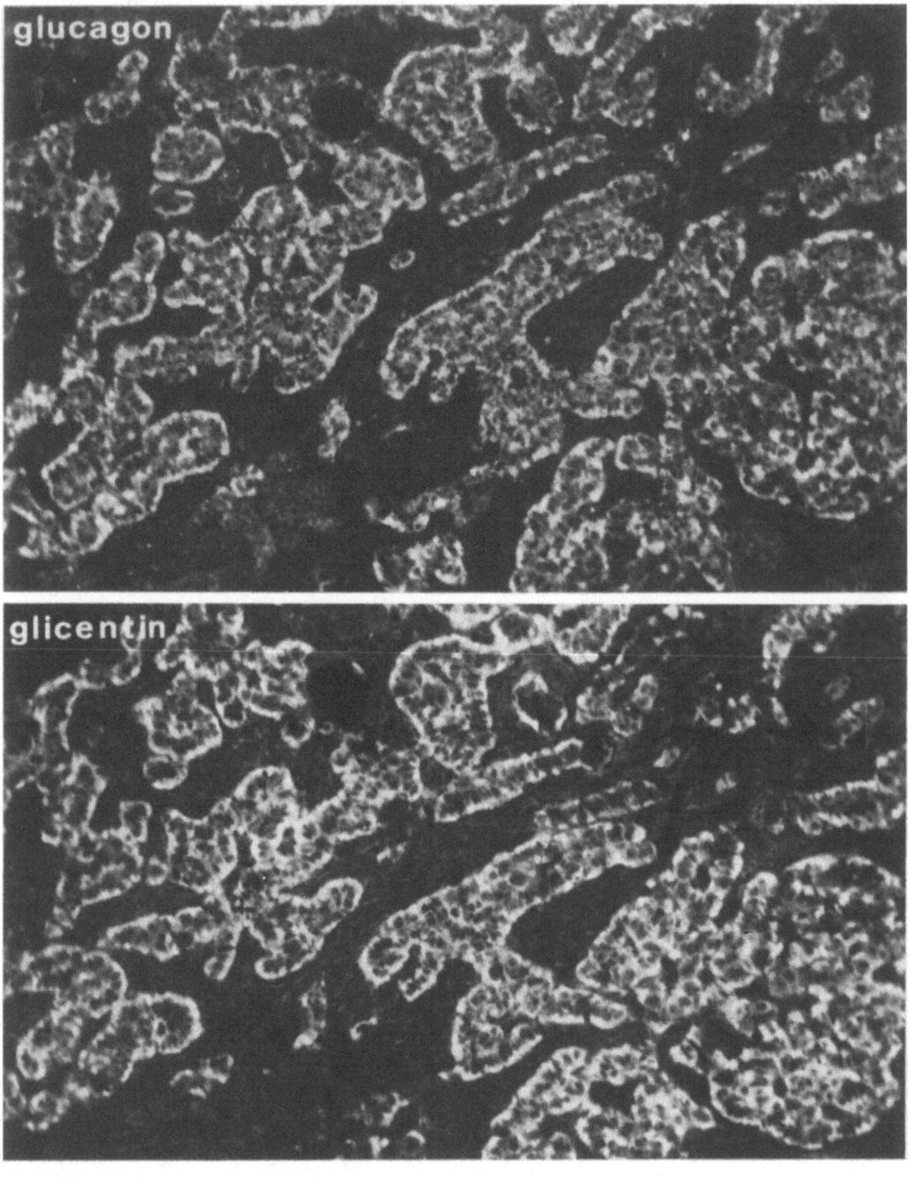


**Fig. 10.** Light micrographs of consecutive sections taken through a pancreatic A-cell tumor not accompanied by the glucagonoma syndrome. In this tumor, abundant glucagon-immunofluorescent cells are detectable and are superposable with an equally abundant number of glicentin-immunofluorescent cells. BORDI et al. (1981) ×185

(2) the distribution pattern of the A-cells among the islets, as well as in the entire pancreas, also appears fully understood in the normal adult human – their characterization in the diseased pancreas is beginning; (3) the role of these patterns, and of A-cell communication with neighboring islet cells, remains, on the other hand, undetermined; (4) the existence of an extrapancreatic A-cell is definitely estab-

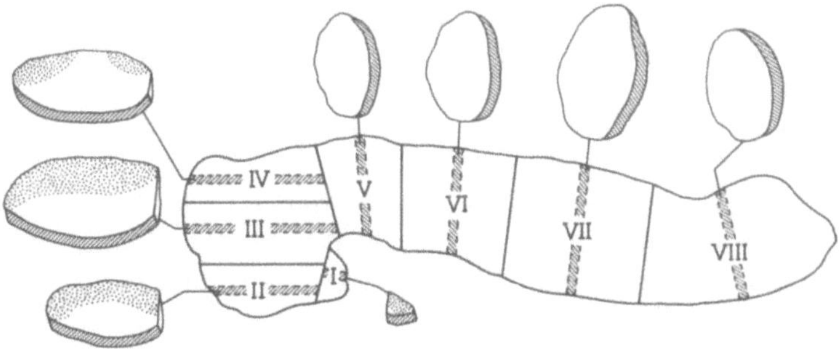


**Fig. 11.** Drawing of the shape of a human pancreas with the delimitation of the eight parts (I–VIII) sampled. In each part, the 5-mm thick slice, fixed and processed for light microscope immunofluorescence, is also represented. In the slices from parts I–IV, the stippled areas correspond to the pancreatic polypeptide-rich region. MALAISSE-LAGAE et al. (1979 b)

lished (not yet in the adult human stomach), as are the points of similarity and dissimilarity of this cell type with the intestinal L-cell regarding their morphology and stored polypeptides.

## F. Appendix

In view of the crucial importance, for comparative purposes, of unified approach in the estimation of endocrine cell populations in the human pancreas, we felt it is not out of the context of the present chapter to detail the technique used in collecting the data shown in Table 1 and Fig. 5, and to encourage prospective workers to comply with a similar approach.

### I. Sampling of the Pancreas

Pancreas are sampled in eight parts as shown in Fig. 11 (ORCI et al. 1978 b). The volume of each part is measured by immersing the sample in a graduated cylinder filled with saline and reading the increase in the level of the fluid; a 5-mm thick slice of tissue is then taken in the center of each sample and fixed for 24 h in Bouin's fluid.

### II. Immunofluorescence Technique

After dehydration through increasing concentrations of ethanol, the fixed slices are embedded in paraffin and cut serially into 5-μm thick sections. Consecutive serial sections are put individually on glass slides and one of each is stained with one of the following substances:

1. Hemalum–eosin (HE);
2. Guinea pig anti-insulin antiserum (Dr. P. H. WRIGHT, Indianapolis, Indiana), 1:100 or 1:200 dilution;
3. Rabbit anti-glucagon antiserum 15 K (Dr. R. H. UNGER, Dallas, Texas), diluted 1:20, or rabbit anti-glucagon antiserum K 5563 (Dr. L. HEDING, Copenhagen,

Denmark) absorbed with human albumin, diluted 1:100; or rabbit anti-glucagon antiserum M107/8 (Dr. R.A. DONALD, Christchurch, New Zealand) diluted 1:100;
4. Rabbit anti-pancreatic polypeptide antiserum (Dr. R.E. CHANCE, Indianapolis, Indiana), 1:100 or 1:200 dilution;
5. Rabbit anti-somatostatin (synthetic) antiserum (Dr. M.P. DUBOIS, Nouzilly, France) absorbed with human albumin, 1:100 or 1:200 dilution; or rabbit anti-somatostatin (synthetic) antiserum (Dr. S. ITO, Niigata, Japan) absorbed with human albumin, diluted 1:200.

Dilutions are carried out with phosphate-buffered saline (PBS) and bound antisera are revealed by the indirect immunofluorescence technique (COONS et al. 1955) as follows. After an incubation of 2 h at room temperature with each antiserum, the sections are rinsed in PBS, exposed to anti-rabbit or anti-guinea pig IgG (1:20 dilution) labeled with fluorescein isothiocyanate (Pasteur Institute, Paris) for 1 h, rinsed in PBS and counterstained with Evans blue (0.01% in PBS). Each section is coverslipped in 25% glycerin, diluted in PBS, and finally observed for fluorescence in a Leitz Orthoplan microscope (Leitz, Wetzlar, West Germany) equipped with a Ploemopak condensor. The specificity of each immunofluorescent staining is tested by control experiments in which sections are incubated with the respective antisera adsorbed with their respective antigens. No staining is observed under such conditions.

## III. Quantitative Evaluation

Four steps are performed:

1. Determination of the volume (μl) of the eight parts (I–VIII) of the gland obtained during sampling (see Sect. I of this Appendix).

2. Determination of the volume (μl) of the PP-rich region on anteroposterior, 5-μm thick sections of parts I–IV of the pancreatic head stained with anti-PP antiserum; the contour of the PP-rich region is outlined and its surface expressed as a percentage of the total surface of the section (both surfaces are measured by planimetry). The percentage obtained then multiplies the volume of parts I–IV determined under 1.

3. Determination of the volume density $V_v$ of the glandular (endocrine + exocrine) tissue in hemalum-eosin-stained sections of the PP-rich and glucagon-rich regions. This is performed by the point-counting method (WEIBEL 1969) at a low magnification (×72) to subtract from the total volume density the volume occupied by nonglandular structures (e.g., fat, large vessels, connective septa). The volume (μl) of the glandular tissue in each region is obtained by multiplying the respective values of the volume density $V_v$ by the volume (μl) of these regions obtained according to 1. and 2.

4. Determination of the volume density $V_v$ of each immunofluorescent cell type within glandular tissue in three lobules of the PP-rich region of the pancreas and in three lobules of the glucagon-rich region. This measurement is performed by the point-counting method at high magnification (×500). The expression of the volumes of immunofluorescent cells on a "per lobule" basis proved to be a more real-

istic expression of the endocrine cell content in the human pancreas than the "per islet" basis used in previous work on the rodent pancreas (ORCI et al. 1976b; STEFAN et al. 1978; BAETENS et al. 1979) since in the human PP-rich lobe, endocrine cells are grouped in ill-defined colonies. Moreover, the "per lobule" quantification allows one to take into account fluorescent cells dispersed outside the islets.

Point-counting is performed directly in the fluorescence microscope equipped with a drawtube through which the lattice of the test grid is projected onto the fluorescent image seen in the eyepiece. The original formula for point-counting (WEIBEL 1969) was modified (WEIBEL 1973) to take into account the overestimation of volume density of fluorescent cells due to the superposition of positive and negative structures within the thickness of the section.

Thus, for each antiserum:

$$V_v \text{ (volume density)} = \frac{P\text{ (cells)}}{P\text{ (glandular tissue)}} - \frac{T}{2} \times \frac{I\text{ (cells)}}{2P\text{ (glandular tissue)} \times d}$$

where

$P$ (cells) = number of intersection points of the test grid over immunofluorescent cells

$P$ (glandular tissue) = number of intersection points of the test grid over glandular tissue

$T$ = thickness of the section (5 μm)

$I$ (cells) = number of lines (vertical and horizontal) of the test grid intersected by fluorescent cells or cell clusters (correction for superposition)

$d$ = distance between horizontal and vertical lines of the test grid

The volume (μl) of each immunofluorescent cell population is then obtained by multiplying the respective values of volume density $V_v$ by the volume of glandular tissue calculated as detailed under 2 and 3 (Table 1).

The immunofluorescent evaluation of endocrine cell populations requires a sufficient degree of preservation of cellular structures and of antigenicity of the hormones. In our experience, we found that all pancreas fixed within 6 h after death, and most glands sampled 12 h postmortem, were suitable in both respects. The storage of the body in a cold room allows the extension of these delays for up to 24 h.

*Acknowledgments.* The work reported in this review was supported by the Swiss National Science Foundation, grant 3.668.80 and the Sandoz Foundation, Basle, Switzerland. The help of Ms. M. SIDLER-ANSERMET for photography and of Ms. I. BERNARD for typing is gratefully acknowledged.

## References

Alumets J, Håkanson R, O'Dorisio T, Sjölund K, Sundler F (1978) Is GIP a glucagon cell constituent? Histochemistry 58:253–257

Alumets J, Falkmer S, Grimelius L, Håkanson R, Ljungberg O, Sundler F, Wilander E (1980) Immunocytochemical demonstration of enkephalin and β-endorphin in endocrine tumours of the rectum. Acta Pathol Microbiol Scand [A]88:103–109

Baetens D, Rufener C, Srikant BC, Dobbs R, Unger RH, Orci L (1976) Identification of glucagon-producing cells (A-cells) in dog gastric mucosa. J Cell Biol 69:455–464

Baetens D, Stefan Y, Ravazzola M, Malaisse-Lagae F, Coleman DL, Orci L (1978) Alteration of islet cell populations in spontaneously diabetic mice. Diabetes 27:1–7

Baetens D, Malaisse-Lagae F, Perrelet A, Orci L (1979) Endocrine pancreas: three dimensional reconstruction shows two types of islets of Langerhans. Science 206:1323–1325

Baum J, Simons BF, Unger RH, Madison LL (1962) Localization of glucagon in the alpha cells in the pancreatic islet by immunofluorescent technics. Diabetes 11:371–374

Bordi C, Ravazzola M (1979) Endocrine cells in the intestinal metaplasia of gastric mucosa. Am J Pathol 96:391–398

Bordi C, Gabrielli M, Missale G (1978) Pathological changes of endocrine cells in chronic atrophic gastritis. An ultrastructural study on peroral gastric biopsy specimens. Arch Pathol Lab Med 102:129–135

Bordi C, Ravazzola M, Baetens D, Gorden P, Unger RH, Orci L (1979) A study of glucagonomas by light and electron microscopy and immunofluorescence. Diabetes 28:925–936

Bordi C, Ravazzola M, Orci L (1981) The morphology of glucagonomas. In: Unger RH, Orci L (eds) Glucagon. Physiology, pathophysiology and morphology of the pancreatic A cells. Elsevier, Amsterdam Oxford New York, p 399

Bussolati G, Capella C, Vassallo G, Solcia E (1971) Histochemical and ultrastructural studies on pancreatic A-cells. Evidence for glucagon and non-glucagon components of the α-granule. Diabetologia 7:181–188

Carpentier JL, Malaisse-Lagae F, Muller W, Orci L (1977) Glucagon release from rat pancreatic islets: a combined morphological and functional approach. J Clin Invest 60:1174–1182

Conlon JM (1980) The glucagon-like peptide. Order out of chaos? Diabetologia 18:85–88

Conlon JM (1981) Molecular forms of the glucagon-like polypeptides (IRG and GLI) in tissues and plasma. In: Unger RH, Orci L (eds) Glucagon. Physiology, pathophysiology and morphology of the pancreatic A-cells. Elsevier, Amsterdam Oxford New York, p 55

Coons AH, Leduc EH, Connolly JM (1955) Studies on antibody production. I. A method for the histochemical demonstration of specific antibody and its application to a study of the hyperimmune rabbit. J Exp Med 102:49–63

Deamer DW (1977) The relation of membrane ultrastructure to membrane function. In: Jamieson GA, Robinson DM (eds) Mammalian cell membranes, vol 4. Butterworth, London, p 1

Fiocca R, Capella C, Buffa R, Fontana P, Solcia E, Hage E, Chance RE, Moody AJ (1980) Glucagon-, glicentin-, and pancreatic polypeptide-like immunoreactivities in rectal carcinoids and related colorectal cells. Am J Pathol 100:81–92

Gepts W, De Mey J, Marichal-Pipeleers M (1977) Hyperplasia of pancreatic polypeptide cells in the pancreas of juvenile diabetics. Diabetologia 13:27–34

Grimelius L, Capella C, Buffa R, Polak JM, Pearse AGE, Solcia E (1976) Cytochemical and ultrastructural differentiation of enteroglucagon and pancreatic-type glucagon cells of the gastrointestinal tract. Virchows Arch [Cell Pathol] 20:217–228

Grube D, Voigt KH, Weber E (1978 a) Pancreatic glucagon cells contain endorphin-like immunoreactivity. Histochemistry 59:75–79

Grube D, Maier V, Raptis S, Schlegel W (1978 b) Immunoreactivity of the endocrine pancreas. Evidence for the presence of cholecystokinin-pancreozymin. Histochemistry 56:13–35

Helmstaedter V, Feurle GE, Forssmann WG (1977) Relationship of glucagon-somatostatin and gastrin-somatostatin cells in the stomach of the monkey. Cell Tissue Res 177:29–46

Hertzberg EL, Lawrence TS, Gilula NB (1981) Gap junctional communication. Ann Rev Physiol 43:479–491
Howell SL, Hellerström C, Tyhurst M (1974) Intracellular transport and storage of newly synthesized proteins in the guinea pig pancreatic A-cell. Horm Metab Res 6:267–271
Ito S, Iwanaga T, Kusumoto Y, Sudo N, Sano M, Suzuki T, Shibata A (1981) Is glucagon present in the human gastric fundus? Horm Metab Res 13:419–422
Korànyi F, Péterfy F, Szabò J, Török A, Guoth M, Tamàs G (1981) Evidence for transformation of glucagon-like immunoreactivity of gut into pancreatic glucagon in vivo. Diabetes 30:792–794
Lange R (1973) Histochemistry of the islet cells. In: Graumann W, Neumann K (eds) Handbuch der Histochemie, vol VIII, Suppl I. Fischer, Stuttgart, p 1
Larsson LI, Moody AJ (1980) Glicentin and gastric inhibitory polypeptide immunoreactivity in endocrine cells of the gut and pancreas. J Histochem Cytochem 28:925–933
Like AA, Orci L (1972) Embryogenesis of the human pancreatic islets: a light and electronmicroscopical study. Diabetes [Suppl 2] 21:511–534
Lokich J, Anderson N, Rossini A, Hadley W, Federman M, Legg M (1980) Pancreatic alpha cell tumors: case report and review of the literature. Cancer 45:2675–2683
Malaisse-Lagae F, Orci L, Perrelet A (1979a) Anatomic and hormonal markers of the ventral primordium in the human pancreas? N Engl J Med 300:436
Malaisse-Lagae F, Stefan Y, Cox J, Perrelet A, Orci L (1979b) Identification of a lobe in the adult human pancreas rich in pancreatic polypeptide. Diabetologia 17:361–365
Meda P, Kohen E, Kohen C, Orci L (1981) Couplage hétérocellulaire dans les cultures de cellules endocrines pancréatiques. C R Acad Sci [D] (Paris) 293:607–610
Meda P, Kohen E, Kohen C, Rabinovitch A, Orci L (1982) Direct communication of homologous and heterologous endocrine islet cells in culture. J Cell Biol 92:221–226
Michaels RL, Sheridan JD (1981) Islets of Langerhans: dye coupling among immunocytochemically distinct cell types. Science 204:862–865
Munger BL (1972) The histology, cytochemistry and ultrastructure of pancreatic islet A-cells. In: Lefebvre PJ, Unger RH (eds) Glucagon. Pergamon, Oxford New York, p 7
Orci L (1974) A portrait of the pancreatic B-cell. Diabetologia 10:163–187
Orci L (1977) Morphologic events underlying the secretion of peptide hormones. In: James VHT (ed) Endocrinology. Proc V th Intern Congr Endocrinol, Hamburg 1976. Excerpta Medica, Amsterdam London New York, p 7
Orci L (1982) Macro- and microdomains in the endocrine pancreas. Banting Lecture 1981. Diabetes 31:538–565
Orci L, Perrelet A (1977) Morphology of membrane systems in pancreatic islets. In: Volk BW, Wellmann KF (eds) The diabetic pancreas. Plenum, New York, p 171
Orci L, Perrelet A (1979) La microarchitecture des îlots de Langerhans. Pour Sci 22:30–44
Orci L, Perrelet A (1981) The morphology of the A-cell. In: Unger RH, Orci L (eds) Glucagon. Physiology, pathophysiology and morphology of the pancreatic A-cells. Elsevier, Amsterdam Oxford New York, p 1
Orci L, Unger RH (1975) Functional subdivision of islets of Langerhans and possible role of D-cells. Lancet 2:1243–1244
Orci L, Pictet R, Forssmann WG, Renold AE, Rouiller C (1968) Structural evidence for glucagon producing cells in the intestinal mucosa of the rat. Diabetologia 4:56–67
Orci L, Stauffacher W, Rufener C, Lambert AE, Rouiller C, Renold AE (1971) Acid phosphatase activity in secretory granules of pancreatic A-cells of normal rats. Diabetes 20:385–388
Orci L, Like AA, Amherdt M, Blondel B, Kanazawa Y, Martin EB, Lambert AE, Wollheim CB, Renold AE (1973) Monolayer cell cultures of neonatal rat pancreas: an ultrastructural and biochemical study of functioning endocrine cells. J Ultrastruct Res 43:270–297
Orci L, Amherdt M, Malaisse-Lagae F, Perrelet A, Dulin WE, Gerritsen GC, Malaisse WJ, Renold AE (1974) Morphological characterization of membrane systems in A- and B-cells of the Chinese hamster. Diabetologia 10:529–539
Orci L, Malaisse-Lagae F, Amherdt M, Ravazzola M, Weisswange A, Dobbs R, Perrelet A, Unger RH (1975a) Cell contacts in human islets of Langerhans. J Clin Endocrinol Metab 41:841–844

Orci L, Malaisse-Lagae F, Ravazzola M, Rouiller D, Renold AE, Perrelet A, Unger RH (1975b) A morphological basis for intercellular communication between A- and B-cells in the endocrine pancreas. J Clin Invest 56:1066–1070
Orci L, Baetens D, Ravazzola M, Stefan Y, Malaisse-Lagae F (1976a) Pancreatic polypeptide and glucagon: non-random distribution in pancreatic islets. Life Sci 19:1811–1816
Orci L, Baetens D, Rufener C, Amherdt M, Ravazzola M, Studer P, Malaisse-Lagae F, Unger RH (1976b) Hypertrophy and hyperplasia of somatostatin-containing D-cells in diabetes. Proc Natl Acad Sci USA 73:1338–1342
Orci L, Perrelet A, Gorden P (1978a) Less-understood aspects of the morphology of insulin secretion and binding. Recent Prog Horm Res 34:95
Orci L, Malaisse-Lagae F, Baetens D, Perrelet A (1978b) Pancreatic polypeptide-rich regions in human pancreas. Lancet 2:1200–1201
Palade E (1975) Intracellular aspects of the process of protein synthesis. Science 189:247–253
Patzelt C, Tager HS, Carroll RJ, Steiner DF (1979) Identification and processing of proglucagon in pancreatic islets. Nature 282:260–266
Polak JM, Pears AGE, Grimelius L, Marks V (1975) Gastrointestinal apudosis in obese hyperglycaemic mice. Virchows Arch [Cell Pathol] 19:135–150
Ravazzola M, Orci L (1979) Anatomie immunocytochimique des granules α dans le pancréas endocrine humain. C R Acad Sci [D] (Paris) 289:1161–1163
Ravazzola M, Orci L (1980a) Transformation of glicentin-containing L-cells into glucagon-containing cells by enzymatic digestion. Diabetes 29:156–158
Ravazzola M, Orci L (1980b) Glucagon and glicentin immunoreactivity are topologically segregated in the α-granule of the human pancreatic A cell. Nature 284:66–67
Ravazzola M, Siperstein A, Moody AJ, Sundby F, Jacobsen H, Orci L (1979a) Glicentin: a precursor of glucagon? Life Sci 25:287–290
Ravazzola M, Siperstein A, Moody AJ, Sundby F, Jacobsen H, Orci L (1979b) Glicentin immunoreactive cells: their relationship to glucagon-producing cells. Endocrinology 105:499–508
Ravazzola M, Unger RH, Orci L (1981a) Demonstration of glucagon in the stomach of human fetuses. Diabetes 30:879–882
Ravazzola M, Orci L, Perrelet A, Unger RH (1981b) Immunocytochemical quantitation of glicentin and glucagon during maturation of A-cell secretory granules. Diabetologia 21:319 (Abstract)
Roth J, Bendayan M, Orci L (1978) Ultrastructural localization of intracellular antigens by the use of protein A-gold complex. J Histochem Cytochem 26:1074–1081
Ruttman E, Klöppel G, Bommer G, Kiehn M, Heitz PhU (1980) Pancreatic glucagonoma with and without syndrome. Immunocytochemical study of 5 tumour cases and review of the literature. Virchows Arch [Pathol Anat] 388:51–67
Smith PH, Merchant FW, Johnson DG, Fujimoto WJ, Williams RH (1977) Immunocytochemical localization of a gastric inhibitory polypeptide-like material within A-cells of the endocrine pancreas. Am J Anat 189:585–590
Solcia E, Capella C, Vassallo G, Buffa R (1975) Endocrine cells of the gastric mucosa. Int Rev Cytol 42:223–286
Stefan Y, Malaisse-Lagae F, Yoon JW, Notkins AL, Orci L (1978) Virus-induced diabetes in mice: a quantitative evaluation of islet cell populations by immunofluorescence technique. Diabetologia 15:395–401
Stefan Y, Orci L, Malaisse-Lagae F, Perrelet A, Patel Y, Unger RH (1982a) Quantitation of endocrine cell content in the pancreas of non diabetic and diabetic humans. Diabetes 31:694–700
Stefan Y, Ravazzola M, Grasso S, Perrelet A, Orci L (1982b) Glicentin precedes glucagon in the developing human pancreas. Endocrinology 110:2189–2191
Sutherland EW, de Duve C (1948) Origin and distribution of the hyperglycemic-glycogenolytic factor of the pancreas. J Biol Chem 175:663–674
Tager HS (1981) Biosynthesis of glucagon. In: Unger RH, Orci L (eds) Glucagon. Physiology, pathophysiology and morphology of the pancreatic A-cells. Elsevier, Amsterdam Oxford New York, p 39

Thim L, Moody AJ (1981) The primary structure of porcine glicentin (proglucagon). Regul Peptides 2:139–150
Unger RH, Orci L (1981 a) Glucagon and the A-cell. Physiology and pathophysiology (first of two parts). N Engl J Med 304:1518–1524
Unger RH, Orci L (1981 b) Glucagon and the A-cell. Physiology and pathophysiology (second of two parts). N Engl J Med 304:1575–1580
Weibel ER (1969) Stereological principles for morphometry in electron microscopic cytology. Int Rev Cytol 26:235–302
Weibel ER (1973) Stereological techniques for electron microscopic morphometry. In: Hayat MA (ed) Principles and technics of electron microscopy, vol 3. Van Nostrand Reinhold, New York, p 237
Wellmann KF, Volk BW (1977) Histology, cell types and functional correlation of islets of Langerhans. In: Volk BW, Wellmann KF (eds) The diabetic pancreas. Plenum, New York, p 99
Wilander E, Portela-Gomes G, Grimelius L, Lundqvist G, Skoog V (1977) Enteroglucagon and substance P-like immunoreactivity in argentaffin and argyrophil rectal carcinoids. Virchows Arch [Cell Pathol] 25:117–124
Wilander E, Grimelius L, Portela-Gomes G, Lundqvist G, Skoog V, Westermark P (1979) Substance P- and enteroglucagon-like immunoreactivity in argentaffin and argyrophil midgut carcinoid tumours. Scand J Gastroenterol [Suppl 53] 14:19–25

CHAPTER 5

# Ontogeny and Phylogeny of the Glucagon Cell

S. FALKMER and S. VAN NOORDEN

## A. Introduction

Since an earlier review on the phylogeny and ontogeny of glucagon production (FALKMER and MARQUES 1972), published in a monograph on glucagon a decade ago, research on the gastroenteropancreatic (GEP) neuroendocrine system has undergone an explosively rapid progress, making several statements in that review obsolete and even false. The reason is that, to a great extent, they were based on observations made by means of silver staining techniques (GRIMELIUS and WILANDER 1980) and that during the 1970s immunohistochemical (IHC) methods – immunofluorescence and/or the PAP (peroxidase–antiperoxidase) procedures – and radioimmunoassays (RIA) have been developed, permitting observations that have revolutionised our view on the histophysiology and evolution of the GEP neuroendocrine system, including the glucagon-producing cells (cf. VAN NOORDEN and POLAK 1979). For instance, a major breakthrough in the research front was made in 1975 when it was found that the hypothalamic peptide, somatostatin, occurred not only in neuronal cells, but also in endocrine cells in the antral mucosa and the pancreatic islets (POLAK et al. 1975). Thus, the D-cells (formerly $\alpha_1$-cells) previously suspected of producing gastrin, were now shown to be not only the storage site, but also the place for actual production of somatostatin (cf. NOE 1981 a, b). Since then, practically all the GEP hormonal peptides have been demonstrated to have this dual distribution (cf. VAN NOORDEN and FALKMER 1980; FALKMER et al. 1981 a) and thus to form the brain–gut axis of the diffuse neuroendocrine system (VAN NOORDEN and POLAK 1979; DOCKRAY and GREGORY 1980). Glucagon is one of the latest additions to this group, having recently been localised in several areas of the mammalian (including human) brain (cf. TAGER et al. 1980; SANDERS et al. 1981; see also Chap. 33).

Another major advance during these last few years has been the elucidation of the amino acid sequences of the pancreatic glucagon of several species (SUNDBY 1976). Even more spectacular is the determination of the tertiary molecular structure of pancreatic glucagon (BLUNDELL 1979) and the clarification of the molecular and biosynthetic interrelationship between pancreatic glucagon and glicentin (enteroglucagon, proglucagon) (LUND et al. 1980; SHIELDS et al. 1981; THIM and MOODY 1981). These topics are reviewed in detail in Chaps. 6, 7, 10 and 11.

Glucagon shares amino acid sequences with some other GEP neurohormonal peptides, in particular with secretin, VIP (vasoactive intestinal peptide) and GIP (glucose-dependent insulin-releasing peptide or gastric inhibitory peptide); together these four neurohormonal peptides make up a hormone family, presumed to have developed from an ancestral molecule (DOCKRAY 1977, 1979 a, b; VAN NOORDEN and POLAK 1979). This is the reason why this chapter is not strictly limited to the cells producing pancreatic glucagon and glicentin, but also covers some other aspects of those producing other peptides of this hormone family and its affiliations. It is thus evident that there is urgent need for an updated review of the phylogeny and ontogeny of the glucagon-producing cell. In the present survey some new, still unpublished, observations from our laboratories are also given.

## B. Material, Methods, Nomenclature

Of course, a study of the phylogeny of any part of the GEP neuroendocrine system must be limited to extant species and, thus, the true course of evolution can only be guessed at from fragmentary information (cf. VAN NOORDEN and POLAK 1979). It became a matter of excitement when in the 1970 s it was discovered that immunological methods, using antisera raised against mammalian or avian GEP hormonal peptides, could be used to detect antigenically similar substances in poikilothermic vertebrates and even in invertebrates (cf. FALKMER 1980 a).

However, because of changes in structure and function of GEP neurohormonal peptides over the millions of years of evolution (cf. DOCKRAY 1977, 1979 a, b), caution must be exercised in stating that a certain peptide is present in an animal merely on the basis of an immunochemical reaction with a heterologous antiserum (cf. VAN NOORDEN and FALKMER 1980). For full identification, IHC evidence should be added to that obtained from extraction, chromatography and RIA, followed by isolation, purification and molecular sequencing of the peptide. Biological assays should be performed in the same species, and perhaps also in other species, and thus some idea of the function of the peptide might be attained. Unfortunately, this battery of tests cannot usually be applied at present, so that most of the statements made in this chapter are based on the use of one or two investigative methods only. Nevertheless, enough information is now available on animals from widely separated groups for an attempt to be made at an integrated concept of evolutionary progress. So far, the only GEP neurohormonal peptide in a low order animal that has been completely investigated according to this criterion is the insulin produced by the cyclostome, *Myxine glutinosa* (EMDIN 1981; FALKMER and EMDIN 1981). Details of the various procedures used, the key species chosen and the tissues investigated, have previously been given elsewhere (cf. VAN NOORDEN and FALKMER 1980; FALKMER et al. 1981; EMDIN 1981). When possible, the nomenclature principles proposed by CONLON (1980) have been followed.

## C. Prokaryotes, Eukaryote Protozoa, Coelenterates

A simple evolutionary tree (Fig. 1) shows the progression of animal forms from the simplest unicellular protozoa to the coelenterates, with two layers of cells, and thence to the more complicated groups. An accepted theory divides the animal kingdom above the coelenterates into two evolution lines, based on embryological criteria: the protostomian and the deuterostomian, the latter containing the vertebrates (cf. FALKMER et al. 1973).

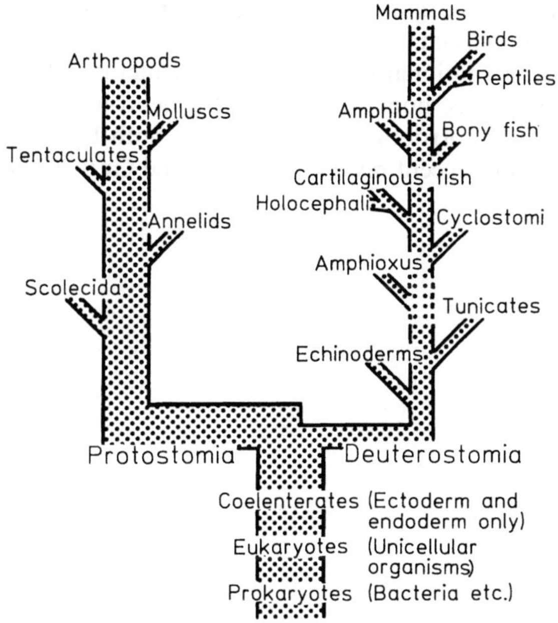


**Fig. 1.** Simplified evolutionary tree, following GROBBEN's classification (cf. FALKMER et al. 1973) of the animals above the coelenterates into two evolutionary lines, protostomian and deuterostomian, distinguished by their embryology. The overwhelming majority of the multicellular organisms belong to the protostomian lines; only some of the major groups have been indicated

As to the occurrence of catecholamines and GEP neurohormonal peptides in organisms below the coelenterates, there have recently appeared some reports, claiming that adrenaline/noradrenaline (cf. COUPLAND 1979) and peptides identical with, or closely related to, insulin, somatostatin, human chorionic gonadotropin (HCG), adrenocorticotropic hormone (ACTH), and neurotensin, have been found by means of RIA and/or biological tests in extracts of species representing unicellular eukaryotes (fungi and ciliated protozoa) and even prokaryotes (bacteria) (LEROITH et al. 1980, 1981 a, b; BHATNAGAR and CARRAWAY 1981). Similarly, it has been claimed that some unicellular organisms contain receptors not only for catecholamines, but also for GEP neurohormones (cf. CSABA 1981). The support for the hypothesis that catecholamine and GEP neurohormonal peptides already functioned as messenger signals at this evolutionary level (more than $2 \times 10^9$ years ago) comes from the existence of humoral signalling mechanisms for cell–cell communication by slime mould, yeast and myxobacteria, from known effects in protozoa and bacteria of vertebrate hormones (catecholamines, insulin, opiates) that can be inhibited by specific receptor-blocking agents, and from the conservation of the biologically important regions of the hormonal molecule (cf. CSABA 1981; LEROITH et al. 1981 b). However, no IHC observations or ultrastructural studies supporting these claims that catecholamines and GEP neurohormonal peptides occur in prokaryotes and eukaryotes have been given (COUPLAND 1979). Moreover, the hormonal content found in the extracts has often been extraordinarily low. In any case, no work on glucagon and/or glicentin in subcoelenterian organisms seems to have been published.

It is not until the evolutionary stage of the coelenterates that the presence of immunoreactive substances, homologous to GEP neurohormonal peptides, has been found by means of corroborative results from both IHC examinations and RIA. As summarised in a recent review (FALKMER and GRIMMELIKHUIJZEN 1981),

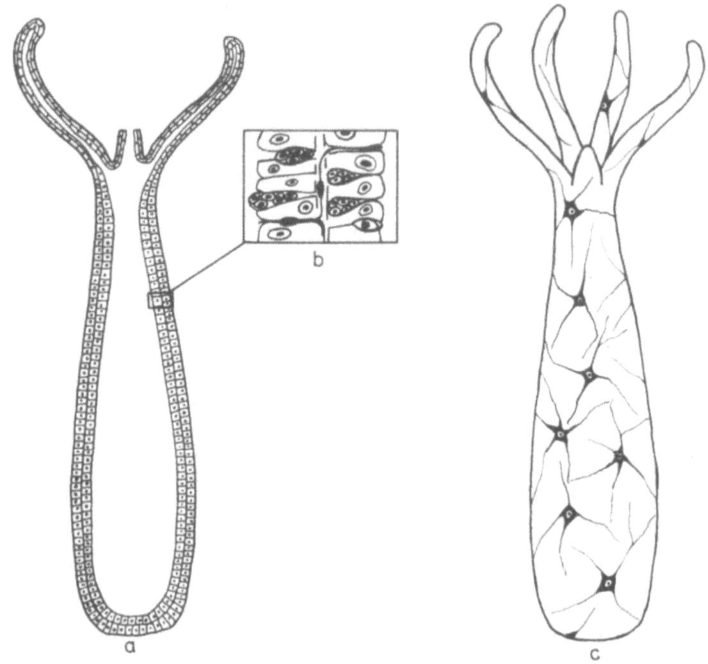


**Fig. 2a–c.** Schematic illustrations (partly after GRIMMELIKHUIJZEN and SCHALLER 1979) of the freshwater coelenterate *Hydra attenuata*, showing (**a**) its general structure with a foot disk (*bottom*), a mouth opening, surrounded by tentacles (*top*), and a large gastric cavity (*centre*). The body wall (**b**) consists of an ectodermal and an endodermal cell layer, separated by the thin acellular mesogloea, containing collagen and a fine net of nerve cells (**c**) densest in the mouth and foot regions (not indicated). The nerve cells (deliberately exaggerated in size and diminished in number) possess a cilium, they form synapses with other nerve cells, nematocytes and epitheliomuscular cells, and contain neurosecretory granules of the same type as in peptidergic nerves of higher animals. This is the most primitive nervous system in the animal kingdom

---

**Fig. 3.** Simplified schematic illustration (after FALKMER and GRIMMELIKHUIJZEN 1981 and GRIMMELIKHUIJZEN 1983) of the IHC distribution of some of the neurohormonal immunoreactive neurons in *Hydra attenuata*. The gastrin/CCK-like neurons are concentrated in the mouth region. The neurotensin-like (NT) nerve cells are more widely distributed with a condensation in an ectodermal ring above the foot plate and in a circular area along the bases of the tentacles, outside the gastrin/CCK cells in the mouth region. The neurons, showing substance P immunoreactivity (SP), are localised in the foot plate and in the tentacles. Bombesin-like and oxytocin-like immunoreactivities partly coexist and show a similar distribution pattern as SP, whereas that of the cardiac excitatory peptide FMRF-amide is more like that of NT. No neurohormone immunoreactivity has been observed in the ectodermal or endodermal epithelial cells. A high molecular weight glucagon-immunoreactive peptide has been extracted from *Hydra* at a concentration of 1 pmol/g wet weight, but its IHC distribution pattern is still unknown. (C.J.P. GRIMMELIKHUIJZEN and J.J. HOLST 1981, unpublished work) ▶

the fine network of neurofibrils in *Hydra attenuata*, situated in the mesogloea between the ectoderm and the endoderm of this primitive multicellular organism (Fig. 2), obviously contains peptides, related to cholecystokinin (CCK), substance P and neurotensin, with a distinct localisation to certain regions, characteristic for each of the three immunoreactive peptides (Fig. 3; GRIMMELIKHUIJZEN et al. 1980, 1981 a, b). Similar observations have been made in another coelenterate, the sea anemone, *Tealia felina*, (GRIMMELIKHUIJZEN et al. 1980). The hormonal contents, determined by RIA of acetic acid or methanol extracts of *Hydra*, were at least 5, 1, and 300 fmol/g wet weight for CCK, substance P and neurotensin, respectively. More recently, a bombesin-like immunoreactive peptide has also been

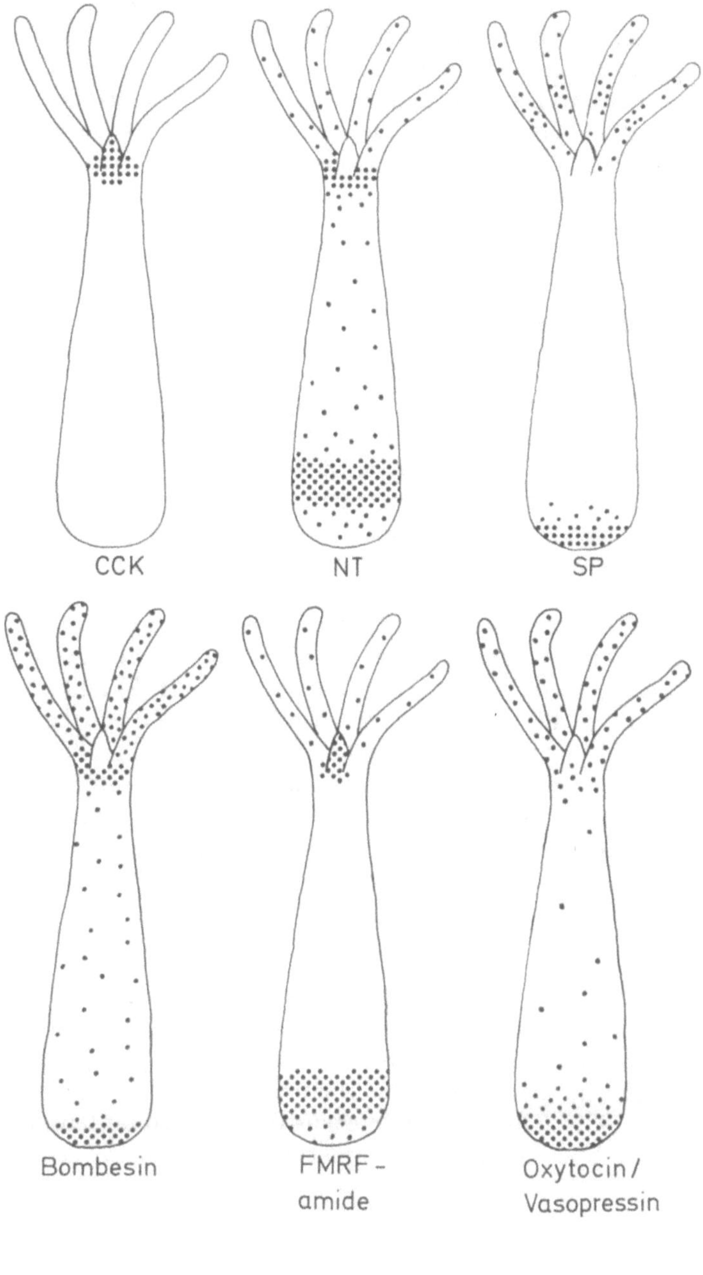

detected in *Hydra* at a concentration of at least 0.2 pmol/g wet weight, showing an IHC distribution pattern similar to that of substance P (GRIMMELIKHUIJZEN et al. 1981c). Also other hormonal peptides seen to exist in this primitive nervous system (Fig. 3), partly coexisting in the same neurones (GRIMMELIKHUIJZEN 1983). Preliminary observations, where the IHC support is still lacking, indicate that some glucagon-like and somatostatin-like immunoreactive peptides might also occur in this nervous system at concentrations of 1 and 800 pmol/g wet weight, respectively; these peptides seemed from their elution patterns after gel filtration to have a higher molecular weight than their mammalian counterparts (C.J.P. GRIMMELIKHUIJZEN and J.J. HOLST 1981, unpublished work).

Until now, none of the "classical" neurotransmitters, like the biogenic amines and acetylcholine, have been unambiguously demonstrated in *Hydra attenuata*; it was tempting, therefore, to speculate that peptide hormones were the first in evolution to act as neurotransmitters and that only later did the classical transmitters emerge (FALKMER and GRIMMELIKHUIJZEN 1981). However, by several histochemical techniques, catecholamines have been observed in the nervous system of both *Hydra* and other coelenterates in other laboratories (cf. COUPLAND 1979; FUJITA et al. 1980), consequently, this speculation remains to be further substantiated by correlated IHC and chemical assay procedures.

## D. Protostomian Invertebrates

About 14 phyla and some 95% of all animal species are found in the groups belonging to this evolutionary line; nevertheless, only fragmentary information is available as regards the diffuse neuroendocrine system in these animals (cf. VAN NOORDEN and POLAK 1979; VAN NOORDEN and FALKMER 1980). In general, it seems that protostomian invertebrates, when compared with vertebrates, possess few endocrine organs, whereas neurosecretory cells are numerous (cf. JOOSSE 1979), indicating that many physiological processes are regulated by neurohormones (SCHOT et al. 1981).

Although isolated reports occur about the presence of a hyperglycaemic factor in homogenates of the suprapharyngeal ganglia of an annelid, the common earthworm, *Lumbricus terrestris* (MCVAY LAWRENCE et al. 1972) and a neuronal localisation by IHC of PP (pancreatic polypeptide) and VIP in the same species (SUNDLER et al. 1977), it is not until the evolutionary level of molluscs and arthropods that at least some species have been studied with correlated IHC, RIA, and electron microscopic (EM) procedures in order to trace the appearance of neuroendocrine cells, showing GEP neurohormonal immunoreactivity in general, and glucagon/glicentin immunoreactivity in particular.

### I. Molluscs

The molluscan central nervous system, with its convenient "giant" neurosecretory neurons, has been an object of study for neurophysiologists for many years (cf. JOOSSE 1979). Recently, it has been possible to investigate the system for the presence of GEP neurohormonal peptides different from, or akin to, vertebrate ones (cf. SCHOT et al. 1981). The most comprehensive studies so far have been made in two gastropod molluscs, the freshwater or terrestrial snails, *Lymnaea stagnalis*, and

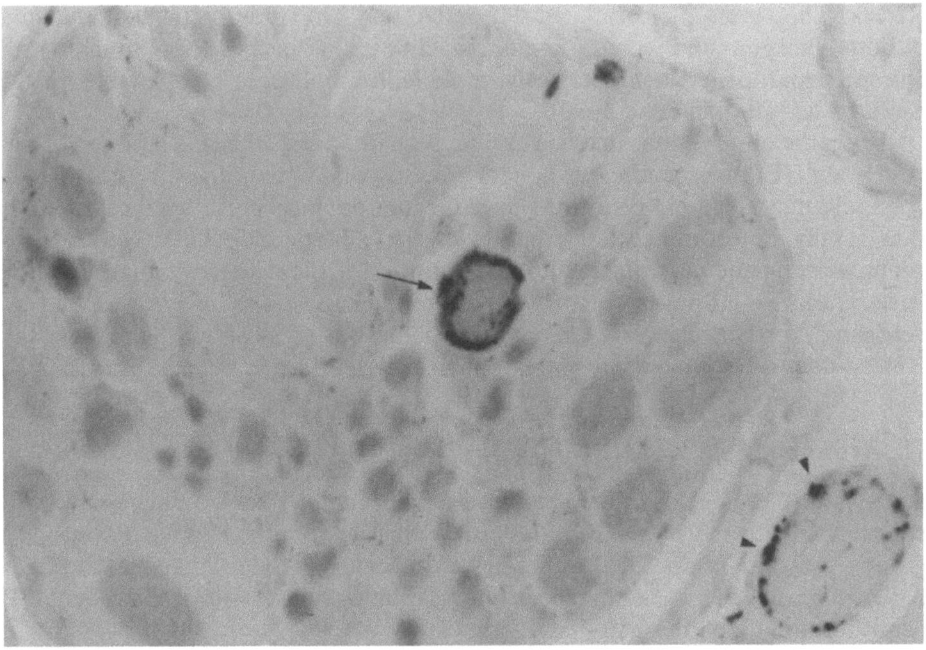

**Fig. 4.** Medium power photomicrograph of the visceral ganglion in a gastropod mollusc, the freshwater snail, *Lymnaea stagnalis*, freeze-dried and fixed in *p*-benzoquinone, paraffin embedded, cut and immunostained for glucagon by the PAP technique, using an antibody to pig glucagon. The *arrow* indicates an immunostained ganglion cell. The *arrowheads* indicate glucagon-immunoreactive fibres in the periphery of the intestinal nerve, suggesting that the nerve is a neurohaemal area whence the glucagon-like material is released into the blood. SCHOT et al. (1981) ×400

*Achatina fulica*, where the results of broad IHC investigations have recently been published (SCHOT et al. 1981; VAN NOORDEN et al. 1980).

Although immunostaining with an antibody to the related GEP hormonal peptide, secretin, showed some positively reactive fibres in the peripheral nerve supply to the intestine in the giant African snail, *Achatina fulica* (VAN NOORDEN and FALKMER 1980; VAN NOORDEN et al. 1980), neuronal glucagon-like immunoreactivity has, so far, been shown only in IHC-reactive perikarya and/or nerve tracts in the nervous system of the pond snail, *Lymnaea stagnalis*, when antibodies to glucagon, secretin, VIP, and even GIP, were used in addition to antisera raised against a dozen other mammalian neurohormonal peptides (Fig. 4; SCHOT et al. 1981). As with several other peptides demonstrated in this species (SCHOT et al. 1981), the various antisera identified completely separate sets of neurons so that, although the identification at present rests only on IHC tests, there is strong evidence that these two molluscs, and presumably others too, utilise a wide range of peptides to regulate their bodily functions. However, it must be remembered that until the molecular structures of the various invertebrate peptides have been analysed, little of use can be said about their relationships.

There appears to be a general belief that only a few cells of endocrine type occur in the alimentary tract of molluscs (FRITSCH et al. 1976; PLISETSKAYA et al. 1978).

Recently, however, in a correlated IHC, RIA, and EM study, significant amounts of both glucagon- and secretin-immunoreactive endocrine cells were found in the gut and hepatopancreas, respectively, of a bivalve mollusc, *Cardium* (or *Cerastoderma*) *edule* (BANKS et al. 1980).

From the observations made in molluscs it can – against the background of the vertebrate GEP neuroendocrine system – be speculated that digestive tract endocrine cells may represent a more advanced regulatory mechanism than neurosecretion. It may, however, be simply that the digestive endocrine cells developed more prominence in the vertebrates and invertebrates of the deuterostomian line (see Sects. E and F). However, since only a few molluscan species have as yet been investigated, further revelations may again change our ideas about the neuroendocrine system of molluscs and protostomian invertebrates in general.

## II. Arthropods

Reports have been given of the occurrence of glucagon-like or hyperglycaemic/glycogenolytic neurohormonal peptides in several species of the two main arthropod groups, crustaceans and insects. Studies of insects in both larval and adult stages have also provided some ontogenetic information on glucagon-like and other GEP neurohormonal peptides (EL-SALHY 1981).

### 1. Crustaceans

The decapod crustaceans (shrimps, crayfish, crabs, lobsters etc.) possess a most diversified neuroendocrine organ in the eyestalk, composed of neurosecretory cells and their neurohormonal storage and release organ, the sinus gland (cf. KLEINHOLZ and KELLER 1979). Like the vertebrate pituitary gland, its neurohormonal substances are vital in the regulation of a great variety of physiological functions.

Among the many eyestalk neurohormonal peptides is the hyperglycaemic hormone (HGH), known since 1944 (cf. KLEINHOLZ and KELLER 1979). Owing to its hyperglycaemic and glycogenolytic activities, HGH has been supposed to be a glucagon-like substance in crustaceans, homologous to the glucagon-like peptides isolated from the central nervous system of insects (see Sect. D.II.2). When, however, the amino acid composition of HGH from two kinds of crabs (*Cancer magister* and *Carcinus maenas*) was settled, the data obtained did not suggest any similarity between crustacean HGH and glucagon (cf. KELLER and WUNDERER 1978). The *Carcinus* hormone molecule was found to contain 57 amino acid residues, and a minimum molecular weight of 6,726 daltons was calculated (KELLER and WUNDERER 1978). Using antisera raised against *Carcinus* HGH (JAROS and KELLER 1979) and *Astacus* HGH (VAN HERP and VAN BUGGENUM 1979), respectively, IHC identifications of the HGH producing and storing cells have also been made in two of the decapod crustacean species. So far, no clear-cut evidence has been obtained that glucagon/glicentin cells occur in the central nervous system of crustaceans.

In the crustacean digestive tract, however, it has been claimed that a glucagon-like immunoreactive peptide is present. By using two antisera raised against mammalian glicentin (one cross-reacting with "high molecular GLI" only; the other with "all GLI fractions"), a biologically and immunologically glucagon-like

sialoglycopeptide was isolated from acid–alcohol extracts of the midintestinal gland of the crayfish, *Astacus fluviatilis* (MAIER et al. 1975). No corroborative IHC or EM investigations of the cellular site for this glucagon/glicentin production in the crayfish seem to have been published yet.

## 2. Insects

In the insects, however, there are more solid data as to the occurrence of glucagon/glicentin-like peptides, both in the central nervous system and in the digestive tract (KRAMER 1983). Here, also, IHC evidence has been obtained that glucagon/glicentin, like most other GEP neurohormonal peptides, has not only its phylogenetic but also its ontogenetic origin in the nervous system (EL-SALHY 1981). Some of these statements are, however, still controversial.

It was not until 1961, i.e. 17 years after the discovery of the HGH in the eyestalk of decapod crustaceans, that hyperglycaemic activity was found also in the neurosecretory system of insects (cf. BOUNIAS 1979), notably in the corpora cardiaca (TAGER et al. 1975, 1976; ZIEGLER 1979). In the best studied insects, the hormonal substance was reported to function similarly to glucagon in vertebrates: it increased both the glucose and trehalose levels in the circulating haemolymph in vivo, it stimulated glycogenolysis and the activation of phosphorylase in the fat body in vitro, and it seemed to be a low molecular weight polypeptide (cf. TAGER et al. 1975, 1976; BOUNIAS 1979). By means of gel filtration and RIA, its hormonal nature was demonstrated by the fact that a glucagon-like peptide was found to be present in the haemolymph from larvae and pupae of the tobacco hornworm, *Manduca sexta* (KRAMER et al. 1980). In a brief report at a recent "brain–gut axis" symposium, it was stated that glucagon-immunoreactive dendrites of neurons could be identified in the brain of the honey bee, *Apis mellifera* (MAIER et al. 1981). In contrast, other reports claim that no glucagon-immunoreactive nerve cells or nerve fibres occur in the insect brain (YUI et al. 1980), and that the hyperglycaemic factor isolated differs in several important respects biochemically and physiologically from glucagon (ZIEGLER 1979). No such comprehensive correlated IHC and RIA investigations as those recently made for insulin-immunoreactive (DUVE and THORPE 1979; DUVE et al. 1979) and pancreatic polypeptide (PP)-immunoreactive (DUVE and THORPE 1980; DUVE et al. 1981) cells in the brain of the blowfly, *Calliphora vomitoria*, have yet been published as regards neuronal glucagon cells in any insect (KRAMER 1983).

Some aspects on the ontogeny of the neuronal glucagon-immunoreactive cells in insects were supposed to have been obtained when the central nervous system and the gut of the fifth instar larva of the hoverfly, *Eristalis aeneus*, were systematically investigated by means of IHC, using a great number of antisera, raised against mammalian GEP neurohormonal peptides (EL-SALHY et al. 1980; EL-SALHY 1981). Whereas glucagon-immunoreactive nerve cells (but no nerve fibres) – in addition to insulin-, somatostatin-, PP-, secretin-, gastrin/CCK/caerulein-, substance P-, enkephalin-, and endorphin-immunoreactive neurons and/or nerve fibres – were present in the central nervous system, no GEP-immunoreactive cells at all were found in various parts of the alimentary tract. These observations were interpreted as a support for the hypothesis that glucagon cells belong to those GEP neuroendocrine elements that appear as neuronal structures of the brain before they begin to inhabit the digestive tract (FALKMER et al. 1981a).

However, recent comprehensive ultrastructural studies in insects have shown that, in several lepidopteran species, endocrine cells or basally granulated cells are sparsely scattered throughout the midgut, both in larval/pupal stages and in the adult (ENDO and NISHIITSUTSUJI-UWO 1981). Greater numbers of such cells were found in the midgut caeca of the cockroach, *Periplaneta americana*. They were tentatively correlated with the human gastrointestinal endocrine cell classification on the basis of their shape and size. One cell type was said to resemble the human glicentin-producing L-cell (NISHIITSUTSUJI-UWO and ENDO 1981). Among these presumed endocrine cells were some which showed glicentin-like immunoreactivity (IWANAGA et al. 1981). Moreover, in the fifth instar larva of *Manduca sexta*, acid–ethanol extracts of the midgut were found to contain a glicentin-like immunoreactive peptide with a high molecular weight (approximately 15,000 daltons); it showed no reactivity with antisera specific for 3,500 daltons glucagon (TAGER and KRAMER 1980). The reasons for the failure to detect immunoreactive gut endocrine cells of the hoverfly larva (EL-SALHY et al. 1980) remain to be investigated; reasonably, it is a matter of differences in the specificities of the antisera used and those of the structure of the large glicentin molecule (LARSSON and MOODY 1980). There may also be significant differences between the Diptera, the Lepidoptera, and the Dictyoptera.

The main physiological role of the insect glucagon/glicentin-like neurohormonal peptide (or peptides) is still poorly understood (cf. ZIEGLER 1979). It is, however, well established that trehalose and glycogen levels are modulated by insulin-like and glucagon-like peptides (cf. KRAMER et al. 1980) and that ox/pig glucagon administration evokes hyperglycaemia and (to a lesser extent) hypertrehalosaemia in insects (BOUNIAS 1979). Therefore, it would not be surprising if the insect glucagon/glicentin-like peptides were soon found to be involved in regulating mechanisms in carbohydrate and lipid metabolism in a similar way to mammalian glucagon.

## E. Deuterostomian Invertebrates

### I. Echinoderms, Hemichordates

There seems to be no recent information on the presence of GEP neurohormonal regulatory peptides in these two groups, but in a recent paper WELSCH and DILLY (1980) described the ultrastructure of the hemichordate alimentary tract. Although there were apparently no endocrine cells in the mucosa, there were, between the epithelial cells, nerve endings that contained several types of secretory granules, larger than classical adrenergic or cholinergic vesicles and reminiscent of peptidergic nerves. It could be of great interest to know whether these nerve endings contain substances similar to GEP neurohormonal peptides. If so, they might, as the authors suggested, represent an early stage in the derivation of gut endocrine cells, if these did indeed evolve from neurosecretory neurons. Echinoderms also have intraepithelial alimentary tract nerve endings. Here, however, insulin cells are known to occur (WILSON and FALKMER 1965).

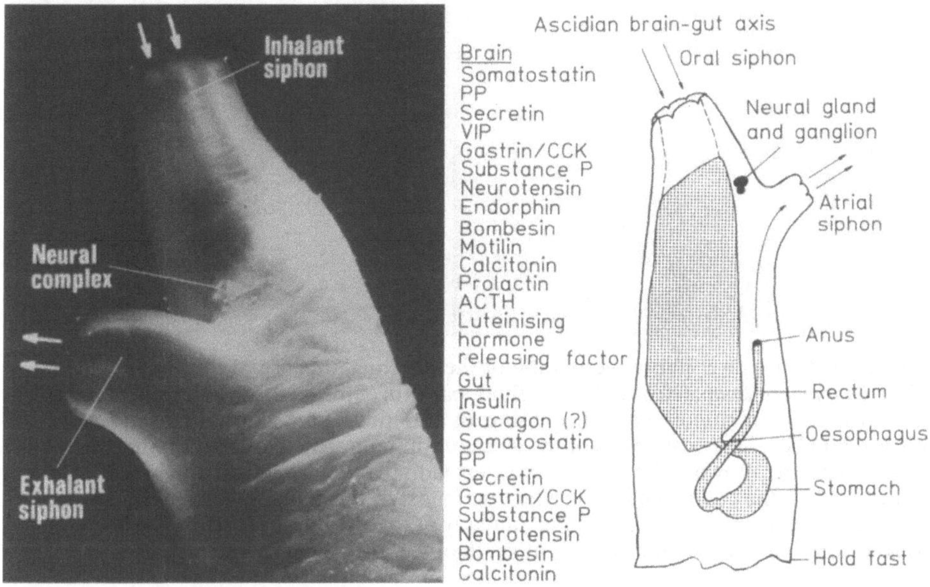


a b

**Fig. 5.** **a** Gross photograph of the neural complex (neural ganglion and neural gland) in a tunicate, the sea squirt *Ciona intestinalis*. The neural complex is situated between the inlet and outlet siphons as a whitish spot, measuring scarcely more than 1 mm in diameter, even in large specimens. **b** Schematic representation of our present knowledge of the tunicate GEP neuroendocrine system, particularly its brain–gut axis. So far, no glucagon-like immunoreactive cells have been discovered in these sedentary feeders

## II. Protochordates (Tunicates, Amphioxus)

Considerably more attention has been paid to the GEP neuroendocrine system of the protochordates, both to tunicates (urochordates) and to the cephalochordate amphioxus. The favoured position of these animals as possible representatives of the early vertebrates is justified by the number and distribution of their GEP neurohormonal peptides. A variety of such peptides is identified by IHC and, in some cases, by RIA in the central and peripheral nervous system and also in a mixed population of apparently endocrine cells dispersed along the alimentary tract (Figs. 5 and 6). In this, the protochordates bear more resemblance to their putative vertebrate descendants than to their presumed invertebrate forebears (cf. BEVIS and THORNDYKE 1981).

Deuterostomian invertebrate GEP neuroendocrine cells belonging to the glucagon family have not as yet been widely reported, but glucagon-like immunoreactivity has been localised by IHC to some alimentary tract endocrine cells in *Branchiostoma lanceolatum* (the amphioxus) (Fig. 6; VAN NOORDEN and PEARSE 1976; REINECKE 1981) and has also been shown by RIA of extracts of whole animals as reported in the previous glucagon phylogeny review (FALKMER and MARQUES 1972).

A VIP-like substance has been shown by IHC in the neural ganglion (Fig. 5) of the sea squirt, *Ciona intestinalis* (Ascidiacaea), as has a secretin-like substance

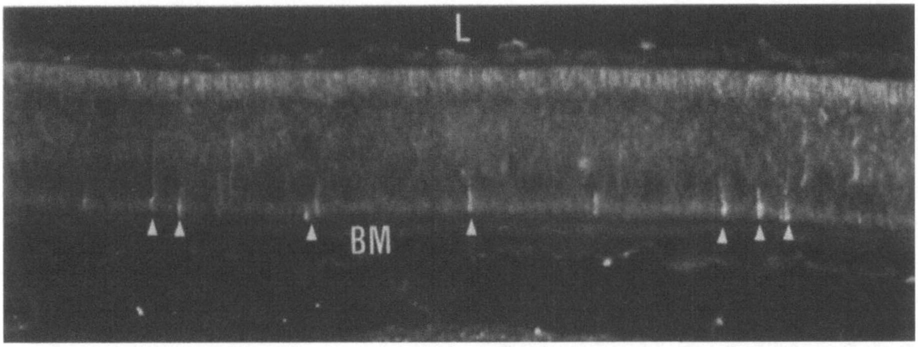


**Fig. 6.** Medium power photomicrograph of the intestinal mucosa of the amphioxus, *Branchiostoma lanceolatum*, fixed in Bouin's fluid, paraffin embedded, cut and stained for glucagon by indirect immunofluorescence, using an antibody to pig glucagon (as in Fig. 4). *Arrowheads* indicate the basal parts of the glucagon-immunoreactive cells that (in consecutive sections) can be shown to be of open type, extending all the way from the basement membrane (*BM*) to the gut lumen (*L*). The fact that most of the immunoreactive material occurs near the basal membrane indicates that, already at this evolutionary stage, the hormone release is directed towards the vessels rather than towards the gut lumen. ×344

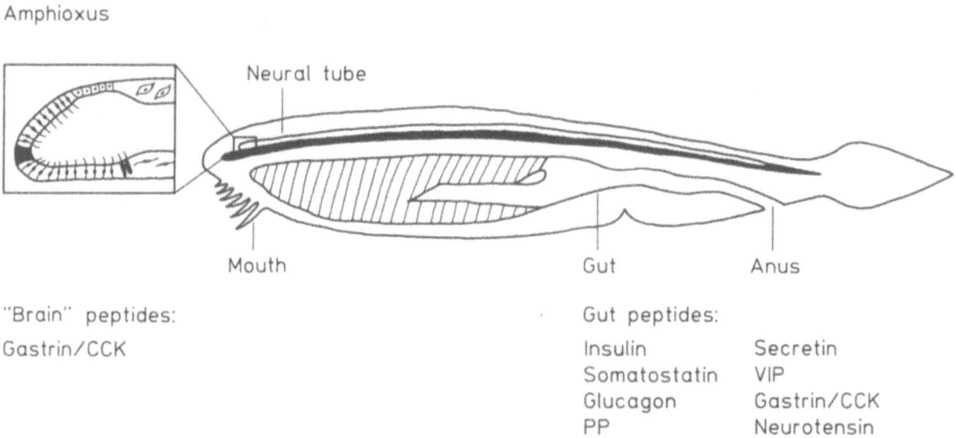


**Fig. 7.** The brain–gut axis in the amphioxus, schematically illustrated in a similar manner as for the tunicates in Fig. 5b. In this faster-moving representative of the protochordates, glucagon-immunoreactive cells occur, indicating a regulatory role for glucagon in eliciting a quick supply of energy

(FRITSCH et al. 1981). A secretin-like immunoreaction is also found in cells of the stomach of another ascidian, *Styela clava* (BEVIS and THORNDYKE 1979), but the secretory cells of the stomach epithelium in this tunicate species respond to infusions, not of secretin or glucagon, but of CCK (BEVIS and THORNDYKE 1981). The central nervous system of the amphioxus has still not been investigated systematically by IHC/RIA for the occurrence of GEP neurohormonal peptides. Recently, however, a discrete population of gastrin/CCK-immunoreactive cells has been reported to occur in its cerebral vesicle (Fig. 7; THORNDYKE 1982).

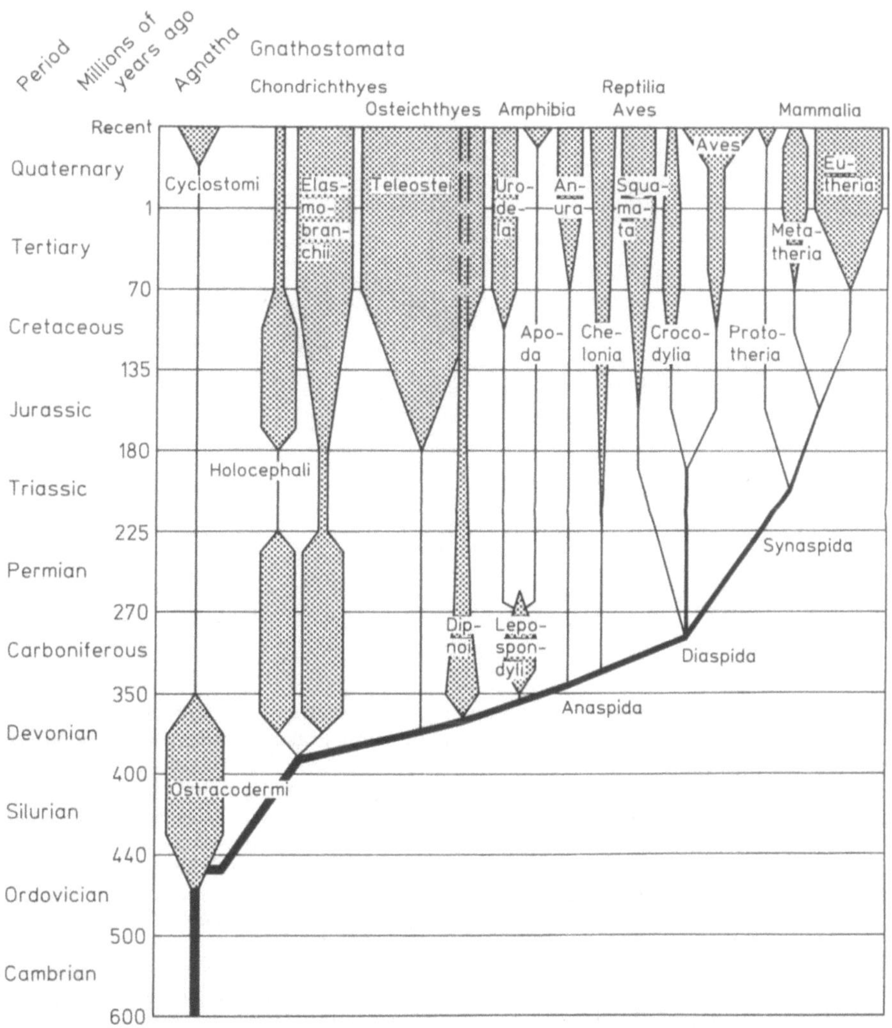


**Fig. 8.** Simplified schematic outline of the phylogeny of the vertebrates. For a study of the stepwise evolution of the GEP neuroendocrine system in general, and of the endocrine pancreas in particular see Fig. 9; pivotal key species are found among the extant cyclostomes (a two-hormone islet organ), the holocephalan cartilaginous fish (a three-hormone endocrine pancreas), and the elasmobranchian cartilaginous fish (a four-hormone endocrine pancreas – as in higher vertebrates). As a rule, the teleostean bony fish are of more recent phylogenetic age and, in some cases, seem to be even more highly specialised in their GEP neuroendocrine system than some tetrapods (amphibians, reptiles, birds, mammals)

If the function of glucagon in the protochordates is to promote glucose availability, as it is in the islets of higher chordates, perhaps the sedentary feeders such as *Ciona* and *Styela*, do not have the same need for a quick supply of energy, elicited via glucagon release, as the faster-moving amphioxus. Since *Branchiostoma* "glucagon" has only been studied by IHC, using pancreatic glucagon antisera

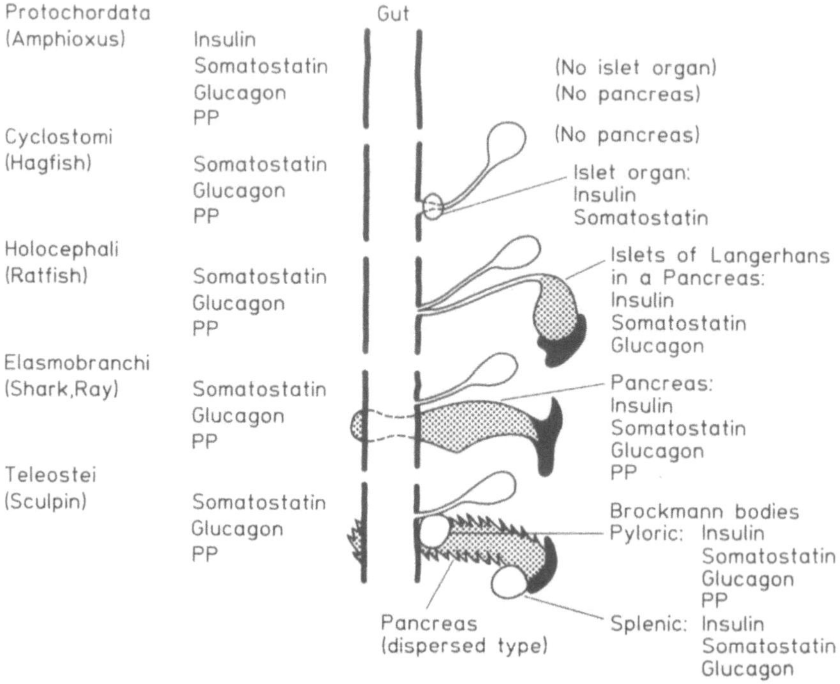


**Fig. 9.** Schematic outline, giving a summarising survey of the stepwise evolution of the endocrine pancreas, presumed to occur from the disseminated endocrine cells of open type, present in the digestive tract mucosa of the protochordates. The first cells to leave the gut to form a separate islet organ are the insulin cells, followed by the somatostatin cells; in the hagfish, most of the somatostatin cells remain in the bile duct mucosa as cells of closed and open type. The glucagon cells start to invade the endocrine pancreas at the evolutionary stage of the holocephalan cartilaginous fish. Already at the level of elasmobranchian cartilaginous fish, the islets of Langerhans have become a four-hormone endocrine organ. A further development occurs in some teleost fish, (such as the daddy sculpin, *Cottus scorpius*) and mammals, where a kind of inverse interrelationship between glucagon and PP cells is indicated by the finding of a concentration of PP cells, at the expense of the glucagon cells, in those parts of the endocrine pancreas situated most closely to the gastrointestinal tract, derived from the ventral anlage of pancreas

(VAN NOORDEN and PEARSE 1976; REINECKE 1981), but not chemically characterised, this consideration is entirely speculative until more is known of the physiology of the protochordates, including their peptide receptors and target tissues.

## F. Vertebrates

In the vertebrates, the GEP neuroendocrine system in general, and the endocrine pancreas in particular, show a stepwise evolution, in several respects well correlated with the phylogenetic history of these animals (cf. VAN NOORDEN and FALKMER 1980; FALKMER et al. 1981a). As schematically illustrated in Fig. 8, the first vertebrates, the agnathan ostracoderms, appeared some 500–600 million years ago and were the predominant vertebrates during the Silurian and Devonian eras. The

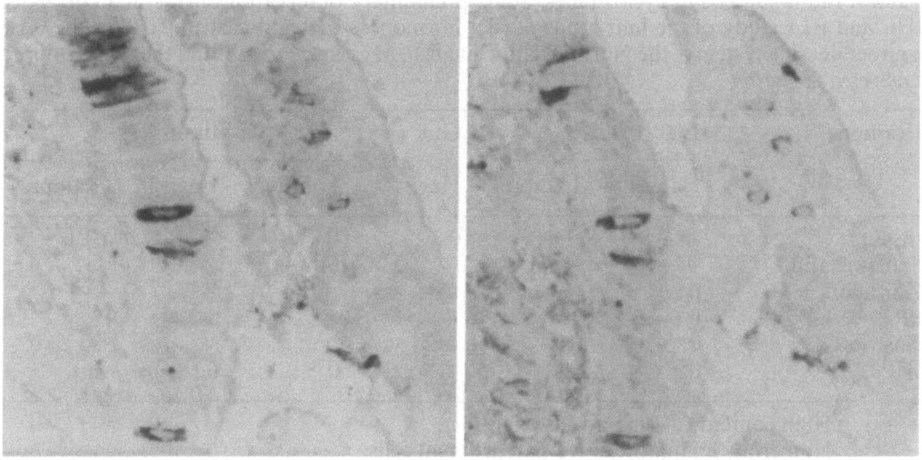

a b

**Fig. 10 a, b.** Medium power photomicrographs of adjacent serial sections of a part of the gut mucosa of the river lamprey, *Lampetra fluviatilis*. The specimen was fixed in methanol-free formaldehyde, embedded in Araldite, cut at 1 μm, and incubated with antisera raised against porcine pancreatic-type glucagon (**a**) and bovine pancreatic polypeptide (BPP) (**b**), respectively. The immunoreactive mucosa cells were visualised by the PAP procedure. The anti-glucagon antiserum was preabsorbed with BPP and the anti-BPP antiserum with porcine glucagon. It is evident that the glucagon-immunoreactive cells also react with antisera to BPP (see Fig. 14). The immunoreactive cells are of open type. They showed no immunoreactivity with antisera raised against porcine glicentin. S. FALKMER and S. VAN NOORDEN (1981, unpublished work) ×320

cyclostomes (hagfish, lamprey) are the only extant representatives of these original vertebrates. Among the gnathostomes, the holocephalan (ratfish, rabbit fish, elephant fish) and elasmobranchian (sharks, rays) cartilaginous fish are the earliest appearing groups. In particular, the few extant representatives of the holocephalan fish, have attracted great interest and are often looked upon as "living fossils" (cf. FALKMER et al. 1981a). The teleostean bony fish are of more recent age; in some cases they seem to be even more highly specialised in their GEP neuroendocrine system than some tetrapods (amphibians, reptiles, birds, mammals). In the cyclostomes, the endocrine pancreas is a separate two-hormone islet organ, in the holocephalans a three-hormone microorgan, and from the elasmobranchs up to and including mammals, a four-hormone parenchyma (Fig. 9; cf. FALKMER et al. 1981 a).

## I. Agnatha (Cyclostomes; Jawless Fish)

As described in some detail in previous reviews (VAN NOORDEN and FALKMER 1980; FALKMER et al. 1981 a), the first endocrine pancreas in evolution appears in the cyclostomes. Here, it is a separate, grossly visible islet organ, budding out from the bile duct (hagfish) or the anterior gut (lamprey), free from exocrine, acinar pancreatic parenchyma, and consisting of insulin cells and somatostatin cells only. The glucagon cells, which also react with antisera to PP, remain as endocrine cells of open type in the gut mucosa (Fig. 10); analogous observations have been made in islet cells in amphibia (see Sect. F.II.3.a).

**Table 1.** Brain-gut axis in cyclostomes. Some preliminary observations made in a correlated IHC and RIA study of the four kinds of islet hormones, GIP and gastrin/CCK in the GEP neuroendocrine system of the North Atlantic hagfish, *Myxine glutinosa*, and the river lamprey, *Lampetra fluviatilis*

| Hormone | Islet organ | | Gut | | Brain | |
|---|---|---|---|---|---|---|
| | Hagfish | Lamprey | Hagfish | Lamprey | Hagfish | Lamprey |
| Insulin | +++ | +++ | 0 | 0 | + | (+) |
| Somatostatin | ++ | ++ | + | (0) | + | + |
| Glucagon | 0 | 0 | + | + | + | (+) |
| PP | 0 | (+) | + | + | + | + |
| Gastrin/CCK | 0 | (+) | + | + | + | + |
| GIP | 0 | n.t. | + | 0 | 0 | n.t. |

+++ High content (~1 mg/g wet weight)
Numerous immunoreactive parenchymal cells
++ Moderate content (~1 μg/g wet weight)
Moderate numbers of immunoreactive cells
+ Low contents (~1 ng/g wet weight)
Few immunoreactive parenchymal cells
(+) (0) Divergent results obtained, so far
0 No convincing evidence obtained for the presence of the GEP neurohormone cell looked for
n.t. Not tested, so far

As in the amphioxus, (VAN NOORDEN and PEARSE 1976), the glucagon-immunoreactive gut endocrine cells of the hagfish and the lamprey also show gastrin-like immunoreactivity (cf. VAN NOORDEN and FALKMER 1980), perhaps indicating the presence in such cells of a "prosecgastrin", a hypothetical single parent ancestor molecule from which both the secretin/glucagon/VIP/GIP family and the gastrin/CCK/caerulein family may have arisen (WEINSTEIN 1972). Recent observations on tunicates, based on combined IHC and experimental investigations, showed secretin-immunoreactive cells, but a gastric secretory response only to CCK, not to secretin (BEVIS and THORNDYKE 1981). However, in immunochemical studies of GEP neurohormonal peptides in lamprey gut and brain, results have been obtained indicating that small secretin/glucagon-like and gastrin/CCK-like peptides are produced separately (HOLMQUIST et al. 1979). A reinvestigation of the brain–gut axis in cyclostomes has, therefore, recently been started in our laboratories, where correlated IHC and RIA studies of the occurrence of GEP neurohormonal peptides in the brain, the gut, and the islet organ of the North Atlantic hagfish, *Myxine glutinosa*, and the river lamprey, *Lampetra fluviatilis*, are being made, using a wide range of antisera. Some preliminary results have been summarised in Table 1, as regards glucagon and the other three islet hormones, GIP, and gastrin/CCK. Subject to confirmation by the results of more extensive studies, including EM examination, it seems that the glucagon cells were present in the brain as neuronal cells, even in the earliest vertebrates.

As described in the previous glucagon phylogeny/ontogeny review (FALKMER and MARQUES 1972), administration of mammalian glucagon does not seem to affect the blood glucose level in the hagfish. No other physiological function is known for glucagon in myxinoid cyclostomes. Its absence from the hagfish islet parenchyma has, so far, not been disproved; the unclassified hagfish islet parenchymal cells illustrated in the previous review were later shown to be islet somatostatin cells (FALKMER et al. 1977, 1978; VAN NOORDEN et al. 1977), and the glucagon-like immunoreactivity observed in extracts of the cranial portion of the islet organ of the river lamprey (ZELNIK et al. 1977) could well be due to contamination from the tissues of the anterior intestine (as stated by ZELNIK et al., themselves). The ontogeny of the glucagon cells in the brain and gut of the cyclostomes is still essentially unknown.

## II. Gnathostomata

Common to the GEP neuroendocrine system of all gnathostomian vertebrates are the facts that their endocrine pancreas is closely associated with the exocrine, acinar, pancreatic parenchyma and that the glucagon cells form a substantial portion of their islet tissue (VAN NOORDEN and FALKMER 1980; FALKMER et al. 1981 a). This statement applies also to those teleostean bony fish species equipped with the giant accumulations of islet parenchymal cells, called Brockmann bodies or "principal islets," as detailed in a previous review (FALKMER and ÖSTBERG 1977). It has earlier been speculated that the seemingly concomitant evolutionary appearance of glucagon cells in the islet parenchyma on one hand and a separate gland for the zymogen exocrine pancreatic cells on the other hand might indicate a mutual regulation of islet glucagon cells and the pancreatic acinar parenchyma (cf. FALKMER and PATENT 1972). However, this evolutionary coincidence still remains essentially unexplained.

### 1. Cartilaginous Fish

As schematically illustrated in a recent review (FALKMER et al. 1981 a), and in Fig. 9, there are already marked differences in the gross anatomy of the pancreatic glands of the holocephalan and elasmobranchian cartilaginous fish. Additional differences are found with regard to the number of islet hormones in the islet parenchyma (see Sect. F) and to the glucagon/glicentin cells.

#### a) Holocephalan Fish

Although the occurrence of glucagon or other GEP neurohormonal cells in the brain of the two main species of holocephalan cartilaginous fish that can be obtained in the northern hemisphere; the rabbit fish, *Chimaera monstrosa* (from the Atlantic); and the ratfish, *Hydrolagus colliei* (from the Pacific), is still unknown, there are some recent observations of great evolutionary interest on the glucagon/glicentin cells in their endocrine pancreas. Here, as detailed and illustrated in a previous review (FALKMER and ÖSTBERG 1977), about half a dozen ultrastructurally more or less well defined types of parenchymal cells have been identified. Among these are the so-called X-cells, constituting at least 50% of the holocephalan islet cells. They are obviously unique to this group of animals. Their function is unknown. By subsequent IHC investigations (FALKMER et al. 1981 a) it has now been possible to show that the number of cell types is reduced to four and that the previous guess that the X-cells (and the amphiphil intermediate X-A-cells) were related to the glucagon cells (cf. FALKMER and ÖSTBERG 1977) was correct; the X-cells have now been found to contain an immunoreactant similar to the $NH_2$ terminal sequence of glucagon (STEFAN et al. 1981). Such an immunoreactivity was common to the X-cells and the glucagon cells of the islet parenchyma and was also found in isolated cells of the epithelium of the long pancreatic duct and in endocrine cells of the gut mucosa. After proteolytic treatment of the cut sections, the X-cells also became immunoreactive to COOH terminal glucagon antisera, like the islet glucagon cells (STEFAN et al. 1981). Thus, these X-cells in the holocephalan islet parenchyma and those in the pancreatic duct epithelium and gut mucosa obviously behave like the so-called L-cells of the mammalian digestive mucosa, containing glicentin and perhaps also other glucagon-like immunoreactive (GLI) peptides; the mammalian cells also respond to mild proteolytic digestion by showing immunoreactivity with antisera raised against the COOH terminal part of the glucagon molecule (RAVAZZOLA and ORCI 1980 a). These data suggested that the holocephalan X-cells are precursor forms of the islet glucagon cells and represent the pancreatic counterparts of the intestinal glicentin cells of mammals; they were also compatible with the hypothesis that COOH terminal tryptic fragments of glucagon derive from larger glucagon-related peptides and they emphasize the common origin of glucagon and glicentin cells (cf. STEFAN et al. 1981).

The fact that PP cells are absent from the holocephalan islet parenchyma, thus making it a three-hormone islet organ, but occur in the gut mucosa (Figs. 9 and 11), has been interpreted as support for the hypothesis that the islet PP cells are phylogenetically youngest among the four kinds of islet hormone cells and that they appear in the pancreas first at an evolutionary stage when the gland is closely apposed to the digestive tract, as in the elasmobranchian cartilaginous fish (VAN NOORDEN and FALKMER 1980; FALKMER et al. 1981 a). In its lack of PP cells and in its rich supply of glucagon and glucagon-related cells, the holocephalan endocrine pancreas is similar to the cellular composition of the Brockmann body near the spleen in some highly developed teleost bony fish, such as the daddy sculpin, *Cottus scorpius*, (Fig. 9; STEFAN and FALKMER 1980; FALKMER et al. 1981 a) and that of the splenic parts of the mammalian pancreas, derived from the embryonal dorsal anlage (cf. FALKMER et al. 1981 b; RAHIER et al. 1981). Whether or not this anatomical–ontogenetic interrelationship between glucagon and PP cells has any functional aspects, remains to be investigated. A schematic representation of the evolution of the vertebrate endocrine pancreas is given in Fig. 9, based on the results of correlated IHC and RIA studies performed in our laboratories (cf. VAN NOORDEN and FALKMER 1980; FALKMER 1981 a).

### b) Elasmobranchian Fish

Already at the time of the previous glucagon review (FALKMER and MARQUES 1972) circumstantial evidence indicated the presence of glucagon-producing cells in the endocrine pancreas of sharks and rays. Since then, proofs have been obtained that glucagon cells are present in the elasmobranchian pancreas, both by IHC (FALKMER et al. 1981 a; SEKINE and YUI 1981), by EM (KOBAYASHI and ALI 1981) and by direct isolation of the hormone and its chemical characterisation (SUNDBY 1976). As to the general light microscopic structure, it has been known for more

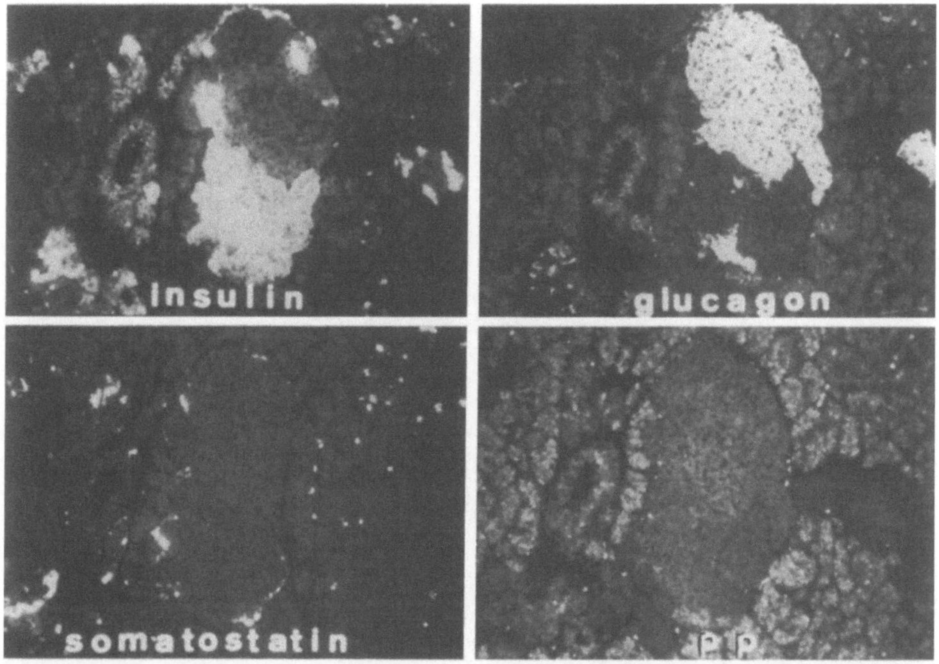


a

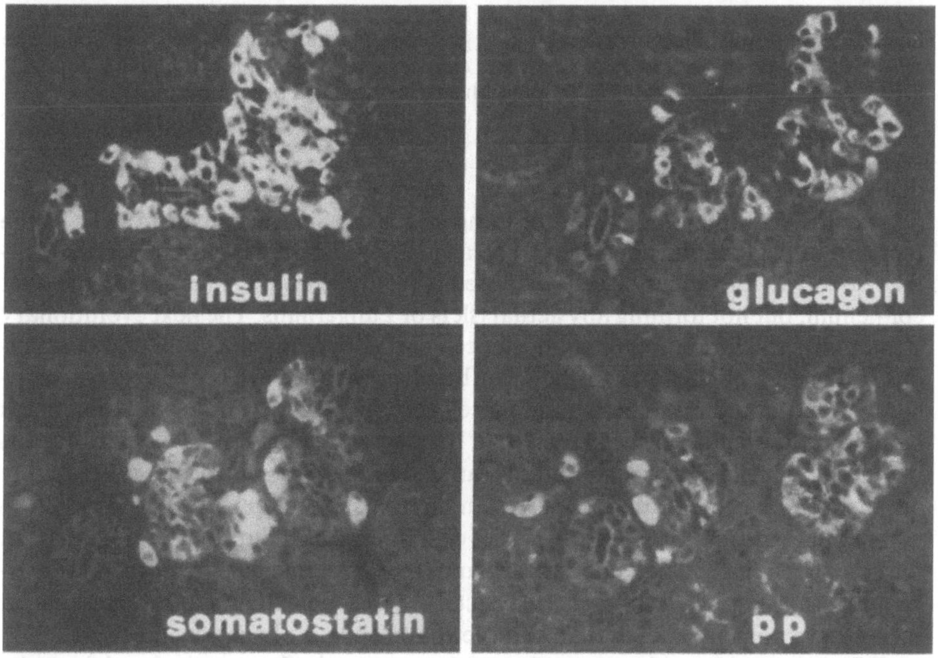


b

**Fig. 11 a, b.** Two series of immunofluorescence medium power photomicrographs of consecutive serial sections of the endocrine pancreas in the holocephalan cartilaginous fish, *Chimaera monstrosa* (the Atlantic rabbit fish) (**a**) and in the elasmobranchian one, *Raja radiata* (the starry ray) (**b**). The sections have been incubated with antisera raised against the four kinds of islet hormones of mammalian origin. The two sets of photomicrographs demonstrate that the islet parenchyma evolves from a three-hormone organ in the holocephalan fish (**a**) to a four-hormone organ in the elasmobranchian (**b**). Although both types of endocrine pancreas have attained the evolutionary stage where distinct islets of Langerhans are being formed, all four kinds of islet parenchymal cells can still be found as endocrine cells of closed type in the duct epithelium. S. FALKMER and Y. STEFAN (1977–1978, unpublished work) ×250

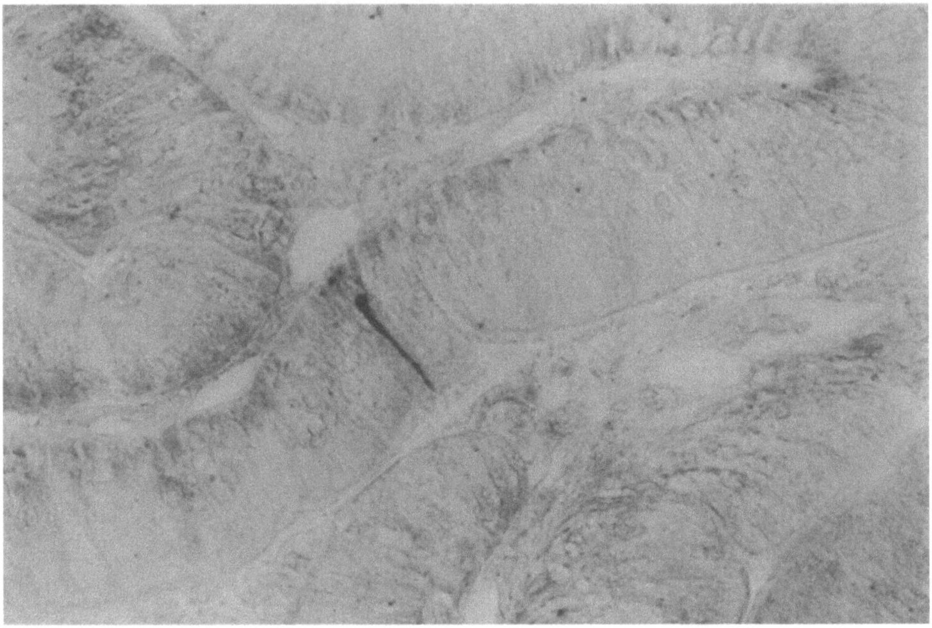

**Fig. 12.** High power photomicrograph of a glucagon/glicentin-immunoreactive endocrine cell of open type in the gut mucosa of the shark, *Squalus acanthias* (the spiny dogfish). S. FALKMER and M. REINECKE (1981, unpublished work) PAP procedure; ×375

than 40 years that the elasmobranchian endocrine pancreas offers interesting aspects of the evolution of the islets of Langerhans (cf. SEKINE and YUI 1981). In the most primitive type of endocrine pancreas in these fish, the islet parenchyma consists essentially of endocrine cells of closed type, immediately surrounding the duct epithelium, but not forming any actual islets of Langerhans. The next evolutionary step implies the formation of solid islet cell cords, budding out from and contiguous to the duct epithelium. In the most highly developed types of endocrine pancreas in sharks and rays, these budding cell cords form typical islets of Langerhans and only a few endocrine cells remain in the duct epithelium. As shown in Fig. 11, the glucagon cells in an elasmobranchian endocrine pancreas of the more highly developed type occur both in the islets and in the duct epithelium. The same statement applies also to the three other islet cell types. It is interesting to note that, in the compact holocephalan pancreatic gland, the endocrine pancreas forms large islets (Fig. 11), despite all the facts (see Sect. F.II.1.a) indicating that it represents an earlier developmental stage than the islet parenchyma of sharks and rays.

When the amino acid composition of the pancreatic glucagon isolated from the spiny dogfish, *Squalus acanthias*, was analysed (SUNDBY 1976), it was found to have 29 amino acid residues; it had two lysine and two methionine residues where pig glucagon had only one of each. In contrast, shark glucagon lacked one leucine and one glutamic acid residue, compared with pig glucagon. Whereas the shark glucagon reacted strongly with "nonspecific antiglucagon sera," it showed only a weak reaction with "specific antiglucagon sera," indicating that at least one of the differences in the amino acid composition of pig and shark glucagon is to be found in the COOH terminal part of the molecule (SUNDBY 1976).

As in the holocephalan cartilaginous fish (STEFAN et al. 1981), endocrine cells of open type, showing glucagon/glicentin-like immunoreactivity, were found by IHC in the gut mucosa of elasmobranchian cartilaginous fish (Fig. 12). At the same time, cells immunoreactive to antisera raised against another GEP neurohormonal peptide, belonging to the same family as glucagon, were discovered in these fish; astonishingly great numbers of VIP-producing cells of endocrine type were found in the colorectal parts of the gut of sharks and in particular, rays (FALKMER et al. 1980a; FOUCHEREAU-PERON et al. 1980).

So far, it has not been established whether or not any glucagon production and/or glucagon-immunoreactive cells occur in the elasmobranchian central nervous system. Since the previous review, some observations have been published on the ontogeny of the glucagon cells in these fish, using the almost ideal conditions offered by nature in this subphylum. In the shark, *Scyliorhinus stellaris* (the larger spotted dogfish), the gastric mucosa of embryos aged from 5 months to a few days prior to hatching was studied by histological/histochemical methods (TAGLIAFIERRO and FARALDI 1976). Four kinds of argentaffin and/or argyrophil cells were already present at early developmental stages. Whether any of these cells were glucagon/glicentin cells was not investigated by IHC.

## 2. Bony Fish

Ever since the early reports in the 1950s of glucagon-producing islet parenchymal cells in the Brockmann bodies of teleost bony fish (cf. FALKMER and MARQUES 1972), these anatomical structures have been widely used as an excellent source for studies of the molecular events in the proglucagon/glucagon biosynthesis (cf. LUND et al. 1980; NOE 1981a, b). Best suited for such studies are those fish (of convenient size and availability) whose Brockmann bodies are devoid of strands of exocrine parenchyma; already in the 1920s it was known that the "principal islets" of the anglerfish, *Lophius piscatorius*, and the daddy sculpin, *Cottus scorpius*, offered histologically "pure islet tissue," giving exceptionally high yields of insulin (cf. FALKMER 1961) and, later on, of other islet hormones, as well (FALKMER et al. 1977, 1978, 1981a).

As recently reported (FALKMER et al. 1981a), RIA of acid–ethanol extracts of the pyloric and the splenic Brockmann bodies (Fig. 9) of the daddy sculpin, *Cottus scorpius*, using two different anti-glucagon sera: one reacting with glucagon 11–15, glicentin, and avian glucagon, and one reacting with glucagon 24–29, not at all with glicentin, and only to some extent with avian glucagon, gave an almost tenfold higher concentration (119–126 μg equiv./g wet weight) with the 11–15 antiserum than with the 24–29 one (8–22 μg equiv./g wet weight). This observation shows that, even when dealing with glucagon, great care must be exercised in interpreting GEP neurohormonal peptide concentrations based on heterologous RIA of tissue extracts (cf. EMDIN 1981). Whether or not the observations can be interpreted as indicating that the sculpin glucagon cells of the islet parenchyma, like those of the ratfish (STEFAN et al. 1981), contain glicentin-immunoreactive peptides, and have more resemblance to glicentin than to pancreatic glucagon, remains to be investigated. The amino acid composition of sculpin islet glucagon is still unknown. That of anglerfish glucagon, however, has been published (SUNDBY 1976); like pig and shark glucagon it has 29 residues; the $NH_2$ terminal histidine is the same as in pig glucagon but, as COOH terminus, it has a lysine residue instead of a threonine. As detailed in the previous glucagon review (FALKMER and MARQUES 1972), the function of glucagon in bony fish seems to be the same as in higher vertebrates. The fine structural characteristics of both the islet glucagon cells (KLEIN and VAN NOORDEN 1980; STEFAN and FALKMER 1980) and the glucagon/glicentin cells in the gut mucosa (ROMBOUT et al. 1979) of some teleost bony fish have been given in recent, comprehensive reports. However, it is essentially unknown whether glucagon/glicentin cells also occur in the brain of bony fish.

No major advances as regards the distribution and histophysiology of glucagon cells (or the GEP neuroendocrine system as a whole) in lungfishes (Dipnoi) or fringe-fins (Crossopterygii) seem to have been made since the previous glucagon review (FALKMER and MARQUES 1972). The same applies to fish belonging to the two smaller groups of the subclass Actinopterygii (rayfins), where the teleost fish represent the largest proportion; sturgeons and paddle-fishes (Chondrostei) and gars and bowfins (Holostei). Neither has any new information on ontogeny of the glucagon cells in bony fish appeared since 1972.

The teleostean islet parenchyma differs from that of the jawless and cartilaginous fish in being conspicuously well innervated. One of the putative peptide neurotransmitters in a gobiid fish, *Gillichthys mirabilis*, has been identified as a VIP-like peptide by IHC (VAN NOORDEN and PATENT 1980). When, however, RIA of extracts from sculpin Brockmann bodies were made, no VIP-immunoreactive substances were found (unpublished observations, March 1981, made together with Dr. JAN FAHRENKRUG, Department of Clinical Chemistry, Copenhagen County Hospital, Glostrup, Denmark). This could be due to species differences (see Sect. F.II.3.b) or the teleostean islet neurotransmitter may be so-called VIP-oid in nature, whereas the gut peptide may well be "true" VIP (FOUCHEREAU-PERON et al. 1980).

### 3. Tetrapods

Among the tetrapods, the glucagon cells of the GEP neuroendocrine system are of greatest phylogenetic interest in the amphibia and reptiles (FALKMER and MARQUES 1972; FALKMER and ÖSTBERG 1977). Their distribution and histophysiology in adult birds and mammals are well known and will be dealt with in other parts of the present monograph. In the latter two subclasses, however, some interesting observations have been made with regard to the ontogeny of the glucagon/glicentin cells.

The tetrapod pancreas is uniform in containing a mixture of endocrine and exocrine tissue, but the anatomical details of the endocrine elements vary considerably, even between members of the same family (FALKMER and ÖSTBERG 1977; BONNER-WEIR and WEIR 1979; EPPLE et al. 1980). There seems also to be a rather uniform pattern as to the general organization of the other parts of the GEP neuroendocrine system: the brain, the gastrointestinal canal, and the respiratory tract (cf. BUCHAN 1981).

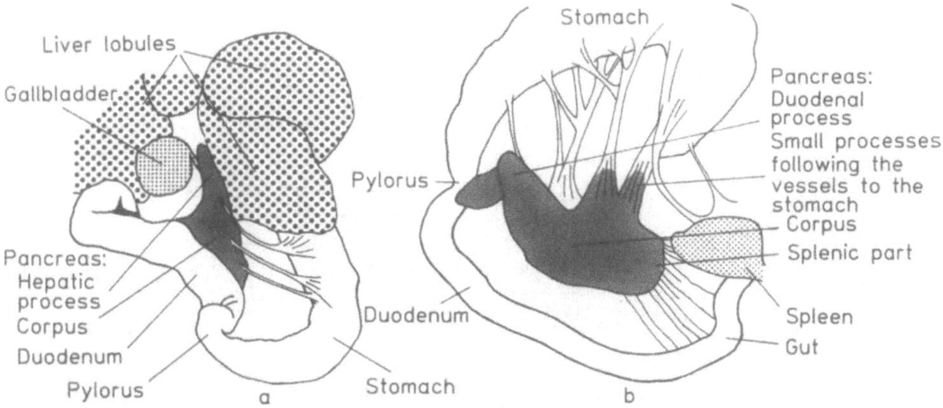


**Fig. 13 a, b.** The gross anatomy of the GEP organs in an anuran amphibian, the marsh frog, *Rana ridibunda*, schematically outlined as seen when the peritoneal cavity is opened and the organs are in situ (**a**), and after lifting up the stomach and duodenum (**b**) to show the major parts of the pancreatic gland. It is of the compact type, ubiquitous in the tetrapods. Despite the presence of distinct processes of the pancreas to the liver (**a**), the duodenum, the stomach (following the vessels), and the spleen (**b**), no marked regional differences were found with regard to glucagon islet cells or BPP-immunoreactive cells in this particular species (S. FALKMER 1977–1978, unpublished work)

### a) Amphibians

As described in the previous review (FALKMER and MARQUES 1972), the comparative anatomy of the endocrine pancreas and its glucagon cells supports the claims that Amphibia is a diphyletic subclass, where Urodela (or Caudata – salamanders and newts) and Apoda (or Gymnophiona – legless forms) form one main group and the Anura (or Salentia – frogs and toads) the other. The anuran amphibians are best known; they represent one of the classical major research objects in the comparative endocrinology of the GEP neurohormonal peptides (FALKMER and ÖSTBERG 1977).

In general, the amphibian pancreas is of the compact mammalian type, in some species with processes towards the liver, duodenum, spleen, and stomach (Fig. 13). Often, endocrine elements appear in ribbon-like formations (Fig. 14) rather than as regular small round islets, or even as scattered cells. An accumulation of islet tissue near the spleen has been reported only in the toad, *Bufo arenarum* (cf. EPPLE et al. 1980), but sampling errors may account for its apparent absence in some groups (BONNER-WEIR and WEIR 1979). Where islets or islet-like structure are present, as in the anuran amphibians, the islet cells are arranged conventionally with insulin cells at the centre, surrounded by glucagon and somatostatin cells with PP cells in the periphery (Fig. 14). In the urodeles (e.g. *Salamandra salamandra*, *Ambystoma mexicana*), endocrine cells are present singly or in small clusters with no distinct distribution pattern of cell types (BUCHAN 1981).

The presence of four types of endocrine islet cells in amphibia has been reported both by IHC and RIA (EPPLE et al. 1980; FALKMER 1980 b) and recently also by EM (TOMITA and POLLOCK 1981). As in the lamprey gut mucosa (Fig. 10), immunostaining with an antibody directed to bovine PP (BPP) often revealed the

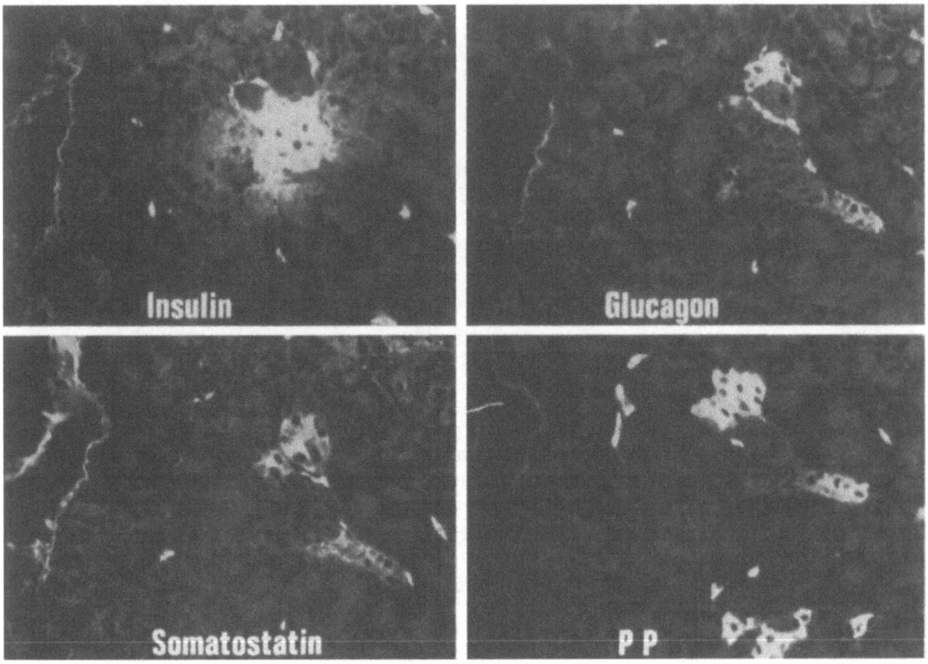


**Fig. 14.** Same type of islet hormone picture as in Fig. 11, consisting of medium power photomicrographs of consecutive serial sections of the hepatic process (Fig. 13a) of the endocrine pancreas in the marsh frog, *Rana ridibunda*, incubated with antisera raised against the four kinds of islet hormone of mammalian origin (S. FALKMER 1977–1978, unpublished work). The amphibian pancreatic endocrine cells form islet-like structures of rather irregular outline, sometimes with ribbon-like processes (as shown in the figure). The insulin cells occur in the central parts of the islets, surrounded by glucagon and somatostatin cells with PP cells in the periphery. Amphibian PP cells were found to be immunoreactive with bovine PP (BPP) antisera only, but not with antisera raised against avian or human PP; they have still not been identified ultrastructurally. As shown, there are islet cells displaying both glucagon and BPP immunoreactivity; this phenomenon, seen both in adult and embryonal amphibian islet parenchyma, is still unexplained (cf. Fig. 10). ×200

same islet cells as the antibody directed to glucagon (Fig. 14) throughout the amphibia, although staining separate cells in the islets of fish, reptiles, and mammals (BUCHAN 1981). Whether this is due to immunological cross-reactivity, or whether two peptides are produced in the same cell, cannot be resolved until amphibian glucagon has been isolated and sequenced. It should be noted that in all the vertebrates the glucagon/glicentin-immunoreactive cells of the gut frequently also stain with the antibody to BPP, despite a separate staining in the islets, so it may be that the glicentin precursor and PP have similar antigenic determinants, not shared by pancreatic glucagon after its split from the larger precursor molecule (RAVAZZOLA and ORCI 1980b). Antibodies directed to avian and human PP did not stain amphibian glucagon cells in the pancreas (BUCHAN 1981). Glucagon cells were not confined to islets, but were also scattered through the exocrine tissue, singly or in clumps; no concentration of glucagon or PP cells to specific regions or processes of the gland seems to exist, at least not generally (Fig. 13; TOMITA and POLLOCK 1981).

In a comprehensive IHC and EM study of the GEP neuroendocrine system in amphibians (BUCHAN 1981), only one species, *Cynops hongkongensis*, showed four ultrastructural types of pancreatic endocrine cells, and the fourth cell type was not immunostained by any of the peptide hormone antibodies tested. In these EM studies it was, however, found that the ultrastructural characteristics of the pancreatic glucagon cells and, indeed, that of the other islet and gut peptide-secreting endocrine cells, of both the amphibia and the reptiles, showed too much species variation for any useful statement to be made about their appearance (BUCHAN 1981). The use of histological staining methods, and preferably of IHC, is thus a sine qua non for the identification of particular endocrine cells in amphibian and reptilian species. By means of the thin/semithin sectioning technique the ultrastructure of the BPP immunoreactive amphibian islet cells have, however, recently been shown in the bullfrog, *Rana catesbeiana* (TOMITA and POLLOCK 1981).

Glucagon-immunoreactive cells were found in the fundus of the stomach of representatives of both anurans and urodeles, but their distribution along the rest of the gut varied markedly from species to species (BUCHAN 1981; EL-SALHY et al. 1981; GIRAUD and YEOMANS 1981). In no case did the gut glucagon cells react with an antibody to pig glicentin. This probably reflects species differences in the structure of the large glicentin molecule (LARSSON and MOODY 1980).

There are some recent reports on the brain of anuran amphibians where both IHC and EM investigations have been made of the occurrence of neuronal cells storing GEP neurohormonal peptides, notably gastrin/CCK (DOERR-SCHOTT and GARAUD 1981), substance P (INAGAKI et al. 1981) and endorphins/enkephalin (DOERR-SCHOTT et al. 1981). So far, no glucagon-immunoreactive cells seem to have been discovered in amphibian brain.

The function of glucagon in amphibians seem to be similar to that in birds and mammals (FARRAR and FRYE 1979). Seasonal variations have been observed in the serum insulin and glucagon concentrations in a frog (SCHLAGHECKE and BLÜM 1981); high serum glucagon concentrations (and low insulin levels) occurred during fasting in winter when the animal depends on its endogenous stores of nutrients for energy supply; high serum glucagon levels were thought to promote their conversion to substrates that can be metabolised, notably glucose, fatty acids, and ketone bodies. The glucagon concentration in the pancreas, however, showed no such seasonal variations.

Recently, some ontogenetic aspects of the amphibian glucagon cells have also been obtained. In an IHC study of embryos and young larvae of the leopard frog, *Rana pipiens*, (KAUNG 1981), it was observed that the first islet cells that could be detected were the insulin cells at somite stage 22 when the embryo is about 8 mm long. Glucagon and PP cells were first identified at somite stage 24. Somatostatin cells appeared later on. Thus, in this species, both similarities and differences were observed between phylogeny and ontogeny as regards the evolution of the islet parenchyma. Interestingly, glucagon and PP were also reported to be present in the same cells during early development; at this stage the endocrine cells were found to be widely scattered in the exocrine parenchyma and not arranged in islet-like structures.

b) Reptiles

As accounted for in a previous review (FALKMER and ÖSTBERG 1977), a wealth of new information on glucagon/glicentin cells – and the whole GEP neuroendocrine system – in most of the three main extant subclasses of the reptiles: the Chelonia (or Testudines – turtles and tortoises), Squamata (snakes and lizards), and Crocodylia (alligators and crocodiles) (Fig. 8), has appeared since the 1972 glucagon ontogeny/phylogeny overview (FALKMER and MARQUES 1972). In addition during the last 5 years, there have been some major advances; nevertheless, the reptiles still belong to the least investigated of the vertebrates, resulting in a relative paucity of information about their GEP neuroendocrine system. The reptiles are phylogenetically important; they constitute the first wholly terrestrial subclass of vertebrates; they have overcome the major environmental barriers that faced the amphibians; being a transitional subclass, they contain representatives that can be considered primitive (turtles) and those that can be regarded as modern (crocodiles) (Fig. 8) with respect to their GEP neuroendocrine system. As in their phylogenetic successors, the birds (Fig. 8), the glucagon cells are numerous in the endocrine pancreas and obviously play an important physiological role (FALKMER and MARQUES 1972; FALKMER and ÖSTBERG 1977).

*α) Turtles, Lizards, Snakes.* In the turtles and in most of the lizards, the pancreatic islets seem to be of the mammalian type, scattered through the exocrine parenchyma, with insulin cells at the centre and glucagon cells at the periphery (BUCHAN 1981; EL-SALHY and GRIMELIUS 1981). However, in some of the lizards (Varanidae) the endocrine tissue is concentrated into a single giant islet, reminiscent of a Brockmann body in teleost bony fish, and in the Serpentes (snakes) there is a tendency for the islet tissue to be concentrated in the splenic portion of the pancreas, or even within the spleen (e.g. *Crotalus atrox*) (FALKMER and ÖSTBERG 1977; BUCHAN 1981). In the snakes, particularly the vipers, and in some lizards, the glucagon cells are found in the centre of the islets, surrounded by insulin and somatostatin cells (BUCHAN 1981; EL-SALHY and GRIMELIUS 1981). The reversal of the usual arrangement of glucagon and insulin cells in the islets is also occasionally found in mammals, e.g. in the horse (cf. FALKMER and ÖSTBERG 1977), and it has been shown that in such islets the arterial blood supply enters the islet at the centre, instead of at the periphery as usual (cf. FUJITA et al. 1976, 1981). It is not recorded whether this arrangement is present in reptiles with central islet glucagon cells. The tendency of the glucagon cells to follow and surround vessels penetrating the islet parenchyma is seen also in the Brockmann bodies of teleost bony fish, such as the daddy sculpin, *Cottus scorpius* (cf. FALKMER 1961; FALKMER and ÖSTBERG 1977; STEFAN and FALKMER 1980).

In the snapping turtle, *Chelydra serpentia*, PP cells were not found in the splenic lobe, but only in the head of the pancreas (BUCHAN 1981) as in mammals (see Sect. F.II.1.a). Analogous observations have been made in some lizards, but not in others (EL-SALHY and GRIMELIUS 1981), illustrating the statement that marked species differences can occur in the structure of the GEP neuroendocrine system even within one and the same family of tetrapods. Where reptilian PP cells have been immunostained, they were identified by antibodies to both BPP and avian PP and were separate from the glucagon cells; the snake glucagon cells did not stain with the PP antibody (BUCHAN 1981). In a comprehensive correlated IHC and EM study of the four kinds of islet parenchymal cell in an iguanid lizard, *Anolis carolinensis*, it was recently observed that without IHC staining, the qualitative ultrastructural characteristics of the PP cell secretory granules were inadequate to permit certain identification of the PP cell, especially with regard to the somatostatin cell (RHOTEN and HALL 1981). Similar difficulties appeared in the ultrastructural identification of the PP cells in the teleost bony fish, *Cottus scorpius*, (STEFAN and FALKMER 1980).

*β) Crocodiles.* In the pancreas of the most highly developed reptiles, the crocodiles, endocrine cells occur as islets of ductular type or as single cells, as shown in the American alligator, *Alligator mississippiensis*; in the caiman, *Caiman crocodilus* large islets were found to be present which could be immunostained for all the four islet hormones (BUCHAN 1981). Ultrastructurally, all the four types of islet parenchymal cells were observed in these two species, showing a characteristic fine structure of the secretory granules (TITLBACH 1981). In addition, enterochromaffin cells seemed to be present. In other reptilian species, however, the ultrastructure of the pancreatic glucagon cells varied from group to group, so that no useful general statement could be made (BUCHAN 1981). It is interesting that the crocodiles resemble the birds, supposedly derived from the same ancestral stock (Fig. 8), in that the PP cells were detected by antibody to avian PP, but not to BPP (BUCHAN 1981).

As in the teleost bony fish islet parenchyma (VAN NOORDEN and PATENT 1980), the islet glucagon cells of snakes and crocodiles are particularly well innervated (cf. FALKMER and ÖSTBERG 1977), and one of the putative peptide neurotransmitters has been identified as a VIP-like peptide by IHC (FUJI et al. 1980; BUCHAN 1981). EM observations in snake islets confirmed the occurrence of numerous nerve terminals in the pericapillary spaces, possessing both large endocrine-type granules and small synaptic-type vesicles (FUJI et al. 1980). Seasonal variations in the concentrations of immunoreactive glucagon in the pancreas of a lizard have been observed by heterologous RIA; high values (particularly in the cranial parts of the gland) were found during starvation, and low concentrations during the food intake period (DUPÉ-GODET and ADJOVI 1981). These observations conform to some extent with those made in amphibians (see Sect. F.II.3.a). The physiological role of glucagon in reptiles seems to be similar to that in higher vertebrates (cf. FALKMER and ÖSTBERG 1977).

As to the occurrence of glucagon/glicentin cells in the gastrointestinal tract mucosa of reptiles, it is now well established by some IHC studies that glucagon-(pancreatic type)-immunoreactive cells of both closed and open types occur in the fundus region of the stomach in representatives of all the main reptilian families investigated (BUCHAN 1981; EL-SALHY 1981). The distribution of glicentin-type-immunoreactive endocrine cells in the gastrointestinal tract showed, however, great species variations (BUCHAN 1981); again a feature shared with the amphibians (see Sect. F.II.3.a). Whether or not glucagon cells also occur in the reptilian central nervous system, is essentially unknown.

### c) Birds

The detailed accounts of the avian endocrine pancreas given in the previous reviews (FALKMER and MARQUES 1972; FALKMER and ÖSTBERG 1977) are still valid (SITBON and MIALHE 1980). As before, most information available stems from the results of investigations in fowls (EPPLE et al. 1980). Most conspicuous are the facts that in birds the glucagon cells are the predominant islet cell type, and that total pancreatectomy induces a fatal hypoglycaemia, indicating that glucagon is indispensable for the maintenance of life in birds.

The compact pancreas contains four lobes: a ventral, a dorsal, a third lobe, and a splenic lobe. The islet parenchyma is to some extent compartmentalised into the classical, large, "dark" A-islets, consisting mainly of glucagon and somatostatin cells, and the small, "light," B-islets, where insulin and somatostatin cells predominate. The A-islets are located exclusively to the third and the splenic lobes, whereas the B-islets are evenly distributed in all four lobes. The PP cells are mainly scattered in the exocrine parenchyma, but occur also in the periphery of the islets. It should be emphasised that there is no complete segregation of glucagon and insulin cells; still the rule holds true that "the A-cells in the B-islets are not as frequent as the B-cells in the A-islets" (FALKMER and ÖSTBERG 1977). Widely divergent opinions still exist as to whether the avian islet parenchyma is abundantly equipped with nerves or whether it completely lacks innervation (cf. FALKMER and ÖSTBERG 1977; EPPLE et al. 1980).

The small splenic lobe can consist of almost equal amounts of endocrine and exocrine parenchyma. This fact and the predilection for A-islets to occur in this lobe imply that high yields of glucagon can be obtained here. This concentration of glucagon cells in the splenic lobe has even been regarded as a compact, well-vascularised endocrine organ of its own – comparable to the splenic Brockmann body in teleost bony fish (Fig. 9) and its homologue in some reptiles – with direct access for its hormones to the portal circulation; there, they can affect intermediary metabolism, particularly through their first-reached target organ, the liver (BONNER-WEIR and WEIR 1979). In a situation of this kind, the putative local effects of glucagon on the exocrine, acinar parenchyma (cf. HENDERSON et al. 1981) must be negligible; actually, it has even been suggested that any functional association between glucagon and the exocrine pancreas is by chance only (EPPLE et al. 1980).

One of the main functions of avian glucagon concerns the regulation of lipid metabolism (SITBON and MIALHE 1980). Avian adipose tissue is particularly sensitive to glucagon (see Chap. 19); unlike in mammals, insulin apparently has no antilipolytic effect. As stated initially, the glycaemic state of birds, too, is far more dependent on glucagon than on insulin (HAZELWOOD 1976, EPPLE et al. 1980; SITBON and MIALHE 1980).

When the amino acid composition and sequences of glucagon molecules began to become known (see Chap. 1), it seemed as if the glucagon molecule had undergone only minor changes during evolution (SUNDBY et al. 1974; SUNDBY 1976); turkey and chicken glucagon are both almost identical with mammalian glucagon, differing from it by one residue at position 28 only; duck glucagon has just one further substitution at position 16 (MARKUSSEN et al. 1972; HAZELWOOD 1973, 1976).

Pancreatic glucagon content is 4–8 times higher in birds than in mammals (HAZELWOOD 1976), and the importance of this hormone in avian physiology is also exemplified by the early embryonic appearance of glucagon cells in the chick pancreas (SWENNE and LUNDQVIST 1980). They are recognisable by IHC at the third day of incubation (DIETERLEN-LIEVRE and BEAUPAIN 1976) and are functioning by the fifth day (BENZO and STEARNS 1976).

Also as regards the glucagon/glicentin cells in the avian gastrointestinal mucosa, the previous detailed accounts are still valid, both in adult birds and in fetuses (FALKMER and MARQUES 1972; FALKMER and ÖSTBERG 1977). Birds are well equipped with gut glucagon/glicentin cells and they appear early during embryogenesis. In the chicken, glucagon cells occur in the pylorus and ileum and are numerous in the proventriculus; the pylorus mucosa seems to be a particularly important endocrine organ in this bird (RAWDON and ANDREW 1980; KRAMER and ANDREW 1981; ANDREW et al. 1981). So far, the localisation of glucagon-like neurohormonal peptides in the brain of birds is essentially unknown.

d) Mammals

It is beyond the scope of this chapter to give an account of the histophysiology of the glucagon/glicentin cells in mammals. Relevant parts of their mammalian ontogeny and phylogeny have already been covered in the previous reviews (FALKMER and MARQUES 1972; FALKMER and ÖSTBERG 1977) and most of the statements made there are still valid. Thus, there have been no major advances in the phylogeny of glucagon/glicentin cells based on studies in mammals during the last 5 years. There is still no recent information about the Prototheria (egg-laying mammals) and the Metatheria (marsupials) (Fig. 8), and of the 24 subgroups of Eutheria (with genuine placenta), still only about one-third have been studied with regard to at least some aspects of their GEP neuroendocrine system (FALKMER and ÖSTBERG 1977; MCDONALD and WARING 1979).

In one respect, however, studies in mammals have contributed to a real breakthrough in the research front as regards the evolution of the glucagon/glicentin cells: the discovery that these cells, too, form part of the GEP neurohormonal brain–gut axis (see Sect. A). In addition, there is a recent report that significant amounts of immunoreactive glucagon were detected by RIA in acid–ethanol extracts of gastrointestinal tract, salivary glands, thymus, thyroid, and adrenal glands of rats; glicentin appeared in gut mucosa, and in salivary and adrenal glands (PEREZ-CASTILLO and BLÁZQUEZ 1980). This report indicates an unexpectedly wide distribution of glucagon/glicentin-like peptides that could be explained by a common neuroectodermal origin (cf. GOULD 1978; PEARSE 1980). The putative glucagon/glicentin-immunoreactive cells in the rat salivary glands, thymus, thyroid, and adrenal glands remain, however, to be visualised by IHC and EM. In the rat brain, "gut-type glucagon"-immunoreactive nerve fibres (LORÉN et al. 1979) and nerve cells (DORN et al. 1980; TAGER et al. 1980) have been described, being numerous in certain areas of hypothalamus and thalamus, particularly in the paraventricular and supraoptical nuclei, but absent from neocortex and hippocampus. They did not react with antisera specific for pancreatic type glucagon. Glucagon-like immunoreactive substances have also been described in the brains of dogs (CONLON et al. 1979; SASAKI et al. 1980) and humans (SANDERS et al. 1981), even in human fetuses (KORÁNYI et al. 1981). As in fetal pancreas (TUNG et al. 1980), the immunoreactive substances are of high molecular weight, compatible with a biosynthesis of glucagon from a larger precursor glicentin molecule (THIM and MOODY

1981). A comparison has recently been made of the various kinds of glucagon-like material discovered in the brain of different mammals (TOMINAGA et al. 1981).

The importance of being alert for artifacts and problems in methodology before reporting the findings of well-known, well-characterised GEP neurohormonal peptides in unusual places has, however, recently been emphasised (ENG and YALOW 1981; YALOW et al. 1981). It is obvious that it is justified to conclude that research on the presence of glucagon-like peptides in the vertebrate nervous system is still in its infancy. Nevertheless, a summarising survey in tabular form of the brain–gut axis for glucagon and the other three islet hormones, for the glucagon/secretin family, for the gastrin/CCK family and for the five best known of the rest of the GEP neurohormonal peptides, is given in Fig. 15.

The ontogeny of glucagon/glicentin cells has been the subject of some recent studies in rat pancreas (FUJI 1979), human pancreas (VAN ASSCHE and AERTS 1979) and the human intestine (BUCHAN et al. 1981). The glucagon/glicentin cells are among the first to appear in both species; being present earlier than insulin and PP cells. Their functional role during fetal life is not well understood (VAN ASSCHE and AERTS 1979). Of particular interest for the ontogeny of islet glucagon cells are some recent IHC studies in embryonic mice where evidence has been obtained that glucagon cells arise from transformation in situ of cells that transiently express a catecholaminergic phenotype (TEITELMAN et al. 1981). Observations of this kind suggest that other GEP peptidergic cells may also arise from aminergic precursors.

As to the ontogeny of the glucagon cells in humans, it seems that some maturation processes continue in the pancreatic islet parenchyma, even during the first few months of postnatal life (RAHIER et al. 1981). Histopathological pictures that previously were considered almost pathognomonic for so-called nesidioblastosis in persistent neonatal hypoglycaemia with hyperinsulinism, must now be included in the normal structure of the endocrine pancreas in neonates and young infants (cf. FALKMER et al. 1981 b; RAHIER et al. 1981).

## G. Discussion and Summarising Conclusions

The incomplete information given in this very patchy survey of the glucagon-producing cell in the phylogenetic series will serve as evidence of the vast world we still have to discover about the structure and function of glucagon in all classes of animal. It goes without saying that what we broadly class as "glucagon" cannot be assumed to be identical with, or even closely related to, mammalian glucagon in structure and function until it has been sequenced and its function in its parent animal determined. The most we can say at present, in defence of our statement in

---

**Fig. 15.** Tabular survey of definite and possible occurrence of GEP neurohormonal peptides ▶ in the nervous system (NS) and in the alimentary tract mucosa (AT) of coelenterates and four kinds of protostomian invertebrates (to the left of the *thick vertical line*) and of deuterostomian protochordates and all the major subclasses of vertebrates (to the right of the *thick vertical line*). The tabular survey has been subdivided into three major parts: the four islet hormones; the members of the secretin and gastrin/CCK families; and the best known of the rest of the GEP neurohormonal peptides. *Shaded areas*, established localisation or information from more than one method; *question marks*, incomplete information, so far; possible localisation; *blank areas*, no studies made or negative information

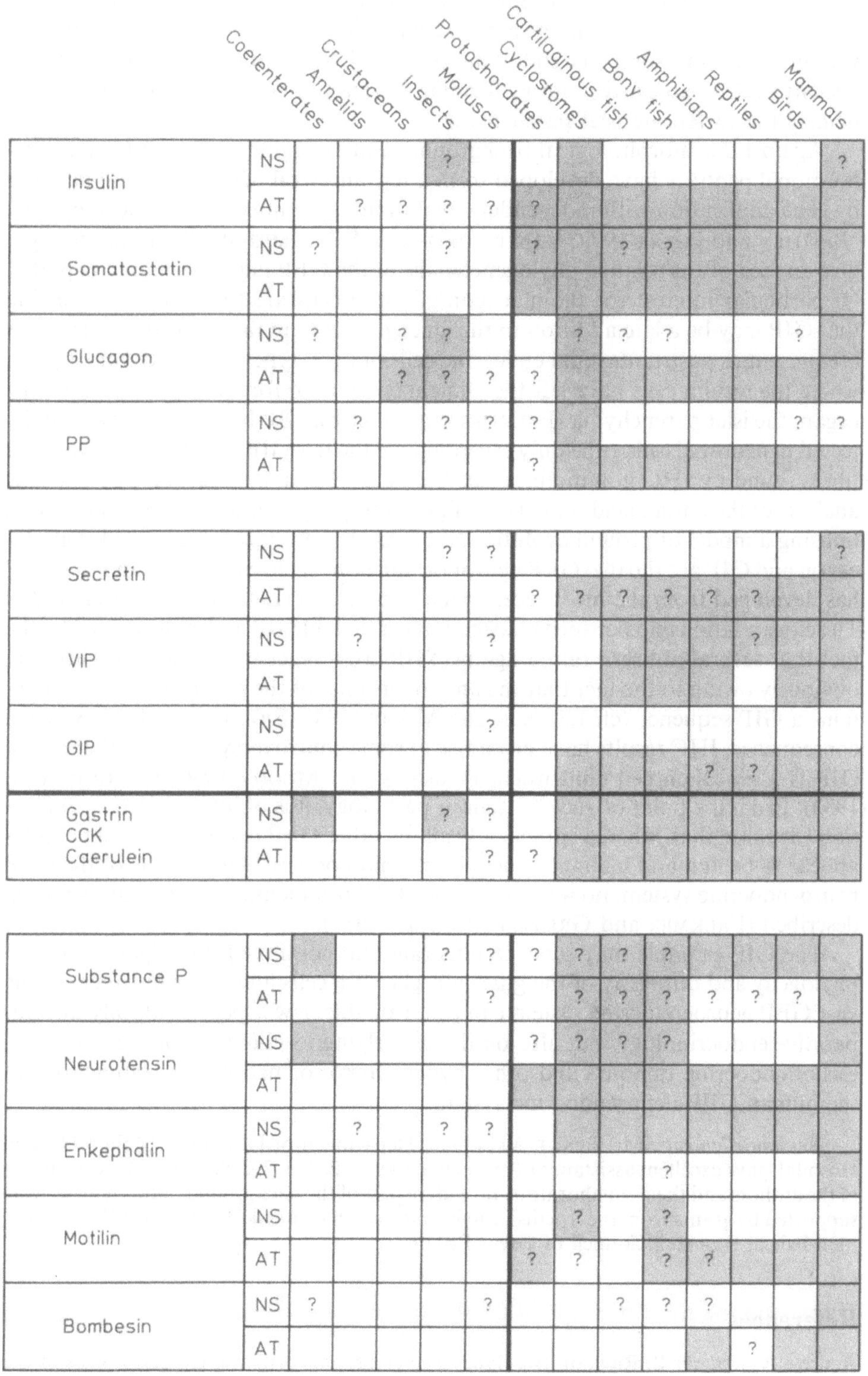
Coelenterates
Annelids
Crustaceans
Insects
Molluscs
Protochordates
Cyclostomes
Cartilaginous fish
Bony fish
Amphibians
Reptiles
Birds
Mammals
Insulin
Somatostatin
Glucagon
PP
Secretin
VIP
GIP
Gastrin
CCK
Caerulein
Substance P
Neurotensin
Enkephalin
Motilin
Bombesin
NS
AT

Sect. A that an integrated concept can be reached, is that glucagon/glicentin-like molecules obviously occur in the central nervous systems of some protostomian invertebrates and in some gut endocrine cells in higher forms; lastly, the glucagon/glicentin cells also participate in forming a separate compact endocrine organ, namely the vertebrate islet parenchyma.

Against this morphological background, and assuming that most GEP neurohormonal peptides have developed from a few ancestral neurohormonal peptides by gene duplication with independent evolution of the mutated products (cf. VAN NOORDEN and POLAK 1979; VAN NOORDEN and FALKMER 1980), figure 15 might give an idea of the relative phylogenetic age of the GEP neurohormonal peptides. Of particular interest for the glucagon/glicentin cells are the features indicating that GIP may be a late addition to the glucagon/secretin family. GIP-storing cells are present as gastrointestinal endocrine cells of open type in the vertebrates only, where the insulin cells have left the alimentary tract to form a separate endocrine organ, the islet parenchyma (FALKMER et al. 1980b). GIP has, so far, not been detected in neuronal cells – the only exception are the few GIP-immunoreactive nerve fibres found by IHC in a mollusc (see Sect. D.I). From a structural phylogenetic analysis of the amino acid sequences of glicentin/proglucagon and GIP molecules, utilising a model of protein evolution, it has been theoretically proposed that glucagon and GIP are the two GEP neurohormones most closely related and that GIP has developed from the most recent gene duplication (LEHMAN and GURD 1980). The close relationship between glucagon/glicentin and GIP is also illustrated by the fact that several antisera raised against GIP cross-react with glucagon/glicentin, obviously owing to the fact that the amino terminal of the glicentin molecule contains a GIP sequence (cf. LARSSON and MOODY 1980; SJÖLUND et al. 1983). As a consequence, IHC results have given rise to some controversy as to whether or not GIP is a glucagon cell constituent (LARSSON and MOODY 1980; SJÖLUND et al. 1983). From the point of view of tumour pathology, it is of interest in this connection to realise that, whereas practically all the other GEP neurohormonal peptides appear in benign and malignant neoplasms and nodular hyperplasias of the GEP neuroendocrine system, no well-established GIP-producing tumours have yet been described (FALKMER and GRIMMELIKHUIJZEN 1981).

The GIP example may serve to illustrate that continued investigations of the phylogeny and ontogeny of the glucagon/glicentin cells and other members of the vast GEP neuroendocrine system can give valuable new aspects, not only in comparative endocrinology, but also on the underlying mechanisms of the pathogenesis of endocrine tumours and other hyper-, hypo- or neoplastic dysfunctions of the human GEP neuroendocrine system.

*Acknowledgments.* Mr. BENGT WÄNGELIN, Department of Pathology, Malmö General Hospital, gave excellent assistance in producing Figs. 1, 2, 3, 5, 7, 8, 9, and 13. Investigations of the authors and their collaborators, providing part of the background of this review, were supported by grants from the Swedish Medical Research Council (project No 12X-718) and the Medical Research Council of Great Britain.

## References

Andrew A, Kramer B, Rawdon BB (1982) The embryonic origin of endocrine cells of the gastro-intestinal tract. Gen Comp Endocr 47:249–265

Banks I, Sloan JM, Buchanan KD (1980) Glucagon-like (GLI) and secretin-like immunoreactivity (SLI) within the digestive system of the bivalve mollusc *Cerastoderma edule*. Regul Peptides [Suppl 1] 1:7

Benzo CA, Stearns SR (1976) Functional differentiation of the chick endocrine pancreas. II. The alpha cells and glucagon. Am J Anat 147:273–280

Bevis PJR, Thorndyke MC (1979) A cytochemical and immunofluorescence study of endocrine cells in the gut of the ascidian *Styela clava*. Cell Tissue Res 199:139–144

Bevis PJR, Thorndyke MC (1981) Stimulation of gastric enzyme secretion by porcine cholecystokinin in the ascidian, *Styela clava*. Gen Comp Endocrinol 45:458–464

Bhatnagar YM, Carraway RE (1981) Bacterial peptides with C-terminal similarities to bovine neurotensin. Peptides 2:51–59

Blundell T (1979) Conformation and molecular biology of polypeptide hormones. II. Glucagon. Trends Biochem Sci 4:80–83

Bonner-Weir S, Weir GC (1979) The organization of the endocrine pancreas: a hypothetical unifying view of the phylogenetic differences. Gen Comp Endocrinol 38:28–37

Bounias M (1979) Relations doses/effets de l'insuline et du glucagon sur la glycémie de l'abeille. CR Soc Biol (Paris) 173:1031–1036

Buchan AMJ (1981) Gut hormones from amphibia to man. PhD thesis. University of London

Buchan AMJ, Bryant MG, Polak JM, Gregor M, Ghatei MA, Bloom SR (1981) Development of regulatory peptides in the human fetal intestine. In: Bloom SR, Polak JM (eds) Gut hormones, 2nd edn. Churchill Livingstone, Edinburgh, pp 119–124

Conlon JM (1980) The glucagon-like peptides – order out of chaos? Diabetologia 18:85–88

Conlon JM, Samson WK, Dobbs RE, Orci L, Unger RH (1979) Glucagon-like polypeptides in canine brain. Diabetes 28:700–702

Coupland RE (1979) Catecholamines. In: Barrington EJW (ed) Hormones and evolution. Academic Press, London New York, pp 309–340

Csaba G (1981) Ontogeny and phylogeny of hormone receptors. Monogr Dev Biol 15:1–172

Dieterlen-Lièvre F, Beaupain D (1976) Immunocytological study of endocrine pancreas ontogeny in the chick embryo; normal development and pancreatic potentialities in the early splanchnopleura. In: Grillo TAI, Leibson L, Epple A (eds) The evolution of pancreatic islets. Pergamon Press, Oxford New York, pp 37–50

Dockray GJ (1977) Molecular evolution of gut hormones: application of comparative studies on the regulation of digestion. Gastroenterology 72:344–358

Dockray GJ (1979a) Comparative biochemistry and physiology of gut hormones. Annu Rev Physiol 41:83–95

Dockray GJ (1979b) Evolutionary relationships of the gut hormones. Fed Proc 38:2295–2301

Dockray GJ, Gregory RA (1980) Relations between neuropeptides and gut hormones. Proc R Soc Lond [Biol] 210:151–164

Doerr-Schott J, Garaud J-C (1981) Ultrastructural identification of gastrin-like immunoreactive nerve fibres in the brain of *Xenopus laevis* by means of colloidal gold or ferritin immunocytochemical methods. Cell Tissue Res 216:581–589

Doerr-Schott J, Dubois MP, Lichte C (1981) Immunohistochemical localisation of substances reactive to antisera against $\alpha$- and $\beta$-endorphin and met-enkephalin in the brain of *Rana temporaria* L. Cell Tissue Res 217:79–92

Dorn A, Bernstein H-G, Hahn HJ, Kostmann G, Ziegler M (1980) Regional distribution of glucagon-like immunoreactive material in the brain of rats and sand rats – an immunohistochemical investigation. Acta Histochem 66:269–272

Dupé-Godet M, Adjovi Y (1981) Seasonal-variations of immunoreactive glucagon contents in pancreatic extracts of a sahelian lizard *(Varanus exanthematicus)*. Comp Biochem Physiol 69A:31–42

Duve H, Thorpe A (1979) Immunofluorescent localization of insulin-like material in the median neurosecretory cells of the blowfly, *Calliphora vomitoria* (Diptera). Cell Tissue Res 200:187–191

Duve H, Thorpe A (1980) Localization of pancreatic polypeptide (PP)-like immunoreactive material in neurons of the brain of the blowfly, *Calliphora erythrocephala* (Diptera). Cell Tissue Res 210:101–109

Duve H, Thorpe A, Lazarus NR (1979) Isolation of material displaying insulin-like immunological and biological activity from the brain of the blowfly, *Calliphora vomitoria*. Biochem J 184:221–227
Duve H, Thorpe A, Neville R, Lazarus NR (1981) Isolation and partial characterization of pancreatic polypeptide-like material in the brain of the blowfly *Calliphora vomitoria*. Biochem J 197:767–770
El-Salhy M (1981) On the phylogeny of the gastro-entero-pancreatic (GEP) neuroendocrine system. MD thesis. University of Uppsala. (Acta Univ Ups 385:1–39)
El-Salhy M, Grimelius L (1981) Histological and immunohistochemical studies of the endocrine pancreas of lizards. Histochemistry 72:237–247
El-Salhy M, Abou-El-Ela R, Falkmer S, Grimelius L, Wilander E (1980) Immunohistochemical evidence of gastro-entero-pancreatic neurohormonal peptides of vertebrate type in the nervous system of the larva of a dipteran insect, the hoverfly, *Eristalis aeneus*. Regul Peptides 1:187–204
El-Salhy M, Grimelius L, Wilander E, Abu-Sinna G, Lundqvist G (1981) Histological and immunohistochemical studies of the endocrine cells of the gastrointestinal mucosa of the toad *(Bufo regularis)*. Histochemistry 71:53–65
Emdin SO (1981) Myxine insulin. Amino-acid sequence, three dimensional structure, biosynthesis, release, physiological role, receptor binding affinity, and biological activity. MD thesis. Umeå University, Sweden (N S 66:1–159)
Endo Y, Nishiitsutsuji-Uwo J (1981) Gut endocrine cells in insects: the ultrastructure of the gut endocrine cells of the lepidopterous species. Biomed Res 2:270–280
Eng J, Yalow RS (1981) Evidence against extra-pancreatic insulin synthesis. Proc Natl Acad Sci USA 78:4576–4578
Epple A, Brinn JE, Young Y (1980) Evolution of pancreatic islet function. In: Pang PKT, Epple A (eds) Evolution of vertebrate endocrine systems. Texas Tech Press, Lubbock, pp 269–321
Falkmer S (1961) Experimental diabetes research in fish. On the morphology and physiology of the endocrine pancreatic tissue of the marine teleost *Cottus scorpius* with special reference to the role of glutathione in the mechanism of alloxan diabetes using a modified nitroprusside method. Acta Endocrinol [Suppl 59] 37:1–122
Falkmer S (1980a) Some recent aspects on the phylogeny of the GEP paraneurons. Biomed Res [Suppl 1] 1:17–23
Falkmer S (1980b) Evolution of pancreatic hormones. In: Ishii S, Hirano T, Wada M (eds) Hormones, adaptation and evolution. Japan Sci Soc Press, Tokyo/Springer, Berlin Heidelberg New York, pp 295–299
Falkmer S, Emdin SO (1981) Insulin evolution. In: Dodson G, Glusker JP, Sayre D (eds) Structural studies of molecules of biological interest. Oxford University Press, Oxford, pp 420–440
Falkmer S, Grimmelikhuijzen CJP (1981) Phyletic aspects of the tumor pathology of the liver and the gastro-entero-pancreatic (GEP) neuroendocrine system (carcinoid-islet-cell tumors). In: Dawe CJ et al (eds) Phyletic approaches to cancer. Japan Sci Soc Press, Tokyo
Falkmer S, Marques M (1972) Phylogeny and ontogeny of glucagon production. In: Lefebvre PJ, Unger RH (eds) Glucagon: molecular physiology, clinical and therapeutic implications. Pergamon Press, Oxford New York, pp 343–361
Falkmer S, Östberg Y (1977) Comparative morphology of pancreatic islets in animals. In: Volk BW, Wellmann KF (eds) The diabetic pancreas. Plenum Press, New York, pp 15–59
Falkmer S, Patent GJ (1972) Comparative and embryological aspects of the pancreatic islets. In: Steiner DF, Freinkel N (eds) The endocrine pancreas. Handbook of Physiology, vol 1. Williams and Wilkins, Baltimore, pp 1–23
Falkmer S, Emdin SO, Havu N, Lundgren G, Marques M, Östberg Y, Steiner DF, Thomas NW (1973) Insulin in invertebrates and cyclostomes. Am Zool 13:625–638
Falkmer S, Elde RP, Hellerström C, Petersson B, Efendić S, Fohlman J, Siljevall J-B (1977) Some phylogenetic aspects on the occurrence of somatostatin in the gastro-enteropancreatic endocrine system. A histological and immunocytochemical study, combined with quantitative radioimmunological assays of tissue extracts. Arch Histol Jpn 40:99–117

Falkmer S, Elde RP, Hellerström C, Petersson B (1978) Phylogenetic aspects of somatostatin in the gastro-entero-pancreatic (GEP) endocrine system. Metabolism 27:1193–1196
Falkmer S, Fahrenkrug J, Alumets J, Håkanson R, Sundler F (1980a) Vasoactive intestinal polypeptide (VIP) in epithelial cells of the gut mucosa of an elasmobranchian cartilaginous fish, the ray. Endocrinol Jpn [Suppl] 27:31–35
Falkmer S, Ebert R, Arnold R, Creutzfeldt W (1980b) Some phylogenetical aspects on the entero-insular axis with particular regard to the appearance of the gastric inhibitory peptide. Front Horm Res 7:1–6
Falkmer S, Carraway RE, El-Salhy M, Emdin SO, Grimelius L, Rehfeld JF, Reinecke M, Schwartz TFW (1981a) Phylogeny of the gastroenteropancreatic neuroendocrine system. A review with special reference to the occurrence of CCK-like and neurotensin-like polypeptides in lower vertebrates and invertebrates. UCLA Forum Med Sci 23:21–42
Falkmer S, Søvik O, Vidnes J (1981b) Immunohistochemical, morphometric, and clinical studies of the pancreatic islets in infants with persistent neonatal hypoglycemia of familial type with hyperinsulinism and nesidioblastosis. Acta Biol Med Ger 40:39–54
Farrar ES, Frye BE (1979) A comparison of adrenalin and glucagon effects on carbohydrate levels of larval and adult *Rana pipiens*. Gen Comp Endocrinol 39:372–380
Fouchereau-Peron M, Laburthe M, Besson J, Rosselin G, LeGal Y (1980) Characterization of the vasoactive intestinal polypeptide (VIP) in the gut of fishes. Comp Biochem Physiol 65:489–492
Fritsch HAR, Van Noorden S, Pearse AGE (1976) Cytochemical and immunofluorescence investigations on insulin-like producing cells in the intestine of *Mytilus edulis* L (Bivalvia). Cell Tissue Res 177:407–413
Fritsch HAR, Van Noorden S, Pearse AGE (1981) Gastrointestinal and neurohormonal peptides in the alimentary tract and cerebral complex of *Ciona intestinalis* (Ascidiacaea). Their relevance to the evolution of the diffuse neuroendocrine system. Cell Tissue Res 223:369–402
Fujii S (1979) Development of pancreatic endocrine cells in the rat fetus. Arch Histol Jpn 42:467–479
Fujii S, Kobayashi I, Fujita T, Yanaihara N (1980) VIP-immunoreactive nerves in the pancreas of the snake, *Elaphe quadrivirgata* (Boie) – Another model for insular neurosecretion. Biomed Res 1:180–184
Fujita T, Yanatori Y, Murakemi T (1976) Insulo-acinar axis, its vascular basis and its functional and morphological changes caused by CCK-PZ and caerulein. In: Fujita T (ed) Endocrine gut and pancreas. Elsevier, Amsterdam Oxford New York, pp 347–357
Fujita T, Kobayashi S, Yui R, Iwanaga T (1980) Evolution of neurons and paraneurons. In: Ishii S, Hirano T, Wada M (eds) Hormones, adaptation and evolution. Japan Sci Soc Press, Tokyo/Springer, Berlin Heidelberg New York, pp 35–43
Fujita T, Kobayashi S, Fuji S, Iwanaga T, Serizawa Y (1981) Langerhans islets as the neuro-paraneuronal control center of the exocrine pancreas. UCLA Forum Med Sci 23:231–242
Giraud AS, Yeomans ND (1981) Fine structure of the gastric mucous and endocrine cells of the toad, *Bufo marinus*. Cell Tissue Res 218:663–668
Gould RP (1978) The APUD cell system. Recent Adv Histopathol 10:1–22
Grimelius L, Wilander E (1980) Silver stains in the study of endocrine cells of the gut and pancreas. Invest Cell Pathol 3:3–12
Grimmelikhuijzen CJP (1983) Coexistence of neuropeptides in Hydra. Neuroscience (in press)
Grimmelikhuijzen CJP, Schaller HC (1979) Hydra as a model organism for the study of morphogenesis. Trends Biochem Sci 4:265–267
Grimmelikhuijzen CJP, Sundler F, Rehfeld JF (1980) Gastrin/CCK-like immunoreactivity in the nervous system of coelenterates. Histochemistry 69:61–68
Grimmelikhuijzen CJP, Balfe A, Emson PC, Powell D, Sundler F (1981a) Substance P-like immunoreactivity in the nervous system of Hydra. Histochemistry 71:325–333
Grimmelikhuijzen CJP, Carraway RE, Rökaens Å, Sundler F (1981b) Neurotensin-like immunoreactivity in the nervous system of Hydra. Histochemistry 72:199–209
Grimmelikhuijzen CJP, Dockray GJ, Yanaihara N (1981c) Bombesin-like immunoreactivity in the nervous system of Hydra. Histochemistry 73:171–180

Hazelwood RL (1973) The avian endocrine pancreas. Am Zool 13:699–709
Hazelwood RL (1976) Three avian pancreatic hormones. In: Grillo TAI, Leibson L, Epple A (eds) The evolution of pancreatic islets. Pergamon Press, Oxford New York, pp 321–334
Henderson JR, Daniel PM, Fraser PA (1981) The pancreas as a single organ: the influence of the endocrine upon the exocrine part of the gland. Gut 22:158–167
Holmquist AL, Dockray GJ, Rosenquist GL, Walsh JH (1979) Immunochemical characterization of cholecystokinin-like peptides in lamprey gut and brain. Gen Comp Endocrinol 37:474–481
Inagaki S, Senba E, Shiosaka S, Takagi H, Kawai Y, Takatsuki K, Sakanaka M, Matsuzaki T, Tohyama M (1981) Regional distribution of substance P-like immunoreactivity in the frog brain and spinal cord: Immunohistochemical analysis. J Comp Neurol 201:243–254
Iwanaga T, Fujita T, Nishiitsutsuji-Uwo J, Endo Y (1981) Immunohistochemical demonstration of PP-, somatostatin-, enteroglucagon- and VIP-like immunoreactivities in the cockroach midgut. Biomed Res 2:202–207
Jaros PP, Keller R (1979) Immunocytochemical identification of hyperglycemic hormone-producing cells in the eyestalk of *Carcinus maenas*. Cell Tissue Res 204:379–385
Joosse J (1979) Evolutionary aspects of the endocrine system and of the hormonal control of reproduction in molluscs. In: Barrington EJW (ed) Hormones and evolution. Academic Press, London New York, pp 119–157
Kaung HC (1981) Immunocytochemical localization of pancreatic endocrine cells in frog embryos and young larvae. Gen Comp Endocrinol 45:204–211
Keller R, Wunderer G (1978) Purification and amino acid composition of the neurosecretory hyperglycemic hormone from the sinus gland of the shore crab, *Carcinus maenas*. Gen Comp Endocrinol 34:328–335
Klein C, Van Noorden S (1980) Pancreatic polypeptide (PP)- and glucagon cells in the pancreatic islet of *Xiphophorus helleri* H (Teleostei). Correlated immunohistochemistry and electron microscopy. Cell Tissue Res 205:187–198
Kleinholz LH, Keller R (1979) Endocrine control in crustacea. In: Barrington EJW (ed) Hormones and evolution. Academic Press, London New York, pp 159–213
Kobayashi K, Ali SS (1981) Cell types of the endocrine pancreas of the shark, *Scyliorhinus stellaris*, as revealed by correlative light and electron microscopy. Cell Tissue Res 215:475–490
Korányi L, Burger ZS, Peterfy F, Gouth J, Tamás G Jr (1981) Glucagon-like immunoreactivity in the brain of human fetuses. Diabetologia 21:293
Kramer B, Andrew A (1981) Further investigations into the source of pancreatic endocrine cell types. Gen Comp Endocrinol 44:279–287
Kramer KJ (1983) Vertebrate hormones in insects. In: Kerkut GA, Gilbert LI (eds) Endocrinology. (Comprehensive insect physiology, biochemistry and pharmacology, vol 7/I) Pergamon Press, Oxford (in press)
Kramer KJ, Tager HS, Chilos CN (1980) Insulin-like and glucagon-like peptides in insect hemolymph. Insect Biochem 10:179–182
Larsson L-I, Moody AJ (1980) Glicentin and gastric inhibitory polypeptide immunoreactivity in endocrine cells of the gut and pancreas. J Histochem Cytochem 28:925–933
Lehman LD, Gurd RS (1980) Primary structure of ancestral pancreatic and gut hormones and their phylogenetic relationships. Clin Res 28:800 A
LeRoith D, Shiloach J, Roth J, Lesniak MA (1980) Evolutionary origins of vertebrate hormones: substances similar to mammalian insulins are native to unicellular eukaryotes *(Tetrahymena/Neurospora)*. Proc Natl Acad Sci USA 77:6184–6188
LeRoith D, Lesniak MA, Roth J (1981 a) Insulin in insects and annelids. Diabetes 30:70–76
LeRoith D, Shiloach J, Roth J, Lesniak MA (1981 b) Insulin or a closely related molecule is native to *Escherichia coli*. J Biol Chem 256:6533–6536
Lorén I, Alumets J, Håkanson R, Sundler F, Thorell J (1979) Gut-type glucagon immunoreactivity in nerves of the rat brain. Histochemistry 61:335–341
Lund PK, Goodman RH, Jacobs JW, Habener JH (1980) Glucagon precursors identified by immunoprecipitation of products of cell-free translation of messenger RNA. Diabetes 29:583–586

Maier V, Kroder A, Groner E, Keller R, Pfeiffer EF (1975) Glucagon-like activity (GLI) in the intestine of *Porcus domesticus* and *Astacus fluviatilis*. Acta Endocrinol [Suppl] 193:41

Maier V, Grube D, Steiner R, Greischel A, Pfeiffer EF (1981) Glucagon and insulin in the brain of the honeybee (Abstr). Int Symp "Brain-gut axis: the new frontier". Florence, June 29/30–July 1, 1981, p 119

Markussen J, Frandsen E, Heding LG, Sundby F (1972) Turkey glucagon: crystallization, amino-acid composition and immunology. Horm Metab Res 4:360–363

McDonald IR, Waring H (1979) Hormones of marsupials and monotremes. In: Barrington EJW (ed) Hormones and evolution. Academic Press, London New York, pp 873–916

McVay Lawrence J, Craig JV, Clough D (1972) The presence of a hyperglycemic factor in the suprapharyngeal ganglia of *Lumbricus terrestris*. Gen Comp Endocrinol 18:260–267

Nishiitsutsuji-Uwo J, Endo Y (1981) Gut endocrine cells in insects: the ultrastructure of the endocrine cells in the cockroach midgut. Biomed Res 2:30–44

Noe BD (1981 a) Inhibition of islet prohormone to hormone conversion by incorporation of arginine and lysine analogs. J Biol Chem 256:4940–4946

Noe BD (1981 b) Characterization of proinsulin- and proglucagon-converting activities in isolated islet secretory granules. J Cell Biol 90:312–318

Pearse AGE (1980) APUD: concept, tumours, molecular markers and amyloid. Mikroskopie 36:257–269

Perez-Castillo A, Blázquez E (1980) Tissue distribution of glucagon, glucagon-like immunoreactivity, and insulin in the rat. Am J Physiol 238:E 258–E 266

Plisetskaya E, Kazakov VK, Solbitskaya L, Leibson LG (1978) Insulin-producing cells in the gut of freshwater bivalve molluscs *Anodonta cygnea* and *Unio pictorum* and the role of insulin in the regulation of their carbohydrate metabolism. Gen Comp Endocrinol 35:133–145

Polak JM, Pearse AGE, Grimelius L, Bloom SR, Arimura A (1975) Growth-hormone release-inhibiting hormone in gastrointestinal and pancreatic D-cells. Lancet 1:1220–1225

Rahier J, Wallon J, Henquin J-C (1981) The cell populations in the endocrine pancreas of human neonates and infants. Diabetologia 20:540–546

Ravazzola M, Orci L (1980 a) Transformation of glicentin-containing L-cells into glucagon-containing cells by enzymatic digestion. Diabetes 29:156–158

Ravazzola M, Orci L (1980 b) A pancreatic polypeptide (PP)-like immunoreactant is present in the glicentin-containing cell of the cat intestine. Histochemistry 67:221–224

Rawdon BB, Andrew A (1980) An immunocytochemical study of endocrine cells in the chick gut. Regul Peptides [Suppl 1] 1:S 89

Reinecke M (1981) Immunohistochemical localization of polypeptide hormones in endocrine cells of the digestive tract of *Branchiostoma lanceolatum*. Cell Tissue Res 219:445–456

Rhoten WB, Hall CE (1981) Four hormones in the pancreas of the lizard, *Anolis carolinensis*. Anat Rec 199:89–97

Rombout JHWM, Rademakers LHPM, Van Hees JP (1979) Pancreatic endocrine cells of *Barbus conchonius* (Teleostei, Cyprinidae), and their relation to the enteroendocrine cells. Cell Tissue Res 203:9–23

Sanders DJ, Zahedi-Asl S, Marr AP, Perry EK, Perry RH (1981) Glucagon in human brain: Distribution in controls and patients with Alzheimer disease (Abstr). Int Symp "Brain-gut axis: the new frontier". Florence, June 29/30–July 1, 1981, p 160

Sasaki H, Ebitani I, Tominaga M, Yamatani K, Yawata Y, Hara M (1980) Glucagon-like substance in the canine brain. Endocrinol Jpn [Suppl] 27:135–140

Schlaghecke R, Blüm V (1981) Seasonal variation in insulin and glucagon concentrations of *Rana esculenta* (L). Gen Comp Endocrinol 43:479–483

Schot LPC, Boer HH, Swaab DF, Van Noorden S (1981) Immunocytochemical demonstration of peptidergic neurons in the central nervous system of the pond snail *Lymnaea stagnalis* with antisera raised to biologically active peptides of vertebrates. Cell Tissue Res 216:273–291

Sekine Y, Yui R (1981) Immunohistochemical study of the pancreatic endocrine cells of the ray, *Dasyatis akajei*. Arch Histol Jpn 44:95–101

Shields D, Warren TG, Roth SE, Brenner MJ (1981) Cell-free synthesis and processing of multiple precursors to glucagon. Nature 289:511–514
Sitbon G, Mialhe P (1980) The endocrine pancreas of birds. J Physiol (Paris) 76:5–24
Sjölund K, Ekelund M, Håkanson R, Moody AJ, Sundler F (1983) Gastric inhibitory peptide-like immunoreactivity in glucagon and glicentin cells: properties and origin. An immunocytochemical study using several antisera. J Histochem Cytochem 31 (in press)
Stefan Y, Falkmer S (1980) Identification of four endocrine cell types in the pancreas of *Cottus scorpius* (Teleostei) by immunofluorescence and electron microscopy. Gen Comp Endocrinol 42:171–178
Stefan Y, Ravazzola M, Orci L (1981) Primitive islets contain two populations of cells with differing glucagon immunoreactivity. Diabetes 30:192–195
Sundby F (1976) Species variations in the primary structure of glucagon. Metabolism [Suppl 1] 25:1319–1321
Sundby F, Markussen J, Danho W (1974) Camel glucagon: isolation, crystallization and amino acid composition. Horm Metab Res 6:425
Sundler F, Håkanson R, Alumets J, Walles B (1977) Neuronal localization of pancreatic polypeptide (PP) and vasoactive intestinal peptide (VIP) in the earthworm *(Lumbricus terrestris)*. Brain Res Bull 2:61–65
Swenne I, Lundqvist G (1980) Islet structure and pancreatic hormone content of the developing chick embryo. Gen Comp Endocrinol 41:190–198
Tager HS, Kramer KJ (1980) Insect glucagon-like peptides – Evidence for a high molecular-weight form in midgut from *Manduca sexta* (L). Insect Biochem 10:617–619
Tager HS, Markese J, Spiers RD, Kramer KJ (1975) Glucagon-like immunoreactivity in insect corpus cardiacum. Nature 254:707–708
Tager HS, Markese J, Kramer KJ, Speirs RD, Childs CN (1976) Glucagon-like and insulin-like hormones of the insect neurosecretory system. Biochem J 156:515–520
Tager HS, Hohenbok M, Markese J, Dinerste RJ (1980) Identification and localization of glucagon-related peptides in rat brain. Proc Natl Acad Sci USA 77:6229–6233
Tagliafierro G, Faraldi G (1976) Differentiation of the endocrine cells of the gastric epithelium of *Scyliorhinus stellaris*. Boll Zool 43:75–85
Teitelman G, Joh TH, Reis DJ (1981) Transformation of catecholaminergic precursors into glucagon (A) cells in mouse embryonic pancreas. Proc Natl Acad Sci USA 78:5225–5229
Thim L, Moody AJ (1981) The primary structure of porcine glicentin (proglucagon). Regul Peptides 2:139–150
Thorndyke MC (1982) Cholecystokinin (CCK) gastrin-like immunoreactive neurones in the cerebral ganglion of the protochordate ascidians *Styela clava* and *Ascidiella aspersa*. Regul Peptides 3:281–288
Titlbach M (1981) Light and electron microscopic study of crocodile islets of Langerhans. Z Mikrosk Anat Forsch 95:401–427
Tominaga M, Ebitani I, Marubashi S, Kaminura T, Katagiri T, Sasaki H (1981) Species difference of glucagon-like materials in the brain. Life Sci 29:1577–1581
Tomita T, Pollock HG (1981) Four pancreatic endocrine cells in the bullfrog *(Rana catesbeiana)*. Gen Comp Endocrinol 45:355–363
Tung AK, Ruse JL, Cockburn E (1980) Glucagon from the bovine fetal pancreas: chromatographic and electrophoretic characterizations of high molecular weight immunoreactive species. Can J Biochem 58:707–714
Van Assche FA, Aerts L (1979) The fetal endocrine pancreas. Contrib Gynecol Obstet 5:44–57
Van Herp F, Van Buggenum HJM (1979) Immunocytochemical localization of hyperglycemic hormone (HGH) in the neurosecretory system of the eyestalk of the crayfish *Astacus leptodactylus*. Experientia 35:1527–1528
Van Noorden S, Falkmer S (1980) Gut-islet endocrinology – some evolutionary aspects. Invest Cell Pathol 3:21–35
Van Noorden S, Patent GJ (1980) Vasoactive intestinal polypeptide-like immunoreactivity in nerves of the pancreatic islet of the teleost fish, *Gillichthys mirabilis*. Cell Tissue Res 212:139–146

Van Noorden S, Pearse AGE (1976) The localization of immunoreactivity to insulin, glucagon and gastrin in the gut of Amphioxus *(Branchiostoma lanceolatum)*. In: Grillo TAI, Leibson L, Epple A (eds) The evolution of pancreatic islets. Pergamon Press, Oxford New York, pp 163–178

Van Noorden S, Polak JM (1979) Hormones of the alimentary tract. In: Barrington EJW (ed) Hormones and evolution. Academic Press, London New York, pp 791–828

Van Noorden S, Östberg Y, Pearse AGE (1977) Localization of somatostatin-like immunoreactivity in the pancreatic islets of the hagfish, *Myxine glutinosa*, and the lamprey, *Lampetra fluviatilis*. Cell Tissue Res 177:281–285

Van Noorden S, Fritsch HAR, Grillo TAI, Polak JM, Pearse AGE (1980) Immunohistochemical staining for vertebrate peptides in the nervous system of a gastropod mollusc. Gen Comp Endocrinol 40:375–376

Weinstein B (1972) A generalized homology correlation for various hormones and proteins. Experientia 28:1517–1522

Welsch U, Dilly PN (1980) Elektronenmikroskopische Beobachtungen am Epithel des Verdauungstraktes der Hemichordaten. Ein Beitrag zur Evolution des Darmtraktes niederer Deuterostomier (Hemi- und Protochordaten) und seiner nervösen und hormonalen Steuerung. Zool Jahrb Anat 104:25–39

Wilson S, Falkmer S (1965) Starfish insulin. Can J Biochem 43:1615–1624

Yalow RS, Eng J, Maldow RL (1981) Peptide hormones in strange places – are they there? (Abstr). Int Symp "Brain-gut axis: the new frontier". Florence, June 29/30 – July 1, 1981, p 197

Yui R, Fujita T, Ito S (1980) Insulin, gastrin, pancreatic polypeptide-like immunoreactive neurons in the brain of the silkworm, *Bombyx mori.* Biomed Res 1:42–46

Zelnik PR, Hornsey DJ, Hardisty MW (1977) Insulin and glucagon-like immunoreactivity in the river lamprey *(Lampetra fluviatilis)*. Gen Comp Endocrinol 33:53–60

Ziegler R (1979) Hyperglycaemic factor from the corpora cardiaca of *Manduca sexta* (L) (Lepidoptera: Sphingidae). Gen Comp Endocrinol 39:350–357

CHAPTER 6

# The Biosynthesis of Glucagon

C. HELLERSTRÖM

## A. Introduction

Glucagon or peptides with glucagon-like immunoreactivity have been detected in several locations throughout the mammalian body, including pancreatic islets, salivary glands, stomach, intestine, and the central nervous system (see Chap. 33). The study of the biosynthesis of such compounds from radioactively labeled precursors has so far been almost entirely confined to the pancreatic islets, which contain a relatively high proportion of glucagon-producing cells. Important information has also been obtained by extraction and chemical characterization of glucagon-like material from various sources. As will be seen, the combined data from these two approaches has now shed considerable light on the mechanism of glucagon biosynthesis. It is the purpose of this chapter to summarize the present knowledge on the biosynthesis of glucagon as it has emerged from studies of fish, birds, and mammals. In addition, some general aspects of the formation and processing of peptide hormones will be given since recent developments in this field are of such fundamental importance for the correct understanding of glucagon biosynthesis. Excellent reviews on glucagon biosynthesis have recently been published by TAGER et al. (1980), TAGER (1981), and NOE et al. (1981).

## B. General Aspects of the Formation, Intracellular Conversion, and Storage of Peptide Hormones

In the last 15 years or so, there has been an almost explosive progress in the understanding of the molecular events which lead to the formation of mature secretory peptides. The starting point of this development was the discovery by STEINER and OYER (1967) and STEINER et al. (1967) that insulin is synthesized as a larger precursor which is modified to biologically active insulin by limited intracellular proteolysis. Further milestones are represented by the discovery of BLOBEL and SABATINI (1971) and MILSTEIN et al. (1972) that the coordinate biosynthesis and segregation of secretory proteins requires that the growing polypeptide chain has a leading sequence of hydrophobic amino acid residues which directs the formation of a functional ribosome–membrane junction. The development of methods for analyses of the structure and organization of specific genes and of their associated sequences in genomic DNA as well as the mapping of their transcription products has also greatly contributed to our present understanding of peptide hormone formation. A brief outline of the various steps in the process will be given in the following paragraphs.

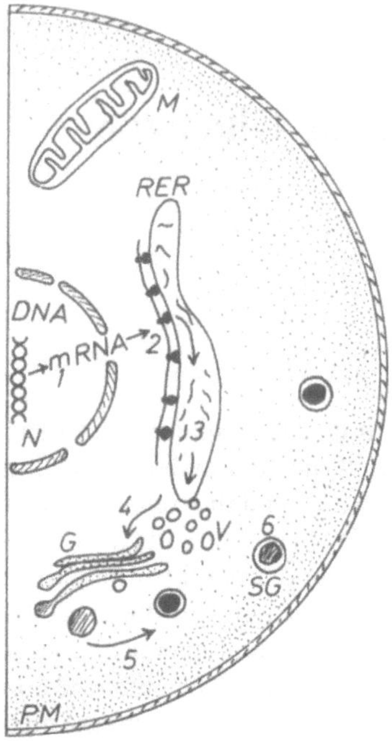


**Fig. 1.** An outline of the various steps in the biosynthesis and storage of proteins in a secretory cell. The cell is delimited by a plasma membrane (*PM*) and contains a nucleus (*N*) which is surrounded by an envelope, penetrated by pores. In the cytoplasm mitochondria (*M*), rough endoplasmic reticulum (*RER*), smooth vesicles (*V*), a Golgi complex (*G*), and secretory granules (*SG*) have been depicted. The genes encoding the secretory protein (or proteins) are transcribed to mRNA in the nucleus (*1*). After translocation into the cytoplasm, the mRNA is translated in the polysomes attached to the membranes of the RER (*2*), and the elongating polypeptide chain segregated in the cisternal space of the RER (*3*). From the RER, the newly synthesized protein (prohormone) is transported to the Golgi complex, possibly via a vesicular system (*4*). In the Golgi complex, the peptide is concentrated in condensing vacuoles and finally packed into mature secretory granules surrounded by a smooth membrane (*5*). Proteolytic cleavage of the prosecretory peptide takes place in both the Golgi complex and during storage in the secretory granule (*6*). After PALADE (1975)

As can be seen in Fig. 1 the formation, processing, and storage of peptide hormones in secretory cells can be subdivided in certain well-defined steps. The flow of information from the specific genes coding for the secretory products is generally depicted as follows:

$$\text{DNA} \xrightarrow{\text{transcription}} \text{RNA} \xrightarrow{\text{translation}} \text{protein.}$$

Of the various classes of RNA, only messenger RNA (mRNA) is translated into proteins. Recent evidence suggests that eukaryotic genes contain nucleotide sequences (designated "introns" or "intervening sequences") which are not found as complementary sequences in the mature mRNA (CRICK 1979). The entire gene including these regions is thought to be first transcribed into larger, primary forms of mRNA which are then modified to mature mRNA through a process designated as RNA splicing. Further posttranscriptional processing of mRNA is effected by

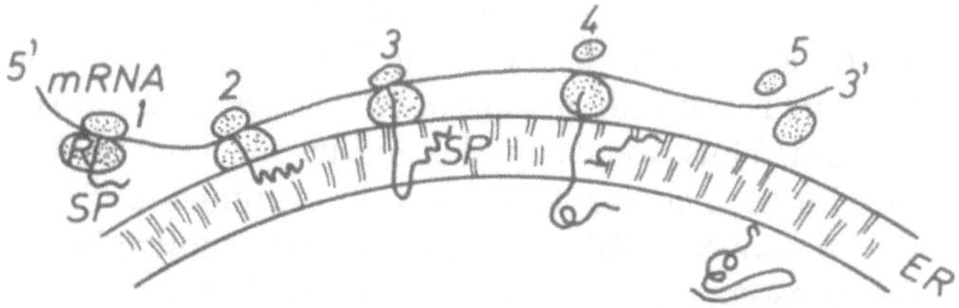


**Fig. 2.** Model for the transfer of presecretory proteins across the membrane of the endoplasmic reticulum. After the signal peptide (*SP*) has emerged from the large subunit of the ribosome (*R*) (*step 1*) it permeates into the membrane of the endoplasmic reticulum (*ER*) and the ribosome becomes associated with the membrane, possibly by interaction with a receptor (*step 2*). Elongation of the presecretory protein forces the peptide chain through the membrane (*step 3*), and the signal peptide is removed by signal peptidase (*step 4*). After termination of translation, the completed protein is released into the cisternal space, the ribosome–membrane junction is disrupted, and the ribosomal subunits dissociated from the mRNA (*step 5*). JACKSON and BLOBEL (1980), STEINER et al. (1980)

the addition at the 5′ end (Fig. 2) of a cap, consisting of a methylated guanosine residue which may be of importance for efficient translation. At the 3′ end of the mRNA (see Fig. 2), a string of 40–200 adenosine nucleotides is added as a so-called poly(A) tail. There is evidence to suggest that this addition may stabilize the mRNA and thus prolong the survival of the message.

The assembly of amino acids into the full-length polypeptide occurs on ribosomes attached to the membranes of the rough endoplasmic reticulum (RER). By this process, the newly synthesized protein also becomes segregated into the cisternae of the RER. However, with a few exceptions, translation of the mRNA leads to the production of a peptide chain, which is considerably larger than that forming the mature hormone and which represents a precursor form of the exportable product. It has recently been shown that the first step in the production of the precursor is the formation at the $NH_2$ terminal end of an extension with a length of 15–30 amino acid residues. This leading sequence has been designated "signal peptide," and its biosynthesis is thought to be initiated on free cytoplasmic ribosomes. However, after the signal peptide region emerges from the large ribosomal unit, it embeds itself in the RER membrane and, in this way, effects the formation of a ribosome–membrane junction (Fig. 2). This latter process is thought to reflect the fact that the signal peptide is composed of a central core of hydrophobic amino acid residues, which penetrates the lipid bilayer of the RER membrane and in this way may not only attach the ribosome, but also facilitate the vectorial growth of the peptide chain into the interior of the cisternal space of the RER. At this early stage of its formation, the hormonal precursor is designated a "prehormone" or "preprohormone." As elongation continues, however, the signal peptide is split off enzymatically by signal peptidases and released into the cisternal space. This co-translational processing of the protein is very rapid, which means that cellular contents of preprohormones are extremely low.

When translation of the mRNA is completed, the secretory protein is released in its entirety into the cisternae of the RER. After the removal of the signal peptide, the remaining part is in many, but not all, cases still larger than the mature hormone and therefore designated "prohormone." The reason for the existence of these larger precursors is not fully understood, but several tentative explanations have been put forward. Thus, some peptides, like insulin, contain disulfide bridges which may require for their correct formation the spontaneous folding of a single chain propeptide. Moreover, a full extension of a polypeptide between its site of formation on the ribosomal subunit and the cisternal space would take a chain length of at least 78 amino acid residues (BLOBEL and DOBBERSTEIN 1975). It is also possible that at least some peptide hormone precursors reflect their evolutionary origin from larger primordial proteins (NIALL 1980).

Full biologic activity of a hormone is achieved by a limited enzymatic proteolysis of its precursor and is effected intracellularly in the Golgi region or in the secretory granules. In other situations, activation may occur extracellularly as, for example, in the case of zymogens. Intracellular cleavage of the prohormone is carried out by a variety of proteases having tryptic and carboxypeptidase B-like specificity. Indeed, the same prohormone can be processed differently in different cell types as exemplified by the ACTH (adrenocorticotropic hormone) precursor which can give rise to either ACTH and $\beta$-LPH ($\beta$-lipotropic hormone) in the anterior pituitary or CLIP (corticotropin-like intermediate lobe peptide), MSH (melanocyte-stimulating hormone), and $\beta$-endorphin in the intermediate pituitary lobe (KRIEGER and LIOTTA 1979).

Intracellular translocation of the newly formed prohormone from the cisternae of the RER to the Golgi complex occurs via a vesicular system, presumably derived from transitional elements of the RER cisternae. This is an energy-dependent process which is blocked by inhibitors of oxidative phosphorylation and may be the functional equivalent of a lock along the channels used for intracellular transport (PALADE 1975). Translocation of the prohormone to the Golgi complex is a rapid event, occurring within only 20 min or so of the beginning of peptide biosynthesis.

The Golgi complex provides both the enzymatic apparatus for the conversion of the prohormone and the membranes that form the envelope of the secretory granules. The activities of the Golgi region therefore are of great importance for posttranslational processing and the storage of the newly synthesized hormonal peptide. In addition, lysosomes are formed in the Golgi region and there is evidence to indicate that, at least under certain conditions, they may degrade considerable quantities of secretory proteins.

Secretory cells with an episodic need for release of peptides require an intracellular store of these compounds. This is effected by the formation of granules consisting of a central core of secretory material surrounded by a smooth membrane. In some cell types, like the pancreatic B-cell, the proteolytic processing seems to continue also within the secretory granule. Thus, a limited amount of proinsulin is secreted from the B-cell along with insulin and C-peptide and all these three products can be readily detected in the circulation. In other cells, like those of the exocrine pancreas, the precursor is stored in the granules without further processing, which takes place extracellularly. It should also be mentioned that some cells with a more continuous secretion lack a storage form of their export products. Examples of such cells are those producing nerve growth factor or immunoglobulins.

# C. Biosynthesis of Glucagon

## I. Formation of Preproglucagon

For the reasons given, the molecular size of pancreatic glucagon makes it likely that the biosynthesis of this peptide proceeds via larger precursors. Confirmation of this hypothesis has, indeed, been obtained in several recent reports although the precise molecular weight and amino acid composition of the precursors are still matters of debate. In this section, the different steps in glucagon biosynthesis will be analyzed according to the sequence of events that occurs in the cellular formation of the hormone.

The structure of the gene or genes coding for preproglucagon mRNA is still unknown. However, a poly(A) RNA has been extracted from the fairly large quantities of glucagon-producing cells which can be obtained from the principal islets of anglerfish *(Lophius americanus)*. After translation of this material in a cell-free wheat germ system, several peptides with glucagon immunoreactivity could be detected. Thus, LUND et al. (1980) identified one peptide with a molecular weight of 14,500 daltons and another of 12,500 daltons. SHIELDS et al. (1981), in a similar study, identified three different proteins with glucagon immunoreactivity, two of which had an apparent molecular weight of 14,000 daltons and a third of 16,000 daltons. All these peptides were taken to represent the primary translation products of glucagon mRNA and were therefore tentatively designated preproglucagon. More recent studies, indeed, support this notion in that two mRNAs have been identified by the cloning of separate complementary DNAs encoding each of two glucagon precursors (LUND et al. 1981).

Further identification of the glucagon-immunoreactive peptides was attempted in experiments in which anglerfish islet mRNA was translated in a cell-free wheat germ system supplemented with dog pancreas microsomal membranes (LUND et al. 1980; SHIELDS et al. 1981). The presence of signal peptidase in the membrane preparation could be expected to convert a putative preproglucagon to proglucagon by splitting off an $NH_2$ terminal amino acid sequence corresponding to the signal peptide. Indeed, evidence in support of this notion was found in that both the 14,500 daltons protein and the 16,000 daltons protein were converted to slightly smaller peptides. Somewhat surprisingly the 12,500 daltons protein of LUND et al. (1980) and the 14,000 daltons protein of SHIELDS et al. (1981) were both processed to products with higher molecular weights than the nascent proteins. Although this observation would be consistent with cotranslational removal of a signal peptide and addition of core sugars to the growing polypeptides, evidence of sugar residues has so far not been found in the proglucagon molecule (PATZELT et al. 1979; SHIELDS et al. 1981). It is nevertheless of considerable interest that the different precursor molecules seemed to be subjected to different modes of posttranslational processing. If these observations are confirmed, they add to the complexity of the early steps of glucagon biosynthesis and will be an interesting area of future studies.

Both LUND et al. (1980) and SHIELDS et al. (1981) obtained further information on the properties of the glucagon-related proteins by showing that addition of microsomal membranes to the translation products protected these products from limited proteolysis with trypsin or chymotrypsin. This finding suggests that the nascent proteins are segregated within the microsomal vesicles during their cotranslational processing. Indeed, when the microsomal membranes were lysed before addition of the enzymes, conversion of the proteins proceeded uninterrupted. These important observations suggest that the primary translation products of glucagon, like other preprohormones are not only degraded by proteases associated with the RER, but are also able to traverse the membrane system and become segregated from the extracisternal space.

While these observations all refer to glucagon-related peptides translated from mRNA obtained from fish islets, little is so far known of the corresponding process in mammals. This, of course, reflects the inherent difficulties of obtaining sufficient mammalian glucagon-producing islet cells for purification of mRNA. An alternative approach, in which incubation of intact rat islet cells with radioactive amino acids was followed by electrophoretic separation of labeled proteins, did not produce convincing evidence of the formation of preproglucagon, although in the same system the biosynthesis of preproinsulin was readily demonstrated (PATZELT et al. 1979). On the basis of the molecular weight of proglucagon-like peptides in the rat, it was predicted that mammalian preproglucagon would have a molecular weight of about 21,000 daltons (PATZELT et al. 1979). Preliminary evidence for the existence of such a precursor in mammals has been obtained by translating RNA extracted from a human pancreatic endocrine tumor (PATZELT et al. 1980). Owing to the abnormal nature of many tumor proteins, confirmation of these observations using normal islet mRNA seems warranted.

## II. Formation and Conversion of Proglucagon

When studies on the mechanism of glucagon biosynthesis were first launched a little more than a decade ago, they were all directed towards a search for larger precursors and their conversion to the native hormone. This approach was inspired not only by the discovery, some years before, of proinsulin, but also of the previous finding of large glucagon-like immunoreactivity (GLI) in extracts of plasma and pancreas of dogs and humans (VALVERDE et al. 1970; RIGOPOULOU et al. 1970). At that time, the heterogeneous nature of GLI could not be ascertained and its identity with a putative glucagon precursor was not excluded.

NOE and BAUER (1971), in experiments on anglerfish islets, were the first to observe the incorporation of $^{3}$H-labeled tryptophan (present in all known species of glucagon, but not in insulin) into glucagon-related proteins with a considerably higher molecular weight than native glucagon. Tryptic digestion of this material resulted in the generation of a radioactive component with the molecular weight, immunologic characteristics, and electrophoretic mobility of glucagon. These data therefore strongly suggested that a larger peptide, tentatively designated proglucagon, might serve as a precursor in the formation of glucagon.

In further studies of newly synthesized anglerfish islet proteins, NOE and BAUER (1975) demonstrated that two glucagon-immunoreactive molecules with approximate molecular weights 12,000 and 9,000 daltons were being synthesized during a 20-min pulse incubation with radioactive tryptophan. During a following chase incubation with nonradioactive tryptophan, there was a gradual appearance of radioactivity in glucagon-immunoreactive peptides with the approximate size of glucagon and a concomitant disappearance of radioactivity in the 12,000 and 9,000 daltons peptides. In a more recent study, NOE et al. (1977a, b) found an additional, smaller intermediate with a molecular weight of 4,900 daltons which was converted into glucagon. This latter work also indicated that proglucagon was rapidly segregated into the RER cisternae and that cleavage to smaller peptides continued during maintenance in secretory granules. Altogether, the data were taken to suggest a stepwise conversion process, involving several intermediates between proglucagon and glucagon.

An attempt to purify and characterize anglerfish proglucagon was made by TRAKATELLIS et al. (1975). A large glucagon-immunoreactive protein, synthesized by anglerfish islets incubated with tryptophan $^3$H was purified and subjected to amino acid analyses after acid hydrolysis. This material was found to contain 78 amino acid residues, corresponding to a molecular weight of about 9,100 daltons. However, NOE and BAUER (1975) presented compelling evidence to support the view that the 78 amino acid peptide proposed as anglerfish proglucagon by TRAKATELLIS et al. (1975) is not the primary conversion product of preproglucagon, but rather an intermediate in the metabolic conversion of proglucagon to glucagon. This intermediate, indeed, would be identical with the 9,000 daltons material that was rapidly synthesized in the studies of NOE and BAUER (1975).

The avian endocrine pancreas is comprised of small, B-cell-rich, islets, and larger islets mainly composed of A-cells (CLARA 1924; NAGELSCHMIDT 1939; HELLMAN and HELLERSTRÖM 1960). The former islet type has been called "light" and the latter "dark." The dark islets are localized in the splenic part of the pancreas which therefore contains a particularly high concentration of glucagon-producing cells. In attempts to study the mechanism of glucagon biosynthesis, islets from this pancreatic region have been isolated, either by treatment with collagenase or by microdissection, and incubated with radioactive amino acids. By this means, TUNG and ZEREGA (1971) and TUNG (1973, 1974) demonstrated that tryptophan $^3$H was incorporated into glucagon-immunoreactive proteins with widely different molecular weights. While one of these components was of the same molecular weight as glucagon, two others were considerably larger, i.e., with estimated sizes of 6,000–9,000 and 69,000 daltons, respectively. Tryptic hydrolysates of the larger biosynthetic component contained radioactive peptides identifiable chromatographically with those obtained from a corresponding hydrolysate of bovine–porcine glucagon. However, little evidence was presented for a precursor–product relationship between any of these components and glucagon.

In further studies of glucagon biosynthesis by pigeon islets, O'CONNOR and LAZARUS (1976) also demonstrated incorporation of tryptophan $^3$H into both large glucagon-related proteins and proteins of the same molecular weight as glucagon. In these studies, however, the molecular weight of the large component corresponded to 20,000 daltons and, again, no precursor–product relationship was evident between this molecular species and glucagon.

Studies of glucagon biosynthesis in mammals were initiated by HELLERSTRÖM et al. (1972), who utilized for this purpose isolated pancreatic islets of guinea pigs. Previous studies suggested that islets of this species were particularly suitable for studies of glucagon formation since they can be isolated without much damage to the A-cells and can be converted into A-cell-rich islets by elimination of the B-cells with streptozotocin (PETERSSON et al. 1970; HOWELL et al. 1971). Gel chromatography on Sephadex G-100 at pH 8.8 of extracts of guinea pig islets incubated for up to 17 h with tryptophan $^{3}$H resolved two distinct peaks of radioactivity, one migrating with the void volume and another in a region corresponding to a molecular weight of about 18,000 daltons. Despite the relatively long incubation period, only little radioactivity became incorporated into proteins migrating with the marker glucagon. Subsequent work indicated that the 18,000 daltons protein probably comprised a dimer of a protein eluting in the 9,000 daltons region when exposed to an acid pH (HELLERSTRÖM et al. 1974). This 9,000 daltons protein coeluted with a peak of glucagon-like immunoreactivity. The absence of labeled proteins eluting with the glucagon marker, even after a 17-h incubation with tryptophan $^{3}$H, appeared puzzling and suggested either than glucagon was being synthesized only slowly or that it might be rapidly released into the medium after its synthesis. However, investigation of the media after incubation of islets in the presence of tryptophan $^{3}$H failed to reveal any labeled glucagon. Experiments with A-cell-rich islets nevertheless indicated that the source of larger, radioactive proteins were the A-cells and not the B-cells. In addition, these larger proteins seemed to be localized in the secretory granules of the A-cells, as evidenced both by subcellular fractionation of islets (HELLERSTRÖM et al. 1974) and electron microscopic autoradiography (HOWELL et al. 1974a, b). Incubation of the islets in the presence of other amino acids (leucine, tyrosine) instead of tryptophan, although demonstrating the presence of labeled proinsulin and insulin, also failed to provide evidence of radioactivity in the glucagon region of the column eluates (HELLERSTRÖM et al. 1974). It was therefore concluded that, in the conditions of these in vitro experiments, little or no newly synthesized glucagon accumulated in the islets during incubations for up to 17 h. Another protein with an estimated molecular weight of 9,000 daltons which contained tryptophan $^{3}$H and coeluted with a peak of glucagon-like immunoreactivity was, however, consistently recovered from the columns.

Further studies of the labeling pattern of glucagon-related islet proteins indicated that a period of at least 2 days in tissue culture in the presence of tryptophan $^{3}$H was required before the islets showed any significant accumulation of proteins coeluting with the marker glucagon (Fig. 3; HELLERSTRÖM et al. 1974; ÖSTENSON et al. 1980). Immune binding experiments identified a substantial part of these newly synthesized proteins as glucagon (Fig. 4). Labeled material which bound to antiglucagon antibodies also occurred in the 9,000 daltons region and in the void volume, although to a smaller extent. The results of pulse-chase experiments, however, provided no convincing evidence for a conversion of these larger proteins into glucagon. It was still possible, though, that the heterogeneity of a putative precursor pool, as evidenced both by the immune binding and by sodium dodecyl sulfate (SDS)–polyacrylamide gel electrophoresis, might have masked a conversion process. Essentially similar findings were made in corresponding studies of isolated mouse islets (ÖSTENSON et al. 1980). It was concluded that while glucagon is being synthesized in isolated mammalian islets in vitro, this process apparently occurs at a relatively slow rate.

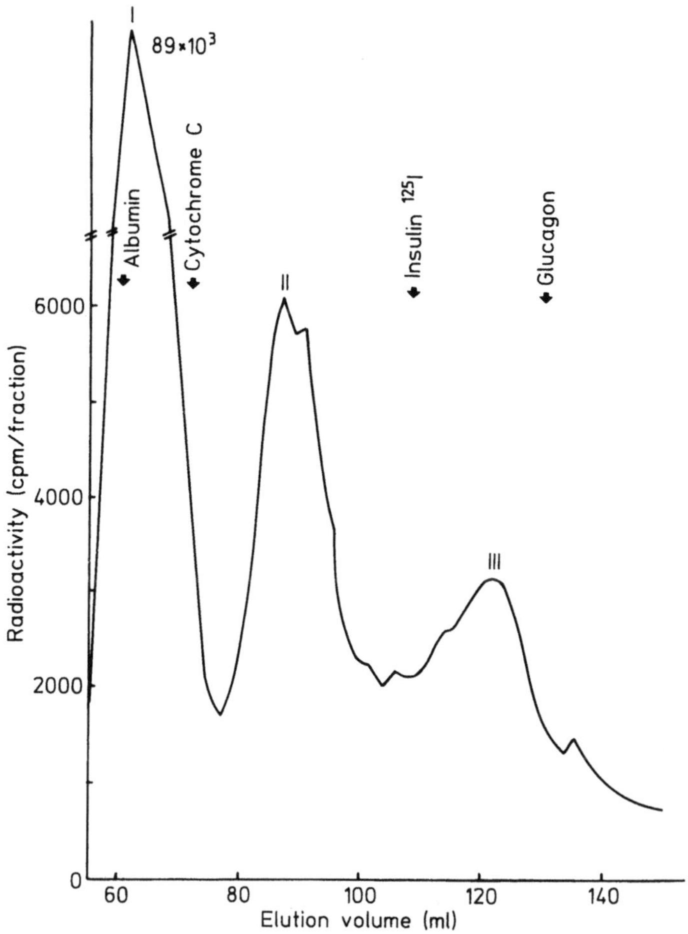


**Fig. 3.** Gel chromatograph of proteins extracted from isolated guinea pig islets maintained in tissue culture and labeled with tryptophan $^3H$. Islets were cultured for 3 days in a medium supplemented with 40 µCi/ml tryptophan $^3H$ and 16.7 m*M* glucose. Islets (230 in number) were then homogenized, extracted with acid–ethanol, and subjected to gel chromatography on a Sephadex G-50 fine column in 1 *M* acetic acid and with 1 mg porcine glucagon as carrier protein. The elution positions of albumin, cytochrome C, insulin $^{125}I$, and porcine glucagon are indicated by *arrows*

Studies of glucagon biosynthesis in the rat were first accomplished by O'CONNOR et al. (1973). Since the A-cells are located at the periphery of the rat islets, the potentially injurious collagenase treatment was avoided and the experiments were undertaken in a recirculated perfused rat pancreas preparation. In agreement with the observations of HELLERSTRÖM et al. (1972, 1974) and ÖSTENSON et al. (1980), an early labeling of a 9,000 daltons peptide which coeluted with a small amount of glucagon immunoreactivity was observed. With incubations longer than 4 h, labeling appeared in a second peak with migration characteristics identical with those of native rat glucagon. This newly synthesized protein was identified as glucagon by demonstration of specific binding and dissociation behavior with gluca-

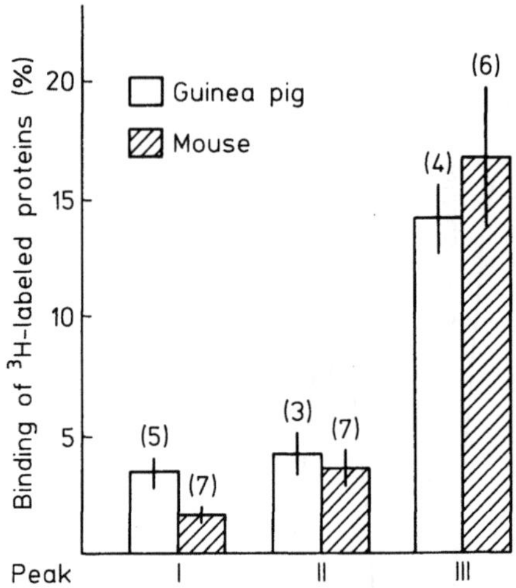


**Fig. 4.** Binding of proteins labeled with tryptophan $^3H$ to Sepharose-coupled anti-glucagon serum. Extracts of isolated guinea pig or mouse islets, cultured in media containing tryptophan $^3H$, were fractionated in 1 *M* acetic acid on a Sephadex G-50 column. Fractions corresponding to peaks I, II, or III in Fig. 3 were pooled, freeze-dried, and subsequently redissolved prior to assay of the immune binding capacity as described by ÖSTENSON et al. (1980). The immune binding was calculated as the radioactivity bound to antiporcine glucagon rabbit antibodies (K 44, Novo Research Institut, Bagsvaerd, Denmark), expressed as a percentage of the total radioactivity in each pool of fractions. Values were corrected for nonspecific binding exhibited by normal serum. Numbers of experiments are given within parentheses

gon antibodies. However, no evidence for conversion of the 9,000 daltons protein to glucagon was presented.

Using fluorography of labeled proteins separated by SDS slab-gel electrophoresis, immunoprecipitation, and tryptic peptide analysis, PATZELT et al. (1979, 1980) studied glucagon-related proteins synthesized in pulse-chase experiments with isolated rat islets. With these sensitive methods, they identified a family of peptides which were assigned the role of precursors and intermediates in the sequential formation of glucagon. The most rapidly formed of these glucagon-immunoreactive peptides had a molecular weight of 18,000 daltons. The rapid disappearance of this component in the pulse-chase experiments and its structural similarities both to glucagon and another peptide appearing during the chase period were consistent with its proposed role as proglucagon. It was noteworthy that the 18,000 daltons protein underwent a small posttranslational increase in molecular weight before it disappeared after a chase period of 30–120 min. The nature of this modification was not identified and several attempts to verify the involvement of a glycosylation process failed.

The fate of the proglucagon was further analyzed in the pulse-chase experiments. After conversion of proglucagon to its slightly larger form, a labeled component with a molecular weight of about 10,000 daltons was first split off, followed by the appearance of label in native glucagon. The 10,000 daltons component lacked glucagon immunoreactivity and was therefore believed to be an end product in the cleavage of the precursor. Its relationship to glucagon was con-

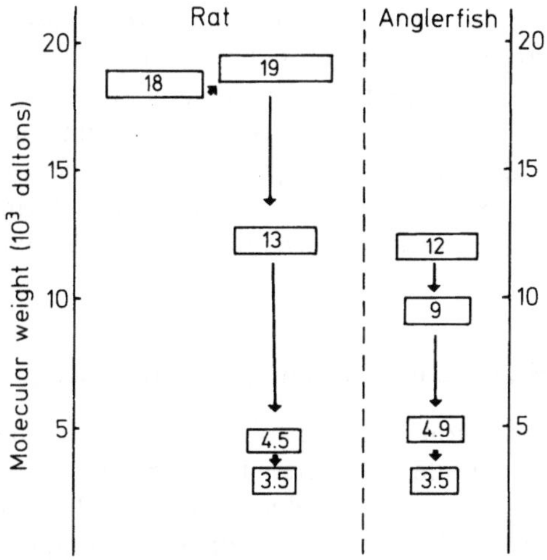


**Fig. 5.** Proposed scheme for the conversion of proglucagon to glucagon in the rat and anglerfish. Numbers in *boxes* represent approximate molecular weights ($\times 10^3$ daltons) of the various peptides. After PATZELT et al. (1979) and NOE et al. (1975, 1981)

firmed in extensive tryptic peptide analyses. Small amounts of glucagon-related intermediates of 13,000 and 4,500 daltons were also identified, but it was not ascertained whether these were true intermediates or side products.

The elegant studies of PATZELT et al. (1979, 1980) have so far provided the most conclusive evidence of a stepwise conversion of larger, glucagon-related peptides into glucagon in mammalian islets. A similar sequential cleavage of a glucagon precursor was previously demonstrated in the islets of the anglerfish (NOE and BAUER 1975) and the molecular weights of the two sets of glucagon-immunoreactive peptides emerging in the process are compared in Fig. 5. Although the end products, native glucagon, seem to be almost identical, there are still apparent differences between both prohormones and intermediates. Two features are particularly noteworthy in the formation of rat glucagon. One is the large size of the proglucagon molecule in relation to glucagon, suggesting that the biologically active, mature hormone comprises less than 20% of the proglucagon molecule. By comparison, insulin represents approximately 70% and parathyroid hormone 90% of their respective prohormones. The second interesting feature is the posttranslational increase in molecular weight at the initial modification step of proglucagon. As pointed out in Sect. C.I, a corresponding change was observed by LUND et al. (1980) and SHIELDS et al. (1981) in their studies of glucagon precursors synthesized by anglerfish mRNA in a cell-free wheat germ system. The nature of this alteration is so far unknown, but its occurrence in both fish and mammals suggests that it might represent a more general mechanism.

There is so far only one report dealing with glucagon biosynthesis in humans (NOE et al. 1975). Gel filtration of extracts of isolated human islets incubated in the presence of tryptophan $^3H$ indicated that radioactivity became incorporated into at least two proteins, the larger of which had the approximate size of proinsulin, and the smaller coeluted with glucagon. Radioimmunoassay and polyacrylamide gel electrophoresis confirmed the identity of the smaller protein with glucagon, whereas partial tryptic digestion of the larger molecular species yielded labeled products having charge and immunologic characteristics indistinguishable from porcine glucagon. These observations are consistent with the notion that, in human islets, also glucagon biosynthesis proceeds via larger precursors containing covalently bound glucagon. The results also suggest that formation of native glucagon may be a more rapid process in the human A-cells than in those of the rat or guinea pig.

## III. Structure and Storage of Proglucagon and Glucagon

Immunologic characterization and determination of the amino acid sequences of purified glucagon-like peptides extracted from various tissues has contributed much to our present understanding of glucagon biosynthesis and the mutual relationships between different glucagon-immunoreactive compounds. Also, immunocytochemical localization of such compounds in the pancreas and gut has clarified the question of what is stored and in what form in the cells of origin. An extensive review of this field is given in Chap. 4 and, therefore, only a summary of data relevant to glucagon biosynthesis will be presented here.

Early studies of extracts of intestine and pancreas indicated the presence in these tissues of numerous glucagon-related peptides of widely varying sizes (RIGOPOULOU et al. 1970; VALVERDE et al. 1970; MOODY et al. 1970). In addition, antisera directed towards pancreatic glucagon reacted differently with these compounds which added to the confusion and hampered relevant structural comparisons between the different glucagon-related peptides. The situation changed, however, when TAGER and STEINER (1973) were able to isolate from commercial preparations of glucagon and structurally define a peptide, which was found to consist of the whole sequence of glucagon plus eight additional amino acid residues attached to the COOH terminal end of the molecule. With a molecular weight of 4,500 daltons, it was suggested that this peptide represented a fragment of a larger precursor of glucagon. At about the same time it became evident that antibodies raised against glucagon could be subdivided into two groups, depending on whether they were directed against an immunodeterminant in the $NH_2$ terminal and central portion (residues 2–23) or the COOH terminal portion (residues 24–29) of the molecule (ASSAN and SLUSHER 1972). The former group was designated "nonspecific" and the latter "pancreatic glucagon specific." The peptide of TAGER and STEINER (1973) reacted with the $NH_2$ terminal-specific but not the COOH terminal-specific antibodies, showing that the COOH terminal extension blocked the antigenic site located at this end of the molecule. Thus, the reaction pattern of the two groups of antibodies provided a powerful tool for structural analyses of glucagon-related polypeptides occurring in intestinal and pancreatic extracts.

Analyses of glucagon-immunoreactive peptides isolated from the gut indicated that one of these molecular species, subsequently purified and designated glucagon-like immunoreactivity-1 (GLI-1) or glicentin, contained the full homologous sequence of glucagon (MOODY et al. 1977). Immunocytochemical studies further suggested that a peptide located together with glucagon in the pancreatic A-cell was immunologically related to glicentin (RAVAZZOLA et al. 1979; RAVAZZOLA and

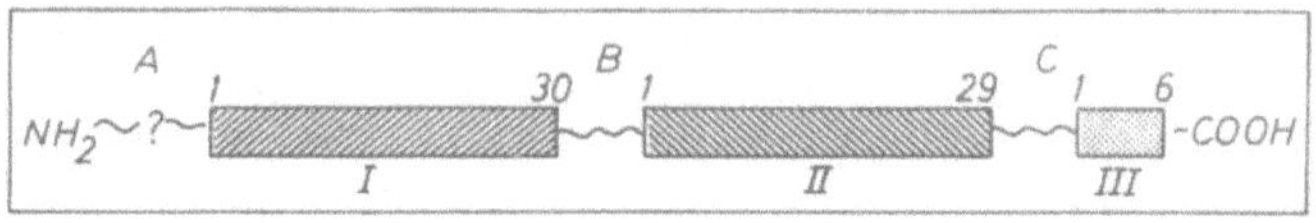


**Fig. 6.** Schematic representation of the proglucagon molecule. *I*: Glicentin-related pancreatic peptide (30 amino acid residues); *II*: glucagon (29 amino acid residues); *III*: COOH-terminal hexapeptide; *A*: $NH_2$ terminal extension which may be of different length in different species and which presumably contains a cleavage site(?); *B* and *C*: Lys-Arg sequences linking the three main parts of the molecule together and representing cleavage sites in the postsynthetic modification of the molecule. After THIM and MOODY (1981)

ORCI 1980a). TAGER and MARKESE (1979), in their comparative investigations of intestinal and pancreatic glucagon-like peptides, confirmed and extended these observations and concluded that at least two large glucagon-like peptides originating in the intestine are counterparts of identical pancreatic peptides, but that differences in their posttranslational modifications result in altered amounts of these peptides in the two tissues. A schematic representation of the proposed structural interrelationships between these various glucagon-related compounds is given in Fig. 6.

The relationship between glicentin and pancreatic glucagon precursors was further elucidated when MOODY and his colleagues (THIM and MOODY 1981; MOODY et al. 1981) determined the primary structure of glicentin and investigated the glucagon-like and glicentin-like peptides in extracts and perfusates of the porcine pancreas. According to these observations, glicentin is composed of 69 amino acid residues and has a molecular weight of 8,100 daltons. The amino acid sequence 1–30 of glicentin represents a previously purified peptide designated glicentin-related pancreatic peptide (GRPP; MOODY et al. 1980). The next 29 amino acid residues (sequence 33–61) form the full sequence of glucagon and the sequence 64–69 is a COOH terminal hexapeptide. GRPP, glucagon, and the hexapeptide are linked by two Lys-Arg pairs (31–32, 62–63) which presumably represent the cleavage sites in the posttranslational processing of the peptide (Fig. 6). Except for inversion of two amino acids, the COOH terminal octapeptide of the intact glicentin molecule is identical to that described by TAGER and STEINER (1973). On the basis of these observations, THIM and MOODY (1981) concluded that glicentin fulfills the structural requirements for its identification as proglucagon.

In agreement with these conclusions, perfusion of porcine pancreas showed that glucagon and a peptide with glicentin-like immunoreactivity and intermediate in size between glicentin and glucagon were secreted in approximately equimolar amounts upon stimulation with arginine (MOODY et al. 1981). Small amounts of a larger peptide, very similar to glicentin, were also secreted. These observations were taken to indicate that the intermediate, glicentin-related peptide represented a major fragment formed during processing of proglucagon. The small amounts of the larger, glicentin-like peptide secreted synchronously with glucagon might then represent proglucagon.

The observations on glucagon-related peptides extracted or released from the pancreas have been extended in a series of elegant immunocytochemical investigations with the aim of elucidating the cellular origin and the storage form of the different compounds. In these studies, as in those referred to previously, the availability of an antiserum reacting only with the nonglucagon part of the glicentin molecule has been of great significance (MOODY et al. 1977; RAVAZZOLA et al. 1979). With the aid of this antiserum, glicentin was specifically identified in the glucagon cells of both the pancreas and of the gastric mucosa and also in the L-cell of the gut (RAVAZZOLA and ORCI 1980 b). In the pancreatic islets, the same A-cells were positive for both glicentin and glucagon, except in the guinea pig islets in which only glicentin-positive cells were demonstrated. Analyses of the human pancreas with electron microscopic immunocytochemistry indicated that glicentin and glucagon were, indeed, located in the same secretory granules of the pancreatic A-cell and that in the individual granule glucagon occupied the central, most electron-dense, part of the granule, while glicentin was preferentially located in the granule periphery (RAVAZZOLA and ORCI 1980 a). This intriguing observation not only sheds light upon the structural inhomogeneity of the human pancreatic α-granule, but also explains the observation by MOODY et al. (1981) of a cosecretion of glicentin-like compounds and glucagon from the perfused pancreas. It is also of interest in this context that the intestinal L-cells which normally bind only antiglicentin or $NH_2$ terminal-specific anti-glucagon sera could be made to react also with COOH terminal-specific anti-glucagon serum by enzymatic treatment with trypsin and carboxypeptidase B (RAVAZZOLA and ORCI 1980 b). Thus, by this treatment the full complement of antiglucagon antigenic sites present in the pancreatic and gastric A-cells could also be demonstrated in the intestinal L-cell. The combined data suggest that, in all these cells, a common glucagon precursor is being synthesized. However, the L-cell probably lacks the enzymatic apparatus necessary for degrading the molecule sufficiently to expose its COOH terminal antigenic site.

## D. Concluding Remarks

The review presented in this chapter indicates that our present understanding of the mechanism of glucagon biosynthesis is mainly based on results of studies performed in the last 3–4 years. It can now be concluded that glucagon, like insulin and somatostatin (TAGER et al. 1979 a; NOE et al. 1981), is indeed synthesized via a larger precursor, which is processed to glucagon through a stepwise proteolytic degradation. There is furthermore little doubt that an intermediate, most probably identical with glicentin, is stored together with glucagon in the secretory granules of the pancreatic A-cell and is secreted along with the native hormone. Recent studies also suggest a continuous conversion in the granules of glicentin into glucagon (RAVAZZOLA et al. 1981), which conforms to early suggestions of a slow formation of the native hormone and the presence of a glucagon precursor in the granule fraction of islet homogenates (HELLERSTRÖM et al. 1974). The homology of gut glicentin with the corresponding storage product of the pancreatic α-granules, moreover, has clarified the hitherto obscure relationship between large, glucagon-related polypeptides identified in extracts of pancreas and intestine. Finally, the determination of the complete amino acid sequence of porcine glicentin provides a reasonable basis for predicting the enzymatic requirements for postsynthetic cleavage of this molecule.

It is, however, also clear that a number of questions regarding the mechanism of glucagon biosynthesis remains to be answered. For example, marked disagreements as to the size of preproglucagon, proglucagon, and their smaller cleavage products still exist. Although these discrepancies may reflect species differences or methodological differences, it is nevertheless noteworthy that, in some studies of glucagon biosynthesis in the rat, glucagon-related peptides of a molecular weight corresponding to that of glicentin have been found (O'CONNOR and LAZARUS 1973), whereas in other investigations the occurrence of such peptides was apparently lacking (PATZELT et al. 1979). It is also worthy of note in this context that knowledge of glucagon biosynthesis in extrapancreatic sites is only fragmentary. One recent report indicates that the mRNA encoding the glucagon precursors from the intestine of the anglerfish is similar in size to the corresponding mRNA identified in the islets (LUND et al. 1981). However, nothing is so far known of the structure and proteolytic cleavage of the translation products.

The observation that glicentin is stored in the secretory granules of the intestinal L-cells without being converted to glucagon also calls for further studies. Is glicentin biologically active in itself or is it secreted and, as suggested by the recent observations of KORÁNYI et al. (1981), degraded into active fragments in the extracellular space? The possible significance of the $NH_2$ terminal, nonglucagon-containing portion of the molecule, is of particular interest in this context.

The regulation of glucagon biosynthesis at the cellular level is another area for future research. While some studies suggest that exposure in vitro of pancreatic A-cells to high glucose concentrations leads to increased rates of glucagon biosynthesis (HELLERSTRÖM et al. 1974; ÖSTENSON et al. 1980) other observations suggest either an inhibitory action of glucose (TUNG and ZEREGA 1971; O'CONNOR and LAZARUS 1976) or no effect at all (PATZELT et al. 1979). At any rate, there is nothing to indicate that a rapid glucose action like that observed for insulin biosynthesis (STEINER et al. 1972) is of importance for the control of glucagon biosynthesis. Possibly, other substrates like amino acids may be of significance in this context and perhaps also more slowly acting feedback systems involving the cellular store of the hormone.

Finally, very little is known about the control of glucagon biosynthesis in various pathologic states. The glucagon excess in clinical diabetes mellitus (DOBBS and UNGER 1979; UNGER and ORCI 1981) would involve an increased rate of glucagon biosynthesis and/or conversion of glucagon precursors. Indeed, experimental evidence of enhanced glucagon biosynthesis has been presented in pancreatic islets from C57BL/KsJ misty diabetic mice (LEITER et al. 1979). This A-cell dysfunction was observed before the appearance of frank diabetes and was taken to suggest that glucagon could play an early role in the events that led to the overt disease in this strain of mice. Another area of interest in this context would be to look for structurally abnormal glucagons, particularly in infants with neonatal hypoglycemia. A deletion or substitution of one or several amino acids in the receptor-binding part of the glucagon molecule would theoretically lead to deficient biologic activity. Although no such cases have so far been described, it is of note that an abnormal variant of insulin has been discovered in a patient with diabetes mellitus (TAGER et al. 1979b). By the same token, a deficient proteolytic processing of proglucagon would cause a change in the biologic activity of the secreted product, which might affect glucose homeostasis. Indeed, hyperproinsulinemia, possibly due to a genetic defect in the cleavage of proinsulin, has been observed in a few patients, who nevertheless exhibited normal glucose tolerances (RUBENSTEIN et al. 1977).

*Acknowledgments.* The survey of the literature for this chapter was concluded in September 1981. The work reported from the author's laboratory was supported by the Swedish Medical Research Council (12X-109; 12X-2297), the Swedish Diabetes Association, the Medical Faculty of the University of Uppsala, the Swedish Association for Medical Research, and Stiftelsen Claes Groschinskys Minnesfond. The typing of the manuscript by KERSTIN CLAESSON and AGNETA SNELLMAN is gratefully acknowledged.

## References

Assan R, Slusher N (1972) Structure/function and structure/immunoreactivity relationship of the glucagon molecule and related synthetic peptides. Diabetes 21:843–855

Blobel G, Dobberstein B (1975) Transfer of proteins across membranes. I. Presence of proteolytically processed and unprocessed nascent immunoglobulin light chains on membrane and ribosomes of murine myeloma. J Cell Biol 67:835–851

Blobel G, Sabatini DD (1971) Ribosome membrane interaction in eukaryotic cells. In: Manson LA (ed) Biomembranes, vol 2. Plenum, New York London, p 193

Clara M (1924) Das Pankreas der Vögel. Anat Anz 57:257–265

Crick F (1979) Split genes and RNA splicing. Science 204:264–271

Dobbs RE, Unger RH (1979) Glucagon and somatostatin. In: Freinkel N (ed) Contemporary metabolism. Plenum, New York London, p 307

Hellerström C, Howell SL, Edwards JC, Andersson A (1972) An investigation of glucagon biosynthesis in isolated pancreatic islets of guinea-pigs. FEBS Lett 27:97–101

Hellerström C, Howell SL, Edwards JC, Andersson A, Östenson C-G (1974) Glucagon biosynthesis in isolated pancreatic islets of guinea-pigs. Biochem J 140:13–21

Hellman B, Hellerström C (1960) The islets of Langerhans in ducks and chickens with special reference to the argyrophil reaction. Z Zellforsch 52:278–290

Howell SL, Edwards JC, Whitfield M (1971) Preparation of B-cell deficient guinea-pig islets of Langerhans. Horm Metab Res 3:37–43

Howell SL, Hellerström C, Whitfield M (1974a) Radioautographic localization of labelled proteins after incubation of guinea-pig islets of Langerhans with $^{3}$H tryptophan. Biochem J 140:22–23

Howell SL, Hellerström C, Tyhurst M (1974b) Intracellular transport and storage of newly synthesized proteins in the guinea-pig pancreatic A-cell. Horm Metab Res 6:267–271

Jackson RC, Blobel G (1980) Post-translational processing of full-length presecretory proteins with canine pancreatic signal peptidase. Ann NY Acad Sci 343:391–405

Korányi L, Peterfy F, Szabó J, Török A, Guóth M, Tamás GY Jr (1981) Evidence for transformation of glucagon-like immunoreactivity of gut into pancreatic glucagon in vivo. Diabetes 30:792–794

Krieger DT, Liotta AS (1979) Pituitary hormones in brain: where, how and why? Science 205:366–372

Leiter E, Coleman DL, Eppig JJ (1979) Endocrine pancreatic cells of postnatal "diabetes" (DB) mice in cell culture. In Vitro 15:507–521

Lund PK, Goodman RH, Jacobs JW, Habener JF (1980) Glucagon precursors identified by immunoprecipitation of products of cell-free translation of messenger RNA. Diabetes 29:583–586

Lund PK, Goodman RH, Habener JF (1981) Intestinal glucagon mRNA identified by hybridization to a cloned islet cDNA encoding a precursor. Biochim Biophys Res Comm 100:1659–1666

Milstein C, Brownlee GG, Harrison TM, Matheus MB (1972) A possible precursor of immunoglobulin light chains. Nature 239:117–120

Moody AJ, Markussen J, Sundby F, Steenstrup C, Schaich Fries A (1970) The insulin releasing activity of extracts of the porcine intestinal tract. In: Falkmer S, Hellman B, Täljedal IB (eds) The structure and metabolism of the pancreatic islets. Pergamon, Oxford, p 469
Moody AJ, Jacobsen H, Sundby F, Frandsen EK, Baetens D, Orci L (1977) Heterogeneity of gut glucagon-like immunoreactivity. In: Foà PP, Bajaj JS, Foà N (eds) Glucagon: its role in physiology and clinical medicine. Springer, Berlin Heidelberg New York, p 129
Moody AJ, Thim L, Holst JJ (1980) Porcine pancreatic glicentine-related peptide. Diabetologia 19:300
Moody AJ, Holst JJ, Thim L, Lindkaer Jensen S (1981) Relationship of glicentine to proglucagon and glucagon in the porcine pancreas. Nature 289:514–516
Nagelschmidt L (1939) Untersuchungen über die Langerhansschen Inseln der Bauchspeicheldrüse bei den Vögeln. Z Mikros Anat Forsch 45:200–232
Niall HD (1980) The evolution of peptide hormones. In: Cumming JA, Funder JW, Mendelsohn FAO (eds) Endocrinology 1980. Elsevier/North Holland, Amsterdam, p 13
Noe BD, Bauer GE (1971) Evidence for glucagon biosynthesis involving a protein intermediate in islets of the anglerfish *(Lophius americanus)*. Endocrinology 89:642–651
Noe BD, Bauer GE (1975) Evidence for sequential metabolic cleavage of proglucagon to glucagon in glucagon biosynthesis. Endocrinology 97:868–877
Noe BD, Bauer GE, Steffes MW, Sutherland DEK, Najarian JS (1975) Glucagon biosynthesis in human pancreatic islets: Preliminary evidence for a biosynthetic intermediate. Horm Metab Res 7:314–322
Noe BD, Baste CA, Bauer GE (1977 a) Studies on proinsulin and proglucagon biosynthesis and conversion at the subcellular level. I. Fractionation procedure and characterization of the subcellular fractions. J Cell Biol 74:578–588
Noe BD, Baste CA, Bauer GE (1977 b) Studies on proinsulin and proglucagon biosynthesis and conversion at the subcellular level. II. Distribution of radioactive hormones and hormone precursors in subcellular fractions after pulse and pulse-chase incubation of islet tissue. J Cell Biol 74:589–604
Noe BD, Fletcher DJ, Bauer GE (1981) Biosynthesis of glucagon and somatostatin. In: Cooperstein SJ, Watkins DT (eds) Biochemistry, physiology and pathology of the islets of Langerhans. Academic Press, New York, p 189
O'Connor K, Lazarus NR (1976) Studies on the biosynthesis of pancreatic glucagon in the pigeon *(Columba livia)*. Biochem J 156:279–288
O'Connor KJ, Gay A, Lazarus NR (1973) The biosynthesis of glucagon in perfused rat pancreas. Biochem J 134:473–480
Östenson C-G, Andersson A, Eriksson U, Hellerström C (1980) Glucagon biosynthesis in isolated pancreatic islets of mice and guinea-pigs. Diab Metab 6:141–149
Palade G (1975) Intracellular aspects of the process of protein synthesis. Science 189:347–358
Patzelt C, Tager HS, Caroll RJ, Steiner DF (1979) Identification and processing of proglucagon in pancreatic islets. Nature 282:260–266
Patzelt C, Chan SJ, Quinn PS, Carroll RJ, Tager HS, Steiner DF (1980) Biosynthetic precursors of glucagon: identification of glucagon and proglucagon. In: Waldhäusl WK (ed) Diabetes 1979. Excerpta Medica, Amsterdam, p 119
Petersson B, Hellerström C, Gunnarsson R (1970) Structure and metabolism of the pancreatic islets in streptozotocin treated guinea-pigs. Horm Metab Res 2:313–317
Ravazzola M, Orci L (1980 a) Glucagon and glicentin immunoreactivity are topologically segregated in the alpha granule of the human pancreatic A-cell. Nature 284:66–67
Ravazzola M, Orci L (1980 b) Transformation of glicentin-containing L-cells into glucagon-containing cells by enzymatic digestion. Diabetes 29:156–158
Ravazzola M, Siperstein A, Moody AJ, Sundby IF, Jacobsen H, Orci L (1979) Glicentin immunoreactive cells: their relationship to glucagon-producing cells. Endocrinology 105:499–508
Ravazzola M, Orci L, Perrelet A, Unger RH (1981) Immunocytochemical quantitation of glicentin and glucagon during maturation of A-cell secretory granules. Diabetologia 21:319

Rigopoulou D, Valverde I, Marco J, Faloona G, Unger RH (1970) Large glucagon immunoreactivity in extracts of pancreas. J Biol Chem 245:496–501

Rubenstein AH, Horwitz DL, Jaspan JB, Mako ME, Blix PM, Kuzuya H (1977) Circulating proinsulin and C-peptide. In: Bajaj JS (ed) Diabetes. Excerpta Medica, Amsterdam, p 134

Schields D, Warren TG, Roth SE, Brenner MJ (1981) Cell-free synthesis and processing of multiple precursors to glucagon. Nature 289:511–514

Steiner DF, Oyer PE (1967) The biosynthesis of insulin and a probable precursor of insulin by a human islet cell adenoma. Proc Natl Acad Sci USA 57:473–480

Steiner DF, Cunningham DD, Spigelman L, Aten B (1967) Insulin biosynthesis: evidence for a precursor. Science 157:697–700

Steiner DF, Kemmler W, Clark JL, Oyer PE, Rubenstein AH (1972) The biosynthesis of insulin. In: Steiner DF, Freinkel N (eds) Endocrine pancreas. American Physiological Society, Washington (Handbook of physiology, sect 7, vol 1, p 175)

Steiner DF, Quinn PS, Chan SJ, Marsh J, Tager HS (1980) Processing mechanisms in the biosynthesis of proteins. Ann NY Acad Sci 343:1–16

Tager HS (1981) The biosynthesis and molecular forms of glucagon and glucagon-like immunoreactivity. In: Unger RH, Orci L (eds) Glucagon: physiology, pathophysiology and morphology of the pancreatic A-cells. Elsevier/North Holland, Amsterdam London New York, p 39

Tager HS, Markese J (1979) Intestinal and pancreatic glucagon-like peptides. J Biol Chem 254:2229–2233

Tager HS, Steiner DF (1973) Isolation of a glucagon-containing peptide: primary structure of a possible fragment of proglucagon. Proc Natl Acad Sci USA 70:2321–2325

Tager HS, Patzelt C, Chan S-J, Quinn PS, Steiner DF (1979a) The biosynthesis and conversion of islet cell hormone precursors: a brief review. Biol Cell 36:127–136

Tager HS, Given B, Baldwin D, Mako M, Markese J, Rubenstein A, Olefsky J, Kobayashi M, Kolterman O, Poucher R (1979b) A structurally abnormal insulin causing human diabetes. Nature 281:122–125

Tager HS, Patzelt C, Assoian RK, Chan SJ, Duguid JR, Steiner DF (1980) Biosynthesis of islet cell hormones. Ann NY Acad Sci 343:133–147

Thim L, Moody AJ (1981) The primary structure of porcine glicentine. Regul Peptides 2:139–150

Trakatellis AG, Tada K, Yamaji K, Gardiki-Kouidou P (1975) Isolation and partial characterization of anglerfish proglucagon. Biochemistry 14:1508–1575

Tung AK (1973) Biosynthesis of avian glucagon: evidence for a possible high molecular weight biosynthetic intermediate. Horm Metab Res 5:416–424

Tung AK (1974) Glucagon biosynthesis in avian pancreatic islets: evidence for medium-sized biosynthetic intermediates. Can J Biochem 52:1081–1086

Tung AK, Zerega F (1971) Biosynthesis of glucagon in isolated pigeon islets. Biochem Biophys Res Commun 45:387–390

Unger RH, Orci L (1981) Glucagon and the A-cell. Physiology and pathophysiology. N Engl J Med 304:1575–1580

Valverde I, Rigopoulou D, Marco J, Faloona GR, Unger RH (1970) Molecular size of extractable glucagon and glucagon-like immunoreactivity (GLI) in plasma. Diabetes 19:624–629

CHAPTER 7

# Glucagon, Glicentin, and Related Peptides

A. J. Moody and L. Thim

## A. Introduction

The observations that the canine and human intestine contain glucagon-like immunoreactivity (GLI) (Unger et al. 1961, 1968), that the gastrointestinal tract of the dog contains hyperglycemic factors (Makman and Sutherland 1964), and that oral glucose increases plasma GLI in humans (Samols et al. 1965) laid the foundation for the concept that the intestine contains peptides which differ from glucagon, but which have certain immunochemical and biologic properties in common with glucagon. It is now possible, some 20 years after the first demonstration of GLI in the gastrointestinal tract (Unger et al. 1961), to describe the chemistry and immunochemistry of porcine gut glucagon-like immunoreactants (GLIs) and to identify some of their biologic effects. As will be seen, the gut GLIs contain the full sequence of glucagon and share some immunochemical and biologic properties with glucagon, but their spectrum of biologic effects and possible biologic role overlap only slightly with those of glucagon itself. The first of the gut GLIs to be isolated in a highly purified form was the porcine peptide, glicentin or gut GLI-1 (Sundby et al. 1976), and this chapter describes the structure of glicentin and one other gut GLI, their relationship to other peptides, and the structure–function relationships of the gut GLIs.

Glicentin was partially sequenced soon after isolation (Jacobsen et al. 1977) and found to have a COOH terminal decapeptide which closely resembled that of the possible fragment of porcine proglucagon isolated by Tager and Steiner (1973). With this information in mind, together with the finding that an anti-glicentin serum detected a glicentin-like component in the pancreatic A-cell, it was proposed that glicentin could contain the full sequence of glucagon (Moody et al. 1977) and might be proglucagon (Moody et al. 1978; Ravazzola et al. 1979a). Proof that glicentin contains the sequence of glucagon has been obtained by sequencing the molecule (Thim and Moody 1981). Porcine glicentin contains 69 amino acid residues (not 100 residues as first reported by Sundby et al. 1976), and glucagon forms the sequence 33–61. The $NH_2$ terminal portion (1–30) comprises the sequence of the porcine glicentin-related pancreatic peptide (GRPP) whose presence in, and secretion by, the porcine pancreas has been documented (Moody et al. 1981). GRPP has been isolated from the porcine pancreas and fully sequenced (Thim and Moody, 1982). Glicentin, an intestinal peptide, thus contains the sequence of glucagon, and the two main fragments of glicentin – GRPP and glucagon – are found in, and secreted by, the pancreas.

## B. The Immunochemistry of Gut GLIs

The antigenicity of glucagon and the radioimmunoassay (RIA) of glucagon are dealt with elsewhere in this volume (see Chaps. 9 and 10), but a brief description of anti-glucagon sera is included in this chapter, since their use has been crucial in the development of our knowledge of the gut GLIs. Unmodified glucagon mainly evokes the formation of two classes of antibodies in animals. The most common type of glucagon antiserum reacts with both glucagon and materials in extracts of the intestinal tract, whereas a few antisera react only with glucagon. ASSAN and SLUSHER (1972) pointed out that the glucagon antibodies that react with gut extracts are directed towards the $NH_2$ terminus and midportion of the glucagon molecule. In contrast, the other class of antisera that react only with glucagon are directed towards the COOH terminal end of the glucagon molecule.

These findings were extended by HEDING et al. (1976), who showed that the COOH terminal immunoreactant of glucagon lies in the sequence 24–29, and YANAIHARA et al. (1979) who proposed that the $NH_2$ terminal immunoreactant includes the sequence 11–16. A similar location of the $NH_2$ terminal immunoreactant has also been proposed on the basis of a comparison of the sequences of glucagon, secretin, vasoactive intestinal peptide (VIP), and gastric inhibitory peptide (GIP) (MOODY 1980). These immunoreactants are not the only ones present in the glucagon molecule (see Chap. 9) and it is thus difficult rigidly to define the specificity of anti-glucagon sera and the material each antiserum actually measures in a RIA. In this chapter, the two classes of anti-glucagon sera will be referred to as $NH_2$ terminal-specific and COOH terminal-specific. The $NH_2$ terminal antisera are defined as those which react with glucagon and gut peptides in a more or less equivalent manner. The COOH terminal antisera are those which react with glucagon, but not with crude extracts of intestinal tissue or purified gut GLIs such as glicentin. These antisera are considered as measuring (in a RIA) or detecting (in immunocytochemistry) $NH_2$ terminal and COOH terminal GLIs, respectively (see Table 1).

The term gut GLIs is used to describe peptides of enteral origin which react with $NH_2$ terminal-specific anti-glucagon sera, but not with COOH terminal-specific anti-glucagon sera (MOODY 1980). Other terms in current use for this class of peptides include: gut glucagon-like immunoreactivity (gut GLI), enteroglucagon, and glucagonoid. Provided that the terms used by the various laboratories are defined (and this definition revised from time to time as the assays used are modified) we feel that there is little to choose between "immunoreactant," which implies a

**Table 1.** The possible immunoreactive sites of glucagon and the specificity of anti-glucagon sera

| | | |
|---|---|---|
| Antiserum type | $NH_2$ terminal-specific | COOH terminal-specific |
| Immunoreactant | 11 12 13 14 15 16<br>-Ser-Lys-Tyr-Leu-Asp-Ser- | 24 25 26 27 28 29<br>-Gln-Trp-Leu-Met-Asn-Thr |
| Reacts with | Mammalian glucagon<br>Avian glucagon<br>Gut GLI | Mammalian glucagon |

concrete compound with a specific characteristic, and "immunoreactivity," which implies a more or less ill-defined material possessing immunoreactivity. The term enteroglucagon, although widely used, is, however, misleading. It is our opinion that the name glucagon should be reserved for the 29 amino acid residue peptide whose sequence is known for many animal species, and the prefixes entero-, gastro- or, for that matter, plasma- reserved for those molecules which are considered to be identical with glucagon derived from the pancreas. As we will show, the gastrointestinal tract of several animal species contains glucagon, especially that of the dog in which a well-defined gastric glucagon exists (as reviewed in detail in Chap. 33), but the bulk of the gut GLI differs chemically, biologically, and immunochemically from glucagon.

## C. The Chemistry of Crude Gut GLIs

### I. Distribution

UNGER et al. (1961, 1966) laid the basis for the studies of gut GLI by showing that GLI was present in extracts of intestinal tissue of rats, dogs, and humans. Material that reacts only with $NH_2$ terminal-specific anti-glucagon sera has been found in the gastrointestinal tract of a single insect species (TAGER and KRAMER 1980) and all the animal species (see MOODY 1972, for review) so far studied.

The distribution of gut GLI in the gastrointestinal tract of mammals has been studied in detail. Low concentrations are found in the duodenum and jejunum, with the tissue concentration progressively rising to a maximum in the lower ileum or colon in humans (BLOOM and POLAK 1978), primates (BRYANT and BLOOM 1979), pigs (HOLST 1977), and the rat (PEREZ-CASTILLO and BLAZQUEZ 1980). The concentration of gut GLI in the gastrointestinal tract is lower than the concentration of glucagon in the pancreas; for example, in the rat the maximum (ileal) concentration of gut GLI is 0.4 μg/g wet weight, whereas the glucagon content of the pancreas is 6 μg/g wet weight (PEREZ-CASTILLO and BLAZQUEZ 1980). Since the total weight of the gastrointestinal tract is greater than that of the pancreas, the total amount of gut GLI in each animal is approximately equal to the total glucagon.

### II. Size

Gut GLI was shown to be heterogeneous in size by VALVERDE et al. (1968) who found that gel filtration of crude extracts of dog jejunum separated the gut GLIs into two fractions: peak I, with a molecular weight of about 7,000 daltons; and peak II, with a molecular weight close to 3,500 daltons. Throughout this text we will refer to gut GLIs with molecular weights close to 7,000 and 4,000 daltons as peak I and peak II, respectively.

In porcine gastrointestinal tissues, gut GLI (in forms similar to VALVERDE'S peak I and peak II) is mainly located in the mucosa of the small intestine and colon (HOLST 1977). Material indistinguishable from glucagon (on the basis of its chemical, immunochemical, and biologic characteristics) was found in small amounts in the cardiac gland region of the stomach, duodenum, and small intestine, and in significant amounts (12% of the total gut GLI) in the colon.

The rabbit small intestine contains gut GLI which can be resolved into two main peaks, with approximate molecular weights of 12,000 and 8,000 daltons, and a minor peak with a molecular weight of approximately 3,500 daltons (TAGER and MARKESE 1979). Duck gut GLI consists of a single molecular species, with a molecular weight of between 6,000 and 13,000 daltons (KRUG and MIAHLE 1971).

The midgut of the only insect so far examined (the tobacco hornworm *Manduca sexta*) contains small amounts of gut GLI which, after gel filtration, separate into a minor component (molecular weight greater than 15,000 daltons) and a major component (molecular weight approximately 10,000 daltons) (TAGER and KRAMER 1980).

It can be concluded that, in mammals, gut GLI extracted from the intestine can be separated into two main size classes: peak I, which forms the majority of the gut GLI, with a molecular weight between 10,000 and 7,000 daltons, and peak II, with a molecular weight of some 3,500 daltons. In birds and the single insect species so far examined, the bulk of the gut GLI has a high molecular weight.

### III. Charge

A new facet, that of isoelectric pH (pI), as determined by isoelectric focusing, was added to the characterization of gut GLIs by MARKUSSEN and SUNDBY (1970), who found that the gut GLI extracted from the small intestines of several mammalian species was heterogeneous with respect to pI. Most of the peak I gut GLI of porcine tissue had a pI close to 7.3, whereas the peak II gut GLIs were basic, with pI > 9.0. These observations foreshowed the isolation of glicentin, the main peak I gut GLI with pI of 6.9 (SUNDBY et al. 1976), and oxyntomodulin, a peak II gut GLI with a very basic pI (BATAILLE et al. 1981 b).

The range of pIs possessed by gut GLIs is reflected in the heterogeneity of gut GLIs separated by ion exchange chromatography, a heterogeneity displayed by porcine (MARKUSSEN and SUNDBY 1970) as well as by canine (HORIGOME et al. 1977) gut GLIs. It is considered likely that the gut GLIs in the mammalian L-cell in vivo exist in relatively few molecular forms (perhaps with sizes close to molecular weights of 8,000 and 3,500 daltons) and that the plethora of gut GLIs found in extracts of the intestine is the result of enzymatic and chemical modifications of these stem peptides during extraction.

## D. The Chemistry of Glicentin, GRPP, and Oxyntomodulin

### I. Isolation of Glicentin

Many attempts were made during the 1970s to isolate the gut GLIs from extracts of intestinal tissue (for reviews, see SUNDBY and MOODY 1980; MOODY 1972). One of the more promising approaches, that utilizing immunoabsorption, led to the isolation of small amounts of peak I gut GLI (MURPHY et al. 1973). This material was found to contain three $NH_2$ terminal amino acids, indicating that the peptide was not pure, and the amount isolated was insufficient for further purification and chemical analysis.

In the mid-1970s, the isolation of gut GLIs in our own laboratory was intensified when anti-glucagon sera (prepared by L. G. HEDING) became available that reacted with gut extracts without dilution effects and thus permitted a quantitative study of recoveries during fractionation. This programm resulted in the isolation from porcine small intestine of a peptide called gut GLI-1, the major form of porcine gut GLI. This peptide was characterized by its amino acid composition, its apparent molecular weight (11,625 daltons), its $NH_2$ terminal and COOH terminal sequences, and its approximately molar reactivity with $NH_2$ terminal-specific anti-glucagon sera, combined with its lack of reactivity with COOH terminal-specific anti-glucagon sera (SUNDBY et al. 1976).

Inspired by the observation that gut GLI-1 contained 100 amino acid residues (as judged by a combination of molecular composition and apparent molecular weight), the name glicentin was coined (MOODY et al. 1978) for this peptide (*GLI* = glucagon-like immunoreactant, *cent* = 100). As will be seen, our interpretation of the chemical data was at fault. We feel, however, that the name glicentin should be kept for this peptide and its homologs in other species since the name has already been used extensively in the literature.

## II. Sequence of Glicentin

In 1976, MOODY et al. proposed that glicentin contained the full sequence of glucagon and that gut GLIs and glucagon have the same primordial precursor. This proposal was based on observations that glicentin contained the amino acids comprising the sequence of glucagon and that an anti-glicentin serum, which did not react with glucagon, bound to the A-cells in the rat pancreas.

This proposal was strengthened by the demonstration that the COOH terminal sequence of glicentin consists of the sequence glucagon 27–29 extended at its COOH terminal by a basic octapeptide (JACOBSEN et al. 1977). This COOH terminal undecapeptide differs only slightly from the homologous sequence of a proposed fragment of proglucagon isolated by TAGER and STEINER (1973). The hypothesis that glicentin contains the sequence of glucagon was further developed to include the concepts that the COOH terminal octapeptide of glicentin "masks" the immunoreactant present in glucagon 24–29 (MOODY et al. 1977) and that a glicentin-like molecule is a biosynthetic precursor of glucagon in the porcine pancreas (MOODY et al. 1978; RAVAZZOLA et al. 1979a).

Proof that glicentin contains the sequence of glucagon has now been obtained by determining the full sequence of porcine glicentin (THIM and MOODY 1981). The recently established secondary structure of glicentin is shown in Fig. 1 and the amino acid composition in Table 2. The glicentin molecule contains 69 amino acid residues, not the 100 previously reported (SUNDBY et al. 1976).

## III. Relationship of Glicentin to Other Peptides

The sequence of porcine glucagon forms the sequence glicentin 33–61, and the sequence glicentin 1–30 corresponds to the sequence (THIM and MOODY, 1982) of the glicentin-related pancreatic peptide (GRPP) which is present in, and secreted by, the porcine pancreas (MOODY et al. 1981). The Lys-Arg sequences at the ends of

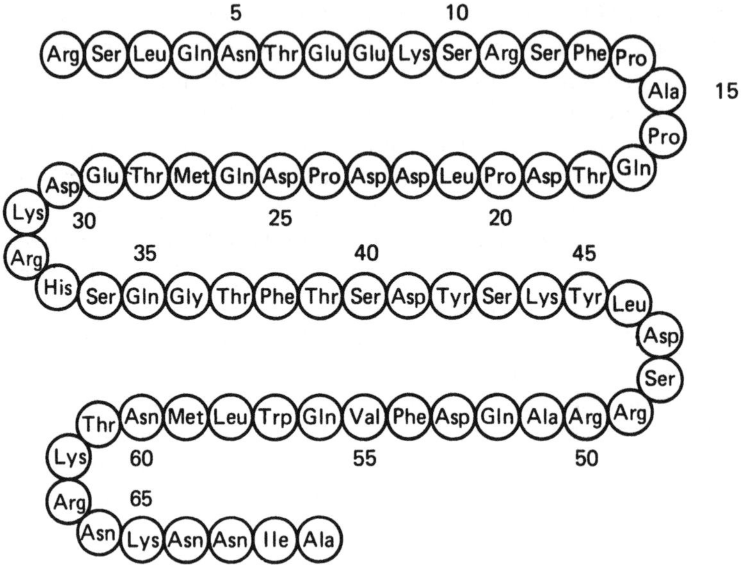


**Fig. 1.** The amino acid sequence of porcine glicentin. Thim and Moody (1981)

**Table 2.** The amino acid composition of high pressure liquid chromatographically purified glicentin (A) compared with the previously published composition (B)

| Amino acid | Amino acid residues/molecule | |
|---|---|---|
| | A[a] | B[b] |
| Asx | 13 | 16 |
| Thr | 6 | 8 |
| Ser | 7 | 10 |
| Glx | 9 | 14 |
| Pro | 4 | 6 |
| Gly | 1 | 3 |
| Ala | 3 | 5 |
| Val | 1 | 2 |
| Met | 2 | 3 |
| Ile | 1 | 2 |
| Leu | 4 | 6 |
| Tyr | 2 | 3 |
| Phe | 3 | 4 |
| Lys | 5 | 7 |
| His | 1 | 2 |
| Trp | 1 | 1 |
| Arg | 6 | 8 |
| Total | 69 | 100 |

[a] Data from Thim and Moody (1981)
[b] Data from Sundby et al. (1976)

**Table 3.** Sequence homologies[a] between glicentin (33–69) and related porcine peptides

| | |
|---|---|
| Glicentin[b] | HSQGTFTSDYSKYLDSRRAQDFVQWLMNTKRNKNNIA |
| Proglucagon fragment[c] | HSQGTFTSDYSKYLDSRRAQDFVQWLMNTKRNNKNIA |
| GIP[d] | YAEGTFISDYSIAMDKIRQQDFVNWLLAQKGKKSDWKHNITQ |
| Secretin[e] | HSDGTFTSELSRLRDSARLQRLLQGLV ($NH_2$) |
| VIP[f] | HSDAVFTDNYTRLRKQMAVKKYLNSILN ($NH_2$) |

[a] The residues homologous with the sequence glicentin 33–69 are underlined
[b] THIM and MOODY (1981)
[c] TAGER and STEINER (1973)
[d] JÖRNVALL et al. (1981)
[e] MUTT et al. (1972)
[f] MUTT and SAID (1974)

the glucagon moiety in glicentin indicate that glicentin fulfills the structural requirements for being a glucagon precursor (this point is discussed in detail in Sect. G and in Chap. 6).

The sequence of the glucagon moiety of glicentin has homologies to the gut peptides secretin, VIP, and GIP if the His-33 (homologous to glucagon 1) is aligned with the $NH_2$ terminal amino acids of these peptides. These homologies, which perhaps indicate a common primordial ancestor, are shown in Table 3. Less striking homologies can be found between glicentin and other peptides if the peptides are aligned irrespective of the relative positions of their $NH_2$ terminal amino acids, and these homologies are summarized in Table 4. Five of the six residues in glicentin 5–10 and α-endorphin 5–10 are identical, and so are three of the four residues in the sequences glicentin 65–68 and β-endorphin 19–22. GIP 37–40 has three residues in four in common with glicentin 65–68.

Glicentin 30–33 (one of the putative postsynthetic cleavage sites) is identical to β-lipotrophin 58–60, one of the known cleavage sites of β-lipotrophin. Bovine pancreatic polypeptide 14–20 is related to glicentin 24–30, and somatostatin 10–13 is identical to glicentin 37–40. Finally, an extensive homology exists between glicentin 30–38 and part of the sequence of chymodenin (an enzyme-releasing intestinal peptide).

These homologies, characteristically 3–4 residues of the 69 found in the sequence of glicentin, indicate that some degree of cross-reactivity could exist between glicentin and these peptides in immunochemical systems. Cross-reactivity has been described between glicentin and GIP in studies of the immunocytochemical characteristics of the rat pancreatic A-cell (LARSSON and MOODY 1980).

## IV. Structural Analysis of Glicentin

The amino acid sequence of glicentin (Fig. 1) has been analyzed for helical regions, β-pleated sheets, and β-turns in accordance with the guidelines laid down for the prediction of peptide conformation by CHOU and FASMAN (1974). According to this analysis, the GRPP moiety of the glicentin molecule in solution could contain a

**Table 4.** Sequence similarities[a] between glicentin and other peptides

| | |
|---|---|
| | 1 5 10 15 |
| Glicentin[b] | R S L Q N T E E K S R S F A … |
| | 5 10 |
| α-Endorphin[c] | … M T S E K S Q … |
| | 20 25 30 |
| Glicentin[b] | … P L D D P D Q M T E D K R H S Q G T F … |
| | 58 60 |
| β-Lipotrophin[d] | … D K R … |
| | 13 17 |
| BPP[e] | … T P E Q M … |
| Chymodenin[f] | B B R R A Z G T F |
| | 61 65 69 |
| Glicentin[b] | … W K R M K N N I A |
| | 19 23 |
| β-Endorphin[g] | … K N A I I … |
| | 37 42 |
| GIP[h] | … K H N I W Q |

[a] The homologies between glicentin and other peptides are underlined
[b] THIM and MOODY (1981)
[c] LING et al. (1976)
[d] LI (1969)
[e] CHANCE et al. (1974)
[f] ADELSON et al. (1980)
[g] LI and CHUNG (1976)
[h] JÖRNVALL et al. (1981)

short helical section in the $NH_2$ terminal portion and two β-turns in the mid- to COOH terminal portion. The structure of the glucagon moiety is similar to that predicted for glucagon itself (CHOU and FASMAN 1975), but it could be altered by the COOH terminal octapeptide of glicentin which has a high β-turn potential. This effect of the very similar COOH terminal hexapeptide of the possible proglucagon fragment isolated by TAGER and STEINER (1973) was predicted by CHOU and FASMAN (1975). The COOH terminal hexapeptide could prevent the sequence glicentin 51–59 from forming the helical structure predicted for the homologous sequence glucagon 19–27 (CHOU and FASMAN 1975) which is important for the binding of glucagon to its receptor on the liver plasma membrane (FRANDSEN et al. 1981; see also Chap. 13). The hexapeptide could also either mask or alter the structure of the sequence glicentin 56–61, which is homologous with the COOH terminal immunoreactant glucagon 24–29.

## V. Isolation of GRPP

Porcine pancreatic extracts have been shown to contain glicentin-related immunoreactants by means of the RIA for glicentin (MOODY et al. 1981). The main component of these glicentin-like peptides has a molecular weight between that of

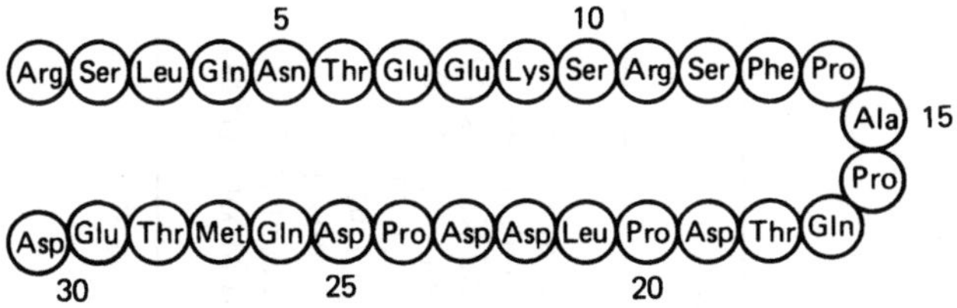


**Fig. 2.** The amino acid sequence of porcine GRPP. THIM and MOODY (1982)

glicentin and glucagon, as judged by gel filtration, and pI of 4.0 (MOODY et al. 1981). A few μg of GRPP were isolated in 1980 and the partial sequence established (MOODY et al. 1980). It has recently been possible to isolate about 10 mg GRPP from 140 kg porcine pancreas and establish the full structure of the molecule (THIM and MOODY 1982).

## VI. Chemistry of GRPP

GRPP consists of 30 amino acid residues and the sequence (Fig. 2) is identical to the sequence of glicentin 1–30 (THIM and MOODY 1982). Owing to the lack of tyrosine and tryptophan residues, GRPP has a low absorption at 280 nm.

## VII. Relationship of GRPP to Other Peptides

The sequence similarities between GRPP (identical to glicentin 1–30) and other peptides are depicted in Table 4. GRPP is linked to the glucagon part of glicentin via a Lys-Arg sequence. A similar dibasic sequence is found in other known biosynthetic precursors of peptide hormones. Table 5, which is analogous to that published by PRADAYROL et al. (1980), compares the dibasic region sequences of a number of peptide hormone precursors. The hydrophobic fragmental constants (REKKER 1977), have been summarized for the three amino acid residues on either side of the dibasic cleavage center. As can be seen from these calculations, the dibasic regions consist mainly of nonhydrophobic amino acid residues. As suggested by PRADAYROL et al. (1980), these characteristics could account for the specificity of the cleavage enzyme (or enzymes). However, the glucagon sequence from residue 14 to 21 (Leu-Asp-Ser-Arg-Arg-Ala-Glu-Asp) also precedes a dibasic sequence surrounded largely by nonhydrophobic residues, and so far no cleavage in vivo of this dibasic site has been reported. A comparison of the linear arrangements of glicentin (GRPP-Lys-Arg-Glucagon-Lys-Arg-hexapeptide) and proinsulin (B-chain-Arg-Arg-C-peptide-Lys-Arg-A-chain) indicates that the sites of postsynthetic enzymatic cleavage also depend, at least in the A- and B-cells of the pancreas, upon the chain length of the peptides between the cleavage points, inasmuch as the B chain of proinsulin and GRPP both contain 30 residues and the C-peptide of proinsulin and glucagon both have 29.

Although the C-peptide region of proinsulin was at first considered to play a role in the postsynthetic folding of proinsulin, it is now thought that it either covers the biologically active site of insulin during synthesis (SNELL and SMYTH 1975) or simply functions as a spacer which enlarges the proinsulin molecule sufficiently for

**Table 5.** The dibasic region of porcine peptide hormone precursors

| Precursor | Residues | Sequence (postsynthetic cleavage site indicated by box) | | | HFC[a] | Reference |
|---|---|---|---|---|---|---|
| Proinsulin | 28–35 | Pro–Lys–Ala– | Arg–Arg | –Glu–Ala–Gln | +1.48 | Chance et al. (1968) |
| Proinsulin | 59–66 | Pro–Pro–Gln– | Lys–Arg | –Gly–Ile–Val | +4.38 | Chance et al. (1968) |
| Gastrin 34 | 13–20 | Asp–Leu–Ala– | Lys–Lys | –Gln–Gly–Pro | +2.42 | Harris and Kenner (1978) |
| Somatostatin 28 | 10–17 | Pro–Arg–Glu– | Arg–Lys | –Ala–Gly–Cys | +1.49 | Pradayrol et al. (1980) |
| β-Lipotropic hormone | 36–43 | Ala–Ala–Glu– | Lys–Lys | –Asp–Glu–Gly | +0.90 | Gráf et al. (1971) |
| β-Lipotropic hormone | 56–63 | Pro–Lys–Asp– | Lys–Arg | –Tyr–Gly–Gly | +3.21 | Gráf et al. (1971) |
| Proparathyrin | 2– 9 | Pro–Ile–Lys– | Lys–Arg | –Ser–Val–Ser | +4.98 | Chu et al. (1975) |
| Proglucagon (glicentin) | 59–66 | Met–Asn–Thr– | Lys–Arg | –Asn–Lys–Asn | −1.81 | Jacobsen et al. (1977) |
| Proglucagon (glicentin) | 28–35 | Thr–Glu–Asp– | Lys–Arg | –His–Ser–Gln | −1.21 | Thim and Moody (1981) |

[a] HFC: hydrophobic fragmental constant (Rekker 1977) calculated for the three residues on either side of the pair of basic amino acids

it to span the space between the ribosomes and the rough endoplasmic reticulum (PATZELT et al. 1978). Owing to the lack of disulfide bridges in glucagon, it is unlikely that GRPP play any role in the postsynthetic folding of glucagon. By analogy with the C-peptide of proinsulin, GRPP may thus act as a simple "transport peptide" or may be a biologically active peptide. On balance, it is most likely that GRPP has biologic activities since other fragments of hormonal precursors such as C-peptide (DRYBURGH et al. 1980) and the fragments of proopiocorticotrophin are known to have biologic effects.

### VIII. Chemistry of Oxyntomodulin

A peak II gut GLI has been isolated from extracts of the porcine small intestine by BATAILLE et al. (1981 a). The biologic effects of this peptide have been partially characterized (see Sect. J). By finger printing techniques and partial sequence analysis the peptide was found to consist of glucagon extended at the carboxyl terminal and by the carboxyl terminal octapeptide of glicentin, i.e. is glicentin 33–69 (BATAILLE et al., 1982a).

## E. The Immunochemistry of Glicentin

### I. Reaction of Glicentin with Anti-Glucagon Sera

Glicentin has full molar reactivity with $NH_2$ terminal-specific anti-glucagon sera but does not react with COOH terminal-specific anti-glucagon sera (SUNDBY et al. 1976). It has been proposed that the COOH-terminal octapeptide of glicentin masked the immunoreactant homologous with glucagon 26–29 (MOODY et al. 1976). Synthetic glicentin 51–69 does not react with COOH terminal-specific anti-glucagon sera such as OAL-123 and 30K (YANAIHARA et al. 1979). Trypsin and carboxypeptidase B liberate COOH terminal GLI from glicentin (HOLST 1980) and from the glicentin-like material in the L-cell of the lower small intestine (RAVAZZOLA and ORCI 1980a). These findings, together with the findings of the predictive structural analysis of glicentin (Sect. D.IV) support this proposal. The observation that synthetic glicentin 51–69 in which the lysine residues (glicentin 62 and 65) were protected had significant COOH terminal GLI (YANAIHARA et al. 1979) suggests that the lysine residues in the COOH terminal octapeptide of glicentin are important in the masking of the COOH terminal immunoreactant in the glucagon moiety of glicentin.

### II. Anti-Glicentin Sera

Anti-glicentin sera were raised in five rabbits by immunizing them with glicentin coupled to human serum albumin with glutaraldehyde. The antisera were tested for their ability to bind glucagon $^{125}I$, GIP $^{125}I$, and glicentin $^{125}I$. One serum, R 64, was selected for further study because it had a high affinity and capacity for glicentin $^{125}I$ and no detectable capacity for glucagon $^{125}I$ or GIP $^{125}I$. This antiserum has been used as the basis for a RIA measuring immunoreactive glicentin and in the study of the glicentin cell (see Sect. F).

**Table 6.** Reactivity of porcine peptides with anti-glicentin serum R64

| Peptide[a] | Immunoreactive glicentin (mol %) |
|---|---|
| GIP | 0.2 |
| Glucagon | 0.005 |
| CCK | 0.005 |
| VIP | 0.005 |
| Secretin | 0.005 |

[a] Source of peptides:
GIP natural, porcine, Novo Research Institute, Bagsvaerd, Denmark
Glucagon natural, porcine, Novo Industry, Bagsvaerd, Denmark
CCK, VIP, Secretin } natural, porcine, produced by V. Mutt

## III. Radioimmunoassay for Glicentin

The RIA for immunoreactive glicentin is based on R 64, standards of highly purified porcine glicentin, and glicentin $^{125}$I prepared by a modification of the technique used for the iodination of glucagon (Jørgensen and Larsen 1972). The details of the assay are given in Moody et al. (1981). The molar reactivities of VIP, GIP, glucagon, and secretin in the RIA for immunoreactive glicentin are given in Table 6.

On the basis of studies with peptides liberated from glicentin by digestion with *Armillarea mellea* protease, we previously suggested that R 64 reacted with the $NH_2$ terminal portion of glicentin between the Lys in position 9 and the $NH_2$ terminal His of the glucagon moiety (Moody et al. 1979). The R 64 immunoreactant probably lies between glicentin 10 and 30, since GRPP itself has full molar reactivity with this antiserum.

## IV. Distribution of Immunoreactive Glicentin

The distribution of immunoreactive glicentin in the porcine small intestine is very similar to that of $NH_2$ terminal GLI (Moody et al. 1978), indicating that the majority of porcine gut GLI reacts with both R 64 and $NH_2$ terminal-specific anti-glucagon sera, i.e., it is similar to glicentin itself. Separation of porcine gut GLI by ion exchange chromatography showed however that, while the bulk of gut GLI was congruent with immunoreactive glicentin, a considerable percentage (10%–20%) of the gut GLI was not associated with immunoreactive glicentin (Moody et al. 1978). This material may well be related to oxyntomodulin, a peak II gut GLI recently isolated by Bataille et al. (1981 a). Preliminary studies have been carried out on the levels of immunoreactive glicentin in pig plasma after a mixed feed, and it has been found that fasting levels are significantly increased after feeding (MOODY 1980). More detailed studies of the levels of plasma immunoreactive glicentin in the pig are in progress.

Extracts of rat (LARSSON and MOODY 1980) and human pancreas (A. J. MOODY 1981, unpublished work) and human ileum (GHATEI and BLOOM 1981) do not contain significant amounts of immunoreactive glicentin. This species specificity of R 64 when used in RIA is at variance with the findings that R 64 reacts with material in the A- and L-cells of all mammalian species so far studied by immunocytochemical techniques. This discrepancy could be the result of different antibodies reacting with immunoreactive glicentin in the two techniques or to histologic fixation inducing a structural change in glicentin-related molecules (see LARSSON and MOODY 1980 for a further discussion).

## F. Glicentin Cells

### I. Intestinal Gut GLI Cells

Although the ultrastructure of the glucagon-producing cells is dealt with elsewhere in this volume, it is useful to consider the glicentin cells in this chapter since the study of the distribution of glicentin-like immunoreactivity at the cellular level has been crucial to the development of the concept of glicentin's relationship to glucagon.

In 1968 ORCI et al. reported a study of the gastroenteropancreatic tissues of the rat with electron microscopic techniques. Taking into account that glucagon-like immunoreactivity and glucagon-like biologic activity had at that time been detected in the gastrointestinal tract, the authors searched for cells that resembled – at the ultrastructural level – the A-cells of the pancreas. They concluded that one candidate endocrine cell with electron-dense particles could be the cell containing glucagon-like material. This cell was found at all levels of the gastrointestinal tract and had granules with a mean size of 250 nm which resembled those of the pancreatic A-cell. POLAK et al. (1971) demonstrated, with indirect immunofluorescent techniques, the presence of glucagon-like immunoreactivity in cells of the gastrointestinal tract of the dog, thus confirming the suggestion by ORCI et al. that such cells existed. These findings were confirmed by LARSSON et al. (1975).

ITO and KOBAYASHI (1976) drew attention to the fact that the anti-glucagon sera used in earlier studies had not been characterized, and used anti-glucagon sera that were either specific for the COOH terminal portion of glucagon or reacted with glucagon and with extracts of the gastrointestinal tract (i.e., were $NH_2$ terminal specific). In this way, ITO and KOBAYASHI were able to show that, in the dog, COOH terminal GLIs were found in the pancreas and in the deeper portions of the fundic gastric glands, but that the COOH terminal GLIs were found not only in the pancreas and gastric glands, but also in the jejunum, ileum, and colon. GRIMELIUS et al. (1976) studied the correlation between the ultrastructurally defined intestinal L-cell (L = large granules) and the gut GLI cell, as detected by immunocytochemistry. They confirmed the presence in the dog of the gastric A-cell, and concluded that the intestinal L-cell contained gut GLI.

## II. The Glicentin Cell

### 1. Animal Tissue

The cellular localization of glicentin-like immunoreactivity was reported in a preliminary form by Moody et al. (1976) when it was stated that an anti-glicentin serum reacted with the A-cell of the pancreas as well as with the L-cell of the intestine. These findings have been fully described by Ravazzola et al. (1979 b), who found that the pancreatic A-cell and cells in the intestinal mucosa of rats, dogs, pigs, and guinea pigs contained glicentin-like immunoreactive material. In the species examined, gastric glicentin-positive cells were found solely in the dog oxyntic mucosa. The distribution of the glicentin cells in the intestine was similar in all species studied, being extremely rare in the upper small intestine, but increasing in frequency in the lower small intestine, to reach a maximum in the terminal ileum and colon. This distribution of glicentin cells is similar to the distribution of immunoreactive glicentin in extracts of the segments of porcine intestine (Moody et al. 1978). The cells were in contact with the intestinal lumen, triangular in shape, and appeared to be preferentially located in the crypts of Lieberkühn. Studies at the ultrastructural level revealed that, in the pancreas, the same cells were stained with an anti-glucagon and an anti-glicentin serum, and that the staining was essentially restricted to the secretory granules. Similar congruent staining with anti-glucagon and anti-glicentin serum was found in the dog gastric A-cells. In the dog ileum, the use of serial semithin sections showed that glicentin-like immunoreactive material was clearly located in cells which, on an ultrastructural basis, were L-cells. The glicentin-like material was located in the secretory granules of the L-cells.

The distribution of glicentin-like immunoreactivity in the rat, dog, and cat was studied by Larsson and Moody (1980). The twin location of glicentin-like immunoreactivity in the pancreatic A-cell and in the L-cell of the intestinal mucosa was confirmed, and the presence of glicentin-like material in the gastric A-cell of the cat was observed. Glicentin-like immunoreactivity was found in the tissues of the rat from the 18th day of gestation.

### 2. Normal Human Tissue

Garaud et al. (1980) described the presence of glicentin-like material in human intestinal L-cells and in human pancreatic A-cells. No gastric glicentin-containing cells were demonstrated in the adult material used. Glicentin-positive cells were rare in the duodenum and most frequent in the ileum and colon. Cells which contained COOH terminal glucagon-like material as well as glicentin-like material were also described in the ileum/colon.

Moxey et al. (1982) have also shown that the colonic mucosa contains cells which stain with both a COOH terminal-specific anti-glucagon serum and a glicentin antiserum, but that the colon also contains cells which stain only with the anti-glicentin serum. The demonstration of COOH terminal GLI in the cells of the human colon is in accordance with the previous demonstration (Knudsen et al. 1975) that the human colon contains cells which stain with both COOH and $NH_2$ terminal-specific anti-glucagon sera. The presence of COOH terminal GLI in extracts of porcine intestine (Holst 1977), together with the findings of COOH terminal

GLI in fixed cells indicate that the intestine is a potential source of peptides with COOH terminal GLI.

The ontogeny of glicentin cells in the human has been studied by LEDUQUE et al. (1982) who showed that glicentin and glucagon cells were found in the oxyntic mucosa of the fetus and that they were of the closed type. Glicentin cells were found also in the intestine, appearing in the ileum at the 8th gestational week and in the colon at the 12th gestational week. Glicentin cells, congruent with glucagon cells, were found in the pancreas from the 8th week and were rare in the lower posterior part of the head.

### 3. Pathologic Human Tissue

Glicentin-like material has been found in cells of tumors removed from subjects with a well-defined glucagonoma syndrome (BORDI et al. 1979). The stained cells were not of an appearance similar to normal pancreatic A-cells. The presence of glicentin-positive cells in small glucagon adenomas was also reported in the same study. In this case, the glicentin-positive cells were very much like the normal A-cell.

FIOCCA et al. (1980) found that the cells of three nonargentaffin rectal carcinoids contained glicentin-like, glucagon-like, and bovine pancreatic polypeptide-(BPP)-like immunoreactivities. In many cells, glicentin-like and BPP-like materials were present simultaneously. This presence of glucagon-like material in rectal carcinoids has also been reported by WILANDER et al. (1977).

## III. Fine Structure of the A- and L-cell Secretory Granules

The secretory granules of the A- and L-cells display characteristic differences at the ultrastructural level. As was first noted by LIKE (1967), the A-cell secretory granule consists of an electron-dense core surrounded by a "halo" of electron-translucent material, whereas this characteristic halo is absent in the dog L-cell (GRIMELIUS et al. 1976). The difference in electron density between the core and halo of the pancreatic A-cell was shown by BUSSOLATI et al. (1971) to agree with the distribution of glucagon-like material within the granules, since glucagon-like material is restricted to the electron-dense core of the granule.

In the secretory granules of the human A-cell, glucagon-like material is restricted to the granule core, whereas glicentin-like material is located in the halo of the granule (RAVAZZOLA and ORCI 1980b; GARAUD et al. 1980). This segregation of glucagon- and glicentin-like material was not found in the L-cell. The distribution of glicentin-like material in the A- and L-cell granules is similar to that of the material that reacts with Grimelius's silver stain which is found throughout the L-cell granule (GRIMELIUS et al. 1976) but which is restricted to the halo of the A-cell granule (BUSSOLATI et al. 1971).

In summary, the L-cell of all mammalian species so far studied contains glicentin-like material which is probably evenly distributed throughout the secretory granules. Glicentin-like material is essentially restricted to the halo of the secretory granules of the pancreatic A-cells and probably consists of GRPP. Although $NH_2$ terminal GLI has been demonstrated by immunocytochemical techniques in the gastroenteropancreatic system of nonmammalian species (see STEFAN et al. 1981), glicentin-like material has not been detected in tissue from holocephalian *(Hydrolagus colliei)* or teleostian fish *(Cottus scorpius)* (STEFAN et al. 1981).

## G. Glicentin and Glucagon Biosynthesis

Glucagon biosynthesis is considered elsewhere in this volume (Chap. 6), and this section is restricted to a consideration of how glicentin could be related to glucagon biosynthesis. The elucidation of the full sequence of glicentin (THIM and MOODY 1981), in which the glucagon moiety is bracketed by Lys-Arg sequences, the location of immunoreactive glicentin in the halo of the pancreatic A-cell secretory granules, and the isolation of GRPP from the pancreas all support the concept that a glicentin-like molecule is porcine proglucagon. The way in which a glicentin-containing preproglucagon could be converted into glucagon, GRPP, and the hexapeptide is shown in Fig. 3.

The simultaneous storage of two or more fragments of a hormonal precursor is known in the case of the pancreatic B-cell (TAGER et al. 1972), the anterior pituitary (WEBER et al. 1979), and the pancreatic polypeptide cell (SCHWARTZ et al. 1980). The synchronous and approximately equimolar secretion of GRPP and glucagon by the perfused porcine pancreas (Fig. 4) is also commensurate with the concept that GRPP and glucagon are derived from a common precursor.

Studies of the glucagon precursor or the kinetic aspects of glucagon biosynthesis have not yet been carried out in the pig. Comparison between the concept that glicentin is proglucagon and the results of experimental studies in nonporcine tis-

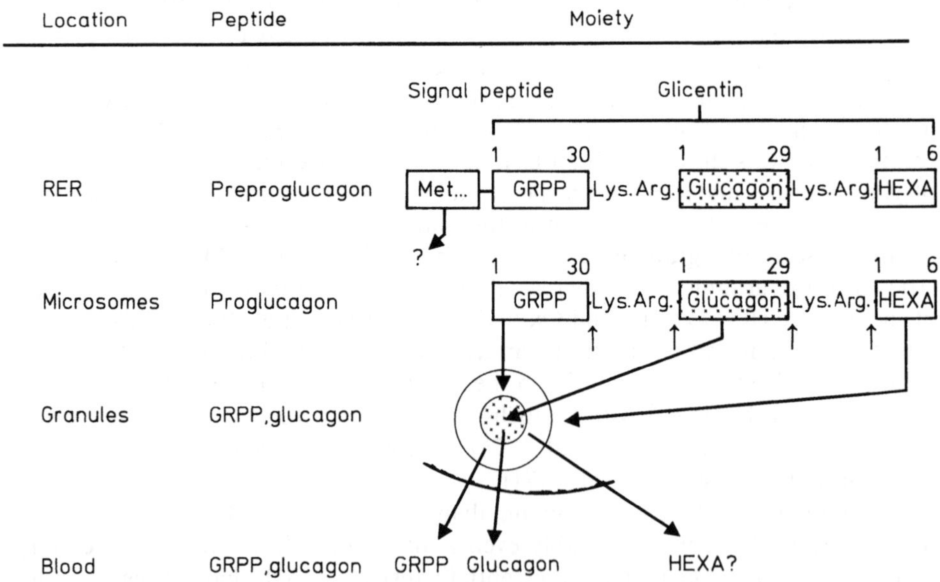


**Fig. 3.** A proposed conversion of porcine proglucagon into glucagon, GRPP and hexapeptide (HEXA). RER = rough endoplasmic reticulum

sues reveals that, although none of the available evidence negates the role of glicentin as proglucagon, there are some discrepancies which indicate that the scheme shown in Fig. 3 requires modification.

PATZELT et al. (1979) found that rat islets synthesize a large "proglucagon" molecule of molecular weight 18,000 daltons, which is rapidly converted into two fragments. One of these fragments, molecular weight 10,000 daltons, is devoid of GLI, whereas the other, molecular weight 4,500 daltons contains GLI and is slowly converted into glucagon. The former fragment could be considered equivalent to porcine GRPP, but has some three times the molecular weight.

A study of the cell-free translation of anglerfish glucagon mRNA (SHIELDS et al. 1981) indicates that the mRNA generates three forms of primary translation products: two with an approximate molecular weight of 14,000 daltons and the third with a molecular weight of 16,000 daltons. In a similar study, LUND et al. (1980) found that the anglerfish preproglucagon had a molecular weight of 14,500 daltons, which was converted by microsomal membranes into a stable peptide with a molecular weight of 12,500 daltons. The consensus of the studies with anglerfish mRNA are that the preproglucagon in the anglerfish has a molecular weight of 14,000–16,000 daltons, which is converted into a stable intermediate with a slightly lower molecular weight (see also NOE et al. 1977). The partial sequence of one anglerfish preproglucagon was established by Lund et al. in 1981 and fully determined in 1982 (LUND et al. 1982). Anglerfish preproglucagon is a 124 amino acid peptide containing the following sequences: a signal peptide, a 30 amino acid se-

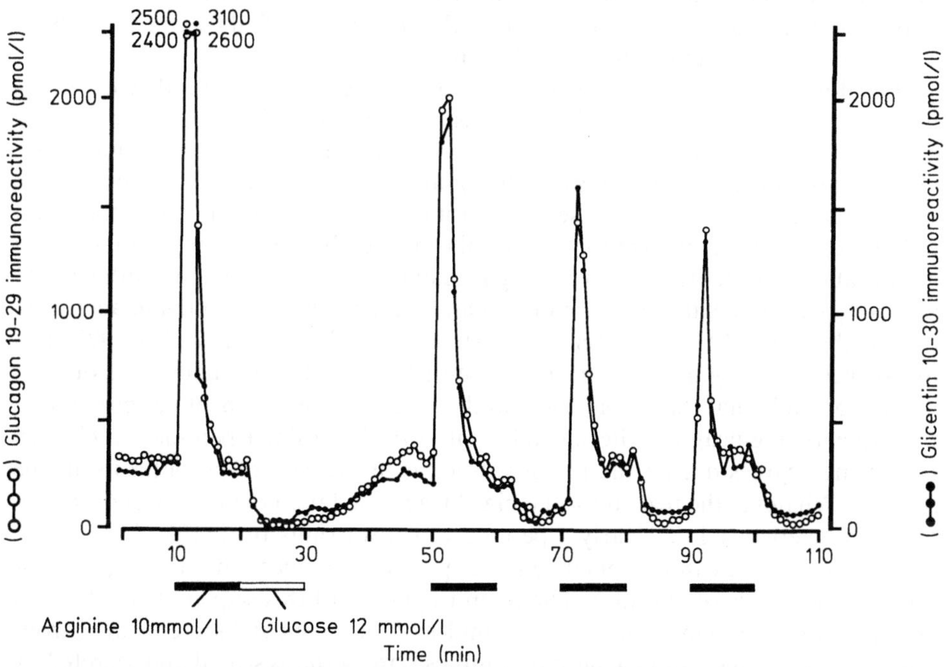


**Fig. 4.** The release of glucagon-like (*open circles*) and glicentin-like (*full circles*) immunoreactivity by a perfused porcine pancreas. MOODY et al. (1981)

quence with some homology to porcine GRPP, glucagon, a pentapeptide and a 34 amino acid sequence which has homologies to glucagon as well as to GIP and VIP. Two basic amino acids are located between the GRPP-like peptide and glucagon, glucagon and the pentapeptide, and the pentapeptide and the carboxy terminal 34 amino acid peptide. The size of the molecule indicates that it contains some 120 amino acid residues. The glucagon moiety contains the sequence of anglerfish glucagon bracketed by Lys-Arg sequences. Since the calculated molecular weight of glicentin is 8,180 daltons, there is clearly a major discrepancy between the proglucagons identified in the anglerfish and rat, which have molecular weights of 12,000–14,000 daltons, and the glicentin/proglucagon shown in Fig. 4.

This discrepancy can be resolved in one of two ways. Porcine proglucagon could have much the same size as the proglucagon proposed in other species and, during storage in the secretory granules, one or more sequences devoid of immunoreactive glicentin and GLI are removed to yield GRPP, glucagon, and hexapeptide, plus the putative fragments of proglucagon. The alternative is that the stable intermediates identified as proglucagon in cell-free systems are unstable in the A-cell, being rapidly shortened to a glicentin-size, stable form which in turn is split into GRPP, glucagon, and hexapeptide for storage in the secretory granules. These alternatives differ only in that, in the first, a 12,000 daltons peptide is considered to be the stable intracellular glucagon precursor, whereas in the second this precursor is considered as being unstable. A functional difference is that the first alternative suggests that the A-cell granule contains (and secretes) an as yet unidentified 4,000 daltons fragment of proglucagon. In the second alternative this fragment (or fragments) is removed from the proglucagon molecule before its storage in the A-cell granule. A third possibility is that the unidentified fragment has been cleaved off from glicentin during isolation.

Ox, dog, rat, and turkey pancreata contain a glucagon-like peptide with an estimated molecular weight of 8,200 daltons (O'CONNOR and LAZARUS 1976). The immunochemical and biologic properties of a crude preparation (24 μg equiv. GLI/mg protein) containing this big glucagon include reactivity with some COOH terminal-specific anti-glucagon sera, the ability to activate adenylate cyclase and to bind to the glucagon receptor of liver plasma membranes. These characteristics appear at first mutually exclusive: if big glucagon contained glucagon with COOH and $NH_2$ terminal extensions then it would be expected not to activate adenylate cyclase (because of the $NH_2$ terminal extension) and not react with COOH terminal-specific anti-glucagon sera (because of the COOH terminal extension).

It is possible that the preparation studied included several peptides containing the sequence of glucagon. Alternatively, it may be that a single peptide was cleaved to varying degrees during the investigation, as in the case of glicentin itself (see Sect. D). Whatever the reason for the big glucagon so far studied having contradictory characteristics, it is clearly a peptide which warrants further study.

In summary, it can be concluded that porcine proglucagon either is, or contains, the sequence of glicentin. The main fragments of this sequence (GRPP and glucagon) are stored in the secretory granules of the A-cell and secreted synchronously. Other products of glucagon biosynthesis are perhaps stored and secreted by the A-cell, and their isolation and characterization would provide valuable information concerning the role of the secretory products of the A-cell in mammals. The observations that $NH_2$ terminal GLI is found in the central nervous system (CNS) of the dog (CONLON et al. 1979) and rat (LOREN et al. 1979; TAGER et al. 1980) indicate that either glucagon or other products of its biosynthesis could be involved in the function of the CNS.

# H. Circulating Gut GLIs

## I. Radioimmunoassay of Gut GLI

Development of information concerning the immunochemistry of glucagon led to the concept that gut GLIs reacted only with $NH_2$ terminal-specific anti-glucagon sera, whereas glucagon reacted both with $NH_2$ terminal-specific and COOH terminal-specific anti-glucagon sera (see Sect. B). This in turn led to the introduction (HEDING 1971) of a differential RIA for gut GLIs. In this system, an $NH_2$ terminal-specific anti-glucagon serum is used to measure the total total GLI in a sample, and a COOH terminal specific anti-glucagon serum is used to measure the pancreatic contribution to the total GLI. The difference between $NH_2$ terminal and COOH terminal GLI is taken to represent the gut GLI in the sample. This approach provides an adequate measure of gut GLI during an oral or intraluminal challenge, and permits a simultaneous measurement of COOH terminal GLI (pancreatic GLI). Many laboratories measure only the total circulating $NH_2$ terminal GLI. Providing that it is borne in mind that variations in the GLI of pancreatic origin will alter the precise amount of gut GLI in each sample, this approach provides an adequate measure of gut GLI.

## II. Factors Controlling Gut GLI Release

The pioneer studies of UNGER et al. (1968) showed that intraluminal glucose loads stimulated the secretion of gut GLI by the dog lower small intestine. Such studies have also been carried out in the rat and cat (FRAME 1976, 1977). In the rat, glucose in the lower small intestine caused the release of gut GLI. In the cat, the contribution of the colon was studied separately, and it was found that only the colon released significant amounts of gut GLI into the colonic venous circulation. The release of gut GLI in response to glucose by the isolated dog ileum has the same anomeric specificity for D-glucose as the islet B- and A-cells in that the $\alpha$-anomer of glucose is a more potent releaser of gut GLI than the $\beta$-anomer (MATSUYAMA et al. 1979). This finding indicates that the intestinal L-cell may have stereospecific glucose receptors.

Triglyceride-rich meals and intraduodenal infusion of triglycerides cause large increases in plasma gut GLI in dogs (BÖTTGER et al. 1973). Triglycerides are perhaps a weaker stimulant of gut GLI release in humans than calorifically equivalent amounts of glucose, whether administered in the form of a meal (BÖTTGER et al. 1973) or an intraduodenal infusion (HOLST et al. 1976).

The effects of nonnutrients on gut GLI release have been studied in some detail. In dogs, glucose-induced gut GLI release can be inhibited by parenteral infusion of somatostatin (SAKURAI et al. 1975). Furthermore, the increase in gut GLI evoked in dogs by a mixed meal can be inhibited by the simultaneous oral administration of somatostatin (SCHUSDZIARRA et al. 1979). In humans, somatostatin infusion is also able to inhibit glucose-induced gut GLI release (MARCO et al. 1977).

Bombesin infusion in the dog decreases blood glucose, increases basal levels of plasma gut GLI, and increases the gut GLI response to intraduodenal glucose (MATSUYAMA et al. 1980). Motilin infusion in humans, during the period immediately following a mixed breakfast, did not alter plasma levels of gut GLI (CHRISTOFIDES et al. 1979). It should be noted that the mixed meal used (containing approximately 3,000 kJ) did not cause any significant change in plasma gut GLI. Pancreatectomy in dogs causes a marked increase in plasma gut GLI (see, for instance, MATSUYAMA and FOÀ 1974; YOSHIDA and KONDO 1980; WIDER et al. 1976) and a marked increase in the secretion of gut GLI into the venous drainage of dog ileum after the intraluminal infusion of glucose (WIDER et al. 1976).

Similar elevated levels of gut GLI in basal plasma are observed in pancreatectomized humans (KARESEN et al. 1980) and in insulin-dependent diabetics (MATSUYAMA et al. 1975; HEDING and MUNKGAARD RASMUSSEN 1972). These elevated levels of gut GLI are reversed, in part, in the dog by insulin administration (MATSUYAMA and FOÀ 1974) and are normalized in the diabetic human by adequate insulin treatment (MATSUYAMA et al. 1975).

These findings of elevated gut GLI in depancreatized dogs and humans and in insulin-dependent diabetics indicate that a pancreatic factor associated with insulin secretion can inhibit gut GLI release. The hypersecretion of gut GLIs could also be caused by a dysfunction of the upper small intestine which permits nutrients to pass into the small intestine. The observations that the elevated levels of gut GLI can be reduced by insulin in dogs (MATSUYAMA and FOÀ 1975) and in diabetic humans (MATSUYAMA et al. 1975) indicate that insulin is the lacking factor, although it is not known whether the insulin acts directly on the L-cell or via a modification of intestinal function.

### III. Circulating Forms of Gut GLI

Surprisingly little is known about the molecular forms of gut GLI that are released into the mesenteric circulation and thereafter appear in the posthepatic circulation. The isolated, perfused rat small intestine released gut GLI into serosal secretion (O'CONNOR et al. 1979) which was heterogeneous with respect to size and immunoreactivity. Although some $NH_2$ terminal GLI with a molecular weight close to 12,000 daltons was detected in the serosal secretion, the bulk of the GLI was smaller in size and consisted of a mixture of peptides with $NH_2$ terminal GLI alone and with both $NH_2$ terminal and COOH terminal GLI.

On the contrary, VALVERDE et al. (1979) showed that gut GLI with a size similar to the peak I found in extracts of the dog small intestine was the only GLI form to increase after intraduodenal glucose. In humans, ingestion of a mixed meal also causes increases only in peak I gut GLI in the peripheral circulation (GHATEI and BLOOM 1981). This disagreement between findings in vitro and in vivo may be caused by the different techniques used, by the isolated perfused intestine having split the larger gut GLIs into smaller molecules during secretion – a process which is theoretically possible and which has been partially achieved by enzymatic degradation in vitro (TAGER and MARKESE 1979; HOLST 1980) on histologic specimens

**Table 7.** Fasting and stimulated levels of plasma gut GLI and total GLI[a] in normal humans

| Load[c] | Parameter | | Plasma GLI | | Reference |
|---|---|---|---|---|---|
| | Total/Gut GLI | | Fast-ing | Maximum[b] | |
| OGTT | | Gut GLI | 109 | 217 (60) | HEDING (1971) |
| OGTT | | Gut GLI | 138 | 268 (60) | HEDING and MUNKGAARD RASMUSSEN (1972) |
| OGTT | Total | | 433 | 578 (45) | MARCO et al. (1972) |
| OGTT | Total | | 89 | 126 (45) | VANCE et al. (1972) |
| Nil | | Gut GLI | 37 | | CZYZYK et al. (1975) |
| OGTT | Total | | 34 | 40 (60) | BREUER et al. (1975) |
| OGTT | | Gut GLI | 86 | 115 (60) | BOTHA et al. (1976) |
| Meal | | Gut GLI | 28 | 45 (180) | BESTERMAN et al. (1978) |
| OGTT | | Gut GLI | 25 | 35 (45) | HORNNES et al. (1981) |
| Meal (3,000 kJ) | | Gut GLI | 33 | 38 (70) | GHATEI and BLOOM (1981) |
| Meal (8,820 kJ) | | Gut GLI | 36 | 95 (70) | GHATEI and BLOOM (1981) |

[a] The GLI values are in pmol equiv./l plasma
[b] The time, in min, at which the maximum values were measured is in parentheses
[c] OGTT: oral glucose tolerance test; Meal mixed meal

(RAVAZZOLA and ORCI 1980a) and by incubated hepatocytes (THIEDEN et al. 1981) – or by the liver modifying the forms of gut GLI present in the mesenteric vein.

If it is assumed that the gut GLIs secreted into the mesenteric veins are released in much the same balance of molecular forms as found in extracts of intestine, i.e., approximately 70% peak I (molecular weight ~ glicentin); 30% peak II (molecular weight ~ glucagon) then the forms of gut GLI appearing in the posthepatic circulation will reflect this pattern, as modified by the relative hepatic clearance of the two forms. In the pig, the half-life of glucagon is 3.4 min, whereas that of the gut GLI in a crude extract of porcine intestine was 15.9 min (TANAKA et al. 1979). If the hepatic clearances of peak I and peak II gut GLIs bear approximately the same relationship to each other as the half-lives of total gut GLI and glucagon, then it is likely that only in circumstances of extremely high secretion rates of gut GLI will detectable changes in peak II gut GLI be observed in the peripheral circulation. When considering reported levels of gut GLI in humans, one must therefore apply the following riders to the clinical data:

a) The material in the peripheral circulation is probably a glicentin-like molecule with a half-life greater than that of glucagon.
b) Highly significant amounts of peak II gut GLI may well have been present in the portal vein but have been removed by the liver.

## IV. Circulating Gut GLI Levels in Adult Humans

In normal humans, fasting gut GLI or total $NH_2$ terminal GLI is in the range 15–90 pmol/l, although higher values have been recorded. The plasma gut GLI is only slightly increased by small (50–75 g) oral glucose loads or by a light (3,000 kJ) mixed meal (Table 7). A large mixed meal (8,800 kj) causes a marked increase in plasma gut GLI (GHATEI and BLOOM 1981).

**Table 8.** Fasting and stimulated levels of plasma gut GLI and total GLI[a] in patients

| Disease[c] | Load[d] | Parameter | Fasting GLI | | Reference |
|---|---|---|---|---|---|
| | | Total/gut GLI | Fasting | Maximum[b] | |
| C | Meal | gut GLI | 97 | 263 (180) | Besterman et al. (1978) |
| Dgn | OGTT | gut GLI | 28 | 38 (45–60) | Hornnes et al. (1981) |
| Dgo | OGTT | gut GLI | 23 | 25 (45) | Hornnes et al. (1981) |
| Da | OGTT | gut GLI | 176 | 300 (60) | Heding and Munkgaard Rasmussen (1972) |
| Db | OGTT | gut GLI | 242 | 591 (60) | Heding and Munkgaard Rasmussen (1972) |
| G | OGTT | Total | 50 | 312 (60) | Bloom et al. (1972) |
| Gd | OGTT | Total | 50 | 598 (60) | Bloom et al. (1972) |
| G | OGTT | Total | 30 | 202 (45) | Breuer et al. (1975) |
| G | OGTT | gut GLI | 130 | 300 (30) | Shima et al. (1975) |
| G | OGTT | Total | 92 | 300 (30) | Vance et al. (1972) |
| G | OGTT | Total | 490 | 1590 (30) | Marco et al. (1972) |
| P | OGTT | gut GLI | 86 | 462 (15–60) | Botha et al. (1976) |
| S | Meal | gut GLI | 45 | 150 (150) | Holst et al. (1979) |
| Vt | OGTT | Total | 28 | 401 (50) | Russell et al. (1974) |
| Vs | OGTT | Total | 39 | 411 (50) | Russell et al. (1974) |

[a] The GLI values are in pmol equiv./l plasma
[b] The time, in min, at which the maximum values were measured is in parentheses
[c] C celiac disease;
Dgn gestational diabetics, normal weight;
Dgo gestational diabetics, overweight;
Da diabetics with IRI response to glucose load;
Db diabetics without IRI response to glucose load;
G gastrectomized subjects;
Gd gastrectomized subjects with dumping syndrome;
P patients with pancreatitis;
S subjects after jejunoileal shunt for obesity;
Vs patients after selective vagotomy;
Vt patients after total vagotomy
[d] Meal mixed meal; OGTT oral glucose tolerance test

In pregnant women, basal levels of gut GLI are variously reported as being normal (Hornnes et al. 1981) or elevated (Ghatei and Bloom 1981). In normal pregnant women, an oral glucose load or a mixed meal does not cause an increase in plasma gut GLI (Hornnes et al. 1981; Ghatei and Bloom 1981). In gestational diabetics, the gut GLI response to an oral glucose load is abolished in women of normal weight and impaired in overweight subjects (Hornnes et al. 1981). These changes in the response of plasma gut GLI to oral nutrients are normalized postpartum.

Pathologic conditions, such as celiac disease, which are associated with an accelerated or elevated passage of nutrients into the small intestine are associated with an increased response of plasma gut GLI to oral nutrients (Table 8). This association indicates that the amount of nutrients reaching the lower small intestine is of major importance in controlling the release of gut GLI. This direct effect of nutrients on gut GLI release is, however, probably modulated by hormonal and neural influences in such states as pregnancy and diabetes.

### V. Circulating Gut GLI Levels in the Perinatal Period

During the immediate postnatal period, the newborn has to adapt from receiving a steady flow of nutrients across the placenta to receiving an intermittent supply by mouth. The study of the gut hormone response to this dramatic alteration in the mode of nutrient supply may well provide evidence for the role of gut GLIs.

In term babies suffering from respiratory distress, the first feed of breast milk caused marked increases in gastrin and gut GLI, although GIP was unchanged (AYNSLEY-GREEN et al. 1977). A similar study was carried out in preterm infants in whom it was found that breast milk did not cause any alteration in the level of these hormones in arterial plasma (LUCAS et al. 1978). These observations indicate that functional maturation of the L-cells occurs around term. In the neonatal infant, a dextrose feed did not cause any significant increase in plasma gut GLI (AYNSLEY-GREEN et al. 1979). It is not known whether this is because the dextrose feed was totally absorbed in the upper small intestine, or whether the neonatal L-cell responds only to triglyceride.

Plasma levels of gut GLI are above adult levels in both preterm and term infants at birth, and these levels increase further during the immediate postnatal period (LUCAS et al. 1980). The maximum postnatal basal level of gut GLI occurred on day 6 of postnatal life in preterm infants and the maximum postprandial response was greatest 24 days after birth (LUCAS et al. 1980). This increase in the postnatal response was not caused by an increase in the maximum level of gut GLI, which remained unchanged, but was caused by a steady decrease in basal level. The time after feeding at which maximum gut GLI concentrations were observed also changed during the immediate postnatal period from 60 min at 2.5 days to 30 min at 24 days.

In the human during the first 6 weeks or so of extrauterine life, circulating levels of plasma gut GLI are very high and are further increased on feeding (perhaps only in response to the triglyceride component of the feed). These findings suggest that gut GLIs have a role in the developmental changes that occur in the neonate, and it has been proposed that gut GLIs in some way influence the maturation of the intestinal mucosa (LUCAS et al. 1980).

## J. Effects of Gut GLIs

### I. Postulated Effects

A wide range of effects has been ascribed to gut GLIs based on the circumstantial evidence that postprandial increases in circulating levels of gut GLI are associated with such events as an increase in insulin release or an inhibition of gastric acid secretion. SAMOLS et al. (1965) proposed that the increases in what is now known as gut GLI after an oral load could account for the increase in insulin release observed after oral administration of glucose. The observation that gut GLI is elevated in a single patient with reactive hypoglycemia (REHFELD and HEDING 1970) led to the proposal that gut GLI may be a pathogenic factor in reactive hypoglycemia.

In humans, pentagastrin-stimulated gastric acid secretion is strongly, and equally, inhibited by intrajejunal infusion of triglycerides, hypertonic glucose, and hypertonic sodium chloride (HOLST et al. 1976). Gut GLI levels were increased markedly after glucose administration and to a lesser extent after triglyceride administration, and the authors concluded that gut GLI could act as a glucose- and triglyceride-dependent enterogastrone in humans. It was also observed that neither glucose nor insulin levels were altered during the triglyceride infusion, and it was suggested that gut GLIs in humans are devoid of glycogenolytic and insulin-releasing effects in the presence of normoglycemia.

A single patient has so far been described with a renal tumor rich in $NH_2$ terminal GLI (GLEESON et al. 1971). This tumor was associated with elevated plasma levels of $NH_2$ terminal GLI, severe intestinal stasis, and small intestinal mucosal hypertrophy. These abnormalities disappeared following resection of the tumor. Extracts of the tumor contained large amounts (150 μg equiv./g wet weight) of $NH_2$ terminal GLI (GLEESON et al. 1971) with a molecular weight (as judged by gel filtration) of 7,000 daltons (BLOOM 1972). These findings of tumor-associated intestinal changes were taken to indicate (BLOOM 1972) that gut GLI could control the motor function and development of the intestine. This, and other observations, led BARROWMAN (1975) to suggest that gut GLI could be an "enterotrophin," a proposal that is supported by the perinatal changes in plasma gut GLIs found in infants (see Sect. H).

## II. Effects of Partially Purified Gut GLIs

Despite the clinical observations which indicate that gut GLIs may control intestinal development and function, studies with partially purified gut GLIs have been concentrated on the study of the material's glucagon-like metabolic effects. This concentration on what may turn out to be minor effects of gut GLI can be largely attributed to the fact that the assays for glucagon-like effects are quantitative and specific and require relatively little material.

The attribution of effects of crude gut extracts to the gut GLIs contained in the extracts is somewhat hazardous because of the possibility that biologically active peptides other than gut GLIs are present in the extracts. This difficulty was circumvented by GUTMAN et al. (1973), who showed that crude fractions of the rat intestine containing peak I and peak II gut GLIs possessed lipolytic activity and insulin-releasing activity in vitro, but only the peak II activities were suppressed by an $NH_2$ terminal-specific antiglucagon serum.

Partial purification of gut GLIs has been achieved by specific immunoabsorption. Porcine peak I gut GLI, purified by MURPHY et al. (1973) was devoid of lipolytic activity (LANGSLOW 1973). Canine peak I and II gut GLIs were found to have lipolytic activities which were suppressed by anti-glucagon serum (HORIGOME et al. 1977).

Porcine peak II gut GLI has been shown to be insulin releasing in the anesthetized dog (OHNEDA et al. 1976) although these findings were not confirmed by TA-

NAKA et al. (1977) who studied the effects of purified gut GLIs on the immunoreactive insulin (IRI) release of the perfused rat pancreas. TANAKA et al. (1977) did observe that extracts of the porcine fundus contained a small amount of GLI ($NH_2$ terminal, COOH terminal) and that affinity-purified fundic GLI possessed insulin-releasing activity. It is not clear whether this material was similar to the gastric glucagon found in cats and dogs, or whether it was a form of gut GLI.

An interesting study of the possible biologic effects of gut GLIs was carried out by BATAILLE et al. (1974), who used cell-free preparations of plasma membranes in order to study the effects of a crude preparation of porcine peak II gut GLI. It was shown that the crude preparation inhibited the binding of glucagon $^{125}I$ to its receptors on liver and adipose tissue plasma membranes, and induced a commensurate increase in adenylate cyclase. A similar demonstration of porcine peak II gut GLI binding to porcine hepatic glucagon receptors has been carried out by HOLST (1977). In all cases where crude gut GLI has been shown to exert glucagon-like metabolic effects, it has done so with less than its immunometric potency. (See Chap. 12 for further discussion.)

Studies of crude mammalian gut GLI thus support the concept that mammalian peak I is probably devoid of glucagon-like metabolic effects, but that the peak II is a weak glucagon agonist. Avian gut GLI with a size corresponding to that of the mammalian peak I material, does not stimulate the lipolysis of chicken (LANGSLOW 1973) or duck adipocytes (KRUG and MIAHLE 1977). Crude duck gut GLI inhibits the lipolytic effects of low molecular weight peptides present in the duck intestine (KRUG and MIAHLE 1977).

## III. Effects of Pure Gut GLIs

### 1. Synthetic Peptides

The sequences glicentin 33–69 and 37–69 have been synthesized (YANAIHARA et al. 1979; KANEKO et al. 1979) and their immunochemical (YANAIHARA et al. 1979) and biologic (KANEKO et al. 1980) characteristics studied. Synthetic glicentin 33–69 (which corresponds to the sequence of oxyntomodulin) and glicentin 37–39 (which corresponds to des-(1–4)-glucagon extended by the octapeptide) both bound to the glucagon receptor of rat hepatic plasma membranes with a molar potency of some 10%, relative to glucagon. Both glicentin fragments stimulated the adenylate cyclase of hepatic plasma membranes. Glicentin 33–69 had a molar potency of some 10% relative to glucagon, and exerted a maximum effect 80% of that of glucagon, whereas glicentin 37–69 was a very weak activator of adenylate cyclase and increased it to only 15% of the extent observed with glucagon. Both glicentin fragments inhibited, at a 100-fold molar excess, the submaximal stimulation of adenylate cyclase induced by glucagon. These findings indicate that glicentin 33–69 is a weak glucagon agonist, which may under certain circumstances be a glucagon antagonist, whereas glicentin 37–69 is a weak glucagon antagonist. The observation that glicentin 33–36 (homologous with glucagon 1–4) is not obligatory for the binding of the glicentin fragments to the glucagon receptor is in agreement with the finding that deletion of the first four amino acids from glucagon does not abolish its binding to the glucagon receptor (FRANDSEN et al. 1981). The overall characteristics of glicentin 33–69, i.e., its possession of $NH_2$ terminal GLI but lack of COOH terminal GLI, its being a weak glucagon agonist, its molecular weight, and its basic character are very similar to those of oxyntomodulin.

### 2. Oxyntomodulin

This peak II gut GLI binds to glucagon receptors and activates the adenylate cyclase of hepatic plasma membranes with molar potencies of 20% and 10%, respectively (Bataille et al. 1982b) relative to glucagon. The peptide has been termed oxyntomodulin because of the finding that it stimulates the formation of adenylate cyclase by plasma membrane fractions of the stomach oxyntic mucosa with a molar potency 20 times that of glucagon (Bataille et al. 1981 a).

### 3. Glicentin

Glicentin does not bind to the glucagon receptor of pig (Holst 1977) or rat (E. K.-Frandsen 1981, personal communication) hepatic plasma membranes, nor does glicentin $^{125}I$ bind to porcine hepatic plasma membranes (Holst 1977). The addition of glicentin to rat hepatocytes caused activation of adenylate cyclase and the release of glucose (Thieden et al. 1981). These effects were small (glicentin having 1% of the effect of molar equivalents of glucagon) and were probably caused by the release of glucagon-like peptides from glicentin by enzymatic digestion during the incubation. These findings indicate that, in complex systems, glicentin could be converted into peptides that have biologic effects other than those of glicentin itself.

Long-term administration of glicentin to rats (15 µg, three times daily for 10 days) caused a significant increase in the pancreatic content of somatostatin-like immunoreactivity (Yamada et al. 1980). There were no effects of glicentin on the pancreatic content of insulin and glucagon. The acute intravenous injection of glicentin into mice (0.7 nmol/kg) had no effect on the basal level of plasma IRI, but significantly decreased glucose-induced IRI release (Ahren and Lundquist 1980). The plasma levels of glicentin were not measured in these studies, but were probably higher than the levels observed in pigs after feeding, which reached a maximum of 160 pmol/l (Moody 1980). It is possible that this inhibitory effect of glicentin is only of pharmacologic importance, or was not a direct effect of low concentrations of glicentin on the pancreatic B-cell.

Glicentin has been infused at low doses (370 pmol $kg^{-1}$ $h^{-1}$) into unanesthetized rats with a permanent gastric fistula and found to inhibit gastrin-stimulated (25 µg $kg^{-1}$ $h^{-1}$) gastric acid release (Kirkegaard et al. 1982). It was not possible to measure the circulating levels of glicentin in the rats. The total doses infused were small, so it was unlikely that very high plasma concentrations were achieved. This observation that glicentin has an inhibitory effect on gastric acid secretion is the first experimental indication that glicentin, a peak I gut GLI, may have a role in the control of the function of the gastrointestinal tract. Taken together with the finding of oxyntomodulin's effect on the cyclase levels in preparations of rat gastric mucosa (see Sect. J.III.2), this indicates that the search for the biologic effects of gut GLI should be concentrated on the gastrointestinal tract, rather than on the glucagon-sensitive tissues, such as the liver and adipose tissue.

# K. Discussion

## I. Clinical Significance of Circulating Gut GLIs

The measurement by RIA of plasma gut GLIs has provided a background for studies with pure materials, and has indicated the type of effect gut GLIs may have in the whole organism. The findings that plasma gut GLIs are elevated in the basal state in the human infant, and that there is an increase in plasma gut GLI in response to a normal feed, indicate that gut GLIs are associated with (or are a causal agent of) the adaptive changes that occur in the intestinal mucosa in the immediate perinatal period. This concept that gut GLIs modify the development and function of the gastrointestinal tract is supported by the observation of intestinal stasis and mucosal hypertrophy in the single patient with a tumor secreting only $NH_2$ terminal GLI so far reported (see Sect. J).

Measurement of the peripheral level of gut GLIs can have clinical significance. As has been described by BESTERMAN et al. (1978), the establishment of a profile of the response of as many gut hormones as possible to a standard oral load of mixed nutrients provides a valid insight into the functioning of the intestinal tract as a whole and, in particular, the degree to which the nutrients pass unabsorbed from the upper small intestine into the lower small intestine. Measurement of gut GLIs alone is of less value, since one cannot discriminate between hypersecretion of gut GLIs caused by hyperactive L-cells and that due to an inappropriate passage of nutrients into the lower intestine.

There remain some questions concerning the plasma gut GLIs – particularly in humans – which warrant investigation. The first is that of the molecular form of gut GLIs in the mesenteric circulation draining the intestine compared with that of the gut GLIs in the peripheral circulation. The small peak II gut GLIs, exemplified by oxyntomodulin (glicentin 33–69) can have a spectrum of biologic effects which include that of being a weak glucagon agonist as well as (in the case of oxyntomodulin) that of stimulating the adenylate cyclase of gastric mucosa, with possible effects on such parameters as the secretion of gastric acid and pepsin. In contrast, the described biologic effects of glicentin overlap only slightly with those of glucagon. The net effect of the secretion of gut GLIs into the mesenteric vein in response to intraluminal nutrient will thus depend on: (a) the proportion of the two main forms of gut GLIs released; and (b) the proportion of the two forms which pass through the liver.

The second question is that of the colonic gut GLIs, which are present in large amounts, but are unlikely to be released into the bloodstream in response to the intracolonic presence of nutrients. Perhaps at this level of the intestine, the gut GLIs have a paracrine role, a concept which is supported by the observations (BUCHAN and POLAK 1980) that L-cells are frequently furnished with processes which increase the possibility of cell–cell transmission of gut GLIs within the mucosa itself.

## II. Structure–Function Relationships of Gut GLIs

The elucidation of the structure of glicentin (THIM and MOODY 1981) has provided the basis for a structural explanation of some of the biologic and immunochemical characteristics of the gut GLIs. The predictive structural analysis presented in Sect. D.IV indicates that, in the glicentin molecule, the sequence homologous with glucagon 24–29 may be involved in $\beta$-bend formation, resulting in the possible masking of the 24–29 immunoreactant by the COOH terminal octapeptide. The predictive analysis of the structure of glicentin does not indicate that any marked change will occur in the structure homologous with glucagon 22–26, that part of the glucagon molecule which plays a major role in the binding of glucagon to its receptor (FRANDSEN et al. 1981). The reduced affinity of synthetic glicentin 33–69 and of oxyntomodulin for the glucagon receptor indicates that some modification of the receptor-binding portion of the glucagon moiety has been induced by the presence of the COOH terminal extension. Whether this modification also extends to the $NH_2$ terminal portion of glucagon moiety, which is mainly responsible for the transduction of receptor occupancy into cyclase activation, cannot yet be decided.

The total failure of glicentin to bind to hepatic glucagon receptors is indicative of the $NH_2$ terminal moiety of glicentin (the GRPP moiety) inducing an alteration in the affinity of the glucagon 22–26 moiety for the glucagon receptor. This could be achieved, either by an induced change in the conformation of the region, or by the $NH_2$ terminal portion of glicentin masking possible contact between glucagon 22–26 and the glucagon receptor. The degree to which the gut GLIs possess glucagon-like biologic properties (in particular their reaction with the glucagon receptor) is thus a function of the extent to which the extensions to the glucagon sequence alter, first, the affinity of glucagon 22–26 for the receptor, and second, the conformation of the molecule around the sequence glucagon 1–4. The part of the gut GLI molecule that might be responsible for the specific effect of the gut GLIs remains to be identified.

The presence of the two Lys-Arg sequences in glicentin at either end of the glucagon moiety raises the question of whether glicentin is in fact stable in vivo in blood and the extracellular fluid, or whether it is broken down into smaller peptides each with its own specific biologic effects.

Although the human intestinal L-cell contains COOH terminal GLI (see Sect. F.II) there is no convincing evidence that the human lower intestine releases significant amounts of COOH terminal GLI into the circulation after pancreatectomy (B. TRONIER 1981, personal communication). This observation implies that either gut GLI is released into the circulation with a masked COOH terminal immunoreactant or that the COOH terminal GLI released is removed by the liver. Glicentin possesses the structural requirements for being proglucagon, and the presence of two fragments of glicentin in the A-cell (GRPP and glucagon) supports the contention that glicentin (or a molecule containing glicentin) is proglucagon.

The observation (STEFAN et al. 1981) that the pancreatic tissue of a primitive fish (the ratfish, *Hydrolagus colliei*) contains two types of A-like cells, one containing only $NH_2$ terminal GLI, the other containing $NH_2$ terminal and COOH terminal GLI (i.e., similar to the mammalian A-cell), identified a possible stage in the development of the primitive intestinal L-cell into the mammalian and avian A-cell which contains the same material that is cleaved after synthesis into the $NH_2$ terminal GRPP, the midportion glucagon and, possibly, the COOH terminal hexapeptide. One cannot help wondering whether this development of the L-cell into the A-cell occurred only in the pancreas and in the gastric mucosa of the cat and dog, or whether it occurs in all adult higher mammals. Such a partial development could account for the findings of COOH terminal GLI in the intestinal mucosa of many animal species, the detection of COOH terminal GLI in intestinal cells which are similar to L-cells, and the detection of COOH terminal GLI in the human colon and in colonic cancer cells.

## III. Role of Gut GLIs

The small amount of information concerning the biologic effects of purified gut GLIs does not permit the definition of the role of these peptides in the whole organism, but it does permit an identification of those areas of the functioning of the body as a whole in which the gut GLIs may play a modifying role. On balance, the peak II gut GLIs such as oxyntomodulin are unlikely to have a marked effect on tissues known to be sensitive to the metabolic effects of glucagon. This conclusion is reached on the basis of the demonstrated low molar potency of such peptides as glucagon agonists, and doubts about the presence of large amounts of small gut GLIs in the peripheral circulation.

It is more likely that the small gut GLIs, like the larger gut GLIs, exert their effect via specific gut GLI receptors on target cells (although these receptors may have some affinity for glucagon, in much the same way as glucagon receptors have some affinity for gut GLIs) and that these target cells are located in the gastrointestinal tract or the pancreas. The large, peak I, gut GLIs probably do not interact in the intact form with glucagon receptors to exert glucagon-like metabolic effects, but exert their effects (such as the inhibition of gastrin-stimulated gastric acid secretion) via specific receptors. It cannot be decided at present whether the peak I and II gut GLIs have discrete receptors, have some receptors in common (for the glucagon moiety of the molecule), or have a single common receptor.

In conclusion, the role of the gut GLIs in the whole organism remains to be established. There are grounds for proposing that the primary target organs of the gut GLIs are in the gastroenteropancreatic system and that the peptides probably act as modifiers of the system's function, as inducers of the system's postnatal development and maturation and, possibly, as weak glucagon agonists. The complete delineation of the role of the gut GLIs requires detailed investigation of the biologic activities of highly purified gut GLIs, establishment of the molecular forms of gut GLIs in the mesenteric and peripheral circulation, and studies of the way in which the effects of the gut GLIs are integrated with the effects of other regulatory peptides.

## References

Adelson JN, Nelback ME, Chang R, Glaser CB, Yates GB (1980) Chymodenin: between factor and hormone. In: Jerzy Glass GB (ed) Gastrointestineal hormones. Raven, New York, pp 387–396

Ahren B, Lundquist I (1980) Effects of glicentin on insulin secretion. Horm Metab Res 12:582–586

Assan R, Slusher N (1972) Structure/function and structure/immunoreactivity relationships of the glucagon molecule and related synthetic peptides. Diabetes 21:843–855

Aynsley-Green A, Bloom SR, Williamson DH, Turner RC (1977) Endocrine and metabolic response in the human newborn to first feed of breast milk. Arch Dis Child 52:291–295
Aynsley-Green A, Lucas A, Bloom SR (1979) The effect of freeds of differing composition on entero-insulin hormone secretion in the first hours of life in human neonates. Acta Paediatr Scand 68:265–270
Barrowman JA (1975) The trophic action of gastro-intestinal hormones. Digestion 12:92–104
Bataille D, Freychet P, Rosselin G (1974) Interactions of glucagon, gut glucagon, vasoactive intestinal polypeptide and secretin with liver and fat cell membranes: binding to specific sites and stimulation of adenylate cyclase. Endocrinology 95:713–721
Bataille D, Gespach C, Coudray AM, Rosselin G (1981a) Enteroglucagon: a specific effect on gastric glands isolated from the rat fundus. Evidence for an "oxyntomodulin" action. Biosci Rep 1:151–155
Bataille D, Gespach C, Tatemoto K, Marie JC, Caudray AM, Rosselin G, Mutt V (1981b) Bioactive enteroglucagon (oxyntomodulin): Present knowledge on its chemical structure and its biological activities. Peptides 2 Suppl 2:41–44
Bataille D, Coudray AM, Carqvist M, Rosselin G, Mutt V (1982a) Isolation of glucagon-37 (bioactive enteroglucagon/oxyntomodulin) from porcine jejuno-ileum. Isolation of the peptide. FEBS Lett 146:73–79
Bataille D, Tatemoto K, Gespach C, Jörnwall H, Rosselin G, Mutt V (1982b) Isolation of glucagon-37 (bioactive enteroglucagon/Oxyntomodulin) from porcine jejuno-ileum. Characterization of the peptide. FEBS Lett 146:79–85
Besterman HS, Sarsen DL, Johnston DI, Stewart JS, Guerin S, Bloom SR, Blackburn AM, Patel HR, Modigliani R, Mallinson CN (1978) Gut hormone profile in coeliac disease. Lancet 1:785–788
Bloom SR (1972) An enteroglucagon tumour. Gut 13:520–523
Bloom SR, Polak JM (1978) Gut hormone overview. In: Bloom SR (ed) Gut hormones. Churchill Livingstone, Edinburgh London, pp 3–18
Bloom SR, Roysten CMS, Thompson JPS (1972) Enteroglucagon release in the dumping syndrome. Lancet 2:789–791
Böttger I, Dobbs R, Faloona GR, Unger RH (1973) The effects of triglyceride absorption upon glucagon, insulin and gut glucagon-like immunoreactivity. J Clin Invest 52:2532–2541
Bordi C, Ravazzola M, Baetens D, Gorden P, Unger RH, Orci L (1979) A study of glucagonomas by light and electron microscopy and immunofluorescence. Diabetes 28:925–936
Botha JL, Vinik AJ, Brown JC (1976) Gastric inhibitory polypeptide (GIP) in chronic pancreatitis. J Clin Endocrinol Metab 42:791–797
Breuer RI, Zuckerman L, Hanck TW, Green W, O'Hara P, Lawrence AM, Foà PP, Matsuyama T (1975) Gastric operations and glucagon homeostasis. II. Glucagon and secretin. Gastroenterology 69:598–606
Bryant MG, Bloom SR (1979) Distribution of the gut hormones in the primate intestinal tract. Gut 20:653–659
Buchan AMJ, Polak JM (1980) The classification of the human gastroenteropancreatic endocrine cells. Invest Cell Pathol 3:51–71
Bussolati G, Capella C, Vassallo G, Solcia E (1971) Histochemical and ultrastructural studies on pancreatic A cells. Evidence for glucagon and non-glucagon components of the α granule. Diabetologia 7:181–188
Chance RE, Ellis RM, Bromer WW (1968) Porcine proinsulin: characterization and amino acid sequence. Science 161:165–167
Chance RE, Johnsen ML, Koppenberger JI (1974) Polypeptide from bovine, ovine, human and porcine pancreases. US Patent No. 3842063
Chou PY, Fasman GD (1974) Prediction of protein conformation. Biochemistry 14:222–245, Chou PY, Fasman GD (1975) The conformation of glucagon: predictions and consequences. Biochemistry 14:2536–2541
Christofides ND, Modlin IM, Fitzpatrick ML, Bloom SR (1979) Effect of motilin on the rate of gastric emptying and gut hormone release during breakfast. Gastroenterology 76:903–907

Chu LLH, Huang W-Y, Littledike ET, Hamilton JW, Cohn DV (1975) Porcine proparathyroid hormone. Identification, biosynthesis, and partial amino acid sequence. Biochemistry 14:3631–3635

Conlon JM, Samsen WK, Dobbs RE, Orci L, Unger RH (1979) Glucagon-like polypeptides in bovine brain. Diabetes 28:700–702

Czyzyk A, Heding LG, Malczewski B, Miedzinska E (1975) The effect of phenformin upon the plasma pancreatic and gut glucagon-like immunoreactivity in diabetics. Diabetologia 11:129–133

Dryburgh JR, Hampton SM, Marks V (1980) Endocrine pancreatic control of the release of gastric inhibitory polypeptide. A possible physiological role for C-peptide. Diabetologia 19:397–401

Fiocca R, Capella C, Buffa R, Fontana R, Solcia E, Hage E, Chance RE, Moody AJ (1980) Glucagon-, glicentin- and pancreatic polypeptide-like immunoreactivities in rectal carcinoids and related colorectal cells. Am J Pathol 100:81–92

Frame CM (1976) The contribution of the distal gastrointestinal tract to glucagon-like immunoreactivity secretion in the rat. Proc Soc Exp Med 152:667–670

Frame CM (1977) Regional release of glucagon-like immunoreactivity from the intestine of the cat. Horm Metab Res 9:117–120

Frandsen EK, Grønvald FC, Heding LG, Johansen NL, Lundt BF, Moody AJ, Markusen J, Vølund Aa (1981) Glucagon: structure-function relationships investigated by sequence deletions. Hoppe Seylers Z Physiol Chem 362:665–677

Garaud JC, Eloy R, Moody AJ, Stock C, Grenier JF (1980) Glucagon- and glicentin-immunoreactive cells in the human digestive tract. Cell Tissue Res 213:121–136

Ghatei MA, Bloom SR (1981) Enteroglucagon in man. In: Bloom SR, Polak JM (eds) Gut hormones. Churchill Livingstone, Edinburgh London, pp 332–338

Gleeson MH, Bloom SR, Polak JM, Henry K, Dowling RH (1971) Endocrine tumour in kidney affecting small bowel structure, motility and absorbtive function. Gut 12:773–782

Gráf L, Barát E, Cseh G, Sajgó M (1971) Amino acid sequence of porcine β-lipotropic hormone. Biochim Biophys Acta 229:276–278

Grimelius L, Capella C, Buffa R, Polak JM, Pearse AGE, Solcia E (1976) Cytochemical and ultrastructural differentiation of enteroglucagon and pancreatic-type glucagon cells of the gastrointestinal tract. Virchows Arch [Cell Pathol] 20:217–228

Gutman RA, Fink G, Voyles N, Selawry H, Penhos JC, Lepp A, Recant L (1973) Specific biologic effects of intestinal glucagon-like materials. J Clin Invest 52:1165–1175

Harris JI, Kenner EW (1978) In: Dayhoff (ed) Atlas of protein sequence and structure, vol 5, suppl 3. National Biomedical Research Foundation, Silver Spring, p 160

Heding LG (1971) Radioimmunological determination of pancreatic and gut glucagon in plasma. Diabetologia 7:10–19

Heding LG, Munkgaard Rasmussen S (1972) Determination of pancreatic and gut glucagon-like immunoreactivity (GLI) in normal and diabetic subjects. Diabetologia 8:408–411

Heding LG, Frandsen EK, Jacobsen H (1976) Glucagon. Structure-function relationship: immunologic. Metabolism 25:1327–1329

Holst JJ (1977) Extraction, gel filtration pattern, and receptor binding of porcine gastrointestinal glucagon-like immunoreactivity. Diabetologia 13:159–169

Holst JJ (1980) Evidence that glicentin contains the entire sequence of glucagon. Biochem J 187:337–343

Holst JJ, Christiansen J, Kühl K (1976) The enteroglucagon response to intrajejunal infusion of glucose, triglycerides and sodium chloride and its relation to jejunal inhibition of gastric acid. Scand J Gastroenterol 11:297–304

Holst JJ, Sorensen TIA, Andersen AN, Stadil F, Andersen B, Lauritsen KB, Klein HC (1979) Plasma enteroglucagon after jejuno-ileal bypass with 3:1 or 1:3 jejuno-ileo ratio. Scand J Gastroenterol 14:205–207

Horigome K, Ohneda A, Maruhama Y, Abe R, Hai Y (1977) Heterogeneity of extractable gut glucagon-like immunoreactivity (GLI) and its lipolytic activity. Horm Metab Res 9:370–374

Hornnes S, Kühl C, Lauritsen KB (1981) Gastrointestinal insulinotrophic hormones in normal and gestational-diabetic pregnancy: response to oral glucose. Diabetes 30:504–509
Ito S, Kobayashi S (1976) Immunohistochemical demonstration of glucagon and GLI-containing cells in the canine gut and pancreas. Arch Histol Jpn 39:193–202
Jacobsen H, Demandt A, Moody AJ, Sundby F (1977) Sequence analysis of porcine gut GLI-1. Biochim Biophys Acta 493:452–459
Jørgensen KH, Larsen UD (1972) Purification of 125-I glucagon in anion exchange chromatography. Horm Metab Res 4:223–224
Jörnvall H, Carlquist M, Kwaak S, Otte SC, McIntosh CHS, Brown JC, Mutt V (1981) Amino acid sequence and heterogenicity of gastric inhibitory polypeptide (GIP). FEBS Lett 123:205–216
Kåresen R, Tronier B, Aune S (1980) Immunoreactive glucagon and insulin C-peptide in man after resection of the pancreas and total pancreatectomy. Am J Surg 140:272–276
Kaneko T, Cheng PY, Toda G, Oka H, Yanaihara N, Yanaihara C, Mihara S, Nishida T, Kaise N, Shin S, Imagawa K (1979) Biological and binding actions of synthetic possible C-terminal fragments of glicentin in rat liver plasma membrane. In: Miyoshi A (ed) Gut peptides, secretion, function and clinical aspects. Elsevier North Holland, Amsterdam Oxford New York, pp 157–161
Kirkegaard P, Moody AJ, Holst JJ, Loud FB, Skov Olsen P, Christiansen J (1982) Glicentin inhibits gastric acid secretion in the rat. Nature 297:156–157
Knudsen JB, Holst JJ, Asness S, Johansen A (1975) Identification of cells with pancreatic-type and gut-type glucagon immunoreactivity in the human colon. Acta Pathol Microbiol Scand [A] 83:741–743
Krug E, Mialhe P (1971) Pancreatic and intestinal glucagon in the duck. Horm Metab Res 3:24–27
Krug E, Miahle P (1977) A possible role for gut GLI: an inhibitor of lipolysis. Horm Metab Res 6:465–469
Langslow DR (1973) The action of gut glucagon-like immunoreactivity and other intestinal hormones on lipolysis in chicken adipocytes. Horm Metab Res 5:428–432
Larsson L-I, Moody AJ (1980) Glicentin and gastric inhibitory polypeptide immunoreactivity in endocrine cells of the gut and pancreas. J Histochem Cytochem 28:925–933
Larsson L-I, Holst J, Håkanson R, Sundler F (1975) Distribution and properties of glucagon immunoreactivity in the digestive tract of various mammals: an immunohistochemical and immunochemical study. Histochemistry 44:281–290
Leduque P, Moody AJ, Dubois PM (1982) Ontogeny of immunoreactive glicentin in the human gastro-intestinal tract and endocrine pancreas. Regul Peptides 4:261–274
Li CH (1969) β-lipotrophin, a new pituitary hormone. In: La specificité zoologique des hormones hypophysaires et leur activités. Colloq Int Cent Natl Rech Sci 117:93–101
Li CH, Chung D (1976) Isolation and structure of an unatriacontapeptide with opiate activity from camel pituitary glands. Proc Natl Acad Sci USA 73:1145–1148
Like AA (1967) The ultrastructure of the secretory cells of the islets of Langerhans in man. Lab Invest 16:937–951
Ling N, Burgus R, Guillemin R (1976) Isolation, primary structure and synthesis of alpha-endorphin and gamma-endorphin, two peptides of hypothalamic-hypophyseal origin with morphinometic activity. Proc Natl Acad Sci USA 73:3942–3946
Lorén I, Alumets J, Håkanson R, Sundler F, Thorell J (1979) Gut-type glucagon immunoreactivity in nerves of the rat brain. Histochemistry 61:335–340
Lucas A, Bloom SR, Aynsley-Green A (1978) Metabolic and endocrine events at the time of the first feed of human milk in preterm and term infants. Arch Dis Child 53:731–736
Lucas A, Adrian TE, Christofides N, Bloom SR, Aynsley-Geen A (1980) Plasma motilin, gastrin and enteroglucagon and feeding in the human newborn. Arch Dis Child 55:673–677
Lund PK, Goodman RH, Jacobs JW, Habener JF (1980) Glucagon precursors identified by immunoprecipitation of products of cell-free translation of messenger RNA. Diabetes 29:583–586
Lund PK, Goodman RH, Habener JF (1981) Pancreatic preproglucagons are encoded by two separate mRNAs. J Biol Chem 256:6515–6518

Lund PK, Goodmann RH, Dee PC, Habener JF (1982) Pancreatic preproglucagon CDNA contains two glucagon-related coding sequences arranged in tandem. PNAS 79:345–349
Makman MH, Sutherland EW (1964) Use of liver adenyl cyclase for assay of glucagon in human gastrointestinal tract and pancreas. Endocrinology 75:127–134
Marco J, Baroja IM, Diaz-Ferros M, Villanueva ML, Valverde I (1972) Relationship between insulin and gut glucagon-like immunoreactivity (GLI) in normal and gastrectomised subjects. J Clin Endocrinol Metab 34:188–191
Marco J, Hedo JA, Villanueva ML (1977) Inhibition of intestinal glucagon-like immunoreactivity (GLI) secretion by somatostation in man. J Clin Endocrinol Metab 44:695–698
Markussen J, Sundby F (1970) Separation and characterization of glucagon-like immunoreactive components from gut extracts by electrofocusing. In: Peeters H (ed) Protides of the biological fluids. Pergamon, Oxford New York, pp 471–474
Matsuyama T, Foà PP (1974) Plasma glucose, insulin, pancreatic and enteroglucagon levels in normal and depancreatized dogs. Proc Soc Exp Biol Med 124:97–102
Matsuyama T, Hoffman WH, Dunbar JC, Foà NL, Foà PP (1975) Glucose, insulin, pancreatic glucagon and glucagon-like immunoreactive materials in the plasma of normal and diabetic children. Effect of initial insulin treatment. Horm Metab Res 7:452–456
Matsuyama T, Tanaka R, Shima K, Tarui S (1979) Anomeric specificity in the response of gut glucagon-like immunoreactive materials to glucose. Horm Metab Res 11:214–216
Matsuyama T, Namba M, Nonaka K, Tarui S, Tanaka R, Shima K (1980) Decrease in blood glucose and release of gut glucagon-like immunoreactive materials by bombesin infusion in the dog. Endocrinol Jpn SR 1:115–119
Moody AJ (1972) Gastrointestinal glucagon-like immunoreactivity. In: Lefebvre PJ, Unger RH (eds) Glucagon, molecular physiology, clinical and therapeutic implications. Pergamon, Oxford New York, pp 319–341
Moody AJ (1980) Gut glucagon-like immunoreactants (GLIs) and other enteric glucagon-like peptides. J Clin Pathol [Suppl 8] 33:58–62
Moody AJ, Frandsen EK, Jacobsen H, Sundby F, Orci L (1976) The structural and immunologic relationship between gut GLI and glucagon. Metabolism [Suppl 1] 25:1336–1338
Moody AJ, Jacobsen H, Sundby F, Frandsen EK, Baetens D, Orci L (1977) Heterogeneity of gut glucagon-like immunoreactants (GLIs). In: Foà PP, Bajaj JS, Foà NL (eds) Glucagon: its role in physiology and clinical medicine. Springer, Berlin Heidelberg New York, pp 129–135
Moody AJ, Jacobsen H, Sundby F (1978) Gastric glucagon and gut glucagon-like immunoreactants. In: Bloom SR (ed) Gut hormones. Churchill Livingstone, Edinburgh London, pp 369–378
Moody AJ, Frandsen EK, Jacobsen H, Sundby F (1979) Speculations on the structure and function of gut GLIs. In: Rosselin G, Fromageot P, Bonfils S (eds) Hormone receptors in digestion and nutrition. Elsevier/North Holland, Amsterdam Oxford New York, pp 55–64
Moody AJ, Thim L, Holst JJ, Lindkaer-Jensen S (1980) Porcine pancreatic glicentin-related peptide. Diabetologia 19:300
Moody AJ, Holst JJ, Thim L, Lindkaer-Jensen S (1981) Relationship of glicentin to proglucagon and glucagon in the porcine pancreas. Nature 289:514–516
Moxey PC, Helmstaeder V, Moody AJ, Garaud JC, Forssman WG (1982) Glucagon and glicentin immunoreactive cells in human colon. Cell Tissue Res 221:483–491
Murphy RF, Buchanan KD, Elmore DT (1973) Isolation of glucagon-like immunoreactivity of gut by affinity chromatography on anti-glucagon antibodies coupled to Sepharose 4B. Biochim Biophys Acta 303:118–127
Mutt V, Said SI (1974) Structure of porcine vasoactive intestinal octosapeptide. Eur J Biochem 42:581–589
Mutt V, Jorpes JE, Magnusson S (1972) Structure of porcine secretin. The amino acid sequence. Eur J Biochem 29:199–204
Noe BD, Baste CA, Bauer GE (1977) Studies on proinsulin and proglucagon biosynthesis and conversion at the subcellular level. II. Distribution of radioactive peptide hormones

and hormone precursors in subcellular fractions after pulse and pulse-chase incubation of islet tissue. J Cell Biol 74:589–604
O'Connor FA, Conlon JM, Buchanan K, Murphy RF (1979) The use of the perfused rat intestine to characterise the immunoreactivity released into serosal secretions following stimulation by glucose. Horm Metab Res 11:19–23
O'Connor KJ, Lazarus NR (1976) The purification and biological properties of pancreatic big glucagon. Biochem J 156:265–277
Ohneda A, Horigome K, Kai Y, Itabashi H, Ishii S, Yamagata S (1976) Purification of canine gut glucagon-like immunoreactivity (GLI) and its insulin releasing activity. Horm Metab Res 8:170–174
Orci L, Pictet R, Forssmann WG, Renold AE, Rouiller C (1968) Structural evidence for glucagon producing cells in the intestinal mucosa of the rat. Diabetologia 4:56–67
Patzelt C, Chan SJ, Duguid J, Hortin G, Keim P, Heinrikson RL, Steiner DF (1978) Biosynthesis of polypeptide hormones in intact and cell free systems. FEBS Proc Meet 47:69–78
Patzelt C, Tager HS, Carroll RJ, Steiner DF (1979) Indentification and processing of proglucagon in pancreatic islets. Nature 282:260–266
Perez-Castillo A, Blazquez E (1980) Tissue distribution of glucagon, glucagon-like immunoreactivity, and insulin in the rat. Am J Physiol 238:258–266
Polak JM, Bloom SR, Coulling J, Pearse AGE (1971) Immunofluorescent localization of enteroglucagon cells in the gastrointestinal tract of the dog. Gut 12:311–318
Pradayrol L, Jörvall H, Mutt V, Ribet A (1980) N-terminally extended somatostatin: the primary structure of somatostatin-28. FEBS Lett 109:55–58
Ravazzola M, Orci L (1980a) Transformation of glicentin-containing L cells into glucagon-containing cells by enzymatic digestion. Diabetes 29:156–158
Ravazzola M, Orci L (1980b) Glucagon and glicentin immunoreactivity are topologically segregated in the a-granule of the human pancreatic cell. Nature 284:66–67
Ravazzola M, Siperstein A, Moody AJ, Sundby F, Jacobsen H, Orci L (1979a) Glicentin: a precursor of glucagon? Life Sci 25:287–290
Ravazzola M, Siperstein A, Moody AJ, Sundby F, Jacobsen H, Orci L (1979b) Glicentin immunoreactive cells: their relationship to glucagon-producing cells. Endocrinology 105:499–508
Rehfeld JF, Heding LG (1970) Increased release of gut glucagon in reactive hypoglycaemia. Br Med J 2:706–707
Rekker RE (1977) The hydrophobic fragmental constant. Elsevier/North Holland, Amsterdam Oxford New York, p 300
Russell RCG, Thompson JPS, Bloom SR (1974) The effect of truncal and selective vagotomy on the release of pancreatic glucagon, insulin and enteroglucagon. Br J Surg 61:821–824
Sakurai H, Dobbs RE, Unger RH (1975) The effect of somatostatin on the response of GLI to the intraduodenal administration of glucose, protein and fat. Diabetologia 11:427–430
Samols E, Tyler J, Marri G, Marks V (1965) Stimulation of glucagon secretion by oral glucose. Lancet 2:1257–1259
Schusdziarra V, Rouiller D, Unger RH (1979) Oral administration of somatostatin reduces postprandial plasma triglycerides, gastrin and gut glucagon-like immunoreactivity. Life Sci 24:1595–1600
Schwartz TW, Gingerich RL, Tager HS (1980) Biosynthesis of pancreatic polypeptide. Identification of a precursor and a co-synthesised product. J Biol Chem 255:11494–11498
Shields D, Warren TG, Roth SE, Brenner MJ (1981) Cell-free synthesis and processing of multiple precursors to glucagon. Nature 289:511–514
Shima K, Sawazaki N, Morishita S, Tarui S, Nishihawa M (1975) Effect of phenformin on the response of plasma intestinal glucose-like immunoreactivity to oral glucose in gastrectomized subjects. Proc Soc Exp Biol Med 150:232–236
Snell CR, Smyth DG (1975) Proinsulin: a proposed threedimensional structure. J Biol Chem 250:6291–6295

Stefan Y, Ravazzola M, Orci L (1981) Primative islets contain two populations of cells with differing glucagon immunoreactants. Diabetes 30:192–195
Sundby F, Moody AJ (1980) Gut glucagon-like immunoreactants (gut GLIs): isolation, structure and possible role. In: Jerzy Glass GB (ed) Gastrointestinal hormones. Raven, New York, pp 307–313
Sundby F, Jacobsen H, Moody AJ (1976) Purification and characterisation of a protein from porcine gut with glucagon-like immunoreactivity. Horm Metab Res 8:366–371
Tager HS, Kramer KJ (1980) Insect glucagon-like peptides: evidence for a high molecular weight form in midgut from *Manduca sexta (L)*. Insect Biochem 10:617–619
Tager HS, Markese J (1979) Intestinal and pancreatic glucagon-like peptides. Evidence for identity of higher molecular weight forms. J Biol Chem 254:2229–2233
Tager HS, Steiner DF (1973) Isolation of a glucagon containing peptide. Primary structure of a possible fragment of proglucagon. Proc Natl Acad Sci USA 70:2321–2325
Tager HS, Emdin SO, Clark JL, Steiner DF (1972) Studies on the conversion of proinsulin to insulin. II. Evidence for a chymotrypsin-like cleavage in the connecting peptide region of insulin precursors in the rat. J Biol Chem 248:3476–3482
Tager HS, Hohenbohen M, Markese J, Dinerstein RJ (1980) Identification and localisation of glucagon related peptides in the rat brain. Proc Natl Acad Sci USA 77:6229–6233
Tanaka R, Matsuyama T, Shima K, Sawazaki N, Tarui S, Kumahara Y (1977) Insulin releasing activity of gastrointestinal glucagon-like immunoreactive materials in perfused rat pancreas. Endocrinol Jpn 6:575–579
Tanaka R, Matsuyama T, Shima K, Sawazaki N, Tarui S, Kamahara Y (1979) Half life of gastrointestinal glucagon-like immunoreactive materials. Endocrinol Jpn 26:59–63
Thieden H, Holst JJ, Dich J, Moody AJ, Sundby F (1981) Effect of highly purified porcine gut glucagon-like immunoreactivity (glicentin) on glucose release from isolated rat hepatocytes. Biochim Biophys Acta 675:163–170
Thim L, Moody AJ (1981) The primary structure of porcine glicentin (proglucagon). Regul Peptides 2:139–150
Thim L, Moody AJ (1982) Porcine glicentin related pancreatic peptide (GRPP): purification and chemical characterization of a glicentin-related pancreatic peptide (proglucagon fragment). Biochim Biophys Acta 703:134–141
Unger RH, Eisentraut AM, Sims K, McCall MS, Madison LL (1961) Site of origin of glucagon in dogs and humans. Clin Res 9:53–60
Unger RH, Ketterer H, Eisentraut AM (1966) Distribution of immunoassayable glucagon in gastrointestinal tissues. Metabolism 15:865–867
Unger RH, Ohneda A, Valverde I, Eisentraut AM, Exton J (1968) Characterization of the response of circulating glucagon-like immunoreactivity to intraduodenal and intravenous administration of glucose. J Clin Invest 47:48–65
Valverde I, Rigopoulou D, Exton J, Ohneda A, Eisentraut A, Unger RH (1968) Demonstration and characterization of a second fraction of glucagon-like immunoreactivity in jejunal extracts. Am J Med Sci 255:415–420
Valverde I, Ghiglione M, Matesanz R, Casado S (1979) Chromatographic pattern of gut glucagon-like immunoreactivity (GLI) in plasma before and during glucose absorption. Horm Metab Res 11:343–346
Vance JE, Stoll RW, Farin BL, Williams RH (1972) Exaggerated intestinal glucagon and insulin responses in human subjects. Metabolism 21:405–412
Weber E, Martin R, Voigt KH (1979) Corticotrophin/β-endorphin precursor: concomitant storage of its fragments in the secondary granules of the anterior pituitary corticotrophin/endorphin cells. Life Sci 25:1111–1118
Wider M, Marsuyama T, Dunbar JC, Foà PP (1976) Elevated gut glucagon-like immunoreactive material in human and experimental diabetes and its suppression by somatostatin. Metabolism [Suppl 1] 25:1487–1489
Wilander E, Portela-Gomes G, Grimelius L, Lundquist G, Skoog V (1977) Enteroglucagon and substance P-like immunoreactivity in argentaffin and argophil rectal carbinoids. Virchows Arch [Cell Pathol] 25:117–124

Yamada T, Solomon TE, Petersen H, Levin SR, Lewin K, Walsh JH, Grossman MI (1980) Effects of gastrointestinal polypeptides on hormone of endocrine pancreas in the rat. Am J Physiol 238 (Gastrointest Liver Physiol 1): G 526–G 530
Yanaihara N, Yanaihara C, Nishida T, Hiraiwa T, Mihara S, Sakagami M, Ozaki J, Imagawa K, Shin S (1979) Synthesis of glicentin- and proglucagon-related peptides and their immunological properties. In: Rosselin G, Fromageot P, Bonfils S (eds) Hormone receptors in digestion and nutrition. Elsevier/North Holland, Amsterdam Oxford New York, pp 65–68
Yoshida T, Kondo M (1980) Effect of acetylcholine on the secretion of gut glucagon immunoreactivity and gut glucagon-like immunoreactivity in pancreatectomised dogs. Endocrinol Jpn 27:33–38

# Production and Assay of Glucagon

CHAPTER 8

# Glucagon Preparations

M. PINGEL, B. SKELBAEK-PEDERSEN, and J. BRANGE

## A. Introduction

Discovered in 1923 by KIMBALL and MURLIN, glucagon is now a well-defined substance. Its chemistry and structure have been reviewed in detail in the present volume by BROMER (Chap. 1) and BLUNDELL (Chap. 3) respectively. As emphasized by BROMER (Chap. 1), the primary structures of porcine, bovine, and human glucagon are identical, but different primary structures have been reported for guinea pig, avian, and piscine glucagons. The present chapter will summarize the production of glucagon, its pharmaceutical preparations, the physicochemical and bioassay methods available for its measurement, and the factors affecting its stability and timing of action.

## B. Production

Glucagon can be produced synthetically (WÜNSCH 1967; LUNDT et al. 1979; see also Chap. 2). Since, however, the primary structures of porcine, bovine, and human glucagon are identical, synthetic glucagon presumably does not afford clinical advantages over nonsynthetic porcine and bovine glucagon, and, since until now the need for glucagon has not exceeded the amount that can be produced as a by-product of insulin production, glucagon for clinical use will probably continue to be produced from animal pancreata originating, generally from beef and pork.

The pancreata are removed at the slaughterhouse, freed of attached fat, and immediately placed in containers for deep-freezing to prevent bacterial and enzymatic decomposition. The tightly packed glands are then transported deep-frozen to the insulin factories. The procedures for collection, storage, and transport have proved to be of great importance in increasing insulin and glucagon yields from the pancreas.

The manufacture of glucagon from pancreas to crystals is illustrated in Fig. 1. The deep-frozen blocks of pancreas are minced by machine and extracted with acid, and aqueous ethyl alcohol. The acid reaction has the effect of inhibiting the action of the proteolytic enzymes. After centrifugation to remove suspended fat material, the acid–alcohol extract is neutralized and the precipitate is removed by filtration and discarded. The supernatant liquid is then acidified and the alcohol is removed by vacuum evaporation. The proteins in the concentrate are salted out, usually by means of sodium chloride at weak acid pH. The isolated proteins are dissolved in dilute acid, and from this solution insulin and some other proteins, including glucagon, are precipitated by adjusting the pH to the isoelectric point of

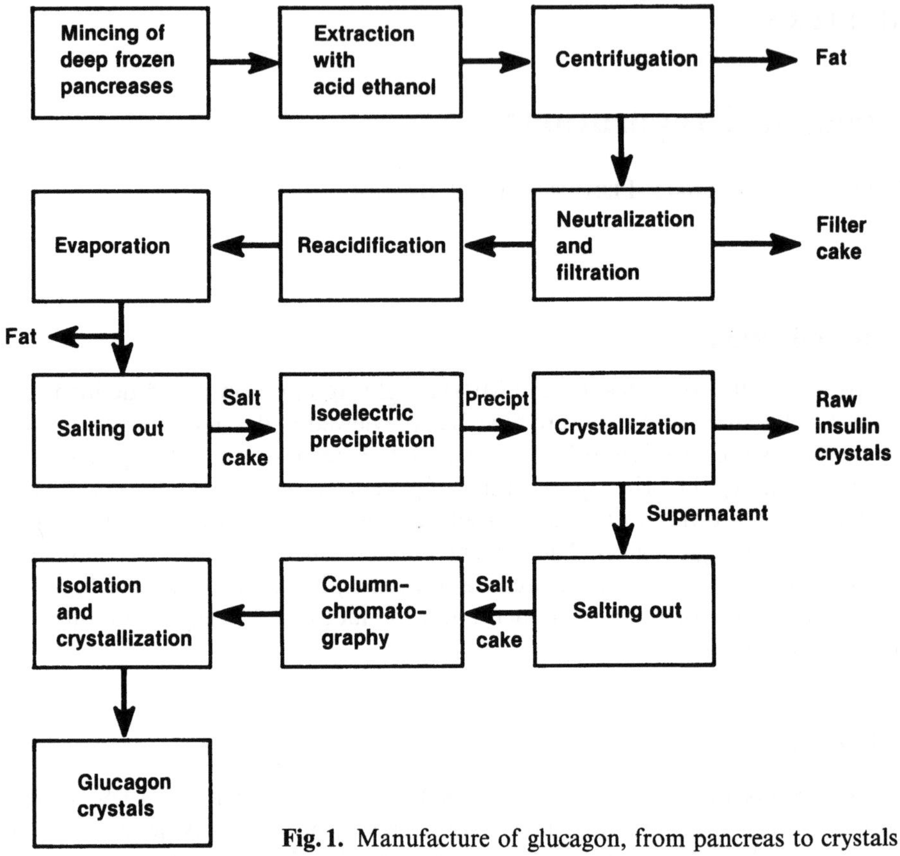


**Fig. 1.** Manufacture of glucagon, from pancreas to crystals

insulin. The precipitate is dissolved and the insulin is crystallized as rhombohedral crystals from a citrate buffer containing medium. The glucagon-containing supernatant liquid from this first crystallization of insulin is salted out by addition of sodium chloride. The precipitate is redissolved in dilute hydrochloric acid and purified several times by column chromatography on Sephadex gel. The purified glucagon is isolated by isoelectric precipitation and salting out. The glucagon salt cake is dissolved in dilute hydrochloric acid and the glucagon crystallizes at 4 °C overnight after the addition of dilute sodium hydroxide to pH 6. The glucagon precipitates as rhombic dodecahedral crystals. The pharmacopoeia (*BP* 1980; *USP* 1980) standards for glucagon are given in Table 1.

## C. Pharmaceutical Preparations

The rhombic dodecahedral glucagon crystals prepared from beef and pork pancreata are used for the preparation of pharmaceutical glucagon preparations. Glucagon preparations can be divided into the rapidly acting acid glucagon solutions (glucagon injection 1 mg and glucagon injection 10 mg) and the more slowly acting neutral glucagon suspensions (zinc glucagon 10 mg and zinc protamine glucagon 10 mg).

**Table 1.** The pharmacopeia (*BP* 1980; *USP* 1980) standards for glucagon

| | *BP* 1980 | *USP* 1980 |
|---|---|---|
| Species | Porcine or bovine | Porcine or bovine |
| Identification | A. Causes a rise of blood glucose<br>B. Ability to crystallize as rhombic, dodecahedral crystals | Property of increasing the concentration of glucose in blood |
| Light absorption | $E^{0.025\%}_{1\,cm}$ = 0.52–0.62 measured at the maximum at about 276 nm | |
| Related peptides | Content of monodesamidoglucagon ≦20%<br>Content of other related peptides (e.g., insulin) ≦0.2% | |
| Zinc content | ≦ 0.15% | ≦ 0.05% |
| Nitrogen content | 16.0%–18.5% | 16.0%–18.5% |
| Loss on drying | ≦10% | ≦10.0% |
| Potency | 80%–125% of stated potency; fiducial limits 64%–156% of stated potency | |

**Table 2.** Composition of the lyophilized content in one vial of the glucagon preparations and the volume of solvent normally used to dissolve/suspend the lyophilized content

| Glucagon preparation | Glucagon (IU) | Lactose (mg) | Volume of solvent normally used (ml) |
|---|---|---|---|
| Glucagon injection 1 mg | 1 | 107 or 49 | 1 |
| Glucagon injection 10 mg | 10 | 140 | 10 |
| Zinc glucagon 10 mg[a] | 10 | 125 | 10 |
| Zinc protamine glucagon 10 mg[a] | 10 | 125 | 5 |

[a] Contains in addition 3.4 mg $CH_3COONa$; $3H_2O$ and 0.5 mg $ZnCl_2$

Glucagon injections 1 mg and 10 mg are prepared by lyophilization in vials of an acid solution of glucagon, hydrochloric acid, and lactose. To make the preparations ready for injection the lyophilized content is redissolved in the accompanying solvent. Zinc glucagon 10 mg and zinc protamine glucagon 10 mg are prepared by lyophilization in vials of a suspension of rhombic dodecahedral zinc glucagon crystals in a neutral solution of zinc chloride, sodium acetate, and lactose. The suspended zinc glucagon crystals contain about 0.9% zinc, equivalent to 1 mol zinc to 2 mol glucagon. To make the preparations ready for injection, the lyophilized content is resuspended in the accompanying solvent. The only difference between zinc glucagon and zinc protamine glucagon is that the solution used for resuspending the lyophilized preparation in order to obtain zinc protamine glucagon contains in addition 0.05% protamine. The composition of the glucagon preparations is given in Tables 2 and 3.

**Table 3.** Composition of the dissolved/suspended glucagon preparations ready for injection

| Glucagon preparation | Retarding substance | Physical state of glucagon | pH | Buffer | Isotonicum | Preservative |
|---|---|---|---|---|---|---|
| Glucagon injection 1 mg | None | Dissolved | ~3 | None | 10% Lactose or 5% Lactose + 1.6% glycerin | None or 0.2% phenol |
| Glucagon injection 10 mg | None | Dissolved | ~3 | None | 1.4% Lactose + 2% glycerin or 1.4% Lactose + 1.6% glycerin | 0.1% Methylparaben + 0.015% propylparaben or 0.2% phenol |
| Zinc glucagon 10 mg | 0.005% zinc | Crystalline | 6–7 | 0.136% $CH_3COONa$; $3H_2O$ | 1.25% Lactose + 2% glycerin | 0.1% Methylparaben |
| Zinc protamine glucagon 10 mg | 0.015% zinc + 0.05% protamine | Crystalline | 6–7 | 0.136% $CH_3COONa$; $3H_2O$ | 2.5% Lactose + 1.6% glycerin | 0.1% Methylparaben |

The glucagon preparations are analyzed by several analytic methods to establish the purity, identity, composition, and the biologic activity of the preparation (*BP* 1980; *USP* 1980). The insulin content of the glucagon used for the pharmaceutical preparations is now in the range 0.01%–0.02%, as determined by radioimmunoassay according to HEDING (1972).

## D. Assays

### I. Physicochemical Methods

Methods dependent on the physical and chemical properties of glucagon are useful with purified preparations, but require confirmation and correlation with other methods, including bioassay. Such procedures are applicable to the determination of the concentration of crystalline glucagon in solution, to the estimation of purity and, in some instances, to the separation of glucagon from complex mixtures. Most of the methods have not been developed as quantitative assay procedures, although all may be suitable for such development.

#### 1. Ultraviolet Absorption

The concentration of crystalline glucagon in aqueous solution may readily be obtained by measurement of the ultraviolet (UV) absorption at the absorption maximum 278 nm, followed by comparison with the $E_{1\,\text{cm}}^{0.1\%}$ value of 2.38 (GRATZER et al. 1967). By another photometric method of analysis (*BP* 1980) the absorption at 244 nm of an acid (pH 1–2) dilution of the sample is subtracted from that of a basic (pH ~ 13) dilution. The difference in molar absorptivity of these solutions at 244 nm is 21,900; $\Delta E_{1\,\text{cm}}^{0.1\%} = 6.30$. This difference in absorption is due to a shift of the absorption bands of un-ionized tyrosine at 223 and 275 nm to 240 and 295 nm, respectively, upon ionization. This results in a maximum of the difference spectrum at 244 nm (WETLAUFER 1962). As glucagon contains two tyrosine groups, the difference in molar absorptivity corresponds to a difference of 10,950 for each tyrosine group. This is in good agreement with the results of HERMANS (1962), who for a group of hemoproteins found a mean difference in molar absorptivity of 11,000 for each tyrosine group. The advantage of this procedure is that the absorption of other UV-absorbing compounds, with the absorption at 244 nm independent of pH, will be eliminated.

#### 2. Electrophoretic Methods

Starch electrophoresis has been employed by STAUB et al. (1955) to provide data on purity of glucagon. LIGHT and SIMPSON (1956) have suggested that paper electrophoretic separation of insulin and glucagon may be useful in determining small amounts of glucagon in insulin. LOCHNER et al. (1964) described a cellulose acetate electrophoresis method that is useful in preparing small amounts of glucagon from tissues and in helping to identify glucagon in crude extracts. Agar gel electrophoresis has been employed by ZIEGLER and LIPPMANN (1968) for the separation of insulin and glucagon. SUNDBY and MARKUSSEN (1971) used isoelectric focusing for isolation and analysis of glucagon from rat pancreata. Polyacrylamide gel electrophoresis has been used by BROMER et al. (1971, 1972) for analysis of crystalline bovine and porcine glucagon; by SUNDBY and MARKUSSEN (1972), SUNDBY et al. (1972, 1974), MARKUSSEN et al. (1972), and POLLOCK and KIMMEL (1975) for analysis of glucagon from several species, and by CALAM and STORRING (1975) to study the heterogeneity of glucagon and changes in composition resulting from thermal degradation.

### 3. Chromatographic Methods

Cation exchange chromatography on Amberlite IRC 50 in urea-containing buffers has been used by COLE (1960) for purification and analysis of glucagon samples. WIESEL et al. (1963) demonstrated that glucagon and insulin can be separated from serum proteins by descending chromatography. HRUBY and GROGINSKY (1971) described the purification of glucagon by partition chromatography on Sephadex gel. BROMER et al. (1972) used cation exchange chromatography on DEAE-cellulose for purification of crystalline bovine glucagon and monodesamidoglucagon. EDVINSSON et al. (1972) have investigated thin layer chromatography of histidylpeptides, including glucagon, by using their o-phthalaldehyde-induced fluorescence to detect the compounds. POLLOCK and KIMMEL (1975) used a combination of cation and anion exchange chromatography on CM-Sephadex and on QAE-Sephadex for isolation of chicken glucagon.

### 4. High Pressure Liquid Chromatography

High pressure liquid chromatography (HPLC) has been used by BURGUS and RIVIER (1976) for the separation of small peptides, including glucagon and peptide analogs, which are closely related in structure. FISCHER et al. (1978) described a procedure for rapid separation of insulin, glucagon, and somatostatin by HPLC on glycophase-coated, controlled-porosity glass. Reverse-phase HPLC has been applied to the analysis of peptides and proteins, including glucagon, by the addition of hydrophilic or hydrophobic ion-pairing reagents to the mobile phase (HANCOCK et al. 1978). BATAILLE et al. (1979) used a solvent system containing trifluoroacetic acid diethylamine for analysis of glucagon, vasoactive intestinal peptide (VIP), and secretin by HPLC. BIEMOND et al. (1979) used gradient elution HPLC for the quantitative determination of glucagon and other polypeptides. Thirty-two hormonal polypeptides and nine proteins, including glucagon, have been used by O'HARE and NICE (1979) to evaluate the potential of HPLC on alkylsilane-bonded silica for separating and recovering biologically active compounds of this type. MEEK (1980) has investigated the retention time of several peptides, including glucagon, in reverse-phase HPLC.

## II. Bioassay Methods

Several in vivo and in vitro bioassay methods for glucagon have been developed. In this section, only the in vivo bioassays based on the hyperglycemic effect of glucagon will be mentioned. The in vitro bioassays for glucagon have been reviewed by LUYCKX and LEFEBVRE (1976). The classical means of detection and assay of glucagon is based on the injection of the sample into an experimental animal and the subsequent measurement of the rise in blood glucose. The methods differ primarily in the species of animal employed, in the experimental design, and in the procedure for handling the data.

ROWLINSON and LESFORD (1951) have proposed an assay based on the hyperglycemic response of glucagon in rabbits. Based on their results, a potency determination of glucagon necessitates the use of about 300 rabbits to obtain a confidence interval ($P=0.95$) of about $\pm 25\%$. Pretreatment of rabbits with cortisone has been used by HELMER and ROOT (1954) and by LAZARUS et al. (1957) in connection with the demonstration of a glucagon content in insulin preparations. VOLK and LAZARUS (1959) have suggested that cortisone pretreatment of rabbits could be used in an assay of glucagon and, finally, TARDING et al. (1969 a) have developed such an assay where a confidence interval ($P=0.95$) of about $\pm 20\%$ can be obtained by employing 32 rabbits.

An assay in anesthetized cats has been described by STAUB and BEHRENS (1954) where the test substance and the glucagon standard are administered by intravenous injection. The assay is performed according to a crossover design with 2 h intervals and a confidence interval ($P=0.95$) of about $\pm 20\%$ necessitates the use of 16 cats. The use of chickens (BEEKMAN 1958) and of cortisone-primed, adrenalectomized rats (HOLT et al. 1956) in a glucagon assay has also been suggested. ANNABLE et al. (1974) describe the use of a twin crossover assay of glucagon in mice in connection with bioassays of the first international standard for glucagon.

The chief virtues of the in vivo bioassay methods lie in the simplicity of the procedures. The difficulties encountered with such methods include individual variation, lack of specificity, slowness, and the need for large numbers of animals. Despite these objections, the only two methods employed at present for bioassay of glucagon intended for parenteral use in humans are the two in vivo bioassay methods described in the pharmacopoeias (*USP* 1980; *BP* 1980). One is a slight modification of the method of STAUB and BEHRENS (1954), in which fasted, anesthetized cats are injected at two dose levels in a modified Latin square design. The other (TARDING et al. 1969 a) uses fasted cortisone-pretreated rabbits in a twin crossover design somewhat similar to the classical insulin assay. Both bioassays are based on the four-point (high dose–low dose), parallel-line approach, giving confidence intervals ($P=0.95$) in the range $\pm 20\%$ when 16–32 animals are used.

## E. Stability

The stability of the first international standard for glucagon has been investigated by CALAM and STORRING (1975) using polyacrylamide gel electrophoresis. They found that thermal degradation of the freeze-dried standard at 37° and 45 °C resulted in some deamidation as well as production of an unidentified component which is electrophoretically distinct. The standard is prepared somewhat similarly to the pharmaceutical glucagon preparations and contains a freeze-dried mixture of 1.5 mg glucagon, 5 mg lactose, 0.25 mg sodium chloride, and sufficient hydrochloric acid to give a pH $\sim 3$.

**Table 4.** Biologic potency of glucagon injection 1 mg, glucagon injection 10 mg, and zinc glucagon 10 mg after storage of the lyophilized preparation for different periods at 4°, 15°, 25°, or 37°C. The figures stated are the potencies obtained, expressed as percentages of initial biologic potency with 95% confidence limits in brackets

| Temperature (°C) | Glucagon preparation | Potency (%) 1 year | 2 years | 3 years | 4 years | 5 years |
|---|---|---|---|---|---|---|
| 4 | Glucagon injection 1 mg | 110 ( 88–137) | 90 ( 75–107) | 103 ( 78–136) | 99 (82–120)<br>96 (73–127) | 103 ( 83–128) |
| | Glucagon injection 10 mg | | 103 ( 79–135) | 100 ( 80–124) | 88 (74–105)<br>90 (67–119)<br>112 (85–149) | |
| | Zinc glucagon 10 mg | | 101 ( 88–116) | 83 ( 63–108)<br>93 ( 76–114) | 97 (80–117)<br>116 (88–154) | 112 ( 94–134)<br>106 ( 82–137) |
| 15 | Glucagon injection 1 mg | 115 ( 97–137) | 83 ( 60–112) | 83 ( 65–103)<br>98 ( 84–114)<br>109 ( 92–130) | 104 (82–132) | 75 ( 60– 94) |
| | Glucagon injection 10 mg | | | 97 ( 73–129)<br>81 ( 68– 97) | 107 (86–133)<br>90 (75–109)<br>118 (91–154) | |
| | Zinc glucagon 10 mg | | | 95 ( 78–116) | 95 (78–116) | 99 ( 79–125)<br>115 (101–132) |
| 25 | Glucagon injection 1 mg | 97 ( 79–118) | 94 ( 70–127)<br>115 ( 89–150) | 63 ( 53– 75)<br>95 ( 80–113)<br>84 ( 69–101) | | |
| | Glucagon injection 10 mg | 94 ( 75–118) | 121 (105–140) | 93 ( 82–105)<br>84 ( 67–105)<br>98 ( 81–119) | 100 (81–123)<br>101 (75–134) | |
| | Zinc glucagon 10 mg | | 95 ( 71–127)<br>97 ( 78–121) | 116 ( 90–151)<br>127 (100–163) | 105 (82–135)<br>123 (95–161) | 44 ( 38– 51)<br>102 ( 85–124) |
| 37 | Glucagon injection 1 mg | 74 ( 58– 94)<br>100 ( 81–123) | 104 ( 89–122) | | | |
| | Glucagon injection 10 mg | 126 (106–149)<br>101 ( 84–121) | | | | |
| | Zinc glucagon 10 mg | 97 ( 74–128) | 26 ( 18– 38) | | | |

**Table 5.** Biologic potency of glucagon injection 1 mg, glucagon injection 10 mg, zinc glucagon 10 mg, and zinc protamine glucagon 10 mg after storage of the lyophilized preparation dissolved/suspended in the accompanying solvent for different periods at 4°, 15°, or 25°C. The figures stated are the potencies obtained, expressed as percentages of initial biologic potency with 95% confidence limits in brackets

| Temperature (°C) | Glucagon preparation | Potency (%) | | | | | |
|---|---|---|---|---|---|---|---|
| | | 1 month | 2 months | 3 months | 4 months | 6 months | 1 year |
| 4 | Glucagon injection 1 mg | 99 (81–121) | 88 (67–113) | 101 (79–127) | | | |
| | Glucagon injection 10 mg | 92 (71–119) | 94 (76–117) | | 79 (62–101) | 70 (52– 95) | |
| | | | | 102 (80–131) | 104 (87–124) | 100 (74–136) | |
| | Zinc glucagon 10 mg | | | | | | 87 (68–111) |
| | Zinc protamine glucagon 10 mg | | | 114 (95–137) | | 104 (83–132) | 89 (73–108) |
| 15 | Glucagon injection 1 mg | | | 87 (74–103) | | | |
| | Glucagon injection 10 mg | | 97 (77–123) | | | | |
| | Zinc glucagon 10 mg | | | 70 (52– 95) | | | |
| | Zinc protamine glucagon 10 mg | | 101 (81–126) | 89 (73–109) | | | |
| 25 | Glucagon injection 1 mg | 99 (81–121) | 82 (64–102) | 74 (60– 90) | | | |
| | Glucagon injection 10 mg | | 63 (49– 81) | | | | |
| | Zinc protamine glucagon 10 mg | 111 (93–133) | | 83 (71– 97) | | | |

Stability studies have been performed on the pharmaceutical glucagon preparations after storage for different periods at 4°, 15°, 25°, and 37 °C using the lyophilized preparations before and after dissolution/suspension in the accompanying solvent. The biologic potency has been determined by using the bioassay based on the hyperglycemic effect of glucagon in fasted, cortisone-pretreated rabbits (TARDING et al. 1969a; *BP* 1980). In Tables 4 and 5 are shown the results of these stability studies. Each result in the tables refers to one batch of glucagon preparation. The results of these bioassays show that the biologic potency of the glucagon preparations complies with the pharmacopeia standards for biologic assay after storage of the lyophilized preparations for at least 5 years at 4 °C, 4–5 years at 15 °C or 2–4 years at 25 °C, or after storage of the redissolved/resuspended preparations for at least 3 months at 4 °C, 2 months at 15 °C, or 1 month at 25 °C.

In addition to the biologic potency, several other parameters have been included in the stability studies. Appearance, including gel formation as an indicator of fibril formation, microscopic appearance, pH, glucagon content determined by the photometric method ($\Delta E$ at 244 nm) as described in Sect. D.I.1, ability to crystallize as rhombic dodecahedral crystals, etc., do not change significantly after storage of the lyophilized preparations for 5 years at 4 °C, 4 years at 15 °C, 2 years at 25 °C, or 0.5–1 year at 37 °C, or after storage of the redissolved/resuspended preparations for 3 months at 4 °C, 2 months at 15 °C, or 1 month at 25 °C.

All pharmaceutical glucagon preparations are prepared as a two-vial system, one with the lyophilized glucagon preparation and one with the solvent, because glucagon in acid solutions/neutral suspensions has only a few months' stability. In the acid solutions, the glucagon will, upon standing, easily change conformation, leading to the formation of fibrils with diminished biologic activity, and undergo deamidation also leading to diminished biologic potency (see Chap. 1). The kinetics of fibril formation have been investigated by BEAVEN et al. (1969) who found the formation of fibrils to be promoted by high glucagon concentration, presence of sodium chloride, and increased temperature (below 30 °C). The fibril formation starts after a long lag phase, which can be eliminated by nucleation of the solution with preformed glucagon fibrils. In neutral suspensions, the glucagon crystals will form aggregates upon standing.

## F. Timing of Action

The timing of action of different glucagon preparations has been studied in rats, rabbits, dogs, normal human subjects, diabetics, and patients with liver glycogen storage diseases.

Using inhibition of liver glycogen deposition, after feeding glucose to rats, as the parameter, SOKAL (1960) demonstrated a significantly prolonged action of zinc glucagon suspension in comparison with glucagon in solution. The effect on the blood glucose concentration in the rats was not included in the study. BROMER and CHANCE (1969) have developed a laboratory method in rabbits for comparing the duration of action of preparations of glucagon. Their data show that zinc glucagon suspension produces a significantly more prolonged hyperglycemia and depression of amino acids than does glucagon in solution. Glucagon in suspension without added zinc provided a longer duration of action than glucagon in solution. The ef-

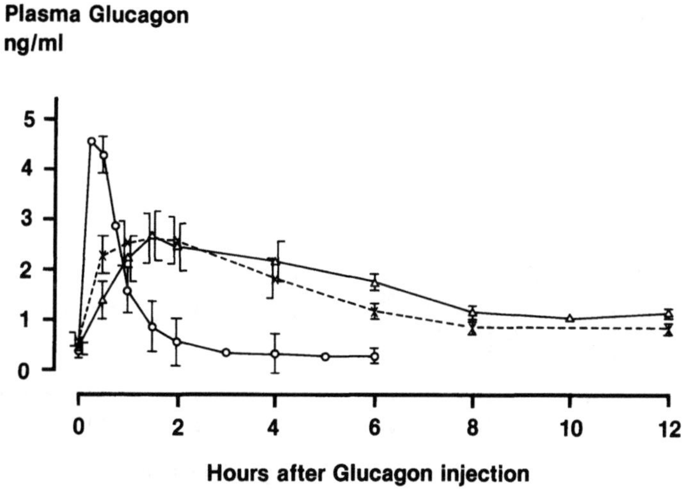


**Fig. 2.** Glucagon concentrations in the plasma of ten human subjects after intramuscular injection of 1 mg glucagon (*circles*), 5 mg zinc glucagon (*triangles*), and 5 mg zinc protamine glucagon (*crosses*). Each point and interval represents the mean ± SEM. (B. TRONIER 1976, unpublished work)

fects of zinc glucagon suspension, zinc protamine glucagon suspension, and glucagon in solution have been investigated by TARDING et al. (1969b) after subcutaneous injection to normal dogs. It was shown that the effect of zinc protamine glucagon suspension on the concentration of glucose, free fatty acids, and insulin in plasma was more prolonged than that of zinc glucagon suspension and glucagon in solution. Zinc glucagon suspension also had a more prolonged action than glucagon in solution.

The effect of zinc glucagon on blood glucose has been investigated by WEINGES (1959, 1960), who administered zinc glucagon suspension to normal human subjects and to diabetics and compared it with the effect of glucagon in solution. No hyperglycemia was observed after administration of zinc glucagon, while glucagon administration resulted in an elevation of the blood glucose. Supression of amino acids in blood was also measured. From these results the duration of effect of zinc glucagon was estimated to be about 9 h compared with 2–3 h for glucagon alone. LOWE et al. (1962) have compared the effect of zinc glucagon suspension and glucagon in solution in patients with liver glycogen storage disease. They found a consistent prolongation of the effect on blood glucose after administration of zinc glucagon, the duration of effect estimated as being about 8 h compared with about 2 h for glucagon. ASSAN and DELAUNAY (1972) have compared the effect of four different glucagon preparations with prolonged action, including zinc glucagon and zinc protamine glucagon with that of glucagon in solution in normal human subjects.

Studies in ten normal human subjects have been performed to investigate the timing of action of glucagon (acid solution) zinc glucagon (neutral suspension) and zinc protamine glucagon (neutral suspension). Figure 2 shows the changes in the plasma glucagon concentration after intramuscular injection of 1 mg glucagon, 5 mg zinc glucagon, and 5 mg zinc protamine glucagon, respectively. The glucagon

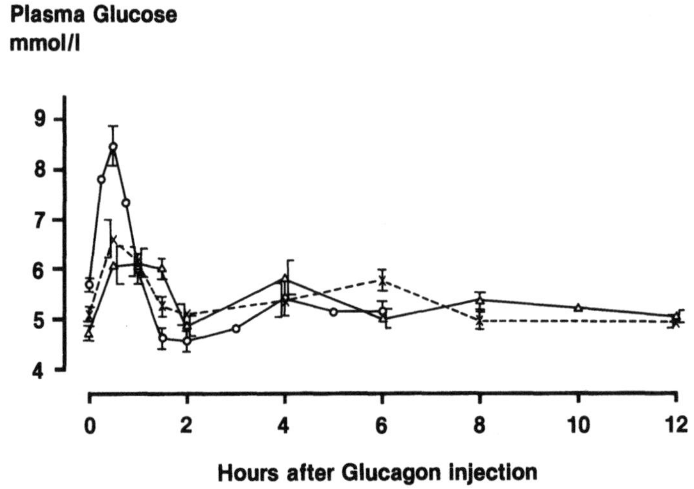


**Fig. 3.** Glucose concentrations in the plasma of ten human subjects after intramuscular injection of 1 mg glucagon (*circles*), 5 mg zinc glucagon (*triangles*), and 5 mg zinc protamine glucagon (*crosses*). Each point and interval represents the mean ± SEM. (B. TRONIER 1976, unpublished work)

concentration was measured by radioimmunoassay, according to HEDING (1971); 15 min after the injection of glucagon, the plasma glucagon concentration increased to about 13 times the initial level. The concentration then fell to the normal level within 2–3 h. After zinc glucagon and zinc protamine glucagon injections, there was a 4- to 7-fold increase in the glucagon concentration 1.5–2 h after the injection, with a slower return to about twice the initial level within 8–12 h. Analyses of variance followed by paired *t*-tests show statistically significant differences (5% level) between glucagon on the one hand and zinc glucagon and zinc protamine glucagon on the other at 30 and 90 min, and 2, 4, and 6 h after the injection.

Figure 3 shows the plasma concentrations of glucose in these experiments. The glucose concentration was analyzed by using the hexokinase/glucose-6-phosphate dehydrogenase method adapted to the Technicon AutoAnalyzer (SCHMIDT 1971). The increases in blood glucose after injection of glucagon, zinc glucagon, and zinc protamine glucagon were maximal at 30, 60, and 30 min, respectively. The peak value after glucagon administration was much higher than the peak value after zinc glucagon and zinc protamine glucagon administration, despite the fact that the last-mentioned were injected in doses five times that of glucagon alone. Analyses of variance followed by paired *t*-tests show statistically significant differences (5% level) between glucagon on the one hand and zinc glucagon and zinc protamine glucagon on the other, 30 and 90 min after the injection. Thus, the duration of action of zinc glucagon suspension and zinc protamine glucagon suspension is indeed prolonged compared with that of glucagon in solution.

## References

Annable L, Bangham DR, Salokangas AA, Storring PL (1974) The first international standard for glucagon. Acta Endocrinol (Copenh) 77:705–714

Assan R, Delaunay J (1972) Activités biologiques comparées de différents glucagons-retard et de glucagon ordinaire. Pathol Biol (Paris) 20:979–984

Bataille D, Besson J, Gespach C, Rosselin G (1979) High performance liquid chromatography of hormonal peptides: use of trifluoroacetic acid-diethylamide for their separation. In: Rosselin G (ed) Horm Recept Dig Nutr. Proc 2nd. Int Symp Horm Recent Dig Tract Physiol. Elsevier, Amsterdam Oxford New York, pp 79–88

Beaven GH, Gratzer WB, Davies HG (1969) Formation and structure of gels and fibrils from glucagon. Eur J Biochem 11:37–42

Beekman BE (1958) A bioassay for glucagon based on the hyperglycaemic response of the fowl. Poultry Sci 37:595–599

Biemond MEF, Sipman WA, Olivié J (1979) Quantitative determination of polypeptides by gradient elution high pressure liquid chromatography. J Liq Chromatogr 2:1407–1435

British Pharmacopoeia (1980) pp 210–211, 611, A 142–A 143

Bromer WW, Chance RE (1969) Zinc glucagon depression of blood amino acids in rabbits. Diabetes 18:748–754

Bromer WW, Boucher ME, Koffenberger JE (1971) Amino acid sequence of bovine glucagon. J Biol Chem 246:2822–2827

Bromer WW, Boucher ME, Patterson JM, Pekar AH, Frank BH (1972) Glucagon structure and function. I. Purification and properties of bovine glucagon and monodesamidoglucagon. J Biol Chem 247:2581–2585

Burgus R, Rivier J (1976) Use of high pressure liquid chromatography in the purification of peptides. In: Loffett A (ed) Peptides. Proc 14th Eur Peptide Symp, Brussels, pp 85–94

Calam DH, Storring PL (1975) The heterogeneity and degradation of glucagon studied by polyacrylamide gel electrophoresis. J Biol Stand 3:263–265

Cole D (1960) Ion exchange chromatography of glucagon in urea-containing buffers. J Biol Chem 235:2300–2301

Edvinsson L, Håkanson R, Rönnberg AL, Sundler F (1972) Separation of histidyl-peptides by thin-layer chromatography acid microspectrofluorometric characterization of their o-phthalaldehyde-induced fluorescence. J Chromatogr 67:81–85

Fischer LJ, Thies RL, Charkowski D (1978) High performance liquid chromatographic separation of insulin, glucagon, and somatostatin. Anal Chem 50:2143–2144

Gratzer WM, Bailey E, Beaven GH (1967) Conformational states of glucagon. Biochem Biophys Res Commun 28:914–919

Hancock WS, Bishop CA, Prestidge RL, Harding DRK (1978) Reversed phase, high-pressure liquid chromatography of peptides and proteins with ion-pairing reagents. Science 200:1168–1170

Heding LG (1971) Radioimmunological determination of pancreatic and gut glucagon in plasma. Diabetologia 7:10–19

Heding LG (1972) Determination of total serum insulin (IRI) in insulin-treated diabetic patients. Diabetologia 8:260–266

Helmer OM, Root M (1954) The effect of ACTH and cortisone on the hyperglycaemic response to glucagon. Endocrinology 54:338–342

Hermans J Jr (1962) Normal and abnormal tyrosine side-chains in various heme proteins. Biochemistry 1:193–196

Holt Cv, Holt Lv, Kröner B, Kühnen J (1956) Metabolic effects of $\alpha$-cell destruction. Ciba Found Coll Endocrinol 9:14–34

Hruby VJ, Groginsky CM (1971) Partition chromatography of glucagon and secretin on sephadex. J Chromatogr 63:423–428

Kimball CP, Murlin JR (1923) Aqueous extracts of pancreas. J Biol Chem 58:337–346

Lazarus SS, Volk BW, Lew H (1957) Cortisone potentiation of the hyperglycaemic action of insulin preparations. J Clin Endocrinol Metab 17:542–551

Light A, Simpson MV (1956) Studies on the biosynthesis of insulin. I. The paper chromatographic isolation of $^{14}C$-labelled insulin from calf pancreas slices. Biochim Biophys Acta 20:251–261

Lochner JDeV, Esterhuizen AC, Unger RH (1964) Separation of human insulin, glucagon and other pancreatic proteins. Diabetes 13:387–394

Lowe CU, Sokal JE, Mosovich LL, Sarcione EJ, Doray BH (1962) Studies in liver glycogen disease. Effects of glucagon and other agents on metabolic pattern and clinical status. Am J Med 33:4–19
Lundt BF, Grønvald FC, Johansen NL, Markussen J (1979) Synthesis of glucagon. Res Disclosure 246–247
Luyckx AS, Lefebvre PJ (1976) Pancreatic hormones. Assay of glucagon and glucagon-like substances. Bioassay and immunoassay of plasma glucagon and glucagon-like substances of gastrointestinal origin. In: Antoniades HN (ed) Hormones in human blood-detection and assay. Harvard University Press, Cambridge, MA London, pp 293–324
Markussen J, Frandsen EK, Heding LG, Sundby F (1972) Turkey glucagon: crystallization, amino acid composition and immunology. Horm Metab Res 4:360–363
Meek JL (1980) Prediction of peptide retention times in high-pressure liquid chromatography on the basis of amino acid composition. Proc Natl Acad Sci USA 77:1632–1636
O'Hare MJ, Nice EC (1979) Hydrophobic high-performance liquid chromatography of hormonal polypeptides and proteins on alkylsilane-bonded silica. J Chromatogr 171:209–226
Pollock HG, Kimmel JR (1975) Chicken glucagon: isolation and amino acid sequence studies. J Biol Chem 250:9377–9380
Rowlinson HR, Lesford JM (1951) An in vivo test for the hyperglycaemic glycogenolytic factor in insulin. J Pharm Pharmacol 3:887–896
Schmidt FH (1971) Methoden der Harn- und Blutzuckerbestimmung. II. Blutzucker. In: Pfeiffer EF (ed) Handbuch des Diabetes mellitus. Lehmanns, Munich, p 938
Sokal JE (1960) The duration of glucagon effect. Endocrinology 67:774–783
Staub A, Behrens OK (1954) The glucagon content of crystalline insulin preparations. J Clin Invest 33:1629–1633
Staub A, Sinn L, Behrens OK (1955) Purification and crystallization of glucagon. J Biol Chem 214:619–632
Sundby F, Markussen J (1971) Isolation, crystallization and amino acid composition of rat glucagon. Horm Metab Res 3:184–187
Sundby F, Markussen J (1972) Rabbit glucagon: isolation, crystallization and amino acid composition. Horm Metab Res 4:56
Sundby F, Frandsen EK, Thomsen J, Kristiansen K, Brunfeldt K (1972) Crystallization and amino acid sequence of duck glucagon. FEBS Lett 26:289–293
Sundby F, Markussen J, Danho W (1974) Camel glucagon: isolation, crystallization and amino acid composition. Horm Metab Res 6:425
Tarding F, Nielsen P, Keiser-Nielsen B, Nielsen AaV (1969a) Biological assay of glucagon in rabbits. Diabetologia 5:146–154
Tarding F, Nielsen P, Pingel M, Vølund Aa (1969b) Biological and chemical properties of two glucagon preparations with prolonged action. Eur J Pharmacol 7:206–210
United States Pharmacopeia XX (1980) pp 350–351
Volk BW, Lazarus SS (1959) Glucagon content of glucagon-insulin mixtures. Diabetes 8:128–131
Weinges KF (1959) Der Einfluß eines protrahiert wirkenden Glucagons auf den Blutzucker, das anorganische Serumphosphat und die Gesamtaminosäuren im Serum. Arch Exp Pathol Pharmakol 237:22–26
Weinges KF (1960) Stoffwechseluntersuchungen mit einem verzögert wirkenden Glukagon an Gesunden und Diabetikern. In: Bergmann JF (ed) Verhandlungen der Deutschen Gesellschaft für innere Medizin, 66. Kongreß, pp 207–212
Wetlaufer DB (1962) Ultraviolet spectra of proteins and amino acids. Adv Protein Chem 17:303–390
Wiesel LL, Positano V, Kologlu Y, Anderson GE (1963) Chromatographic separation of glucagon and insulin from serum by resinimpregnated paper. Proc Soc Exp Biol Med 112:515–518
Wünsch E (1967) Die Totalsynthese des Pankreas-Hormones Glucagon. Z Naturforsch 22b:1269–1276
Ziegler M, Lippmann HG (1968) Isolation of pure insulin and glucagon by means of agar-gel electrophoresis. Acta Biol Med Ger 21:733–738

CHAPTER 9

# The Immunogenicity of Glucagon

L. G. HEDING

## A. Introduction

The role which glucagon plays in the regulation of carbohydrate and fat metabolism is still not fully understood. Apart from early childhood, where glucagon seems to play a vital role in preventing hypoglycemia, it remains an enigma to what extent glucagon participates in maintaining normoglycemia in health and disease. Thus, in insulin-treated diabetics, insulin-induced hypoglycemia has been described as being defective, absent, or near normal. The role which glucagon may play in the development and maintenance of hyperglycemia in diabetes likewise remains a controversy, as some investigations report hyperglucagonemia while others find normal glucagon levels. A number of these differences may be apparent and caused by the use of glucagon antibodies with poorly defined specificity. Such antibodies when used in glucagon radioimmunoassays may yield different basal, as well as stimulated, levels of glucagon-like immunoreactivity (GLI).

It has now become widely accepted to distinguish between two essentially different types of glucagon antibodies, namely: (1) $NH_2$ terminal-specific or nonspecific gut GLI-specific, and (2) the COOH terminal-specific or pancreatic glucagon specific antibodies. Even when classified as (1) or (2), different antibodies belonging to the same class can give widely different levels of basal glucagon concentrations in plasma from normal persons, indicating that the specificity is not sufficiently well characterized and that different fragments are present in plasma which react more or less with the antibodies used. Another characteristic which has remained an obstacle to the use of glucagon antibodies in a radioimmunoassay (RIA) is the affinity which determines the sensitivity and detection limit of the RIA, provided a highly purified glucagon $^{125}I$ is used (JØRGENSEN and LARSEN 1972).

Although the first glucagon RIA was described by UNGER et al. in 1959, the assay has not been perfected, i.e., the RIAs do not only determine the intact glucagon molecule, but in addition, to varying degrees, fragments having an immunoreactive site corresponding to the site on the antibody molecule. The aim of this chapter is to give a review of recent literature and my own results on the production of glucagon antibodies in order to reach a conclusion as to which methods are the most suitable for the preparation and characterization of glucagon antibodies.

## B. Immunogenicity of Glucagon and Glucagon Fragments

### I. Species Used for Immunization

The rabbit has been the most widely used animal for the production of glucagon antibodies, and comparing rabbits, hens, pigeons, sheep, and guinea pigs immunized with the same glucagon preparation, the rabbit was clearly superior (HEDING 1972). In contrast to other mammalian glucagons, e.g., porcine, bovine, rabbit, which have the same composition, guinea pig glucagon was found to differ considerably (see Chap. 6). This difference has probably been the rationale behind the many attempts to produce glucagon antibodies in guinea pigs, taking into account the successful outcome of using this animal species for the production of insulin antibodies. However, the binding capacity was found to be similar in rabbits and guinea pigs by GREY et al. (1970) and GOLDFINE and RYAN (1969). CUATRECASAS et al. (1971) claimed to have produced high affinity antibodies in guinea pigs, but as no standard curve was shown the results were difficult to compare with others. The same holds true for the work by SENYK et al. (1971, 1972) where the antibodies produced in guinea pigs were not characterized according to capacity and affinity, but in detail regarding the cellular immunity found in the immunized animals.

Recently FLATT and SWANSTON-FLATT (1979, 1980) have described the production of glucagon antibodies in guinea pigs in similar amounts and of similar affinity as in rabbits. Owing to the higher sensitivity (mean value twice as high) of the guinea pig sera when used for RIA, the authors concluded that the guinea pig should be selected in preference to the rabbit. However, this statement is not justified by the data since the sensitivity (defined as the percentage fall in blood count caused by 125 pg glucagon) was 63.4% (range 54%–75%), for the guinea pig antibodies and 33.8% (range 2%–58%) for the rabbit antibodies which is hardly statistically significant. Furthermore, the antibodies produced reacted with gut GLI to varying degrees, rendering the procedure less attractive (FLATT and SWANSTON-FLATT 1980). Apart from the obvious practical advantages of rabbits over guinea pigs, such as facilitated blood sampling (ear blood versus cardiac puncture) and access to taking larger blood samples (30 ml/month versus maximum volumes of 10 ml/month), the resulting glucagon antibodies have shown improved qualities and almost all the commonly used antibodies are of rabbit origin (HEDING 1971; ASSAN et al. 1971; HOLST and AASTED 1974; VON SCHENCK 1977; ALFORD et al. 1977; ZIEGLER et al. 1980). It should be mentioned that large differences can be observed between different breeds of rabbits (HEDING 1972). Chickens, whose glucagon differs by one amino acid, were also found to be inferior to rabbits (HEDING 1972; TAGER et al. 1977).

### II. Immunogen

#### 1. Glucagon

Until a few years ago, porcine, and bovine glucagon were mostly used coupled to high molecular weight proteins to induce antibodies in animals. To enhance the immunogenicity, complete and incomplete adjuvant are used to emulsify the glucagon solution prior to its injection, in order to prolong the time which the organism is exposed to the immunogenic stimulus. Excellent high affinity and high capacity antisera have been obtained with this approach (WOROBECK et al. 1967; FROHMAN et al. 1970; TAGER et al. 1977). The use of glucagon alone was abandoned more than 10 years ago, owing to the low immunogenicity of the intact glucagon molecule.

Knowing the many technical manipulations performed with glucagon, such as coupling and use of adjuvant, which are necessary to raise glucagon antibodies, it is surprising that some insulin-treated diabetics have been reported to form glucagon antibodies. The insulins used had not been purified by several chromatographic procedures and contained, among other pancreatic peptides, glucagon (1–150 ppm, 1 ppm = 1 ng/mg insulin) (BLOOM et al. 1979; VILLALPANDO and DRASH 1979). It should be noticed that the frequency of glucagon antibodies increased with increasing duration of treatment from about 12% to 24% (VILLALPANDO and DRASH 1979), while BLOOM et al. (1979) reported that 7% of the diabetics treated with crystalline insulin had glucagon antibodies although none of the patients on purified monocomponent insulin developed glucagon antibodies. The explanations for the observed immunogenicity of glucagon in such low concentrations are probably manifold: the HLA constitution of the patients could play an important role, the daily treatment year in year out seems to have a cumulative effect as very few patients have antibodies in the first year or two, after which time the frequency rises and, in addition, the glucagon is most likely incorporated in the protamine insulin or zinc insulin crystals which thereby act as an adjuvant. It should be stressed that the capacity of these antibodies is usually low.

## 2. Glucagon Fragments

A few glucagon fragments, e.g., glucagon 1–17, 1–21, 1–23, and 18–29 have been available in limited quantities for the characterization of the antibodies (ASSAN and SLUSHER 1972; HEDING et al. 1976; VON SCHENCK 1977). Lately, larger amounts of a COOH terminal fragment (a 1:3 mixture of glucagon 18–29 and 19–29) were prepared by tryptic digestion of glucagon and used for the preparation of glucagon antibodies (IMAGAWA et al. 1979). The antibodies raised had a reasonable capacity and one of the five antisera had an acceptable affinity. The main reason for using this fragment was to ensure formation of antibodies directed exclusively towards the COOH terminus which was partly achieved (see Sect. C.II). The same was attempted using a synthetic glucagon 15–29, but the resulting antibody gave a standard curve of low sensitivity (OHNEDA et al. 1979). Immunization of groups of ten rabbits with synthetic fragments (1–15 and 1–21) coupled to albumin in order to raise $NH_2$ terminal-directed antibodies resulted in the formation of low titers in all but two rabbits receiving glucagon 1–15 and 1–21, even after 1 year of immunization. In contrast, glucagon 16–29 gave high titers of antibodies in five of ten rabbits (HEDING and MARKUSSEN to be published a). The standard curves for an antiserum raised against glucagon 1–15 and 16–29 are shown in Fig. 1. For comparison are shown in Fig. 2 two reference standard curves for antisera K 5563 and K 4023 raised against glucagon coupled to albumin. The antiserum K 6048 gave standard curves with the synthetic fragments 1–15 and 1–21 as well as with glicentin which were similar to that of porcine glucagon.

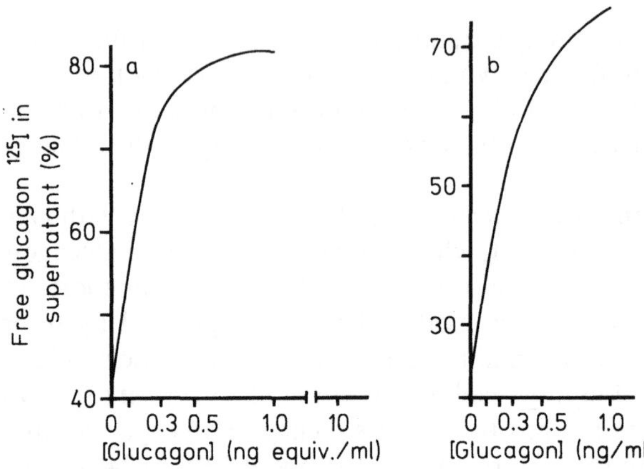


**Fig. 1a, b.** Standard curves with **a** anti-glucagon 1–15 serum K 6248, dilution 1:10,000–1:30,000 and **b** anti-glucagon serum K 6251, dilution (1:60,000–1:180,000)

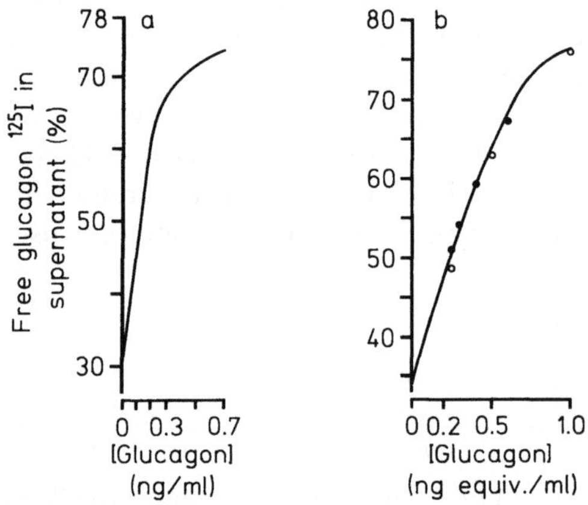


**Fig. 2a, b.** Standard curves with **a** anti-glucagon serum K 5563, dilution 1:20,000–1:60,000 and **b** anti-glucagon serum K 4023, dilution 1:4,000–1:12,000 (*open circles* indicate purified gut GLI; *full circles* indicate crude gut GLI)

In conclusion, glucagon fragments, when used coupled to albumin, can induce extremely useful glucagon antibodies, but the proportion of useful animals in each group was found to be lower than when the whole glucagon molecule was used under otherwise similar conditions (coupling procedure, amount, injection technique, frequency, etc.) (HEDING and MARKUSSEN to be published a).

### 3. Coupling Procedures

The use of adjuvant was introduced from the very first production of glucagon antibodies prepared by UNGER et al. (1959), but quite soon it became evident that the immunogenic properties of glucagon had to be potentiated by other means. Glucagon was alum precipitated (M. ROOT 1965, personal communication), absorbed by polyvinylpyrrolidone (ASSAN et al. 1965) and injected as zinc protamine gluca-

**Table 1.** Coupling of glucagon with proteins

| Reference | Coupling reagent | Coupling protein | Glucagon: protein molar ratio |
|---|---|---|---|
| ASSAN et al. (1965) | Carbodiimide | Albumin | |
| HEDING (1969) | Carbodiimide | Glucagon | |
| GOLDFINE and RYAN (1969) | Carbodiimide | Albumin | |
| GREY et al. (1970) | Diethylmalonimidate | *Limulus* hemocyanin | 19:1 |
| FROHMAN et al. (1970) | Glutaraldehyde | Rabbit albumin | 4.7:1 |
| CUATRECASAS et al. (1971) | Carbodiimide | Poly-L-lysine | |
| HOLST and AASTED (1974) | Carbodiimide | Bovine albumin | 10:1 |
| SPERLING et al. (1974) | Carbodiimide | Bovine thyroglobulin | 30:1 |
| GARAUD et al. (1976) | Glutaraldehyde | Bovine albumin | |
| LUNDQUIST et al. (1976) | Carbodiimide | Rabbit albumin | |
| von SCHENCK (1977) | Glutaraldehyde | Bovine thyroglobulin | 20:1 |
| TAGER et al. (1977) | Difluorodinitrobenzene | Human albumin | 12:1 |
| MCEVOY et al. (1977) | Difluorodinitrobenzene | Keyhole limpet hemocyanin | |
| ALFORD et al. (1977) | Carbodiimide | Albumin | |
| IMAGAWA et al. (1979)[a] | Glutaraldehyde | Bovine albumin | 8:1 |
| OHNEDA et al. (1979)[a] | Glutaraldehyde | Bovine albumin | |
| HEDING et al. (1979) | Difluorodinitrobenzene | Human albumin | 40:1 |
| ZIEGLER et al. (1980) | Difluorodinitrobenzene | Bovine albumin | 9:1 |
| ZIEGLER et al. (1980) | Glutaraldehyde | Bovine albumin | |

[a] Glucagon 19–29

gon (HEDING 1972). The covalent coupling of glucagon to a protein was, however, the real breakthrough in the production of antibodies with higher capacity and affinity.

Table 1 shows a list of some of the most important procedures employing coupling of the glucagon to a protein with higher molecular weight. As can be seen, the most commonly used coupling reagents are carbodiimide (several different types), glutaraldehyde, and difluorodinitrobenzene. The chemical processes leading to the covalent coupling of glucagon to a protein (or to itself) are not clearly understood for any of the three reagents. However, from the affinity of the resulting antibodies it can be concluded that carbodiimide coupling leaves both some $NH_2$ terminal and COOH terminal parts of glucagon free, whereby a spectrum of antibodies as well as mixtures of antibodies are formed (HOLST and AASTED 1974; SPERLING et al. 1974; ALFORD et al. 1977). The antisera K 4023, giving linear dilution curves with gut GLI, and K 964 (Fig. 2), showing less than 0.2% reactivity with gut GLI (HEDING et al. 1976), were both raised against glucagon coupled to albumin with carbodiimide. In general, carbodiimide conjugates gave antisera of good affinity and capacity, but with varying specificity.

Glutaraldehyde was introduced by FROHMAN et al. (1970), but the antisera raised by these authors were not of high sensitivity and reacted with gut GLI. Surprisingly, GARAUD et al. (1976) observed that with glutaraldehyde five rabbits of eight gave antisera directed towards the COOH terminus and postulated that the glutaraldehyde reacted mainly with amino acids at positions 1 and 12 thereby leaving the COOH terminus free. This result was not confirmed by VON SCHENCK

(1977) using a similar technique, who found that antibodies were also produced against the midportion of the glucagon molecule and reacted with gut GLI. Finally, ZIEGLER et al. (1980), who compared the glutaraldehyde coupling with that obtained with difluorodinitrobenzene, could not obtain useful high affinity or specific glucagon antibodies with the glutaraldehyde conjugate. The varying results obtained with glutaraldehyde could easily be explained by minor differences in the technique (pH, temperature, coupling yield – which are not always calculated) or in the variation of the ability of the rabbits to produce glucagon antibodies (genetic constitution).

The introduction of coupling glucagon to albumin with 1,5-difluoro-2,4-dinitrobenzene by TAGER et al. (1977) yielded high capacity antibodies with a low reactivity to gut GLI (<2.5%). Similar results were reported by MCEVOY et al. (1977), who used hemocyanin as the protein. The same authors observed that glucagon coupled to glucagon (called a homopolymer) by difluorodinitrobenzene also yielded antibodies that reacted poorly with gut GLI. This suggests that the coupling procedure involves the $NH_2$ terminus and/or the midportion of the molecule (e.g., the $NH_2$ groups in the $NH_2$ terminal histidine, the lysine (position 12), or the arginines (positions 17 and 18). This hypothesis was investigated by immunizing 40 rabbits with glucagon conjugated by difluorodinitrobenzene to human albumin (HEDING et al. 1980 to be published b). All but three animals gave useful antibodies for RIA and showed less than 1% cross-reactivity with glicentin and crude gut GLI. The final dilutions in RIA were up to 1:90,000. An example is K 5563, which is shown in Fig. 2 and has less than 0.2% cross-reactivity with glicentin. The results of ZIEGLER et al. (1980) show the same trend and confirm that this coupling procedure is superior to the previous regarding capacity and predictability of the specificity.

The use of difluorodinitrobenzene in the coupling of glucagon 16–29 to albumin to produce useful COOH terminal antibodies should also be mentioned (HEDING and MARKUSSEN to be published a), but when fragments are used, the coupling procedure becomes less critical in regard to specificity (IMAGAWA et al. 1979; OHNEDA et al. 1979).

### 4. Mode of Administration, Dose, Frequency, Adjuvant

In rabbits, the intramuscular or especially the subcutaneous injection between the shoulder blades has become the routine mode of administration. Several reports of more exotic and, for the animals very discomforting injection sites, such as the footpads (GREY et al. 1971) or intradermally (VON SCHENCK 1977; TAGER et al. 1977; KORANYI et al. 1977) have not justified these techniques by systematic comparisons with the more conventional routes. In fact, the antisera gained by the footpad, interdermal, or multidose techniques do not appear superior to those obtained by the subcutaneous route. It is therefore considered that these painful techniques are in no way justified.

The dose of glucagon varies considerably from one study to another, but many studies do not give information about the amount of glucagon injected, partly because the ratio between glucagon and the carrier protein used for coupling has not been determined. More than 10 years ago maximum doses of 5–20 mg were frequently used, but recently with the more immunogenic coupled glucagon–albumin polymers, the amount of glucagon injected each time is of the order of 0.1 mg.

The frequency of injections varies from several a week to once monthly, and it is of importance to note that one injection a month is sufficient to maintain a high antibody titer year after year (HEDING 1972). It is also important to know that many studies report that it takes 3–6 months of immunization before a maximum of capacity and affinity is reached.

Adjuvants – Freund's complete the first time and incomplete subsequently – are by now routinely used no matter which immunogen is employed. Preparation of the emulsion is normally done by mixing equal volumes of the immunogen solution with Freund's adjuvant and then mixing thoroughly either in the syringe (in the case of small volumes) or using a blender at 10,000 rpm for about 1 min. The resulting emulsion is stable for days and looks like hand lotion.

## C. Characterization of the Glucagon Antibodies

### I. Affinity and Capacity

These two important antibody characteristics are defined by the law of mass action:

$$[Ab_G] + [G] \rightleftharpoons [Ab_G - G]$$

$$\frac{[Ab_G - G]}{[Ab_G][G]} = K,$$

where $Ab_G$ stands for the concentration of glucagon antibodies and G for the concentration of glucagon. The affinity or equilibrium constant $K$ is a measure of the "attraction" between the antibody and glucagon and determines how far the reaction is "pushed" towards the right in the reaction illustrated. Antibodies with a high affinity yield RIA standard curves with a low detection limit (defined as the smallest amount of glucagon which can be distinguished significantly from the zero value, e.g., with $P < 0.01$). The maximum binding capacity, i.e., the total number of binding sites, determines the dilution in which the antiserum can be used for RIA. In order to calculate $K$ and the maximum binding capacity, a Scatchard plot has to be constructed. Only few authors have performed this tedious task (HOLST and AASTED 1974) probably owing to the fact that the calculated values for the capacity and $K$ are not directly useful for establishing and optimizing RIA. The development of improved immunization techniques, including the introduction of difluorodinitrobenzene for the coupling of glucagon, has resulted in antisera with such a high affinity that the detection limits of the standard curve are of the order of or less than 0.010 ng/ml or 1 pg/tube (FRANDSEN et al. 1981), and the final dilution ranges from 1:60,000 to 1:180,000 (see Fig. 1 and 2).

### II. Specificity

#### 1. Reactivity with Glicentin, Gut GLIs, and Glucagon Analogs

Traditionally, glucagon antibodies are characterized by their reaction with crude gut extracts containing a number of gut GLIs or, more recently, by their reactivity

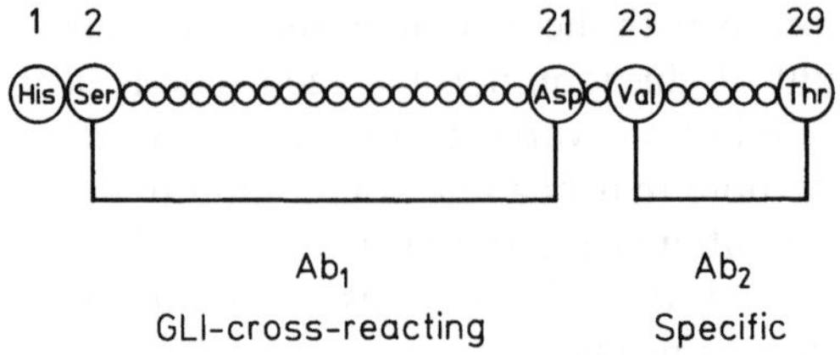


**Fig. 3.** Schematic representation of the two types of glucagon antibodies formed after immunizing with glucagon

**Table 2.** Evidence for the immunoreactive site of $Ab_1$ (gut GLI-reacting) being located within the 2–21 section of glucagon, excluding position 16

$Ab_1$ reacts with:

1. Glucagon fragments 1–21 and 1–23 [a]
2. Turkey glucagon (Asn-28 replaced by Ser) [b]
3. Duck glucagon (Asn-28 replaced by Ser and Ser-16 replaced by Thr) [b]
4. Glucagon modified at Trp-25 by 2-nitrophenylsulfenyl [b]
5. [desHis[1]]Glucagon [b]

$Ab_1$ does not react with:

1. des-(10–15)-Glucagon (1%–7%) [c]
2. des-(16–21)-Glucagon (0.14%) [c]

[a] ASSAN and SLUSHER (1972), HEDING et al. (1979)
[b] HEDING et al. (1976)
[c] FRANDSEN et al. (1981)

with porcine glicentin, a pure gut GLI, isolated and characterized by MOODY and SUNDBY (1979). Glicentin contains the full 1–29 sequence of glucagon and in addition is extended at both the $NH_2$ and COOH termini (see Chap. 7). In general, two types of antibodies can be formed after immunization with glucagon (Fig. 3), namely one type which is directed against $NH_2$ or the midportion of glucagon ($Ab_1$) and reacts with gut GLI and the other which reacts with the COOH terminal part 23–29 of glucagon. The evidence for the antigenic site of $Ab_1$ to be located between 2 and 21, excluding position 16, is shown in Table 2. It appears that changes at positions 1, 16, and 21–29 hardly affect the reactivity, whereas deletion of amino acids 10–15 or 16–21 reduced or abolished the reactivity. Serial dilutions of gut GLI in a RIA which uses only type $Ab_1$ yield linear dilution curves, a prerequisite for the quantitative determination of gut GLI.

The evidence for the immunoreactive site of the so-called specific or COOH terminal antibodies ($Ab_2$) being in the COOH terminal region is listed in Table 3. Most convincing are the results with the glucagon analogs, where changes in the 1–21 section do not affect the reactivity grossly, whereas any change in the 22–29 section leads to virtually total disappearance of reactivity. The reason for the lack of reactivity of glicentin (and possibly other gut GLI) with the $Ab_2$ seems to be the extension of the molecule at the COOH terminus, whereby the immunoreactive site becomes masked (YANAIHARA et al. 1979).

**Table 3.** Evidence for the immunoreactive site of $Ab_2$ (not reacting with gut GLI) being located in the COOH terminal 24–29 section of glucagon

$Ab_2$ reacts with:
1. des-(1–4)-Glucagon
2. des-(10–15)-Glucagon
3. des-(16–21)-Glucagon

$Ab_2$ does not react with:
1. Glucagon 1–21 and 1–23 [a]
2. Turkey glucagon (Asn-28 replaced by Ser) [b]
3. Duck glucagon (Asn-28 replaced by Ser and Ser-16 replaced by Thr) [b]
4. Glucagon modified at Trp-25 by 2-nitrophenylsulfenyl [b]
5. des-(22–26)-Glucagon [c]
6. des-(27–29)-Glucagon [c]

[a] ASSAN and SLUSHER (1972), HEDING et al. (1979)
[b] HEDING et al. (1976)
[c] FRANDSEN et al. (1981)

## 2. Reactivity with Glucagon Fragments

Further characterization of the glucagon antibodies has been done using a number of fragments, e.g., the tryptic glucagon fragments, 1–12, 1–17, and 18–29 and others made by syntheses such as 1–23 and 9–23 (ASSAN et al. 1971; ASSAN and SLUSHER 1973; SENYK et al. 1972; HEDING et al. 1976; VON SCHENCK 1977; TAGER et al. 1977). In general, these fragments have proved less useful than anticipated; e.g., fragment 18–29, which contains the whole COOH terminal immunoreactive site, shows very low (2%–15%) reactivity with COOH terminal antisera such as K964, K5563, 30 K, and E7 (HEDING et al. 1976; VON SCHENCK 1977), probably due to an altered tertiary structure of the 18–29 peptide as compared with the same peptide sequence when located in the intact glucagon molecule. Likewise, the $NH_2$ terminal fragments have to be of a certain size in order to give useful information, and glucagon 1–12 and 1–17 seem to be too small and have low affinity, whereas 1–21 reacts well with the $NH_2$ terminal, but not with the COOH terminal antisera (Fig. 4).

YANAIHARA et al. (1979) prepared a very useful peptide, namely the glicentin fragment 11–37 which contains glucagon 11–29 and in addition the glicentin octapeptide, which is the partial sequence that extends glucagon at the COOH terminus. This peptide was elegantly used to demonstrate that extension of glucagon at the COOH terminus leads to loss of immunoreactivity with the COOH terminal antisera in spite of the fact that the whole immunoreactive site is present. The masking of the site could be due to steric hindrance or changes in the tertiary structure of the site involved in the antibody binding.

In summary, glucagon antibodies still have to be characterized by their reaction with gut GLI and possibly glicentin. Further characterization can be performed using glucagon 1–21 and 11–37 or, even more extensively, by glucagon analogues such as des-(1–4)-, des-(16–21)-, and des-(22–26)-glucagons. This information, however, is not sufficient to predict the values of glucagon that will be obtained by using the characterized antibody in RIA. Thus the antisera K964, K5563, and RC5

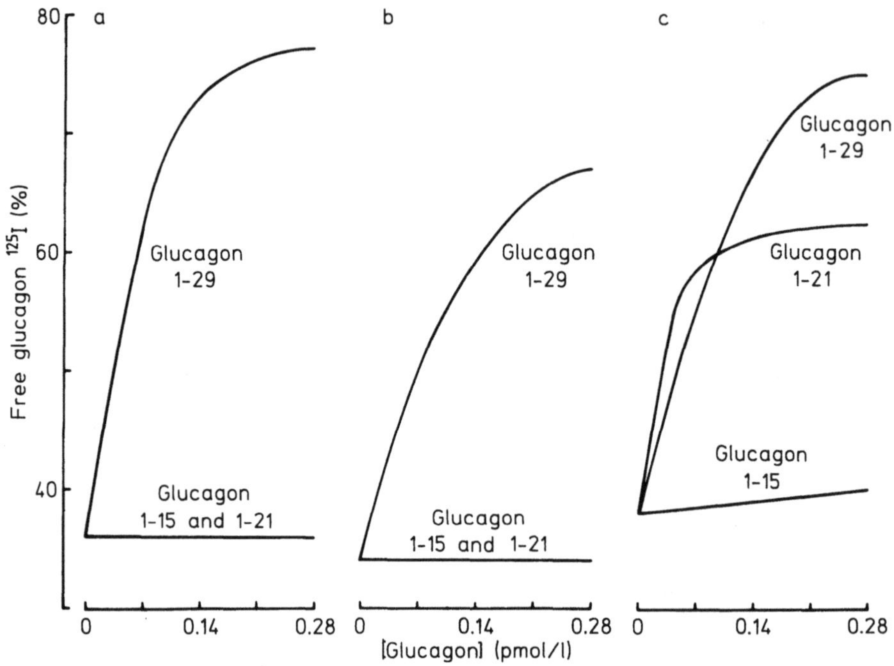


**Fig. 4a–c.** Use of glucagon 1–15 and 1–21 to characterize **a** anti-glucagon serum K 5563, dilution 1:60,000, **b** anti-glucagon serum K 964, dilution 1:18,000, and **c** anti-glucagon serum K 4023, dilution 1:12,000

(HEDING et al. 1976, 1980; ALFORD et al. 1980), which are all COOH terminal specific and have less than 0.2% reactivity to glicentin, give different basal glucagon levels in fasting normal persons. Thus with K 964 0.160 ± 0.04 ng/ml was found, K 5563 gave 0.091 ± 0.013 ng/ml, and RC 5 gave 0.076 ± 0.003 ng/ml (mean ± standard deviations) (TRONIER et al. 1981). Interestingly K 964 and RC 5 gave the same fall in glucagon values after pancreatectomy (10 pg/ml) and the same rise after arginine (~0.200 ng/ml) (TRONIER et al. 1981).

## D. Purification of Mixtures of Glucagon Antibodies

The use of glucagon 16–29 and coupling techniques, which produce a glucagon immunogen that elicits COOH terminal antibodies almost exclusively, have proved very useful. Another way to secure specific antibodies was introduced by HEDING et al. (1979), who purified mixtures of $NH_2$ and COOH terminal antibodies by a solid phase technique. Glucagon 1–21 was bound covalently to AH-Sepharose 4 B (Pharmacia, Uppsala, Sweden) using 1,5-difluoro-2,4-dinitrobenzene. This solid phase Sepharose–glucagon 1–21 was then used to remove glucagon antibodies directed against the $NH_2$ terminal section. It was shown that virtually all antibodies in K 4023 (linear cross-reacting) were removed while the COOH terminal K 964 was virtually unaffected. In an antiserum (K 44) which had significant but low reactivity with gut GLI, the gut GLI-specific antibodies were removed and the K 44 antiserum rendered specific, showing less than 1% reactivity with glicentin. It is anticipated that a solid phase glucagon 16–29 could be used in a similar way to make $NH_2$ terminal-specific antibodies. A prerequisite is that the starting mixture of antibodies contains high concentrations of the desired antibody, and it should be realized that the antibodies to be prepared must not be bound to the solid phase, since no efficient method of recovering glucagon antibodies from the solid phase has been described as yet.

## E. Summary

### I. Production of COOH Terminal-Specific Antibodies

With bovine or porcine glucagon coupled to albumin by difluorodinitrobenzene, many reports now agree that rabbits immunized with the resulting glucagon–albumin immunogen in doses corresponding to 0.1 mg glucagon per injection produce high affinity, high capacity antibodies with COOH terminal specificity. The specificity can be easily checked by using gut GLI, glucagon 1–21 or des-(22–26)-glucagon. Owing to individual differences between rabbits, it is advisable to start with at least ten animals and make a selection after approximately 6 months. The animals can then be maintained for years with a nearly constant concentration of antibodies by one monthly injection of the same immunogen.

The use of the COOH terminal fragments, glucagon 18–29 or 16–29, will consequently yield antibodies directed against that particular part of the glucagon molecule, and no reactivity at all could be detected with gut GLI with such an antiserum which, in addition, showed excellent affinity and higher capacity than previously seen with intact glucagon (Fig. 1). However, the percentage of successful immunization was lower than when the whole glucagon molecule was used. Finally, COOH terminal antibodies can be obtained from mixtures of antibodies by removal of the $NH_2$ terminal antibodies using a solid phase glucagon 1–21 reagent.

### II. Production of $NH_2$ Terminal-Specific Antibodies Giving Linear Dilution Curves with Gut GLI

Owing to the interest in measuring the gut GLI release after various stimuli in health and disease, there is a need for an $NH_2$ terminal antibody that can be used in a RIA to quantitate the gut GLI. Linear dilution curves with gut GLI then have to be obtained. In contrast to the production of COOH terminal antibodies, there is no safe and convenient prescription as to how to obtain this type of antibody. The $NH_2$ terminal antibodies with linear cross-reaction described in the literature have been raised against glucagon coupled to albumin by carbodiimide (HOLST and AASTED 1974; HEDING et al. 1976; ALFORD et al. 1977) or using glucagon fibrils (YANAIHARA et al. 1979). An antiserum showing linear dilution is very rare and occurs only by chance. Most antisera produced with the carbodiimide–glucagon immunogen contain mixtures of $NH_2$ terminal and COOH terminal antibodies or have different affinity towards pancreatic glucagon and gut GLI, rendering them useless in a quantitative RIA of gut GLI. An attempt to produce $NH_2$ terminal-specific antibodies using glucagon 1–15 was not successful as the only acceptable antibodies (in terms of affinity and capacity) did not react with gut GLI to any great extent. A further characterization of the specificity of this type of antisera has to be performed.

The separation of COOH terminal antibodies from $NH_2$ terminal, e.g., by solid phase glucagon 16–29 should also be attempted. Thus, in conclusion, production of $NH_2$ terminal glucagon antibodies having linear dilution curves with gut GLI in RIA still has to be done in the old-fashioned way by immunizing a large number of rabbits, e.g., with glucagon–carbodiimide–albumin, whereby 1 rabbit in 20 may come up with the expected antibody.

*Acknowledgments.* I wish to thank my co-worker Lisbet P. Pedersen for her excellent technical assistance in glucagon immunogenicity which represents more than 10 years of involvement in our many attempts to produce improved and better-defined glucagon antibodies. Thanks to her skill, perseverance, and fine cooperation I have gained the experience with glucagon antibodies which is described as part of this chapter.

## References

Alford FP, Bloom SR, Nabarro JDN (1977) Glucagon levels in normal and diabetic subjects: use of a specific immunoabsorbent for glucagon radioimmunoassay. Diabetologia 13:1–6

Assan R, Slusher N (1972) Structure/function and structure/immunoreactivity relationships of the glucagon molecule and related synthetic peptides. Diabetes 21:843–855

Assan R, Rosselin G, Drouet J, Dolais J, Tchobroutsky G (1965) Glucagon antibodies. Lancet 2:590

Assan R, Tchobroutsky G, Derot M (1971) Glucagon radioimmunoassay: technical problems and recent data. Horm Metab Res [Suppl] 3:82–90

Bloom SR, Barnes AJ, Adrian TE, Polak JM (1979) Autoimmunity in diabetics induced by hormonal contaminants of insulin. Lancet 1:14–17

Cuatrecasas P, Illiano G, Green I (1971) Production of anti-glucagon antibodies in poly-L-lysine "responder" guinea-pigs. Nature New Biol 230:60–61

Flatt PR, Swanston-Flatt SK (1979) Rapid production of potent glucagon antisera in the guinea pig. Horm Metab Res 11:581–582

Flatt PR, Swanston-Flatt SK (1980) Production and evaluation of antibodies to glucagon-albumin conjugates in the guinea pig. Acta Physiol Scand 108:317–319

Frandsen EK, Grønvald FC, Heding LG, Johansen NL, Lundt BF, Moody AJ, Markussen J, Vølund Aa (1981) Glucagon: structure-function relationships investigated by sequence deletions. Hoppe Seylers Z Physiol Chem 362:665–667

Frohman LA, Reichlin M, Sokal JE (1970) Immunologic and biologic properties of antibodies to a glucagon-serum albumin polymer. Endocrinology 87:1055–1061

Garaud JC, Moody AJ, Eloy R, Grenier JF (1976) Unusual specificities of antibodies to glucagon-glutaraldehyde-albumin conjugates. Horm Metab Res 8:241–243

Goldfine ID, Ryan WG (1969) Rapid production of glucagon antibodies. Horm Metab Res 2:47–48

Grey N, McGuigan JE, Kipnis DM (1970) Neutralization of endogenous glucagon by high titer glucagon antiserum. Endocrinology 86:1383–1388

Heding LG (1969) The production of glucagon antibodies in rabbits. Horm Metab Res 1:87–88

Heding LG (1971) Radioimmunological determination of pancreatic and gut glucagon in plasma. Diabetologia 7:10–19

Heding LG (1972) Immunologic properties of pancreatic glucagon: antigenicity and antibody characteristics. In: Lefebvre PJ, Unger RH (eds) Glucagon: molecular physiology, clinical and therapeutic implications. Pergamon, Oxford New York, pp 187–200

Heding LG, Markussen J (to be published a) Production of glucagon antibodies using glucagon (1–15) and (16–29).

Heding LG, Frandsen EK, Jacobsen H (1976) Structure-function relationship: immunologic. Metabolism [Suppl 1] 25:1327–1329

Heding LG, Markussen J, Grønvald FC, Lundt BF, Johansen NL (1979) Preparation of specific glucagon antibodies by removal of gut GLI reacting antibodies with immobilized glucagon fragment 1–21. Acta Endocrinol (Copenh) [Suppl] 227:34–35

Heding LG, Naithani KV, Markussen J (to be published b) Production of C-terminal specific glucagon antibodies.

Holst JJ, Aasted B (1974) Production and evaluation of glucagon antibodies for radioimmunoassay. Acta Endocrinol (Copenh) 77:715–726

Imagawa K, Nishino T, Shin S, Uehata S, Hashimura E, Yanaihara C, Yanaihara N (1979) Production of anti-glucagon sera with a C-terminal fragment of pancreatic glucagon. Endocrinol Jpn 26:123–131

Jørgensen KH, Larsen UD (1972) Purification of $^{125}$I-glucagon by anion exchange chromatography. Horm Metab Res 4:223–224

Koranyi L, Peterfy F, Paksy A, Vargha P (1977) Production of glucagon antibodies by thyroglobulin-zinc glucagon conjugate. Horm Metab Res 9:434–435

Lundqvist G, Edwards J, Wide L (1976) A solid phase radioimmunoassay for pancreatic glucagon. Ups J Med Sci 81:65–69

McEvoy RC, Madson KL, Elde RP (1977) Stimulation of antiglucagon antibodies in rabbits. Horm Metab Res 9:272–274

Moody AJ, Sundby F (1979) The structure and biosynthesis of glucagon. Proceedings of the 10th Congress of the International Diabetes Federation. Vienna, Austria, Sept. 9–14

Ohneda A, Watanabe K, Wakimatsu M, Fujino M (1979) Production of a specific antiserum by synthetic C-terminal fragment of glucagon. Horm Metab Res 11:463–468

Senyk G, Nitecki D, Goodman JW (1971) Immunogenicity of glucagon: Determinants responsible for antibody binding and lymphocyte stimulation. Science 171:407–408

Senyk G, Nitecki DE, Spitler L, Goodman JW (1972) The immune response to glucagon in conjugated form. Immunochemistry 9:97–110

Sperling MA, DeLamater PV, Kazenelson M, Fiser RH, Fisher DA (1974) Development and application of a radioimmunoassay for plasma glucagon. Clin Chem 20:566–570

Tager HS, Hohenboken M, Markese J (1977) High titer glucagon antisera. Endocrinology 100:367–372

Tronier B, Kåresen R, Aune S (1981) Glucagon immunoreactivity in man after pancreas resection and total pancreatectomy measured by different antisera. Horm Metab Res 13:56–57

Unger RH, Eisentraut AM, McCall MS, Keller S, Lanz HC, Madison LL (1959) Glucagon antibodies and their use for immunoassay for glucagon. Proc Soc Exp Biol Med 102:621–623

Villalpando S, Drash A (1979) Circulating glucagon antibodies in children who have insulin-dependent diabetes mellitus. Clinical significance and characterization. Diabetes 28:294–299

Von Schenck H (1977) Production and characterization of an antiserum against pancreatic glucagon. Clin Chem Acta 80:455–463

Worobec R, Locke R, Hall A, Ertl R, Ernst K, Deininger E (1967) Production of antibodies of high binding affinities to glucagon in rabbits. Biochem Biophys Res Commun 29:406–412

Ziegler M, Keilacker H, Woltanski KP, Besch W, Schubert J (1980) Radioligandassays: Methodik und Anwendung. Acta Biol Med Ger 30:305–314

Yanaihara N, Nishino T, Kodaira T, Imagawa K, Nishida T, Mihara S, Yanaihara C (1979) Characterization of antiglucagon sera elicited against a C-terminal fragment of pancreatic glucagon and their use in glucagon radioimmunoassay. In: Symposium on proinsulin, insulin, and C-peptide, Tokushima 1978. Excerpta Medica, Amsterdam London New York, pp 426–431

CHAPTER 10

# Immunoassays for Glucagon

J. W. Ensinck

## A. Introduction

In 1959, Unger et al. first applied the principles of the radioimmunoassay (RIA) developed by Berson et al. (1956) to the measurement of glucagon. The subsequent refinement of this tool has enabled major strides to be made in describing the sources and functions of an array of substances related to glucagon and interpreting their roles in fuel homeostasis. It is now clear that glucagon is not only synthesized and released from the pancreatic A-cell, but that cells within the gastrointestinal tract (A- and L-cells) and brain also contain glucagon and/or peptides that share structural components of the glucagon molecule (molecular weight 3,485 daltons). As is the case for other peptide hormones, glucagon is the product of processing of precursor molecules which are detected within cells and also may be secreted into the circulation (see Chaps. 6, 7, 11). Because of structural homologies, several, if not all of these substances, may cross-react with antibodies generated against pancreatic glucagon. Since the regulation of the secretion of glucagon-related peptides may differ and these peptides do not share similar actions on target cells, in the early years, antisera used in RIA which were unable to distinguish between the different species of glucagon-related peptides inadvertently led to misinterpretations (Heding 1971; Assan and Slusher 1972; Unger 1972).

Advances in our knowledge of the controls of secretion and functions of these substances have depended upon resolution of the physicochemical properties of these related molecular species, coupled with the development of antisera which differentiate pancreatic and gastric glucagon and its precursors from glucagon-like substances distributed throughout the intestine. The latter have been termed gut glucagons or glucagon-like immunoreactants (gut GLIs) (Assan 1973; Sundby and Moody 1980). The relationships between pancreatic and extrapancreatic glucagon and gut GLIs are described in Chaps. 7 and 11. With the availability of discriminant antisera, it is now possible to distinguish glucagons from gut GLIs and to quantify and categorize the multiple forms of peptides that collectively cross-react with nonspecific antisera against glucagon. The 29 amino acid glucagon molecule (Bromer et al. 1972) is schematically shown as $IRG^{3500}$ in Fig. 1. As detailed by Heding in Chap. 9 and discussed in Sect. B.IV.1 of this chapter, immunodeterminants "specific" for pancreatic and gastric glucagon have their loci, involving a variable sequence of amino acids, oriented towards the carboxyl (COOH) terminus, whereas the GLIs share with glucagon immunodeterminants in the central and amino ($NH_2$) terminal regions of the glucagon molecule. Furthermore, complex three-dimensional conformations of the polypeptide are also important in determining specificity of antibody recognition.

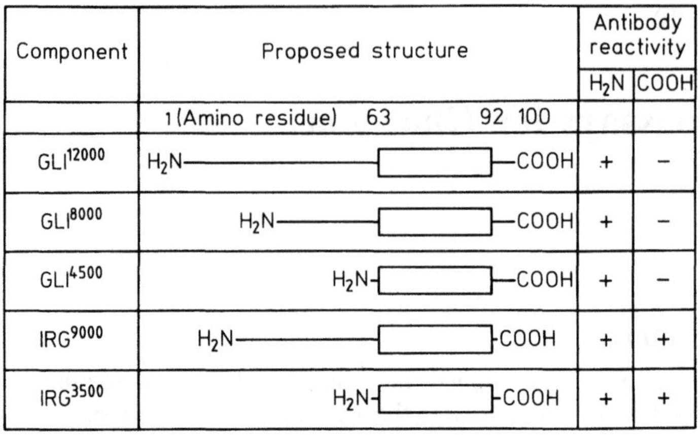


**Fig. 1.** Scheme of proposed relationships amongst glucagon-related peptides. GLI and IRG designate peptides referred to as glucagon-like immunoreactivity and immunoreactive glucagon, respectively. Superscripts refer to approximate molecular weights of the peptides. After CONLON (1980)

Based upon the physical, chemical, and biologic properties and of the known molecular structures and reactivities of antibodies directed to the $NH_2$ or COOH terminal loci of glucagon, a rational classification of the glucagon-related peptides has been outlined by CONLON (1980) (Fig. 1). The proposed notation of glucagon and its biosynthetic precursor as immunoreactive glucagon (IRG) differentiate them from the gut GLIs by molecular size and immunologic specificity, nevertheless it also underscores similarities among these species, reflecting their probable common derivation from a single precursor (glicentin) and subsequent tissue-specific processing (TAGER and STEINER 1973; TAGER and MARKESE 1979; MOODY et al. 1978; PATZELT et al. 1979; HOLST 1980; THIM and MOODY 1981). This nomenclature further illustrates the dilemmas posed by the heterogenity of glucagon-like peptides by RIA in various tissues and in plasma. In this context, it is the intent of this chapter to elaborate upon the methods and practical applications of the RIA for glucagon-related peptides, in particular focusing upon their measurement in plasma.

## B. Method of Radioimmunoassay

### I. Sources of Peptides

Purified crystalline porcine or bovine glucagon is available from several commercial sources (Eli Lilly, Indianapolis, Indiana; Novo Research Institute, Bagsvaerd, Denmark; Sigma Chemical Co., St. Louis, Missouri; CalBioChem, San Diego, California). Porcine, bovine and human glucagons have identical structural sequences (SUNDBY 1976). Except for the guinea pig, all of the mammalian glucagons as yet examined (rat, rabbit, and camel) show homology with porcine glucagon, indicating a strong conservation of primary structure (see also Chap. 1). Therefore, their quantification by RIA can be obtained by comparison with a porcine or bovine glucagon standard. In contrast, avian and fish glucagons so far characterized contain substantial substitutions in the primary sequence with resultant weak cross-reactivity with specific antisera directed to mammalian glucagon; (HEDING et al. 1976); however, all react strongly with nonspecific anti-glucagon sera. To avoid loss in potency it is advisable to store preparations of crystalline glucagon at −20 °C in the presence of a desiccant.

## II. Preparation and Purification of Radiolabeled Ligand

Porcine glucagon labeled with $^{125}I$ is the radioligand most generally used because of the long half-life of the isotope. Monoiodinated glucagon with a specific activity of 100–200 μCi/μg its commercially available (New England Nuclear, Boston, Massachusetts; Amersham, Arlington Heights, Illinois). Alternatively, radioiodination of the peptide may be carried out in the investigator's laboratory. The most commonly employed technique is the chloramine-T oxidation method of GREENWOOD et al. (1963). Although satisfactory iodopeptides are frequently obtained, the reaction may lead to unwanted oxidation of the methionine residue at position 27 which may hinder interaction between the radioligand and a COOH terminal-specific antibody (SHIMA et al. 1975). Furthermore, the product may be unstable and repurification may be necessary prior to repetitive use in the RIA. VON SCHENCK et al. (1976) reported the radiolabeling of glucagon by the lactoperoxidase technique which led to a product with less damage to the peptide. An analogous experience was published by TOWER et al. (1977). VON SCHENCK observed that at pH 10, the enzyme-mediated reaction favors the formation of monoiodotyrosine with 80% of the iodine bound to the tyrosine residue at position 13. Specific activities of 500–600 μCi/μg were achieved. This has led to an enhancement in immunoreactivity and sensitivity of the RIA with COOH terminal directed antisera (VON SCHENCK and JEPPSSON 1977). The alleged improvement in immunogenicity and prolongation of shelf life without need for frequent repurification are strong recommendations for the more widespread application of this technique. The following protocol is modified from VON SCHENCK et al. (1976):

1. 1–2 μl Na $^{125}I$ (100–200 μCi; Amersham, Arlington Heights, Illinois)
2. 50 μl glucagon (800 pmol)
3. 2 μl lactoperoxidase (50 pmol; Sigma, St. Louis, Missouri)
4. 5 μl $H_2O_2$ (48 pmol).

The solvent for all reactants is 0.1 *M* glycine/NaOH buffer, pH 10. After mixing reactants for 60 s, the reaction is stopped by addition of 0.5 ml phosphate buffer (0.5 *M*, pH 7.5) containing 0.05% sodium azide. The solution may be stored in an appropriate buffer containing 0.1% human serum albumin (HSA) or bovine serum albumin (BSA) in 0.02% merthiolate. Usually, the radiolabeled glucagon is purified immediately after the iodination procedure. Several methods have been employed, including gel permeation chromatography (HARRIS et al. 1978), elution from cellulose or charcoal (LAWRENCE 1966; LECLERCQ-MEYER et al. 1970), extraction with organic solvents, (HEDING 1971), and ion exchange chromatography (JORGENSEN and LARSEN 1972). We have routinely used the last method which entails separation of iodoglucagon from radioactive contaminants and unlabeled glucagon on a column (1 × 30 cm) QAE Sephadex A-25 (Pharmacia, Uppsala,

Sweden) in 0.08 *M* Tris, 0.02 *N* HCL, 0.08 *M* NaCl, pH 8.6, containing 1% HSA and 500 kallikrein inactivating units (KIU) aprotinin (Trasylol, FBA Pharmaceuticals, New York) or 0.1 *M* benzamidine hydrochloride (Aldrich Chemicals, Milwaukee, Wisconsin) to attenuate tracer degradation. The radioactivity peaking at an elution volume of 100 ml is collected, diluted, frozen at −20 °C, and subsequently tested for antibody binding. In our hands, the shelf life in dilute solutions is frequently 2–3 months without need for repurification. Although this procedure generally provides satisfactory results, the introduction of reverse-phase high pressure liquid chromatography (SEIDAH et al. 1980) and polyacrylamide gel electrophoresis (LINDE et al. 1980) holds promise of applications of fast, simple, and quantitative procedures for greater purification of the monoiodinated hormone, free of unlabeled peptides, thereby increasing specific activity, improving sensitivity, and prolonging shelf life.

## III. Preparation of Standard

An appropriate amount (1 mg) of crystalline glucagon powder is dissolved in 0.1 *N* NaOH (0.2 ml) and diluted to 10 ng/ml with assay buffer (0.13 *M* borate buffer, pH 8.2, containing 0.5% BSA, Pentex reagent grade, Miles Laboratories, Elkhart, Indiana) and 1:10,000 merthiolate. A variety of buffers between pH 7 and 9 have been used. In most instances the buffer pH is not critical. It should be cautioned, that, occasionally, commercial preparations of BSA contain proteolytic activity, leading to inactivation of glucagon standard or tracer. Batches should be checked by determining tracer damage by either immunoprecipitation or ion exchange chromatography (JORGENSON and LARSEN 1972). After serial dilution of the stock solution to the desired concentration of the working standard, i.e., 10 ng/ml, aliquots are frozen at −20 °C, thawed for each assay, and the residual discarded.

## IV. Production and Characterization of Antisera

### 1. Antigenic Determinants and Coupling Procedures

In the years ensuing after UNGER et al. (1959) described the antigenicity of pancreatic glucagon, a number of methods for developing glucagon antiserum of variable specificity have been published (UNGER et al. 1961; ASSAN et al. 1965; LAWRENCE 1966; SHIMA and FOÀ 1968; HAZZARD et al. 1968; EISENTRAUT et al. 1968b; HEDING 1969, 1971; SENYK et al. 1971, 1972; PEK et al. 1972a; HOLST and AASTED 1974; SPERLING et al. 1974; FLANAGAN et al. 1974; GARAUD et al. 1976; ALFORD et al. 1977; VON SCHENCK 1977; TAGER et al. 1977, IMAGAWA et al. 1979; OHNEDA et al. 1979). In most instances, antisera have been produced in randomly bred rabbits by means of free glucagon, polymers of glucagon, or the peptide covalently conjugated to a number of compounds which modify its haptenic specificity. HEDING has reviewed in Chap. 9 the various factors influencing antibody production and methods for obtaining antisera of high avidity.

A list of the characteristics of the most widely used antisera for RIA are given in Table 1. Those that are highly specific for pancreatic glucagon have usually (but not uniformly) been obtained by a covalent coupling of the intact glucagon mol-

**Table 1.** Characteristics of selected antisera with specificity for glucagon

| Designation | Conjugate[a] | Cross-reaction with gut GLI (%) | Sensitivity (pg) | Reference |
|---|---|---|---|---|
| 30K | | <1 | 2 | EISENTRAUT et al. (1968b) |
| G58 | | <1 | 2 | FALOONA (1972) |
| K47 | G–polymer | <1 | 10 | HEDING (1971) |
| G11 | G–carbodiimide–thyroglobulin | <1 | 2 | SPERLING (1974) |
| K4317 | G–carbodiimide–BSA | <3 | 1 | HOLST and AASTED (1974) |
| K964 | G–polymer | <1 | | HEDING et al. (1976) |
| E7 | G–glutaraldehde–thyroglobulin | <1 | 2 | VON SCHENK (1977) |
| R1 | G–difluorodinitro-benzene–HSA | <2 | < 2 | TAGER et al. (1977) |
| GC5 | G 18–29–glutaral-dehyde–BSA | <1 | < 2 | IMAGAWA et al. (1979) |
| G42 | G 15–29–glutaral-dehyde–BSA | <1 | 2 | OHNEDA et al. (1979) |

[a] G = glucagon; BSA = bovine serum albumin; HSA = human serum albumin

ecule or its COOH terminal fragment to a macromolecule such as albumin or thyroglobulin. Coupling conditions favoring the linkage of the macromolecule to the $NH_2$ terminus of glucagon are generally recommended. Nonetheless, polymers of glucagon and absorption to compounds such as polyvinylpyrrolidone have been advocated (ASSAN et al. 1965; HEDING 1972; HARRIS et al. 1978). Of these, antisera denoted 30K, K47, and K964 with high specificity and sensitivity have been widely distributed and have been the benchmark antisera for numerous studies, establishing the role of glucagon in physiology and in diverse pathologic states. Many antisera have been obtained which cross-react with glucagon and gut GLIs (ASSAN et al. 1965; ASSAN 1973; UNGER et al. 1961; HEDING 1971; HOLST and AASTED 1974). Recently, an antiserum specific for purified glicentin (SUNDBY et al. 1976), denoted R64, has been generated in a rabbit (MOODY et al. 1978). No cross-reaction with glucagon, small basic gut GLI, gastric inhibitory peptide (GIP), cholecystokinin (CCK), or vasoactive intestinal peptide (VIP) was noted and RIA with high sensitivity has recently been reported (MOODY and SUNDBY 1980). Although the successful production of high titer antibody with the required avidity and specificity for glucagon is not guaranteed by any given recipe, the methods outlined later in this section have been selected from the literature based on a reasonable expectancy of obtaining glucagon-specific antiserum that would be applicable for RIA of IRG. The options include the use of the intact molecule or its 19–29 COOH terminal sequence. The advantage of the use of intact glucagon is its ready availability, in contrast to the COOH terminal fragment which necessitates custom synthesis. A disadvantage of the use of glucagon is a higher probability of developing nonspecific antisera whereas the COOH terminal fragment is less likely to induce antibodies cross-reacting with gut GLIs.

a) Method of HOLST and AASTED (1974)

Glucagon, BSA, and ethylcarbodiimide in ratios of 5:6:25 are dissolved in 0.05 *M* phosphate, 0.15 *M* sodium chloride buffer, pH 7.4, at 22 °C. The conjugate may be either stored diluted in buffer at −20 °C or dialyzed against water and lyophilized.

b) Method of TAGER et al. (1977)

Crystalline glucagon (1–5 mg) is dissolved in 0.1–1.0 ml 7 *M* guanidine hydrochloride, 0.1 *M* potassium phosphate, pH 7.2. Difluorodinitrobenzene (DFDNB) (Sigma, St. Louis, Missouri) in purified methanol, 30 mg/ml, is added to the glucagon solution in a ratio of 5:1 (v:v). After 15 min at 22 °C, excess free agent is removed by shaking with cold diethyl ether. The activated peptide is then dissolved in 0.4 *M* sodium borate buffer, pH 10, containing HSA (100 mg/ml). After standing overnight at 22 °C, the mixture is dialyzed against water and lyophilized. The potential benefit of DFDNB is that it ensures a high degree of reaction coupling of the α-amino groups of lysine, histidine, and the two tyrosine residues to albumin, thereby favoring the development of antibodies directed to the COOH terminus.

c) Method of OHNEDA et al. (1979)

A sample of 15 mg COOH terminal glucagon fragment (15–29) is dissolved in 5 ml phosphate buffer (0.2 *M*, pH 7.3), containing 20 mg BSA and 5 ml glutaraldehyde (5%) is added (GARAUD et al. 1976). After 3 h stirring, the solution is dialyzed against distilled water and freeze-dried.

## 2. Immunization Procedure

Of the various empiric methods for immunization, the following schedule is derived from widespread experience in developing high antibody titers for diverse antigens. The lyophilized antigen conjugate (5–10 mg, equivalent to 30–50 mg glucagon or its COOH terminal fragment) is dissolved in 1 ml NaCl (0.15 *M*), emulsified in 1 ml complete Freund's adjuvant, and injected intradermally at several sites (VAITUKAITIS et al. 1971). Booster injections are given every 2 weeks, to a total of four and at intervals of 2 months thereafter. Blood is withdrawn either from ear vein or by cardiac puncture and the sera stored at −20 °C. In general, antisera are stable to repeated thawing and freezing. Another "immunization" technique which has yet to be exploited for glucagon is that of hybridoma technology which leads to a generation of homogeneous antibody (KENNETT et al. 1980). This revolutionary serologic method is based on selection of clones of fused myeloma and spleen lymphocytes which are maintained indefinitely in culture and produce a single immunoglobulin. This offers the potential of "immortalizing" antibodies highly specific to pancreatic glucagon. The application of this method for glucagon would guarantee perpetuation of unlimited quantities of a standardized antibody for this peptide for universal distribution.

**Table 2.** Immunoassay procedure

| Step | Volume (ml) |
|---|---|
| 1. To 12 × 75 mm disposable culture tubes add: | |
| a. Standard glucagon | 0.1 |
| b. Plasma | 0.01–0.1 |
| c. Buffer to total | 1.0 |
| d. Anti-glucagon serum diluted in assay buffer | 0.05 |
| 2. Incubate at 4 °C for 24 h | |
| 3. Add glucagon $^{125}I$ diluted in assay buffer to contain desired concentration (10–20 pg glucagon $^{125}I$, specific activity of 500 μCi/μg) | 0.05 |
| 4. Incubate at 4 °C for 24 h | |
| 5. Add cellulose solution (10% slurry) in assay buffer | 1.0 |
| 6. Mix by vortex instrument and centrifuge at 3000 rpm for 10 min | |
| 7. Decant supernatant into vial and count supernatant or residual cellulose | |

### 3. Characterization of Antisera

The characterization of antisera in terms of affinity, capacity, and specificity is reviewed in detail by HEDING in Chap. 9.

## V. Assay Procedure

Although adhering to the principles of antigen–antibody reactions, techniques of the RIA of glucagon and gut GLIs in different laboratories vary in preparation of samples, specificity of antisera, radioligand concentration, incubation conditions, and separation procedures. However, except for the processing of plasma which is not uniform in most laboratories, assay conditions generally conform to similar formats and an exhaustive discussion of the spectrum of minor modifications is beyond the scope of this chapter. The interested reader may refer to the following authors as illustrative of the different assay conditions: SHIMA and FOÀ (1968), HAZZARD et al. (1968), ASSAN et al. (1969), HEDING (1971), BUCHANAN and MCCARROL (1971) LUYCKX (1972), ASSAN (1973), SPERLING et al. (1974), HOLST and AASTED (1974), VON SCHENCK (1977), TAGER et al. (1977), HARRIS et al. (1978), IMAGAWA (1979), OHNEDA et al. (1979), MOODY and SUNDBY (1980).

The procedure used in our laboratory, is outlined in Table 2. All the agents are diluted in assay buffer (0.13 *M* borate buffer, pH 8.2, containing 0.25% BSA, and 1:10,000 final concentration merthiolate. The buffer also contains either aprotinin (1,000 KIU/ml) or 0.01 *M* benzamidine hydrochloride. It is to be noted that benzamidine may interfere with binding of ligand, with some antisera; therefore, it should be used only after evaluating its suitability with a particular antiserum. The total volume for the reaction is 1.2 ml and the total incubation period is 48 h. Antibody is added to all tubes except those in which the total number of counts of radioligand is to be determined. In our hands, this preincubation step has enhanced the sensitivity of the assay. Routinely we employ 30K antiserum (Diabetes Research Foundation, the University of Texas Health Science Center, Dallas, Texas). Final concentrations will depend upon 30K lot number. The number of counts of radiolabeled glucagon added is determined to give between 25% and 30% for the radioligand bound as a percentage of total counts in the absence of glucagon. A typical assay includes duplicate tubes for (1) nonspecific binding; (2) zero dose; (3) glucagon standards of 2, 5, 10, 15, 20, 30, and 40 pg; (4) control mammalian sera at low and high concentration of endogenous IRG; (5) unknown sample (if plasma, "native" or "extracted"). Each unknown has matching tubes in which the antibody is omitted to correct for nonspecific binding. To measure basal levels of IRG in plasma, the conditions of the RIA must be established to record with confidence of 2 pg/tube (~20 pg/ml plasma). When unextracted or "native" plasma samples are analyzed, we do not exceed 100 µl/tube. As described later in this section, plasma samples are also routinely extracted with polyethyleneglycol (PEG) and additional values of PEG are added to those tubes containing standard when this procedure is used. For separation of the antibody-bound tracer from the free radioligand, we employ the method of NONAKA and FOÀ (1969). Cellulose (Cellulosepulver MN 300, Brinkman Instruments Inc., Westbury, New York) is made up in assay buffer as a 10% slurry and added to the incubation tube. After thorough mixing by a vortex instrument, the supernatant is separated by centrifugation and decanted into a counting vial for measurement of radioactivity in a $\gamma$-counter.

Several alternative methods for separation of the bound from free tracer have been described. They include chromatoelectrophoresis, ion exchange chromatography, salt or ethanol precipitation, absorption on charcoal, and immunoprecipitation (LUYCKX 1972; HARRIS et al. 1979). In general, the techniques provide comparable results (PALMIERI et al. 1971; BUCHANAN and MCCARROL 1971). The advantage of cellulose or charcoal absorption or ethanol precipitation is that they enable rapid handling of large numbers of samples and one or other of these techniques is most commonly used. Chromatoelectrophoresis, while enabling analysis of tracer damage, is time consuming and allows only a limited number of analyses. It is no longer popular. Immunoprecipitation offers the advantage of simplicity and clean separation of free from bound peptides. Nevertheless, it extends the duration of the assay and additional difficulties include possibility of interfering effects of serum on formation of the precipitate as well as the necessity of establishing the optimal concentration of precipitating antibody. Neither does this technique allow the recognition of damage to tracer during incubation with the first antibody.

Data processing can be achieved by manual methods and computer programs. RODBARD (1974) and RODBARD and FRAZIER (1975) have listed several guidelines for reduction of RIA data. Although plotting points from the standard curve against dose on semilogarithmic or logarithmic paper and reading the unknown from these plots can be accomplished by hand, a number of sophisticated computer programs are now available (YANAGISHITA and RODBARD 1978; DAVIS et al. 1980). Most entail the application of the logit-log method for linear transformation of standard curves. A typical standard curve for glucagon with 30 K antiserum is plotted in Fig. 2. Measurement of accuracy, precision, and sensitivity of assays are included in most packages to enable the quantification of the least detectable doses and the intra- and interassay coefficients of variation. The variance at doses giving 50% displacement with outriders and values of standard sera are routinely included. The coefficients of variation for single replicate measurements of glucagon in our laboratory are 8% in the midrange of the standard curve and 14% at levels

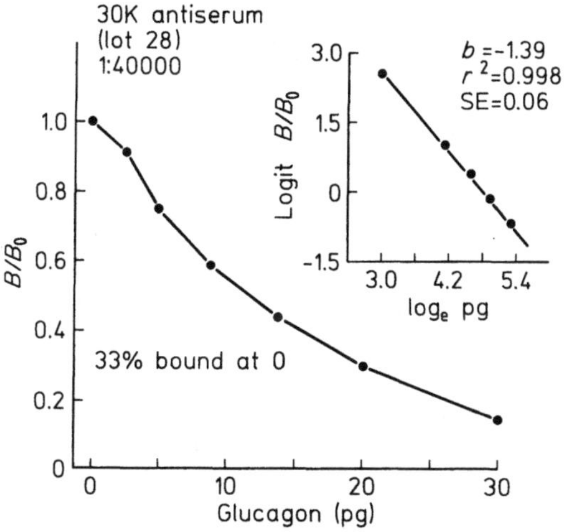


**Fig. 2.** A representative standard curve of immunoassay of glucagon using 30 K antiserum. $B$ = radioligand bound with varying amounts of glucagon, $B_0$ = radioligand bound without added glucagon. *Insert* shows logit-log transformation of data

between 0 and 2 pg/tube. The interested reader is referred to reviews by RODBARD (1974) and RODBARD and FRAZIER (1975) and papers by FAURE et al. (1980), RAAB et al. (1980), and SMIGEL and LAZAR (1981) for theoretical and practical methods for RIA data reduction.

The RIA method for measurement of total gut GLIs and glicentin are detailed by MOODY and SUNDBY (1980) and reviewed in Chap. 7. To quantify total gut GLIs, it is necessary to measure both the contribution of IRG and gut GLIs to the total immunoreactivity. This requires the use of an antiserum that cross-reacts with the midportion and/or the $NH_2$ terminus of glucagon as well as separate assay values obtained for IRG content using a COOH terminus directed antiserum. The GLIs are quantified by subtracting the values of IRG content by RIA with a COOH terminal-specific antiserum. If is important to select an antiserum which has an affinity equivalent for glucagon and gut GLI so that standard curves of glucagon and gut GLI are superimposable. Glucagon $^{125}$I and porcine glucagon standards are used. The mechanics of performing the assay can be carried out by a method similar to that outlined in Table 2. Since not all the structures of the gut GLIs have been identified, it is recommended that the results be expressed as either molar or equiv. GLI rather than pg/ml glucagon. The RIA specific for glicentin, purified from pork intestine, involves a specific antiserum (R 64), highly purified porcine glicentin calibrated against porcine glucagon, and $^{125}$I-labeled glicentin with separation of the antibody-bound ligand by alcohol. The working range is 50–1,500 pg equiv./ml standard glicentin (MOODY and SUNDBY 1980).

## C. Measurement of Glucagon-Related Peptides

The detection and quantification of glucagon-related substances in tissue and circulation by RIA has profoundly influenced understanding of the control of the biosynthesis and release of these peptides and their action on target cells in a number of species. However, it is important to be alert to a number of potential pitfalls in the RIA which may invalidate results. Except for the guinea pig, all mammals share the common primary sequence for glucagon, thus, absolute IRG levels amongst mammalian species are comparable. Owing to weaker affinities of the guinea pig, avian, and amphibian glucagons for antisera generated against porcine glucagon, their comparison with other species are qualitative, although quantitative changes within species are valid. Troubleshooting for emergent and recurrent problems in the glucagon RIA is similar to that for other hormones. Awareness of factors such as lack of sensitivity of the assay system, chemical interference with the antigen–antibody reaction, incubation damage of labeled antigen or antibody, and nonspecificity of antisera may contribute to spurious values (UNGER 1972; MOODY 1972). General approaches to the recognition and solution of these problems are discussed by YALOW and STRAUS (1980) and BRYANT (1978).

Although glucagon has been quantified under a variety of conditions in tissue extracts, and media from perfused organs and tissue culture, its measurement in plasma is most likely to be fraught with difficulties, and precautions must be taken to avoid inclusion of artifacts. Furthermore, since plasma contains species of gut GLI and IRG with different controls of their secretion and metabolic fates, the unwary investigator may be misled by assigning specificity of a determination of a level of IRG or GLI for glucagon in unmodified ("native") plasma. Thus, in the following sections, emphasis is placed on the measurements of glucagon-related peptides in plasma which apply in a qualitative sense in all mammals so far examined. Because most experience has been gained in study of human plasma, discussion will be restricted to observations in humans. Further details of peptide measurements are given in Chap. 11.

## I. Collection and Processing of Plasma

As is the case for several peptide hormones, glucagon-related polypeptides are susceptible to degradation by proteases in plasma. MIRSKY et al. (1958) first showed that glucagon $^{131}$I is degraded by serine proteases, i.e., plasmin and trypsin. EISENTRAUT et al. (1968 a) found that destruction of the radioligand during incubation with plasma in the RIA created spuriously high values, wrongly attributed to glucagon. This artifact can be overcome by including protease inhibitors in the assay buffer (EISENTRAUT et al. 1968 a; ENSINCK et al. 1972). Whether or not significant losses of endogenous hormones occur during plasma collection and storage is moot. NONAKA and FOÀ (1969) and HEDING (1971) reported significant losses of IRG in blood samples that did not contain protease inhibitors. In contrast, EISENTRAUT et al. (1968 a) and HAZZARD et al. (1968) did not observe any disappearance of IRG in unprotected blood. HENDRIKS and BENRAAD (1981) have systematically addressed this question and were unable to demonstrate any loss of IRG in human plasma when blood was collected and left at room temperature for up to 7 h. This is in accord with the generally low levels of serine protease activity in plasma unless their precursors are activated. Nevertheless, as protease activities vary among species (MIRSKY et al. 1959) and inactive precursors may be activated by glass or freezing and thawing, it seems prudent to attempt to minimize any potential losses from proteolysis. Therefore, we recommend that each 1 ml blood be collected into aprotinin, 500 KIU and 1.2 mg EDTA (EISENTRAUT et al. 1968a) or 0.1 ml benzamidine hydrochloride (1 *M*) (ENSINCK et al. 1972). A list of other enzyme inhibitors that may be suitable for glucagon RIA are given by GOODFRIEND and ODYA (1978). Blood is kept in ice until plasma is separated from red cells by centrifugation (ideally in a refrigerated centrifuge) and subsequently frozen at −20 °C. Repeated freezing and thawing of plasma samples are to be avoided.

## II. Contribution of Different Species to Plasma Levels

There is abundant evidence that several components of IRG and gut GLIs occur in mammalian plasma, reflecting the secretion and clearance of glucagon-related peptides from pancreas, stomach, and gut (see Chaps. 11 and 33). By chromatography of total plasma IRG measured by 30K antiserum, VALVERDE et al. (1970b, 1974) were the first to show that, in addition to pancreatic glucagon (3,500 daltons), three components of molecular weight 160,000, 9,000, and 2,000 daltons were present in peripheral venous plasma under basal conditions in normal humans. These observations have subsequently been confirmed (KUKU et al. 1976; PALMER et al. 1978).

Gut GLIs in plasma have been less extensively studied, yet it is clear that they are also heterogeneous and are likely to be the same molecular size as that found in gut extracts (VALVERDE et al. 1970b; JACOBSEN et al. 1977; TAGER and STEINER 1973; TAGER and MARKESE 1979; SUNDBY and MOODY 1980). Control of the secretion and metabolic fate of glucagon-related peptides in normal circumstances and disease states are dealt with in greater detail elsewhere in this volume (see Chaps. 23–38). It is to be emphasized here that levels of IRG or gut GLI in native plasma, either in the basal state or under conditions where these levels are perturbed, should take into account the probability that the various species may have different rates of release and clearance, and therefore the measurement of total IRG or GLI levels may be inexact indicators of changes of the specific substances.

Since most reports of plasma IRG levels have not specified the cross-reactivity among the different components within plasma, comparisons of the values obtained amongst the different antisera in terms of absolute measurement might be predicted to be highly variable. In Table 3 are given levels of IRG in "native" and "extracted" plasma of healthy subjects after overnight fast as measured by several "glucagon-specific" antisera. Because of immunoheterogeneity, it seems most appropriate to express levels of IRG in "native" and "extracted" plasma as "equivalents" to the 3,500 daltons glucagon used as standards in the RIA. It is of some surprise that the mean values of IRG registered by the different antiserum are not more discordant. However, based on the significantly lower values obtained with "extraction" procedures and the distribution of IRG components by gel filtration, it seems likely that the "glucagon-specific" antisera all recognize a fraction with molecular weight ~160,000 daltons and that the concentrations of the component vary amongst individuals. This component has been designated "big plasma glucagon" (BPG) by VALVERDE et al. (1974), "interference factor" by WEIR et al. (1973), "peak A" by KUKU et al. (1976). The identity, origin, and biologic function of BPG

**Table 3.** Plasma levels of immunoreactive glucagon (IRG) in normal humans after an overnight fast

| Anti-glucagon serum (COOH terminal) | Reference | IRG (pg equiv./ml) | | |
|---|---|---|---|---|
| | | Number | Mean | Standard error |
| "Native" plasma | | | | |
| G58 | AQUILAR-PARADA et al. (1969) | 16 | 150 | ±20 |
| G58 | MÜLLER et al. (1970) | 11 | 126 | ±15 |
| G58 | HARRIS et al. (1978) | 59 | 75 | ± 4 |
| 30K | WEIR et al. (1975) | 30 | 106 | ±57 |
| 30K | KUKU et al. (1976) | 32 | 113 | ± 9 |
| 30K | VALVERDE (1977) | 20 | 155 | ±10 |
| G9I | PEK et al. (1972b) | 34 | 84 | ± 5 |
| GI5 | SPERLING et al. (1974) | 10 | 131 | ±31 |
| G42 | OHNEDA et al. (1979) | 10 | 127 | ±39 |
| "Extracted" plasma | | | | |
| K47 | HEDING and RASMUSSEN (1972)[a] | 29 | 330 | ±30 |
| YY89 | KALK et al. (1974)[a] | 7 | 121 | ±24 |
| K814 | CZYZYK et al. (1975)[a] | 17 | 210 | ±12 |
| 30K | WEIR et al. (1975)[b] | 30 | 34 | ±20 |
| 30K | DUDL and ENSINCK (1977)[c] | 34 | 40 | ± 3 |
| RCS5 | ALFORD et al. (1977)[d] | 18 | 24 | ± 3 |

[a] Ethanol extraction
[b] Charcoal absorption
[c] Acetone extraction
[d] Immunoabsorption

have not been resolved. As yet there is no compelling evidence that BPG is a biosynthetic precursor, a polymer of glucagon, or glucagon bound to a protein. UNGER and EISENTRAUT (1967) first noted that globulin solutions interfered with the binding of labeled glucagon to some glucagon antibodies. VON SCHENCK (1977) speculated that BPG may be an immunoglobulin, containing a sequence of amino acids homologous with glucagon. In support of this conjecture, JULLIARD et al. (1980) reported that a high molecular weight immunoreactive $\beta$-endorphin found in human placenta is a fragment of immunoglobulin G. As reviewed by VALVERDE in Chap. 11, BPG levels comprise the highest proportion of total IRG in plasma under basal conditions in normal subjects.

In addition, a species with a molecular weight of 9,000 daltons was also detected in 75% of humans. This component of IRG may be proglucagon (NOE and BAUER 1975). The 3,500 daltons IRG, presumed to represent pancreatic glucagon, accounted for 17% of the total. IRG of approximate molecular weight 2,000 daltons was found in 85% of humans examined. The nature of this component is unknown, but it may represent a glucagon fragment. A substance (or substances) reacting with 30K antiserum with molecular weight between 10 and 20,000 daltons has also been described in several members of a kindred with extremely high levels of total glucagon immunoreactivity (PALMER et al. 1978). Thus, despite use of antisera with specificity for the COOH terminus of glucagon, it is obviously improper to equate total IRG with the 3,500 daltons pancreatic glucagon moiety.

Currently, the only precise method for quantifying the relevant contribution of the different components comprising total IRG in plasma is by physical separation such as gel filtration. This procedure is time consuming and few samples can be conveniently processed. Because BPG contributes variably, but usually represents the greatest amount of total IRG and is unaltered acutely by physiologic signals that modify the secretion of glucagon-related peptides from the A cell, it would be advantageous to have a simple and reliable method to enable separation of BPG from the group of lower molecular weight glucagon-related peptides. WEIR et al. (1973, 1975) have made use of the property of activated charcoal to absorb peptides of small molecular weight. The method entails treatment of plasma with charcoal followed by centrifugation. The residual IRG in the supernatant represents BPG or "interference factor." The contribution of the lower molecular weight IRG species is obtained by subtracting the measurement of BPG from total IRG (Table 3). Unfortunately, this subtraction is cumbersome and the differences may be small and within assay variance, thus diminishing confidence in the measurement of the group of low molecular weight IRG species. ALFORD et al. (1977) have circumvented the measurement of BPG by submitting plasma to affinity chromatography with a nonspecific glucagon antiserum bound to agarose. Since BPG is not bound to the immunoabsorbant, the filtered plasma from each individual is used to derive a standard curve with which plasma not treated by affinity chromatography is compared. Mean levels of IRG of $25 \pm 6$ pg/ml (18 samples) were found in basal human plasma comparable to those measured by the subtraction technique of WEIR et al. (1975) ($34 \pm 20$ pg/ml; 30 samples) although lower than the combined measurements of 9,000, 3,500, and 2,000 daltons IRG components. However, the method is tedious, specialized, and therefore not generally applicable.

Alternative techniques for separating BPG from smaller IRG components make use of the solubility of the latter in organic solvents such as alcohol or acetone. HEDING (1971) has routinely added 96% ethanol to plasma (1.8: 1; v: v). The supernatant obtained after centrifugation is decanted into a bottle and, after desiccation, is reconstituted into assay buffer. Recovery of pancreatic glucagon was reported to be greater than 80%. WALTER et al. (1974) described an analogous procedure using acetone, with mean recoveries of pancreatic glucagon of 70%. Extraction procedures are laborious, however, requiring specialized equipment and more importantly, nonspecific factors interfering in the RIA have been detected in extracts from glucagon-free plasma (J. W. ENSINCK and C. NIST 1978, unpublished work). In this context, it is of interest to note that the highest values for IRG were measured using HEDING's antisera in "extracted" plasma (Table 3).

We have recently developed an alternative method to separate BPG from smaller glucagon-related peptides by precipitation of the former using polyethylene glycol (PEG) at concentrations that selectively precipitate $\gamma$-globulins (CHESEBRO and SVEHAG 1968). PEG 6000 (Eastman Kodak, Rochester, New York) is diluted in distilled water to final concentrations of 26% and 13%. To a volume of plasma is added an equal volume of 26% PEG. Following thorough mixing by a vortex instrument, the solution is kept at 4 °C for at least 1 h with mixing two or

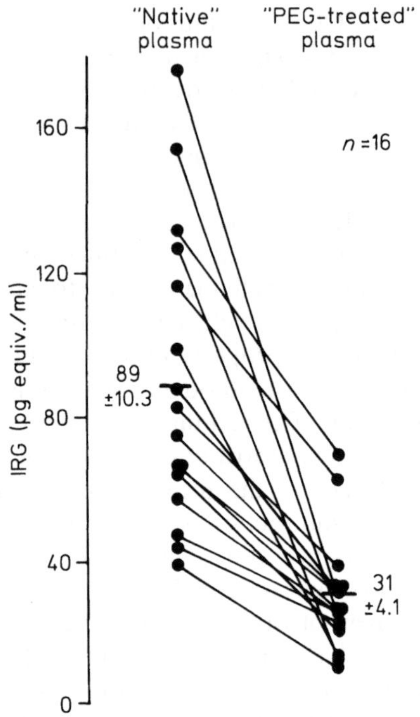


**Fig. 3.** Measurement of levels of immunoreactive glucagon (IRG) equivalents in "native" and polyethyleneglycol- (PEG)-treated plasma from healthy men and women after overnight fast. Numbers refer to mean ± standard error

three times during that period. The solution is centrifuged for 20 min at 3,000 rpm and the supernatant decanted into clean tubes. Volumes of up to 0.4 ml supernatant may be measured in the glucagon RIA using 30K antiserum. Solutions with glucagon standard contain 13% PEG. Recovery of added glucagon to glucagon-free plasma is 90%. Preliminary measurements of total PEG-treated plasma from 16 healthy volunteers are shown in Fig. 3. The mean values of 31 ± 4 pg/ml are similar to those measured by ALFORD et al. (1977). The use of PEG is relatively simple in that it does not necessitate evaporation or special apparatus and artifacts have not as yet been detected. The main requirement is an antiserum in which the affinity is not altered to the presence of 13% PEG. Although this procedure offers an advantage in eliminating BPG, it will not distinguish among the low molecular weight species. In practical terms, however, the major fluctuation in levels occurs with the 3,500 daltons IRG, therefore the procedure enables a more precise measurement of small changes in glucagon.

Levels of gut GLIs in "native" plasma of healthy persons after an overnight fast are displayed in Table 4. Because of the likelihood of multiple forms (VALVERDE et al. 1970a) and the possible cross-reactivity with IRG, the measurement of gut GLIs under basal conditions is likely to be inaccurate (MOODY and SUNDBY 1980). On the other hand, the rise in gut GLI after the ingestion of glucose is considered to be a reliable estimate of the increase in gut GLIs in peripheral blood. Following the intake of food containing protein, however, and in diabetic subjects after oral

**Table 4.** Plasma levels of gut glucagon-like immunoreactivity (GLI) in normal humans after an overnight fast

| Anti-glucagon sera | Reference | GLI [c] (pg equiv./ml) | | |
|---|---|---|---|---|
| | | Number | Mean | Standard error |
| K36 [a] K47 [b] | HEDING and RASMUSSEN (1972) | 29 | 480 | ± 60 |
| 78J [a] 30K [b] | BÖTTGER et al. (1973) | 4 | 600 | ±100 |
| YY57 [a] YY89 [b] | KALK et al. (1974) | 8 | 43 | ± 10 |
| K52 [a] K814 [b] | CZYZYK et al. (1975) | 17 | 210 | ± 12 |
| K57 [a] YY89 [b] | BOTHA et al. (1976) | 5 | 312 | ± 76 |

[a] Nonspecific antiserum
[b] COOH terminal directed antiserum
[c] Measured by subtraction of values with [b] from those with [a]

glucose administration, various species of IRG are released from the gastrointestinal cells and the pancreas, and at present cannot be differentiated from gut GLIs which are secreted from the lower small intestine. MOODY and SUNDBY (1980) have reported that the RIA specific for glicentin has enabled the quantification of this putative precursor of glucagon-related peptides in pig plasma. Fasting levels of 100 pmol/l were found with a 2- to 3-fold increase after feeding. Thus, the wider availability of glicentin-specific antisera in conjunction with pancreatic glucagon-specific antibody should allow the elucidation of the various factors regulating the processing of glucagon precursors to their active and inactive products.

## D. Summary

It is clear from two decades of experience that the RIA for glucagon has been pivotal in probing the role of glucagon in fuel homeostasis. The initial confusion in measuring glucagon has been resolved with the clarification of the relationship of this peptide within a family of related substances that share common immunodeterminants. The development of more sophisticated RIAs linked with techniques for separating molecular species has significantly advanced our understanding of the sources of these peptides in gastrointestinal and pancreatic cells and offers the prospect of opening up new areas for investigation in the control of their secretion and action. It seems probable that with isolation and purification of gut GLIs, such as glicentin, specific RIA will be developed enabling the quantification of all glucagon-related peptides.

## References

Aguilar-Parada E, Eisentraut AM, Unger RH (1969) Pancreatic glucagon secretion in normal and diabetic subjects. Am J Med Sci 257:415–419

Alford FP, Bloom SR, Nabarro JDN (1977) Glucagon levels in normal and diabetic subjects: use of a specific immunoabsorbent for glucagon radioimmunoassay. Diabetologia 13:1–6

Assan R (1973) Gut glucagon. In: Berson SA, Yalow RS (eds) Methods in investigative and diagnostic endocrinology. Part II: Non-pituitary hormones. Elsevier North Holland, Amsterdam London New York, p 888

Assan R, Slusher N (1972) Structure/function and structure/immunoreactivity relationships of the glucagon molecule and related synthetic peptides. Diabetes 21:843–855

Assan R, Rosselin G, Drouet J, Dolais J, Tchobroutsky G (1965) Glucagon antibodies. Lancet 2:590–591

Assan R, Drouet J, Rosselin G, Wünsch E, Schröder E (1969) Étude radio-immunologique de glucagon naturel et synthétique et de peptides synthétiques apparentés. Pathol Biol (Paris) 17:757–762

Berson SA, Yalow RS, Baumann A, Rothschild MA, Newerly K (1956) Insulin $^{131}$I metabolism in human subjects: demonstration of insulin-binding globulin in the circulation of insulin treated subjects. J Clin Invest 35:170–190

Böttger I, Dobbs R, Faloona GR, Unger RH (1973) The effects of triglyceride absorption upon glucagon, insulin, and gut glucagon-like immunoreactivity. J Clin Invest 52:2533–2541

Botha JL, Vinik AI, Brown JC (1976) Gastric inhibitory polypeptide (GIP) in chronic pancreatitis. J Clin Endocrinol Metab 42:791–797

Bromer WW (1972) Chemical and physical properties of pancreatic glucagon. In: Lefèbvre PJ, Unger RH (eds) Glucagon molecular physiology. Clinical and therapeutic implications. Pergamon, Oxford New York Toronto Sydney Braunschweig, p 27

Bryant MG (1978) Plasma artefacts in radioimmunoassay. In: Bloom SR (ed) Gut hormones. Churchill Livingstone, Edinburgh London, p 120

Buchanan KD, McCarrol AM (1971) Comparison of methods of separation of free from bound hormone in the radioimmunoassay of insulin and glucagon. In: Kirkham KE, Hunter WM (eds) Radioimmunoassay methods. ES Livingstone, Edinburgh, p 266

Chesebro B, Svehag SE (1968) Precipitation of human serum proteins by polethyleneglycol. Clin Chim Acta 20:527–529

Conlon JM (1980) The glucagon-like polypeptides – order out of chaos? Diabetologia 18:85–88

Czyzyk A, Heding LG, Malczewski B, Miedzinska E (1975) The effect of phenformin upon the plasma pancreatic and gut glucagon-like immunoreactivity in diabetics. Diabetologia 11:129–133

Davis SE, Munson PJ, Jaffe M, Rodbard D (1980) Radioimmunoassay data processing with a small programmable calculator. J Immunoassay 1:15–25

Dudl RJ, Ensinck JW (1977) Insulin and glucagon relationships during aging in man. Metabolism 26:33–41

Eisentraut AM, Whissen N, Unger RH (1968 a) Incubation damage in the radioimmunoassay for human plasma glucagon and its prevention with "trasylol". Am J Med Sci 225:137–142

Eisentraut AM, Ohneda A, Parada E, Unger RH (1968 b) Immunologic discrimination between pancreatic glucagon and enteric glucagon-like immunoreactivity (GLI) in tissues and plasma (Abstr). Diabetes [Suppl 1] 17:321–322

Ensinck JW, Shepard C, Dudl RJ, Williams RH (1972) Use of benzamidine as a proteolytic inhibitor in the radioimmunoassay of glucagon in plasma. J Clin Endocrinol Metab 35:463–467

Faloona GR (1972) The structure-function relationships of pancreatic glucagon. In: Lefèbvre PJ, Unger RH (eds) Glucagon molecular physiology. Clinical and therapeutic implications. Pergamon, Oxford New York Toronto Sydney Braunschweig, p 201

Faure A, Nemoz C, Claustrat B, Paultre CZ, Site J (1980) Control of routine radioimmunoassays: a computer program for calculation of control charts for precision and accuracy. Comput Programs Biomed 12:105–110

Flanagan RWJ, Buchanan KD, Murphy RF (1974) Specificity of antibodies in radioimmunoassay for glucagon (Abstr). Diabetologia 10:365

Garaud JC, Moody AJ, Eloy R, Grenier JF (1976) Unusual specificities of antibodies to glucagon-glutaraldehyde-albumin conjugates. Horm Metab Res 8:241–243

Goodfriend TL, Odya CE (1978) Bradykinin. In: Jaffe BM, Behrman HR (eds) Methods of hormone radioimmunoassay, 2nd edn. Academic Press, New York London, p 909

Greenwood FC, Hunter WM, Glover JS (1963) The preparation of $^{131}$I-labeled human growth hormone of high specific radioactivity. Biochem J 89:114–123

Harris V, Faloona GR, Unger RH (1978) Glucagon. In: Jaffe BM, Behrman HR (eds) Methods of hormone radioimmunoassay, 2nd edn. Academic Press, New York San Francisco London, p 643

Hazzard WR, Crockford PM, Buchanan KD, Vance JE, Chen R, Williams RH (1968) A double antibody immunoassay for glucagon. Diabetes 17:179–186

Heding LG (1969) The production of glucagon antibodies in rabbits. Horm Metab Res 1:87–88

Heding LG (1971) Radioimmunological determination of pancreatic and gut glucagon in plasma. Diabetologia 7:10–19

Heding LG (1972) Immunological properties of pancreatic glucagon: antigenicity and antibody characteristics. In: Lefèbvre PJ, Unger RH (eds) Glucagon, molecular physiology. Clinical and therapeutic implications. Pergamon, Oxford New York, p 187

Heding LG, Rasmussen SM (1972) Determination of pancreatic and gut glucagon-like immunoreactivity (GLI) in normal and diabetic subjects. Diabetologia 8:408–411

Heding LG, Frandsen EK, Jacobsen H (1976) Structure-function relationship: immunologic. Metabolism [Suppl 1] 25:1327–1329

Hendriks T, Benraad TJ (1981) On the stability of immunoreactive glucagon in plasma samples. Diabetologia 20:553–557

Holst JJ (1978) Extra-pancreatic glucagon. Digestion 17:168–190

Holst JJ (1980) Evidence that glicentin contains the entire sequence of glucagon. Biochem J 187:337–343

Holst JJ, Aasted B (1974) Production and evaluation of glucagon antibodies for radioimmunoassay. Acta Endocrinol (Copenh) 77:715–726

Imagawa K, Nishino T, Shin S, Uehata S, Hashimura E, Yanaihara C, Yanaihara N (1979) Production of anti-glucagon sera with a C-terminal fragment of pancreatic glucagon. Endocrinol Jpn 26:123–131

Jacobsen H, Demandt A, Moody AJ, Sundby F (1977) Sequence analysis of porcine gut GLI-I. Biochim Biophys Acta 493:452–459

Jørgensen KH, Larsen UD (1972) Purification of $^{125}$I glucagon by anion exchange chromotography. Horm Metab Res 4:223–224

Julliard JH, Shibasaki T, Ling N, Guillemin R (1980) High-molecular weight immunoreactive β-endorphin in extracts of human plasma is a fragment of immunoglobulin G. Science 208:183–185

Kalk WJ, Vinik AI, Bank S, Buchanan KD, Keller P, Jackson WPU (1974) Glucagon responses to arginine in chronic pancreatitis. Possible pathogenic significance in diabetes. Diabetes 23:257–263

Kennett RH, McKearn TJ, Bechtol KB (eds) (1980) Monclonal antibodies: hybridomas: a new dimension in biological analyses. Plenum, New York London

Kuku SF, Jaspan JB, Emmanouel DS, Zeidler A, Katz AI, Rubenstein AH (1976) Heterogenity of plasma glucagon. Circulating components in normal subjects in patients with chronic renal failure. J Clin Invest 58:742–750

Lawrence AM (1966) Radioimmunoassayble glucagon levels in man: effects of starvation, hypoglycemia and glucose administration. Proc Natl Acad Sci USA 55:310–320

Leclercq-Meyer V, Miahle P, Malaisse WJ (1970) Une méthode de dosage radioimmunologique du glucagon comportant une séparation par le charbon-dextran. Diabetologia 6:121–129

Linde S, Hansen B, Lernmark A (1980) Stable iodinated polypeptide hormones prepared by polyacrylamide gel electrophoresis. Anal Biochem 107:165–176

Luyckx AS (1972) Immunoassays for glucagon. In: Lefèbvre PJ, Unger RH (eds) Glucagon, molecular physiology. Clinical and therapeutic implications. Pergamon, Oxford New York, p 285

Mirsky IA, Perisutti G, Davis M (1958) The destruction of glucagon, adrenocorticotrophin and somatotrophin by human plasma. J Clin Invest 38:14–20
Mirsky IA, Perisutti G, Davis NC (1959) The destruction of glucagon by the blood plasma from various species. Endocrinology 64:992–1001
Moody AJ (1972) Gastrointestinal glucagon-like immunoreactivity. In: Lefèbvre PJ, Unger RH (eds) Glucagon molecular physiology. Clinical and therapeutic implications. Pergamon, Oxford New York, p 319
Moody AJ, Sundby F (1980) Radioimmunoassay of gut glucagon-like immunoreactants. In: Jerzy Glass GB (ed) Gastrointestinal hormones. Raven, New York, p 831
Moody AJ, Jacobsen H, Sundby F (1978) Gastric glucagon and gut glucagon-like immunoreactants. In: Bloom SR (ed) Gut hormones. Churchill Livingstone, Edinburgh London, p 369
Müller WA, Faloona GR, Aguilar-Parada E, Unger RH (1970) Abnormal alpha-cell function in diabetes. Response to carbohydrate and protein ingestion. N Engl J Med 283:109–115
Nonaka K, Foà PP (1969) A simplified glucagon immunoassay and its use in a study of incubated pancreatic islets. Proc Soc Exp Biol Med 130:330–336
Noe BD, Bauer E (1975) Evidence for sequential metabolic cleavage of pro-glucagon to glucagon in glucagon biosynthesis. Endocrinology 97:868–877
Ohneda K, Watanabe K, Wakimatsu M, Fujino M (1979) Production of a specific antiserum by synthetic C-terminal fragment of glucagon. Horm Metab Res 11:463–468
Palmer JP, Werner PL, Benson JW, Ensinck JW (1978) Dominant inheritance of large molecular weight immunoreactive glucagon. J Clin Invest 61:763–769
Palmieri GMA, Yalow RS, Berson SA (1971) Absorbent techniques for the separation of antibody-bound from free peptide hormones in radioimmunoassay. Horm Metab Res 3:301–305
Patzelt C, Tager HS, Carrol RJ, Steiner DF (1979) Identification and processing of proglucagon in pancreatic islets. Nature 282:260–266
Pek S, Fajans SF, Floyd JC Jr, Knopf RF, Conn JW (1972a) Failure of sulfonylureas to suppress plasma glucagon in man. Diabetes 21:216–223
Pek S, Fajans SS, Floyd JC Jr, Knopf RF, Weissman PN, Conn WJ (1972b) Plasma levels of glucagon in patients with diabetes mellitus (Abstr). Diabetes [Suppl 1] 21:324
Raab GM, Thompson R, McKenzie (1980) Variance function estimation for immunoassays. Comput Programs Biomed 12:111–120
Rodbard D (1974) Statistical quality control and routine data processing for radioimmunoassays (RIA) and immunoradiometric assays (IRMA). Clin Chem 20:1255–1270
Rodbard D, Frazier GR (1975) Statistical analysis of radioligand assay data. Methods Enzymol 37 B:3
Seidah NG, Dennis M, Corvol P, Rochemont J, Chrétien M (1980) The rapid high-performance liquid chromotography purification method of iodinated polypeptide hormones. Anal Biochem 109:185–191
Senyk G, Williams EB, Nitecki D, Goodman JW (1971) The functional dissection of an antigen molecule: specificity of humoral and cellular immune responses to glucagon. J Exp Med 133:1294–1308
Senyk G, Nitecki DE, Spitler L, Goodman JW (1972) The immune response to glucagon in conjugated form. Immunochemistry 9:97–110
Shima K, Foà PP (1968) A double-antibody assay for glucagon. Clin Chim Acta 22:511–520
Shima K, Sawazaki N, Tanaka R, Tarui S, Nishikawa M (1975) Effect of an exposure to chloramine-T on the immunoreactivity of glucagon. Endocrinology 96:1254–1260
Smigel MD, Lazar JD (1981) Improved fitting of radioimmunoassay data by Scatchard analysis. Int J Biomed Comput 12:189–203
Sperling MA, DeLamater PV, Kazenelson M, Fiser RH, Fisher DA (1974) Development and application of a radioimmunoassay for plasma glucagon. Clin Chem 20:566–570
Sundby F (1976) Species variations in the primary structure of glucagon. Metabolism [Suppl 1] 25:1319–1321

Sundby F, Moody AJ (1980) Gut glucagon-like reactants (Gut GLI's): isolation, structure and possible role. In: Jerzy Glass GB (ed) Gastrointestinal hormones. Raven, New York, p 307
Sundby F, Jacobsen H, Moody AJ (1976) Purification and characterization of a protein from porcine gut with glucagon-like immunoreactivity. Horm Metab Res 8:366–371
Tager HS, Markese J (1979) Intestinal and pancreatic glucagon-like peptides. Evidence for identity of higher molecular weight forms. J Biol Chem 254:2229–2233
Tager HS, Steiner DF (1973) Isolation of a glucagon-containing peptide: primary structure of a possible fragment of proglucagon. Proc Natl Acad Sci USA 70:2321–2325
Tager HS, Hohenboken M, Markese J (1977) High titer glucagon antisera. Endocrinology 100:367–372
Thim L, Moody AJ (1981) The primary structure of porcine glicentin (proglucagon). Regul Peptides 2:139–148
Tower BB, Clark BR, Rubin RT (1977) Preparation of $^{125}I$ polypeptide hormones for radioimmunoassay using glucose oxidase with lactoperoxidase. Life Sci 21:959–966
Unger RH (1972) Glucagon and glucagon immunoreactivity in plasma and pancreatic tissues. In: Lefèbvre PJ, Unger RH (eds) Glucagon molecular physiology. Clinical and therapeutic implications. Pergamon, Oxford New York Toronto Sydney Braunschweig, p 205
Unger RH, Eisentraut AM (1967) Études récentes sur la physiologie du glucagon. In: Journées annuelles de diabètologie de l'Hôtel-Dieu. Flammarion, Paris, pp 7–18
Unger RH, Eisentraut A, McCall MS, Keller S, Lanz HC, Madison LL (1959) Glucagon antibodies and their use for immunoassay for glucagon. Proc Soc Exp Biol 102:621–623
Unger RH, Eisentraut AM, McCall MS, Madison LL (1961) Glucagon antibodies and an immunoassay for glucagon. J Clin Invest 40:1280–1289
Vaitukaitis J, Robbins JB, Nieschlag E, Ross ET (1971) A method for producing specific antisera with small doses of immunogen. J Clin Endocrinol Metab 33:988–991
Valverde I (1977) Quantification of plasma glucagon immunoreactive components in normal and hyperglucagonemic states. In: Foà PP, Bajaj JS, Foà NL (eds) Glucagon: its role in physiology and clinical medicine. Springer, Berlin Heidelberg New York, p 77
Valverde I, Rigopoulou D, Marco J, Faloona GR, Unger RH (1970a) Characterization of glucagon-like immunoreactivity (GLI). Diabetes 19:614–623
Valverde I, Rigopoulou D, Marco J, Faloona GR, Unger RH (1970b) Molecular size of extractable glucagon and glucagon-like immunoreactivity (GLI) in plasma. Diabetes 19:624–629
Valverde I, Villanueva ML, Lozano I, Marco J (1974) Presence of glucagon immunoreactivity in the globulin fraction of human plasma (big plasma glucagon). J Clin Endocrinol Metab 39:1090–1098
Von Schenck H (1977) Production and characterization of an antiserum against pancreatic glucagon. Clin Chim Acta 80:455–463
Von Schenck H, Jeppsson JO (1977) Preparation of monoiodotyrosine-13-glucagon. Biochim Biophys Acta 491:503–508
Von Schenck H, Larsson I, Thorell JI (1976) Improved radioiodination of glucagon with the lactoperoxidase method. Influence of pH on iodine substitution. Clin Chim Acta 69:225–232
Walter RM, Dudl RJ, Palmer JP, Ensinck JW (1974) The effect of adrenergic blockade on the glucagon responses to starvation and hypoglycemia in man. J Clin Invest 54:1214–1220
Weir GC, Turner RC, Martin DB (1973) Glucagon radioimmunoassay using antiserum 30K: interference by plasma. Horm Metab Res 5:241–244
Weir GC, Knowlton SD, Martin DB (1975) High molecular weight glucagon-like immunoreactivity in plasma. J Clin Endocrinol Metab 40:296–302
Yalow RS, Straus E (1980) Problems and pitfalls in radioimmunoassay of gastrointestinal hormones. In: Jerzy Glass GB (ed) Gastrointestinal hormone. Raven, New York, p 751
Yanagishita M, Rodbard D (1978) Computer optimization of radioimmunoassays for HCG and estradiol: an experimental evaluation. Anal Biochem 88:1–19

CHAPTER 11

# Heterogeneity of Circulating Glucagon and Glucagon-Like Immunoreactivity

I. VALVERDE

## A. Introduction

Multiple molecular forms of immunoreactive glucagon and glucagon-like immunoreactants have been described in various tissues and in the circulation. Some of these forms, as in the case of many other polypeptide hormones, may be related to glucagon biosynthesis (see Chaps. 6 and 7). At present, it is not clearly established whether the glucagon heterogeneity found in plasma is merely a reflection of that found in pancreatic or extrapancreatic tissues. The heterogeneity of circulating glucagon has been demonstrated by means of plasma gel filtration and subsequent glucagon radioimmunoassay of the eluates with different antisera.

## B. Types of Glucagon Antibodies

As reviewed in detail in Chap. 9, two major types of glucagon antibodies have been recognized in sera of animals (most frequently rabbits) that were immunized with crystalline glucagon. One type was first believed to be specific for glucagon, based upon its weak cross-reactivity with glucagon-like material of intestinal origin (gut GLI) (VALVERDE et al. 1970a). It was later shown that antibodies of this type reacted with a variety of peptides, of larger or smaller size than glucagon, present in pancreas, stomach, intestine, salivary glands, and plasma (RIGOPOULOU et al. 1970; VALVERDE et al. 1974; HOLST 1977; SRIKANT et al. 1977; BHATHENA et al. 1977; LAWRENCE et al. 1977; PEREZ-CASTILLO and BLAZQUEZ 1980). Since these antibodies seem to be directed against the COOH terminus of the glucagon molecule (ASSAN and SLUSHER 1972) they are now referred to as "COOH terminal-specific." Structural studies on glucagon-related peptides suggested that only those which share their COOH terminus with that of glucagon could be bound to this type of antibody (TAGER and MARKESE 1979; MOODY et al. 1977). The other type was traditionally known as GLI cross-reactive or nonspecific, because of its affinity for gut GLI (EISENTRAUT et al. 1968; HEDING 1969). These antibodies were shown to be directed against the $NH_2$ terminal and central regions of the glucagon molecule (ASSAN and SLUSHER 1972; HEDING et al. 1976). The term "$NH_2$ terminal-specific" is used to designate this group of antibodies.

It has been proposed that polypeptides that are recognized by COOH terminal-specific antibodies should be referred to as IRG (immunoreactive glucagon), and polypeptides that react mainly with $NH_2$ terminal-specific antibodies should be referred to as GLI (glucagon-like immunoreactants). To specify these compounds more precisely, their approximate molecular weights should be indicated as superscripts (UNGER 1976a).

The reported data on circulating IRG and GLI components have been obtained using the following anti-glucagon sera.

### I. COOH Terminal-Specific

G 58 and 30 K (Dr. R.H. UNGER, Dallas), RCS 5 (Dr. S.R. BLOOM, London), E 7 (Dr. H. VON SCHENCK, Malmo), 4317 (Dr. J.J. HOLST, Copenhagen), K 964 (Dr. L.G. HEDING, Copenhagen), YY 89 (Dr. K.D. BUCHANAN, Belfast), 2601 (Dr. V. LECLERCQ-MEYER, Brussels), and N6E (Dr. A. OHNEDA, Sendai).

### II. $NH_2$ Terminal-Specific

G 128 and 78 J (Dr. R.H. UNGER, Dallas), R 8 (Dr. M.L. VILLANUEVA, Madrid), YY 92 (Dr. K.D. BUCHANAN, Belfast), AGS (Dr. V. LECLERCQ-MEYER, Brussels), and G 7 (Dr. A. OHNEDA, Sendai).

These antisera were obtained from rabbits immunized with crystalline glucagon, with the exception of antibody N6E, which was raised against the glucagon COOH terminal peptide (15–29).

## C. Plasma COOH Terminal-Specific Antibody-Reacting Components

VALVERDE et al. (1970 b) presented the first information about the molecular size of circulating glucagon. Filtration on Biogel P-10 columns of acid–alcohol extracts of canine plasma, obtained during an amino acid infusion, revealed the existence of two discrete fractions that reacted with a COOH terminal-specific antiserum (G 58). A major one, comprising more than 90% of the recovered immunoreactivity, appeared in the zone of the glucagon $^{131}I$ marker (molecular weight 3,500 daltons) and a minor one in a region of approximately 7,000 daltons, eluting immediately before the insulin $^{131}I$ marker. On an immunoreactive basis, the $IRG^{3500}$ component was equipotent to canine glucagon, in terms of its ability to stimulate glucose release in the perfused rat liver. The $IRG^{7000}$ fraction was thought to correspond to the "large glucagon immunoreactivity" (LGI) of the same molecular weight, described by RIGOPOULOU et al. (1970) in pancreatic extracts, and proposed as a proglucagon candidate. In 1973, WEIR et al. reported the presence of a nonspecific factor in human plasma, which interfered with the glucagon assay. This "interference factor" was not adsorbable by charcoal and represented more than 70% of the 30 K glucagon immunoreactivity. VALVERDE et al. (1974) reported that filtration of untreated human plasma on Biogel P-30 columns yields four glucagon-immunoreactive fractions, as determined with 30 K antiserum: fraction I eluted with the globulins and was termed big plasma glucagon (BPG); fraction II appeared in a molecular weight zone of 9,000 daltons ($IRG^{9000}$); fraction III coeluted with native glucagon ($IRG^{3500}$); and fraction IV appeared immediately before the salt peak ($IRG^{2000}$). WEIR et al. (1975), in further characterization of the "interference factor," found that it corresponded to a glucagon-immunoreactive material with a molecular weight of approximately 160,000 daltons. There is evidence that this material corresponds to BPG (WEIR 1977; VILLANUEVA 1976).

Concerning BPG, it is of note that after treatment with 8 *M* urea/1 *M* acetic acid its initial elution volume was not modified, a finding suggesting that BPG is

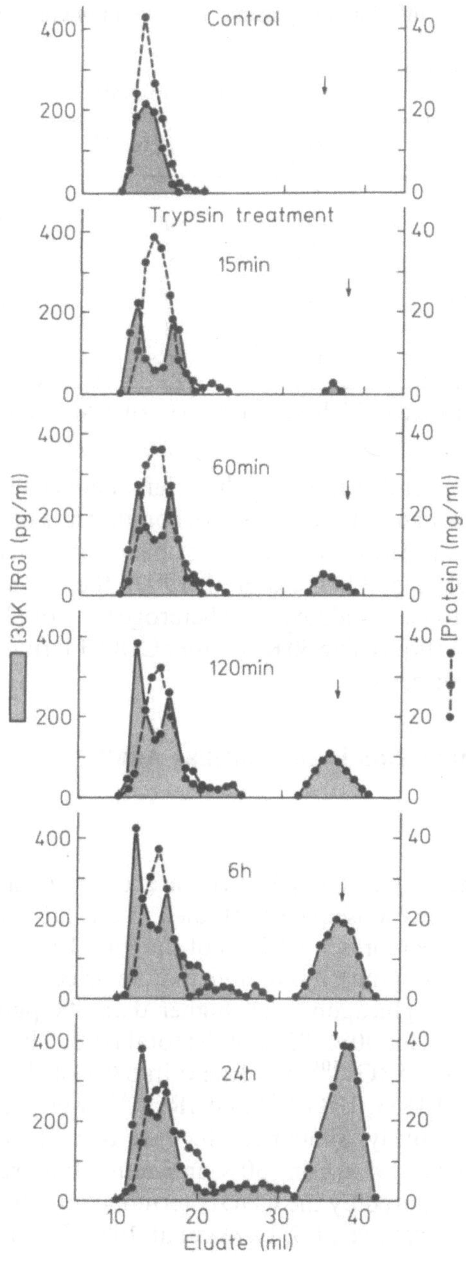


**Fig. 1.** Trypsin treatment of a globulin preparation containing BPG. Filtration on Sephadex G-100. *Arrows* indicate the elution volume of added glucagon $^{125}$I. After VALVERDE et al. (1974)

not a complex of glucagon bound to itself or to another protein by noncovalent bonds. On the other hand, incubation of BPG with trypsin resulted in an increase in the amount of immunoreactivity and generated IRG peptides of smaller size, down to the region of the glucagon marker (Fig. 1). This indicated that BPG contains immunologic glucagon determinants within its structure, which are better ex-

**Table 1.** Effect of arginine infusion on plasma IRG components in 17 normal subjects (Biogel P-30, 30 K)

| | Total plasma | Concentrations (pg/ml) | | | |
|---|---|---|---|---|---|
| | | BPG | $IRG^{9000}$ | $IRG^{3500}$ | $IRG^{2000}$ |
| Basal | | | | | |
| Mean | 241 | 131 | 9 | 23 | 17 |
| Standard error | 25 | 15 | 3 | 5 | 3 |
| Post-arginine | | | | | |
| Mean | 450 | 131 | 7 | 225 | 20 |
| Standard error | 51 | 17 | 3 | 36 | 6 |
| $P$[a] | < 0.01 | n.s. | n.s. | < 0.001 | n.s. |

[a] Student's $t$-test in comparison with basal values; n.s. = not significant

posed to the antibody after disruption of the original molecule or molecules (VALVERDE et al. 1974). Glucagon-immunoreactive components of sizes comparable to that of BPG have been described in extracts of pancreas and stomach (VALVERDE et al. 1974; SRIKANT et al. 1977; TUNG et al. 1980); their exact nature, as well as that of BPG, remains to be elucidated. The heterogeneity of plasma IRG has been confirmed by several authors using 30 K or other COOH terminal-specific antibodies, as will be further discussed.

## I. Plasma IRG Components in the Normal Adult

### 1. Basal State

#### a) Humans

Table 1 contains the data reported by VALVERDE et al. (1974) and those from seven additional normal subjects subsequently studied in our laboratory. In basal samples, the mean BPG value represented 75% of the total IRG, with a wide range in individual values (45–257 pg/ml); it is of note that, in three apparently healthy subjects with fasting plasma glucagon levels higher than 800 pg/ml (not included in Table 1), BPG accounted for 90%–95% of the total IRG. The other fractions were not consistently detected. $IRG^{3500}$ was found in 88% of the subjects and represented 13% of the total IRG; $IRG^{9000}$ and $IRG^{2000}$ were observed in 65% and 71% of the cases, respectively. Similar results have been reported by KUKU et al. (1976a) and RECANT et al. (1976) using 30 K antiserum. It must be emphasized that BPG is not equally recognized by the COOH terminal antisera so far tested (VALVERDE et al. 1974; VON SCHENCK 1977; IWASA et al. 1981; I. VALVERDE 1981, unpublished work).

#### b) Dogs

In a group of 21 adult dogs (Fig. 2), the following concentrations were found: BPG, $76 \pm 18$ pg/ml (mean ± standard error); $IRG^{9000}$, $12 \pm 3$ pg/ml; $IRG^{3500}$, $15 \pm 3$ pg/ml; $IRG^{2000}$, $11 \pm 3$ pg/ml. In two dogs, no BPG was detected. $IRG^{3500}$ was present in 86% of the animals; $IRG^{9000}$ and $IRG^{2000}$ appeared in 67% and 81% of the cases, respectively (VALVERDE and GHIGLIONE 1980).

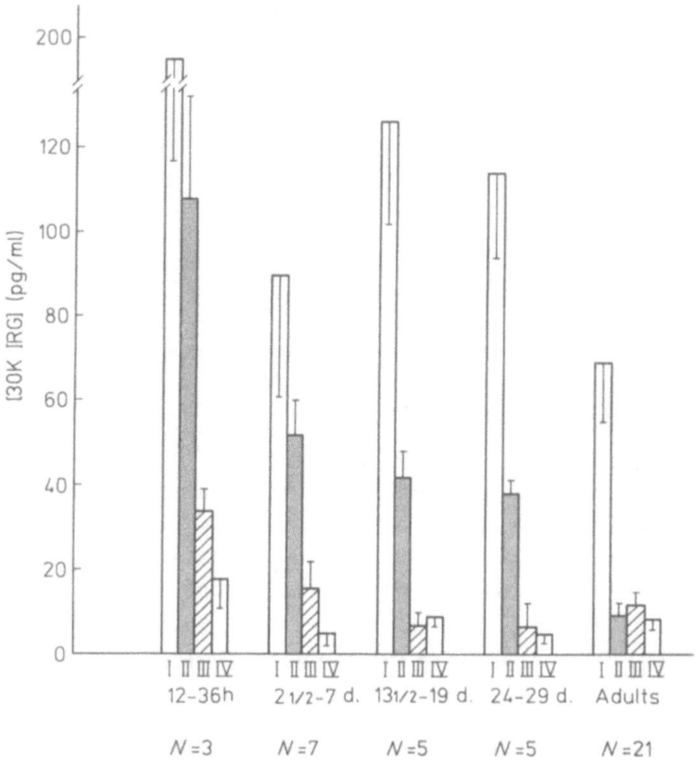


**Fig. 2.** Basal plasma IRG components in dogs of different ages; mean ± standard error. Bio-gel P-30 columns. I = BPG; II = $IRG^{9000}$; III = $IRG^{3500}$; IV = $IRG^{2000}$; $N$ = number of dogs. After VALVERDE and GHIGLIONE (1980)

c) Rats

BPG has also been found in rat peripheral plasma. In a group of six animals, it represented 12% of the total IRG as determined with antiserum 30 K; $IRG^{3500}$ was the predominant component (EMMANOUEL et al. 1976). In a later report from the same laboratory, the following 30 K values were given; BPG 26 ± 3 pg/ml (mean ± standard error); $IRG^{9000}$, 1 ± 8 pg/ml; $IRG^{3500}$, 28 ± 6 pg/ml; $IRG^{2000}$, 5 ± 3 pg/ml (JASPAN et al. 1977).

### 2. After Intravenous Arginine Administration

a) Humans

The infusion of arginine in healthy subjects (Table 1) elicited an increase of $IRG^{3500}$, the other fractions remaining at the basal levels. A similar observation has been reported by KUKU et al. (1976a).

b) Dogs

In adult dogs (Fig. 3), $IRG^{3500}$ is also the only fraction which increases after intravenous arginine administration (VALVERDE and GHIGLIONE 1980).

**Table 2.** Effect of glucose infusion on plasma IRG components in 3 normal subjects (Biogel P-30, 30 K)

| | Total plasma | Concentrations (pg/ml) | | | |
|---|---|---|---|---|---|
| | | BPG | $IRG^{9000}$ | $IRG^{3500}$ | $IRG^{2000}$ |
| Basal | | | | | |
| Mean | 104 | 93 | 14 | 9 | 12 |
| Standard error | 2 | 12 | 4 | 3 | 5 |
| Post-glucose | | | | | |
| Mean | 59 | 75 | | | 7 |
| Standard error | 12 | 12 | | | 4 |

### 3. After Glucose Administration

In three normal subjects (Table 2), $IRG^{3500}$ and $IRG^{9000}$ were not detectable after intravenous glucose infusion. The other fractions remained unchanged (VALVERDE 1977). Oral glucose did not significantly affect the basal values of IRG components (see 30 K values in Fig. 9). KUKU et al. (1976a) reported that, in normal subjects, glucose ingestion was followed by a significant decline of plasma $IRG^{3500}$.

## II. Plasma IRG Components in the Neonatal Period

The contribution of the different plasma IRG components to hyperglucagonemia of early life was investigated in the dog. Compared with adult values (Fig. 2), $IRG^{9000}$ values were extremely high 12 h after birth; they declined progressively with age and at the fourth week of life still remained significantly elevated. $IRG^{3500}$ was higher than in adults during the first week only; BPG was shown to be elevated in the 12- to 36-h period; and $IRG^{2000}$ did not differ from adult values at any time.

Arginine infusion induced a significant increase in both $IRG^{3500}$ and $IRG^{9000}$ in pups, in contrast to adult dogs, in which arginine did not affect $IRG^{9000}$ (Fig. 3). Somatostatin infusion (100 μg/kg for 30 min) reduced $IRG^{3500}$ to undetectable values in four of five pups aged 3–26 days; $IRG^{9000}$ was reduced to less than one-half of the preinfusion values (VALVERDE and MATESANZ 1977; VALVERDE and GHIGLIONE 1980). These immunoassays were performed using 30 K antiserum; the analysis with RCS 5 yielded similar results, except for BPG which was not detected by this antiserum (VALVERDE et al. 1980a).

## III. Plasma IRG in Pathologic States

### 1. Glucagonoma (see also Chap. 43)

VALVERDE et al. (1976a) were the first to report on the IRG components in the plasma of a patient with a glucagonoma. Chromatographic analysis of this plasma revealed the presence of the four recognized IRG components (Fig. 4). Compared with normal values, the levels of $IRG^{3500}$ and $IRG^{9000}$ were elevated. BPG and $IRG^{2000}$ appeared to be in lower proportions than those commonly found in healthy subjects. Treatment of this patient with diaminotriazenoimidazolecar-

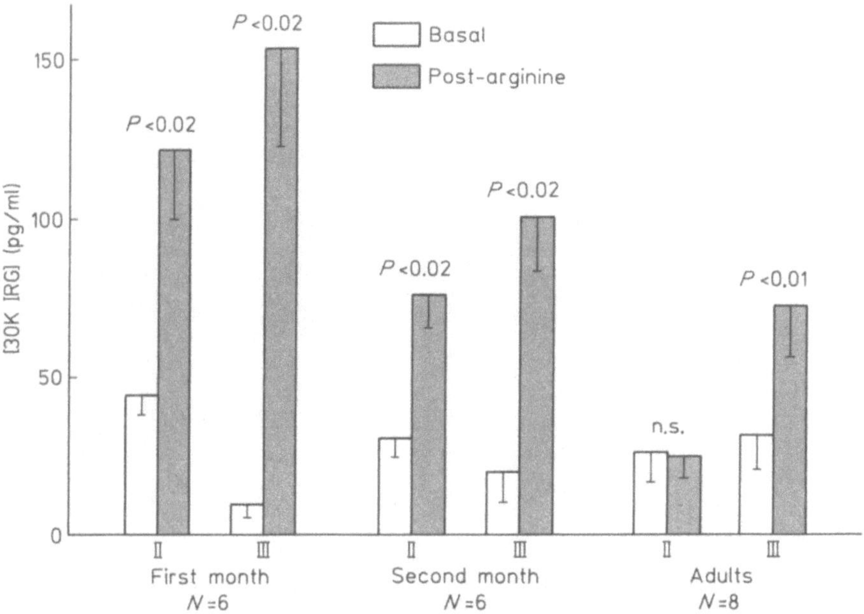


**Fig. 3.** Effect of intravenous arginine on plasma $IRG^{9000}$ (II) and $IRG^{3500}$ (III) components in dogs of different ages; mean ± standard error. First month = 6–12 days; second month = 31–53 days. Biogel P-30 columns. $N$ = number of dogs; n.s. = not significant. After VALVERDE and GHIGLIONE (1980)

boximide resulted in a decrease in both $IRG^{3500}$ and $IRG^{9000}$, while BPG and $IRG^{2000}$ remained unchanged (Fig. 4).

Large quantities of plasma $IRG^{9000}$ accompanying high plasma levels of $IRG^{3500}$ has been a constant finding in all glucagonoma patients thus far reported. In the plasma collected from a tumor-draining vein of two patients, the concentration of $IRG^{3500}$ and $IRG^{9000}$ was higher than that in peripheral plasma (WEIR et al. 1977). JASPAN and RUBENSTEIN (1977) showed that $IRG^{9000}$ and $IRG^{3500}$ obtained from plasma of two glucagonoma patients diluted out in parallel to the glucagon standard, whereas BPG did not. This has been confirmed by us (I. VALVERDE 1977, unpublished work). Conversely, RECANT et al. (1976) reported that, in three cases, dilution of the four fractions resulted in parallel lines to dilutions of standard glucagon. Furthermore they found that all these fractions possessed glucagon-like activity in terms of stimulating cyclic AMP generation in isolated hepatocytes. However, as pointed out by these authors, the latter results must be interpreted with caution as the plasma fraction tested might contain substances interfering in the bioassay system.

There is little information about plasma IRG components in glucagonoma patients after stimulation or inhibition of A-cell secretion. WEIR et al. (1977) reported that in two patients, intravenous administration of arginine was followed by a clearcut increase of $IRG^{3500}$, but there was also an elevation of $IRG^{9000}$. In two patients, oral glucose evoked a paradoxical rise of circulating IRG; this was shown to be due to increments in $IRG^{3500}$ and $IRG^{9000}$. In one case, intravenous glucose reduced $IRG^{3500}$ and $IRG^{9000}$.

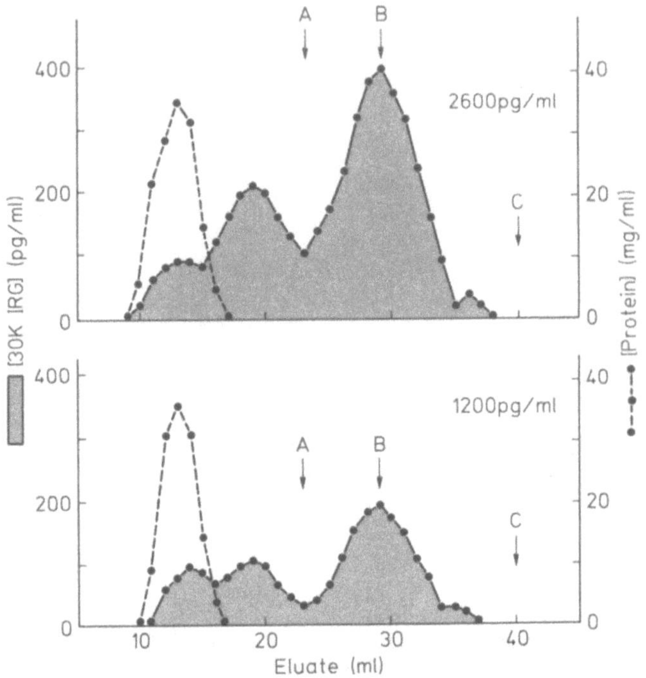


**Fig. 4.** Chromatographic pattern on Biogel P-30 of a glucagonoma patient's plasma (2 ml) before (2,600 pg/ml, total IRG) and after (1,200 pg/ml, total IRG) chemotherapy. *Arrows* indicate the elution volume of added: insulin $^{125}$I (A); glucagon $^{125}$I (B); and Na$^{125}$I (C). After VALVERDE et al. (1976a)

Recently a family with multiple endocrine neoplasia of type I was described (STACPOOLE et al. 1981), in which three members had A-cell pancreatic tumors. The proband had high plasma $IRG^{3500}$ and $IRG^{9000}$ levels. Arginine infusion increased mainly $IRG^{3500}$, but $IRG^{9000}$ was also increased. Intravenous secretin produced a twofold increase in plasma glucagon levels due almost entirely to an increment of $IRG^{3500}$. Somatostatin infusion induced a rapid fall in plasma glucagon levels ($\sim$50% of basal values) which was due to a decrease in $IRG^{3500}$ and $IRG^{9000}$.

All these data on glucagonoma patients were obtained with 30K antiserum. Using antiserum E7, an identical profile has been reported in gel-filtered plasma from an additional glucagonoma patient (VON SCHENCK et al. 1979). Administration of arginine in this patient was followed by a rise in both $IRG^{3500}$ and $IRG^{9000}$, whereas alanine increased only $IRG^{3500}$. After somatostatin infusion, no $IRG^{3500}$ could be detected, but the other fractions remained unaltered.

## 2. Diabetes

### a) Humans

As shown in Fig. 5, gel filtration of basal plasma from three diabetic patients (type I) revealed an increase in $IRG^{3500}$ and also a modest elevation of $IRG^{9000}$,

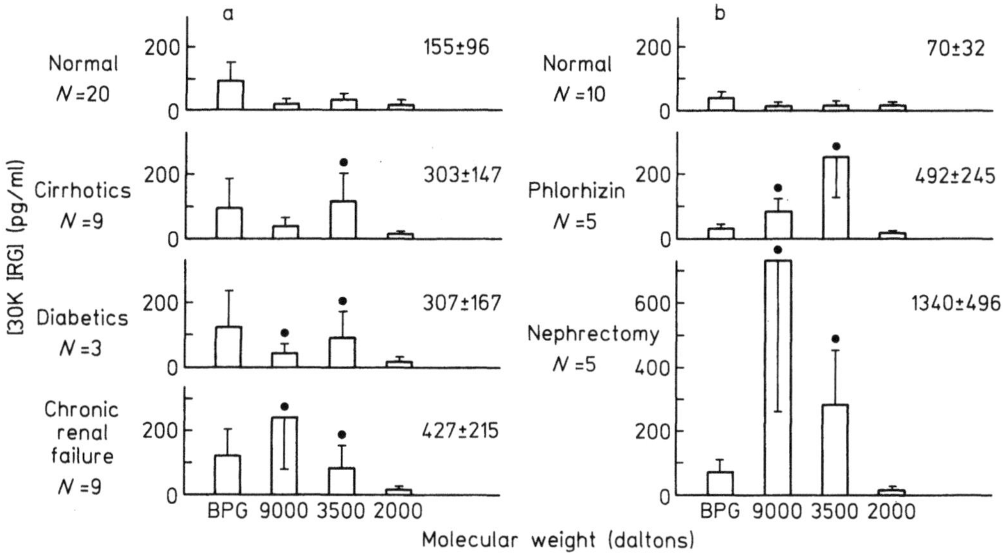


**Fig. 5a, b.** Basal plasma IRG components in humans (**a**) and dogs (**b**) in different hyperglucagonemic states compared with normal subjects. Biogel P-30 columns. Total plasma IRG values are given as mean ± standard deviation. $N$ = number of subjects; *asterisks* indicate $P<0.05$ compared with normal subjects. After VALVERDE and VILLANUEVA (1976)

when compared with normal values (VALVERDE and VILLANUEVA 1976). Arginine infusion in one diabetic resulted in increases of $IRG^{3500}$ and $IRG^{9000}$. Insulin treatment in another patient was shown to reduce both components. No alterations were found in BPG and $IRG^{2000}$ levels in these diabetics (VALVERDE 1977).

KUKU et al. (1976b) have reported that, in diabetic ketoacidosis and diabetic hyperosmolar coma, the marked increases in plasma IRG were due to excess $IRG^{3500}$, although $IRG^{9000}$ may also be present in large amounts. UNGER (1976b) has mentioned that 1,000 pg equiv. BPG isolated from diabetic patients' plasma and perfused into the isolated rat liver had the biologic activity of 5,000 pg crystalline glucagon.

b) Dogs

The chromatographic study of basal plasma samples from four dogs with alloxan-induced diabetes showed an increase in $IRG^{3500}$ and $IRG^{9000}$ levels when compared with normal dogs (VALVERDE 1977).

### 3. Pancreatectomy

a) Dogs

The persistence of high circulating levels of IRG in the plasma of depancreatized dogs led to the discovery of the functioning A-cell in the canine gastric fundus (see Chap. 33). Gel filtration of plasma from pancreatectomized dogs yielded the same four fractions as those found in intact animals (MASHITER et al. 1975; VALVERDE et al. 1975). In a group of seven dogs, $IRG^{3500}$ and $IRG^{9000}$ appeared significantly

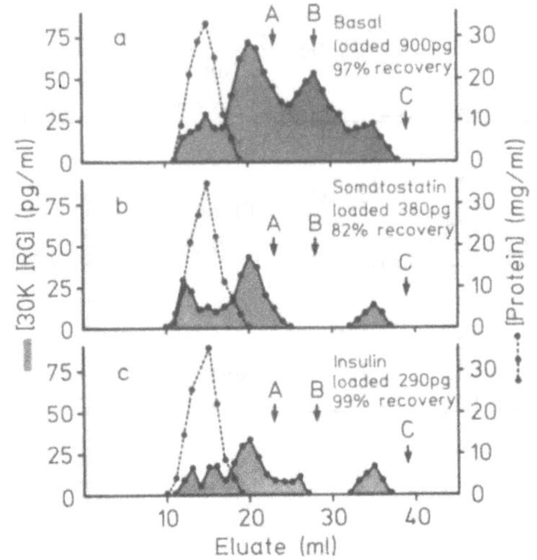


**Fig. 6a–c.** Chromatographic pattern on Biogel P-10 of plasma (4 ml) from a pancreatectomized dog in the basal state (**a**) and during somatostatin (**b**) or insulin infusion (**c**). *Arrows* indicate the elution volume of added: insulin $^{125}I$ (A); glucagon $^{125}I$ (B); and $Na^{125}I$ (C). After VALVERDE et al. (1975)

increased (VALVERDE 1977). Similarly, antiserum RCS 5, which does not react with BPG, recognized fractions $IRG^{9000}$, $IRG^{3500}$, and $IRG^{2000}$ (VALVERDE et al. 1980b). In two pancreatectomized dogs, intravenous arginine elicited an increase of plasma $IRG^{3500}$ and, to a lesser extent of $IRG^{9000}$. Both intravenous insulin and somatostatin (Fig. 6) totally suppressed $IRG^{3500}$ and substantially reduced $IRG^{9000}$ (VALVERDE et al. 1975).

b) Humans

Normal or high levels of total IRG has been the most frequent finding in plasma from pancreatectomized patients. Although A-cells have been identified in the human fundic mucosa (MUÑOZ-BARRAGAN et al. 1977), the presence of plasma $IRG^{3500}$ in pancreatectomized patients has been difficult to establish (see Chaps. 33 and 42). VILLANUEVA et al. (1976) were the first to report the chromatographic analysis of the plasma IRG from one of these patients, by means of 30 K antiserum. They found that most of the total IRG corresponded in size to BPG and the rest to $IRG^{2000}$. No changes in the IRG components were observed after arginine infusion or oral glucose administration.

MÜLLER et al. (1979) studied plasma from five pancreatectomized patients. In plasma from fasting patients, no $IRG^{3500}$ was detected. Most of the IRG eluted as BPG, with minor fractions eluting in the 9,000 and 2,000 zones. After arginine infusion, the same pattern was observed, with the exception of one case in which some $IRG^{3500}$ was detected. BODEN et al. (1980) were able to detect $IRG^{3500}$ in all gel-filtered plasma from six pancreatectomized patients by means of 30 K antiserum; in five of them, RCS 5 antiserum was also employed and all the IRG detected by this antiserum was recovered in the $IRG^{3500}$ zone, whereas with 30 K most of the IRG appeared in the BPG zone. The mean plasma $IRG^{3500}$ value obtained with both antisera was similar, and significantly lower than in five normal controls.

BOTHA et al. (1977) subjected an ethanolic extract of plasma, from a pancreatectomized patient, to polyacrylamide gel electrophoresis. They found two IRG (K 964) components, one migrating with the glucagon marker and the other one behind it, indicating that the latter has either a higher molecular weight or a more basic chemical nature than glucagon. KRAUSE et al. (1980) reported the chromatographic pattern of 30 K IRG present in plasma of four pancreatectomized subjects after oral glucose administration. The resulting peaks were those corresponding to the BPG, $IRG^{9000}$, and $IRG^{2000}$ eluting zones; no peak was clearly detected in the $IRG^{3500}$ region.

c) Pigs

HOLST et al. (1978) reported that gel filtration of plasma from four pancreatectomized pigs assayed with antiserum 4317 revealed only one IRG component with a size larger than glucagon, but smaller than 20,000 daltons. In one animal, a small amount of $IRG^{3500}$ was also observed. In plasma from two normal pigs, most of the total IRG corresponded to $IRG^{3500}$, while the remainder eluted earlier, in the same zone as the IRG present in the pancreatectomized pigs. BPG was not detected.

d) Cats

The chromatographic pattern of the plasma IRG from four depancreatized cats was studied by CHISHOLM et al. (1978) using RCS 5 antiserum. It was found that 40%–90% of the total IRG was $IRG^{9000}$, only 10%–60% corresponded to $IRG^{3500}$. There are no available data on IRG components in basal plasma from normal cats; CHISHOLM et al. (1978) reported that after arginine infusions 90% of the total IRG appeared as $IRG^{3500}$.

e) Rats

SMITH et al. (1978) reported persistent plasma IRG in rats whose pancreas and gastrointestinal tract had been removed. Chromatographic analysis of the IRG present in the plasma from these eviscerated rats revealed that most of it was BPG and the rest $IRG^{2000}$. The source of the persistent plasma IRG in these animals has not been clarified.

### 4. Chronic Hypoglycemia

a) Dogs

Sustained hypoglycemia induced by phlorhizin treatment in five dogs produced a highly significant elevation of both $IRG^{9000}$ and $IRG^{3500}$ in plasma (Fig. 5). During a 30-min somatostatin infusion, there was a reduction of $IRG^{9000}$ and $IRG^{3500}$ values with disappearance rates of 19 and 3 min, respectively (VALVERDE and GHIGLIONE 1980).

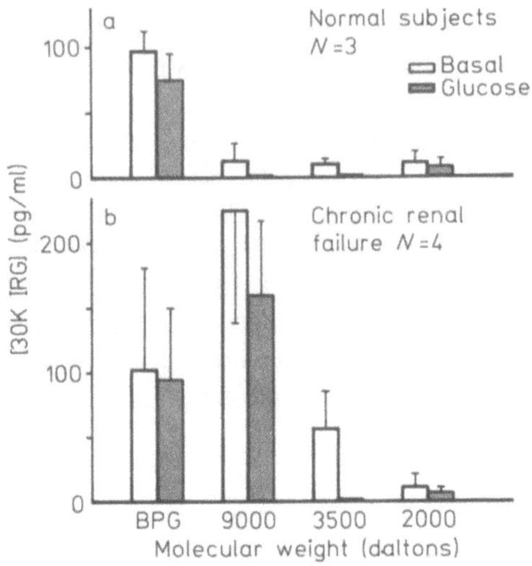


**Fig. 7 a, b.** Effect of 30 min intravenous glucose infusion on plasma IRG components in normal subjects (**a**) and chronic renal failure patients (**b**). Mean ± standard deviation. $N$ = number of subjects. After VALVERDE et al. (1976b)

### 5. Renal Failure

#### a) Humans

The high levels of total plasma IRG in patients undergoing long-term hemodialysis were found to be due to increases in $IRG^{9000}$ and $IRG^{3500}$ (Fig. 5). Intravenous glucose infused for 30 min (Fig. 7) produced a total disappearance of $IRG^{3500}$ from plasma, while $IRG^{9000}$ levels were only slightly reduced (VALVERDE et al. 1976b). KUKU et al. (1976a) showed that the hyperglucagonemia of uremic patients was largely due to an elevation of $IRG^{9000}$, but an increase in $IRG^{3500}$ was also detected. After oral glucose administration, $IRG^{3500}$was markedly suppressed, while $IRG^{9000}$ declined to a lesser extent. Arginine infusion induced only an $IRG^{3500}$ rise. $IRG^{9000}$, isolated from the plasma of one of these patients, when treated with trypsin, was converted to $IRG^{3500}$ and smaller IRG material.

#### b) Dogs

In five dogs, 1 day after bilateral nephrectomy, plasma IRG rose to strikingly high levels. Chromatographic studies of these plasma samples revealed the presence of large amounts of $IRG^{3500}$ and even greater quantities of $IRG^{9000}$ (Fig. 5). Somatostatin infusion reduced the concentrations of both components. In these experiments, the half-lives of $IRG^{3500}$ and $IRG^{9000}$ were increased (twofold and threefold, respectively) in comparison with the values found in intact dogs (MATESANZ et al. 1977).

#### c) Rats

The renal metabolism of IRG components has been studied by EMMANOUEL et al. (1976) using three models of uremic rats, i.e., bilateral nephrectomy, bilateral

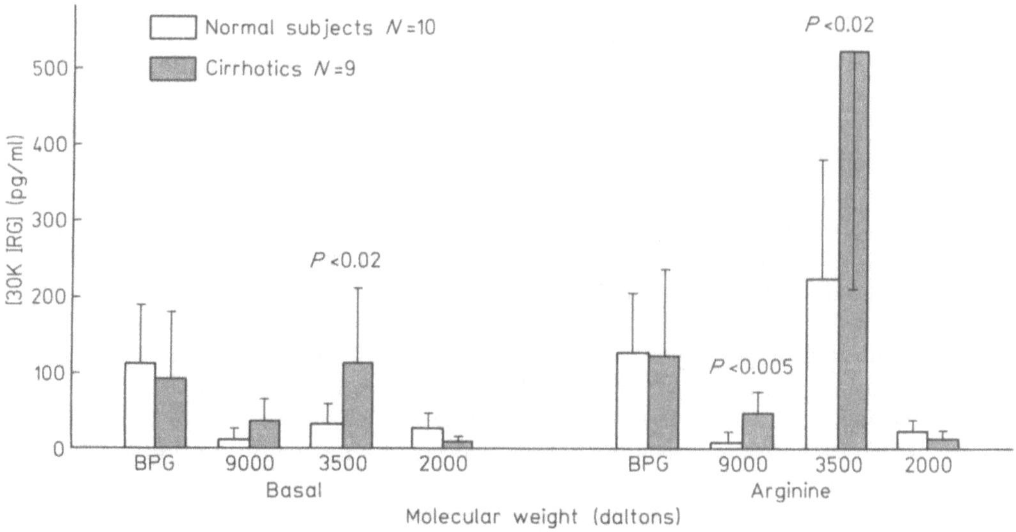


**Fig. 8.** Basal and post-arginine plasma IRG components in cirrhotic patients compared with normal subjects. Mean ± standard deviation. $N$ = number of subjects. After VALVERDE (1977)

ureteral ligation, and urine autoinfusion. Only the first two models developed marked hyperglucagonemia, indicating that uremia per se does not account for the high IRG levels in renal failure. In ligated rats, $IRG^{3500}$ was almost entirely responsible for the hyperglucagonemia, whereas in nephrectomized rats, both $IRG^{9000}$ and $IRG^{3500}$ accounted for the total high IRG values in a ratio of 2:3. The renal metabolism of $IRG^{9000}$ seemed to depend mainly on peritubular uptake whereas $IRG^{3500}$ appeared to be handled by both glomerular filtration and peritubular uptake (EMMANOUEL et al. 1978).

### 6. Cirrhosis of the Liver

In nine hyperglucagonemic patients without surgical portacaval shunts gel filtration of basal plasma samples revealed the presence of the four IRG fractions (Figs. 5 and 8). $IRG^{3500}$ and $IRG^{9000}$ appeared in higher levels than those from control individuals, although for $IRG^{9000}$ the difference was not statistically significant. After intravenous administration of arginine (Fig. 8) only $IRG^{3500}$ increased significantly (VALVERDE 1977).

MCDONALD et al. (1979) have found that in cirrhotic patients with surgical portal systemic shunts $IRG^{3500}$ was always elevated, and $IRG^{9000}$ was readily detected in most plasma samples, being the predominant peak in one subject. Glucose administration, either oral or intravenous, reduced mainly $IRG^{3500}$ plasma levels. The hepatic metabolism of IRG components has been studied by JASPAN et al. (1977) in normal and nephrectomized rats; they measured these fractions on plasma drawn simultaneously from the proximal portal vein and the superior vena cava. According to their results, the liver does not extract IRG components other than $IRG^{3500}$. This was later confirmed in dogs (JASPAN et al. 1981).

### 7. Other Conditions

The plasma IRG components have also been studied in several situations, such as pregnancy, severe obesity, insulinoma, disseminated carcinomatosis, and septicemia. Their relative amounts were not different from those found in normal subjects; $IRG^{3500}$ was increased and $IRG^{9000}$ was occasionally high in some of these patients (JASPAN and RUBENSTEIN 1977). Sporadic cases with very high BPG levels, close to 1,000 pg/ml (30 K), have been found in a few patients and also in healthy subjects. In these individuals, a significant correlation between BPG and immunoglobulin values was not obtained. BPG levels were shown to be unmodified by stimuli of the A-cell (VALVERDE and VILLANUEVA 1976). VON SCHENCK (1978) has provided data suggesting that immunoglobulins react with COOH terminal antisera and may be, at least in part, responsible for the BPG fraction.

PALMER et al. (1978) reported the presence of high values of $IRG^{9000}$ and of a 10,000–20,000 daltons component in nine healthy members of a family. Both fractions increased more than $IRG^{3500}$ after arginine infusion; they were suppressed by oral glucose. The distribution of individuals with high levels of large IRG components in this pedigree was consistent with autosomal dominant inheritance. PALMER et al. (1978) suggested that these components may be precursors of glucagon and may be elevated in this family because of an inherited defect in either their synthesis or degradation. BODEN and OWEN (1977) studied the plasma IRG fractions in nine healthy members of a glucagonoma patient, four of whom had elevated total plasma IRG, suggesting also autosomal dominant inheritance of hyperglucagonemia. In these four relatives, IRG eluted mainly between the void volume and the 9,000 daltons zone. $IRG^{3500}$ was less than 15% of the total immunoreactivity.

Gel filtration of plasma containing a low concentration of total IRG from a child with severe neonatal hypoglycemia and nesidioblastosis revealed the absence of $IRG^{3500}$. All of the immunoreactivity corresponded to BPG (SCHWARTZ et al. 1979). The lack of plasma $IRG^{3500}$ in this infant may have contributed to the development of hypoglycemia.

## IV. Origin of Plasma IRG Components

There is a general belief that plasma $IRG^{3500}$ represents the native hormone secreted by the pancreatic or gastric A-cell. $IRG^{9000}$ seems also to be a secretory product of the A-cell. As can be seen in Table 3, there is a higher concentration of this component in the pancreatic vein than in the peripheral plasma, both in basal state and during the first 5 min of an arginine infusion (0.033 g $min^{-1}$ $kg^{-1}$); in a pancreatectomized dog, $IRG^{9000}$ is also greater in the gastric vein than in systemic blood, in both states (VALVERDE et al. 1980a). These findings and those of WEIR et al. (1977) in tumor-draining veins suggest a cellular origin for $IRG^{9000}$. In the effluent of the perfused dog pancreas, CONLON et al. (1978) did not find any $IRG^{9000}$. However, $IRG^{9000}$ was detected in the medium bathing canine pancreatic pieces incubated with arginine; it represented 8% of the total IRG (VALVERDE et al. 1980a). HOLST et al. (1979) only found $IRG^{3500}$ in the effluent from the isolated perfused porcine pancreas during stimulation of glucagon secretion with

**Table 3.** IRG components in peripheral and organ-draining veins of a normal dog and a pancreatectomized dog in basal state and during an arginine infusion (0.17 g/kg in 5 min; Biogel P-30, 30 K)

| | Vein [a] | Total plasma | Concentrations (pg/ml) | | | |
|---|---|---|---|---|---|---|
| | | | BPG | $IRG^{9000}$ | $IRG^{3500}$ | $IRG^{2000}$ |
| Normal | | | | | | |
| Basal | Pe.v. | 244 | 252 | 12 | 14 | 8 |
| | P.v. | 652 | 240 | 55 | 382 | 10 |
| Arginine [b] | Pe.v. | 320 | 247 | 31 | 68 | 5 |
| | P.v. | 1368 | 310 | 94 | 1030 | 12 |
| Pancreatectomized | | | 310 | | | |
| Basal | Pe.v. | 62 | 41 | | 8 | 7 |
| | G.v. | 268 | 52 | 37 | 102 | 7 |
| Arginine [b] | Pe.v. | 134 | 58 | 29 | 23 | 8 |
| | G.v. | 792 | 84 | 89 | 490 | 9 |

[a] Pe.v. = peripheral vein; P.v. = pancreatic organ-draining vein; G.v. = gastric organ-draining vein

[b] Pool of samples collected at 1.5, 3, and 5 min during the arginine infusion

arginine. Analysis of the effluent of the perfused rat pancreas during stimulation of the A-cell showed also that most of the 30 K IRG released was $IRG^{3500}$ (LECLERCQ-MEYER et al. 1980).

The origin and nature of BPG are not clear. BPG-like components have been reported in extracts of pancreas, stomach, and salivary glands (VALVERDE et al. 1974; SRIKANT et al. 1977; LAWRENCE et al. 1977, TUNG et al. 1980), but none of these large compounds has been fully characterized. CONLON et al. (1978) reported the existence of a similar sized IRG component in plasma-free effluent of isolated dog pancreas after prolonged stimulation with arginine. This component was converted to $IRG^{3500}$ by treatment with 6 *M* guanidium chloride, suggesting that it may be a complex of glucagon and a nonimmunoreactive protein. Although the amount of BPG-sized material secreted by the pancreas was low, this material could contribute significantly to the levels of plasma BPG, which seemed to have a very low rate of removal from the circulation. $IRG^{2000}$ is believed to be a degradation product of glucagon.

## D. Plasma GLI Components

VALVERDE et al. (1970 b) demonstrated the presence of two GLI components (7,000 and 3,500 daltons) in acid–alcohol extracts of plasma obtained from partially pancreatectomized dogs which were being subjected to an intraduodenal glucose load. These two GLI fractions have also been found by OHNEDA et al. (1975) in normal dog plasma under the same conditions. FLANAGAN et al. (1975), using immunoaffinity chromatographic methods, have found a variety of plasma peptides reacting only with an $NH_2$ terminal-specific antibody.

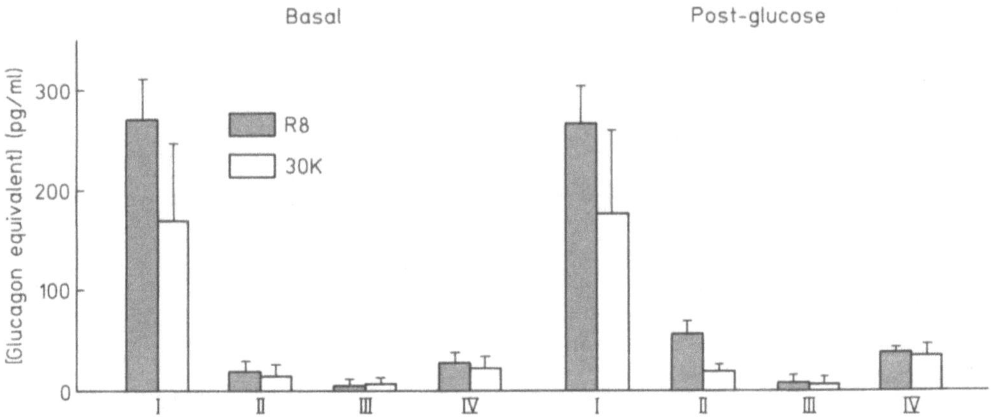


**Fig. 9.** R 8 and 30 K reactivity of the Biogel P-30 plasma fractions from 9 normal subjects before and after oral glucose administration. Mean ± standard error. After VALVERDE et al. (1980 b)

## I. Basal State and During Glucose Absorption

### 1. Normal Humans

Figure 9 shows the amounts of glucagon-immunoreactive fractions found in the plasma of nine normal subjects as determined with both R 8 ($NH_2$ terminal) and 30 K (COOH terminal) antisera. In this study, the plasmas were native and filtered on Biogel P-30 columns. In the basal state, the immunoreactivity recognized by R 8 appeared in the same four molecular weight regions as with 30 K antiserum. Most of the immunoreactivity eluted in the void volume. There was no significant correlation between the individual R 8 and 30 K values of this fraction. After glucose ingestion, fraction II (9,000 daltons) as determined with R 8, appeared significantly increased, while the other fractions remained unchanged. Glucose administration induced no change in 30 K-reacting fractions (VALVERDE et al. 1980 b).

### 2. Normal Dogs

Plasma was obtained from anesthetized normal dogs before and after an intraduodenal glucose load. As for human plasma, both R 8 and 30 K, recognized four immunoreactive fractions (Fig. 10). It is remarkable that, in the post-glucose samples, fraction II appeared to be increased 13-fold as measured with R 8 (VALVERDE et al. 1979). With 30 K, an elevation of fraction II was also demonstrated (78% over the basal level). The 30 K immunoreactivity of post-glucose fraction II represented 13% of that measured with R 8 (VALVERDE et al. 1980 a). After glucose administration, no significant changes in the levels of the other fractions were observed with either antiserum. In interpreting the 30 K rise of fraction II, two possibilities can be considered: (1) a true increase of COOH terminal peptide (or peptides) of approximate molecular weight 9,000 daltons, obviously measured by this antiserum; (2) a possible cross-reactivity of massively released GLI peptides with 30 K antiserum.

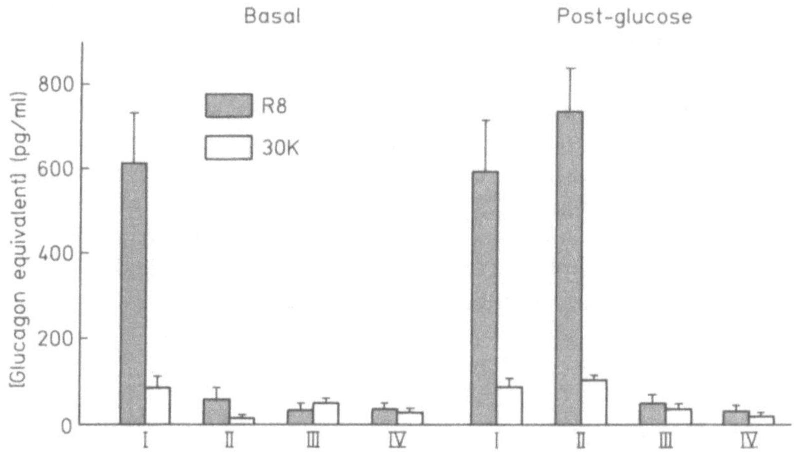


**Fig. 10.** R 8 and 30 K reactivity of the Biogel P-30 plasma fractions from 11 normal dogs before and after intraduodenal glucose administration. Mean ± standard error. After VALVERDE et al. (1980 a)

CONLON et al. (1980) studied the GLI peptides of plasma samples collected from the mesenteric vein of dogs after instillation of nutrients into the ileum. The GLI peptides were isolated by immunoaffinity chromatography and later gel filtered. The predominant GLI peptides were of molecular weight, 8,000–12,000 daltons, with isoelectric point in the range pI = 5–8, but smaller basic (pI > 9) GLI peptides (3,500–5,000 daltons) were also observed.

### 3. Nephrectomized Dogs

In an attempt to investigate the clearance of GLI peptides by the kidney, we compared the plasma GLI responses to an intraduodenal glucose load in normal and nephrectomized dogs. Samples were simultaneously assayed with R 8 and 30 K antisera. The GLI response was calculated as the difference between the post-glucose increments of immunoreactivity measured with these antibodies. It was found that intraduodenal glucose administration induced a rise in GLI fraction II only. This rise was not different in control and nephrectomized animals (ROVIRA et al. 1980). In further studies performed with the isolated perfused rat kidney, we found that a partially purified preparation of gut GLI peak I (7,000–12,000 daltons) was not removed by this system (ROVIRA et al. 1980). These results suggest that the kidney may not play an important role in gut GLI catabolism.

## II. Origin of Plasma GLI Components

O'CONNOR et al. (1979) investigated the GLI components released from the isolated perfused rat intestine during glucose stimulation. Acid–alcohol extracts of the perfusate contained $GLI^{12000}$ and $GLI^{3500}$, as well as $IRG^{3500}$, the smaller size peptides being the predominant species. This is in contrast with our observation in untreated plasma in which only an increment of GLI fraction II (7,000–

12,000 daltons) was detected (VALVERDE et al. 1979). The treatment of the samples before filtration could have modified the original molecular size of the secreted GLI (VALVERDE et al. 1980b).

Chromatography of perfusates from isolated rat pancreas, obtained during stimulation of glucagon secretion by lowering the glucose concentration of the medium, revealed the presence of a GLI component of molecular weight $\geqq$6,000 daltons when using antiserum AGS. A similar sized component was shown to exist in minor concentration in human plasma (LECLERCQ-MEYER et al. 1980). Finally, most of the GLI of basal plasma corresponds to fraction I (>20,000 daltons). This GLI fraction remains unmodified after glucose ingestion. Its nature and origin have not been clarified yet.

## E. Concluding Remarks

There is a vast heterogeneity of circulating glucagon-related peptides. The assessment of "true glucagon" values cannot be achieved in untreated plasma samples with any of the anti-glucagon sera in current use. Plasma gel filtration is the only available technique to distinguish "true glucagon" from the other glucagon-related peptides. There is little doubt that the plasma gel-filtered $IRG^{3500}$ fraction corresponds to the native biologically active hormone. Its concentration quickly changes in response to acute stimulatory or suppresive agents of the A-cell, and it is elevated in situations of known glucagon hypersecretion or impaired catabolism.

Three other IRG components (COOH terminal-specific antibody-reacting peptides) are present in plasma, namely BPG, $IRG^{9000}$, and $IRG^{2000}$. $IRG^{9000}$ is elevated in various chronic hyperglucagonemic states, such as glucagonoma, diabetes, cirrhosis of the liver, the neonatal period, and chronic renal failure. In normal subjects, it is a minor component which responds poorly or not at all to acute stimulatory maneuvers. In neonates as well as in glucagonoma patients, fasting $IRG^{9000}$ levels are elevated and can be further increased by A-cell stimulation. Although its origin and nature has not been clarified, the behavior in plasma of $IRG^{9000}$ is compatible with it being a secretory product of the A-cell which may represent an intermediate product in the biosynthesis of glucagon (see Chap. 6). Its biologic activity, if any, seems to be low. In the kidney it is catabolized by peritubular uptake.

The largest plasma IRG component, BPG, appears as a heterogeneous material of unknown origin. It may represent in part a secretory product of the A-cell since similar-sized material has been detected in the effluent from the perfused canine pancreas. It has been reported that BPG from glucagonoma patients has biologic activity in terms of generating cyclic AMP in rat hepatocytes. Some individuals exhibit unexplained high levels of BPG. The smallest plasma IRG component, $IRG^{2000}$, may correspond to fragments of the glucagon molecule; it is always a minor component of the total immunoreactivity.

Data on circulating GLI components, as assessed by gel filtration of untreated plasma and assay of the eluates with $NH_2$ terminal- and COOH terminal-specific antibodies, indicate that plasma from fasting subjects contains mainly GLI species of high molecular weight (>20,000 daltons) and that during glucose absorption the gut GLI released corresponds to peptides of molecular weight, 7,000–

12,000 daltons. The assessment of total plasma GLI cannot be done by subtracting from the values obtained with an $NH_2$ terminal-specific antibody those measured with a COOH terminal-specific antibody, since GLI fraction I reacts inconsistently with both types of antiserum.

*Acknowledgments.* I wish to express my gratitude to my present and past collaborators who have contributed to the work reviewed here. I am grateful to Dr. J. MARCO, in addition, for revising the manuscript. Finally thanks go to Dr. S. R. BLOOM, Dr. V. LECLERCQ-MEYER, Dr. L. G. HEDING, Dr. R. H. UNGER, and Dr. M. L. VILLANUEVA for their generous supplies of antibodies.

## References

Assan R, Slusher N (1972) Structure/function and structure/immunoreactivity relationships of the glucagon molecule and related synthetic peptides. Diabetes 21:843–855

Bhathena SJ, Smith SS, Voyles NR, Penhos JC, Recant L (1977) Studies on submaxillary gland immunoreactive glucagon. Biochem Biophys Res Commun 74:1574–1581

Boden G, Owen OE (1977) Familial hyperglucagonemia-An autosomal dominant disorder. N Engl J Med 296:534–538

Boden G, Master RW, Rezvani I, Palmer JP, Lobe TE, Owen OE (1980) Glucagon deficiency and hyperaminoacidemia after total pancreatectomy. J Clin Invest 65:706–716

Botha JL, Vinik AI, Child PT, Paul M, Jackson WPU (1977) Pancreatic glucagon-like immunoreactivity in a pancreatectomized patient. Horm Metab Res 9:199–205

Chisholm DJ, Alford FP, Harewood MS, Findlay DM, Gray BN (1978) Nature and biologic activity of "extrapancreatic glucagon": studies in pancreatectomized cats. Metabolism 27:261–273

Conlon JM, Ipp E, Unger RH (1978) The molecular forms of immunoreactive glucagon secreted by the isolated, perfused dog pancreas. Life Sci 23:1655–1658

Conlon JM, Rouiller D, Unger RH (1980) Characterization of the glucagon-like polypeptides released by the dog gut into the circulation. Biochem Soc Trans 8:51–52

Eisentraut A, Ohneda A, Parada E, Unger RH (1968) Immunologic discrimination between pancreatic glucagon and enteric glucagon-like immunoreactivity (GLI) in tissues and plasma. Diabetes 17:321

Emmanouel DS, Jaspan JB, Kuku SF, Rubenstein AH, Katz AI (1976) Pathogenesis and characterization of hyperglucagonemia in the uremic rat. J Clin Invest 58:1266–1272

Emmanouel DS, Jaspan JB, Rubenstein AH, Huen H-J, Fink E, Katz AI (1978) Glucagon metabolism in the rat. Contribution of the kidney to the metabolic clearance rate of the hormone. J Clin Invest 61:6–13

Flanagan RWJ, Trimble ER, Conlon JM, Murphy RF (1975) Purification and characterization of plasma glucagon-like immunoreactivity by immunoaffinity chromatography. Diabetologia 11:342

Heding LG (1969) The production of glucagon antibodies in rabbits. Horm Metab 1:87–88

Heding LG, Frandsen EK, Jacobsen H (1976) Structure-function relationship: immunologic. Metabolism [Suppl 1] 25:1327–1329

Holst JJ (1977) Extraction, gel filtration pattern, and receptor binding of porcine gastrointestinal glucagon-like immunoreactivity. Diabetologia 13:159–169

Holst JJ, Kreutzfeldt M, Holm G, Jensen E, Poulsen JSD, Sparsö B, Sparsö B, Schmidt A (1978) Absence of true pancreatic glucagon but persistence of circulating pancreatic glucagon-like immunoreactivity after pancreatectomy in pigs. Diab Metab 4:74–79

Holst JJ, Von Schenck H, Lindkaer S (1979) Gel filtration pattern of immunoreactive glucagon secreted by the isolated, perfused, porcine pancreas. Scand J Clin Lab Invest 39:47–52

Iwasa S, Kondo K, Ueno H, Wakimasu M, Ohneda A (1981) Enzyme immunoassay for determination of pancreatic glucagon in plasma. Horm Metab Res 13:14–17

Jaspan JB, Rubenstein AH (1977) Circulating glucagon. Plasma profiles and metabolism in health and disease. Diabetes 26:887–902

Jaspan JB, Huen AH-J, Morley CG, Moossa AR, Rubenstein AH (1977) The role of the liver in glucagon metabolism. J Clin Invest 60:421–428
Jaspan JB, Polonsky KS, Lewis M, Pensler J, Pugh W, Moossa AR, Rubenstein AH (1981) Hepatic metabolism of glucagon in the dog. Contribution of liver to the overall metabolic disposal of glucagon. Am J Physiol 240:E233–E244
Krause U, Thiel M, Beyer J, Mangold G, Cordes U (1980) Glucagon secretion in four duodenopancreatectomized patients after arginine and glucose load. Horm Metab Res 12:364–369
Kuku SF, Jaspan JB, Emmanouel DS, Zeidler A, Katz AI, Rubenstein AH (1976a) Heterogeneity of plasma glucagon: circulating components in normal subjects and patients with chronic renal failure. J Clin Invest 58:742–750
Kuku SF, Zeidler A, Emmanouel DS, Katz AI, Rubenstein AH, Levin NW, Tello A (1976b) Heterogeneity of plasma glucagon; patterns in patients with chronic renal failure and diabetes. J Clin Endocrinol Metab 42:173–176
Lawrence AM, Tan S, Hojvat TS, Kirsteins L (1977) Salivary gland hyperglycemic factor. An extrapancreatic source of glucagon-like material. Science 195:70–72
Leclercq-Meyer V, Marchand J, Leclercq R (1980) Studies on the molecular forms of glucagon immunoreactivity (GLI) released by the in vitro perfused rat pancreas. Diabetologia 19:294
Mashiter K, Harding PE, Chou M, Mashiter GD, Stout J, Diamond D, Field JB (1975) Persistent pancreatic glucagon but not insulin response to arginine in pancreatectomized dogs. Endocrinology 96:678–693
Matesanz R, Casado S, Valverde I, Hernando L (1977) Hyperglucagonemia and renal failure. Kidney Int 12:84–85
McDonald TJ, Dupre J, Caussignac Y, Radziuk J, Vliet SV (1979) Hyperglucagonemia in liver cirrhosis with portal-systemic venous anastomoses: responses of plasma glucagon and gastric inhibitory polypeptide to oral or intravenous glucose in cirrhotics with normal or elevated fasting plasma glucose levels. Metabolism 28:300–307
Moody AJ, Jacobsen H, Sundby F, Frandsen EK, Baetens D, Orci L (1977) Heterogeneity of gut glucagon-like immunoreactivity (GLI). In: Foà PP, Bajaj JS, Foà NL (eds) Glucagon: its role in physiology and clinical medicine. Springer, Berlin Heidelberg New York, pp 129–136
Müller WA, Berger M, Suter P, Cüppers HJ, Reiter J, Wyss T, Berchtold P, Schmidt FH, Assal J-P, Renold AE (1979) Glucagon immunoreactivities and amino acid profile in plasma of duodenopancreatectomized patients. J Clin Invest 63:820–827
Muñoz-Barragán L, Rufener C, Srikant CB, Dobbs RE, Shannon WA Jr, Baetens D, Unger RH (1977) Immunocytochemical evidence for glucagon-containing cells in the human stomach. Horm Metab Res 9:37–39
O'Connor FA, Conlon JM, Buchanan KD, Murphy RF (1979) The use of perfused rat intestine to characterise the glucagon-like immunoreactivity released in serosal secretions following stimulation by glucose. Horm Metab Res 11:19–23
Ohneda A, Horigome K, Yanbe A, Ishii S, Itabashi H, Chiba M, Maruhama Y, Yamagata S (1975) Plasma glucagon-like immunoreactivity (GLI) in dogs. Tohoku J Exp Med 115:337–343
Palmer JP, Werner PL, Benson JW, Ensinck JW (1978) Dominant inheritance of large molecular weight immunoreactive glucagon. J Clin Invest 61:763–769
Pérez-Castillo A, Blázquez E (1980) Tissue distribution of glucagon, glucagon-like immunoreactivity, and insulin in the rat. Am J Physiol 238:E258–E266
Recant L, Perrino PV, Bhathena SJ, Danforth DN, Lavine RL (1976) Plasma immunoreactive glucagon fractions in four cases of glucagonoma. Increased "large glucagon-immunoreactivity." Diabetologia 12:319–326
Rigopoulou D, Valverde I, Marco J, Faloona G, Unger RH (1970) Large glucagon immunoreactivity in extracts of pancreas. J Biol Chem 245:496–501
Rovira A, López-Novoa JM, Ghiglione M, Zubiaur M, Pascual JM, Valverde I (1980) Lack of gut-GLI removal by the kidney. Diabetologia 19:311
Schwartz SS, Rich BH, Lucky AW, Strauss FH, Gonen B, Wolfsdorf J, Thorp FW, Burrington JD, Madden JD, Rubenstein AH, Rosenfield RL (1979) Familial nesidioblastosis: severe neonatal hypoglycemia in two families. Pediatrics 95:44–53

Smith SS, Bhathena SJ, Nompleggi D, Penhos JC, Recant L (1978) Studies on persistent circulating immunoreactive glucagon (IRG) and immunoreactive insulin (IRI) found in eviscerated rats with a functional liver. Diabetologia 14:177–184

Srikant CB, McCorkle K, Unger RH (1977) Properties of immunoreactive glucagon fractions of canine stomach and pancreas. J Biol Chem 252:1847–1851

Stacpoole PW, Jaspan J, Kasselberg AG, Halter SA, Polonsky K, Gluck FW, Liljenquist JE, Rabin D (1981) A familial glucagonoma syndrome. Genetic, clinical and biochemical features. Am J Med 70:1017–1026

Tager HS, Markese J (1979) Intestinal and pancreatic glucagon-like polypeptides. Evidence for identity of higher molecular weight forms. J Biol Chem 254:2229–2233

Tung AK, Ruse JL, Cockburn E (1980) Glucagon from the bovine fetal pancreas: chromatographic and electrophoretic characterizations of high molecular weight immunoreactive species. Can J Biochem 58:707–714

Unger RH (1976a) Report of the Nomenclature Committee. Glucagon symposium. Metabolism [Suppl 1] 24:IX

Unger RH (1976b) Discussion to: Distribution and properties of IRGs in animals and man. Metabolism [Suppl 1] 24:1412

Valverde I (1977) Quantification of plasma glucagon immunoreactive components in normal and hyperglucagonemic states. In: Foà PP, Bajaj JS, Foà NL (eds) Glucagon: its role in physiology and clinical medicine. Springer, Berlin Heidelberg New York, pp 77–92

Valverde I, Ghiglione M (1980) Plasma glucagon immunoreactive components. In: Andreani D, Lefèbvre PJ, Marks V (eds) Current views on hypoglycemia and glucagon. Academic Press, London New York, pp 21–26

Valverde I, Matesanz R (1977) High levels of plasma "proglucagon" in early life. Diabetes 26:421

Valverde I, Villanueva ML (1976) Heterogeneity of plasma immunoreactive glucagon. Metabolism [Suppl 1] 24:1393–1395

Valverde I, Rigopoulou D, Marco J, Faloona G, Unger RH (1970a) Characterization of glucagon-like immunoreactivity (GLI). Diabetes 19:614–623

Valverde I, Rigopoulou D, Marco J, Faloona G, Unger RH (1970b) Molecular size of extractable glucagon and glucagon-like immunoreactivity (GLI) in plasma. Diabetes 19:624–629

Valverde I, Villanueva ML, Lozano I, Marco J (1974) Presence of glucagon immunoreactivity in the globulin fraction of human plasma ("big plasma glucagon"). J Clin Endocrinol Metab 39:1090–1098

Valverde I, Dobbs R, Unger RH (1975) Heterogeneity of plasma glucagon immunoreactivity in normal, depancreatized and alloxan-diabetic dogs. Metabolism 24:1021–1028

Valverde I, Lemon HM, Kessinger A, Unger RH (1976a) Distribution of plasma glucagon immunoreactivity in a patient with suspected glucagonoma. J Clin Endocrinol Metab 42:804–808

Valverde I, Matesanz R, Lozano I, Plaza JJ, Casado S (1976b) Evaluation of plasma glucagon immunoreactive fractions in renal failure. Diabetologia 12:423

Valverde I, Ghiglione M, Matesanz R, Casado S (1979) Chromatographic pattern of gut glucagon-like immunoreactivity (GLI) in plasma before and during glucose absorption. Horm Metab Res 11:343–346

Valverde I, Ghiglione M, Calvo G (1980a) Heterogeneity of circulating glucagon and GLI. In: Waldhausl WK (ed) Diabetes 1979. Excerpta Medica, Amsterdam London New York, pp 415–420

Valverde I, Ghiglione M, Pascual JM, Rovira A, Calvo G, Vivanco F (1980b) Glucagon and glucagon-like immunoreactivity in the gastrointestinal tract. In: Cumming IA, Funder JW, Mendelsohn FAO (eds) Endocrinology 1980. Australian Academy of Science, Canberra, pp 554–557

Villanueva ML (1976) PhD thesis. Estudio del glucagón immunorreactivo plasmático. Caracterización del macroglucagón. Complutense University of Madrid

Villanueva ML, Hedo JA, Marco J (1976) Plasma glucagon immunoreactivity in a totally pancreatectomized patient. Diabetologia 12:613–616

Von Schenck H (1977) Production and characterization of an antiserum against pancreatic glucagon. Clin Chim Acta 80:455–463
Von Schenck H (1978) Immunoglobulins interfering in the glucagon radioimmunoassay. Acta Endocrinol (Copenh) [Suppl 219] 88:69
Von Schenck H, Thorell JI, Berg J, Bojs G, Dymling JF, Hallengren B, Ljungberg O, Tibblin S (1979) Metabolic studies and glucagon gel filtration pattern before and after surgery in a case of glucagonoma syndrome. Acta Med Scand 205:155–162
Weir GC (1977) Assesment of glucagon immunoreactivity in plasma. In: Foà PP, Bajaj JS, Foà NL (eds) Glucagon: its role in physiology and clinical medicine. Springer, Berlin Heidelberg New York, pp 65–76
Weir GC, Turner RC, Martin DB (1973) Glucagon radioimmunoassay using antiserum 30K: interference by plasma. Horm Metab Res 5:241–244
Weir GC, Knowlton SD, Martin DB (1975) High molecular weight glucagon-like immunoreactivity in plasma. J Clin Endocrinol Metab 40:296–302
Weir GC, Horton ES, Aoki TT, Slovik D, Jaspan J, Rubenstein AH (1977) Secretion by glucagonomas of a possible glucagon precursor. J Clin Invest 59:325–330

CHAPTER 12

# Radioreceptorassays for Glucagon

J. J. HOLST

## A. Introduction

The interaction of glucagon with its receptor on the plasma membranes of liver cells has been the subject of intensive research (see Chap. 13), and it has been established that the initial step in the effect of glucagon on liver function is specific binding of glucagon to the outside of liver cell membranes. The work of RODBELL et al. (1971) showed that glucagon associated highly specifically with its receptor in a reversible manner, with an equilibrium constant for dissociation of approximately $4 \times 10^{-9}$ *M*. Furthermore, it was found that very small changes in the structure of the glucagon molecule interfered markedly with its ability to activate the adenylate cyclase associated with the receptor on cell membranes. It was therefore conceivable, that a radioreceptorassay for glucagon, i.e., an assay based on the principles of competitive protein binding assays like radioimmunoassays, but utilizing the glucagon receptors as the binding reagent, might throw some light on the many specificity problems associated with the radioimmunologic determination of glucagon in biologic fluids (HOLST 1978 b). Not only are substances with glucagon-like immunoreactivity produced in many tissues outside the pancreatic A-cells (HOLST 1978 b), but also plasma contains peptides with glucagon-like immunoreactivity, which are not derived from the pancreas or differ in size from the originally isolated molecule (see Chap. 11). If any of these substances were capable of binding to the glucagon receptor and possibly activating adenylate cyclase, the radioreceptorassay might more accurately predict the total glucagon-like bioactivity in a given sample, and therefore be a more interesting biologic parameter than even the most restrictive and "specific" radioimmunoassay.

## B. Glucagon Radioreceptorassay Methodology

The binding of glucagon to cell membranes has been studied in numerous systems (see Chap. 13); in this chapter we will only consider systems which have been employed for quantitative or qualitative characterizations of biologic material, i.e., when the systems have been used as assays.

### I. Receptor Preparations

#### 1. Isolated Cells

The use of isolated rat hepatocytes has been described by FREYCHET et al. (1974 b), ROSSELIN et al. (1974 b), and SONNE et al. (1978). In these studies, which were basically directed at the investigation of the properties of glucagon receptor–activity coupling in liver cells, some assays were made of glucagon analog using the principle of competitive protein binding, whereas biologic material as such was not assayed.

Hepatocytes are usually isolated by means of perfusion with collagenase, which might damage the membrane receptors. However, isolated rat hepatocytes are very responsive to glucagon; the concentration needed for half-maximal stimulation of glycogenolysis is about $5 \times 10^{-11}$ $M$ (THIEDEN et al. 1981). The use of isolated hepatocytes offers the opportunity to assay in a very simple manner the biologic activity of glucagon-like substances (THIEDEN et al. 1981), but suffers the same lack of specificity as other bioassays, and it seems that binding studies with intact hepatocytes have no advantage over studies using isolated membranes.

### 2. Isolated Membranes

#### a) Rat Liver Membranes

The thorough investigations of RODBELL et al. (1971) of the binding of glucagon to isolated rat liver cell membranes form the basis for the use of this preparation for receptor assay of glucagon, which was introduced by POHL and CHASE in 1972 and BATAILLE and co-workers (BATAILLE et al. 1973, 1974; ROSSELIN et al. 1974a; FREYCHET 1976). In these studies, it seems that the method of membrane preparation was that originally devised by NEVILLE. This method (NEVILLE 1960, 1968) is based on homogenization of rat liver with the Dounce homogenizer (which is supposed to cause less damage to the membranes than other devices) in hypotonic medium (1 m$M$ $NaHCO_3$) followed by centrifugation and further purification of the membrane-rich pellets by high speed centrifugation in sucrose layers and gradients.

#### b) Pig Liver Membranes

The use of porcine liver cell membranes was introduced by HOLST (1975) in order to avoid problems with differences between species when studying glucagon-like components from porcine gut. It was also felt that there was a need for a method which would require simpler laboratory equipment. The aqueous two-phase polymer method of LESKO et al. (1973) was therefore adapted to pig liver tissue (HOLST 1975). Preliminary experiments with porcine, murine, and human liver had established, that a rather high degree of purification of the membrane fraction was necessary to demonstrate specific binding of glucagon. Thus, none of the fractions from conventional differential centrifugation showed binding activity, probably because of almost instantaneous degradation of glucagon (J. J. HOLST, unpublished work). Pig or rat liver cell membranes isolated by means of the aqueous two-phase polymer method bound glucagon rapidly at room temperature (HOLST 1975). With gelatin included in the incubation medium (in addition to the enzyme inhibitor, aprotinine), degradation of glucagon seemed small. Increases in the amount of membrane material added to the incubation mixtures led to linear increases in specific binding up to at least 70% binding of the labeled hormone added in trace amounts. Thus, tracer degradation did not limit the performance of the assay. Furthermore, the method allowed the processing of large amounts of membrane material from pig liver, and the membrane material, when freeze-dried and stored at −20 °C retained its binding activity for more than 2 years. The method therefore allows a large number of assays to be performed with the same batch of membranes, a fact which minimizes interassay variations.

c) Other Systems

Specific binding of glucagon has been demonstrated: to fat cells of rodents (DESBUQUOIS and LAUDAT 1974; LIVINGSTON et al. 1974; SONNE and GLIEMANN 1977); to myocardial membranes (LEVEY et al. 1975); to human mononuclear cells (GOLDSTEIN et al. 1975; BLECHER and GOLDSTEIN 1977); and to membranes from B-cells originating from insulin-producing tumors of syrian hamsters (GOLDFINE et al. 1972). Some of these systems might be used for the development of receptor assays of an effectiveness and simplicity corresponding to receptor assays described for insulin (OZAKI and KALANT 1977); so far, only B-cell membranes have been used for this purpose; the information obtained was similar to that obtained with liver cell membranes.

## II. Radioactive Glucagon

Most authors have used carefully prepared monoiodinated glucagon (NOTTEY and ROSSELIN 1971; RODBELL et al. 1971; JORGENSEN and LARSEN 1972; DESBUQUOIS 1975b). However, it seems that polyiodination of glucagon, in contrast to other peptides, does not impair its affinity for liver cell membranes; on the contrary, affinity has been reported to increase with the degree of iodination (BROMER et al. 1973; DESBUQUOIS 1975b; HOLST 1975). Similarly, after nitration of the tyrosine residues 10 and 13 (nitro- or dinitroglucagon) or nitration and reduction (amino- or diaminoglucagon), glucagon retains to a very large extent its biologic activity (PATTERSON and BROMER 1973); smaller modification of the tyrosine residues does not seem to interfere with glucagon's binding to its receptor, whereas antibody binding may be importantly influenced (PATTERSON and BROMER 1973; VON SCHENCK et al. 1976). However, most iodoglucagons are prepared by oxidation, and particularly the oxidation of methionine in position 27 may decrease the affinity of glucagon (SONNE and LARSEN 1980; DESBUQUOIS 1975b); gentle oxidation of glucagon, as described by JORGENSEN and LARSEN (1972), or preparation with very small amounts of chloramine-T (ROTH 1975b; HEBER et al. 1978) impairs the glucagon binding to membranes only slightly (SONNE and LARSEN 1980). Iodination by oxidation may be avoided with the use of the so-called Bolton-Hunter reagent, i.e., iodinated 3-(4-hydroxyphenyl)propionic acid *N*-hydroxysuccinimide ester, which has proven useful, e.g., for gastrin–receptor interaction studies (REHFELD 1980). However, this reagent couples to amino groups, and sizable modifications of α- as well as ε-amino groups were associated with extensive loss of biologic activity (EPAND and WHEELER 1975; LANDE et al. 1972; GRANDE et al. 1972); it may therefore be anticipated that glucagon, labeled in this way, will not be useful for receptor studies. Tritiated glucagon has been employed in some studies (LIN et al. 1977), but will probably be of limited value in receptor assays because of the low energy of radiation.

## III. Incubation Conditions

This subject has not been systematically investigated, but experience accumulated in the author's laboratory and elsewhere (Pohl and Chase 1972; Freychet 1976) allows some discussion of the subject. In this connection it must be emphasized that the object of optimizing a radioreceptorassay does not necessarily conform with the determination of parameters for glucagon binding to its receptor under physiologic circumstances.

### 1. Temperature and Incubation Time

As demonstrated by several authors (Rodbell et al. 1971; Holst 1975), glucagon binding proceeds rapidly at 37 °C, but also at room temperature, and because of the simplicity and decreased degradation of tracers (Ozaki and Kalant 1977; Freychet 1976), the lower temperature is preferable for assay purposes. Since sensitivity is an important feature of the assay, any measure to increase it is important. Several authors agree that late addition of labeled hormone is beneficial (Pohl and Chase 1972; Holst 1975; Freychet 1976). Pohl and Chase (1972) recommended 15 min preincubation of sample and receptor and 15 min further incubation with tracer at 22 °C and reported a sensitivity in the low p*M* range; Freychet (1976) recommended 30 min at 30 °C and 2 h at 22 °C, and found a sensitivity of about 60 p*M*, and Holst, working with pig liver membranes, used 30 min and 15 min incubation, at room temperature and obtained a sensitivity of approximately 10–20 p*M*. In this laboratory, we found it essential to rotate the incubation mixture to ensure high precision, and with larger series, the 30- and 15-min intervals were quite impractical (Holst 1975). We therefore studied incubations at 4 °C overnight, and found a marked increase in the binding capacity of the membranes and equal assay sensitivity, in spite of omission of preincubation (Fig. 1); preincubation however, improved the sensitivity further, in agreement with our finding that binding at 4 °C is complete in about 4–7 h and practically irreversible. Incubations overnight at 4 °C are now used routinely in our laboratory.

Another implication of incubation at lower temperatures may turn out to be of considerable importance. Recent research has established that hormone–receptor complexes, rather than dissociating, may be internalized at physiologic temperatures (Gorden et al. 1980). Also glucagon has been reported to become internalized (Barazzone et al. 1979). Internalization means that treatment of binding data according to the law of mass action is no longer justified. However, internalization does not take place at lower temperatures (Gorden et al. 1980), and it may be assumed that binding under such circumstances is fully reversible.

### 2. Reagent Volumes

Another feature of importance for precision as well as sensitivity is the sample volume; Pohl and Chase (1972) and Holst (1975) have found the use of large samples or standard volumes helpful. The former reported the use of 20 μg membrane protein and 4 ml sample, whereas Holst originally used 200 μg membrane protein and 2 ml sample; when used at 4 °C, the amount of membrane protein was reduced to 50 μg.

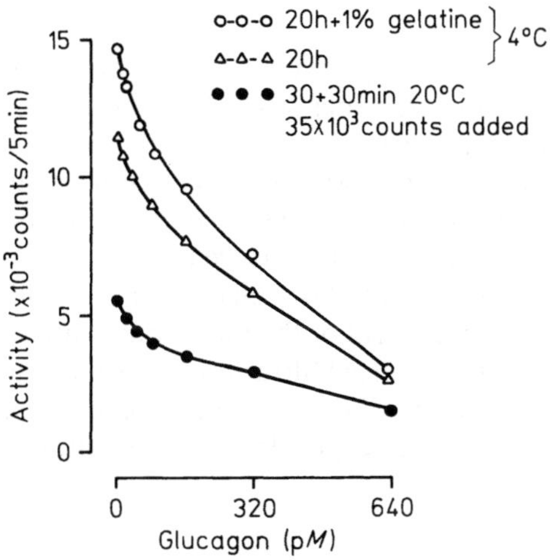


**Fig. 1.** Radioreceptorassay standard curves with pig liver plasma membranes (HOLST 1975) performed with preincubation at 20 °C (30 + 30 min) or without preincubation at 4 °C (with or without the addition of gelatin). After separation by centrifugation, the precipitate, containing the membranes, was counted for radioactivity

### 3. Assay Buffer

Most workers have used balanced salt solutions with a plasma-like composition like Krebs–Ringer solution. However, POHL and CHASE (1972) initially noted that incubation might take place in simpler buffer systems like Tris–HCl, 50 m*M* at pH 7.6. We also found that Tris–HCl titrated to pH 7.5, and the saline–albumin–phosphate buffer (50 m*M*, pH 7.5) which we usually employ for glucagon radioimmunoassay (HOLST 1977b), function as well as balanced salt solutions. HEPES (*N*-2-hydroxyethylpiperazine-*N'*-2-ethonesulfonic acid) buffer may also be used, but whatever buffer is chosen pH should lie in the vicinity of 7.4. Other buffer systems widely used for radioimmunologic studies of glucagon, like the glycine–albumin buffer at pH 8.8 or albumin-containing ammonium bicarbonate at 0.125 *M*, interfere strongly with glucagon binding.

### 4. Degradation

Although much of the degradation of glucagon (or at least labeled glucagon) which is caused by the membrane preparation, can be avoided by relevant measures: maximum reduction of membrane material, temperature reduction, preincubation, etc., most authors have added an inhibitor of protein degradation. The problem has been studied extensively by DESBUQUOIS et al. (1974), who recommended the use of bacitracin, which has also been found beneficial by others (SONNE et al. 1978, FREYCHET 1976). HOLST (1975) used gelatine, the addition of which was also beneficial at 4 °C (Fig. 1). Oxidized insulin B-chain, bacitracin, or aprotinin, could neither substitute the effect of gelatin, nor improve the results by combination with gelatin. Gelatin may be used between 0.25% and 1% (1% solutions will gelatinize at 4 °C without agitation).

### 5. Plasma Effects

Little is known about the effect of plasma and serum on radioreceptorassay for glucagon. In our hands, plasma interferes strongly in the assay, and it has proved impossible to prepare standard curves in plasma (with a very low glucagon content); this is in contrast to radioreceptorassays for insulin (Ozaki and Kalant 1977). Also, plasma extracted with ethanol interfered massively with binding. The use of glucagon radioreceptorassays for direct measurement in plasma-containing samples is therefore not possible at present. Also, changes in salt concentration may interfere with binding, and strict control of the ionic composition of unknown samples is therefore necessary (salt peak in gel filtrations). The concentration of nucleotides in the unknown samples may be of great importance, because of their ability to alter the rate of dissociation of glucagon from the receptors (Frandsen 1979).

### 6. Separation

Separation of free and bound hormone moieties in the radioreceptorassay has been accomplished by filtration through glass filters or cellulose acetate filters (EGWP Milipore), which is relevant for at least microsomal membranes (Desbuquois et al. 1974), but centrifugation in simple laboratory centrifuges is probably the easiest way of separation when employing the larger sheets of membranes obtained with the methods of Neville (1968), Ray (1970), Lesko et al. (1973), or Holst (1975). The precipitate may be washed several times or drained by inversion of the test tube on filter paper, to enhance precision. In the former case, it is practical to employ a device which enables one to suck off a fixed aliquot of the supernatant (preferably from several tubes simultaneously) leaving the tubes in the centrifuge.

## C. Performance and Applications of the Glucagon Radioreceptorassay

As discussed in Sect. B.III, with the receptor assay, detection limits at about 10–60 p*M* have been obtained. This is hardly sufficient for determinations of glucagon-like substances in plasma. Furthermore, as already mentioned, the massive interference of plasma in the assay precludes its use for direct plasma determination. Its use is therefore restricted to assays of material which is purified and fractionated to a degree which will remove the most important interfering substances. The assays have therefore been used mainly to study structure–function relationships of purified glucagon derivatives or glucagon-like material in tissue extracts, subsequent to purification and fractionation by some chromatographic procedure.

### I. Requirements for Ligand–Receptor Interaction

The performance of the radioreceptorassay must be discussed in the light of the following important questions:

1. Does binding of a substance to the glucagon receptor imply that this substance would also, in vivo, have exerted glucagon-like bioactivity? Or, in other

words, are some of the binding sites only acceptors and not true receptors, associated with the effectors of the entire receptor complex? Further, do the structural requirements for binding differ from those required for initiating the biologic activity of glucagon? And finally, does the "spare receptor" phenomenon apply to the glucagon receptors also?

2. What is the specificity of the glucagon receptor, or more precisely, what are the structural requirements of the glucagon receptor for accepting a ligand with sufficient binding energy to interfere in the assay?

### 1. Coupling of Binding to Biologic Activity

Parts of this important issue are discussed in Chaps. 13 and 54. However, since one of the advantages of the radioreceptorassay is its "specificity" compared with radioimmunologic determination of glucagon-like moieties, some relevant findings must be discussed here.

#### a) Spare Receptors

First, JARRET et al. (1971) showed that the ratio between hormone-sensitive and fluoride-sensitive adenylate cyclase could change during the purification of membranes, and if this is so, it would naturally affect the issue. Anyway, their finding is in agreement with the now generally accepted view that the acceptor and the effector (the adenylate cyclase) are separate entities (POHL 1977; LEVITZKI and HELMREICH 1979; SCHRAMM 1979). Therefore, it may be expected that binding can take place without initiating further events. Along the same lines, BIRNBAUMER and POHL (1973) found that 80%–90% of the glucagon-specific sites on liver membranes did not participate in the activation of adenylate cyclase by the hormone, and SONNE et al. (1978) working with isolated hepatocytes, found that maximal stimulation of adenylate cyclase took place when only 10% of the binding sites were occupied by glucagon. However, if the glucagon receptor system thus has the characteristics of a spare receptor system, the beneficial effects of which are a faster response to hormone binding and a greater sensitivity than systems without spare receptors (BAXTER and FUNDER 1979), then these receptors are not spare at all, but all participate somehow in the activation of the effector moiety of the receptor. If this is so, all receptors may be regarded as true receptors, and of equal value in the receptor assay.

#### b) Binding–Bioactivity Relationship

If the measurements with receptor assay were to reflect truly and specifically the glucagon-like bioactivity in an unknown sample, then this implies that binding and activation of the effector component are closely related or parallel events. However, there is evidence that this is not the case (see detailed review in Chap. 13). For instance, in many systems, glucagon shortened by one $NH_2$ terminal amino acid [desHis$^1$]glucagon acts as a weak partial agonist, explained by the finding that its affinity for the binding site in liver cells is greater in molar terms than its potency to activate adenylate cyclase (LIN et al. 1975; BREGMAN and HRUBY 1979). This seems to indicate that the histidyl residue in position 1 is of greater importance for activation of the effector unit than for the affinity of the molecule to the binding unit. Furthermore, in a detailed study of sequence-deleted glucagon analogs, FRANDSEN et al. (1981) found a 5,700-fold difference in potency for binding and cyclase activation for des-(1–4)-glucagon, and BREGMAN and HRUBY (1979) reported that [desHis$^1$]glucagon modified with phenylisothiocyanate in the ε-amino group of the lysine residue was a true glucagon antagonist. These findings contrast markedly to the results obtained with radioreceptorassays of insulin derivatives (FREYCHET et al. 1974a; GLIEMANN and GAMMELTOFT 1974) where affinity for binding was invariably paralleled by biologic activity.

Along these lines, it has been suggested that some naturally occuring glucagon-like molecules might interact with the glucagon receptor, without activation of the effector and thus act as competitive inhibitors of glucagon action, leading to hypoglycemia (REHFELD et al. 1973). It has not been possible to substantiate this theory further (HOLST 1977a), and on the whole it seems that further studies are needed to unravel why binding and activation are affected differently by certain modifications in some systems, but not in others (FRANDSEN et al. 1981; SONNE et al. 1978; DESBUQUOIS 1975a; WRIGHT et al. 1980).

If the biologic activity of glucagon is judged by its ability to promote glycogenolysis, the relation between binding and biologic activity becomes even more complicated, because, as shown by BYUS et al. (1976), activation of all the protein kinase present in liver cells is not necessary for maximal stimulation of glycogenolysis and, as emphasized by THIEDEN et al. (1981), a considerable glycogenolysis may be elicited by glucagon at concentrations which do not measurably activate adenylate cyclase. It follows that the correlation between binding of glucagon derivatives and their biologic activity may be determined only with difficulty.

c) Receptor Heterogeneity

It has been shown by some authors (FREYCHET 1976; SONNE et al. 1978) that rat liver glucagon receptors seem to be heterogeneous with respect to binding capacity and affinity as judged from nonlinear Scatchard plots. Similar findings were made in this laboratory with rat liver membranes isolated by the method of LESKO et al. (1973; J. J. HOLST, unpublished work), whereas pig liver membranes repeatedly yielded linear Scatchard plots (HOLST 1975). The importance of this possible receptor heterogeneity for the results of radioreceptorassays is difficult to ascertain, but it may be assumed that, because of the relatively low affinity of the large capacity class of receptors (in this laboratory $K_a$ was determined as about $4 \times 10^8\ M^{-1}$) it will influence the results but little if the samples to be assayed are diluted so as to give concentrations which do not exceed $10^{-9}\ M$. The possibility that low affinity binding might represent binding to nonreceptor material must also be considered (CUATRECASAS and HOLLENBERG 1975).

## 2. Specificity of the Receptor

a) Related Peptides

The glucagon receptor shows no affinity for the related gastrointestinal peptides like secretin, vasoactive intestinal polypeptide (VIP), or gastric inhibitory polypeptide (GIP), and neither is binding influenced by other gastrointestinal peptides: in-

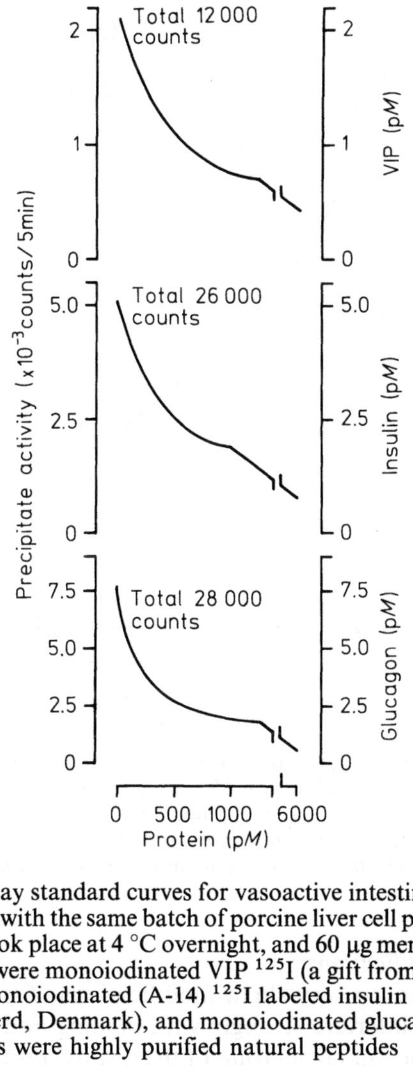


**Fig. 2.** Radioreceptorassay standard curves for vasoactive intestinal peptide (VIP), insulin, and glucagon performed with the same batch of porcine liver cell plasma membranes (HOLST 1975). All incubations took place at 4 °C overnight, and 60 µg membrane protein was added to all tubes. The tracers were monoiodinated VIP $^{125}I$ (a gift from Dr. J. FAHRENKRUG, Copenhagen, Denmark), monoiodinated (A-14) $^{125}I$ labeled insulin (a gift from the Novo Research Institute, Bagsvaerd, Denmark), and monoiodinated glucagon $^{125}I$ (JORGENSEN and LARSEN 1972). Standards were highly purified natural peptides

sulin, gastrins, cholecystokinins, etc. (HOLST 1975). In this connection, it is of interest that the same receptor preparation nevertheless may possess several receptors; thus, the porcine membranes used by HOLST (1975) may be used for insulin, glucagon, and VIP receptor studies (Fig. 2), whereas intact glicentin (see Sect.-C.II.1) apparently has no receptor of its own (Fig. 3).

b) Glucagon Derivatives

In the following discussion of the results of studies with modified glucagons, no distinction will be made between binding and biologic activity, since most studies have been concerned with biologic activity, e.g., activation of adenylate cyclase. For detailed data, see Chap. 1.

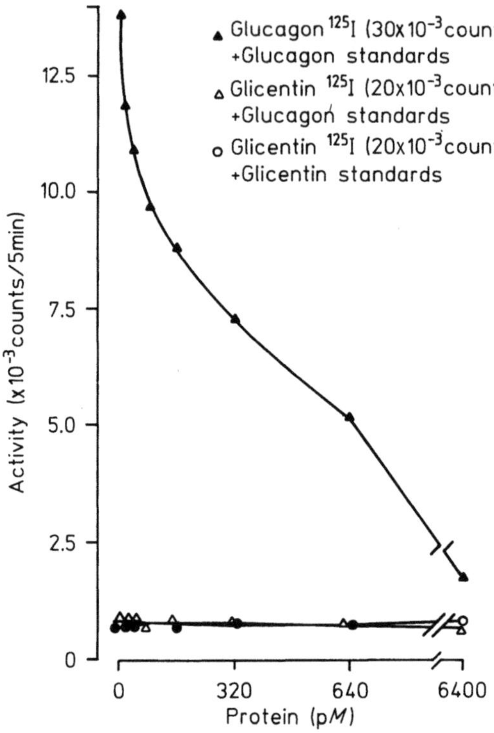


**Fig. 3.** Lack of binding of intact glicentin (THIM and MOODY 1981) to porcine liver cell plasma membranes. Monoiodinated glucagon $^{125}I$ or glicentin $^{125}I$ (gifts from the Novo Research Institute, Bagsvaerd, Denmark) were incubated overnight at 4 °C with 60 μg membrane protein from the same batch of receptors and standards as indicated, and separated by centrifugation. The precipitates with the membranes were counted for radioactivity

*α) $NH_2$ Terminal Modifications*

As already mentioned, even very slight modifications of or loss of the $NH_2$ terminal histidine residue interfere strongly with binding and bioactivity (LIN et al. 1975); it appears that the imidazol group of the histidyl residue is essential (EPAND et al. 1973), whereas minor substitutions at the α-amino group are less important. However, more extensive modifications like carbamylation or trinitrophenylation of the α-amino group lead to complete loss of bioactivity (GRANDE et al. 1972; EPAND and WHEELER 1975).

*β) COOH Terminal Modifications*

Blocking of all four carboxyl groups of glucagon leads to complete loss of activity (EPAND and EPAND 1972), and each of the four groups seem to contribute importantly to the binding to the receptor (WHEELER et al. 1974); however, the analog obtained by extension of the molecule at the COOH terminal carboxyl group with taurine still retains 10% of the bioactivity of intact glucagon (WHEELER et al. 1974). Furthermore, the tryptophan residue in position 25 may be modified by nitrophenylsulfenyl chloride without loss of activity (EPAND and COTE 1976), in agreement with the finding by many authors that the COOH terminal sequence 22–29 is not essential to biologic activity (ASSAN and SLUSHER 1972; EPAND and GREY 1973; WRIGHT et al. 1978).

Indeed, EPAND and COTE (1976) found that two molecules of glucagon covalently coupled to form a dimer through the tryptophan residue at position 25, i.e., dinitrophenyldisulfenyl(glucagon)$_2$, activate adenylate cyclase from rat liver to an extent comparable to the native hormone. This is in agreement with the results of KANEKO et al. (1979) that a synthetic derivative of glucagon, glucagon 1–29 extended at the COOH terminal sequence by a further eight amino acids, corresponding to the sequence of the COOH terminal part of porcine glicentin (THIM and MOODY 1981), stimulated rat liver plasma membrane adenylate cyclase with a potency corresponding to approximately 10% of glucagon's; deletion of the first four $NH_2$-terminal amino acids of the glucagon sequence almost abolished the activity of the derivative, whereas both derivatives were reported to compete with labeled glucagon for binding to liver cell membranes with less difference in affinity than the difference in bioactivity, a finding which seems to be at some variance with the reports discussed in Sect. C.I.2.b.$\alpha$ concerning the deleterious effects of modifications of even single amino acids at the $NH_2$ terminal sequence.

It may thus be concluded that the $NH_2$ terminal sequence of the glucagon molecule is indeed of extreme importance for binding as well as bioactivity and that the $NH_2$ terminus must be free and exposed to allow the molecule to couple with the receptor with appreciable affinity. Changes in the COOH terminus are much less important, and extensions with single or several amino acids do not severely impair the activity of the molecule. The radioreceptorassay for glucagon might therefore be expected to pick up molecular forms of glucagon with exposed $NH_2$ termini and preserved sequence up to residue 21–23 only, but might easily pick up derivatives with COOH terminal extensions.

## II. Applications

### 1. Assay of Gut Peptides

The main purposes of employing a radioreceptorassay for glucagon would be:

1. To determine whether radioimmunologic determinations of glucagon might underestimate the quantity of substances capable of interacting with the glucagon receptor, with subsequent consequences for the postreceptor functions of the tissue in question – although early bioassay studies might indicate that this would be the case (MAKMAN et al. 1958), the structural requirements of the glucagon receptor, as discussed in Sect. C.I.2.b.$\alpha$, make it unlikely that such molecules would escape detection with a region-specific assay for the $NH_2$ terminal immunodeterminant of the glucagon molecule (HOLST 1980). The "excess" glucagon-like bioactivity, if present (SOKAL 1972), therefore most probably reflects the poor specificity of most bioassay systems.

2. To determine the relationship between the large number of substances with glucagon-like immunoreactivity found in the circulation (HOLST 1981; see also Chap. 11) or in tissues (HOLST 1978b) and their possible interaction with the glucagon receptor (HOLST 1977a), and of course to facilitate further studies of the

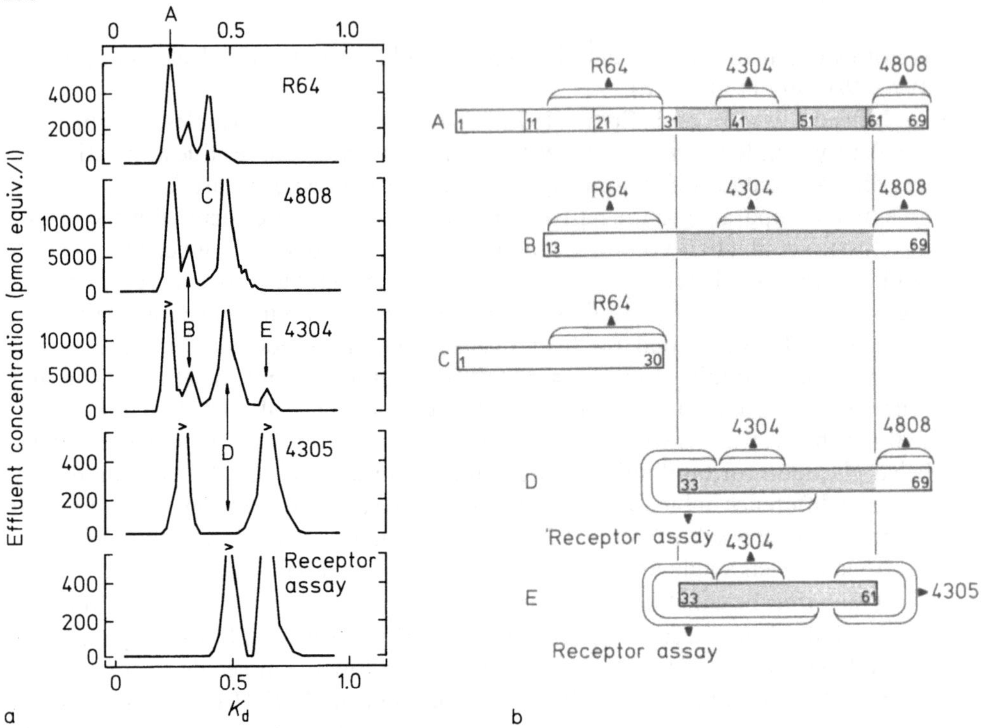


**Fig. 4 a, b.** Application of radioreceptorassay. Analysis of chemical structure by combination of gel filtration, region-specific radioimmunoassays, and radioreceptorassay. An acid–ethanol extract of porcine ileum mucosa (HOLST 1977b) was gel filtered on Sephadex G-50 in 0.5 *M* acetic acid, and the eluted fractions subjected to radioreceptorassay and radioimmunoassay after freeze-drying. **a** shows measured effluent concentrations plotted against $K_d$, the distribution coefficient (the quotient of elution volume, less the void volume, divided by the accessible inner volume of the gel). R 64, 4808, 4303, and 4305 designate antisera, the region specificities of which, relative to the sequence of the glicentin molecule (THIM and MOODY 1981) of which the *bar* A is a diagrammatic representation, are indicated in **b**. Thus, neither R 64 (MOODY et al. 1981) nor 4808 (a gift from Professor N. YANAIHARA, Shizuoka, Japan) bind to regions of the glicentin molecule which are part of the glucagon sequence (*stippled* region), whereas 4304 and 4305 bind exclusively to the glucagon part of the sequence, and 4305 only after exposure of the COOH terminus. Likewise the receptor assay seems to require an exposed $NH_2$ terminus. By comparison of the elution positions and the binding characteristics of the various components of the elution diagrams, it is possible to identify in the extract at least five components, which bear some relation to glicentin: A, a component which is similar to intact glicentin (THIM and MOODY 1981); B, probably a fragment of glicentin with $NH_2$ terminal deletions as suggested in **b**; C, a component which lacks the glucagon sequence and may be related to the similar pancreatic component (MOODY et al. 1981); D, the so-called peak II GLI or basic GLI (HOLST 1977a, b) which seems to consist of the glucagon sequence with a COOH terminal extension; and finally E, which may represent intact glucagon. Tracers for the R 64 and 4808 assays were glicentin $^{125}$I while glucagon $^{125}$I was used for the remaining assays

structure–activity relationship between glucagon and its receptor, as detailed in Chap. 13.

Of the gut peptides with glucagon-like immunoreactivity, the porcine forms have been most extensively investigated (see Chaps. 7, 33; HOLST 1978b), and a detailed analysis of the relationship between immunoreactive and receptor-active forms has been published (HOLST 1977b). The main results of this study are illustrated in Fig. 4. It appears that extracts of gut mucosa contain at least four different components which possess the glucagon sequence 6–15 (or more probably 9–15), whereas only two components possess an exposed sequence corresponding to the 19–29 sequence of the glucagon molecule; the smaller of these, which represents a very small fraction of the total amount, may correspond to intact glucagon as judged from its elution position and the fact that it is recognized in equal quantities by receptor and immunoassays for the free 19–29 sequence and the 6–15 sequence.

The dominating immunoreactive peak, which undoubtedly represents intact glicentin (HOLST 1977a, 1978b) as judged by its p*I* (isoelectric pH) and elution position ($K_d = 0.25$) (Fig. 4) interacts with neither COOH terminal glucagon antisera nor receptors, in agreement with the structural requirements for receptor binding. Similar results were published by MURPHY et al. (1973). The smaller peak, however, which represents what has been called peak II (HOLST 1977a, 1978a, b) or "basic GLI" ($K_d = 0.49$) interacts strongly with the glucagon receptor. Similar findings have been presented by BATAILLE et al. (1973, 1974).

These results have two important implications. Taken together, the chromatographic, immunologic, and receptor assay studies, would indicate that the chemical structure of "peak II" or "basic GLI" correspond to the 1–25 (at least) sequence of the glucagon molecule (with the $NH_2$ terminal sequence exposed, as illustrated by the receptor binding; the only tenfold difference between immunopotency and receptor potency indicates that the glucagon sequence is intact, probably up to residue 25). The lack of COOH terminal immunoreactivity shows that this sequence must be masked by $NH_2$ terminal extensions or deletions, but the elution position indicates that it is extended, and as judged from its pI, by a number of basic residues (HOLST 1977a). In agreement with this, the molecule does not possess an antigenic site corresponding to the nearest $NH_2$ terminal extensions in the glicentin molecule (THIM and MOODY 1981), as indicated by the lack of reactivity with antiserum R64. Finally, the $NH_2$ terminal extension does indeed seem to correspond to the $NH_2$ terminal extension of the glicentin molecule as judged from its immunoreactivity (the antiserum 4808 assay results in Fig. 4). The hypothesis that the sequence of the peak II molecule is indeed identical to the sequence of the last 37 amino acids of glicentin is further supported by immunologic as well as receptor assay studies of a synthetic peptide with this sequence (KANEKO et al. 1979; HOLST 1982). The analysis described here, illustrates how immunologic and receptor assay studies may be used in the elucidation of chemical structures (HOLST 1980, 1982).

The second implication of the assay results is that "basic GLI" if secreted from the gut may indeed exert glucagon-like effects on the liver (HOLST 1977a, 1978a); since blood enzymes may be capable of exposing the glucagon 19–29 sequence of the glicentin molecule, the gut may indeed generate intact glucagon, which can not be distinguished from the pancreatic counterpart.

## 2. Assay of Pancreatic Peptides

Glucagon radioreceptorassay of pancreatic extracts will demonstrate only one component (MOODY et al. 1981; HOLST 1977a, 1983). This seems to be in agreement with the structural requirements of the glucagon receptor, when it is taken into consideration that the large GLI discovered by RIGOPOULOU et al. (1970) has COOH terminal glucagon-like immunoreactivity (indicating the presence of an exposed glucagon 19–29 sequence), but is larger than glucagon, for which reason it follows that the molecule must be extended at the $NH_2$ terminus. In agreement with this point of view, RIGOPOULOU et al. (1970) found that the component lacked glucagon-like bioactivity. Similar results were obtained by O'CONNOR and LAZARUS (1976).

### 3. Glucagonomas

The presence of particularly high concentrations of a component with higher molecular weight than, but immunoreactivity similar to, glucagon in plasma from patients with glucagon-producing tumours has led to speculations about the bioactivity of this component, because of the sometimes surprisingly inconspicuous effects of hyperglucagonemia on glucose homeostasis in these patients. However, as demonstrated recently (HOLST 1983), the high molecular weight component present in the tumor extract is devoid of receptor activity and, by the same reasoning as discussed in Sect. C.I.1, may be assumed not to exhibit glucagon-like bioactivity. Also, in a case of intractable hypoglycemia in an infant, radioreceptor analysis of an extract of resected pancreas showed that the defects in the extractable glucagon were not responsible for the pathologic condition (RANCON et al. 1974).

## References

Assan R, Slusher N (1972) Structure/function and structure/immunoreactivity relationships of the glucagon molecule and related synthetic peptides. Diabetes 21:843–855
Barazzone P, Carpentier J-L, Gorden P, Canivet B (1979) $^{125}$I-glucagon: direct demonstration of binding, internalization and lysosomal association in isolated rat hepatocytes. Excerpta Medica Int Congr Ser 481:16
Bataille DP, Freychet P, Kitabgi PE, Rosselin GE (1973) Gut glucagon: a common receptor site with pancreatic glucagon in liver cell plasma membranes. FEBS Lett 30:215–218
Bataille D, Freychet P, Rosselin G (1974) Interactions of glucagon, gut glucagon, vasoactive intestinal polypeptide and secretin with liver and fat cell plasma membranes: binding to specific sites and stimulation of adenylate cyclase. Endocrinology 95:713–721
Baxter JD, Funder JW (1979) Hormone receptors. N Engl J Med 301:1149–1161
Birnbaumer L, Pohl SL (1973) Relation of glucagon-specific binding sites to glucagon-dependent stimulation of adenylyl cyclase activity in plasma membranes of rat liver. J Biol Chem 248:2056–2061
Blecher M, Goldstein S (1977) Hormone receptors. VI. On the nature of the binding of glucagon and insulin to human circulating mononuclear leukocytes. Mol Cell Endocrinol 8:301–315
Bregman MD, Hruby VJ (1979) Synthesis and isolation of a glucagon antagonist. FEBS Lett 101:191–194
Bromer WW, Boucher ME, Patterson JM (1973) Glucagon structure and function. II. Increased activity of iodoglucagon. Biochem Biophys Res Commun 53:134–139
Byus CV, Hayes JS, Brendel K, Russell DH (1976) Correlation between cAMP, activation of cAMP-dependent protein kinase(s) and rate of glycogenolysis in isolated rat hepatocytes. Life Sci 19:329–336

Cuatrecasas P, Hollenberg MD (1975) Binding of insulin and other hormones to non-receptor materials: saturability, specificity and apparent "negative cooperativity". Biochem Biophys Res Commun 62:31–41
Desbuquois B (1975a) Acetylglucagon: preparation and characterization. Eur J Biochem 60:335–347
Desbuquois B (1975b) Iodoglucagon: preparation and characterization. Eur J Biochem 53:569–580
Desbuquois B, Laudat M-H (1974) Glucagon-receptor interactions in fat cell membranes. Mol Cell Endocrinol 1:355–370
Desbuquois B, Krug F, Cuatrecasas P (1974) Inhibitors of glucagon inactivation. Effect on glucagon-receptor interactions and glucagon-stimulated adenylate cyclase activity in liver cell membranes. Biochim Biophys Acta 343:101–120
Epand RM, Cote TE (1976) Conformational and biological properties of a covalently linked dimer of glucagon. Biochim Biophys Acta 453:365–373
Epand RM, Epand RF (1972) Carboxyl group modification in glucagon. Biochim Biophys Acta 285:176–180
Epand RM, Grey V (1973) Conformational and biological properties of partial sequences of glucagon. Can J Physiol Pharmacol 51:243–248
Epand RM, Wheeler GE (1975) The effects of the trinitrophenylation of the amino groups of glucagon on its conformational properties and on its ability to activate rat liver adenylyl cyclase. Biochim Biophys Acta 393:236–246
Epand RM, Epand RF, Grey V (1973) The essential role of the imidazole group of glucagon in its biological function. Arch Biochem Biophys 154:132–136
Frandsen EK (1979) The function of guanyl nucleotides in the glucagon mediated activation of hepatic adenylate cyclase. In: Rosselin G, Fromageot P, Bonfils S (eds) Hormone receptors in digestion and nutrition. Elsevier/North Holland Amsterdam Oxford New York, pp 313–318
Frandsen EK, Gronvald FC, Heding LG, Johansen NL, Lundt BF, Moody AJ, Markussen J, Volund A (1981) Glucagon: structure-function relationships investigated by sequence deletions. Hoppe Seylers Z Physical Chem 362:665–678
Freychet P (1976) Interactions of polypeptide hormones with cell membrane specific receptors: studies with insulin and glucagon. Diabetologia 12:83–100
Freychet P, Brandenburg D, Wollmer A (1974a) Receptor-binding assay of chemically modified insulins. Comparison with in vitro and in vivo bioassays. Diabetologia 10:1–5
Freychet P, Rosselin G, Rancon F, Fouchereau M, Broer Y (1974b) Interactions of insulin and glucagon with isolated rat liver cells. I. Binding of the hormone to specific receptors. Horm Metab Res [Suppl] 5:72–78
Gliemann J, Gammeltoft S (1974) The biological activity and the binding affinity of modified insulins determined on isolated rat fat cells. Diabetologia 10:105–114
Goldfine ID, Roth J, Birnbaumer L (1972) Glucagon receptors in $\beta$-cells. Binding of $^{125}I$-glucagon and activation of adenylate cyclase. J Biol Chem 247:1211–1218
Goldstein S, Blecher M, Binder R, Perrino PV, Recant L (1975) Hormone receptors 5. Binding of glucagon and insulin to human circulating mononuclear cells in diabetes mellitus. Endocrinol Res Commun 2:367–376
Gorden Ph, Carpentier J-L, Freychet P, Orci L (1980) Internalization of polypeptide hormones. Mechanism, intracellular localization and significance. Diabetologia 18:263–274
Grande F, Grisolia S, Diederich D (1972) On the biological and chemical reactivity of carbamylated glucagon. Proc Soc Exp Biol 139:855–860
Heber D, Odell WD, Schedewie H, Wolfsen AR (1978) Improved iodination of peptides for radioimmunoassay and membrane radioreceptor assay. Clin Chem 24:796–799
Holst JJ (1975) A radioreceptor-assay for glucagon: binding of enteroglucagon to liver plasma membranes. Diabetologia 11:211–219
Holst JJ (1977a) Interactions of hepatocyte membrane receptors with pancreatic and gut glucagon. In: Foà PP, Bajaj JS, Foà NL (eds) Glucagon: its role in physiology and clinical medicine. Springer, Berlin Heidelberg New York, pp 287–303
Holst JJ (1977b) Extraction, gel filtration pattern and receptor binding of porcine gastrointestinal glucagon-like immunoreactivity. Diabetologia 13:159–169

Holst JJ (1978a) Physiology of enteric glucagon-like substances. In: Bloom SR (ed) Gut hormones 1st ed. Churchill Livingstone, Edinburgh London, pp 383–386
Holst JJ (1978b) Extrapancreatic glucagons. Digestion 17:168–190
Holst JJ (1980) Evidence that glicentin contains the entire sequence of glucagon. Biochem J 187:337–343
Holst JJ (1981) Pattern of glucagon release. In: Bloom SR, Polak J (eds) Gut hormones 2nd edn. Churchill Livingstone, Edinburgh London, pp 325–331
Holst JJ (1982) Evidence that enteroglucagon(II) is identical with the C-terminal sequence (residues 33–69) of glicentin. Biochem J 207:381–388
Holst JJ (1982) Molecular heterogeneity of glucagon produced and secreted by glucagonomas. Diabetologia, in press
Jarett L, Reuter M, McKeel DW, Smith RM (1971) Loss of adenyl cyclase hormone receptors during purification of fat cell plasma membranes. Endocrinology 89:1186–1190
Jorgensen KH, Larsen UD (1972) Purification of $^{125}$I-glucagon by anion exchange chromatography. Horm Metab Res 4:223–224
Kaneko T, Cheng PY, Toda G, Oka H, Oda T, Yanaihara N, Yanaihara C, Mihara S, Nishida T, Kaise N, Shin S, Imagawa K (1979) Biological and binding activities of synthetic possible C-terminal fragments of glicentin in rat liver plasma membranes. In: Miyoshi A (ed) Gut peptides. Kodansha, Tokyo, pp 157–161
Lande S, Gorman R, Bitensky M (1972) Selectively blocked and deshistidine-glucagons: preparation and effects on hepatic adenylate cyclase activity. Endocrinology 90:597–604
Lesko L, Donlon M, Marinetti GV, Hare JD (1973) A rapid method for the isolation of rat liver plasma membranes using an aqueous two-phase polymer system. Biochim Biophys Acta 311:173–179
Levey GS, Fletcher MA, Klein I (1975) Glucagon and adenylate cyclase: binding studies and requirements for activation. Adv Cyclic Nucleotide Res 5:53–65
Levitzki A, Helmreich JM (1979) Hormone receptor-adenylate cyclase interactions. FEBS Lett 101:213–219
Lin MC, Wright DE, Hruby VJ, Rodbell M (1975) Structure-function relationships in glucagon: properties of highly purified deshis$^{1}$-, monoiodo, and [Des-Asn$^{28}$, Thr$^{29}$] (homoserine lactone$^{27}$)-glucagon. Biochemistry 14:1559–1563
Lin MC, Nicosia S, Lad PM, Rodbell M (1977) Effects of GTP on binding of [$^{3}$H] glucagon to receptors in rat hepatic plasma membranes. J Biol Chem 252:2790–2792
Livingston JN, Cuatrecasas P, Lockwood DH (1974) Studies on glucagon resistance in large rat adipocytes: 125-I-labelled glucagon binding and lipolytic capacity. J Lipid Res 15:26–32
Makman MH, Makman RS, Sutherland EW (1958) Presence of a glucagon-like material in blood of man and dog. J Biol Chem 233:894–899
Moody AJ, Holst JJ, Thim L, Jensen SL (1981) Relationship of glicentin to proglucagon and glucagon in the porcine pancreas. Nature 289:514–516
Murphy RF, Buchanan KD, Elmore DT (1973) Isolation of glucagon-like immunoreactivity of gut by affinity chromatography on anti-glucagon antibodies coupled to Sepharose 4B. Biochim Biophys Acta 333:118–127
Neville DM (1960) The isolation of a cell membrane fraction from rat liver. J Biophys Biochem Cytol 8:413–422
Neville DM (1968) Isolation of an organ specific protein antigen from cell-surface membrane of rat liver. Biochim Biophys Acta 154:540–552
Nottey JJ, Rosselin G (1971) Monoiodoglucagon: préparation, isolement, identification, contrôle radio-immunologique. C R Acad Sci [D] (Paris) 273:2118–2121
O'Connor KJ, Lazarus NR (1976) The purification and biological properties of pancreatic big glucagon. Biochem J 156:265–277
Ozaki S, Kalant N (1977) A radioreceptor assay for serum insulin. J Lab Clin Med 90:686–699
Patterson JM, Bromer WW (1973) Glucagon structure and function. Preparation and characterization of nitroglucagon and aminoglucagon. J Biol Chem 248:8337–8342
Pohl SL (1977) The glucagon receptor and its relationship to adenylate cyclase. Fed Proc 36:2115–2118

Pohl SL, Chase LR (1972) A radioreceptor assay for glucagon using purified plasma membranes from rat liver. Excerpta Medica, Int Congr Ser 256:201

Rancon F, Laburthe M, Rosselin G, Freychet P (1974) Untractable hypoglycemia in an infant: studies on pancreas insulin and glucagon. Horm Metab Res 6:443–447

Ray RK (1970) A modified method for the isolation of the plasma membrane from rat liver. Biochim Biophys Acta 196:1–9

Rehfeld JF (1980) $NH_2$-terminal monoiodination of hexadecapeptide gastrin: a simple procedure for preparation of 125-I-gastrin for radioimmunoassays and receptor studies. Clin Chim Acta 101:271–275

Rehfeld JF, Heding LG, Holst JJ (1973) Increased gut glucagon release as pathogenic factor in reactive hypoglycaemia? Lancet 1:116–118

Rigopoulou D, Valverde I, Marco J, Faloona G, Unger RH (1970) Large glucagon immunoreactivity in extracts of pancreas. J Biol Chem 245:496–501

Rodbell M, Krans MJ, Pohl SL, Birnbaumer L (1971) The glucagon-sensitive adenyl cyclase system in plasma membranes of rat liver. III. Binding of glucagon: method of assay and specificity. J Biol Chem 246:1861–1871

Rosselin G, Freychet P, Bataille D, Kitabgi P (1974a) Polypeptide hormone-receptor interactions. A new approach to the study of pancreatic and gut glucagons. Isr J Med Sci 10:1314–1323

Rosselin G, Freychet P, Fouchereau M, Rancon F, Broer Y (1974b) Interactions of insulin and glucagon with isolated rat liver cells. II. Dynamic changes in the cyclic AMP induced by hormones. Horm Metab Res [Suppl] 5:78–86

Roth J (1975b) Methods for assessing immunological and biological properties of iodinated peptide hormones. Methods Enzymol 37:223–232

Schramm M (1979) Transfer of a glucagon receptor from liver membranes to a foreign adenylate cyclase by a membrane fusion procedure. Proc Natl Acad Sci USA 76:1174–1178

Sokal JE (1972) Bioassays for glucagon. In: Lefebvre PJ, Unger RH (eds) Glucagon. Molecular physiology, clinical and therapeutic implications. Pergamon, Oxford New York, pp 275–284

Sonne O, Gliemann J (1977) Receptor binding of glucagon and adenosine 3′:5′-monophosphate accumulation in isolated rat fat cells. Biochim Biophys Acta 499:259–272

Sonne O, Larsen UD (1980) Decreased receptor binding affinity of mono-125-I-glucagon when $Met_{27}$ is oxidized during iodination. Acta Physiol Scand 109:15A

Sonne O, Berg T, Christoffersen T (1978) Binding of 125-I-labelled glucagon and glucagon-stimulated accumulation of adenosine 3′:5′-monophosphate in isolated intact rat hepatocytes. J Biol Chem 253:3203–3210

Thieden HID, Holst JJ, Dich J, Moody AJ, Sundby F (1981) Effect of highly purified porcine gut glucagon-like immunoreactivity (Glicentin) on glucose release from isolated rat hepatocytes. Biochim Biophys Acta 675:

Thim L, Moody AJ (1981) The primary structure of porcine glicentin (proglucagon). Regul Peptides 2:139–150

Von Schenck H, Larsson I, Thorell JI (1976) Improved radioiodination of glucagon with the lactoperoxidase method. Influence of pH on iodine substitution. Clin Chim Acta 69:225–232

Wheeler GE, Epand RM, Barrett D (1974) Non-equivalence of the carboxyl groups of glucagon in the carbodiimide-promoted reaction with nucleophiles and the role of carboxyl groups in the ability of glucagon to stimulate adenyl cyclase of rat liver. Biochim Biophys Acta 372:440–449

Wright DE, Hruby VJ, Rodbell M (1978) A reassessment of structure-function relationships in glucagon. J Biol Chem 253:6338–6340

Wright DE, Hruby VJ, Rodbell M (1980) Preparation and properties of glucagon analogues prepared by semi-synthesis from CNBr-glucagon. Biochim Biophys Acta 631:49–58

# Actions of Glucagon

CHAPTER 13

# The Actions of Glucagon at Its Receptor: Regulation of Adenylate Cyclase

M. RODBELL

## A. Introduction

The "second messenger" concept of hormone action, first proposed nearly 20 years ago, states that hormones generally act by changing the levels of chemical signals produced by effector processes localized at the target cell plasma membrane (SUTHERLAND 1972). These chemical signals govern in turn the activity of key intracellular enzymes that regulate, through a complex series of reactions, numerous metabolic processes in the cell. The concept of second messengers arose from the discovery that cyclic AMP mediates the actions of catecholamines and glucagon on glycogen metabolism in liver; numerous studies subsequently showed that cyclic AMP mediates the actions of a large number of hormones and neurotransmitters and that the latter agents act by either stimulating or inhibiting the nucleotide's production (RODBELL 1978). Although the correlations between cyclic AMP production and hormone action are generally good, it must be emphasized that proof is lacking that regulation of cyclic AMP levels is the single action responsible for all effects of hormones or neurotransmitters. As a model process, however, hormonal regulation of cyclic AMP production has proven to be ideal since it can be observed with isolated preparations of membranes from a variety of sources.

The enzyme that catalyzes the production of cyclic AMP is termed adenylate cyclase (some prefer adenylyl cyclase). The enzyme is present in the plasma membrane of most eukaryotic cells and generally responds in isolated membrane preparations to the same hormones that affect the production of cyclic AMP in the intact cell. Because each system displays unique characteristics with respect to the actions of hormones, whereas other characteristics of adenylate cyclase systems are very similar, it has been apparent for some time that the receptors are likely to be separate entities from the enzyme. There is now ample proof for this assertion based on intensive investigations in numerous laboratories. Emphasis has gradually changed in recent years from studies of receptors to study of the process, termed either "transduction" or "coupling," that is responsible for mediating hormone action at the receptor into alterations in adenylate cyclase activity. Based initially on findings with the glucagon-sensitive system in rat liver plasma membranes that GTP is essential for glucagon action (RODBELL et al. 1971 b), it is now clear that transduction and the actions of GTP are synonymous processes carried out by a distinct class of proteins from both hormone receptors and the catalytic unit of adenylate cyclase (RODBELL 1980). Termed nucleotide-regulatory proteins (abbreviated N), these proteins appear to govern a variety of processes in membranes, including those that stimulate adenylate cyclase activity ($N_s$), those that inhibit the

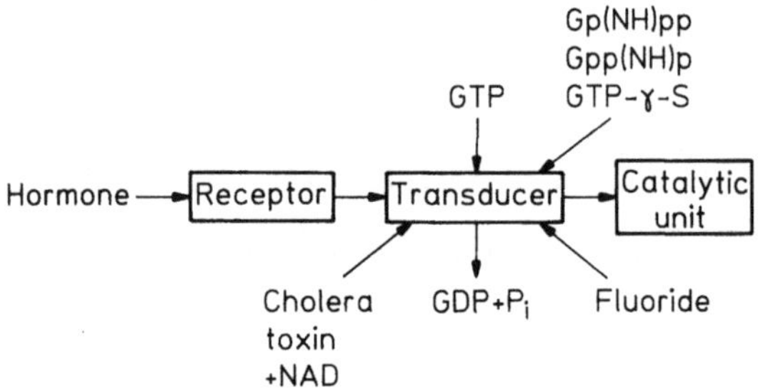


**Fig. 1.** General scheme for "transduction" of hormone action on adenylate cyclase systems. The transducer is a separate component from the receptors and the catalytic unit and is the site of action of guanine nucleotides, cholera toxin, and fluoride ion. It also contains a "GTPase" that converts the highly active guanine nucleotide triphosphates to the less active or inactive GDP. The role of the hormone–receptor complex is to facilitate interaction of GTP, the presumed physiologic activator, with the transducer. GTP-γ-S, guanosine-5′-O-(3-thiotriphosphate); $P_i$, inorganic phosphate

enzyme ($N_i$), and those ($N_x$) that mediate actions of hormones on as yet unidentified processes.

Armed with knowledge that the N units form the structural linkage between receptor (abbreviated R) and the catalytic unit (C), investigators in the field are now in a better position to determine how receptors located on the outer surface of the plasma membrane are able to communicate with and regulate the activity of adenylate cyclase at the inner face of the membrane. In this chapter, emphasis is placed on the characteristics of the glucagon receptor and the N and C units present in the plasma membrane of rat liver membranes, the most intensively studied glucagon-sensitive system. However, in order to provide a broader perspective, discussion of this system is intermingled with current views and status of other hormone-sensitive adenylate cyclase systems which also provide insights into the general mechanism by which the adenylate cyclase system is regulated.

## B. Characteristics of Hormone-Sensitive Adenylate Cyclase Systems

### I. Role of GTP in Hormone Action

Early studies of the glucagon-sensitive system established that the hormone governs the ability of GTP to activate adenylate cyclase in isolated plasma membranes; concerted actions of glucagon and GTP gave rise to the concept illustrated schematically in Fig. 1 that GTP acts at the transduction process (RODBELL 1972a). Hydrolysis of GTP to GDP was postulated because guanine nucleotide analogs, such as guanylylimidodiphosphate [Gpp(NH)p], that are not readily hydrolyzed at the terminal phosphate, cause slowly reversible activation after removal of excess hormone and nucleotide from the incubation medium (RODBELL et al. 1975; SCHRAMM and RODBELL 1975). Since guanine nucleotides and fluoride ion, another stimulant, inhibited each other's actions, it was considered likely that these agents act through a common process. Finally, as discussed further in Sect. B.V, cholera toxin was subsequently shown to enhance the ability of GTP to activate adenylate cyclase in both the absence and presence of hormones.

## II. Specificity for Guanine Nucleotides

Most adenylate cyclase systems that have been investigated using purified plasma membranes require added GTP for significant stimulation of the enzyme to be observed in the presence of hormones. However, there are other factors that can substitute for GTP in some systems; moreover, contamination with mitochondria and other organelles that are potential sources of GTP can often result in activation without added GTP (LONDOS and RODBELL 1975). GTP is a potent activator, half-maximal activation occurring generally with 0.1 $\mu M$ GTP. Commercial preparations of ATP are frequently contaminated with sufficient GTP to cause full activation without added guanine nucleotide. An additional problem stems from the presence of nucleotide phosphotransferases which can convert, for example, GDP to GTP by a variety of nucleotide triphosphates. Thus, apparent activation by UTP, CTP, or ITP seen in some systems may have been due simply to conversion of GDP present in the assay system to the potent activator, GTP. Such conversions have been reported with the glucagon-sensitive system; because activation by GDP in the presence of other nucleotide triphosphates required the presence of glucagon, it was postulated that the transduction process may contain a hormone-sensitive nucleotide phosphotransferase (KIMURA and NAGATA 1979). A simpler explanation, however, is that the hormone increases the reactivity of the N unit with GTP formed from GDP by membrane nucleotide phosphotransferases.

## III. Effects of Adenosine

Another factor related to the mandatory addition of ATP as substrate is that membranes contain nucleotidases that hydrolyze ATP to adenosine (COOPER and LONDOS 1979). Adenosine, at submicromolar concentrations, stimulates adenylate cyclase in membranes through specific, methylxanthine-inhibited receptors; adenosine stimulation, as with hormones, also acts through the GTP transduction process (LONDOS et al. 1979b). If not controlled, either by addition of adenosine deaminase (inosine is inactive) or by addition of methylxanthines, adenylate cyclase systems may display significant activation by GTP with the interpretation that hormones are not necessarily required for nucleotide activation. It has recently been shown that adenosine can also inhibit certain adenylate cyclase systems through receptors that differ from the adenosine-stimulatory receptors primarily in the potency of certain adenosine analogs. As a final complication of the effects of adenosine, the nucleoside also inhibits all adenylate cyclase systems through a process that is distal to hormone receptors and which appears to be located at a site, termed "P-site," which is closely linked to metal ion activation of the enzyme (LONDOS and PRESTON 1977a).

Because many studies have been carried out unmindful of the potential stimulatory and inhibitory effects of adenosine through receptors and P-site, it is likely that some interpretations have been confused by one or more actions of the nucleoside. A notable example is the report (BRAUN and LEVITZKI 1979) that adenosine inhibits through its receptor in turkey erythrocyte membranes the stimulatory effects of catecholamines on adenylate cyclase; there is little question that inhibition was actually achieved through the P-site (LAD et al. 1980b).

## IV. Role of GTPase

Guanine nucleotides having three phosphate groups are generally the most active nucleotides. In some systems, GDP causes modest stimulation of adenylate cyclase [with App(NH)p as substrate to avoid potential phosphorylation of GDP to GTP]. In the liver system, stimulation by GDP requires saturating concentrations of glucagon whereas activation by GTP occurs with minimal occupation of the receptors (KIMURA and NAGATA 1979; RODBELL et al. 1971b; SALOMON et al. 1975). Clearly, on this basis alone, GDP is not the physiologic activator. Guanosine-5′-O-(2-thiodiphosphate) (GDP-$\gamma$-S), which cannot be phosphorylated to yield GTP, is also a modest activator (ECKSTEIN et al. 1979). Such findings indicate that GDP can serve both as inhibitor (at submaximal concentrations of glucagon) and activator (with saturating hormone concentrations); it is essentially a partial agonist on the glucagon-sensitive system and other cyclase systems.

Some cyclase systems, notably that in the turkey erythrocyte, have very different properties from the rat liver system. This is exemplified by GDP serving only as a potent inhibitor of the enzyme either in the absence or presence of hormones (catecholamines or prostaglandins) that stimulate the enzyme through a guanine nucleotide-dependent process. Studies of the turkey system established that hormones stimulate the breakdown of GTP to GDP via a process having all of the characteristics through which the hormones stimulate adenylate cyclase system. From such findings arose the concept of the "GTPase cycle" whereby hormones act through their receptor to stimulate release of bound GDP thus allowing GTP, the activator, to occupy the N unit; GTP is then hydrolyzed to GDP and the enzyme decays to its inhibited state (CASSEL and SELINGER 1978). In this fashion, hormones serve both to enhance the "exchange reaction" and the "GTPase" by allowing GTP access to the site of activation, which is presumably structurally linked to the site of breakdown of GTP to GDP.

Modeling of the comparative actions of Gpp(NH)p and GTP on the liver enzyme in its response to glucagon also gave rise to the concept of a GTPase being responsible for turning down the stimulatory effects of GTP (RENDELL et al. 1977). In this model, it was proposed that glucagon acts through its receptor by promoting access of the enzyme system to a state which displays both high cyclase activity and an "excited" state of GTPase activity. Consistent with this model is the recent finding that glucagon stimulates a GTPase activity in rat liver membranes (KIMURA and SHIMADA 1980). In view of the fact that GDP is not a potent inhibitor of the liver system (its $K_i$ and $K_{act}$ are similar to the $K_{act}$ for GTP), but rather is a partial agonist if the glucagon receptor is saturated, complete turnoff of the enzyme when GDP is formed is not expected as long a saturating concentrations of glucagon are employed. However, under the usual physiologic conditions in which the hormone concentration is submaximal, GDP behaves only as an inhibitor and could play a physiologic role in regulating liver cyclase activity. In this manner, one can visualize the action of glucagon on the adenylate cyclase system as a process oscillating between high and low activities, depending on the amount of receptor occupied and the levels of the N unit occupied by GTP and GDP.

## V. Actions of Cholera Toxin

As evidenced by the initial findings that cholera toxin causes increases in intestinal cyclic AMP production, it is believed that most if not all biologic actions of the toxin are mediated by its effects on adenylate cyclase systems (GILL 1977; MOSS and VAUGHAN 1979). The toxin's actions resemble those of hormones in that, with isolated membrane preparations, activation requires guanine nucleotides. However, unlike hormones the toxin causes sustained activation of the enzyme in the presence of GTP. This sustained action is because the toxin contains in its $A_1$ subunit an ADP-ribosylating component which covalently modifies, in the presence of its substrate NAD, the N unit (CASSEL and PFEUFFER 1978). As discussed later, this enzymatic action has aided considerably in identifying and purifying the N unit. Findings that the toxin inhibits the GTPase activity stimulated by hormones in the turkey erythrocyte system (CASSEL and SELINGER 1977) suggest that its action on cyclase activity relates to this effect. Additionally, it seems alo to promote the exchange reaction as evidenced by findings that toxin treatment enhances the ability of Gpp(NH)p to activate the turkey cyclase system (LAD et al. 1980a).

Treatment of membranes with toxin also generally enhances hormone action, as would be expected from its sustaining effects on GTP action. The toxin also inhibits the stimulatory actions of fluoride, an effect which, as with the inhibitory effects of guanine nucleotides on fluoride action, yielded the suggestive evidence that the N unit is a multisite or multicomponent molecule that mediates that actions of hormones, guanine nucleotides, cholera toxin, and fluoride ion.

## VI. Actions of Guanine Nucleotides on Hormone Receptors

Since GTP acts at an intermediate stage between the hormone receptors and adenylate cyclase, and hormones affect the interaction of GTP with the transducer element, on theoretical grounds GTP should also affect the structure of hormone receptors. In fact, the first evidence for such a reciprocal relationship came from the finding that GTP changes the glucagon receptor from a tight binding, high affinity state to a receptor state having a 6- to 10-fold lower affinity (RODBELL et al. 1971a). When it was found that GTP acts at identical concentrations to alter both the receptor and adenylate cyclase activity, it seemed likely that both effects were mediated by the same transducer process. Such effects of GTP and other guanine nucleotides have been observed on numerous hormone and neurotransmitter receptors, even those that are not demonstrated to be linked to adenylate cyclase systems (RODBELL 1980). The latter formed the basis of the postulate (RODBELL 1980) that there may be distinct classes of N units that mediate the actions of a variety of hormones on processes other than adenylate cyclase. One notable example of such an $N_x$ is the transduction process in rod outer segments in which light activates through rhodopsin a GTP-binding protein (also containing a GTPase activity) which mediates activation of a cyclic GMP phosphodiesterase. Not overlooked is the possibility that this process is analogous to hormone-activated adenylate cyclase systems (SHINOZAWA et al. 1979).

## VII. Effects of GTP on Hormone Binding Versus Action

Concerted activation of adenylate cyclase systems by hormones and GTP generally occurs within seconds, reaching steady state rates in about 1 min. In the presence of GTP, the binding of glucagon and catecholamines to their respective receptors also reach steady state within the same time scale required for attaining steady state rates of cyclase activity in response to these hormones (LIN et al. 1977; LIMBIRD and LEFKOWITZ 1978). These findings have given the strongest evidence that the binding sites measured are functionally involved in activation of adenylate cyclase and that the GTP transduction process is linked to the receptors. Since the tight binding states of the receptors are converted by GTP to the states involved in action, it follows that the former states are precursors of the latter. Because hormones promote interaction of N with GTP, it also follows that the hormones, acting through the tight binding states of the receptors, initiate the events that lead GTP both to alter the affinity states of the receptor and to stimulate adenylate cyclase. By this reasoning, antagonists, though they bind to the tight binding configuration, cannot promote GTP action at the N unit. It has been shown, in fact, that catecholamine antagonists bind to the $\beta$-adrenoceptors with high affinity; GTP does not alter the binding of antagonists to this state (LIMBIRD and LEFKOWITZ 1978).

There remains the important question of whether hormones, in binding to the receptors, promote the formation of complexes with N or stabilize a preformed complex of receptors and N. It is quite likely that both situations are correct and that an equilibrium exists between preformed complexes of receptors and N (R·N) and free R and N units (DELEAN et al. 1980). From a theoretical standpoint, the high affinity state of the receptor is the R·N complex. Evidence that this is correct comes from findings that catecholamine receptors display lower affinities for agonists in membranes lacking N than those membranes of the same cell lines that contain linked N units.

Another important issue is how the GTP-induced low affinity state of the receptor is involved in the activation of adenylate cyclase. One theory suggests that hormones and GTP produce a ternary complex (R·N·GTP) which combines with C and concomitantly forms an activated N(GTP)·C complex with dissociation of R; the latter may represent the low affinity form of the receptor (DELEAN et al. 1980). An alternative possibility is that R·N exceeds the concentration of C and that the ternary complex R·N·GTP has a lower affinity for hormone than the holoenzyme R·N(GTP)·C. In this theory the receptor remains complexed with N during the entire process leading to the activated holoenzyme; the amount of receptor in the low affinity state is a measure of the excess of R·N·GTP over that of C. The holoenzyme has a higher affinity for hormone and reflects the steady state levels of bound hormone during the activation process. Thus, in this theory, three distinct receptor structures with differing affinities for hormones exist in the membrane, depending on the concentration of hormone and GTP. These species and their relative affinities for hormone are:

$$R \cdot N \cong R \cdot \overset{\overset{\textstyle GTP}{\cdot}}{N} \cdot C > R \cdot N \cdot GTP\,.$$

Supporting this theory is the finding that glucagon $^3H$ binds in the absence of GTP to a single species of sites with high affinity; in the presence of GTP, at least two species are observed: 90% with low affinity, the remaining 10% with high affinity (LIN et al. 1977). Interestingly, rat hepatocytes show the identical proportion of low and high affinity states (SONNE et al. 1978). Since the hepatocyte undoubtedly contains GTP, one might suppose that the binding states seen in the intact cell reflect the actions of glucagon and GTP postulated to occur with isolated plasma membranes.

## VIII. Multiple Sites of GTP Action

The possibility that GTP affects hormone binding and adenylate cyclase activity through independent sites or components arose from differences in the characteristics of GTP action on these processes in the glucagon-sensitive cyclase system in rat liver membranes (LAD et al. 1977; IYENGAR et al. 1979). These differences include: (1) potencies of guanine nucleotides; (2) states of reversibility of Gpp(NH)p action on binding and cyclase activation; (3) effects of phospholipases on the two processes; (a) actions of detergents and filipin, a cholesterol binding agent. It was suggested from these findings either that the N unit differs in its characteristics when combined with R or C, or that separate N units are involved in the two processes. Studies comparing the actions of Gpp(NH)p on adenylate cyclase after pretreatment with the nucleotide versus the behavior of the enzyme when subjected to a fresh charge of glucagon and guanine nucleotides (GTP or GDP) were interpreted as evidence for separate sites of GTP action (IYENGAR and BIRNBAUMER 1979). However, it is difficult to distinguish in such kinetic studies whether the different levels of activation are due to different N units or to different species of the enzyme linked with the same N unit. Thus, one could envision the enzyme existing in the membrane in the form of N·C complexes and R·N·C complexes which may have very different properties.

Studies on the adenylate cyclase system in S 49 mouse lymphoma cells have given valuable insights into the question of two or more sites or components of GTP action (ROSS and GILMAN 1980; JOHNSON et al. 1981 a). The wild-type (*wt)* strain contains an adenylate cyclase system that is activated by both prostaglandins and catecholamines, by cholera toxin, fluoride ion, and guanine nucleotides; from labeling experiments with toxin and NAD $^{32}P$, two proteins having molecular weights of 43,000 and 53,000 daltons were detected. In isolated membrane preparations, guanine nucleotides affect both the binding of catecholamines and the activity of cyclase in a manner similar to that discussed in Sect. B.VII for the glucagon-sensitive system. A variant designated $ac^-$ lacks the N protein and displays none of the properties of the *wt* strain. Extracts containing N confer on the $ac^-$ strain all of the properties seen with *wt*. Two other variants have been reported which are especially relevant to the question posed here. One variant, designated *unc*, contains an N unit by several functional criteria, yet the enzyme system is not activated by the hormone; guanine nucleotides also fail to alter receptor affinity. These findings suggest that the receptors are functionally uncoupled from the N unit. Another variant has been discovered recently which behaves diametrically oppositely to *unc*. This variant also contains an adenylate cyclase system that fails to respond to hormones; however, by contrast, it does not respond to the stimulatory effects of Gpp(NH)p, despite the finding that the nucleotide alters the binding of agonist to the $\beta$-adrenoceptor. This strain, therefore, appears to be defective in the linkage of N with the catalytic unit.

The findings reported with the S 49 lymphoma cyclase system also can be interpreted as evidence either for a single N unit mediating both actions of GTP or two N units. However, given recent evidence that the structure of N is modified in the *unc* variant (SCHLEIFER et al. 1980) and the findings (discussed in Sect. B.IX) that the N unit from rabbit liver membranes consists of three subunits, it seems likely that different subunits interact with GTP to regulate receptor and adenylate cyclase structure independently. The interactions of each GTP-binding subunit with R and C may be mediated through different types of lipids and have significantly different affinities and specificities for GTP and various analogs of guanine nucleotides. These putative GTP-binding subunits may also react differently with fluoride ion, phospholipases, detergents, and other agents which appear to act in disparate manner on the coupling of the N unit with R and C.

## IX. Characteristics of the N Unit

From previous discussions of the N unit, it is evident that it acquires very different properties with respect to affinities for guanine nucleotides and susceptibility to deleterious effects of detergents and phospholipases, depending on its linkage to receptors, the catalytic unit, or to both units in the holoenzyme. The N unit exists in human erythrocyte membranes uncomplexed with R or C, which appear to be nonexistent or in very low concentrations with respect to N (NIELSEN et al. 1980; KASLOW et al. 1980; FARFEL et al. 1979). When extracted with detergents or added in its membrane-bound form, the erythrocyte N unit restores to the S 49 lymphoma variant $ac^-$ all of the characteristics of adenylate cyclase displayed by the wild type enzyme. Such findings provided a means of investigating the properties of N in its membrane-bound form before being complexed with the cyclase system. In its free state the N unit displays a much lower affinity for guanine nucleotides than in its coupled forms (R·N, N·C, and R·N·C). Sulfhydryl agents block the ability of N to couple with C functionally, indicating that it contains critical sulfhydryl groups for this function. When extracted from the membrane, it behaves as a water-soluble protein, binds little detergent, and seems to have an asymmetric rod-like structure. In the erythrocyte and probably in all cells, the N unit is associated with the inner aspect of the membrane, but it is not known how it is linked to the membrane. None of its properties suggest that it is an intrinsic membrane protein that spans the membrane bilayer.

A striking feature of N is its ability to couple indiscriminately with receptors and catalytic unit from all species and cell types thus far investigated. In the case

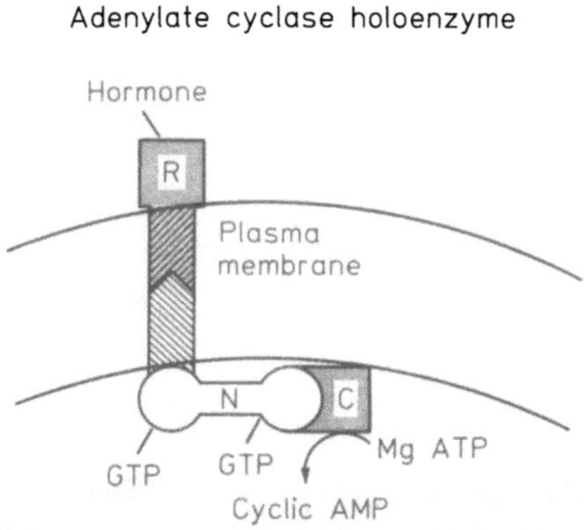


**Fig. 2.** Schematic representation of how the hormone receptor (R), nucleotide regulatory unit (N), and catalytic unit (C) of the adenylate cyclase system may be positioned in the plasma membrane. The N unit is depicted as containing two sites for GTP action, one site linked to the receptor spanning the membrane, the other site linked to the catalytic unit. These sites may be separate protein subunits from each other and a third subunit (the center piece of the dumbbell-shaped molecule) which may serve to "articulate" interaction of the GTP-binding proteins with R and C. See text for further details

of the glucagon receptor, it has been shown that the receptor can be transferred by a fusion procedure to Friend's erythroleukemia cells lacking this receptor with full competency to mediate stimulation by glucagon of the cell's adenylate cyclase system (SCHRAMM 1979). To do so, it must have combined with the N unit in the recipient cell membrane. This seeming lack of specificity of N for different receptor types argues for common structural domains in both R and N units that allow for coupling between these units.

The N unit in rabbit liver plasma membranes has been purified to near homogeneity (NORTHRUP et al. 1980). Its functional size based on hydrodynamic evaluation is about 130,000 daltons (HOWLETT and GILMAN 1980); target size analysis (see Sect. D) also indicates that the functional size of the unit is 130,000 daltons. The purified component yields three protein bands on electrophoresis under denaturing conditions. The apparent molecular weights of these proteins are 35,000, 43,000, and 53,000 daltons. Only the latter two bands are labeled when the N protein is treated with cholera toxin and NAD $^{32}$P. In other systems, it has been shown that the 43,000 and 53,000 daltons proteins are structurally related (HUDSON and JOHNSON 1980). In the turkey system, only the 43,000 daltons protein is necessary to obtain hormone, guanine nucleotide, and fluoride activation of the enzyme. However, it is conceivable that in most systems both cholera toxin-labeled proteins are important and perhaps play different roles in the coupling of receptors and enzyme to the N unit. The role of the 35,000 daltons protein that is not labeled with cholera toxin is unknown. If it turns out not to be a GTP-reactive protein, it might have other roles. As speculation, this subunit may serve to modulate the interactions of the other subunits in their interactions with receptors and adenylate cyclase, as represented schematically in Fig. 2. Note in this scheme that GTP acts through separate subunits, each subunit having a distinct functional role in linkage with receptors and catalytic unit, the central subunit serving to articulate these

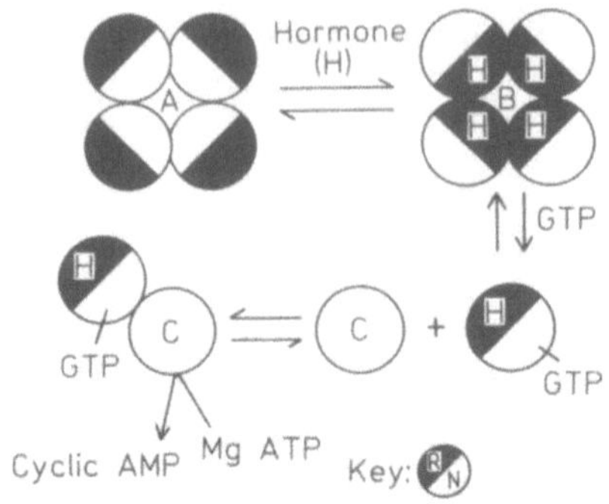


**Fig. 3.** Disaggregation coupling "turnon" cycle as a model for hormone and GTP action on adenylate cyclase systems. The receptor–nucleotide regulatory complex (R·N) is depicted as an oligomeric structure (for illustrative purposes as a tetramer) which is the tight binding form of the receptor (A configuration). Binding of the hormone to structure A converts the oligomer to "monomers" which react with the catalytic unit (C) to form the holoenzyme that converts MgATP to cyclic AMP. See text for further details

functions. Perhaps in this manner it is possible to explain the different characteristics of the N unit when it is linked to receptors, enzyme, or both in the holoenzyme structure.

## X. Independent Complexes of N with R and C

Evidence that the N unit forms independent complexes with receptors stems in part from findings that pretreatment of the liver system with glucagon (trace-labeled with glucagon $^{125}$I) followed by detergent extraction gave, on gel chromatography, a labeled fraction that was separate from a fraction exhibiting activation by guanine nucleotides (WELTON et al. 1977). The hormone-labeled fraction seemed to contain the N unit associated with the glucagon receptor since addition of GTP caused loss of the labeled hormone from this fraction. It was concluded from such studies that the R·N complex is reasonably stable to detergent extraction from the membrane, provided that it contains bound hormone. In similar but more detailed studies of the complex formed between catecholamine $\beta$-receptors and the N unit, it was established that the N unit associated with R is also susceptible to labeling with cholera toxin and NAD $^{32}$P (LIMBIRD et al. 1980). Thus, there is little question that R·N complexes can exist separately from N·C complexes.

The amount of R·N and N·C complexes in membranes may vary. It can be estimated that in liver membranes perhaps 30% of the N units may be precoupled to C. This estimate derives from the findings that treatment with Gpp(NH)p (in the complete absence of other activators) causes at most 30% of the activity that can be generated by combination of the nucleotide with glucagon (WELTON et al. 1977). Similarly, when the enzyme is extracted prior to pretreatment with the nucleotide compared with that pretreated with glucagon and Gpp(NH)p, again about 30% of the activity is susceptible in the extract to the activating effects of Gpp(NH)p or fluoride ion relative to the activity generated by the pretreated enzyme. These findings argue for a form of the N unit that is not complexed with C and which cannot react with C unless it is subjected to the actions of the hormone–receptor complex. This reasoning is part of the argument that N exists predominantly as an R·N complex prior to the actions of hormones and GTP and is a precursor to the holoenzyme, as depicted in Fig. 3 and discussed in Sect. E.

## XI. Characteristics of the Catalytic Unit

The catalytic unit of hormone-sensitive cyclase systems generally utilizes either MgATP or MnATP as substrate with nearly equivalent $K_m$ and $V_{max}$. However, when C is not linked with N, it becomes highly selective for MnATP as substrate (LONDOS et al. 1979a). Thus, C must be linked with N in order to generate cyclic AMP from MgATP, the likely physiologic substrate.

The catalytic unit has not been purified. When extracted with neutral detergents, it exists both in forms that utilize MgATP (N, C) and free forms that utilize MnATP. In the presence of phospholipids the latter can interact with N to acquire regulation by guanine nucleotides and fluoride ion (STRITTMATTER and NEER 1980). Hydrodynamic evaluation of the size of the N·C complex in liver membranes suggests a molecular weight of about 230,000 daltons (STENGEL and HANOUNE 1979). In this form, the enzyme binds detergents and behaves as a protein that has hydrophobic domains appropriate for interaction with membranes. The requirement for phospholipids in the formation of the N·C complex coupled with the fact that the N and C units are likely to be associated with the cytosolic aspect of the membrane, make it likely that the C unit is attached to lipids at the inner aspect of the bilayer.

The structure of the C unit that determines whether the catalytic site utilizes MnATP or MgATP remains unknown. One possibility is that the molecule exists in monomeric and oligomeric forms which have very different substrate specificities and kinetic properties. In the testis, a water-soluble form of adenylate cyclase uses only MnATP as substrate, but it is not known whether it has any other properties in common with the C unit that is associated with membranes; the testis enzyme apparently does not form complexes with N (BRAUN and DODS 1975).

## XII. Regulation by Divalent Cations

Early studies of the adipocyte and cardiac adenylate cyclase systems provided evidence that divalent cations (Mg, Mn) serve as regulatory agents in mediating the stimulatory effects of hormones and act at sites distinct from the catalytic site (BIRNBAUMER 1973; BIRNBAUMER et al. 1969). An alternative theory to explain the stimulatory effects of divalent cations is that the cations complex inhibitory forms of ATP ($ATP^{4-}$, $HATP^{3-}$) that may bind selectively to different states of the enzyme system (DEHAEN 1974; LIN et al. 1975a). Modeling of the rat liver system's kinetics as a function of Mg concentration gave rise to a three-state model that incorporated binding of $HATP^{3-}$ to different guanine nucleotide-activated states of the system (RENDELL et al. 1975). Subsequently, however, independent studies established that the stimulatory effects of Mg and Mn could be explained only if these cations act at sites distinct from the catalytic site; the differences in the stimulatory actions of Mn and Mg could not be accommodated by their ability to form complexes with ATP (LONDOS and PRESTON 1977b). Moreover, it has been shown that the regulation exhibited by divalent cations requires linkage of N with C, findings which suggest that the regulatory sites for divalent cations either reside on the N unit or are expressed at the C unit when it is coupled to N. Given the fact that the enzyme system contains at least three separate components, it is not surprising that divalent cations exert multiple effects on the enzyme system. For example, Mg has been suggested to regulate glucagon binding to its receptor (IYENGAR et al. 1980b). Mn ions cause biphasic stimulatory and inhibitory effects with increasing concentrations on the response of the liver system to glucagon, but not to fluoride ion (BIRNBAUMER et al. 1971). This was an early example of the differences in the processes of activation of adenylate cyclase by hormones and fluoride ion. The inhibitory effects of Mn on hormone action have been referred to as "uncoupling," but it is not clear that there is actual physical dissociation of the receptor from the N unit by the actions of Mn ions (LIMBIRD et al. 1979).

Calcium ion, like Mn and Mg ions, appears to exert multiple effects on the adenylate cyclase system. Calcium ions generally inhibit adenylate cyclase activity in the range of 0.1 m*M* calcium. This inhibitory effect can be observed even when the N unit is uncoupled to C. In some preparations of membrane-bound cyclase, however, calcium ion stimulates activity at concentrations which may be physiologic (0.1 μ*M* or lower). This stimulatory effect may be due to linkage of calmodulin, a calcium-binding protein which appears to regulate a variety of enzymes (TOSCANO et al. 1979). Because of the ubiquitous association of calmodulin with isolated membrane preparations, however, it is difficult to test whether membrane-bound cyclase systems contain this protein as a fundamental regulatory unit. In the case of glucagon action, the addition of calcium ions is not required nor does it appear that calmodulin is involved in regulation as judged from the inability of high concentrations of calcium ion chelators to prevent glucagon action.

## XIII. The Role of Sulfhydryl Groups in Transduction

Regulation of enzymes often involves subtle changes in the reactivity of a few crucial bonds. Sulfhydryl groups have been useful for monitoring changes in structure that accompany regulatory processes because they react with relatively specific reagents such as organic mercurials and tetrathionate, which react reversibly, and with agents such as *N*-ethylmaleimide and iodoacetamide, which react covalently. Glucagon promotes the rate of inactivation of the liver system by iodoacetamide, a reaction that gave the first indication of induced conformational changes in the system (STORM and DOLGINOW 1973). However, the nature of the component, or components, that were functionally altered by iodoacetamide was not established. Treatment of the liver system with glucagon in the presence of labeled sulfhydryl reagents led to apparent selective labeling of a protein having characteristics of a glycoprotein (STORM and CHASE 1975). To date, this material has not been characterized further. As noted previously, the N unit contains crucial sulfhydryl groups necessary for coupling and activation of adenylate cyclase. Additionally, sulfhydryl groups appear to be involved in the binding of glucagon and in the reactivity of the catalytic unit with substrate. A recent report suggests that catecholamines enhance the reactivity of sulfhydryl groups associated with the N unit (WILLIAMS and LEFKOWITZ 1977).

The effects of organic mercurials on adenylate cyclase systems have given interesting new insights into the possible structures of the regulatory components (R, N) (LIN et al. 1980). Organic mercurials also inhibit adenylate cyclase but, unlike iodoacetamide, their effects on activity can be completely reversed by addition of sulfhydryl reducing agents such as dithiothreitol (DTT). When liver, adipocytes, and HeLa cell cyclase systems were treated with organic mercurials at low temperature followed by the addition of DTT, cyclase activity increased unexpectedly to levels approaching that observed with addition of hormones, fluoride, or guanine nucleotides. The stimulatory effect did not require the presence of receptors since it was observed even in HeLa cells lacking receptors. Although the exact mechanism of mercurial action remains unknown, it was suggested that the mercurials, because of their hydrophobic properties, promote disaggregation or dissociation of protein subunits involved in the interactions between N and C. Consistent with this hypothesis was the finding that a hydrophilic sulfhydryl agent, tetrathionate, inhibited adenylate cyclase activity reversibly, but did not cause stimulation of activity after reversal by addition of DTT. These findings with mercurials lend support to the hypothesis that the regulatory units may exist in oligomeric forms that cannot react with the catalytic unit unless dispersed to small or monomeric units (see Fig. 3 and Sect. E).

## XIV. The Role of Membrane Lipids

Treatment of liver membranes with phospholipase $A_2$ results both in losses in glucagon-stimulated activity and reduction in the ability of guanine nucleotides to modify the glucagon receptor (LAD et al. 1979; POHL et al. 1971). Phospholipase C from *Bacillus cereus*, which selectively hydrolyzes acidic phospholipids, also caused diminished glucagon action and reduction in the affinity of the receptor for glucagon (RUBALCAVA and RODBELL 1973). This alone could not explain the loss in glucagon action since even saturating concentrations of the hormone failed to act. Concomitantly, there was also a reduction in the ability of GTP to affect the receptor. These findings are consistent with the view that the linkage between R and N is crucial for hormone action; apparently, acidic phospholipids are involved in this linkage. An unconfirmed report suggests that phosphatidylserine specifically restores glucagon activation of adenylate cyclase in solubilized preparations (LEVEY 1971).

There are many ways of viewing the role of lipids in a complex membrane-bound process. As mentioned previously, lipids may be involved in the associations between R and N, and between N and C. They may also have less specific functions in serving as a matrix within which the various components can move to form associations. In the case of glucagon, an amphipathic molecule, lipids may even be involved in the binding of the hormone to the receptor.

Lateral movements of surface proteins are influenced by the transition in lipid structure that takes place, depending on the type of lipids and the transition temperatures necessary to change the structure from liquid crystal to gel-like states. Studies of the effects of temperature on the responses of liver cyclase to glucagon, guanine nucleotides, and fluoride have been used to monitor the possible relationship between lipid phase changes and activation by the various effectors (HOUSLAY et al. 1976a, b; HOUSLAY and PALMER 1978). A distinctive break in Arrhenius plots of activity versus temperature occurred at 28.6 °C only when the enzyme was activated by glucagon. [des-His[1]]Glucagon, a partial activator of the cyclase system, activated at the same inflection point. Substitution of membrane lipids with synthetic phospholipids exhibiting different phase changes in artifical membranes yielded corresponding shifts in the temperature at which the breaks in Arrhenius plots occur. Electron spin resonance (ESR) changes were also monitored with the various lipid substitutions and correlated well with phase changes being related to breaks in Arrhenius plots. It was suggested that perhaps the phase changes occurred only in the outer lipid leaflet of the membrane because such changes were not observed in pure lipid multilamellar membranes. The latter finding, however, has been challenged (LIVINGSTONE and SCHACHTER 1980). Nonetheless, the data presented are at least consistent with the likelihood that the receptor spans the membrane bilayer and is thus potentially influenced by lipid structure, whereas the N and C units may not be intrinsic membrane proteins as defined by proteins that span the bilayer.

In sum, lipids appear to play important roles both in the structure and coupling of the cyclase components and possibly in the lateral movements of the receptor and associated components within the plane of the membrane. Because of the drastic changes in affinity of the receptor for glucagon when selective lipids are hydrolyzed, it must also be considered that lipids may be involved in the ordered arrangement of receptors and N units, possibly in the form of oligomers (Fig. 3).

## XV. Desensitization of Glucagon Action

Desensitization (refractoriness, tachyphylaxis) is a classical phenomenon in pharmacology and endocrinology. It refers to loss of drug or hormone response after exposure of cells to these agents, often at nonphysiologic concentrations (LEFKOWITZ et al. 1980). Decreased glucagon responses have been observed in the liver after injection of glucagon in whole animals, incubation with slices or tissues, and with cultured cells containing the glucagon receptor (PLAS and NUNEZ 1975; BLAZQUEZ et al. 1976). Following a 4-h exposure of a cloned hepatic cell line (RL-PR-C) to glucagon, the cells became refractory to glucagon, but not to catecholamine action on cyclic AMP production (REILLY et al. 1980; REILLY and BLECHER 1981). This selective loss suggested that desensitization is somehow related to alterations specific to the glucagon receptor. The loss in response was not related, however, to loss in the number of glucagon receptors. Nor was the loss in response due to enhanced destruction of cyclic AMP by phosphodiesterase, egress of cyclic AMP from the cells, or of alterations in the catalytic unit which responded normally in isolated membrane preparations to fluoride ion. "Uncoupling" of the receptor was invoked as a possible mechanism. Others have suggested that desensitization may result from phosphorylation by ATP (BOCKAERT et al. 1976) or by GTP (EZRA and SALOMON 1980). Using isolated liver plasma membranes, some investigators have reported that desensitization results from a combination of high concentrations of hormone, Mg ions, and GTP (IYENGAR et al. 1980b). Such findings imply some involvement of the N component in the desensitization process. However, the problem of how such relationships lead to selective losses in hormone response remains unresolved.

## XVI. Relationship Between Glucagon Binding and Action

Because the N unit plays a crucial intermediate role in the processing of glucagon action on adenylate cyclase, a simple relationship between receptor occupation and levels of cyclic AMP production need not be expected. Under optimal conditions for activation by glucagon and GTP, comparison of receptor occupancy and hormone-stimulated activity follows a hyperbolic relationship in the rat liver system (RODBELL et al. 1974). As little as 10% occupation by glucagon yielded nearly 80% of the maximal activity generated. It has been argued that the majority of glucagon-binding site are not true receptors in the sense that they are not functionally coupled to adenylate cyclase (BIRNBAUMER and POHL 1973). An alternative explanation, however, is that the "signal" generated by the binding of glucagon and GTP to the R·N complex is in large excess over that required to produce the activated state of the catalytic unit. As a theory, this postulate has the testable feature of a concentration dependency of the activation process on the levels of R, N, and C. However, not only concentrations, but also competency for action must be demonstrated for the R and N units. Recent "fusion" techniques have been described which may serve to determine how many of the glucagon receptors in liver membranes are capable of reacting with N units in recipient cells or membranes (SCHRAMM 1979).

Another factor that influences the relationship between hormone binding and action is the breakdown of GTP to GDP during transduction. Because such hydrolysis attenuates the response, the degree of amplification (ratio of action to receptor occupancy) is reduced relative to the amplification that occurs without hydrolysis. Thus, after cholera toxin action, which reduces GTPase activity, or when Gpp(NH)p and other nonhydrolyzable analogs are employed, amplification of response as a function of receptor occupation is enhanced. This generality may account for the report that destruction of over 90% of the $\beta$-receptors in the turkey erythrocyte did not lead to losses in the amount of activity generated by combinations of catecholamines and Gpp(NH)p (TOLKOVSKY and LEVITZKI 1978).

In addition to concentration of regulatory units (R, N), the structure of the R·N complexes may be crucial to the degree of amplification. For example, an ordered arrangement of interacting subunits of R and N may confer special amplifying properties with regard to the binding and actions of both hormones and guanine nucleotides. A specialized arrangement of vasopressin receptors in kidney membranes has been offered as an explanation for the large degree of amplification of the response of adenylate cyclase to vasopressin (BERGMAN and HECHTER 1978). In the theory presented in Fig. 3, the R·N complex is envisioned as an ordered complex of interacting subunits which may be the tight binding, high affinity form of the glucagon receptor observed in the absence of guanine nucleotides; the low affinity form is the "disaggregated" or monomeric complex of R·N that is proposed to be the "signal" responsible for activation of adenylate cyclase.

The reason for discussing the problem of relating receptor occupancy to hormone action is that studies have been reported in which no obvious parameter is responsible for the observed relationships. For example, glucagon receptors in rat liver membranes are reduced approximately threefold in animals chronically treated with glucagon (SRIKANT et al. 1977). Yet, the dose response relationships and the maximal cyclase activities seemed not to change in response to glucagon; in fact, greater production of cyclic AMP occurred with fewer available receptors. It would appear in this case that receptor concentration alone cannot be responsible for these observations. Other parameters, such as the sensitivity of glucagon binding to GTP action, might have revealed structural changes in the R·N complex.

Another example of an apparent lack of correspondance between the number of glucagon-binding sites and the levels of cyclic AMP production is the report that treatment of rats with carbon tetrachloride resulted within 24–48 h in decreased amounts of glucagon receptors and decreased responses of adenylate cyclase to glucagon (MOURELLE and RUBALCAVA 1981). However, the report also indicated that fluoride responsiveness and even basal activity were adversely affected. It would appear from the latter findings that alterations in both R and N units might have resulted from the treatment with carbon tetrachloride.

## C. The Glucagon Receptor

Purification of the glucagon receptor has not been accomplished. One problem in purification is the difficulty of monitoring the receptor after it is extracted from the membrane with detergents. As discussed previously, the high affinity form of the receptor appears to involve combinations of R and N. Dissolution of this complex by detergents apparently leads to very low affinity forms of the glucagon receptor. An added complication is that glucagon binds significantly to detergents and phospholipids because of its amphipathic nature, thus making it difficult to distinguish true receptor binding from interactions with lipids and detergents. Because of these problems, early attempts to identify the glucagon receptor in detergent extracts must be looked at with some reservation. Perhaps a better means of purifying the receptor is to attempt purification of the R·N complex. Not only might such complexes retain the high affinity form of the receptor, but the receptor could be monitored both by the binding of labeled hormone and by the effects of guanine nucleotides on the binding.

Tagging the glucagon receptor by a photoaffinity analog (azidoglucagon) has been reported (BREGMAN and LEVY 1977). However, the tagged analog had no reported biologic activity, making it difficult to assess the biologic significance of the reported labeling of a 23,000 daltons protein. In a different approach, glucagon $^{125}$I previously bound to the receptor was attached covalently by treating the membranes with a heterobifunctional reagent, hydroxysuccinimidyl-*p*-azidobenzoate (JOHNSON et al. 1981 b). The cross-linked material behaved on SDS–PAGE gels as a 58,000 daltons protein. To test for specificity, unlabeled glucagon added prior to cross-linking of labeled hormone abolished appearance of the labeled protein. Perhaps the most convincing argument that the material represents the receptor was the finding that guanine nucleotides reduced both the amount of bound hormone and the amount of material bound to the membrane that could be cross-linked. By these criteria, it appears that the glucagon receptor contains a 58,000 daltons protein. It is also possible that there are additional, nonlabeled proteins in the receptor molecule. The bifunctional agent only links those receptor components to glucagon through accessible groups on receptors and hormone. Nonetheless, the technique of cross-linking subsequent to binding of the hormone has decided advantages over the use of previously derivatized analogs of glucagon which may be devoid of biologic activity or have such reduced affinity for the receptor that their use is severely limited.

With the caveat already expressed that identification of the glucagon receptor requires more than binding of labeled hormones as criterion, the hydrodynamic size of the glucagon receptor extracted from liver membranes is about 100,000 daltons (BLECHER et al. 1973). Target analysis studies (see Sect. D) also suggested that the functional size of the receptor in the activation of adenylate cyclase is about 100,000 daltons. Target analysis of the turkey erythrocyte β-adrenoceptor yielded sizes of about 90,000 daltons (NIELSEN et al. 1981). On the other hand, photoaffinity tagged β-receptors under denaturing conditions suggest that the receptor is composed of proteins ranging in size from 33,000 to 53,000 daltons (STROSBERG et al. 1980). Possibly the β-adrenoceptor is composed of two or more subunits. Similarly, if one assumes that size of the photoaffinity tagged glucagon receptor and that obtained from target analysis represent the receptor, it would appear that the receptor for glucagon may also be composed of subunits.

## D. Target Analysis of the Glucagon-Sensitive Adenylate Cyclase System

METCALFE and associates were the first to employ irradiation by high energy particles (target analysis) to probe the size of the glucagon-sensitive cyclase system in rat liver membranes (HOUSLAY et al. 1977, 1980). The theory behind the use of target analysis has been reviewed (KEMPNER and SCHLEGEL 1979). In essence, ionization due to electron bombardment occurring at random along the chain of a protein (or any macromolecule) causes sufficient energy to be transmitted along the molecule's backbone to result in multiple lesions. In the case of enzymes, catalytic function is destroyed. In the case of receptor function, if the binding of hormone depends on the integrity of the entire molecule, then the binding function is similarly destroyed. By relating the resultant exponential decay to the amount of radiant energy applied, it is possible to determine the functional mass of the material from the slope of the decay curve.

Since the technique yields only the functional sizes of the target, not its composition, it is necessary to test the sizes given by the crude adenylate cyclase system with as many functional parameters as possible. Thus, MnATP can be used as substrate in order to determine the functional size of the catalytic unit, irrespective of its union with the N unit. Activation by Gpp(NH)p or fluoride ion theoretically measures the functional size of N·C. If the receptor is structurally linked to the N·C complex during the formation of the holoenzyme, then target analysis should give the functional sizes of all three components present in the holoenzyme. Moreover, the system can be irradiated prior to exposure to effectors. Theoretically, this gives the size of the "ground state" enzyme system before it has been converted to its activated states, whereas activation prior to irradiation yields the sizes of the activated targets.

When all of these parameters were employed to investigate the liver system, of critical importance was the finding that the mass of the enzyme system increased with increasing regulatory complexity (SCHLEGEL et al. 1979). Thus, the smallest size was that obtained with MnATP as substrate (100,000 daltons) and is suggested to represent the functional size of the catalytic unit. Activation by fluoride ion or by Gpp(NH)p yielded a size that averaged 240,000 daltons. This is not significantly different from the reported size of the solubilized N·C unit obtained from rat liver membranes. Note that the increase in mass over that of the catalytic unit is about 120,000 daltons, which is close to the 130,000 daltons size reported for the hydrodynamic size of N from rabbit liver. When preactivated by glucagon and GTP, the size of the target increased to 340,000 daltons; the increase (about 100,000 daltons) was suggested to represent the glucagon receptor attached to the activated holoenzyme. Similar sizes and increments in size were reported for the turkey erythrocyte cyclase system when examined by target analysis in identical fashion (NIELSEN et al. 1981).

Of particular interest was the finding that the ground state target was considerably larger in size (greater than $10^6$ daltons) than any of the activated targets. Binding of glucagon $^{125}I$ to the GTP-affected receptor (R·N complex) also revealed a larger target than anticipated from the size of R in the presumed holoenzyme. From such findings it was suggested that the ground state target consisted of aggregates or oligomers of R·N units that are precursors of the R·N complex that forms the holoenzyme when linked with C (Fig. 3). Although conjectural, this assumption for the structure of the ground state could at least provide an explanation of how a signal, in this case a macromolecular signal in the form of R·N monomers, may form the basis of the concerted actions of hormone and GTP on the adenylate cyclase system.

MARTIN and associates have also reported data on the target analysis of the glucagon-sensitive system in rat liver membranes (MARTIN et al. 1979, 1980). Using freeze-dried preparations for irradiation [the only similarities in target size were those reported for the enzyme preactivated with fluoride ion or Gpp(NH)p], they reported target sizes of about 240,000 daltons. However, they did not investigate the ground state target sizes of the enzyme system or the size of the enzyme with MnATP as substrate. Nonetheless, based on findings that the target size of the enzyme was larger with glucagon than with guanine nucleotides or fluoride ion, it was speculated that the 240,000 daltons structure represents the catalytic unit and that the actions of fluoride and Gpp(NH)p are to liberate the N unit from an inactive N·C complex. This theory does not fit, however, with all available data which indicate that the activated state of the enzyme formed in the presence of Gpp(NH)p or fluoride is a complex of N and C units. Unfortunately, because of the differences in assay conditions, pretreatment conditions, and the use of freeze-dried preparations for irradiation, it is not possible to compare their other data with the data reported by RODBELL's group using frozen preparations. This emphasizes the caveat in the use of target analysis as a probe for size and possible structure of complex enzyme systems; without independent knowledge of the composition of the target, interpretations are necessarily speculative.

## E. A Model for Glucagon Action

As might be anticipated from the complex structure of adenylate cyclase systems, views on the mechanism of hormone action on these systems vary considerably.

None of the theories disagrees on the complex formed between N and C being essential for activation by guanine nucleotides. The major issue is the role of hormone receptors in the process. Two prominent theories (DELEAN et al. 1980; TOLKOVSKY and LEVITZKI 1978) have in common the idea that the receptor preexists in an uncoupled form from N; the hormone–receptor complex serves to "activate" N by converting it to a GTP-reactive form. In the process of forming the active N·C complex, the receptor dissociates and presumably can be used again. This basic idea has the principal virtue of allowing theoretically a single receptor molecule to active numerous N (and thus C) molecules in the process of hormone action.

An alternative theory is presented in Fig. 3 (RODBELL 1980). In this theory, the receptor is coupled to N through the process leading to the active form of the holoenzyme. This is supported from the target analysis studies discussed in Sect. D. The main role of the receptor in this process is to regulate the ability of N to interact with GTP, as in the models already discussed. Different from these models, however, is the concept that receptors serve to maintain the N unit in an unreactive form; the role of the hormone is to release this inhibitory constraint. Based primarily on the interpretation that the large ground state structure is an oligomer of R·N units, it is suggested in Fig. 3 that the complex is present in an oligomeric structure (tetramer for illustrative purposes only) which is converted from an "inactive" (A) to an active (B) structure; only the latter is reactive with GTP. Disaggregation of the oligomer to monomeric units, as discussed previously, is suggested to be the primary "message" which converted C to its activated form.

Not essential to the "disaggregation" model of hormone action, but an interesting outgrowth of postulating oligomers of R·N is the possibility that the R and N units in the oligomeric structure are heterologous; i.e., different types of receptors (for stimulatory and inhibitory hormones) and N units (stimulatory and inhibitory) might be intercalated into aggregate structures. Depending on the relative concentrations of receptors for a particular hormone or set of hormones, disaggregation of such putative complexes by the concerted actions of hormones (or neurotransmitters) and GTP might give rise to levels of R·N monomers which, when combined with C, yield levels of cyclic AMP commensurate with how much stimulatory and inhibitory monomeric R·N was produced during disaggregation.

Indirect supportive evidence that the glucagon receptors take different structures in the membrane when occupied by hormone alone, compared with when the R·N complex is occupied by both glucagon and GTP, stems from studies with [thioTrp$^{25}$]glucagon. Conversion of glucagon to [thioTrp$^{25}$]glucagon does not modify the affinity or actions of the molecule on liver adenylate cyclase (WRIGHT and RODBELL 1980a). However, oxidation of the thiol group results in formation of a disulfide-linked "dimer" of glucagon. The dimer is fully active and shows identical affinity and $K_{act}$ as long as the activator GTP is present. However, when GTP is absent, binding of the dimer shows an affinity one-fourth that of the monomer. This is clear-cut evidence that there is a structural change in the receptor when it transits from the tight binding to the low affinity form during the concerted actions of glucagon and GTP. Possibly because of the putative oligomeric form of the receptor in the tight binding configuration, the glucagon molecule binds poorly with the large glucagon appendage linked to the sulfhydryl group. Consistent with the postulate that the receptor (R·N complex) disaggregates in the presence of GTP, the "monomer" form of the receptor involved in activation of adenylate cyclase may allow the dimer to interact without the large appendage restricting the binding of the hormone. In this manner, one can explain the identical potencies of native glucagon and the dimer (on a molar basis) during activation.

Heterologous R·N complexes could explain findings with multireceptor adipocyte adenylate cyclase system that, whereas combinations of maximally stimulating concentrations of lipolytic hormones did not yield additive levels of cyclic AMP production, submaximal concentrations of such combinations produced synergistic levels (BIRNBAUMER and RODBELL 1969). Synergism implies that different receptors interact during the process of promoting cyclic AMP production. As another example of heterologous receptor interactions, there are reports that agonists for one type of receptor modify the binding of an agonist to a different receptor in the same membrane (WATANABE et al. 1978).

It has been suggested recently that elements of the cytoskeleton, notably in this case microtubulin, may be linked in some fashion to $\beta$-adrenoceptors in frog erythrocytes (CHERKSEY et al. 1980). The basis of this contention is that colchicine modifies the interactions of fluorescent tagged propranolol to the membranes. From these findings, it was speculated that catecholamine agonists may act by disrupting the putative linkage, allowing the receptor to take an active form. Such linkages, if they exist, could be an explanation for the large target sizes of the ground state structures observed in liver and adipocyte membranes. Moreover, it has been reported that agents that disrupt cytoskeletal structures stimulate the production of cyclic AMP in certain types of cultured cells (HAGMAN and FISHMAN 1980; KENNEDY and INSEL 1979). Thus, it is possible that the cytoskeleton is attached to domains of the plasma membrane that contain receptors and other components of the adenylate cyclase system. The attractiveness of this idea is that it brings into focus the possible structural relationship of the plasma membrane to other elements of the cell which together impinge on the possible regulation of cellular growth and differentiation.

## F. The Glucagon "Message"

In an earlier review of glucagon action on adenylate cyclase (RODBELL 1972b), there was speculation on the structural features of the glucagon molecule that might yield information on the mode of transaction between hormone and receptor. At that time it was known that glucagon does not adopt a definitive ordered structure in dilute aqueous solutions at physiologic pH. However, based on the types and sequence of amino acid residues in the molecule, there was a predicted helical structure encompassing residues 5–16 and another between residues 17–29. It was also known that glucagon acquires structure when interacted with a variety of compounds, including organic solvents, lipids, detergents, and at suitable concentrations, with itself. These findings indicated that the environment plays an important role in determining the structure of the hormone. From the biologic standpoint, however, the most important question is the nature of the transaction between glucagon and its receptor that gives rise to a primary action on the cell.

In the ensuing years there have been notable advances in knowledge of the structure of the glucagon molecule (for review, see BLUNDELL and HUMBEL 1980; and also Chap. 3) both in its crystalline form and from studies of various analogs prepared by either total chemical synthesis, by modification of various residues, or by studying fragments prepared either synthetically or by enzymic treatment. One goal of the latter approach is to prepare glucagon analogs that bind to the receptor, but which do not produce an action. Drug and hormone antagonists have played a key role in investigations of a number of receptors, both for clinical and investigative purposes. From early knowledge that removal of the $NH_2$ terminal histidine residue resulted in loss of activity (RODBELL et al. 1971c), efforts were made to modify this region of the molecule so as to retain a high affinity form of the molecule. Subsequently, it was found that [desHis$^1$]glucagon is a partial agonist, having an affinity for the receptor about 1/20 that of native glucagon (LIN et al. 1975b). Because partial agonists can be both inhibitors and antagonists, depending on their relative concentrations to that of the fully active hormone, this finding pointed up the difficulties of obtaining a potent, pure antagonist form of glucagon. Promising new leads have been obtained, however, from findings that introduction of hydrophobic groups at the $\alpha$- and $\varepsilon$-amino groups (Lys-12) and neutralization of the positive charges result in glucagon antagonists, the most potent being [homoArg$^{12}$]glucagon (BREGMAN et al. 1980). Since amidination of the lysine residue preserves biologic potency of glucagon (WRIGHT and RODBELL 1980b), it is evident that the $\alpha$-amino group plays a distinctive role in biologic activity of this hormone. Although other investigators (EPAND et al. 1981) concluded that modifications of any type resulted in severe losses both in potency and biologic activity, modification at the $NH_2$ terminal region resulted in greater losses in biologic activity compared with glucagon binding to its receptor. This is exemplified by $N^{\alpha}$-trinitrophenylglucagon which behaves as a pure antagonist when tested on the liver adenylate cyclase system.

The latter findings are further testament to the importance of the $NH_2$ terminal region in the biologic actions of glucagon. The most dramatic example of this is the finding that glucagon 1–6 is a partial agonist on the liver cyclase system, although with affinities of about five orders of magnitude less than that of the native hormone (WRIGHT and RODBELL 1979). Unlike glucagon, this derivative does not form a helical structure. It would appear, therefore, that the potential helix-forming regions of glucagon have little to do with action at the receptor, but are clearly essential for recognition and action at physiologic concentrations of the hormone. The fact that glucagon 1–6 is recognized by the receptor suggests, however, that both binding and action at the receptor does occur even in the absence of the large, hydrophobic appendage and may even contain specific "recognition" units. In this regard, it is of interest that vasoactive intestinal peptide and secretin, hormones having homologous structures with glucagon in the $NH_2$ terminal regions, fail to activate or compete effectively with glucagon at its receptor in liver membranes. Accordingly, the seemingly slight differences in the residues of the first six amino acids of these hormones (vasoactive intestinal peptide and secretin contain glutamine rather than aspartate in position 3) may be partially responsible for receptor specificity. As has been stated, however, recognition at physiologic concentrations must involve the large appendages to the first six residues. Based on the large affinity losses when removed, their importance seems more in ensuring that the hormone is tightly "anchored" to the receptor during activation.

An interesting question with regard to the molecular architecture of the glucagon receptor is whether the hydrophilic "action" region (1–5) and the "anchoring" regions (7–29) of glucagon bind to the same molecular chain or to separate chains or subunits that collectively make up the receptor. The possibility that the glucagon receptor exists in aggregate form is raised by the findings that oxidation of [thioTrp$^2$]glucagon to the disulfide-linked dimer (Trp-*S*-glucagon)$_2$ (WRIGHT and RODBELL 1980 a) leads to reduced affinity of the latter in binding to the receptor. It was suggested that the reduction in affinity is due to steric hindrance of the bulky attached glucagon, possibly because the receptor is in an oligomeric structure (see Sect. D). Of interest was the finding that GTP converted the receptor into a form that bound both the monomer and dimer forms of [thioTrp$^2$]glucagon with equal affinity and which also resulted in equivalent potencies on adenylate cyclase activation. The latter is consistent with a proposed model of glucagon action which suggests that the receptor goes from an oligomeric structure to a monomeric structure during activation of adenylate cyclase by glucagon and GTP (see Sect. E).

There are two other reasons for suggesting that the glucagon receptor is composed of two or more components (or subunits). First, target analysis (discussed in Sect. D) suggests that the glucagon receptor has a molecular weight in its "active" state (promoted by the binding of hormone and GTP to their respective protein components) of about 100,000 daltons. This value contrasts with 58,000 daltons for the affinity-labeled receptor (JOHNSON et al. 1981 b). This difference suggests that the receptor may be comprised of two or more subunits. The second reason is based on recent findings with synthetic analogs of glucagon in which various sets of residues are deleted (FRANDSEN et al. 1981). Despite the presence of residues 1–6 in all of the analogs, essentially complete loss occurred in both action (on liver cyclase) and binding to the receptor; affinities were reduced to 0.003% of glucagon.

One interpretation of these findings is that the shortened configuration does not permit the proper orientation of the molecule at the receptor. This seems unlikely because residues 1–6 do not require the presence of the other residues for action. An alternative possibility is that the receptor is composed of two subunits; one subunit reacts with the hydrophilic 1–6 region of glucagon to give action, the other subunit reacts with the helix-forming residues to cause anchoring of the molecule. The failure of the synthetic shortened analogs to act might result from a "frame-shift" in the analogs which does not permit proper alignment of the molecules with the receptor subunits. The attractiveness of a two-chained receptor having separate action and binding functions is that correct alignment of both chains with the hormone is required for physiologic functions of the hormone. Additionally, one might consider that each chain is independently regulated (by GTP, for example) and synthesized. In this manner, the glucagon receptor may have structurally common structures in one subunit (most likely that which is involved in action and which recognizes the hydrophilic regions) with the receptors for secretin and VIP (and other homologous hormones), but markedly different structures in the other subunit that is responsible for tight binding of the hormone. Natural selection pressures may have modified the latter subunit protein, thus giving rise to different types of receptors for hormones such as glucagon, secretin, and VIP.

## G. Summary and Conclusions

A remarkable surge in knowledge of hormone-sensitive adenylate cyclase systems has occurred within the past decade. Given the low concentrations of these systems in cells and the difficulties involved in dealing with membrane-bound enzyme systems, it is all the more remarkable that, within this relatively brief period, some understanding of both the functional structural characteristics of the cyclase system has been reached. The major conclusion is that hormone receptors, rather than controlling directly the activity of adenylate cyclase, act through an intermediate transduction process which involves the participation of GTP, divalent cations, and adenosine (or some other natural substance that reacts with the P-site). Thus far, the evidence on the structure of this transduction element suggests that it is a multisubunit protein with at least three components, two of which appear to be reactive towards cholera toxin and which may contain sites, both for GTP binding and action as well as hydrolysis of the nucleotide. The GTPase element appears to play a crucial role in the dynamics of the hormone–GTP regulatory process. The precise relationship between the glucagon receptor (for example), the N and C units, and the structural organization of these units are not fully understood. The thesis that the R·N exists as oligomers will require testing by techniques that can reveal this possibility in a more direct fashion. What appears to be emerging is that the cytoskeletal elements within the cell are connected somehow to the plasma membrane and possibly to the adenylate system (and other regulatory processes in the membrane). If so, then future research may reveal that hormones act by promoting a propagated wave of events proceeding from the receptor to the N unit and possibly specific cytoskeletal elements that control, not only adenylate cyclase activity, but also other regulatory processes that lead in concert to the physiologic events associated with glucagon action.

Finally, the question of the primary action of glucagon at its receptor is still unresolved. However, the findings that the metal ion sites and the P-site are linked and regulated by glucagon, even in the absence of GTP (Londos and Preston 1977 a) argues either for a control mechanism independent of that proposed to link glucagon binding to GTP action, or that the initial event resulting from glucagon binding to its receptor is promotion of divalent cation binding to sites that alter the ability of guanine nucleotides to activate adenylate cyclase. In either case, it would appear that the N unit is required in the overall regulatory process. It remains to be seen whether other regulatory components are involved in the complex events leading to glucagon action on its target cells.

## References

Bergman RN, Hechter O (1978) Neurohypophyseal hormone-responsive renal adenylate cyclase. IV. Random-hit matrix model for coupling in a hormone-sensitive adenylate cyclase system. J Biol Chem 253:3238–3250

Birnbaumer L (1973) Hormone-sensitive adenylyl cyclases: useful models for studying hormone receptor functions in cell-free systems. Biochim Biophys Acta 300:129–158

Birnbaumer L, Pohl SL (1973) Relation of glucagon-specific binding sites to glucagon-dependent stimulation of adenylyl cyclase activity in plasma membranes of rat liver. J Biol Chem 248:2056–2061

Birnbaumer L, Rodbell M (1969) Adenyl cyclase in fat cells. II. Hormone receptors. J Biol Chem 244:3477–3482
Birnbaumer L, Pohl SL, Rodbell M (1969) Adenyl cyclase in fat cells. I. Properties and the effects of adrenocorticotropin and fluoride. J Biol Chem 244:3468–3476
Birnbaumer L, Pohl SL, Rodbell M (1971) The glucagon-sensitive adenyl cyclase system in plasma membranes of rat liver II. Comparison between glucagon and fluoride-stimula activities. J Biol Chem 246:1857–1862
Blazquez E, Rubalcava B, Montesano R, Orci L, Unger RH (1976) Development of insulin and glucagon binding and the adenylate cyclase response in liver membranes of the prenatal, postnatal, and adult rat; evidence of glucagon "resistance". Endocrinology 98:1014–1023
Blecher M, Giorgio NA, Johnson CB (1973) Hormone receptors: properties of glucagon-binding proteins isolated from liver plasma membranes. J Biol Chem 249:428–437
Blundell TL, Humbel RE (1980) Hormone families: pancreatic hormones and homologous growth factors. Nature 287:771–777
Bockaert J, Hunzicker-Dunn M, Birnbaumer L (1976) Hormone-stimulated desensitization of hormone-dependent adenylyl cyclase. J Biol Chem 251:2653–2663
Braun S, Levitzki A (1979) Adenosine receptor permanently coupled to turkey erythrocyte adenylate cyclase. Biochemistry 18:2134–2138
Braun T, Dods RF (1975) Development of a $Mn^{2+}$-sensitive "soluble" adenylate cyclase in rat testis. Proc Natl Acad Sci USA 72:1097–1101
Bregman MD, Levy D (1977) Labeling of glucagon binding components in hepatocyte plasma membrane. Biochem Biophys Res Commun 78:584–590
Bregman MD, Trivedi D, Hruby VJ (1980) Glucagon amino groups: evaluation of modifications leading to antagonism and agonism. J Biol Chem 255:11725–11731
Cassel D, Pfeuffer T (1978) Mechanism of cholera toxin action: covalent modification of the guanyl nucleotide-binding protein of the adenylate cyclase system. Proc Natl Acad Sci USA 75:2669–2673
Cassel D, Selinger Z (1977) Mechanism of adenylate cyclase activation by cholera toxin: inhibition of GTP hydrolysis at the regulatory site. Proc Natl Acad Sci USA 74:3307–3311
Cassel D, Selinger Z (1978) Mechanism of adenylate cyclase activation through the B-adrenergic receptor: catecholamine-induced displacement of bound GDP by GTP. Proc Natl Acad Sci USA 75:4155–4159
Cherksey BD, Zadunaisky JA, Murphy RB (1980) Cytoskeletal constraint of the β-adrenergic receptor in frog erythrocyte membranes. Proc Natl Acad Sci USA 77:6401–6405
Cooper DMF, Londos C (1979) Evaluation of the effects of adenosine on hepatic and adipocyte adenylate cyclase under conditions where adenosine is not generated endogenously. J Cyclic Nucleotide Res 5:289–302
DeHaen C (1974) A new kinetic analysis of the effects of hormones and fluoride ion. J Biol Chem 249:2756–2762
DeLean A, Stadel JM, Lefkowitz RJ (1980) A ternary complex model explains the agonist-specific binding properties of the adenylate cyclase-coupled β-adrenergic receptor. J Biol Chem 255:7108–7117
Eckstein F, Cassel D, Levkovitz H, Lowe M, Selinger Z (1979) Guanosine 5′-O-(2-thiodiphosphate): an inhibitor of adenylate cyclase stimulation by guanine nucleotides and fluoride ions. J Biol Chem 254:9829–9834
Epand RM, Rosselin G, Hoa DHB, Cote TE, Laburthe M (1981) Structural requirements for glucagon receptor binding and activation of adenylate cyclase in liver. J Biol Chem 256:1128–1132
Ezra E, Salomon Y (1980) Mechanisms of desensitization of adenylate cyclase by lutropin: GTP-dependent uncoupling of the receptor. J Biol Chem 255:653–658
Farfel Z, Kaslow HR, Bourne HR (1979) A regulatory component of adenylate cyclase is located on the inner surface of human erythrocyte membranes. Biochem Biophys Res Commun 90:1237–1241
Frandsen EK, Gronvald FC, Heding LG, Johansen NL, Lundt BF, Moody AJ, Markussen J, Volund A (1981) Glucagon: structure-function relationships investigated by sequence deletions. Hoppe-Seyler's Z Physiol Chem 362:665–677
Gill M (1977) Mechanism of action of cholera toxin. Adv Cyclic Nucleotide Res 8:85–118

Hagmann J, Fishman PH (1980) Modulation of adenylate cyclase in intact macrophages by microtubules. Opposing actions of colchicine and chemotactic factor. J Biol Chem 255:2659–2662

Houslay MD, Palmer RW (1978) Changes in the form of Arrhenius plots of the activity of glucagon-stimulated adenylate cyclase and other hamster liver plasma membrane enzymes occuring on hibernation. Biochem J 174:909–919

Houslay MD, Hesketh TR, Smith GA, Warren GB, Metcalfe JC (1976a) The lipid environment of the glucagon receptor regulates adenylate cyclase activity. Biochim Biophys Acta 436:495–504

Houslay MD, Metcalfe JC, Warren GB, Hesketh TR, Smith GA (1976b) The glucagon receptor of rat liver plasma membrane can couple to adenylate cyclase without activating it. Biochim Biophys Acta 436:489–494

Houslay MD, Ellory JC, Smith GA, Hesketh TR, Stein JM, Warren GB, Metcalfe JC (1977) Exchange of partners in glucagon receptor adenylate cyclase complexes: physical evidence for the independent, mobile receptor model. Biochim Biophys Acta 467:208–219

Houslay MD, Dipple I, Elliott KR (1980) Guanosine 5′-triphosphate and guanosine 5′-(beta gammo-imido) triphosphate effect a collision coupling mechanism between the glucagon receptor and catalytic unit of adenylate cyclase. Biochem J 186:649–658

Howlett AC, Gilman AF (1980) Hydrodynamic properties of the regulatory component of adenylate cyclase. J Biol Chem 255:2861–2866

Hudson TH, Johnson GL (1980) Peptide mapping of adenylate cyclase regulatory proteins that are cholera toxin substrates. J Biol Chem 255:7480–7486

Iyengar R, Birnbaumer L (1979) Coupling of the glucagon receptor to adenylyl cyclase by GDP: evidence for two levels of regulation of adenylyl cyclase. Proc Natl Acad Sci USA 76:3180–3193

Iyengar R, Swartz TL, Birnbaumer L (1979) Coupling of glucagon receptor to adenylyl cyclase: requirement of a receptor-related guanyl nucleotide binding site for coupling of receptor to enzyme. J Biol Chem 254:1119–1123

Iyengar R, Mintz PW, Swartz TL, Birnbaumer L (1980b) Divalent cation-induced desensitization of glucagon-stimulatable adenylyl cyclase in rat liver plasma membrane. GTP-dependent stimulation by glucagon. J Biol Chem 255:11875–11882

Johnson GL, Coffino P, Bourne HR (1981a) Somatic genetic analysis of hormone action. In: Jacobs S, Cuatrecasas P (eds) Membrane receptors, series B, vol 11. Chapman and Hall, London New York, p 173

Johnson GL, Macandrew VI, Pilch PF (1981b) Identification of the glucagon receptor in rat liver membranes by photoaffinity crosslinking. Proc Natl Acad Sci USA 78:875–878

Kaslow HR, Johnson GL, Brothers VM, Bourne HR (1980) A regulatory component of adenylate cyclase from human erythrocyte membranes. J Biol Chem 255:3736–3741

Kempner ES, Schlegel W (1979) Size determination of enzymes by radiation inactivation. Anal Biochem 92:2–10

Kennedy MS, Insel PA (1979) Inhibitors of microtubule assembly enhance beta-adrenergic and prostaglandin $E_1$-stimulated cyclic AMP accumulation in S 49 lymphoma cells. Mol Pharmacol 16:215–223

Kimura N, Nagata N (1979) Mechanism of glucagon stimulation of adenylate cyclase in the presence of GDP in rat liver plasma membranes. J Biol Chem 254:3451–3457

Kimura N, Shimada N (1980) Glucagon-stimulated GTP hydrolysis in rat liver plasma membranes. FEBS Lett 117:172–174

Lad PM, Welton AF, Rodbell M (1977) Evidence for distinct guanine nucleotide sites in the regulation of the glucagon receptor and of adenylate cyclase activity. J Biol Chem 252:5942–5946

Lad PM, Preston MS, Welton AF, Nielsen TB, Rodbell M (1979) Effects of phospholipase $A_2$ and filipin on the activation of adenylate cyclase. Biochim Biophys Acta 551:368–381

Lad PM, Nielsen TB, Lin MC, Cooper DMF, Preston MS, Rodbell M (1980a) Toward a unifying hypothesis for the effects of cholera toxin catalysed ADP-ribosylation in diverse adenylate cyclase systems. In: Smulson EF, Sugimura T (eds) Novel ADP-ribosylations of regulatory enzymes and proteins. Elsevier North Holland, Amsterdam Oxford New York, p 381

Lad PM, Nielsen TB, Londos C, Preston MS, Rodbell M (1980b) Independent mechanisms of adenosine activation and inhibition of the turkey erythrocyte adenylate cyclase system. J Biol Chem 255:10841–10846
Lefkowitz RJ, Wessels MR, Stadel JM (1980) Hormones, receptors, and cyclic AMP: their role in target cell refractoriness. Curr Top Cell Regul 17:205–230
Levey GS (1971) Restoration of glucagon responsiveness of solubilized myocardial adenyl cyclase by phosphatidylserine. Biochem Biophys Res Commun 43:108–113
Limbird LE, Lefkowitz RJ (1978) Agonist-induced increase in apparent beta-adrenergic receptor size. Proc Natl Acad Sci USA 75:228–232
Limbird LE, Hickey AR, Lefkowitz RJ (1979) Unique uncoupling of the frog erythrocyte adenylate cyclase system by manganese. Loss of hormone and guanine nucleotide-sensitive enzyme activities without loss of nucleotide-sensitive, high affinity agonist binding. J Biol Chem 254:2677–2683
Limbird LE, Gill DM, Lefkowitz RJ (1980) Agonist-promoted coupling of the beta-adrenergic receptor with the guanine nucleotide regulatory protein of the adenylate cyclase system. Proc Natl Acad Sci USA 77:775–779
Lin MC, Salomon Y, Rendell M, Rodbell M (1975a) The hepatic adenylate cyclase system. II. Substrate binding and utilization and the effects of magnesium ion and pH. J Biol Chem 250:4246–4252
Lin MC, Wright DE, Hruby VJ, Rodbell M (1975b) Structure-function relationships in glucagon: properties of highly purified des-his-, mono-iodo, and (des-asn$^{28}$, thr$^{29}$) (homoserine lactone$^{27}$)-glucagon. Biochemistry 14:1559–1563
Lin MC, Nicosia S, Lad PM, Rodbell M (1977) Effects of GTP on binding of $^{3}$H-glucagon to receptors in rat hepatic plasma membranes. J Biol Chem 252:2790–2792
Lin MC, Cooper DMF, Rodbell M (1980) Selective effects of organic mercurials on the GTP-regulatory proteins of adenylate cyclase systems. J Biol Chem 255:7250–7254
Livingstone CJ, Schachter D (1980) Lipid dynamics and lipid-protein interactions in rat hepatocyte plasma membranes. J Biol Chem 255:10902–10908
Londos C, Preston MS (1977a) Regulation by glucagon and divalent cations of inhibition of hepatic adenylate cyclase by adenosine. J Biol Chem 252:5951–5956
Londos C, Preston MS (1977b) Activation of the hepatic adenylate cyclase system by divalent cations: a reassessment. J Biol Chem 252:5957–5961
Londos C, Rodbell M (1975) Multiple inhibitory and activating effects of nucleotides and magnesium on adrenal adenylate cyclase. J Biol Chem 250:3459–3465
Londos C, Lad PM, Nielsen TB, Rodbell M (1979a) Solubilization and conversion of hepatic adenylate cyclase to a form requiring MnATP as substrate. J Supramol Struct 10:31–37
Londos C, Wolff J, Cooper DMF (1979b) Action of adenosine on adenylate cyclase. In: Baer HP, Drummond GI (eds) Physiological and regulatory functions of adenosine and adenine nucleotides. Raven, New York, p 271
Martin BR, Stein JM, Kennedy EL, Doberska CA, Metcalfe JC (1979) Transient complexes: a new structural model for the activation of adenylate cyclase by hormone receptors. Biochem J 184:253–260
Martin BR, Stein JM, Kennedy EL, Doberska CA (1980) The effect of fluoride on the state of aggregation of adenylate cyclase in rat liver plasma membranes. Biochem J 188:136–140
Moss J, Vaughan M (1979) Activation of adenylate cyclase by choleragen. Annu Rev Biochem 48:581–600
Mourelle M, Rubalcava B (1981) Regeneration of the liver after carbon tetrachloride: differences in adenylate cyclase and pancreatic hormone receptors. J Biol Chem 256:1656–1660
Nielsen TB, Lad PM, Preston MS, Rodbell M (1980) Characteristics of the guanine nucleotide regulatory component of adenylate cyclase in human erythrocyte membranes. Biochim Biophys Acta 629:143–155
Nielsen TB, Lad PM, Preston MS, Kempner E, Schlegel W, Rodbell M (1981) Structure of the turkey erythrocyte adenylate cyclase system. Proc Natl Acad Sci USA 78:722–726

Northrup JK, Sternweis PC, Smigel MD, Schleifer LS, Ross MR, Gilman AG (1980) Purification of the regulatory component of adenylate cyclase. Proc Natl Acad Sci USA 77:6516–6520

Plas C, Nunez J (1975) Glycogenolytic response to glucagon of cultured fetal hepatocytes. Refractoriness following exposure to glucagon. J Biol Chem 250:5304–5311

Pohl SL, Krans HMJ, Kozyreff V, Birnbaumer L, Rodbell M (1971) The glucagon sensitive adenyl cyclase system in plasma membranes of rat liver. VI. Evidence for a role of membrane lipids. J Biol Chem 246:4447–4454

Reilly TM, Blecher M (1981) Restoration of glucagon responsiveness in spontaneously transformed rat hepatocytes (RL-PR-C) by fusion with normal progenitor cells and rat liver plasma membranes. Proc Natl Acad Sci USA 78:182–186

Reilly T, Beckner S, Blecher M (1980) Uncoupling of the glucagon receptor adenylate cyclase system by glucagon in cloned differentiated rat hepatocytes. J Receptor Res 1:277–311

Rendell M, Salomon Y, Lin MC, Rodbell M, Berman M (1975) The hepatic adenylate cyclase system. III. A mathematical model for the steady state kinetics of catalysis and nucleotide regulation. J Biol Chem 250:4253–4260

Rendell MS, Rodbell M, Berman M (1977) Activation of hepatic adenylate cyclase by guanyl nucleotides: modeling of the transient kinetics suggests an "excited" state of GTPase is a control component of the system. J Biol Chem 252:7909–7912

Rodbell M (1972a) Cell surface receptor sites. In: Goldberger R (ed) Current topics in biochemistry. Academic, London New York, p 187

Rodbell M (1972b) Regulation of glucagon action at it receptors. In: Lefèbvre PJ, Unger RH (eds) Glucagon. Pergamon, Oxford New York, p 61

Rodbell M (1978) The role of nucleotide regulatory components in the coupling of hormone receptors and adenylate cyclase. In: Folco G, Paoletti R (eds) Molecular biology and pharmacology of cyclic nucleotides. Elsevier/North Holland Biomedical, Amsterdam Oxford New York, p 1

Rodbell M (1980) The role of hormone receptors and GTP-regulatory proteins in membrane transduction. Nature 284:17–22

Rodbell M, Krans HMJ, Pohl SL, Birnbaumer L (1971a) The glucagon-sensitive adenyl cyclase system in plasma membranes of rat liver. IV. Effects of gluanyl nucleotides on binding of $^{125}$I-glucagon. J Biol Chem 246:1872–1876

Rodbell M, Birnbaumer L, Pohl SL, Krans HMJ (1971b) The glucagon-sensitive adenyl cyclase system in plasma membranes of rat liver. V. An obligatory role of guanyl nucleotides in glucagon action. J Biol Chem 246:1877–1882

Rodbell M, Birnbaumer L, Pohl SL, Sundby F (1971c) The reaction of glucagon with its receptor: evidence for discrete regions of activity and binding in the glucagon molecule. Proc Natl Acad Sci USA 68:900–913

Rodbell M, Lin MC, Salomon Y (1974) Evidence for interdependent action of glucagon and nucleotides on the hepatic adenylate cyclase system. J Biol Chem 249:59–65

Rodbell M, Lin MC, Salomon Y, Londos C, Harwood JP, Martin BR, Rendell M, Berman M (1975) Role of adenine and guanine nucleotides in the activity and response of adenylate cyclase systems to hormones: evidence for multisite transition states. Adv Cyclic Nucleotide Res 5:3–29

Ross EM (1981) Physical separation of the catalytic and regulatory proteins of hepatic adenylate cyclase. J Biol Chem 1949–1953

Ross EM, Gilman AG (1980) Biochemical properties of hormone-sensitive adenylate cyclase. Annu Rev Biochem 49:533–564

Rubalcava B, Rodbell M (1973) The role of acidic phospholipids in glucagon action on rat liver adenylate cyclase. J Biol Chem 248:3831–3837

Salomon Y, Lin MC, Londos C, Rendell M, Rodbell M (1975) The hepatic adenylate cyclase system. I. Evidence for transition states and structural requirements for guanine nucleotide activation. J Biol Chem 250:4239–4245

Schlegel W, Kempner ES, Rodbell M (1979) Activation of adenylate cyclase in hepatic membranes involves interactions of the catalytic unit with multimeric complexes of regulatory proteins. J Biol Chem 254:5168–5176

Schleifer LS, Garrison JC, Sternweis PC, Northup JK, Gilman AG (1980) The regulatory component of adenylate cyclase from uncoupled S49 lymphoma cells differs in charge from the wild type protein. J Biol Chem 255:2641–2644

Schramm M (1979) Transfer of glucagon receptor from liver membranes to a foreign adenylate cyclase by a membrane fusion procedure. Proc Natl Acad Sci USA 76:1174–1178

Schramm M, Rodbell M (1975) A persistent active state of the adenylate cyclase system produced by the combined actions of isoproterenol and guanylyl-imidodiphosphate in frog erythrocyte membranes. J Biol Chem 250:2232–2237

Shinozawa T, Sen I, Wheeler G, Bitensky M (1979) Predictive value of the analogy between hormone-sensitive adenylate cyclase and light-sensitive photoreceptor cyclic GMP phosphodiesterase: a specific role for a light-sensitive GTPase as a component in the activation sequence. J Supramol Struct 10:185–190

Sonne O, Berg T, Christoffersen T (1978) Binding of $^{125}$I-labeled glucagon and glucagon-stimulated accumulation of adenosine 3′5′-monophosphate in isolated intact rat hepatocytes: evidence for receptor heterogeneity. J Biol Chem 253:3203–3210

Srikant CB, Freeman D, McCorkle K, Unger RH (1977) Binding and biologic activity of glucagon in liver cell membranes of chronically hyperglucagonemic rats. J Biol Chem 252:7434–7436

Stengel D, Hanoune J (1979) Solubilization and physical characterization of the adenylate cyclase from rat liver plasma membranes. Eur J Biochem 102:21–34

Storm DR, Chase RA (1975) Exploitation of hormone-induced conformational changes to label selectively a component of rat liver plasma membranes. J Biol Chem 250:2539–2545

Storm DR, Dolginow YD (1973) Glucagon stimulation of adenylate cyclase sulfhydryl reactivity: evidence for hormone-induced conformational changes. J Biol Chem 248:5208–5210

Strittmatter S, Neer EJ (1980) Properties of the separated catalytic and regulatory units of brain adenylate cyclase. Proc Natl Acad Sci USA 77:6344–6348

Strosberg AD, Vauquelin G, Durieu O, Klutchko C, Bottari S, Andre C (1980) Towards the chemical and functional characterization of the β-adrenergic receptor: a review. Trends Biochem Sci 5:11–14

Sutherland EW (1972) Studies on the mechanism of hormone action. Science 177:401–408

Tolkovsky AM, Levitzki A (1978) Mode of coupling between the β-adrenergic receptor and adenylate cyclase in turkey erythrocytes. Biochemistry 17:3795–3810

Toscano WA, Westcott KR, Laporte DC, Storm DR (1979) Evidence for a dissociable protein subunit required for calmodulin stimulation of brain adenylate cyclase. Proc Natl Acad Sci USA 76:5582–5586

Watanabe AM, McConnaughey MM, Strawbridge RA, Fleming JW, Jones LR, Besch HR (1978) Muscarinic cholinergic receptor modulation of β-adrenergic receptor affinity for catecholamines. J Biol Chem 253:4833–4836

Welton AF, Lad PM, Newby AC, Yamamura H, Nicosia N, Rodbell M (1977) Solubilization and separation of the glucagon receptor and adenylate cyclase in guanine nucleotide-sensitive states. J Biol Chem 252:5947–5950

Williams LT, Lefkowitz RJ (1977) Slowly reversible binding of catecholamine to a nucleotide-sensitive state of the β-adrenergic receptor. J Biol Chem 252:7207–7213

Wright DE, Rodbell M (1979) Glucagon 1–6 binds to the glucagon receptor and activates adenylate cyclase. J Biol Chem 254:268–269

Wright DE, Rodbell M (1980a) Preparation of 2-thioltryptophan-glucagon and (tryptophan-S-glucagon)$_2$: differences in binding to the glucagon receptor in the hepatic adenylate cyclase system. J Biol Chem 255:10884–10887

Wright DE, Rodbell M (1980b) Properties of amidinated glucagons. Eur J Biochem 111:11–16

CHAPTER 14

# Glucagon and Liver Glycogen Metabolism

W. STALMANS

## A. Glucagon, Glycogenolysis, and Glucose Output

The stimulatory effect of glucagon on the conversion of liver glycogen to blood glucose was the first documented property of glucagon, the "hyperglycemic glycogenolytic factor" discovered as a contaminant in some commercial preparations of insulin. SUTHERLAND (1950) has reviewed the early evidence that indicated the hormonal nature of glucagon, distinct from insulin, and that established the liver as the target for the hormone's action. By the analysis of serial blood and liver samples from anesthetized dogs, CAHILL et al. (1957) were able to link directly the glucagon-induced hyperglycemia with hepatic glycogenolysis. The effects of glucagon to increase both glucose output and glycogen breakdown in various isolated liver preparations may be well known; as experimental sophistication has grown over the years, such effects have been demonstrated in rabbit liver slices (SUTHERLAND 1950), isolated perfused rat liver (SOKAL et al. 1964), isolated hepatocytes (GARRISON and HAYNES 1973; SEGLEN 1973; WAGLE and INGEBRETSEN 1973), and liver cells in primary culture (GERSCHENSON and CASANELLO 1968; WALKER and GRINDLE 1977). Numerous reports have stated similar observations, usually with liver preparations from rats or rabbits, but occasionally also from mice (ASSIMACOPOULOS-JEANNET et al. 1973; MÜLLER et al. 1976).

In work with isolated liver preparations, glucose production is usually measured as an index of glycogenolysis. Rightly or wrongly, the two parameters have become so closely linked that data on glucose output are often simply described as "glycogenolysis." This is a potentially dangerous simplification, since glucose can also be produced by gluconeogenesis, which is also stimulated by glucagon (see Chap. 15). However, GARRISON and HAYNES (1973) found that, in the presence as well as in the absence of glucagon, isolated hepatocytes from fed rats produced about ten times more glucose than did cells from fasted animals, provided that no gluconeogenic precursors were included. Furthermore, glucose production by hepatocytes from well-fed rats was not influenced by the presence of gluconeogenic substrates, whether or not glucagon was added (WAGLE 1975). HUTSON et al. (1976) have taken care to correlate directly glucose release with glycogenolysis; they observed that glucose release by hepatocytes from fed rats was mostly or entirely attributable to glycogenolysis, stimulated by hormones or not. In this chapter, it will therefore be accepted that glucose output stems from glycogenolysis when glycogen-rich livers have been studied.

In the following sections, the emphasis will be on the mechanism at the enzymic level whereby glucagon stimulates glycogenolysis and inhibits glycogen synthesis

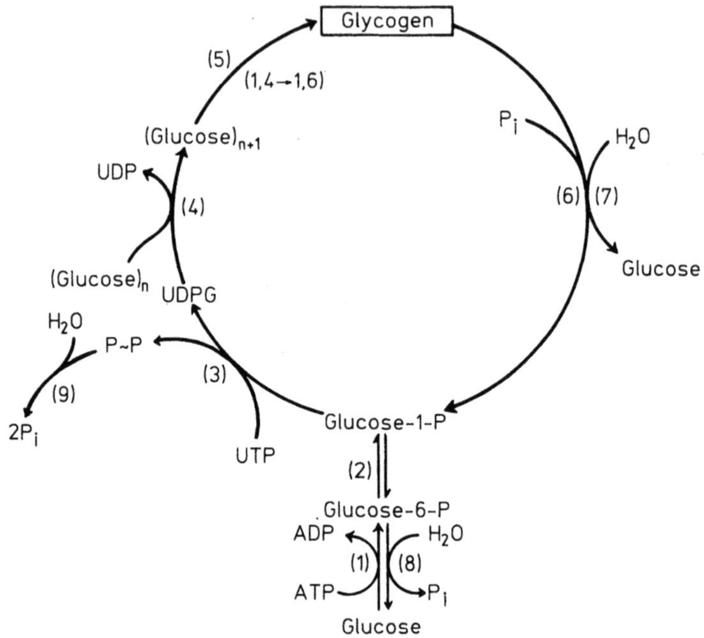


**Fig. 1.** Metabolic pathways for synthesis and degradation of glycogen in the liver. *1* Glucokinase and hexokinase; *2* phosphoglucomutase; *3* UDPG-pyrophosphorylase; *4* glycogen synthase; *5* branching enzyme; *6* glycogen phosphorylase; *7* amylo-1,6-glucosidase; acid maltase (in lysosomes); *8* glucose-6-phosphatase; *9* inorganic pyrophosphatase

(Sect. C). The physiologic expression and modulation of glucagon-induced glycogenolysis is considered in Sect. D. A discussion of how glucagon acts requires a brief introduction to the enzymology and regulation of hepatic glycogen metabolism (Sect. B).

## B. The Enzymes Involved in Hepatic Glycogen Metabolism

This section is intended for the reader who is unfamiliar with the enzymology of glycogen metabolism. References have been indicated only if not available in specific reviews (STALMANS 1976; KREBS and BEAVO 1979; DE WULF et al. 1980).

### I. The General Pathways

Figure 1 summarizes the metabolic steps by which the liver converts glucose to glycogen and vice versa. One may observe that the enzymic equipment of the hepatocyte differs in two important aspects from other cells. A specific glucokinase, which phosphorylates glucose with low affinity, but high maximal velocity, allows the liver to provide sufficient substrate for glycogen synthesis when glucose is abundant in the postprandial period. The presence of glucose-6-phosphatase is essential to the function of the hepatic glycogen stores as a source of glucose for the maintenance of the level of glycemia. The pathway of glycogen synthesis is irreversible

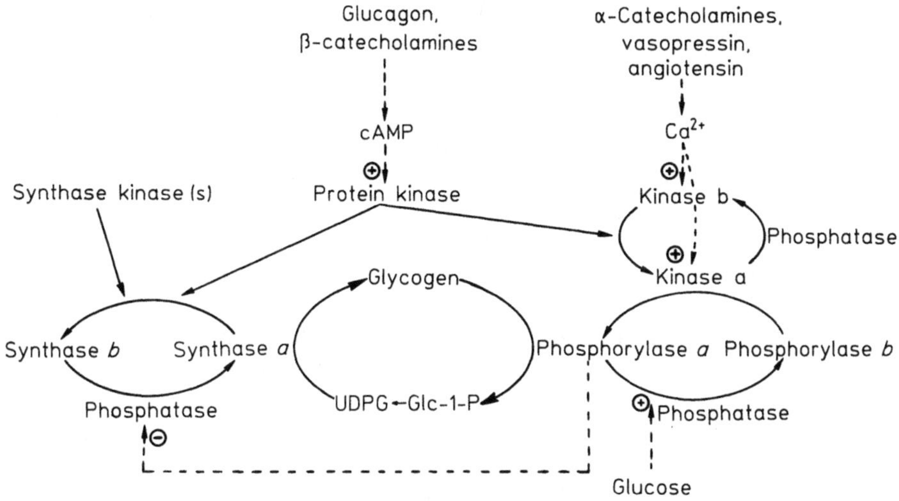


**Fig. 2.** Regulation of the key enzymes in hepatic glycogen metabolism. See the text for explanation. Some regulatory mechanisms have been omitted for the sake of clarity. Glc-1-P: glucose-1-phosphate

because of the reactions catalyzed by glycogen synthase, and by uridine diphosphate glucose (UDPG)-pyrophosphorylase coupled with inorganic pyrophosphatase. Glycogenolysis is also irreversible in vivo because the substrate: product ratio of the phosphorylase reaction, which is readily reversible in vitro, is always kept at approximately 500: 1.

The phosphorolytic degradation of glycogen is completed at branch points by the action of amylo-1,6-glucosidase. Another hydrolytic enzyme, acid maltase, is capable of complete conversion of glycogen to glucose in the lysosomes; the quantitative importance and the possible regulation of the latter pathway remain currently unknown. α-Amylase is abundant in the liver, and its action would cause the fragmentation of glycogen to oligosaccharides; it appears, however, that the latter enzyme is not normally involved in hepatic glycogenolysis (Sect. C.I.2).

## II. The Regulated Enzymes

Figure 2 may serve as a guide through this short account. The rate-limiting steps in the conversion of glucose to glycogen and in glycogen breakdown are catalyzed by glycogen synthase and by phosphorylase, respectively. Either enzyme can be phosphorylated in specific seryl residues; however, in that process phosphorylase is activated (*b*→*a* conversion) whereas glycogen synthase is inactivated (*a*→*b* conversion). The rates of glycogen synthesis and of glycogenolysis are determined by the concentrations of glycogen synthase *a* and of phosphorylase *a*, respectively (see also Sect. C).

### 1. Protein Kinases

The catalytic efficiency of phosphorylase kinase can be increased either by "activation," i.e., phosphorylation by the cAMP-dependent protein kinase, or non-covalently, through "stimulation" of either enzyme form by $Ca^{2+}$. Evidence will be presented in Sect. C that glucagon causes the activation of phosphorylase mainly or exclusively by the former mechanism. So-called cAMP-independent glycogenolytic agents (Fig. 2), which mobilize stored calcium, use the latter mechanism to activate phosphorylase without covalent modification of phosphorylase kinase.

A number of different seryl residues in liver glycogen synthase, a dimer, can be phosphorylated by several distinct protein kinases (JETT and SODERLING 1979). It may be well known that glycogen synthase can be inactivated by the cAMP-dependent protein kinase. However, the potential importance of other kinases is illustrated by the fact that the synthase can be substantially inactivated in mutant cells that lack cAMP-dependent protein kinase (KASLOW 1980). The inactivation of hepatic glycogen synthase caused by cAMP-independent hormones could occur via phosphorylation by phosphorylase kinase (see Sect. C) or by a specific $Ca^{2+}$-dependent synthase kinase (PAYNE and SODERLING 1980). Still other synthase kinases independent of both cAMP and $Ca^{2+}$, have been purified from liver (DOPERÉ et al. 1980a; ITARTE et al. 1981). They inactivate glycogen synthase by introducing one to four phosphate groups per subunit, and their regulation is currently unknown.

### 2. Protein Phosphatases

Protein phosphatases reverse the action of kinases on three enzymes: phosphorylase, glycogen synthase, and phosphorylase kinase. The strong inhibition that phosphorylase *a* exerts on glycogen synthase phosphatase in the liver (Fig. 2) is one argument among others in favor of a single multispecific protein phosphatase. Whether or not the phosphatases involved in hepatic glycogen metabolism are identical is, however, a debated question. A special problem certainly arises in the case of multiply phosphorylated glycogen synthase, which requires at least two distinct protein components for significant $b \rightarrow a$ conversion (DOPERÉ et al. 1980b). For pragmatic reasons also, we will further designate the protein phosphatases as separate enzymes.

Glycogen storage in the liver occurs physiologically after food intake. Both an increased blood glucose level and an altered insulin: glucagon ratio may contribute to the final effect. Glucose binds to phosphorylase *a*, and the resulting complex is a better substrate for phosphorylase phosphatase (Fig. 2; see also Sect. D.II.1). The activation of glycogen synthase occurs in a second phase, when and if phosphorylase *a*, the inhibitor of synthase phosphatase, has been almost completely converted to the *b* form. Insulin appears to cooperate by negative effects at the kinase level, but the number of documented possibilities is rather bewildering (see STALMANS and VAN DE WERVE 1981 for a recent review).

The possible role of a heat-stable protein inhibitor of phosphorylase phosphatase in the glycogenolytic effect of glucagon is discussed in Sect. C.II.2.

## C. The Enzymic Mechanism of Glucagon Action

Glucagon stimulates hepatic glycogenolysis by favoring the activation of glycogen phosphorylase; it also inhibits glycogen synthesis by causing the inactivation of glycogen synthase.

## I. Glucagon Causes Glycogenolysis via Phosphorylase *a*

### 1. Facts

SUTHERLAND and CORI (1951) set out to determine which reaction, in the sequence from glycogen to glucose, is the target for glucagon and epinephrine. Since the hormones increased the concentration of glucose-1-phosphate and glucose-6-phosphate as well as glucose in liver slices, the obvious change appeared to be a facilitation of the phosphorylase reaction (see Fig. 1). The latter interpretation was substantiated by direct assay of the enzyme. SUTHERLAND and CORI concluded that the glycogenolytic agents increase the concentration of active phosphorylase, and that the enhanced velocity of the first reaction causes the subequent nonlimiting conversions to proceed at an increased rate. Subsequent work by SUTHERLAND and coworkers elucidated the mechanism of interconversion of the two forms of phosphorylase. Inactivation was shown to occur by dephosphorylation of the enzyme (WOSILAIT and SUTHERLAND 1956). Evidence was presented for reactivation by a kinase (RALL et al. 1956); incubation of liver slices with $^{32}$P-labeled inorganic phosphate ($P_i$) resulted in the incorporation of $^{32}$P into phosphorylase, and the labeling of the enzyme was manifold increased in the presence of glucagon. The close correlation between phosphorylation and activation of glycogen phosphorylase has recently been confirmed in isolated hepatocytes incubated with increasing concentrations of the hormone (GARRISON 1978; GARRISON et al. 1979).

Since its discovery, the activation of hepatic phosphorylase by glucagon has been observed in work involving liver slices (CORNBLATH 1955, INGEBRETSEN et al. 1974), perfused liver (LEVINE 1965; WEINTRAUB et al. 1969; JACOB and DIEM 1974; WALLI et al. 1974; SAITOH and UI 1975), isolated hepatocytes (CHERRINGTON and EXTON 1976; VANDENHEEDE et al. 1976; VAN DE WERVE et al. 1977a), and intact animals (CAHILL et al. 1957; BISHOP and LARNER 1967; DE WULF and HERS 1968; CURNOW et al. 1975; HEMS et al. 1975; SHIMAZU and AMAKAWA 1975; OKAJIMA and UI 1976, VAN DE WERVE et al. 1977b; NEWMAN and ARMSTRONG 1978).

### 2. Questions

It has been a frequent matter of concern whether the conversion of phosphorylase *b* to the *a* form explains adequately the glucagon-induced glycogenolysis. The experiments by WEINTRAUB et al. (1969) with the isolated perfused rat liver (Fig. 3) illustrate the relationship between glucose output and activity of phosphorylase, as influenced by glucagon. Figure 3 also illustrates a common anomaly that generates a first type of objection to the role of phosphorylase *a*; it is clear indeed that about 40% of phosphorylase would be in the active state in the metabolic situation where the liver neither produces nor takes up glucose. That is an unrealistically high estimate, which is not compatible with the slow turnover of glycogen in the steady state. Such reasoning led SOKAL (1966) to propose that, in the basal condition, most glycogen molecules in the liver would not be exposed to enzymic action, and that glucagon might then induce glycogenolysis by facilitating the contact of glycogen particles with active phosphorylase already present.

The conclusion that "the increase in the total amount of phosphorylase *a* must be a relatively unimportant side effect of the hormone" (SOKAL 1966) created some

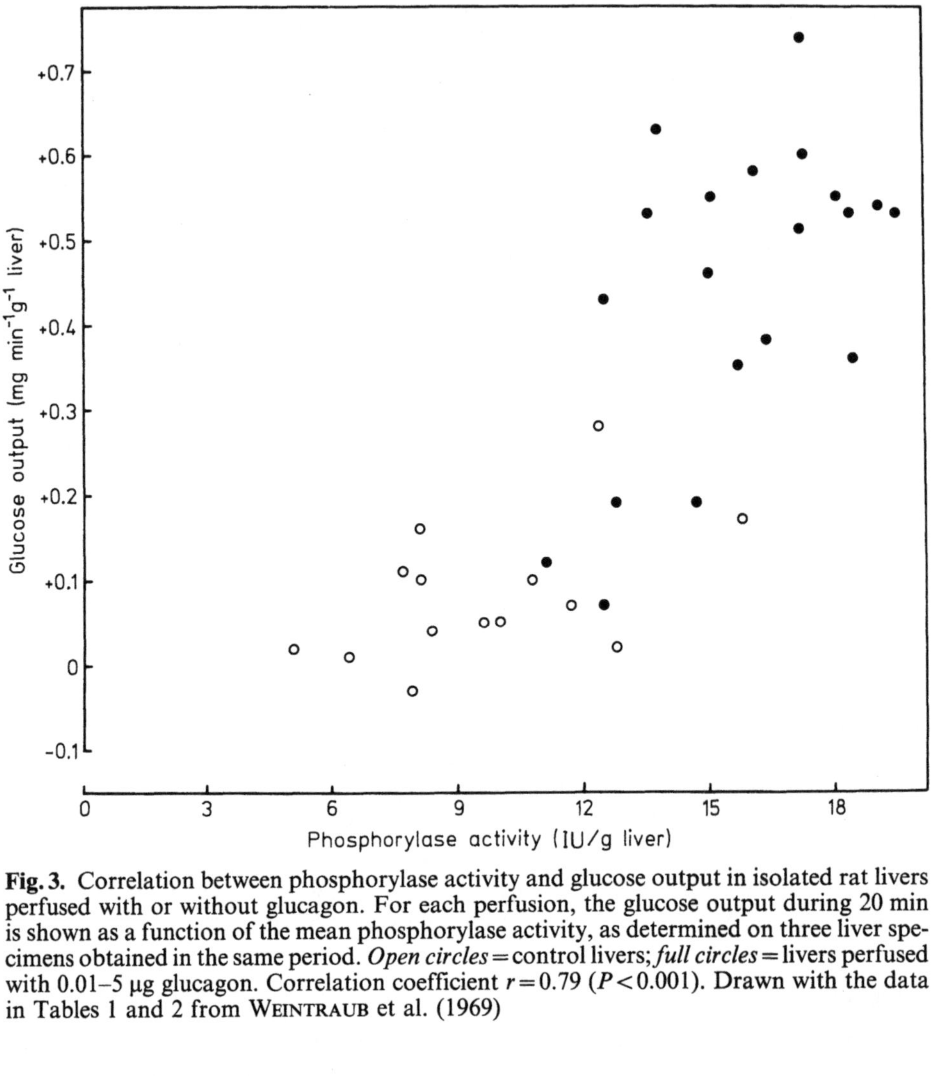


**Fig. 3.** Correlation between phosphorylase activity and glucose output in isolated rat livers perfused with or without glucagon. For each perfusion, the glucose output during 20 min is shown as a function of the mean phosphorylase activity, as determined on three liver specimens obtained in the same period. *Open circles* = control livers; *full circles* = livers perfused with 0.01–5 μg glucagon. Correlation coefficient $r = 0.79$ ($P < 0.001$). Drawn with the data in Tables 1 and 2 from WEINTRAUB et al. (1969)

animosity (ROBISON et al. 1971); but it has required several more years before the various sources of error have become fully evident (see discussion by STALMANS 1976). A major factor is the assay of phosphorylase *a*, which is routinely performed in the nonphysiologic direction of glycogen synthesis, in conditions where the *b* form can acquire significant activity which, as a rule, is erroneously attributed to phosphorylase *a*; this results in a frequent overestimation of basal phosphorylase *a* (STALMANS and HERS 1975). Another factor that may contribute is an artifactual activation of phosphorylase by the general procedures used to obtain and prepare liver samples for enzyme assay. Quick-freezing of the liver is mandatory, preferably in situ to avoid hypoxic activation of phosphorylase (STALMANS et al. 1974; SHARMA et al. 1980). The decapitation of unanesthetized animals leads within a few seconds to an important activation of phosphorylase (STALMANS et al. 1974), which does not involve an activation of phosphorylase kinase (VAN DE WERVE et al. 1977 a), and may be triggered by a stimulation of the splanchnic nerve (SHIMAZU and AMAKAWA 1975).

Another, and quite opposite, type of objection questions the *exclusive* role of phosphorylase *a* in glucagon-induced glycogenolysis. The argument stems to a major extent from observations on human hepatic glycogen-storage diseases. The hyperglycemic response to injected glucagon has some diagnostic value in the latter conditions. One understands easily that the response can be normal when amylo-1,6-glucosidase is deficient, and that it is essentially negative when glucose-6-phosphatase or phosphorylase is lacking (HOWELL 1978). It is difficult to understand, however, why a deficiency of phosphorylase kinase should yield an approximately normal hyperglycemic response, unless either phosphorylase *b* or a nonphosphorolytic pathway were stimulated by glucagon. It has been postulated that 5′-AMP, produced by the breakdown of glucagon-generated cAMP, could stimulate phosphorylase *b* so as to make it catalytically efficient (DE BARSY and LEDERER 1980). The latter proposal has not been substantiated; in physiologic conditions, even high concentrations of 5′-AMP are unable to raise the activity of liver phosphorylase *b* to more than 1% of phosphorylase *a* (STALMANS and GEVERS 1981).

The hypothesis that a nonphosphorolytic, possibly intralysosomal, glycogenolysis may be quantitatively important in some conditions has recently received renewed support (DEVOS and HERS 1980). Another hydrolytic pathway, catalyzed by α-amylase, was active in liver slices and in early preparations of liver cells, but it was insensitive to glucagon (RUTTER and BROSEMER 1961); subsequent work revealed that the α-amylolytic mechanism should be considered as an artifact, since it occurs only in conditions of important cellular damage (RUTTER et al. 1961). Further, glucagon did not alter the measured activities of amylo-1,6-glucosidase, acid maltase, and glucoamylase, in conditions where phosphorylase was activated (ROSENFELD et al. 1971). Of considerable importance are the observations by MALTHUS et al. (1980) on gsd/gsd rats, which have a genetically determined deficiency of liver phosphorylase kinase. These animals do not activate phosphorylase in response to glucagon, and they do not mobilize any glycogen during a 24-h period of starvation. The latter results imply that glucagon-induced hepatic glycogenolysis depends on conversion of phosphorylase *b* to *a*. They also indicate that the human homolog of the disease (see previous paragraph) must be a less complete deficiency of phosphorylase kinase, which can still respond to pharmacologic amounts of glucagon.

## II. Glycogenolysis Mediated by cAMP

It is hardly necessary to recall that the initial events in the hepatic glycogenolytic action of glucagon involve binding of the hormone to the plasma membrane receptor and activation of adenylate cyclase (see Chap. 13). SUTHERLAND (1971) has described in a lively personal account how cAMP was discovered as the "second messenger" of hormones in liver glycogen breakdown. The following observations may be cited among the ample evidence that cAMP mediates the glycogenolytic effect of glucagon: (1) The rise in cAMP precedes the expected result; e.g., in the perfused rat liver, it precedes an increase in glucose output by 0.5 min (EXTON et al. 1971); (2) when injected in vivo or perfused through the isolated liver, cAMP

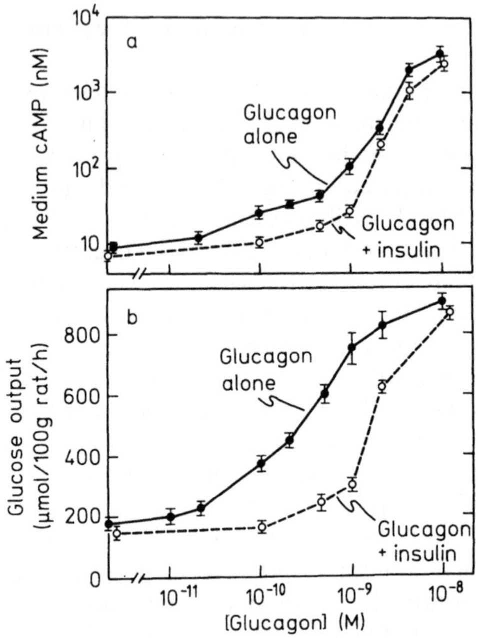


**Fig. 4 a, b.** Relationship between cAMP release (**a**) and glucose production (**b**) by rat livers perfused with glucagon and insulin. Livers from fed rats were perfused for 1 h with recirculating medium. Glucagon was infused at a constant rate to produce the final concentrations indicated on the abscissa, assuming no degradation. When present, insulin was infused at 2.5 mIU/min. Glucose production was measured over the 1 h perfusion, and cAMP in the medium was determined at the end of the experiment. EXTON et al. (1972 b)

mimics the effects of glucagon on hepatic glucose output, glycogenolysis, and activity of glycogen phosphorylase (NORTHROP and PARKS 1964; LEVINE 1965; BERGEN et al. 1966); (3) the concentration of cAMP determines the rate of glycogenolysis more directly than does the concentration of glucagon. For instance, cAMP levels and glucose output are well correlated over a range of glucagon concentrations, whether or not theophylline (an inhibitor of cAMP phosphodiesterase) was added (EXTON et al. 1971).

Along the same lines, Fig. 4 illustrates the relationship between cAMP production and glucose output by isolated livers, perfused with various concentrations of glucagon in the presence or absence of insulin (EXTON et al. 1972 b); what is represented in Fig. 4 a is not the intrahepatic concentration of cAMP, but the extracellular accumulation of the nucleotide. Irrespective of the exact meaning of the extrahepatic accumulation of cAMP, it appears as an exquisitely sensitive index of the intracellular expression of glucagon action.

## 1. Activation of the Glycogenolytic Cascade

In 1964, MAKMAN and SUTHERLAND described a feasible bioassay for glucagon. The hormone was added to a particulate liver fraction capable of producing cAMP; after heating, the amount of generated nucleotide was measured by its abil-

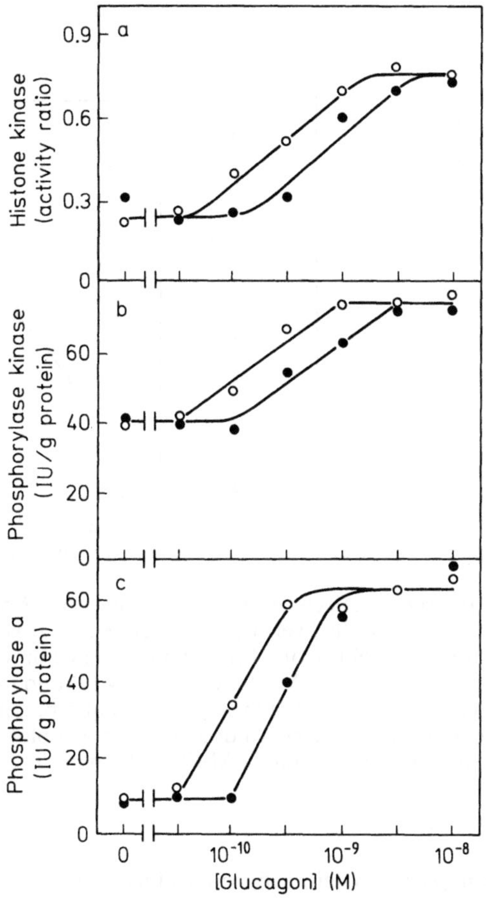


**Fig. 5a–c.** Effects of glucagon and insulin on histone kinase (**a**), phosphorylase kinase (**b**), and phosphorylase *a* (**c**). Isolated hepatocytes from fed rats were incubated for 2 min in the presence of the glucagon concentrations indicated, with *full circles* or without *open circles* 10 n*M* insulin. The activity ratio of histone kinase was determined −/+cAMP. VAN DE WERVE et al. (1977a)

ity to activate phosphorylase in a crude liver system in the presence of MgATP. Though the procedure is now obsolete, it was a remarkable piece of evidence for a hormone-induced cascade of reactions at a time when cAMP-dependent protein kinase was as yet undiscovered and only circumstantial evidence could be adduced for a covalent modification of phosphorylase kinase.

a) Dissociation of cAMP-Dependent Protein Kinase

The inactive protein kinase consists of catalytic (C) and regulatory (R) subunits. It dissociates subsequent to the binding of cAMP to the R subunit, whereby the activity of the C subunit is no longer restrained. The stoichiometry of the reaction has lately been adapted, since there are two binding sites for cAMP in the R polypeptide (RANNELS and CORBIN 1980):

$$C_2R_2 + 4\,cAMP \rightleftharpoons C_2R_2.cAMP_4 \rightleftharpoons 2\,C + R_2.cAMP_4$$

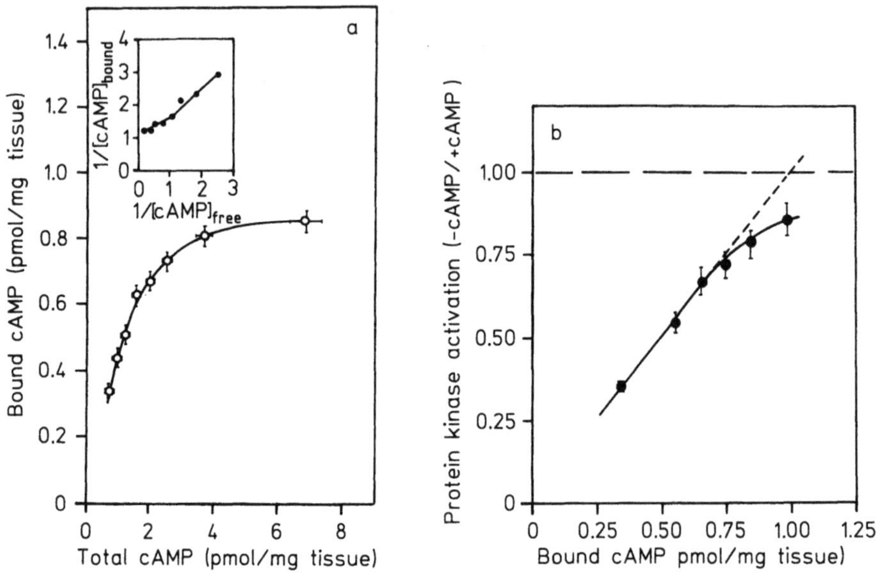


**Fig. 6a, b.** Relationship between the effects of glucagon on total cAMP, protein-bound cAMP, and activation of protein kinase in the liver. Fed rats were killed at various times after the intraperitoneal administration of glucagon (4 μg/g body weight) or saline. Protein-bound cAMP in the cytosol, and either total cAMP (**a**) or the activity ratio (−/+ cAMP) of protein kinase (**b**), were determined in the same livers. Each point represents the mean ± standard error, for 10 or 13 animals. The *inset* is a double-reciprocal plot based on the data in **a**. The extrapolated maximal value for bound cAMP is 0.95 pmol/mg. SCHWOCH and HILZ (1977)

The activation of hepatic protein kinase by glucagon is now well documented from studies in vivo (TAKEDA and OHGA 1973; SUDILOVSKY 1974; VAN DE WERVE et al. 1977 b) and on isolated hepatocytes (BYUS et al. 1976; CHERRINGTON and EXTON 1976; VANDENHEEDE et al. 1976; VAN DE WERVE et al. 1977 a). The dose dependency of the hormone effect on isolated liver cells is illustrated in Fig. 5 a.

An important question concerns the relationship between the level of cAMP and the degree of activation of protein kinase. It is a common finding that the concentration of cAMP can rise 30- to 60-fold when adenylate cyclase is maximally stimulated by glucagon, in vivo (TAKEDA and OHGA 1973; SUDILOVSKY 1974; OKAJIMA and UI 1976) or in isolated liver preparations (EXTON et al. 1971; SIDDLE et al. 1973; CHRISTOFFERSEN and BERG 1974). However, only a slight fraction of those maximal cAMP levels is required for a significant activation of protein kinase. The latter point has been firmly established by SCHWOCH and HILZ (1977); their fundamental observations are illustrated in Fig. 6. It appears that a glucagon-induced rise in total hepatic cAMP from 0.7 pmol/mg tissue (basal value) to 1.7 pmol/mg results in a half-maximal increase in the amount of protein-bound cAMP (Fig. 6 a). It is likely that the protein-bound nucleotide as determined by SCHWOCH and HILZ (1977) corresponds specifically to the R.cAMP complex [for a discussion of cAMP-binding proteins, see SUGDEN and CORBIN (1976)]. The latter assumption is further validated by the data in Fig. 6 b, which show that there is a good linear relation between the amount of bound cAMP and the degree of protein kinase activation; the deviation at higher values is reconsidered in the next paragraph.

There are two types of cAMP-dependent protein kinase, which possess different R subunits. They are present in about equal amounts in rat liver. The dissociated "type II" protein kinase has a tendency to reassociate rapidly in homogenates when the concentration of cAMP decreases (CORBIN et al. 1975; SUGDEN and CORBIN 1976). High salt concentrations inhibit the reassociation of the type II kinase, but provoke some dissociation of the type I enzyme. The assay of cAMP-dependent protein kinases presents therefore some tantalizing problems that have not been fully solved at present (CHERRINGTON et al. 1976; PALMER et al. 1980). When an excess of glucagon causes the complete activation of protein kinase, one arrives of course at the conclusion that both types of the enzyme have been fully dissociated (CHERRINGTON et al. 1976; SCHWOCH 1978). The interesting question is what happens when glucagon achieves a partial activation of protein kinase. SCHWOCH (1978) has investigated this question by combining direct assays of the enzymes in liver cytosol at different salt concentrations with separation of the two isoenzymes on diethylaminoethylcellulose. He concluded that the biphasic decrease in hepatic protein kinase activity, after the initial complete activation by injected glucagon, is explained by an ordered reassociation of type II protein kinase prior to the type I enzyme. This phenomenon may also be responsible for the changes in slope (Fig. 6a, *inset*, and Fig. 6b) which occur when bound cAMP reaches 0.7 pmol/mg tissue (SCHWOCH and HILZ 1977).

### b) Activation of Phosphorylase Kinase

Liver phosphorylase kinase has proved a most difficult enzyme. That activation of the enzyme occurs by cAMP-dependent protein kinase, and inactivation by a protein phosphatase, had long been suspected by analogy with what is known to happen in muscle. VANDENHEEDE et al. (1977) have finally obtained evidence for the latter interconversion in crude preparations from rat liver. More recently, the native enzyme (molecular weight $= 1.3 \times 10^6$ daltons) has been extensively purified in a state where both stimulation by $Ca^{2+}$ and a 30-fold activation by protein kinase are retained (VANDENHEEDE et al. 1979). The liver enzyme is highly susceptible to proteolytic attack; small forms of phosphorylase kinase that have been purified (35,000 and 80,000 daltons) do not respond to protein kinase, and result almost certainly from proteolytic degradation (CHRISMAN et al. 1980).

SHIMAZU and AMAKAWA (1975) were the first to report an increase in liver phosphorylase kinase activity after glucagon administration; a threefold increase had occurred 1 min after intraportal administration of the hormone to anesthetized rabbits. In vivo effects of one and a half and threefold were also observed by VAN DE WERVE et al. (1977b) and by DOORNEWEERD et al. (1981), respectively. As illustrated in Fig. 5, glucagon also activates phosphorylase kinase in isolated hepatocytes, and the latter process is followed by conversion of phosphorylase *b* to *a*. Again, the magnitude of the glucagon effect on phosphorylase kinase did not exceed twofold, whether the hepatocytes were prepared from rats (VANDENHEEDE et al. 1976; VAN DE WERVE et al. 1977a; CHAN et al. 1979b) or rabbits (PROOST et al. 1979). By taking advantage of a specific inhibitor of cAMP-dependent protein kinase, VANDENHEEDE et al. (1976) have substantiated the inference that glucagon-induced activation of phosphorylase kinase is mediated by the protein kinase.

The rather small effects of glucagon on liver phosphorylase kinase remain a puzzling problem. The activity of the enzyme before glucagon administration appears high; even after administration of insulin in vitro or in vivo, it did not drop below 50% of the maximal observed value (VAN DE WERVE et al. 1977a, b; see Fig. 5). DOORNEWEERD et al. (1981) have improved the specific assay of phosphorylase kinase *a*, by choosing conditions where the nonactivated enzyme form is maximally inhibited. Yet, even then the measured effect of glucagon on phosphorylase kinase is limited to 3-fold, whereas the activation of the enzyme in a broken-cell preparation can be over 30-fold (DOORNEWEERD et al. 1981).

c) The Cascade Increases Sensitivity to Glucagon

The existence of a "cascade" of reactions between the hormone and the release of glucose-1-phosphate (Fig. 2) has led to some cogitation on the specific advantages that such a system might confer on the cell. BOWNESS (1966) has stressed the notion of "amplification"; only a minute quantity of hormone is required for the generation of a vast amount of glucose. FISCHER et al. (1971) have pointed out that the specific arrangement of the glycogenolytic cascade provides a means of linking several metabolic controls. Indeed, phosphorylase kinase transmits the effect of cAMP, but it brings glycogenolysis under the control of $Ca^{2+}$ as well. I would like to summarize here some evidence that the cascade of reactions increases the *sensitivity* of glycogenolysis to glucagon; i.e., a half-maximal increase in cAMP accumulation requires considerably more glucagon than does a half-maximal increase in glucose output.

One should, however, interpret the values in the literature with caution. Indeed, half-maximal activation of phosphorylase for instance has been observed in the presence of 200 n*M* glucagon (INGEBRETSEN et al. 1974) as well as 0.1 n*M* (Fig. 5). Part of the latter discrepancy may result from the use of liver slices rather than hepatocytes; however, GARRISON and HAYNES (1973) have made pertinent observations on the importance of glucagon degradation. They found that glucose output by isolated hepatocytes, when measured over 30 min, was half-maximally stimulated by 30 n*M* glucagon; the $A_{0.5}$ value was reduced to 1 n*M* by cutting the incubation time to one-half; and it came down further to 50 p*M* by the simultaneous inclusion of an efficient inhibitor of glucagon degradation. One can therefore only compare data obtained in highly standardized conditions, preferably within the same laboratory.

EXTON et al. (1971) found that the $A_{0.5}$ value of glucagon for cAMP accumulation in the perfused rat liver was 10 n*M*, whereas the value for glucose production was only 1 n*M*. It has already been pointed out that only a small fraction of the cAMP generated by an excess of glucagon is required for full activation of protein kinase. Still further increase in sensitivity to glucagon occurs between protein kinase and phosphorylase (or glycogenolysis). With isolated liver cells, CHERRINGTON et al. (1976) observed half-maximal activation of protein kinase at 0.8 n*M* glucagon; but the latter hormone concentration caused a maximal stimulation of glucose output (HUTSON et al. 1976). A similar shift in the effective hormone concentration was noted by BYUS et al. (1976), and the same is true if one compares activation of protein kinase with activation of phosphorylase (Fig. 5). One may conclude that most of the adenylate cyclase response of the liver requires pharmacologic concentrations of glucagon; and that a major result of the ensuing cascade of reactions is to bring the ultimate glycogenolytic response under the control of physiologic concentrations of the hormone.

### 2. An Effect of cAMP on Phosphorylase Phosphatase

It is possible to extract from liver and muscle two polypeptides that inhibit phosphorylase phosphatase. The isolation of these inhibitory proteins is facilitated by their remarkable ability to survive boiling and treatment with trichloroacetic acid (GORIS et al. 1978; FOULKES and COHEN 1979). Among these proteins, the "type 1" inhibitor from liver is especially relevant to the present discussion, since it becomes inhibitory upon phosphorylation, which is catalyzed by cAMP-dependent protein kinase. By the latter mechanism, glucagon is expected to inhibit phosphorylase phosphatase concertedly with the activation of phosphorylase kinase (GORIS et al. 1978).

This additional action of glucagon remains, however, hypothetical at present, since two essential features have still to be checked: (1) glucagon must elicit the phosphorylation of type 1 inhibitor in the liver – such an effect has been demonstrated in skeletal muscle in response to epinephrine (FOULKES and COHEN 1979); and (2) the phosphorylation of the inhibitor by glucagon must result in a decreased phosphorylase phosphatase activity. At present no such evidence is available, but several groups have reported their failure to detect an influence of glucagon treatment on phosphorylase phosphatase activity (SHIMAZU and AMAKAWA 1975; LALOUX et al. 1978; NUTTALL and GILBOE 1980; MILLER et al. 1981). It may be recalled here that a substantial part of the hepatic phosphorylase phosphatase activity is associated with the glycogen particles, and that the latter enzyme is insensitive to the protein inhibitors (GORIS et al. 1978; LALOUX et al. 1978), apparently because of the presence of a "deinhibitor" protein (DEFREYN et al. 1977). It has also been suggested that inhibition of the cytosolic phosphorylase phosphatase by the "type 2" inhibitor requires previous proteolytic modification of the phosphatase, and that heating is essential for the inhibitory property of the type 2 protein to be revealed (JETT and HERS 1981). Evidently, further research is required to establish or disprove an action of glucagon on the activity of phosphorylase phosphatase.

## III. Can Glucagon Act in a cAMP-Independent Way?

There are persistent claims that glucagon does not act exclusively by the cAMP-dependent pathway described in Sect. C.II. They are based on observations that low glucagon concentrations, or some glucagon derivatives, can activate phosphorylase and cause glycogenolysis without *noticeable* effect on cyclic AMP or cAMP-dependent protein kinase. Some reports are discounted by an obvious methodological error: one should not compare glucose output over an extended time period with cAMP or enzyme activities at the end of that experimental period, when most of the hormone may have been degraded – especially at the low concentrations that matter here. However, this objection does not apply to the work of

OKAJIMA and UI (1976), who administered glucagon to rats in a continuous intravenous infusion. These authors observed drastic changes in hepatic glycogen content and in the activities of phosphorylase and of glycogen synthase (see also Sect. C.IV) by glucagon concentrations that caused no measurable increase in hepatic cAMP, or in the release of cAMP into the hepatic vein; the sensitivity of the extracellular release of cAMP has been pointed out (see Fig. 4). BIRNBAUM and FAIN (1977), using isolated hepatocytes, have also avoided the pitfall by reducing the incubation time to 1 min. They observed that glucagon at 1–3 n*M* activated phosphorylase without measurable rise in cAMP or in protein kinase activity. At 30 n*M* glucagon, however, the classical pathway appeared to operate. Two other reports that challenge the cAMP mechanism are based on work with "capped" derivatives of glucagon (COTE and EPAND 1979) or of [desHis$^1$]glucagon (KHAN et al. 1980). Although the latter derivatives are unable to stimulate adenylate cyclase, they increase glucose release by whole liver or by hepatocytes. Furthermore, when combined with glucagon these compounds inhibit the accumulation of cAMP, but enhance the glycogenolytic response; the latter observation seems to refute the suggestion (KHAN et al. 1980) that the glycogenolytic effect of the capped derivatives might be explained by a minor contaminant, or by some free hormone if the liver were able to remove the protective cap.

With more or less reservation, all these authors have suggested the possibility of a cAMP-independent activation of phosphorylase by glucagon. One reservation is that the available methodology might not allow detection of a weak signal prior to amplification by the glycogenolytic cascade (BIRNBAUM and FAIN 1977). The assay of cAMP-dependent protein kinase also presents inherent difficulties (see Sect. C.II.1.a). CHERRINGTON et al. (1977) have emphasized that, with their methods, glucagon concentrations as low as 50 p*M* are sufficient to elicit small but reproducible and concerted effects on cyclic AMP, protein kinase, and phosphorylase. However, some results with glucagon derivatives (see previous paragraph) cannot be explained by an inadequate methodology. The possibility that glucagon may in part act in a cAMP-independent way can therefore not be dismissed.

Established cAMP-independent glycogenolytic agents, i.e., α-adrenergic agonists, as well as vasopressin and angiotensin, activate liver phosphorylase by increasing the cytosolic concentration of $Ca^{2+}$, which in turn should stimulate phosphorylase kinase (reviewed by DE WULF et al. 1980). Glucagon also affects the movement of $Ca^{2+}$; it promotes the exchange between intracellular and extracellular calcium (FRIEDMANN and PARK 1968; PILKIS et al. 1975; KEPPENS et al. 1977) and enhances the retention of calcium by mitochondria (HUGHES and BARRITT 1978). The literature on the latter aspect is rapidly expanding (see PRPIĆ and BYGRAVE 1980; ANDIA-WALTENBAUGH et al. 1981; and Chap. 50). The relevance of calcium movements to the glycogenolysis induced by low glucagon concentrations remains questionable. Indeed, ASSIMACOPOULOS-JEANNET et al. (1977) noted that glucose output by isolated hepatocytes was stimulated 2- to 3-fold by 100 p*M* glucagon, which was, however, insufficient to enhance the exchange of isotopic calcium. Also, the important factor is expected to be the concentration of free $Ca^{2+}$ in the cytosolic compartment. MURPHY et al. (1980) have developed a method to determine specifically the latter parameter. They report a fourfold increase by norepinephrine, but no effect of 1 n*M* glucagon.

Another way to investigate the possible role of $Ca^{2+}$ in the glycogenolytic action of glucagon is by incubation of hepatocytes with EGTA, which chelates the extracellular Ca, but ultimately depletes the intracellular Ca stores as well. It was initially reported that in those conditions the effect of a high glucagon concentration (0.3 $\mu M$) on glucose output was halved (POINTER et al. 1976). The rate of activation of phosphorylase was also considerably slowed during the initial minutes after glucagon administration (KEPPENS et al. 1977). However, in other studies EGTA interfered minimally or not at all with phosphorylase activation (ASSIMACOPOULOS-JEANNET et al. 1977; CHAN and EXTON 1977; HUE et al. 1978; GARRISON et al. 1979). The interpretation is also hindered by the possibility that prolonged incubation of cells with EGTA may also reduce the cAMP response to glucagon (POINTER et al. 1976).

One arrives at the conclusion that part of the glycogenolytic effect of glucagon may occur by a cAMP-independent mechanism, the nature of which has not been established.

## IV. Glucagon also Causes Inactivation of Glycogen Synthase

BERTHET et al. (1956) observed that glucagon drastically inhibited the incorporation of labeled glucose into glycogen in rabbit liver slices. However, this observation was difficult to interpret at a time when glycogen synthase had not yet been discovered. The explanation is that the mechanism of glucagon action is not limited to a stimulation of glycogenolysis; the hormone also inhibits glycogen synthesis by promoting the conversion of glycogen synthase *a* to *b* (Fig. 2). The latter effect has been demonstrated in vivo (BISHOP and LARNER 1967; DE WULF and HERS 1968; VAN DEN BERGHE et al. 1970; CURNOW et al. 1975; OKAJIMA and UI 1976; NEWMAN and ARMSTRONG 1978) and with incubated liver cells (HUTSON et al. 1976; MASSAGUÉ and GUINOVART 1977; HUE et al. 1978; DE WULF et al. 1980). There are solid arguments in favor of cAMP as the mediator of this glucagon effect. Intravenous injection of cAMP causes a complete inactivation of glycogen synthase in the liver within 1 min, i.e., less time than glucagon requires (DE WULF and HERS 1968). The same effect of cyclic nucleotides has been obtained with the isolated perfused liver (GLINSMANN and HERN 1969). Also, the rise in hepatic cAMP after glucagon injection precedes the inactivation of glycogen synthase (VAN DEN BERGHE et al. 1970).

There is good evidence that the effect of glucagon on glycogen synthase results from a dual action: (1) the hormone stimulates one or more kinases that inactivate the synthase; and (2) it inhibits reactivation by glycogen synthase phosphatase. Glucagon activates cAMP-dependent protein kinase (Sect. C.II.1.a), which also phosphorylates and inactivates glycogen synthase (Sect. B.II). It is not known at present whether the inactivation of the synthase by glucagon occurs exclusively through the action of cAMP-dependent protein kinase. Of special interest among the other kinases that can phosphorylate glycogen synthase (Sect. B.II) is hepatic phosphorylase kinase (SODERLING et al. 1979; VANDENHEEDE et al. 1979). The latter enzyme is also activated in response to glucagon (Sect. C.II.1.b), and may thus participate in the phosphorylation of the synthase. It is currently unknown, however, to what extent phosphorylation by phosphorylase kinase affects the activity of hepatic glycogen synthase.

We (STALMANS and HERS 1973) have proposed an additional mechanism for the control of glycogen synthase by glucagon (see Fig. 2). It is based on the strong inhibition that phosphorylase *a* exerts on glycogen synthase phosphatase: by increasing the hepatic concentration of phosphorylase *a*, glucagon is expected to inhibit the reconversion of synthase *b* to *a*. The evidence in favor of such an action of glucagon is briefly discussed here. When dilute liver extracts from glucagon-treated animals are incubated with purified glycogen synthase *b* from muscle or liver, the initial rate of synthase *a* formation is very low; however, the activity of synthase phosphatase increases manifold when the extracts have been preincubated (BISHOP 1970; STALMANS et al. 1971). This "activation" of synthase phosphatase is entirely explained by the conversion of phosphorylase *a* to the noninhibitory *b* form during the preincubation of the extract, since the same activation could be obtained by the addition of antibodies against phosphorylase *a* (STALMANS et al. 1971). These observations have recently been confirmed and extended by MILLER et al. (1981). They increased the level of phosphorylase *a* in the isolated liver by perfusion, either with glucagon or with vasopressin, a glycogenolytic agent that does not act through cAMP. An inverse relationship was noted between the concentration of phosphorylase *a* and the measured synthase phosphatase activity; further, in the presence of antibodies against phosphorylase, the same high synthase phosphatase activity was recorded in all conditions.

GILBOE and NUTTALL (1978) have described a glucagon-induced decrease in the activity of synthase phosphatase which they do not attribute to a change in the concentration of phosphorylase *a*. The latter interpretation is a debated matter (LALOUX and HERS 1979; NUTTALL and GILBOE 1980).

## D. Expression and Modulation of Glucagon Action

### I. Regulation of Hepatic Glycogen Metabolism by Glucagon

Major physiologic stimuli for an increase in circulating glucagon are brief starvation, protein feeding, and severe exercise (FELIG et al. 1979). Hepatic glycogen metabolism is affected by glucagon in all these conditions.

When food is withdrawn from well-fed rats, there is initially a precipitous fall in the rate of hepatic glycogen synthesis, prior to the onset of glycogen degradation. Glucagon appears to be involved in the suppression of hepatic glycogenesis, since the latter change could be partially reversed by injected anti-glucagon serum (SHIKAMA et al. 1980). After withdrawal of food, liver glycogen is entirely depleted during the ensuing day. SEITZ et al. (1977) have exhaustively investigated the hormonal and metabolic changes during this period. As illustrated in Fig. 7, the serum glucagon concentration rises threefold, with essentially no change in circulating insulin. Associated with the increased serum glucagon is a progressive, almost twofold rise in the hepatic level of cAMP (Fig. 7). Such a rise in total cAMP is known to cause a large increase in protein-bound cyclic nucleotide and in protein kinase activity (Fig. 6). I need hardly recall the relationship between the latter change and the activation of phosphorylase (Fig. 5), and between the level of phosphorylase *a* and glucose production (Sect. C.I).

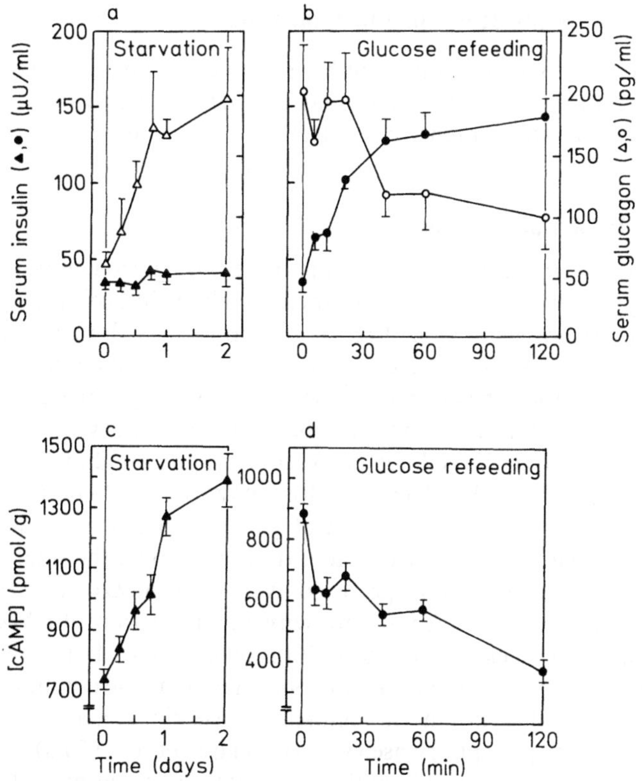


**Fig. 7 a–d.** Changes in plasma glucagon and insulin (**a, b**), and in hepatic cAMP (**c, d**), after food withdrawal and after glucose refeeding. Rats were fed a protein-free, carbohydrate-rich diet for 3 days prior to starvation (**a, c**) in order to lower the serum insulin concentration to the starved level. Normally fed animals were starved for 2 days (**b, d**); at 0 min, glucose was given ad libitum for 1 h. Data are mean ± standard error, for 6–8 rats. SEITZ et al. (1977)

When fasted rats are refed glucose, there is a rapid increase in serum insulin and a slower decrease in circulating glucagon; these hormonal changes are associated with a progressive decrease in the hepatic concentration of cAMP (Fig. 7), and a rapid accumulation of liver glycogen (not illustrated). In contrast, feeding a high protein diet causes a mild increase in circulating insulin, now associated with a severalfold rise in plasma glucagon (TIEDGEN and SEITZ 1980). The latter changes result in a rise in hepatic cAMP; glycogen accumulation in the liver is thereby delayed for several hours, and remains quantitatively reduced. These experiments illustrate strikingly the inhibitory effect of glucagon on hepatic glycogen synthesis.

During exercise, there is an increased hepatic production of glucose, mainly as a result of enhanced glycogenolysis. The latter process may be influenced by the increased plasma concentrations of glucagon and of catecholamines, and by a fall in circulating insulin (see Chap. 38). It may be sufficient here to recall that various approaches, using somatostatin (ISSEKUTZ and VRANIC 1980) or anti-glucagon serum (RICHTER et al. 1981), have indicated that glucagon secretion enhances hepatic glycogenolysis in exercising animals.

## II. Modulation of the Response to Glucagon

A choice had to be made among a number of topics that might have found a place under the present heading. Preference has been given to two situations that have been the target of rather intensive research at the enzymic level.

### 1. Effects of Insulin and Glucose

After food intake, there is a rapid decline in hepatic glycogenolysis, followed by the onset of glycogen synthesis; this occurs through the interplay of decreasing glucagon and increasing insulin (see Fig. 7), and an increased glucose concentration in the portal blood. Insulin and glucose are both able to antagonize the glycogenolytic effect of glucagon. The mechanism or mechanisms by which insulin affects hepatic glycogen metabolism have recently been discussed elsewhere (STALMANS and VAN DE WERVE 1981). In the isolated perfused liver, glucose directly inhibits the activation of phosphorylase by glucagon (BUSCHIAZZO et al. 1970; GLINSMANN et al. 1970; JAKOB and DIEM 1974). No effect of glucose concentration has been noted at any stage of the glycogenolytic cascade. It did not affect the small rise of cAMP caused by 0.1 n*M* glucagon in the perfused liver (BUSCHIAZZO et al. 1970). Glucose administration in vivo did not interfere with the activation of cAMP-dependent protein kinase by an excess of glucagon (SUDILOVSKY 1974), but it still caused a significant decrease in the activation of phosphorylase in those conditions (STALMANS et al. 1974). Glucose does not modify phosphorylase kinase activity in hepatocytes, though it inactivates phosphorylase (VAN DE WERVE et al. 1977a). All these negative observations are compatible with a stimulation of the phosphorylase phosphatase reaction by glucose (Sect. B.II).

### 2. Glucocorticoid Hormones

Adrenalectomy causes a decrease in the maximal glycogenolytic capacity of the liver; this is explained by a 30% decrease in total phosphorylase concentration (SCHAEFFER et al. 1969; EXTON et al. 1972a; CHAN et al. 1979a, b).

More important, however, from a pathophysiologic viewpoint, are the observations that adrenalectomy profoundly inhibits the glycogenolytic effect of small glucagon concentrations (EXTON et al. 1972a; SAITOH and UI 1975). The latter change has been translated into a threefold increase in the glucagon concentration required for half-maximal stimulation of glycogenolysis and of phosphorylase activation (CHAN et al. 1979b). Time course studies have revealed that the defect does not reside in the rate or extent of the early glucagon-induced activation of phosphorylase, but specifically in the inability to *maintain* phosphorylase in the *a* form (CHAN et al. 1979b). The anomaly is at a stage after adenylate cyclase, since (1) cAMP production in response to various glucagon concentrations is normal (EXTON et al. 1972a; SAITOH and UI 1975; CHAN et al. 1979b); and (2) the glycogenolytic effect of cAMP is also reduced (SCHAEFFER et al. 1969; EXTON et al. 1972). The lesion appears also to be at a stage after cAMP-dependent protein kinase, which is normally activated by glucagon (CHAN et al. 1979b). Thus, the impaired activation of phosphorylase could result either from an increased phosphorylase phosphatase activity (which, on the whole, is unlikely: Chan et al. 1979b), or from a decreased ability to keep phosphorylase kinase in the activated state. Further research will be required to choose between these possibilities, and to understand more completely the molecular properties of these complex regulatory enzymes.

## References


Andia-Waltenbaugh AM, Tate CA, Friedmann NK (1981) The effect of glucagon on the kinetics of hepatic mitochondrial calcium-uptake. Mol Cell Biochem 36:177–184

Assimacopoulos-Jeannet F, Exton JH, Jeanrenaud B (1973) Control of gluconeogenesis and glycogenolysis in perfused livers of normal mice. Am J Physiol 225:25–32

Assimacopoulos-Jeannet FD, Blackmore PF, Exton JH (1977) Studies on α-adrenergic activation of hepatic glucose output. Studies on role of calcium in α-adrenergic activation of phosphorylase. J Biol Chem 252:2662–2669

Bergen SS, Hilton JG, Van Itallie TB (1966) Glycogenolytic effect of adenosine 3′,5′-monophosphate in the canine liver. Endocrinology 79:1065–1068

Berthet J, Jacques P, Hers HG, de Duve C (1956) Influence de l'insuline et du glucagon sur la synthèse du glycogène hépatique. Biochim Biophys Acta 20:190–200

Birnbaum MJ, Fain JN (1977) Activation of protein kinase and glycogen phosphorylase in isolated rat liver cells by glucagon and catecholamines. J Biol Chem 252:528–535

Bishop JS (1970) Inability of insulin to activate liver glycogen transferase D phosphatase in the diabetic pancreatectomized dog. Biochim Biophys Acta 208:208–218

Bishop JS, Larner J (1967) Rapid activation-inactivation of liver uridine diphosphate glucose-glycogen transferase and phosphorylase by insulin and glucagon in vivo. J Biol Chem 242:1355–1356

Bowness JM (1966) Epinephrine: cascade reactions and glycogenolytic effect. Science 152:1370–1371

Buschiazzo H, Exton JH, Park CR (1970) Effects of glucose on glycogen synthetase, phosphorylase, and glycogen deposition in the perfused rat liver. Proc Natl Acad Sci USA 65:383–387

Byus CV, Hayes JS, Brendel K, Russell DH (1976) Correlation between cAMP, activation of cAMP-dependent protein kinase(s), and rate of glycogenolysis in isolated rat hepatocytes. Life Sci 19:329–335

Cahill GF, Zottu S, Earle AS (1957) In vivo effects of glucagon on hepatic glycogen, phosphorylase, and glucose-6-phosphatase. Endocrinology 60:265–269

Chan TM, Exton JH (1977) α-Adrenergic-mediated accumulation of adenosine 3′:5′monophosphate in calcium-depleted hepatocytes. J Biol Chem 252:8645–8651

Chan TM, Blackmore PF, Steiner KE, Exton JH (1979a) Effects of adrenalectomy on hormone action on hepatic glucose metabolism. Reciprocal change in α- and β-adrenergic activation of hepatic glycogen phosphorylase and calcium mobilization in adrenalectomized rats. J Biol Chem 254:2428–2433

Chan TM, Steiner KE, Exton JH (1979b) Effects of adrenalectomy on hormone action on hepatic glucose metabolism. Impaired glucagon activation of glycogen phosphorylase in hepatocytes from adrenalectomized rats. J Biol Chem 254:11374–11378

Cherrington AD, Exton JH (1976) Studies on the role of cAMP-dependent protein kinase in the actions of glucagon and catecholamines on liver glycogen metabolism. Metabolism 25:1351–1354

Cherrington AD, Assimacopoulos FD, Harper SC, Corbin JD, Park CR, Exton JH (1976) Studies on the α-adrenergic activation of hepatic glucose output. II. Investigation of the roles of adenosine 3′:5′-monophosphate and adenosine 3′:5′-monophosphate-dependent protein kinase in the actions of phenylephrine in isolated hepatocytes. J Biol Chem 251:5209–5218

Cherrington AD, Hundley RF, Dolgin S, Exton JH (1977) Studies of the role of β-adrenergic receptors in the activation of phosphorylase in rat hepatocytes by catecholamines. J Cyclic Nucleotide Res 3:263–273

Chrisman TD, Vandenheede JR, Khandelwal RL, Gella FJ, Upton JD, Krebs EG (1980) Purification and regulatory properties of liver phosphorylase kinase. Adv Enzyme Regul 18:145–159
Christoffersen T, Berg T (1974) Glucagon control of cyclic AMP accumulation in isolated intact rat liver parenchymal cells in vitro. Biochim Biophys Acta 338:408–417
Corbin JD, Keely SL, Park CR (1975) The distribution and dissociation of cyclic adenosine 3′:5′-monophosphate-dependent protein kinases in adipose, cardiac and other tissues. J Biol Chem 250:218–225
Cornblath M (1955) Reactivation of rabbit liver phosphorylase by epinephrine, glucagon and ephedrine. Am J Physiol 183:240–244
Cote TE, Epand RM (1979) $N^{\alpha}$-trinitrophenyl glucagon: an inhibitor of glucagon-stimulated cyclic AMP production and its effects on glycogenolysis. Biochim Biophys Acta 582:295–306
Curnow RT, Rayfield EJ, George DT, Zenser TV, De Rubertis F (1975) Control of hepatic glycogen metabolism in the rhesus monkey: effect of glucose, insulin, and glucagon administration. Am J Physiol 228:80–87
de Barsy T, Lederer B (1980) Type VI glycogenosis: identification of subgroups. In: Burman D, Holton JB, Pennock CA (eds) Inherited disorders of carbohydrate metabolism. MTP Press, Lancaster, pp 369–380
Defreyn G, Goris J, Merlevede W (1977) A deinhibitor protein neutralizing the effect of the protein inhibitors on dog liver phosphorylase phosphatase. FEBS Lett 79:125–128
Devos P, Hers HG (1980) Random, presumably hydrolytic, and lysosomal glycogenolysis in the livers of rats treated with phlorizin and of newborn rats. Biochem J 192:177–181
De Wulf H, Hers HG (1968) The role of glucose, glucagon and glucocorticoids in the regulation of liver glycogen synthesis. Eur J Biochem 6:558–564
De Wulf H, Keppens S, Vandenheede JR, Haustraete F, Proost C, Carton H (1980) Cyclic AMP-independent regulation of liver glycogenolysis. In: Dumont J, Nunez J (eds) Hormones and cell regulation, vol 4. Elsevier/North Holland Biomedical, Amsterdam Oxford New York, pp 47–71
Doorneweerd DD, Gilboe DP, Nuttall FQ (1981) An assay specific for the active form of liver phosphorylase kinase. Anal Biochem 113:271–276
Doperé F, Goris J, Vandenheede JR, Merlevede W (1980a) Cyclic AMP- and $Ca^{2+}$-independent glycogen synthase kinase from dog liver. Biochem Soc Trans 8:526–527
Doperé F, Vanstapel F, Stalmans W (1980b) Glycogen-synthase phosphatase activity in rat liver. Two protein components and their requirement for the activation of different types of substrate. Eur J Biochem 104:137–146
Exton JH, Robison GA, Sutherland EW, Park CR (1971) Studies on the role of adenosine 3′,5′-monophosphate in the hepatic actions of glucagon and catecholamines. J Biol Chem 246:6166–6177
Exton JH, Friedmann N, Wong EH, Brineaux JP, Corbin JD, Park CR (1972a) Interaction of glucocorticoids with glucagon and epinephrine in the control of gluconeogenesis and glycogenolysis in liver and of lipolysis in adipose tissue. J Biol Chem 247:3579–3588
Exton JH, Lewis SB, Ho RJ, Park CR (1972b) The role of cyclic AMP in the control of hepatic glucose production by glucagon and insulin. Adv Cyclic Nucleotide Res 1:91–101
Felig P, Sherwin RS, Soman V, Wahren J, Hendler R, Sacca L, Eigler N, Goldberg D, Walesky M (1979) Hormonal interactions in the regulation of blood glucose. Recent Prog Horm Res 35:501–532
Fischer EH, Heilmeyer LMG, Haschke RH (1971) Phosphorylase and the control of glycogen degradation. Curr Top Cell Regul 4:211–251
Foulkes JG, Cohen P (1979) The hormonal control of glycogen metabolism. Phosphorylation of protein phosphatase inhibitor-1 *in vivo* in response to adrenaline. Eur J Biochem 97:251–256
Friedmann N, Park CR (1968) Early effects of 3′,5′-adenosine monophosphate on the fluxes of calcium and potassium in the perfused liver of normal and adrenalectomized rats. Proc Natl Acad Sci USA 61:504–508

Garrison JC (1978) The effects of glucagon, catecholamines, and the calcium ionophore A23187 on the phosphorylation of rat hepatocyte cytosolic proteins. J Biol Chem 253:7091–7100

Garrison JC, Haynes RC (1973) Hormonal control of glycogenolysis and gluconeogenesis in isolated rat liver cells. J Biol Chem 248:5333–5343

Garrison JC, Borland MK, Florio VA, Twible DA (1979) The role of calcium ion as a mediator of the effects of angiotensin II, catecholamines, and vasopressin on the phosphorylation and activity of enzymes in isolated hepatocytes. J Biol Chem 254:7147–7156

Gerschenson LE, Casanello D (1968) Metabolism of rat liver cells cultured in suspension: insulin and glucagon effects on glycogen level. Biochem Biophys Res Commun 33:584–589

Gilboe DP, Nuttall FQ (1978) In vivo glucose-, glucagon-, and cAMP-induced changes in liver glycogen synthase phosphatase activity. J Biol Chem 253:4078–4081

Glinsmann WH, Hern EP (1969) Inactivation of rat liver glycogen synthetase by 3′:5′-cyclic nucleotides. Biochem Biophys Res Commun 36:931–936

Glinsmann W, Pauk G, Hern E (1970) Control of rat liver glycogen synthetase and phosphorylase activities by glucose. Biochem Biophys Res Commun 39:774–782

Goris J, Defreyn G, Vandenheede JR, Merlevede W (1978) Protein inhibitors of dog-liver phosphorylase phosphatase dependent on an independent of protein kinase. Eur J Biochem 91:457–464

Hems DA, Whitton PD, Ma GY (1975) Metabolic actions of vasopressin, glucagon and adrenalin in the intact rat. Biochim Biophys Acta 411:155–164

Howell RR (1978) The glycogen storage diseases. In: Stanbury JB, Wyngaarden JB, Fredrickson DS (eds) The metabolic basis of inherited disease, 4th edn. McGraw-Hill, New York, pp 137–159

Hue L, Felíu JE, Hers HG (1978) Control of gluconeogenesis and of enzymes of glycogen metabolism in isolated rat hepatocytes. A parallel study of the effect of phenylephrine and of glucagon. Biochem J 176:791–797

Hughes BP, Barritt GJ (1978) Effects of glucagon and $N^6O^{2'}$-dibutyryladenosine 3′:5′-cyclic monophosphate on calcium transport in isolated rat liver mitochondria. Biochem J 176:295–304

Hutson NJ, Brumley FT, Assimacopoulos FD, Harper SC, Exton JH (1976) Studies on the α-adrenergic activation of hepatic glucose output. I. Studies on the α-adrenergic activation of phosphorylase and gluconeogenesis and inactivation of glycogen synthase in isolated rat liver parenchymal cells. J Biol Chem 251:5200–5208

Ingebretsen C, Clark JF, Allen DO, Ashmore J (1974) Effect of glucagon, dibutyryl adenosine 3′,5′-cyclic monophosphate and phosphodiesterase inhibitors on rat liver phosphorylase activity and adenosine 3′,5′-cyclic monophosphate levels. Biochem Pharmacol 23:2139–2146

Issekutz B, Vranic M (1980) Role of glucagon in regulation of glucose production in exercising dogs. Am J Physiol 238:E13–E20

Itarte E, Mor MA, Salavert A, Pena JM, Bertomeu JF, Guinovart JJ (1981) Purification and characterization of two cyclic AMP-independent casein/glycogen synthase kinases from rat liver cytosol. Biochim Biophys Acta 658:334–347

Jakob A, Diem S (1974) Activation of glycogenolysis in perfused rat livers by glucagon and metabolic inhibitors. Biochim Biophys Acta 362:469–479

Jett MF, Hers HG (1981) Latent phosphorylase phosphatases from rat liver: relationship with the heat-stable inhibitory protein. Eur J Biochem 118:283–288

Jett MF, Soderling TR (1979) Purification and phosphorylation of rat liver glycogen synthase. J Biol Chem 254:6739–6745

Kaslow HR (1980) Apparent phosphorylation of glycogen synthase in mammalian cells lacking cyclic AMP-dependent protein kinase. FEBS Lett 117:219–223

Keppens S, Vandenheede JR, De Wulf H (1977) On the role of calcium as second messenger in liver for the hormonally induced activation of glycogen phosphorylase. Biochim Biophys Acta 496:448–457

Khan BA, Bregman MD, Nugent CA, Hruby VJ, Brendel K (1980) (Deshistidine[1]) ($N^{\varepsilon}$-phenylthiocarbamoyllysine[12])-glucagon: effects on glycogenolysis in perfused rat liver. Biochem Biophys Res Commun 93:729–736
Krebs EG, Beavo JA (1979) Phosphorylation-dephosphorylation of enzymes. Annu Rev Biochem 48:923–959
Laloux M, Hers HG (1979) The role of phosphorylase in the inhibitory effect of EDTA and ATP on liver glycogen synthase phosphatase. Biochem Biophys Res Commun 86:762–768
Laloux M, Stalmans W, Hers HG (1978) Native and latent forms of liver phosphorylase phosphatase. The non-identity of native phosphorylase phosphatase and synthase phosphatase. Eur J Biochem 92:15–24
Levine RA (1965) Effect of glycogenolytic agents on phosphorylase activity of perfused rat liver. Am J Physiol 208:317–323
Makman MH, Sutherland EW (1964) Use of liver adenyl cyclase for assay of glucagon in human gastro-intestinal tract and pancreas. Endocrinology 75:127–134
Malthus R, Clark DG, Watts C, Sneyd JGT (1980) Glycogen-storage disease in rats, a genetically determined deficiency of liver phosphorylase kinase. Biochem J 188:99–106
Massagué J, Guinovart JJ (1977) Insulin control of rat hepatocyte glycogen synthase and phosphorylase in the absence of glucose. FEBS Lett 82:317–320
Miller TB, Garnache A, Vicalvi JJ (1981) Hormonal regulation of hepatic glycogen synthase phosphatase. J Biol Chem 256:2851–2855
Müller P, Singh A, Orci L, Jeanrenaud B (1976) Secretory processes, carbohydrate and lipid metabolism in isolated mouse hepatocytes. Aspects of regulation by glucagon and insulin. Biochim Biophys Acta 428:480–494
Murphy E, Coll K, Rich TL, Williamson JR (1980) Hormonal effects on calcium homeostasis in isolated hepatocytes. J Biol Chem 255:6600–6608
Newman JD, Armstrong JM (1978) On the activities of glycogen phosphorylase and glycogen synthase in the liver of the rat. Biochim Biophys Acta 544:225–233
Northrop G, Parks RE (1964) Studies on epinephrine and 3′,5′-AMP induced hyperglycemia employing the isolated perfused rat liver preparation. J Pharmacol Exp Ther 145:135–141
Nuttall FQ, Gilboe DP (1980) Liver glycogen synthase phosphatase and phosphorylase phosphatase activities in vitro following glucose and glucagon administration. Arch Biochem Biophys 203:483–486
Okajima F, Ui M (1976) Lack of correlation between hormonal effects on cyclic AMP and glycogenolysis in rat liver. Arch Biochem Biophys 175:549–557
Palmer WK, McPherson JM, Walsh DA (1980) Critical controls in the evaluation of cAMP-dependent protein kinase activity ratios as indices of hormonal action. J Biol Chem 255:2663–2666
Payne EM, Soderling TR (1980) Calmodulin-dependent glycogen synthase kinase. J Biol Chem 255:8054–8056
Pilkis SJ, Claus TH, Johnson RA, Park CR (1975) Hormonal control of cyclic 3′:5′-AMP levels and gluconeogenesis in isolated hepatocytes from fed rats. J Biol Chem 250:6328–6336
Pointer RH, Butcher FR, Fain JN (1976) Studies on the role of cyclic guanosine 3′:5′-monophosphate and extracellular $Ca^{2+}$ in the regulation of glycogenolysis in rat liver cells. J Biol Chem 251:2987–2992
Proost C, Carton H, De Wulf H (1979) The α-adrenergic control of rabbit liver glycogenolysis. Biochem Pharmacol 28:2187–2191
Prpić V, Bygrave FL (1980) On the inter-relationship between glucagon action, the oxidation-reduction state of pyridine nucleotides, and calcium retention by rat liver mitochondria. J Biol Chem 255:6193–6199
Rall TW, Sutherland EW, Wosilait WD (1956) The relationship of epinephrine and glucagon to liver phosphorylase. III. Reactivation of liver phosphorylase in slices and in extracts. J Biol Chem 218:483–495
Rannels SR, Corbin JD (1980) Two different intrachain cAMP binding sites of cAMP-dependent protein kinases. J Biol Chem 255:7085–7088

Richter EA, Galbo H, Holst JJ, Sonne B (1981) Significance of glucagon for insulin secretion and hepatic glycogenolysis during exercise in rats. Horm Metab Res 13:323–326

Robison GA, Butcher RW, Sutherland EW (1971) Cyclic AMP, chap. 5. Academic Press, New York London

Rosenfeld EL, Popova IA, Orlova VS (1971) Action of glucagon on γ-amylase and some other enzymes involved in glycogen breakdown. Biochimie 53:939–940

Rutter WJ, Brosemer RW (1961) Glucose production by isolated rat liver cells. An amylase-oligoglucosidase pathway for glycogen breakdown. J Biol Chem 236:1247–1252

Rutter WJ, Arnold M, Brosemer RW, Miller JA (1961) Liver amylase. II. Physiological role. J Biol Chem 236:1259–1263

Saitoh Y, Ui M (1975) Activation and inactivation of phosphorylase and glycogen synthetase during perfusion of rat liver as influenced by epinephrine, glucagon and hydrocortisone. Biochim Biophys Acta 404:7–17

Schaeffer LD, Chenoweth M, Dunn A (1969) Adrenal corticosteroid involvement in the control of liver glycogen phosphorylase activitiy. Biochim Biophys Acta 192:292–303

Schwoch G (1978) Differential activation of type-I and type-II adenosine 3′:5′-cyclic monophosphate-dependent protein kinases in liver of glucagon-treated rats. Biochem J 170:469–477

Schwoch G, Hilz H (1977) Protein-bound adenosine 3′:5′-monophosphate in liver of glucagon-treated rats. Determination of half-maximal binding in vivo and correlation with protein kinase activation. Eur J Biochem 76:269–276

Seglen PO (1973) Effects of anaerobiosis, glucose, insulin and glucagon on glycogen metabolism in isolated parenchymal rat liver cells. FEBS Lett 36:309–312

Seitz HJ, Müller MJ, Krone W, Tarnowski W (1977) Coordinate control of intermediary metabolism in rat liver by the insulin/glucagon ratio during starvation and after glucose refeeding. Arch Biochem Biophys 183:647–663

Sharma RJ, Rodrigues LM, Whitton PD, Hems DA (1980) Control mechanisms in the acceleration of hepatic glycogen degradation during hypoxia. Biochim Biophys Acta 630:414–424

Shikama H, Yajima M, Ui M (1980) Glycogen metabolism in rat liver during transition from the fed to fasted states. Biochim Biophys Acta 631:278–288

Shimazu T, Amakawa A (1975) Regulation of glycogen metabolism in liver by the autonomic nervous system. VI. Possible mechanism of phosphorylase activation by the splanchnic nerve. Biochim Biophys Acta 385:242–256

Siddle K, Kane-Maguire B, Campbell AK (1973) The effects of glucagon and insulin on adenosine 3′:5′-cyclic monophosphate concentrations in an organ culture of mature rat liver. Biochem J 132:765–773

Soderling TR, Sheorain VS, Ericsson LH (1979) Phosphorylation of glycogen synthase by phosphorylase kinase. Stoichiometry, specificity and site of phosphorylation. FEBS Lett 106:181–184

Sokal JE (1966) Glucagon – an essential hormone. Am J Med 41:331–341

Sokal JE, Sarcione EJ, Henderson AM (1964) Relative potency of glucagon and epinephrine as hepatic glycogenolytic agents: studies with the isolated perfused rat liver. Endocrinology 74:930–938

Stalmans W (1976) The role of the liver in the homeostasis of blood glucose. Curr Top Cell Regul 11:51–97

Stalmans W, Gevers G (1981) The catalytic activity of phosphorylase *b* in the liver. With a note on the assay in the glycogenolytic direction. Biochem J 200:327–336

Stalmans W, Hers HG (1973) Glycogen synthesis from UDPG. In: Boyer PD (ed) The enzymes, 3rd edn, vol 9. Academic, New York London, pp 309–361

Stalmans W, Hers HG (1975) The stimulation of liver phosphorylase *b* by AMP, fluoride and sulfate. A technical note on the specific determination of the *a* and *b* forms of liver glycogen phosphorylase. Eur J Biochem 54:341–350

Stalmans W, van de Werve G (1981) Regulation of glycogen metabolism by insulin. In: Hue L, van de Werve G (eds) Short-term regulation of liver metabolism. Elsevier/North Holland Biomedical, Amsterdam Oxford New York, pp 119–138

Stalmans W, De Wulf H, Hers HG (1971) The control of liver glycogen synthetase phosphatase by phosphorylase. Eur J Biochem 18:582–587
Stalmans W, De Wulf H, Hue L, Hers HG (1974) The sequential inactivation of glycogen phosphorylase and activation of glycogen synthetase in liver after the administration of glucose to mice and rats. The mechanism of the hepatic threshold to glucose. Eur J Biochem 41:127–134
Sudilovsky O (1974) *In vivo* regulation of hepatic protein kinase by adenosine 3′,5′-monophosphate mediated glucagon stimulation. Biochem Biophys Res Commun 58:85–91
Sugden PH, Corbin JD (1976) Adenosine 3′:5′-cyclic monophosphate-binding proteins in bovine and rat tissues. Biochem J 159:423–437
Sutherland EW (1950) The effect of the hyperglycemic factor of the pancreas and of epinephrine on glycogenolysis. Recent Prog Horm Res 5:441–459
Sutherland EW (1971) An introduction. In: Robison GA, Butcher RW, Sutherland EW (eds) Cyclic AMP. Academic Press, New York London, pp 1–16
Sutherland EW, Cori CF (1951) Effect of hyperglycemic-glycogenolytic factor and epinephrine on liver phosphorylase. J Biol Chem 188:531–543
Takeda M, Ohga Y (1973) Adenosine 3′,5′-monophosphate and histone phosphorylation during enzyme induction by glucagon in rat liver. J Biochem 73:621–629
Tiedgen M, Seitz HJ (1980) Dietary control of circadian variations in serum insulin, glucagon and hepatic cyclic AMP. J Nutr 110:876–882
Van den Berghe G, De Wulf H, Hers HG (1970) Concentration of cyclic 3′:5′-adenosine monophosphate and glycogen metabolism in the liver. Eur J Biochem 16:358–362
Vandenheede JR, Keppens S, De Wulf H (1976) The activation of liver phosphorylase *b* kinase by glucagon. FEBS Lett 61:213–217
Vandenheede JR, Keppens S, De Wulf H (1977) Inactivation and reactivation of liver phosphorylase *b* kinase. Biochim Biophys Acta 481:463–470
Vandenheede JR, De Wulf H, Merlevede W (1979) Liver phosphorylase *b* kinase. Cyclic-AMP-mediated activation and properties of the partially purified rat-liver enzyme. Eur J Biochem 101:51–58
van de Werve G, Hue L, Hers HG (1977a) Hormonal and ionic control of the glycogenolytic cascade in rat liver. Biochem J 162:135–142
van de Werve G, Stalmans W, Hers HG (1977b) The effect of insulin on the glycogenolytic cascade and on the activity of glycogen synthase in the liver of anaesthetized rabbits. Biochem J 162:143–146
Wagle SR (1975) Interrelationship of insulin and glucagon ratios on carbohydrate metabolism in isolated hepatocytes containing high glycogen. Biochem Biophys Res Commun 67:1019–1027
Wagle SR, Ingebretsen WR (1973) Stimulation of glycogenolysis by epinephrine and glucagon and its inhibition by insulin in isolated rat liver hepatocytes. Biochem Biophys Res Commun 52:125–129
Walker PR, Grindle MJ (1977) Effects of hormones and serum on glycogen metabolism in adult rat liver parenchymal cell primary cultures. J Cell Physiol 91:181–191
Walli AK, Siebler G, Zepf E, Schimassek H (1974) Glycogen metabolism in isolated perfused rat liver. Hoppe Seylers Z Physiol Chem 355:353–362
Weintraub B, Sarcione EJ, Sokal JE (1969) Effect of glucagon on phosphorylase activity of the isolated perfused liver. Am J Physiol 216:521–526
Wosilait WD, Sutherland EW (1956) The relationship of epinephrine and glucagon to liver phosphorylase. II. Enzymatic inactivation of liver phosphorylase. J Biol Chem 218:469–481

CHAPTER 15

# Glucagon and Gluconeogenesis

T. H. CLAUS, C. R. PARK, and S. J. PILKIS

## A. Introduction

Gluconeogenesis is the process whereby lactate, pyruvate, glycerol, and certain amino acids are converted to glucose and glycogen. The liver is the major site of gluconeogenesis, although the kidney becomes important during prolonged starvation. The most important function of gluconeogenesis is the maintenance of blood glucose levels during times when food intake is restricted and/or glycogen stores are depleted. For example, the rate of gluconeogenesis is enhanced during starvation and in the diabetic state. It is also the means whereby the lactate that is produced by glycolysis in erythrocytes and in exercising muscle is reconverted to glucose. Similarly, it conserves the glycerol that is released during lipolysis in adipose tissue and the alanine produced by amino acid metabolism and glycolysis in muscle. Gluconeogenesis also contributes significantly to the utilization of amino acids, which are either absorbed from the alimentary tract or released during protein breakdown in muscle and other tissues.

Figure 1 shows the sequence of reactions by which lactate, pyruvate, glycerol, and various gluconeogenic amino acids are converted to glucose in the hepatocyte. Amino acids and probably lactate enter the cell by various carrier-mediated plasma membrane transport systems. Lactate and most amino acids are converted to pyruvate in the cytoplasm of the cell. Alanine may also be transaminated to pyruvate in the mitochondria (SWICK et al. 1965; DEROSA and SWICK 1975). It has been suggested that mitochondrial metabolism of alanine per se contributes significantly to glucose production from this amino acid (DEROSA and SWICK 1975; MENDES-MOURÃO et al. 1975; DIETERLE et al. 1978). Pyruvate enters the mitochondria by a transport system and, along with that generated in the mitochondria, is converted to oxaloacetate by pyruvate carboxylase or to acetyl-CoA by pyruvate dehydrogenase. The fate of the mitochondrial oxaloacetate varies in different species, depending in large part on the distribution of the enzyme phosphoenolpyruvate carboxykinase between cytosol and mitochondria (TILGHMAN et al. 1976). In the rat and mouse liver, it is converted to malate and/or asparatate, and these metabolites are transported to the cytosol, where they are reconverted to oxaloacetate. The oxaloacetate is then converted to phosphoenolpyruvate by phosphoenolpyruvate carboxykinase, which is predominantly a cytosolic enzyme in rat and mouse liver. This complicated series of interconversions is necessary because the mitochondrial membrane is presumably impermeable to oxaloacetate (however, see GIMPEL et al. 1973). In pigeon, chicken, and rabbit liver, phosphoenolpyruvate carboxykinase is located predominantly in the mitochondria and oxaloacetate is directly converted

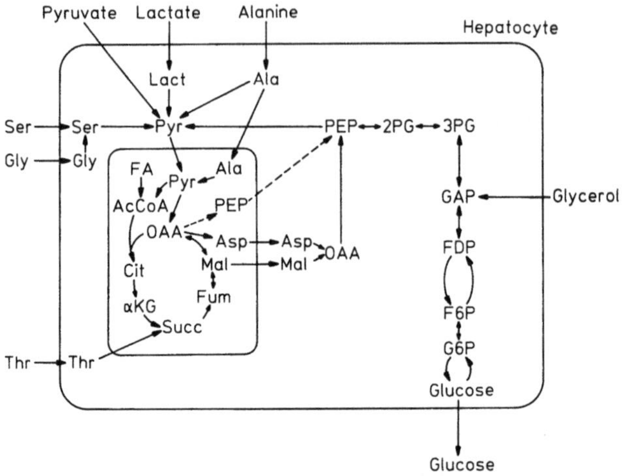


**Fig. 1.** Gluconeogenic pathway in the hepatocyte. The plasma membrane of the hepatocyte is represented by the *large rectangle*. The mitochondrion is depicted by the *small rectangle*. The *broken line* represents the pathway for conversion of oxaloacetate to phosphoenolpyruvate in species that possess mitochondrial phosphoenolpyruvate carboxykinase. Abbreviations: *Lact*, lactate; *Pyr*, pyruvate; *Ala*, alanine; *Ser*, serine; *Gly*, glycine; *FA*; fatty acid; *AcCoA*, acetyl-Coa; *Cit*, citrate; *αKG*, α-ketoglutarate; *Succ*, succinate; *Fum*, fumarate; *Mal*, malate; *oAA*, oxaloacetate; *Asp*, asparatate; *Thr*, threonine; *Pep*, phosphoenolpyruvate; *2 PG*, 2-phosphoglycertate; *3 PG*, 3-phosphoglycerate; *GAP*, glyceraldehyde-3-phosphate; *FDP*, fructose-1,6-bisphosphate; *F6P*, fructose-6-phosphate; *G6P*, glucose-6-phosphate

to phosphoenolpyruvate and then transported to the cytosol. In the human, guinea pig, sheep, and cow liver, the enzyme is distributed about equally between the mitochondria and the cytosol, but the fraction of total phosphoenolpyruvate synthesis at each site is as yet uncertain.

Cytosolic phosphoenolpyruvate can be disposed of by two routes. First, a portion is converted to pyruvate because of the presence of pyruvate kinase. This creates a complicated substrate cycle between pyruvate and phosphoenolpyruvate. Second the remaining phosphoenolpyruvate is converted to fructose-1,6-bisphosphate by the enzymes of the Embden–Meyerhof pathway. Fructose-1,6-bisphosphate is then converted to fructose-6-phosphate by fructose-1,6-bisphosphatase. However, the presence of phosphofructokinase also creates a second substrate cycle between fructose-1,6-bisphosphate and fructose-6-phosphate. The fraction of fructose-6-phosphate not recycled to fructose-1,6-bisphosphate is converted to glucose-6-phosphate by glucose-6-phosphate isomerase. The final step in the pathway is the conversion of glucose-6-phosphate to glucose by glucose-6-phosphatase. However, the presence of hexokinase and glucokinase in liver creates the potential for a third substrate cycle between glucose and glucose-6-phosphate.

Regulation of gluconeogenesis by glucagon occurs on two levels. The first deals with the relatively slow adaptive changes in enzyme activity due to regulation of protein synthesis and/or degradation. For example, glucagon administered in vivo elevates the activity of phosphoenolpyruvate carboxykinase severalfold in 4 h (SHRAGO et al. 1963; RESHEF and HANSON 1972). The injection of dibutyryl cyclic

AMP produces a similar increase in enzyme activity that has been shown to be due to increased de novo synthesis of the enzyme (Reshef and Hanson 1972; Wicks et al. 1972). This topic will not be discussed. The second level is concerned with the minute-to-minute regulation of gluconeogenesis at the level of the hepatocyte, and this will be the central topic of this chapter. The hypothesis will be put forth that glucagon acts by affecting the activity of enzymes involved in the substrate cycles between pyruvate and phosphoenolpyruvate and fructose-1,6-bisphosphate and fructose-6-phosphate. The changes in enzyme activity are brought about by changes in allosteric effectors or by covalent modification of the enzyme or both. A third level of hormonal regulation exists that involves regulation of substrate supply to the liver. While the emphasis will be on the minute-to-minute regulation, it is clear that regulation in vivo occurs on all three levels in an integrated fashion and involves many other hormones besides glucagon (Exton et al. 1970; Exton 1979).

Stimulation of gluconeogenesis by glucagon was first suggested by the observations that the hormone increased urea production in vivo (Tyberghein 1953; Kalant 1954; Rose and Nelson 1956; Salter et al. 1957) and in perfused liver (Miller 1961) and that it stimulated the utilization of lactate by the isolated liver (Schimassek and Mitzkat 1963). More direct evidence was obtained in studies with the perfused liver in which glucagon increased the incorporation of isotope from $^{14}C$-labeled substrate into glucose (Exton and Park 1966, 1968; Garcia et al. 1966). Using the isolated perfused liver system, Exton and Park (1966, 1968) postulated that the effect of this hormone was mediated by an elevation of cyclic AMP levels. The evidence for this was based on the following observations: (1) glucagon caused a rapid elevation of hepatic cyclic AMP; (2) the concentration of hormone necessary to elicit elevation of cyclic AMP levels was the same as that needed to stimulate gluconeogenesis; and (3) addition of exogenous cyclic AMP mimicked the effect of the hormone. These observations are consistent with the concept of cyclic AMP as the second messenger (see Robinson et al. 1971). According to this concept, the hormone, or first messenger, carries the required information to the cell, where it binds to specific hormone receptors. This event in turn stimulates adenylate cyclase to convert ATP to cyclic AMP. This intracellular or second messenger then transfers the information to the cell's enzymatic machinery. In the case of gluconeogenesis, this information transfer is similar to that for glycogen metabolism. That is, cyclic AMP activates a protein kinase, which then catalyzes the phosphorylation of specific enzymes and alters their activity. The resulting changes in enzyme activity lead to a stimulation of gluconeogenesis.

## B. Sites of Action of Glucagon on Hepatic Gluconeogenesis

The major rate-limiting step (or steps) is thought to be in the initial part of the gluconeogenic pathway, since maximal concentrations of lactate, pyruvate, or alanine produce rates of gluconeogenesis that are usually much less than those of substrates that enter at or above the triose phosphate level (Exton and Park 1967; Ross et al. 1967). Analysis of intermediary metabolites from rat livers perfused with high concentrations of lactate, pyruvate, or alanine led to the hypothesis that glucagon and cyclic AMP acted somewhere between pyruvate and phosphoenolpyruvate (EXTON and PARK 1966, 1968; WILLIAMSON et al. 1969a; MALLETTE et al. 1969; PARRILLA et al. 1975, 1976). However, glucagon also appeared to affect reactions between fructose-1,6-bisphosphate and fructose-6-phosphate (EXTON and PARK 1966; WILLIAMSON et al. 1969a; BLAIR et al. 1973; HARRIS 1975; PILKIS et al. 1976b). In addition, the transport of alanine across the plasma membrane was found to be stimulated by glucagon (MALLETTE et al. 1969), but this transport appears unlikely to limit the rate at which alanine is metabolized to glucose (JOSEPH et al. 1978).

The region of the pathway between pyruvate and phosphoenolpyruvate involves many steps, both cytosolic and mitochondrial (Fig. 1). Attempts to localize the affected step by measurement of the total concentration of intermediary metabolites have been unsuccessful, since many of the metabolites exist in both the cytosol and mitochondria. However, a number of mitochondrial steps have been proposed as sites of glucagon action. They include the reactions catalyzed by pyruvate carboxylase and pyruvate dehydrogenase, as well as pyruvate transport into mitochondria and dicarboxylate transport from the mitochondria. Cytosolic sites which have been proposed for hormone action include the enzymes of the pyruvate–phosphoenolpyruvate substrate cycle (phosphoenolpyruvate carboxykinase and pyruvate kinase) and those of the fructose-1,6-bisphosphate–fructose-6-phosphate cycle (fructose-1,6-bisphosphatase and phosphofructokinase). These proposals will be discussed in the following sections.

## I. Action on Mitochondrial Sites

### 1. Pyruvate Carboxylase

Pyruvate carboxylase is an essential enzyme for gluconeogenesis since pyruvate must be converted to oxaloacetate in order for net glucose synthesis to occur. This mitochondrial enzyme requires acetyl-CoA for activity (KEECH and UTTER 1963), and this requirement was the basis for the hypothesis (UTTER et al. 1964) that glucagon stimulates gluconeogenesis by increasing the levels of this effector (STRUCK et al. 1966; GARCIA et al. 1966; WILLIAMSON et al. 1966a–c; SÖLING et al. 1968). The elevation of acetyl-CoA levels could result from an increase in fatty acid oxidation triggered by the activation of lipolysis by a cyclic AMP-sensitive lipase (BEWSHER and ASHMORE 1966). However, DEBEER et al. (1979) could not identify a hormone-sensitive, cyclic AMP-dependent lipase in homogenates of isolated hepatocytes. They proposed that glucagon activated lipolysis by induction of autophagocytosis. In any case, it is clear now that fatty acids and glucagon affect gluconeogenesis by separate mechanisms (for reviews, see EXTON et al. 1970; EXTON 1972; PILKIS et al. 1978b).

Pyruvate carboxylase might be expected to be a phosphoprotein if glucagon directly affects its activity via a cyclic AMP-dependent phosphorylation mechanism, but there is no evidence that the enzyme undergoes phosphorylation (LEITER et al. 1978). However, the activity of this enzyme can be regulated by a number of other factors (for review, see BARRITT et al. 1976) that may be influenced by glucagon. In intact hepatocytes or mitochondria, the activity of pyruvate carboxylase may be limited by the rate of pyruvate entry into mitochondria. This carrier-mediated process (PAPA et al. 1971; PAPA and PARADIES 1974; MOWBRAY 1975; TITHERADGE and COORE 1975, 1976a; HALESTRAP 1978a) has been postulated to be regulated by glucagon. ADAM and HAYNES (1969) found that liver mitochondria isolated from rats injected with glucagon exhibited higher rates of pyruvate carboxylation and oxidation than mitochondria from control rats. The hormone did not alter the activity of pyruvate carboxylase assayed in sonicated extracts of mitochondria, but did increase mitochondrial pyruvate uptake. When the mitochondria were suspended in a medium of low osmolarity, pyruvate uptake increased and the effect of glucagon on pyruvate carboxylation and oxidation was reduced. These results suggested that glucagon increased a reaction common to both carboxylation and oxidation, i.e., pyruvate entry into the mitochondria. GARRISON and HAYNES (1975) obtained similar results with mitochondria from isolated hepatocytes that had been exposed to glucagon. These results ruled out extrahepatic factors as being responsible for the effects. They also found that the glucagon dose-response curves for enhancement of pyruvate carboxylation and gluconeogenesis were identical. The glucagon-induced mitochondrial changes and the increase in the rate of gluconeogenesis were rapid effects, occurring with the same time course, and the increase in pyruvate carboxylation correlated with the hormone-induced changes in cyclic AMP levels.

TITHERADGE and COORE (1976a, b) reported that glucagon treatment of rats had a direct effect on pyruvate transport in isolated mitochondria when transport was driven by oxidation of ascorbate plus tetramethylphenylenediamine. HALESTRAP (1978b) was unable to repeat their observations. He found no differences in the time courses of pyruvate uptake into mitochondria from livers of control or glucagon-treated rats (HALESTRAP 1975, 1978b). Most results suggest that glucagon does not affect pyruvate transport specifically. YAMAZAKI (1975) demonstrated that acute treatment of rats with glucagon caused a stimulation of hepatic mitochondrial respiration as measured by oxygen uptake. In the presence of phosphate and ADP (state 3) or in uncoupled mitochondria, the rates of oxygen uptake were increased using pyruvate–malate, malate–L-glutamate, $\alpha$-ketoglutarate, or succinate as substrates. There was no effect of glucagon on the oxidation of ascorbate in the presence of tetramethylphenylenediamine or when the rate of respiration with NADH- and flavoprotein-linked substrates was measured in the absence of ADP (state 4). These results have been confirmed by TITHERADGE and COORE (1976b), BRYLA et al. (1977), and HALESTRAP (1978b). BRYLA et al. (1977) also observed that mitochondria from glucagon-treated rats showed an increased rate of respiration with $\beta$-hydroxybutyrate as substrate and an increased uptake of ADP. These hormone-induced effects on respiration have been found to persist even after the mitochondria were disrupted by sonication (TITHERADGE et al. 1978). Significant increases in oxygen consumption under state 3 or uncoupled conditions were observed in submitochondrial particles prepared from glucagon-treated rats when either succinate or NADH were the respiratory substrates. No hormone effect was observed when ascorbate plus tetramethylphenylenediamine was used as the substrate. Together, these results indicate that glucagon treatment stimulates mitochondrial function in a general manner rather than just affecting pyruvate transport. Consistent with this notion is the report that glucagon treatment of intact rats stimulated mitochondrial $CO_2$ fixation (pyruvate carboxylation) when alanine was used as substrate (CHAN et al. 1979). Under these conditions, pyruvate was generated inside the mitochondria and there was no need for pyruvate transport.

It has now been shown that mitochondria isolated from livers of rats treated with glucagon display a number of other differences from those of control rats. HAYNES (1976) found that they contained a greater concentration of $Mg^{2+}$ than controls. This was confirmed by HALESTRAP (1978 b) who also found a greater concentration of $K^+$. HAYNES (1976) found no difference in the $K^+$ concentration of control or glucagon-treated mitochondria, but he did observe a greater rate of $K^+$ uptake in glucagon-treated mitochondria. YAMAZAKI et al. (1977) have confirmed the latter observation. Glucagon treatment also results in enhanced rates of $Ca^{2+}$ uptake and prolonged retention of $Ca^{2+}$ (YAMAZAKI 1975; HUGHES and BARRITT 1978; PRPIĆ et al. 1978; TAYLOR et al. 1980). YAMAZAKI and GRAETZ (1977) found that glucagon treatment of rats stimulated the formation of citrulline by isolated liver mitochondria incubated with ornithine, ammonium ion, bicarbonate, phosphate, and an energy source such as succinate which generated intramitochondrial ATP via substrate oxidation. This finding was confirmed in mitochondria isolated from hepatocytes exposed to glucagon (BRYLA et al. 1977; TRIEBWASSER and FREEDLAND 1977) and the increased rate correlated with an increase in ATP content (BRYLA et al. 1977). PRPIĆ et al. (1978), CHAN et al. (1979), and TITHERADGE and HAYNES (1980) observed a higher ATP content in mitochondria from glucagon-treated rats, and a higher ATP: ADP ratio has also been reported (BRYLA et al. 1977; SIESS et al. 1977; TITHERADGE et al. 1979). Glucagon treatment also results in an enhanced mitochondrial ATPase activity in the presence of an uncoupler of oxidative phosphorylation (YAMAZAKI et al. 1977; TITHERADGE et al. 1979; TITHERADGE and HAYNES 1980). The ATPase activity correlated with the glucagon concentration dependency of both gluconeogenesis and pyruvate carboxylation, and the stimulation could be mimicked by addition of cyclic AMP to the hepatocytes (TITHERADGE et al. 1979). Glucagon treatment also caused an increase in the membrane potential (HALESTRAP 1978 b), increased transport of ATP (BRYLA et al. 1977; TITHERADGE and HAYNES 1980), increased $P_i$-ATP exchange (TITHERADGE and HAYNES 1980), and increased succinic dehydrogenase activity (SIESS and WIELAND 1978, 1979).

YAMAZAKI (1975) suggested that the effects of glucagon to stimulate substrate-dependent respiration were consistent with either a general stimulation of mitochondrial transport systems or stimulation of the electron transport chain. TITHERADGE and COORE (1976 b) found that glucagon treatment increased the pH of the mitochondrial matrix, a finding confirmed by HALESTRAP (1978 b). He proposed that glucagon treatment enhanced electron transport chain activity and that this allowed enhanced proton ejection from the mitochondria, thus increasing the pH of the mitochondrial matrix. The increase in hydroxyl ion concentration should stimulate mitochondrial anion transport since hydroxyl ions exchange either directly or indirectly with the various anions. The increase in electron transport chain activity also stimulated the rate of ATP formation within the mitochondria (YAMAZAKI and GRAETZ 1977). Both pyruvate carboxylation and the initial step of citrulline formation (carbamyl phosphate synthesis) are mitochondrial reactions which require ATP as substrate. Thus, the stimulation of oxidative phosphorylation may be necessary to meet the increased demands for ATP during gluconeogenesis and ureagenesis.

The ability of glucagon to stimulate respiration from NADH- and flavoprotein-linked substrates, but not from ascorbate plus tetramethylphenylenediamine, suggested that the hormone activated the respiratory chain at a point prior to the entry of electrons from tetramethylphenylenediamine into the system. This point appeared to be before phosphorylation site III, i.e., between cytochrome *b* and *c* (YAMAZAKI 1975; TITHERADGE et al. 1978; HALESTRAP 1978c). HALESTRAP (1978c) measured the cytochrome spectra under uncoupled conditions in the presence of succinate and rotenone and found a crossover between cytochrome *c* and $c_1$ when mitochondria from glucagon-treated rats were compared with those of control rats. Cytochrome *c* was more reduced and cytochrome $c_1$ more oxidized in mitochondria from glucagon-treated rats and he concluded that glucagon stimulates electron flow between cytochrome $c_1$ and *c*.

The mechanism whereby glucagon stimulates the electron transport chain is uncertain. The fact that the effect persists throughout the time needed to isolate mitochondria, and even after preparation of submitochondrial particles, suggests that the hormone produces a stable, perhaps covalent, modification. Many of the glucagon-induced changes in mitochondrial metabolism have been correlated with changes in cyclic AMP (GARRISON and HAYNES 1975). If the effects of cyclic AMP are mediated through the phosphorylation of specific proteins by the cyclic AMP-dependent protein kinase, it would be predicted that glucagon should enhance the phosphorylation of mitochondrial protein (or proteins). ZAHLTEN et al. (1972) reported that rat liver inner mitochondrial membranes were phosphorylated in vivo and that glucagon increased the phosphorylation. Using isolated hepatocytes, HALESTRAP (1978b, c) reported that inner mitochondrial membrane proteins could be phosphorylated and that one of the proteins had the same molecular weight as cytochrome $c_1$. However, no effect of glucagon on its phosphorylation was reported. If glucagon, via cyclic AMP, acts at the mitochondrial level, one might expect to find a cyclic AMP-dependent protein kinase associated with the mitochondria. However, no reports of the presence of a cyclic AMP-dependent protein kinase in mitochondria have appeared. VARDANIS (1977) found a cyclic AMP-independent protein kinase on the inner membrane of mouse liver mitochondria, but the enzyme was not characterized extensively and it may represent the catalytic subunit of the cyclic AMP-dependent protein kinase. However, since cytochrome $c_1$ is on the outer surface of the inner mitochondrial membrane (RACKER 1970; PAPA 1976), its phosphorylation could be catalyzed by a cyclic AMP-dependent protein kinase located between the inner and outer mitochondrial membranes. The presence of a cyclic AMP-dependent protein kinase in mitochondria is suggested by the claim that addition of cyclic AMP and ATP to isolated mitochondria can mimic the effects of glucagon (HALESTRAP 1978d). However, HALESTRAP et al. (1980) have recently reported that they could find no evidence for phosphorylation of a respiratory chain component.

The intramitochondrial concentration of glutamate may also be a regulator of pyruvate carboxylase activity. The enzyme is inhibited by glutamate, but the high concentration required ($K_i > 5$ m*M*) was thought to preclude its physiologic significance (SCRUTTON and WHITE 1974). UI et al. (1973a, b) reported that glucagon dramatically lowered the intracellular level of glutamate and α-ketoglutarate. These effects were attributed to glucagon activation of either α-ketoglutarate dehydrogenase or succinic thiokinase. SIESS et al. (1977) confirmed the glucagon-induced decrease in α-ketoglutarate and glutamate, and they reported that the intramitochondrial glutamate concentration dropped from approximately 15 to about 3 m*M*. Thus, it is possible that glutamate normally restrains the rate of pyruvate carboxylation and that the addition of glucagon removes that restraint.

Calcium ion can be a potent inhibitor of pyruvate carboxylase by competing with $Mg^{2+}$ (WIMHURST and MANCHESTER 1970; McCLURE and LARDY 1971). Pyruvate carboxylation by isolated rat liver mitochondria can also be inhibited by calcium ions (KIMMICH and RASMUSSEN 1969; MÖRIKOFER-ZWEZ et al. 1973; FOLDES and BARRITT 1977). Regulation of gluconeogenesis by changes in $Ca^{2+}$ has been proposed by FRIEDMAN and RASMUSSEN (1970) and by RASMUSSEN (1970). Changes in intramitochondrial $Ca^{2+}$ concentration which lie within the physiologic range would have significant effects on pyruvate carboxylation (FOLDES and BARRITT 1977). Glucagon has been shown to promote the efflux of $Ca^{2+}$ ions from mitochondria, but it is much less effective in doing so than are catecholamines (CHEN et al. 1978; SALZMANN et al. 1978; BABCOCK et al. 1979; BLACKMORE et al. 1979a). However, glucagon and catecholamines are equally effective in promoting pyruvate carboxylation (GARRISON and BORLAND 1979). Thus, the role of $Ca^{2+}$ efflux in promoting pyruvate carboxylation is uncertain.

## 2. Pyruvate Dehydrogenase

Pyruvate that enters mitochondria can either be converted to oxaloacetate by pyruvate carboxylase or be oxidized to acetyl-CoA by pyruvate dehydrogenase (Fig. 1). Since the two enzymes compete for pyruvate, inactivation of pyruvate dehydrogenase would facilitate carboxylation and thus gluconeogenesis. The discovery that the activity of this enzyme complex could be regulated by a phosphorylation–dephosphorylation mechanism (LINN et al. 1969a, b) suggested that it might be a site of glucagon action. ZAHLTEN et al. (1973) found that glucagon addition to hepatocytes from starved rats decreased glucose $^{14}C$ and $^{14}CO_2$ production from 10 m*M* pyruvate $^{14}C$-1 and they suggested that these effects were due to an inhibition of pyruvate dehydrogenase. On the other hand, YAMAZAKI and HAYNES (1975) concluded that pyruvate dehydrogenase was not a primary site of glucagon action since the hormone was still able to stimulate pyruvate carboxylation under conditions where pyruvate dehydrogenase was inhibited. Also, CRABB et al. (1976) found that dichloroacetate, an activator of pyruvate dehydrogenase (WHITEHOUSE and RANDLE 1973; WHITEHOUSE et al. 1974), activated the enzyme, but had minor effects on glucose synthesis from lactate and alanine in isolated hepatocytes. They concluded that the regulation of pyruvate dehydrogenase was only of marginal importance in the control of gluconeogenesis. CLAUS and PILKIS (1977) found that dichloroacetate did not prevent glucagon from inhibiting glucose synthesis from 10 m*M* pyruvate, even though pyruvate dehydrogenase was completely activated. They concluded that the inhibition of pyruvate gluconeogenesis was not due to inhibition of pyruvate dehydrogenase. MAPES and HARRIS (1976), using a different approach, reached the same conclusion. Thus, it now appears that the glucagon-induced decrease in $^{14}CO_2$ production from 10 m*M* pyruvate $^{14}C$-1 that was observed by ZAHLTEN et al. (1973) was due to the concomitant decrease in glucose synthesis. CLAUS and PILKIS (1977) also found that glucagon stimulated glucose synthesis from a variety of substrates, regardless of whether pyruvate dehydrogenase was in the active or inactive state. Thus, it appears that pyruvate dehydrogenase is not involved in the regulation of gluconeogenesis by glucagon.

## II. Action on Extramitochondrial Sites

The enzymes specific for gluconeogenesis have been identified in cytosol of rat liver. They are: phosphoenolpyruvate carboxykinase, which catalyzes the conversion of oxaloacetate to phosphoenolpyruvate; fructose-1,6-bisphosphatase, which converts fructose-1,6-bisphosphate to fructose-6-phosphate; and glucose-6-phosphatase, which hydrolyzes glucose-6-phosphate to free glucose and inorganic phosphate. These three gluconeogenic enzymes are opposed in the cell by three glycolytic enzymes, pyruvate kinase, phosphofructokinase, and hexokinase, respectively. The existence in the same cell of enzymes that catalyze opposing reactions raised the possibility of cycling between the substrates and products of the enzymes (Fig. 1). Since each of the reactions is nonequilibrium, energy would be expended by such cycling. NEWSHOLME and GEVERS (1967) have speculated that the simultaneous operation of opposing reactions would make possible a very sensitive control system in which both the rate and direction of metabolism could be regulated by very small changes in the concentration of effectors of one or more of the enzymes involved in the cycle. Energy expended by such a substrate cycle would not be wasted, but would create a more efficient regulatory system than a simple "on–off" system.

### 1. Phosphoenolpyruvate–Pyruvate Substrate Cycle

Recent studies have provided evidence that the phosphoenolpyruvate–pyruvate cycle is important in the regulation of gluconeogenesis. Earlier work employing crossover plots of metabolic intermediates had suggested that the rate-limiting step (or steps) occurs in this region of the pathway. The cycle is quite complex. The pyruvate produced by the action of pyruvate kinase must be carboxylated to oxaloacetate and then decarboxylated to phosphoenolpyruvate. The route of these conversions is dependent upon the location of phosphoenolpyruvate carboxykinase, which is distributed in various proportions between cytosol and mitochondria, depending on the species (TILGHMAN et al. 1976). This cycle therefore may include both mitochondrial membrane transport systems and a number of enzymes. During the cycle 1 mol ATP and 1 mol GTP are consumed, and 1 mol ATP is generated. Thus, 1 mol energy-rich phosphate is expended per cycle.

The first evidence that some phosphoenolpyruvate is recycled to pyruvate during gluconeogenesis was presented by FRIEDMAN et al. (1971) using perfused rat liver, and by ROGNSTAD and KATZ (1972) using rat kidney cortex segments. Both groups used pyruvate $^{14}C$-2 as substrate and determined the flux through pyruvate kinase by measuring the distribution of radioactivity among carbon atoms of lactate and pyruvate. Both groups found that, in the fasting state, about one-half as much phosphoenolpyruvate was recycled to pyruvate as was converted to glucose. In the fed state, the situation was reversed, the rate of pyruvate kinase flux being four times that of glucose synthesis (FRIEDMAN et al. 1971). With lactate $^{14}C$-2 as substrate in kidney cortex segments, only 30% as much phosphoenolpyruvate was recycled to pyruvate as was converted to glucose (ROGNSTAD and KATZ 1972).

Another approach to measuring flux through pyruvate kinase was devised by ROGNSTAD (1975). Labeled $NaHCO_3$ was used instead of pyruvate $^{14}C$-2 or lactate, since it will incorporate radioactivity into phosphoenolpyruvate prior to pyruvate. The amount of label found in lactate and pyruvate, along with the estimated specific activity of phosphoenolpyruvate, was used to calculate pyruvate kinase flux. In hepatocytes from starved rats incubated with 20 m*M* pyruvate, the flux was estimated to be about 50% of the rate of gluconeogenesis (ROGNSTAD 1975), and glucagon or cyclic AMP inhibited flux by about 45% (ROGNSTAD 1975, 1976). With 20 m*M* lactate as substrate, the flux through pyruvate kinase was estimated to be 7%–23% of the rate of gluconeogenesis in starvation and about 50% in the fed state (KATZ and ROGNSTAD 1976; ROGNSTAD and KATZ 1977). In hepatocytes from fed rats, glucagon caused a dose-dependent decrease in flux (ROGNSTAD and KATZ 1977). Recently, COHEN et al. (1981) have measured flux through pyruvate kinase in hepatocytes from starved rats using $^{13}C$ nuclear magnetic resonance to determine the amount of $^{13}C$ incorporated into the C-2 position of alanine. With 28 m*M* alanine $^{13}C$-3 as substrate, they found that flux through pyruvate kinase was 25% of the gluconeogenic flux.

Flux through pyruvate kinase in isolated hepatocytes can also be measured from the rate of lactate and pyruvate production from dihydroxyacetone (PILKIS et al. 1976a, b; ROGNSTAD 1976). This substrate is first converted to dihydroxyacetone phosphate by glycerol kinase and then to either glucose or to lactate and pyruvate. Pyruvate kinase flux was about equal to the rate of gluconeogenesis in hepatocytes from fed or 24-h starved rats, but was depressed upon further starvation (PILKIS et al. 1976b). In hepatocytes from fed rats, glucagon caused a dose-dependent decrease in lactate production from dihydroxyacetone and a concomitant, quantitatively equivalent increase in glucose synthesis. Glucagon also decreased flux through pyruvate kinase in hepatocytes from starved rats (PILKIS et al. 1976b; ROGNSTAD 1976, 1979).

It is uncertain just how important changes in pyruvate kinase are as a means of regulating gluconeogenesis. All the studies on pyruvate kinase flux have been done with concentrations of physiologic substrates that produce maximum rates of gluconeogenesis or with nonphysiologic substrates. While the glucagon-induced decrease in pyruvate kinase flux was accompanied by an equivalent increase in glucose synthesis from dihydroxyacetone (PILKIS et al. 1976b; CLAUS et al. 1979), that was not the case when 20 m*M* lactate was the substrate. ROGNSTAD and KATZ (1977) reported that glucagon had no significant effect on pyruvate kinase flux in hepatocytes from starved rats, even though it stimulated glucose synthesis from 20 m*M* lactate. In hepatocytes from fed rats, the glucagon-induced decrease in pyruvate kinase flux accounted for about 40% of the increase in glucose synthesis. Regulation of pyruvate kinase flux may be more (or less) important under conditions that mimic those found in vivo. This question can be resolved only by direct measurement of pyruvate kinase flux with physiologic concentrations of lactate and pyruvate as substrate. However, studies on the regulation of the activity of L-type pyruvate kinase provide strong evidence that the enzyme plays a central role in the regulation of gluconeogenesis by glucagon in rat liver.

a) Pyruvate Kinase

There is approximately ten times more pyruvate kinase activity than phosphoenolpyruvate carboxykinase activity in rat liver (Scrutton and Utter 1968). This suggests that pyruvate kinase must be substantially inhibited in order for net gluconeogenesis to occur. This consideration, and the observation that glucagon can affect flux through pyruvate kinase, has fostered a great deal of research on this enzyme. Many studies conducted in various species and tissues suggest that there are three isozymes of pyruvate kinase in mammalian tissues (Seubert and Schoner 1971; Imamura et al. 1972, Ibsen 1977), but glucagon has been found to affect only the L-isozyme. Therefore, the discussion will be limited to the L-isozyme. Discussion of the properties of the M- and K-isozymes can be found elsewhere (for review, see Pilkis et al. 1978b). The L-isozyme, which is the predominant form in liver, exhibits sigmoidal kinetics with regard to its substrate, phosphoenolpyruvate. When the allosteric activator fructose-1,6-bisphosphate is added or the pH is lowered, the enzyme exhibits normal Michaelis–Menten kinetics (for references see Claus and Pilkis 1981). The L-isozyme is also allosterically inhibited by ATP and alanine (Seubert and Schoner 1971; Imamura et al. 1972).

The presence of ATP and alanine can have profound effects on the degree of cooperativity that L-type pyruvate kinase shows as well as on its $S_{0.5}$ for phosphoenolpyruvate. With physiologic concentrations of alanine, ATP, and phosphoenolpyruvate, in vitro studies predict that the enzyme would be completely inhibited unless it were activated by fructose-1,6-bisphosphate (Flory et al. 1974; Van Berkel et al. 1974). It is reasonable to postulate that the active form of pyruvate kinase in vivo is an enzyme–fructose-1,6-bisphosphate complex. The amount of this complex depends on the concentration of free fructose-1,6-bisphosphate in the cytosol. This in turn depends on the amount of fructose-1,6-bisphosphate bound to other enzymes in liver, such as aldolase, fructose-1,6-bisphosphatase, and phosphofructokinase. Sols and Marco (1970) have calculated that the concentration of free fructose-1,6-bisphosphate is very small compared with that which is bound. Only 10% or less of the total fructose-1,6-bisphosphate, at a concentration of 20–50 $\mu M$, is available to pyruvate kinase. Any change in the affinity of pyruvate kinase for fructose-1,6-bisphosphate, or in the binding of the compound to other enzymes, would have a great influence on the activity of pyruvate kinase. Evidence that changes in the level of fructose-1,6-bisphosphate can affect pyruvate kinase activity was shown in hepatocytes incubated with dihydroxyacetone (Claus et al. 1979). In this case, flux through pyruvate kinase was increased and the increase in flux correlated with an elevation in fructose-1,6-bisphosphate levels, but not with phosphoenolpyruvate levels, which were unchanged. Pyruvate kinase was activated and its affinity for phosphoenolpyruvate was increased sevenfold in hepatocyte extracts. Precipitation of the enzyme from extracts with ammonium sulfate removed fructose-1,6-bisphosphate and activation was no longer observed. Thus, under these conditions, fructose-1,6-bisphosphate controls the activity of and flux through pyruvate kinase (Claus et al. 1979).

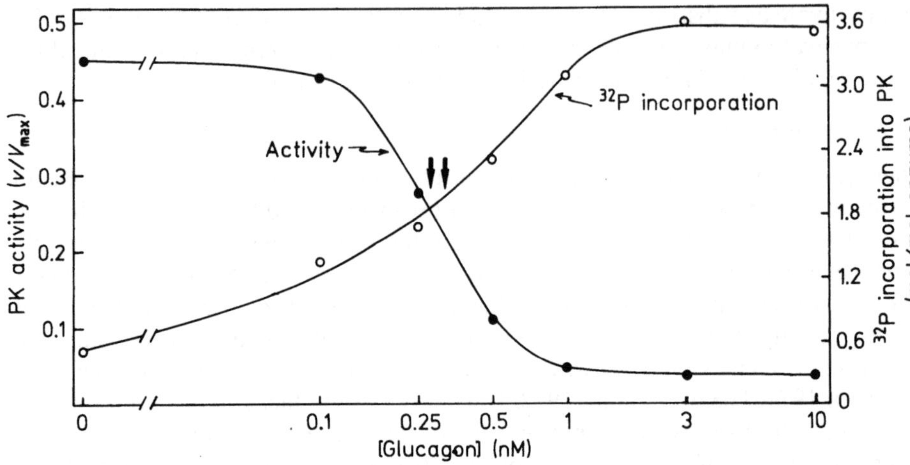


**Fig. 2.** Effect of glucagon concentration on $^{32}$P incorporation into and activity of pyruvate kinase (PK) in isolated hepatocytes. The concentration of glucagon necessary for half-maximal stimulation of $^{32}$P incorporation (0.3 n*M*) and half-maximal inhibition of activity (0.3 n*M*) are given by the *arrows*. Adapted from CLAUS et al. (1980)

TAUNTON et al. (1972, 1974) first reported that glucagon could acutely affect the activity of pyruvate kinase. They showed that injection of glucagon into the portal vein of rats caused a rapid decrease in activity. A number of laboratories now have confirmed this observation. The addition of glucagon to the isolated perfused rat liver (BLAIR et al. 1976) or to isolated hepatocytes (FRIEDRICHS 1976; VAN BERKEL et al. 1976, 1977a, b; RIOU et al. 1976; PILKIS et al. 1976a, b; FELÍU et al. 1976; FOSTER and BLAIR 1978; ISHIBASHI and COTTAM 1978; CHAN and EXTON 1978; GARRISON et al. 1979; CLAUS et al. 1979, 1980) leads to an apparent decrease in the affinity of the enzyme for phosphoenolpyruvate and an increase in the Hill coefficient for phosphoenolpyruvate (FELÍU et al. 1976; VAN BERKEL 1977a). Glucagon treatment also made the enzyme more sensitive to inhibition by ATP and alanine (FELÍU et al. 1976; RIOU et al. 1976; VAN BERKEL et al. 1977a) and less sensitive to activation by fructose-1,6-bisphosphate (RIOU et al. 1976; VAN BERKEL et al. 1977a). RIOU et al. (1978) obtained direct evidence to support the hypothesis that glucagon stimulates the phosphorylation of this enzyme by demonstrating that the hormone stimulates $^{32}$P incorporation into the enzyme in vivo. Glucagon administration increased the number of mol phosphate incorporated per mol enzyme from 0.5 to 1.4. This increase in enzyme-bound phosphate was associated with an inhibition of pyruvate kinase activity and an increase in cyclic AMP. Incubation of rat liver slices (LJUNGSTRÖM and EKMAN 1977) or isolated hepatocytes (ISHIBASHI and COTTAM 1978; GARRISON and BORLAND 1978; GARRISON et al. 1979; CLAUS et al. 1980; STEINER et al. 1980) with $^{32}P_i$ and a high concentration of glucagon also stimulated the incorporation of labeled phosphate into the enzyme by 3- to 6-fold. In isolated hepatocytes, there was a good correlation between the inhibition of pyruvate kinase activity by glucagon and the amount of phosphate incorporated into the enzyme (Fig. 2; CLAUS et al. 1980). Half-maximal effects on both parameters were obtained with 0.3 n*M* glucagon. Half-maximal stimulation of pyruyate kinase phosphorylation in liver slices was obtained with 7 n$M$ glucagon (LJUNGSTRÖM and EKMAN 1977).

The phosphorylation state of pyruvate kinase in intact hepatocytes or in hepatocyte extracts may also be influenced by the concentration of the allosteric effectors of the enzyme. FELÍU et al. (1977) suggested an important role for phosphoenolpyruvate in regulating the phosphorylation state of pyruvate kinase since an inverse relationship existed between its concentration and the activation of pyruvate kinase in rat liver in vivo. CLAUS et al. (1979) found that elevated levels of fructose-1,6-bisphosphate in isolated hepatocytes could influence the ability of glucagon to inactivate pyruvate kinase through a phosphorylation mechanism. When hepatocytes from fasted rats were incubated with 5 m$M$ dihydroxyacetone, the intracellular level of fructose-1,6-bisphosphate rose four-fold and the concentration of glucagon required for half-maximal inhibition increased from 0.3 to 0.8 n$M$. The elevated levels of fructose-1,6-bisphosphate did not affect the activation of the cyclic AMP-dependent protein kinase by glucagon, but it did slow the initial rate of inactivation of pyruvate kinase by the hormone. Elevation of neither alanine nor phosphoenolpyruvate levels altered the ability of glucagon to inactivate pyruvate kinase. However, FOSTER and BLAIR (1978) reported that the inactivation of pyruvate kinase by physiologic glucagon concentrations was enhanced in the presence of alanine. In these studies, the effect of alanine was observed in the presence of dihydroxyacetone, which raises fructose-1,6-bisphosphate levels, and their results may be related to the observation that alanine relieves the effect of fructose-1,6-bisphosphate on the inactivation of the enzyme in vitro (FELÍU et al. 1977; EL-MAGHRABI et al. 1980). The inactivation of rat liver pyruvate kinase in hepatocyte extracts by submaximal concentration of cyclic AMP can also be suppressed by the additions of physiologic concentrations of fructose-1,6-bisphosphate or phosphoenolpyruvate (FELÍU et al. 1977; PILKIS et al. 1978c).

LJUNGSTRÖM et al. (1974) first demonstrated that the purified rat L-isozyme could be phosphorylated by ATP $^{32}$P-$\gamma$ in a reaction catalyzed by cyclic AMP-dependent protein kinase. The activity of the enzyme was decreased by phosphorylation, especially at low phosphoenolpyruvate concentrations. RIOU et al. (1976), PILKIS et al. (1978a), and CLAUS et al. (1979) have confirmed these observations. Under the conditions employed in our laboratory, phosphorylation results in an increase in the $S_{0.5}$ for phosphoenolpyruvate from 0.6 to 1.2 m$M$. Similar results have been obtain by LJUNGSTRÖM et al. (1974, 1976) with the pig liver enzyme. Thus, phosphorylation of the enzyme caused a shift of the phosphoenolpyruvate concentration curve to the right and resulted in an increase in the Hill coefficient (Table 1). The effect of phosphorylation was overcome by saturating concentrations of phosphoenolpyruvate or by fructose-1,6-bisphosphate (LJUNGSTRÖM et al. 1974, 1976; TITANJI et al. 1976; RIOU et al. 1976). Hydrogen ions also overcame the inhibition due to phosphorylation (LJUNGSTRÖM et al. 1976). This is consistent with the observation that fructose-1,6-bisphosphate promoted binding of phosphoenolpyruvate only at high pH, indicating that dissociation of a proton is required for allosteric control (ROZENGURT et al. 1969). Thus, no deviation from Michaelis–Menten kinetics could be demonstrated at pH 6.9, using either the phosphorylated or nonphosphorylated enzyme in the presence or the absence of fructose-1,6-bisphosphate (LJUNGSTRÖM et al. 1976). In contrast, ATP and alanine had

**Table 1.** Effect of phosphorylation on the kinetic properties of the L-type pyruvate kinase

| | L-type | Phosphorylated L-type |
|---|---|---|
| Kinetics with regard to phosphoenylpyruvate (PEP) ($S_{0.5}$) | Sigmoidal (0.6 m$M$) | Sigmoidal (1.2 m$M$) |
| Hill coefficient for PEP | 2.0 | 2.7–2.9 |
| Activation by fructose-1,6-bisphosphate ($K_a$) | 0.04 μ$M$ | 0.3 μ$M$ |
| Inhibition by ATP ($K_{i, apparent}$) | 3.5 m$M$ | 0.8 m$M$ |
| Inhibition by alanine ($K_{i, apparent}$) | 0.6 m$M$ | 0.3 m$M$ |

greater effects on the phosphorylated than on the nonphosphorylated protein (Table 1; LJUNGSTRÖM et al. 1976; RIOU et al. 1976). Phosphorylation of the pig or rat liver enzyme reduced activation by fructose-1,6-bisphosphate (Table 1; LJUNGSTRÖM et al. 1976; RIOU et al. 1976). BERGLUND et al. (1977) reported that alanine increased the rate of phosphorylation of purified pig liver pyruvate kinase by a partially purified preparation of cyclic AMP-dependent protein kinase. Fructose-1,6-bisphosphate and phosphoenolpyruvate had no effect on the phosphorylation.

Recently, EL-MAGHRABI et al. (1980) studied the influence of allosteric effectors on the phosphorylation of purified rat liver pyruvate kinase by a homogeneous preparation of the catalytic subunit of cyclic AMP-dependent protein kinase. When pyruvate kinase that contained less than 1 mol phosphate per mol enzyme was used as substrate, 1–3 μ$M$ fructose-1,6-bisphosphate inhibited the initial rate of phosphorylation by more than 80%. Phosphoenolpyruvate also inhibited the initial rate of phosphorylation, but was not as effective as fructose-1,6-bisphosphate. Alanine had only a slight stimulatory effect on the initial rate of phosphorylation, but it relieved the inhibition by physiologic concentrations of fructose-1,6-bisphosphate and phosphoenolpyruvate. When pyruvate kinase contained 3 mol phosphate per mol enzyme, the influence of allosteric effectors on the phosphorylation of the enzyme was less pronounced; 10- to 100-fold higher concentrations of phosphoenolpyruvate and fructose-1,6-bisphosphate, respectively, were necessary to elicit inhibition similar to the degree seen with the dephosphorylated enzyme. These results are in good agreement with those obtained using hepatocyte extracts (FELÍU et al. 1977; PILKIS et al. 1978c). They suggest that pyruvate kinase is a better substrate for the cyclic AMP-dependent protein kinase when pyruvate kinase is in the inactive form than when it is in the active form.

The kinetic data suggest that enzyme phosphorylation shifts the equilibrium between the active and inactive forms of the enzyme to the inactive form, with the result that binding of substrate and the activator fructose-1,6-bisphosphate is decreased and binding of the inhibitors ATP and alanine is enhanced. EL-MAGHRABI et al. (1981) investigated the effect of phosphorylation on the binding of fructose-1,6-bisphosphate to purified rat liver pyruvate kinase. They found that phosphorylation reduced the $S_{0.5}$ for fructose-1,6-bisphosphate binding by 50%, but had no effect on the cooperativity of binding (Hill coefficient $\eta_H = 2$) (Fig. 3). These results suggest that enzyme phosphorylation caused a conformational change in the enzyme that results in alterations in the interaction between ligand and enzyme. In

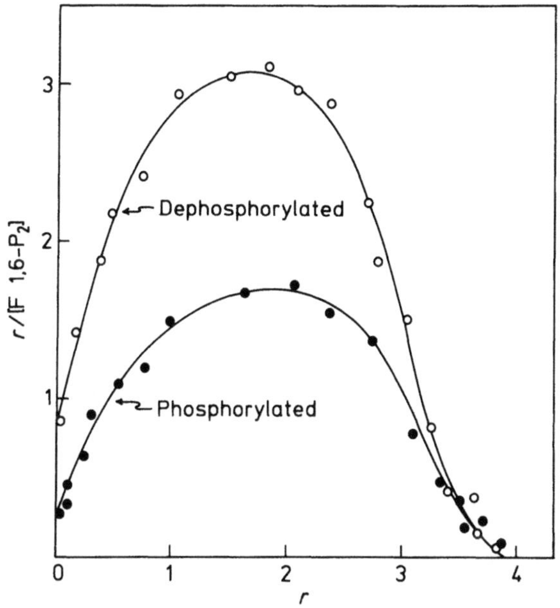


**Fig. 3.** Binding of fructose-1,6-bisphosphate to phosphorylated and dephosphorylated pyruvate kinase. The data are expressed as Scatchard plots. [F-1,6-$P_2$] is the free fructose-1,6-bisphosphate concentration ($\mu M$), and $r$ is the number of mol F-1,6-$P_2$ per mol enzyme. EL-MAGHRABI et al. (1981)

order to investigate this possibility, EL-MAGHRABI et al. (1981) used the technique of ultraviolet difference spectroscopy. Addition of submaximal concentrations of phosphoenolpyruvate or fructose-1,6-bisphosphate induced conformational changes that resulted in perturbations in tryptophan residues of the enzyme. Higher concentrations of fructose-1,6-bisphosphate and phosphoenolpyruvate were necessary to elicit conformational changes with the phosphorylated enzyme than with the nonphosphorylated enzyme (Fig. 4). These data indicate that phosphorylation results in a decrease in the ability of the enzyme to change its conformation in response to substrate and fructose-1,6-bisphosphate.

There have been several reports that attempt to correlate the amount of phosphate incorporated into pyruvate kinase with changes in activity of the enzyme. LJUNGSTRÖM et al. (1974) found that the maximum amount of labeled phosphate that could be incorporated into the pig liver enzyme was 4 mol per mol enzyme. Subsequently, they reported that incorporation of 1.5 mol labeled phosphate per mol enzyme produced only a 25% inhibition of enzyme activity (LJUNGSTRÖM et al. 1976). TITANJI et al. (1976) found that the incorporation by the cyclic AMP-dependent protein kinase of 1.5–2 mol labeled phosphate per mol enzyme was sufficient to inhibit activity of the rat liver enzyme by over 90%. When rat liver slices were incubated with $^{32}P_i$ and glucagon, the incorporation of 3 mol labeled phosphate per mol enzyme caused only a 50% reduction in enzyme activity (LJUNGSTRÖM and EKMAN 1977). Injection of $^{32}P_i$ and glucagon into a rat in vivo caused a 30% reduction in enzyme activity and the incorporation of 1.4 mol labeled phosphate per mol enzyme (RIOU et al. 1978). These different results may be

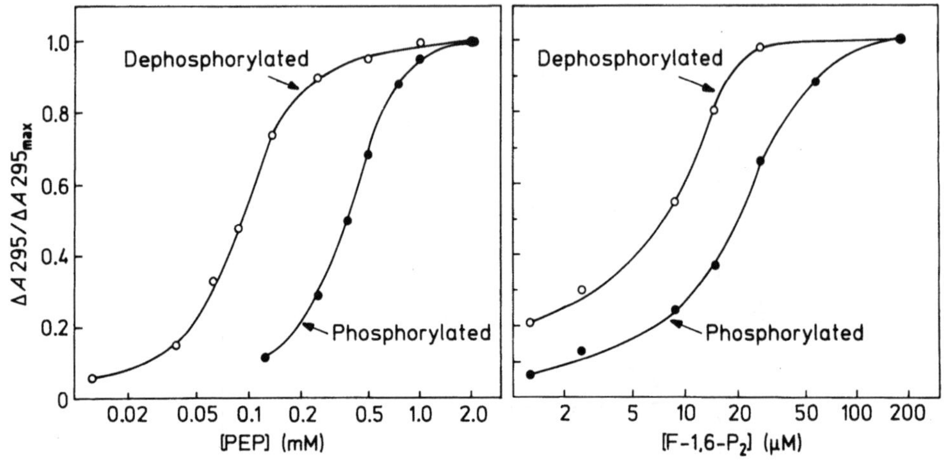


**Fig. 4.** Effect of pyruvate kinase phosphorylation on phosphoenolpyruvate (PEP)- and fructose-1,6-bisphosphate-(F-1,6-$P_2$)-induced absorption differences at 295 nm. $\Delta A\,295_{max}$ represents the maximal change in absorbance induced by 4 m*M* phosphoenolpyruvate or 200 μ*M* fructose-1,6-bisphosphate. EL-MAGHRABI et al. (1981)

due to the fact that only the amount of radioactive phosphate incorporation was determined while the amount of endogenous phosphate in these various preparations was unknown. Recently, EL-MAGHRABI et al. (1980) attempted to correlate changes in activity of the rat liver enzyme with the actual amount of phosphate incorporated into the enzyme by the cyclic AMP-dependent protein kinase. Changes from 90% to 40% of maximal activity occurred when the amount of phosphate increased from 1.5 to 3 mol per mol enzyme. Since there was about 1 mol phosphate per mol enzyme in a site whose phosphorylation was not catalyzed by cyclic AMP-dependent protein kinase (see EL-MAGHRABI et al. 1980 for details), the incorporation by the cyclic AMP-dependent protein kinase of 2 mol phosphate per mol enzyme resulted in essentially maximal inhibition of enzyme activity.

The site of phosphorylation on the enzyme has been determined. HUMBLE et al. (1975) showed that alkali-inactivated pig liver pyruvate kinase and a cyanogen bromide peptide from the same enzyme could be phosphorylated by ATP $^{32}P$-$\gamma$ in the presence of cyclic AMP-dependent protein kinase. They also isolated a peptide from rat liver pyruvate kinase and showed that the minimum structural requirements for phosphorylation were met by the pentapeptide Arg-Arg-Ala-Ser-Val (HJELMQVIST et al. 1974). Evidence from other laboratories suggests that this or very closely related sequences are at the sites of phosphorylation in other proteins that serve as substrates for cyclic AMP-dependent protein kinases (DAILE and CARNEGIE 1974; KEMP 1975; DAILE et al. 1975). The concentration of pentapeptide which gave half-maximal rates of phosphorylation was 0.08 m*M* while that of a hexapeptide (Leu-Arg-Arg-Ser-Val-Ala) was 0.021 m*M* (ZETTERQVIST et al. 1976). On the other hand, PILKIS et al. (1980a) found that the concentration of native pyruvate kinase required for half-maximal rates of phosphorylation was about 17 or 68 μ*M* in terms of its subunit concentration. Thus, the native enzyme was not as good a substrate for the cyclic AMP-dependent protein kinase as is the peptide

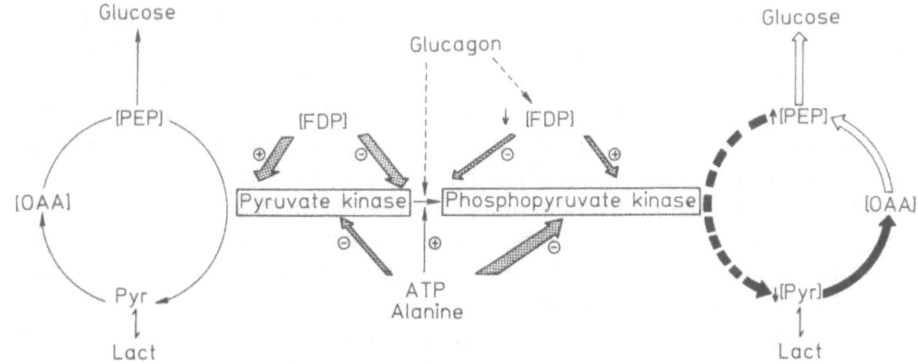


**Fig. 5.** Glucagon effects on pyruvate kinase activity and on flux through the phosphoenolpyruvate–pyruvate substrate cycle. Glucagon stimulates the phosphorylation of pyruvate kinase and lowers the level of fructose-1,6-bisphosphate, thereby decreasing flux through pyruvate kinase. This, along with the glucagon-induced increase in pyruvate carboxylation, leads to stimulation of glucose synthesis. Fructose-1,6-bisphosphate is a more potent activator of the dephosphorylated than of the phosphorylated enzyme. ATP and alanine are more potent inhibitors of the phosphorylated form of pyruvate kinase. Abbreviations as in Fig. 1

analog, a conclusion which differed from that claimed by others (ZETTERQVIST et al. 1976; YEAMAN et al. 1977). In the previous studies, an exact value could not be obtained since the rate of phosphorylation continued to increase, even when micromolar concentrations of pyruvate kinase were used (BERGLUND et al. 1977). PILKIS et al. (1980a) found that enzyme concentrations of up to 100 μ*M* were required to determine $K_m$.

Figure 5 summarizes the effects of glucagon on pyruvate kinase. Addition of the hormone to hepatocytes activates the cyclic AMP-dependent protein kinase which then catalyzes the phosphorylation of pyruvate kinase. Phosphorylation shifts the equilibrium from the active to the inactive form of the enzyme. This form of the enzyme is less active in the presence of low substrate concentrations, is less sensitive to activation by fructose-1,6-bisphosphate, and is more sensitive to inhibition by ATP and alanine. The shift in equilibrium toward the inactive form is aided by the glucagon-induced decrease in fructose-1,6-bisphosphate levels (see Sect. B.II.2). The fall in fructose-1,6-bisphosphate levels not only reduces pyruvate kinase activity directly, but also enhances phosphorylation by making the enzyme a better substrate for cyclic AMP-dependent protein kinase. The combined effects of glucagon on phosphorylation of the enzyme and of fructose-1,6-bisphosphate levels inactive pyruvate kinase and reduce the flux on phosphoenolpyruvate through the enzyme. This allows more of the phosphoenolpyruvate to be converted to glucose. Glucagon also enhances gluconeogenesis by stimulating pyruvate carboxylation as already discussed in Sect. B.I.1.

### b) Phosphoenolpyruvate Carboxykinase

Phosphoenolpyruvate carboxykinase was first discovered in chicken liver mitochondria by UTTER and KURAHASHI (1953). The enzyme has since been found to be present in high concentrations in liver and kidney cortex in all species (UTTER and KOLENBRANDER 1972) where it plays an essential role in gluconeogenesis from three- or four-carbon precursors. Livers of all species studied to date contain both a mitochondrial and a cytosolic form of the enzyme (TILGHMAN et al. 1976). Both forms of the enzyme appear to be a single polypeptide chain of molecular weight ranging from 71,000 daltons for the avian liver enzyme (CHIAO 1976) to possibly 85,000 daltons for the rat liver enzyme (TILGHMAN et al. 1976). However, the two forms of phosphoenolpyruvate carboxykinase have been shown to be distinct proteins on the basis of immunologic and chemical studies (BALLARD and HANSON 1969; BALLARD 1971; DIESTERHAFT et al. 1971).

While there is ample evidence for a dietary and chronic hormonal regulation of phosphoenolpyruvate carboxykinase levels (see TILGHMAN et al. 1976 for review), the evidence that the enzyme is acutely affected by hormones is only indirect. When rat livers were perfused with glucagon or cyclic AMP and the amount of glycolytic intermediates measured, a "crossover" was observed between pyruvate and phosphoenolpyruvate (EXTON and PARK 1969; UI et al. 1973a, b). This crossover was thought to be due to activation of phosphoenolpyruvate carboxykinase. However, these changes appear to be adequately explained by inhibition of pyruvate kinase and activation of pyruvate carboxylation. Attempts to measure a change in the kinetic properties of phosphoenolpyruvate carboxykinase in homogenates of livers treated with glucagon have been unsuccessful. We have also been unable to phosphorylate the purified enzyme with the catalytic subunit of cyclic AMP-dependent protein kinase (T. H. CLAUS 1975, unpublished work). WICKS et al. (1972) also reported that the enzyme was not subject to phosphorylation. Attempts to find metabolic effectors for the enzyme also appear to have been unsuccessful, although it can be postulated that variations in the concentrations of oxaloacetate can regulate the activity of the enzyme. The $S_{0.5}$ for oxaloacetate was originally thought to be too high for the enzyme to participate in gluconeogenesis, but more recent studies have placed the value in the 1–10-$\mu M$ range (BALLARD 1970; WALSH and CHEN 1971; JOMAIN-BAUM et al. 1976), which is about the same as the concentration of oxaloacetate (5–20 $\mu M$) in the cytosol of rat liver (WILLIAMSON et al. 1969b; SIESS et al. 1976). On the other hand, the $S_{0.5}$ for $MnGTP^{2-}$ has been estimated to be 16 $\mu M$ (JOMAIN-BAUM et al. 1976), whereas the GTP concentration in the whole liver is 100–600 $\mu M$ (CHANCE et al. 1965; CLIFFORD et al. 1972). If the nucleotide were evenly distributed in the liver, phosphoenolpyruvate carboxykinase would be nearly saturated with GTP. This would suggest that GTP does not have a regulatory role in phosphoenolpyruvate formation, as was suggested from studies with isolated guinea pig liver mitochondria (ISHIHARAI and KIKUCHI 1968; GARBER and BALLARD 1970).

Another type of control of phosphoenolpyruvate carboxykinase appear to involve metal ions. The enzyme requires two metal ions in order to form phosphoenolpyruvate at maximum rates (HOLTEN and NORDLIE 1965; FOSTER et al. 1967; UTTER and KOLENBRANDER 1972). Magnesium is required in approximately stoichiometric amounts to the nucleotide (GTP or ITP), and micromolar concentrations of a divalent transition metal ion such as $Fe^{2+}$, $Mn^{2+}$, $Co^{2+}$, or $Cd^{2+}$ activate the enzyme (SNOKE et al. 1971). When phosphoenolpyruvate carboxykinase was assayed in rat liver cytosol, the addition of the transition metal ions activated the enzyme 2–3-fold (SNOKE et al. 1971; BENTLE and LARDY 1976; BENTLE et al. 1976). If the cytosol was first incubated with the transition metal ion, even greater effects were observed, and $Fe^{2+}$ was the most effective activator. Evidence has been presented that $Fe^{2+}$ is the natural activator in rat liver cytosol. When rat liver phosphoenolpyruvate carboxykinase was purified to homogeneity, the enzyme lost sensitivity to $Fe^{2+}$, but not to $Mn^{2+}$ stimulation (BENTLE et al. 1976). Addition of rat liver cytosol to the purified enzyme restored the response to $Fe^{2+}$, but not to $Mn^{2+}$ stimulation (BENTLE et al. 1976). Addition of rat liver cytosol to the purified enzyme restored the response to $Fe^{2+}$. This observation prompted the search for and the discovery of a protein that permits $Fe^{2+}$ to activate the purified enzyme 3–4-fold (BENTLE and LARDY 1977). This protein, called phosphoenolpyruvate carboxykinase ferroactivator, has a molecular weight of approximately 100,000 daltons. A subunit molecular weight of 34,600 daltons was obtained by sodium dodecylsulfate electrophoresis. The activity of phosphoenolpyruvate carboxykinase with $Fe^{2+}$, $Mn^{2+}$, $Co^{2+}$, $Cd^{2+}$, $Mg^{2+}$, or $Ca^{2+}$ was not affected by the ferroactivator. These results suggest that the rate of phosphoenolpyruvate synthesis by gluconeogenic tissues may be regulated by the availability of intracellular $Fe^{2+}$ to the ferroactivator and phosphoenolpyruvate carboxykinase. Support for this view comes from studies of the tissue distribution of the ferroactivator and the effects of diabetes and starvation on it (MACDONALD et al. 1978). The highest concentrations of ferroactivator were found in liver, kidney, and erythrocytes, and intermediate levels were found in the heart and pancreas. Except for erythrocytes and heart, the tissue distribution parallels that reported for phosphoenolpyruvate carboxykinase.

## 2. Fructose-6-phosphate–Fructose-1,6-bisphosphate Substrate Cycle

The importance of the fructose-6-phosphate–fructose-1,6-bisphosphate cycle in the regulation of gluconeogenesis from physiologic substrates is less certain than it is for the phosphoenolpyruvate–pyruvate cycle. The rate of gluconeogenesis from three-carbon precursors is not limited directly by reactions in this region of the pathway. There is five times more fructose-1,6-bisphosphatase than phosphofructokinase activity in liver (SCRUTTON and UTTER 1968), so that changes in the flux through phosphofructokinase may not have as large effects on gluconeogenesis as changes in flux through pyruvate kinase. HUE and HERS (1974) have even argued, on the basis of the in vivo concentrations of inhibitors and activators of the enzymes, that little substrate cycling would occur. GARFINKEL et al. (1979) came to the same conclusion using computer modeling techniques. However, phosphofructokinase and fructose-1,6-bisphosphatase activities may regulate gluconeogenesis indirectly by controlling the hepatic level of fructose-1,6-bisphosphate, a potent activator of pyruvate kinase. The addition of glucagon or cyclic AMP to rat hepatocytes or perfused rat liver causes a decrease in the level of fructose-1,6-bisphosphate (WILLIAMSON et al. 1969a; BLAIR et al. 1973; HARRIS 1975; PILKIS et al. 1976b; CLAUS et al. 1979) and thus promotes inactivation of the enzyme, as discussed earlier. The mechanism whereby glucagon lowers fructose-1,6-bisphosphate levels must involve a stimulation of fructose-1,6-bisphosphatase and/or an inhibition of phosphofructokinase activity.

Evidence that both enzymes of this cycle are operative in liver in vivo, and in perfused liver or isolated hepatocytes has been reviewed by KATZ and ROGNSTAD (1976). Effects on this substrate cycle have been studied by following the metabolism of glucose labeled in various positions with tritium and/or $^{14}C$. CLARK et al. (1974a) estimated the rate of substrate cycling in rat liver in vivo by following the metabolism of intraperitoneally injected glucose, $^{3}H$-3, $^{14}C$(ul). They found that the rate of phosphorylation of fructose-6-phosphate was about 60% greater than the in vivo rate of lactate gluconeogenesis in fed rats, but only about 40% of the rate of gluconeogenesis in starved rats. CLARK et al. (1974b) also found that glucagon and cyclic AMP affected the substrate cycle in isolated hepatocytes by inhibiting flux through phosphofructokinase and stimulating flux through fructose-1,6-bisphosphatase. Similar effects were reported by KATZ et al. (1975).

The possibilities for serious quantitative errors in the glucose $^{3}H$, $^{14}C$ method of measuring this substrate cycle in liver have been pointed out (KATZ et al 1975; KATZ and ROGNSTAD 1976). HUE and HERS (1974) have argued that this cycle does not exist in liver and have attributed the production of tritiated water from glucose $^{3}H$-5 to a transaldolase exchange reaction and to the operation of the pentose cycle. In order to avoid these difficulties, ROGNSTAD and KATZ (1976) devised a method that used galactose $^{14}C$-1 instead of tritiated glucose as the tracer substrate. They found, in isolated hepatocytes from starved rats, that the estimated rate of flux through phosphofructokinase ranged from 15% to 40% of the net rate of gluconeogenesis from dihydroxyacetone. Glucagon depressed the rate of flux through phosphofructokinase by as much as 85%, but the decrease in flux accounted for only about 50% of the increase in glucose synthesis. Flux was also depressed by the addition of ethanol or glyoxylate, both of which may cause citrate, an inhibitor of phosphofructokinase, to accumulate. The addition of glycerol instead of dihydroxyacetone also depressed flux through phosphofructokinase. VAN SCHAFTINGEN et al. (1980a) reported that recycling between fructose-6-phosphate and fructose-1,6-bisphosphate occurs in vivo in the livers of fed rats and mice, but not in livers of starved animals unless glucose was first administered. The addition of glucagon abolished the recycling in starved animals that were given glucose. Similar results were reported for hepatocytes from rats (VAN SCHAFTINGEN et al. 1980b). Thus, the weight of evidence supports the existence in rat liver of a fructose-6-phosphate–fructose-1,6-bisphosphate cycle that can be regulated by glucagon and substrates.

a) Phosphofructokinase

Phosphofructokinase is similar to pyruvate kinase in that there appear to be several isozymic forms (TSAI and KEMP 1973). It has been postulated that rabbit heart and muscle contain predominantly a single isozyme (phosphofructokinase A) and the liver and red blood cell predominantly another (phosphofructokinase B). Brain contains a third type, designated phosphofructokinase C.

Liver phosphofructokinase has been purified from pig (MASSEY and DEAL 1973, 1975), rat (DUNAWAY and WEBER 1974; BRAND and SÖLING 1974; PILKIS et al. 1980b; REINHART and LARDY 1980a), sheep (BROCK 1969), chicken (KONO and UYEDA 1971, 1973), and rabbit (RAMAIAH and TEJWANI 1970; KEMP 1971, 1975; MASSEY and DEAL 1975). The rat liver enzyme consists of four apparently identical subunits with a molecular weight of 82,000 daltons (BRAND and SÖLING 1974). The liver enzyme, like that of heart (MANSOUR 1965) and muscle (PAETKAU and LARDY 1967; PAETKAU et al. 1968), tends to form aggregates with molecular weights of the order of several million daltons (BRAND and SÖLING 1974; TRUJILLO and DEAL 1977; REINHART and LARDY 1980b, c). This aggregation is an equilibrium process influenced by enzyme concentration, the presence of allosteric effectors, the oxidation–reduction state of sulfhydryl groups, and temperature (MANSOUR 1972; BLOXHAM and LARDY 1973; RAMAIAH 1974; REINHART and LARDY 1980b, c). The aggregation state of phosphofructokinase may also influence its kinetic behavior. REINHART and LARDY (1980a) observed that the rat liver enzyme gave nonlinear rates of activity when it was diluted, whereas linear rates were obtained when high concentrations of enzyme were used. Evidence that various ligands, including ATP, ADP, and fructose-6-phosphate, affect the quaternary structure of rat liver phosphofructokinase has been reported using enzyme labeled with the fluorescent probe, pyrenebutyric acid (REINHART and LARDY 1980b, c).

Phosphofructokinase from liver exhibits homotropic cooperativity with regard to its substrate, fructose-6-phosphate (BRAND and SÖLING 1974; MASSEY and DEAL 1973, 1975). 5′-Adenosine monophosphate, ADP, and cyclic AMP are allosteric activators of the liver enzyme, and ATP and citrate are allosteric inhibitors (BRAND and SÖLING 1974). KEMP (1971) reported that rabbit liver phosphofructokinase was more sensitive to inhibition by ATP and 2,3-diphosphoglycerate, but less sensitive to inhibition by citrate than the skeletal muscle enzyme. The ATP inhibition of phosphofructokinase decreased markedly as the pH increased from 6.5 to 8.0. Citrate potentiated the inhibitory effect of ATP on hepatic phosphofructokinase from sheep (PASSONEAU and LOWRY 1964) and rats (UNDERWOOD and NEWSHOLME 1965b). Both 5′-AMP and fructose-6-phosphate overcame the increased sensitivity to ATP inhibition induced by the presence of citrate (PASSONEAU and LOWRY 1964; UNDERWOOD and NEWSHOLME 1965b).

Control of phosphofructokinase by enzyme–enzyme interactions has been suggested by UYEDA and LUBY (1974) who reported that chicken liver fructose-1,6-bisphosphatase potentiated ATP inhibition of phosphofructokinase from chicken liver. Other proteins were without effect, but it was not possible to demonstrate that fructose-1,6-bisphosphatase actually interacted with phosphofructokinase. SÖLING et al. (1977) have since shown that fructose-1,6-bisphosphatase removed tightly bound fructose-1,6-bisphosphate from phosphofructokinase. Thus, the inhibition of phosphofructokinase by fructose-1,6-bisphosphatase did not represented enzyme–enzyme interaction, but resulted from removal of fructose-1,6-bisphosphate.

Glucagon has been reported to depress the activity of hepatic phosphofructokinase within minutes after administration to intact rats (TAUNTON et al. 1972, 1974). The effect of glucagon was dose dependent and was independent of changes in protein synthesis. Recently, it has been demonstrated that addition of glucagon to isolated hepatocytes produced an inhibition of phosphofructokinase activity (PILKIS et al. 1979; CASTAÑO et al. 1979; KAGIMOTO and UYEDA 1979; CLAUS et al. 1980; NIETO and CASTAÑO 1980; VAN SCHAFTINGEN et al. 1980b). This inhibition was characterized in crude extracts by a twofold increase in the $S_{0.5}$ for fructose-6-phosphate, but with no change in the activity of the enzyme when assayed under

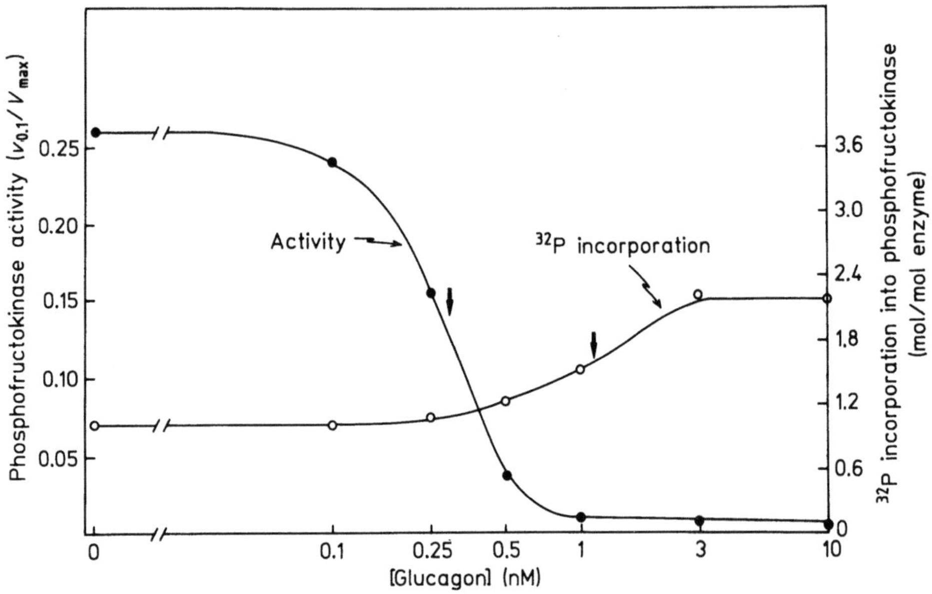


**Fig. 6.** Effect of glucagon concentration on $^{32}$P incorporation into and activity of phosphofructokinase in isolated hepatocytes. The activity of the enzyme is expressed as the activity at 0.1 m*M* fructose-6-phosphate, $v_{0.1}$, divided by the activity at 4 m*M* fructose-6-phosphate plus 1 m*M* AMP, $V_{max}$. The concentration of glucagon necessary for half-maximal stimulation of $^{32}$P incorporation (1 n*M*) and half-maximal inhibition of activity (0.3 n*M*) are given by the *arrows*. Adapted from CLAUS et al. (1980)

$V_{max}$ conditions where enzyme activity is maximal (PILKIS et al. 1979; CASTAÑO et al. 1979; CLAUS et al. 1980; NIETO and CASTAÑO 1980; VAN SCHAFTINGEN et al. 1980b) and by an increased sensitivity of the enzyme to inhibition by ATP (KAGIMOTO and UYEDA 1979). Glucagon stimulates $^{32}$P incorporation into the enzyme in vivo (KAGIMOTO and UYEDA 1979), and the cyclic AMP-dependent protein kinase catalyzes the phosphorylation of purified rat liver phosphofructokinase in vitro (PILKIS et al. 1980b). From these results, it has been postulated that glucagon caused inhibition of the enzyme by stimulating its phosphorylation (CASTAÑO et al. 1979; KAGIMOTO and UYEDA 1979). However, CLAUS et al. (1980) presented several lines of evidence that indicated that phosphorylation was not responsible for the inhibition of phosphofructokinase. First, the increase in phosphorylation in intact cells induced by increasing concentrations of glucagon did not correlate well with the decrease in enzyme activity. In fact, inhibition of the enzyme occurred before any appreciable increase in $^{32}$P incorporation into the enzyme could be detected (Fig. 6). Second, phosphorylation of phosphofructokinase induced by addition of cyclic AMP and MgATP or by addition of MgATP and the catalytic subunit of the cyclic AMP-dependent protein kinase to hepatocyte extracts had no effect on enzyme activity. Third, ammonium sulfate precipitation of the enzyme from extracts of cells incubated with glucagon abolished the hormone effect and shifted the substrate concentration curve to the right. VAN SCHAFTINGEN et al. (1980b) found that partial purification of phosphofructokinase also abolished the glucagon

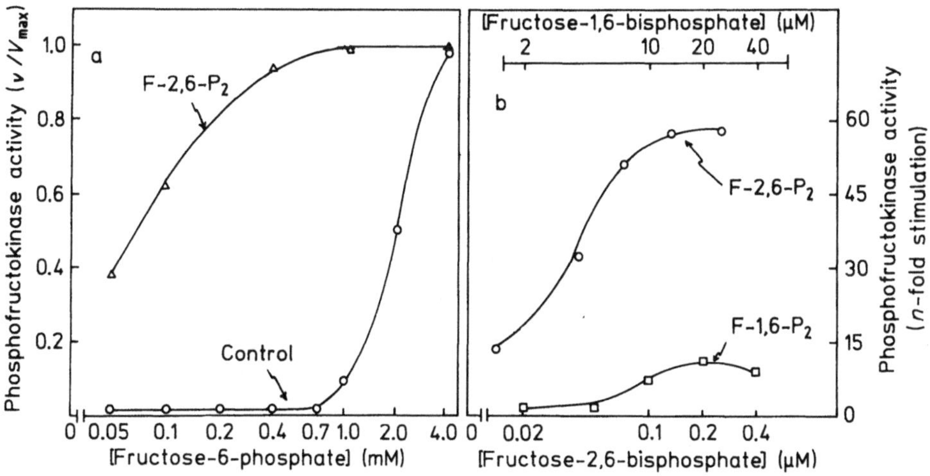


**Fig. 7 a, b.** Effect of fructose-2,6-bisphosphate (F-2,6-$P_2$) and fructose-1,6-bisphosphate (F-1,6-$P_2$) on rat liver phosphofructokinase activity. **a** effect of 0.13 μ*M* fructose-2,6-bisphosphate on the fructose-6-phosphate concentration dependency; **b** concentration dependency of fructose-2,6-bisphosphate on phosphofructokinase activation. Adapted from PILKIS et al. (1981 c)

effect. These results suggested that the inactivation of phosphofructokinase by glucagon was due to a change in the level of an allosteric effector.

Proof that an effector was responsible for the glucagon effect came from mixing experiments (CLAUS et al. 1980). When a phosphofructokinase-free extract from untreated hepatocytes was recombined with ammonium sulfate-treated phosphofructokinase from either untreated or glucagon-treated cells, the substrate concentration curve was shifted to the left and the enzyme had the same kinetic properties as the enzyme in crude extracts of untreated cells. Recombination of a phosphofructokinase-free extract from glucagon-treated cells with phosphofructokinase from either untreated or glucagon-treated cells resulted in an enzyme that had the same properties as the enzyme in crude extracts of glucagon-treated cells. In other words, the hormone effect was conferred on the enzyme by the properties of the phosphofructokinase-free extract from glucagon-treated cells, rather than by the enzyme itself. Identical results have been reported by VAN SCHAFTINGEN et al. (1980 b).

Recently, a number of groups have reported the existence in hepatocyte extracts of an activator of phosphofructokinase (FURUYA and UYEDA 1980 a, b; VAN SCHAFTINGEN and HERS 1980; VAN SCHAFTINGEN et al. 1980 a–c; CLAUS et al. 1980, 1981 b; RICHARDS and UYEDA 1980; PILKIS et al. 1981 b, c). This activator, which is acid labile and alkali stable, has been identified as fructose-2,6-bisphosphate (VAN SCHAFTINGEN and HERS 1980; VAN SCHAFTINGEN et al. 1980 c; CLAUS et al. 1981 b; PILKIS et al. 1981 c) and it has been synthesized chemically (VAN SCHAFTINGEN and HERS 1980; PILKIS et al. 1981 c). The extreme acid lability of this compound is due to the presence of the phosphate on the hemiketalic hydroxyl group at C-2. Acid hydrolysis yields equal molar amounts of fructose-6-phosphate and inorganic phosphate (VAN SCHAFTINGEN et al. 1980 c; PILKIS et al. 1981 c).

**Table 2.** Synergistic effects of fructose-2,6-bisphosphate and AMP on rat liver phosphofructokinase activity (PILKIS et al. 1982)

| Additions | $v/V_{max}$ [a] |
|---|---|
| None | 0.014 |
| 50 $\mu M$ AMP | 0.066 |
| 0.01 $\mu M$ fructose-2,6-bisphosphate | 0.074 |
| 50 $\mu M$ AMP + 0.01 $\mu M$ fructose-2,6-bisphosphate | 0.213 (0.126) |

[a] Results are expressed as the activity with 0.4 m$M$ fructose-6-phosphate ($v$) divided by that with 4 m$M$ fructose-6-phosphate plus 1 m$M$ AMP ($V_{max}$). Number in parenthesis is the value expected from the sum of the effects of AMP and fructose-2,6-bisphosphate

**Table 3.** Level of fructose-2,6-bisphosphate[a] in hepatocytes from fed rats and in whole liver from fed and starved rats

| Hepatocytes | | Whole liver | |
|---|---|---|---|
| Basal | 10 n$M$ Glucagon | Fed | 18-h Starved |
| Fructose-2,6-bisphosphate (nmol/g liver) | | | |
| 11±3 (4) | 1±0.3 (4) | 8±1 (3) | 0.3±0.03 (3) |

[a] Amount of fructose-2,6-bisphosphate was determined as described by PILKIS et al. (1982). Numbers in parentheses are the numbers of preparation tested

Fructose-2,6-bisphosphate is a potent allosteric activator of phosphofructokinase with a $K_a$ of 0.05 $\mu M$, and it is 50–100 times more effective than fructose-1,6-bisphosphate in activating the enzyme (Fig. 7; VAN SCHAFTINGEN and HERS 1980; VAN SCHAFTINGEN et al. 1980c; PILKIS et al. 1981c). Fructose-2,6-bisphosphate also overcomes ATP inhibition of the enzyme and acts synergistically with AMP to activate the enzyme (Table 2; FURUYA and UYEDA 1980a, b; PILKIS et al. 1981d, 1982). Addition of a maximal concentration of glucagon to hepatocytes results in a 90% reduction in the level of the compound (VAN SCHAFTINGEN et al. 1980b; CLAUS et al. 1981b; PILKIS et al. 1981d, 1982; HUE et al. 1981), and starvation also dramatically lowers the level of this activator (Table 3; PILKIS et al. 1982). These reductions probably account for the decrease in flux through phosphofructokinase that is noted in starvation and upon glucagon addition (ROGNSTAD and KATZ 1976; VAN SCHAFTINGEN et al. 1980b).

## b) Fructose-1,6-bisphosphatase

The highest activities of fructose-1,6-bisphosphatase are found in liver and kidney, where high rates of gluconeogenesis occur (SCRUTTON and UTTER 1968). Most studies on the regulatory and kinetic properties of hepatic fructose-1,6-bisphosphatase have been done with the rabbit and rat enzyme (see HORECKER et al. 1975 and PILKIS et al. 1978 b for reviews). This cytosolic enzyme is subject to a multiplicity of controls, including allosteric inhibition by AMP (TAKETA and POGELL 1963, 1965; MENDICINO and VARSARKELY 1963; UNDERWOOD and NEWSHOLME 1965 a; ROSENBERG et al. 1973; DATTA et al. 1974; NIMMO and TIPTON 1975; TEJWANI et al. 1976; RIOU et al. 1977), substrate inhibition by fructose-1,6-bisphosphate (TAKETA and POGELL 1965; NAKASHIMA et al. 1970), and activation by histidine (POGELL et al. 1968), fatty acids (CARLSON et al. 1973), and various chelators. With the exception of AMP, it is unlikely that these effectors have any physiologic role in regulating enzyme activity. Many of the early studies were done with enzyme that had a pH optimum of about 9. TRANIELLO et al. (1971) showed that this was not the native enzyme, but that it arose from proteolytic cleavage of a small peptide (molecular weight 6,000 daltons) from the $NH_2$ terminus of the enzyme subunit during purification. The native enzyme had a pH optimum of less than 8, was more sensitive to AMP inhibition, and had a subunit molecular weight of 35,000 instead of 29,000 daltons. The nature of the proteolytic cleavage has been studied extensively and these studies have been reviewed recently (PILKIS et al. 1978 b). Recent evidence suggests that the subunit molecular weight may be even greater than 35,000 daltons. PILKIS et al. (1980 a) and HUMBLE et al. (1979) have shown that proteolytic digestion of the purified rat liver enzymes also removes a peptide from the COOH terminus that contains the site that is phosphorylated by the cyclic AMP-dependent protein kinase. HOSEY and MARCUS (1981) confirmed this finding. They also reported that the molecular weight of the intact subunit was 41,000 daltons. Trypsin treatment reduced the molecular weight to 36,000–37,000 daltons, and the enzyme could no longer be phosphorylated. CLAUS et al. (1981 a) also found that the molecular weight of the intact subunit was 41,000 daltons while KIDO et al. (1980) recently reported a molecular weight of 39,000 daltons for the subunit of rabbit liver fructose-1,6-bisphosphatase.

POGELL et al. (1968) reported that rabbit liver fructose-1,6-bisphosphatase could be specifically activated by phosphofructokinase. However, SÖLING and KLEINEKE (1976) showed that serum albumin was more effective in this regard than a homogeneous preparation of rat liver phosphofructokinase, and they concluded that the activation by proteins was a nonspecific effect. Thus, it does not appear likely that enzyme–enzyme interaction between fructose-1,6-bisphosphatase and phosphofructokinase plays a significant role in the physiologic regulation of fructose-1,6-bisphosphatase activity.

TAUNTON et al. (1972, 1974) reported that portal vein injection of glucagon into rats increased the activity of fructose-1,6-bisphosphatase by 50%–100% in liver homogenates. The effect of glucagon was observed within 2 min and was maximal 5–10 min after injection of the hormone. Infusion of cyclic AMP produced changes similar to those of glucagon. CHATTERJEE and DATTA (1978) reported that the intravenous administration of glucagon to mice stimulated fructose-1,6-bisphosphatase activity by about 100% within 15 min. Similar results were observed on administration of cyclic AMP. Adrenalectomy largely abolished the stimulation by glucagon, but had no effect on the stimulation by cyclic AMP. The mechanism whereby glucagon affects fructose-1,6-bisphosphatase is unknown, but it has been proposed that glucagon affects enzyme activity by a phosphorylation mechanism (TAUNTON et al. 1974; RIOU et al. 1977; CHATTERJEE and DATTA 1978; PILKIS et al.

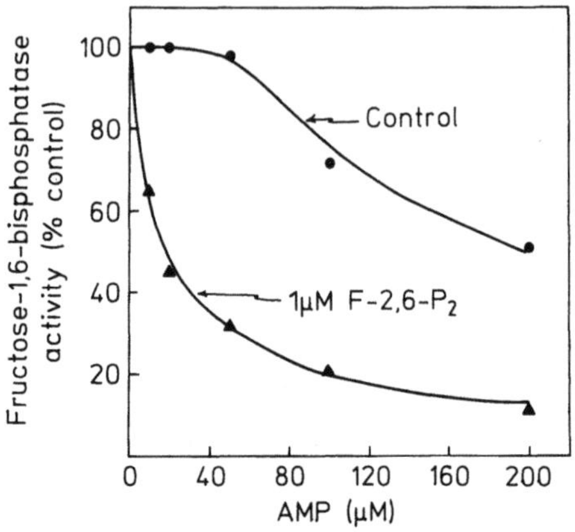


**Fig. 8.** Effect of 1 μ*M* fructose-2,6-bisphosphate (F-2,6-$P_2$) on the inhibition of fructose-1,6-bisphosphatase by AMP. PILKIS et al. (1981 b)

1978 c). Consistent with this idea is the observation that $^{32}$P can be incorporated into the rat liver enzyme in vivo (RIOU et al. 1977). RIOU et al. (1977) did not observe an effect of glucagon, but recently CLAUS et al. (1981 a) showed that the hormone stimulated $^{32}$P incorporation into the enzyme in isolated hepatocytes. However, they did not observe any glucagon-induced activity change. Furthermore, the concentration of glucagon needed for half-maximal stimulation of $^{32}$P incorporation (1 n*M*) was more than three times that needed for half-maximal stimulation of gluconeogenesis (0.3 n*M*). This same high concentration of glucagon was also needed for half-maximal stimulation of phosphofructokinase phosphorylation (CLAUS et al. 1980). In this case, it has been shown that phosphorylation of phosphofructokinase is not responsible for the inhibition of enzyme activity (CLAUS et al. 1980). These results suggest that fructose-1,6-bisphosphatase activity, like that of phosphofructokinase, may be regulated by glucagon-induced changes in one or more allosteric effectors. However, it is also possible that proteolysis may have obscured the real extent of phosphorylation of both enzymes by glucagon.

Since many of the effectors of phosphofructokinase also affect the activity of fructose-1,6-bisphosphatase in a reciprocal manner, PILKIS et al. (1981 a) tested fructose-2,6-bisphosphate for its ability to modulate fructose-1,6-bisphosphatase activity. They found it to be a potent competetive inhibitor, with a $K_i$ of about 0.5 μ*M*. Fructose-2,6-bisphosphate also enhanced inhibition of the enzyme by the allosteric inhibitor AMP (Fig. 8; PILKIS et al. 1981 a, d, 1982). The inhibitory effect of AMP was also potentiated by fructose-1,6-bisphosphate (PONTREMOLI et al. 1968; PILKIS et al. 1981 a, 1982), but much greater concentrations were needed than those required for fructose-2,6-bisphosphate (PILKIS et al. 1981 a, 1982). PILKIS et al. (1981 a, 1982) showed that AMP, fructose-1,6-bisphosphate, and fructose-2,6-bisphosphate each induced a conformational change in the enzyme as revealed by ultraviolet difference spectroscopy, and that both fructose-1,6-bisphosphate and fructose-2,6-bisphosphate potentiated the conformational change brought about by AMP.

While the role of glucagon-induced phosphorylation in modulating the activity of fructose-1,6-bisphosphatase is uncertain, the in vitro phosphorylation of the rat

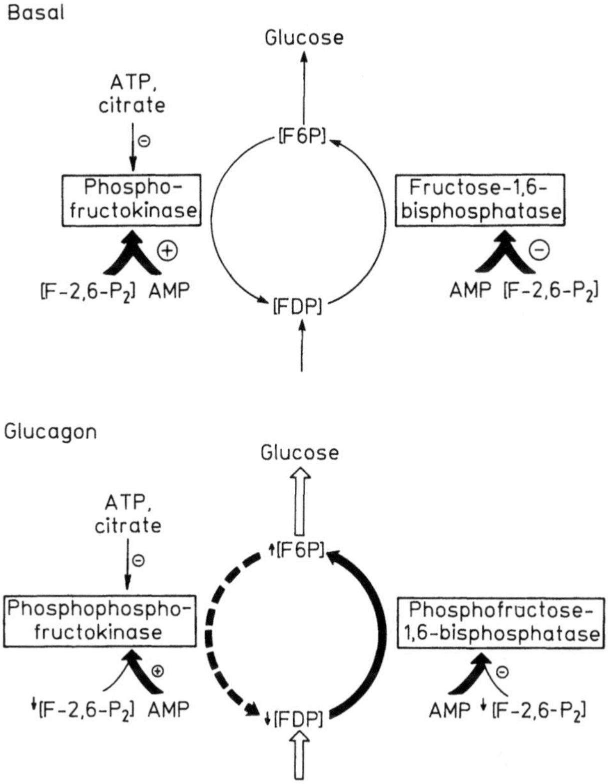


**Fig. 9.** Glucagon effects on phosphofructokinase and fructose-1,6-bisphosphatase activity and on flux through the fructose-6-phosphate–fructose-1,6-bisphosphate cycle. Glucagon stimulates the phosphorylation of both enzymes and lowers the level of fructose-2,6-bisphosphate. The decreased level of fructose-2,6-bisphosphate reduces the synergism with AMP for activation of phosphofructokinase and inhibition of fructose-1,6-bisphosphatase. This leads to a decrease in flux through phosphofructokinase and an increase in flux through fructose-1,6-bisphosphatase and a stimulation of glucose synthesis. The role of phosphorylation of the two enzymes is uncertain

liver enzyme by the cyclic AMP-dependent protein kinase has been characterized. RIOU et al. (1977) showed that 4 mol phosphate were incorporated per mol enzyme. MARCUS and HOSEY (1980) confirmed these results. PILKIS et al. (1980a) compared the phosphorylation of fructose-1,6-bisphosphatase with that of pyruvate kinase. The pH optimum for phosphorylation of fructose-1,6-bisphosphatase was about pH 7 whereas the rate of phosphorylation of pyruvate kinase by the cyclic AMP-dependent protein kinase was still increasing at pH 9. Neither fructose-1,6-bisphosphate nor AMP had any effect on the initial rate of phosphorylation of fructose-1,6-bisphosphatase. In contrast, the allosteric effectors of pyruvate kinase, fructose-1,6-bisphosphate and alanine, inhibited and stimulated, respectively, the rate of phosporylation of that enzyme (EL-MAGHRABI et al. 1980). Fructose-1,6-bisphosphatase was not as good a substrate for the cyclic AMP-dependent protein kinase as pyruvate kinase. The $K_m$ for fructose-1,6-bisphosphatase was threefold greater (58 $\mu M$) than that for pyruvate kinase (17 $\mu M$) while the maximal rate of phosphorylation was about one-third that for pyruvate kinase. These results can be explained by the fact that pyruvate kinase contains two arginine residues on the $NH_2$ terminal side of the phosphorylated serine (HJELMQVIST et al. 1974; EDLUND et al. 1975) whereas fructose-1,6-bisphosphatase contains only one. The sequence around the phosphorylated serine in rat liver fructose-1,6-bisphosphatase has been reported to be either Ser-Arg-Pro-Ser(P)-Leu-Pro-Leu-Pro (PILKIS et al. 1980a) or Ser-Arg-Tyr-Ser(P)-Leu-Pro-Leu-Pro (HUMBLE et al. 1979).

The finding that AMP and fructose-2,6-bisphosphate act synergistically on both phosphofructokinase and fructose-1,6-bisphosphatase and that glucagon dramatically lowers the level of fructose-2,6-bisphosphate seems to offer a possible mechanism for regulating this substrate cycle (Fig. 9). In the absence of hormone, fructose-2,6-bisphosphate and AMP would act together to activate phosphofructokinase and to inhibit fructose-1,6-bisphosphatase, thus limiting the rate of gluconeogenesis by allowing substrate cycling to occur. In the presence of glucagon, the fall in fructose-2,6-bisphosphate concentration would relieve the inhibition of fructose-1,6-bisphosphatase and decrease activation of phosphofructokinase. This would promote flux through fructose-1,6-bisphosphatase, inhibit flux through phosphofructokinase, and promote glucose synthesis. ATP and citrate also inhibit phosphofructokinase. However, glucagon has no effect on the level of ATP, citrate, or AMP (CLARK et al. 1974b). Substantiation of this scheme will require additional studies on the regulation of the level of fructose-2,6-bisphosphate and on flux through these enzymes. The glucagon-induced fall in fructose-2,6-bisphosphate levels suggest that either the synthesis or degradation of this compound may be regulated by a cyclic AMP-dependent phosphorylation mechanism.

## III. Other Possible Sites of Glucagon Action

Glucagon may affect reactions in cellular compartments other than the cytosol and the mitochondria. The hyperglycemic effect of glucagon is associated with ion ($Na^+$, $K^+$, $Ca^{2+}$) redistribution across the plasma membrane (FRIEDMANN and PARK 1968; FRIEDMANN and RASMUSSEN 1970; WILLIAMS et al. 1971; FRIEDMANN 1972) and a concomitant hyperpolarization of the liver cell (FRIEDMANN et al. 1971; SOMLYO et al. 1971; FRIEDMANN and DAMBACH 1973; DAMBACH and FRIEDMANN 1974). Blockage of the hormone-induced $Na^+$ and $K^+$ movement by local anesthetics blocked the hormone-induced $Ca^{2+}$ movement, membrane hyperpolarization, and gluconeogenesis (FRIEDMANN et al. 1971; FRIEDMANN and RASMUSSEN 1970; DAMBACH and FRIEDMANN 1974). Replacement of extracellular $Na^+$ ions with choline also inhibited the gluconeogenic effect of glucagon (FRIEDMANN 1972). It is unknown how changes in membrane potential and ion fluxes regulate gluconeogenesis, but it is possible that they are related to mobilization of intracellular $Ca^{2+}$ since the efflux of $Na^+$ and $K^+$ ions follows the release of intracellular $Ca^{2+}$ (FRIEDMANN and PARK 1968; FRIEDMANN and RASMUSSEN 1970; FRIEDMANN 1972). The effect of glucagon on ion fluxes is probably due to an elevation in cyclic AMP since addition of the nucleotide mimics the effect of the hormone on these parameters (FRIEDMANN 1972; DAMBACH and FRIEDMANN 1974).

Administration of glucagon to intact liver systems induces stable changes in nonmitochondrial (BYGRAVE and TRANTER 1978; WALTENBAUGH and FRIEDMANN

1978; TAYLOR et al. 1979; REINHART and BYGRAVE 1981) as well as mitochondrial (see Sect. B.I.1) $Ca^{2+}$-transport systems. The nonmitochondrial system appears to be confined to the endoplasmic reticulum (BYGRAVE and TRANTER 1978; REINHART and BYGRAVE 1981). Activation of both the mitochondrial and nonmitochondrial systems by glucagon appears to be mediated by cyclic AMP (HUGHES and BARRITT 1978; REINHART and BYGRAVE 1981). The physiologic importance of either system to regulation of gluconeogenesis is unknown, but REINHART and BYGRAVE (1981) have suggested that they may represent a compensatory response to glucagon, lowering the previously elevated concentration of cytoplasmic $Ca^{2+}$ (DEWULF and KEPPENS 1976; BLACKMORE et al. 1978, 1979; CHEN et al. 1978) and hence deactivating previously activated $Ca^{2+}$-dependent enzymes.

The role of $Ca^{2+}$ ions in the regulation of gluconeogenesis is unclear. Removal of $Ca^{2+}$ ions from the medium has generally resulted in a decrease in the rate of lactate gluconeogenesis in isolated hepatocytes (ZAHLTEN et al. 1974; TOLBERT and FAIN 1974; ELLIOT 1976; CLAUS and PILKIS 1976; ROGNSTAD 1976; KNEER et al. 1979) and in perfused rat liver (EXTON et al. 1972), but others have reported no effect of removing $Ca^{2+}$ on basal rates of gluconeogenesis (PILKIS et al. 1975; CHAN and EXTON 1978). Likewise, the effect of high concentrations of glucagon on lactate gluconeogenesis has been reported to be reduced in the absence of $Ca^{2+}$ ions (TOLBERT and FAIN 1974; PILKIS et al. 1975; ROGNSTAD 1976, KNEER et al. 1979) or not affected (EXTON et al. 1972; CHAN and EXTON 1978; SUGANO et al. 1980a). ZAHLTEN et al. (1974) reported that glucagon did not stimulate lactate gluconeogenesis in the presence of $Ca^{2+}$ ions, but did in the absence of $Ca^{2+}$ ions. CLAUS and PILKIS (1976) reported that the stimulation of lactate gluconeogenesis by a low concentration of glucagon was more effective in the absence than in the presence of $Ca^{2+}$ ions. Many of these differences may be explained by differences in experimental conditions. KNEER et al. (1979) reported that $Ca^{2+}$ ions were necessary for glucagon to stimulate gluconeogenesis from lactate, glycerol, sorbitol, and xylitol, but not from dihydroxyacetone or fructose. They concluded that in order for stimulation of gluconeogenesis from reduced substrates to occur $Ca^{2+}$ ions were required for the transfer of reducing equivalents between cellular compartments. Stimulation of gluconeogenesis from the oxidized substrates did not require the transfer of reducing equivalents and was thus unaffected by the absence of $Ca^{2+}$ ions. The results of KNEER et al. (1979) suggest that the ability of glucagon to stimulate gluconeogenesis is dependent upon the redox state of the cytosolic NAD/NADH couple which can be regulated by $Ca^{2+}$ ions. However, the effect of $Ca^{2+}$ on redox state is only manifested when hepatocytes are depleted of $Ca^{2+}$, and there is no evidence yet that $Ca^{2+}$ ions play such a role in the regulation of gluconeogenesis by glucagon under physiologic conditions.

Support for a role of the redox state in modulating the response of gluconeogenesis to glucagon was obtained by MÜLLHOFER et al. (1974), CLARK and JARRETT (1978), and SUGANO et al. (1980b). They found that glucagon or dibutyryl cyclic AMP was able to stimulate gluconeogenesis under reduced conditions (lactate as substrate), but not under oxidized conditions (pyruvate as substrate). MÜLLHOFER et al. (1974) postulated that dibutyryl cyclic AMP stimulated gluconeogenesis from lactate by allowing a higher rate of malate efflux from the mitochondria, and that the excess hydrogen equivalents in the cytosol were recycled into the mitochondria by an enhanced flux through the glycerol phosphate cycle. CLARK and JARRETT (1978) found that the ability of glucagon to stimulate gluconeogenesis from lactate or pyruvate correlated with the glucagon-dependent increase in cyclic AMP. They showed that the activity of the plasma membrane-bound cyclic AMP phosphodiesterase was specifically inhibited by NADH. They proposed that the magnitude of the increase in cyclic AMP and of the stimulation of gluconeogenesis by glucagon was controlled by the activity of the phosphodiesterase through changes in the cellular redox state.

While the efficacy of glucagon may be dependent upon the cytoplasmic redox state, glucagon itself has been shown to alter the redox state. WILLIAMSON et al. (1969a) showed that addition of glucagon to isolated livers perfused with lactate resulted in a rapid reduction of the NAD systems in both the cytosolic and mitochondrial spaces. This has been confirmed by PARILLA et al. (1975) and SIESS et al. (1977). VEECH et al. (1975) found that the injection of glucagon in vivo caused a reduction in the cytoplasm [NADPH]: [NADP] ratio within 3 min while the cytoplasm [NADH]: [NAD] ratio was unaffected. It is not known whether these are primary or secondary effects of the hormone, but the redox state of pyridine nucleotides is generally believed to be under the control of substrate pressure and the phosphate potential (VEECH et al. 1970). Therefore, it is possible that changes in the redox state by glucagon are a response to the hormone-induced altered fluxes of metabolic intermediates that are in turn brought about by the cyclic AMP-dependent phosphorylation of key enzymes in the various metabolic pathways.

## C. Summary and Overview

It is the thesis of this chapter that glucagon stimulates gluconeogenesis by increasing the rate of phosphoenolpyruvate production and decreasing the rate of its disposal by pyruvate kinase. The hormone affects these processes by modulating flux through pyruvate carboxylase, pyruvate kinase, phosphoenolpyruvate carboxykinase, fructose-1,6-bisphosphatase, and phosphofructokinase. Glucagon, via cyclic AMP and the cyclic AMP-dependent protein kinase, affects the activity of pyruvate kinase by altering the phosphorylation state of the enzyme. Glucagon also stimulates the phosphorylation of phosphofructokinase and fructose-1,6-bisphosphatase, but the activity of these enzymes appears to be regulated predominantly by the level of fructose-2,6-bisphosphate instead of by phosphorylation. Pyruvate carboxylase and phosphoenolpyruvate carboxykinase are not phosphorylated, but their activities are altered by changes in the level of substrates and other metabolites which affect the enzymes. Alterations in flux through pyruvate carboxylase, pyruvate kinase, and phosphoenolpyruvate carboxykinase lead directly to a crossover between phosphoenolpyruvate and pyruvate, the rate-limiting portion of the pathway. Alterations in flux through fructose-1,6-bisphosphatase and phosphofructokinase, enzymes which are not located in the rate-limiting portion of the pathway, can affect the activity of and flux through pyruvate kinase by altering the level of fructose-1,6-bisphosphate, a potent activator of the enzyme. This metabolite not only affects pyruvate kinase activity directly, but also influences the phosphorylation state of the enzyme. Thus, fructose-1,6-bisphosphate may serve as a link between the fructose-6-phosphate–fructose-1,6-bisphosphate substrate cycle and the phosphoenolpyruvate–pyruvate substrate cycle.

The importance of pyruvate carboxylase to gluconeogenesis is underscored by the observation that biotin deficiency reduced activity of this enzyme by 90% and gluconeogenesis by over 50% (SIESS et al. 1978). The finding that glucagon was unable to stimulate gluconeogenesis in biotin-deficient rats indicates that increased rates of flux through pyruvate carboxylase are also involved in hormonal regulation of gluconeogenesis. Indeed, GARRISON and HAYNES (1976) showed a direct correlation between the rate of mitochondrial $CO_2$ fixation (pyruvate carboxylation) and the increase in gluconeogenesis caused by glucagon addition to isolated hepatocytes. Glucagon affects flux through pyruvate carboxylase indirectly. Glucagon, via cyclic AMP, appears to stimulate the respiratory chain (YAMAZAKI 1975) by activating cytochrome $c_1$ (HALESTRAP 1978c), but the exact mechanism is unknown. Activation of the respiratory chain stimulates proton efflux from the mitochondrial and mitochondrial ATP production. The rise in intramitochondrial pH promotes anion transport by mitochondria. The increases in both pyruvate and ATP levels, and perhaps the decrease in glutamate levels, stimulate pyruvate carboxylation and gluconeogenesis.

Since pyruvate carboxylation and net glucose synthesis are enhanced by glucagon, the hormone must also increase flux through phosphoenolpyruvate carboxykinase, but there is no evidence for any acute effect on the activity of the enzyme. A ferroprotein activator of this enzyme has been identified (BENTLE and LARDY 1977), but no acute effects of hormones on its activity have been found. Flux through phosphoenolpyruvate carboxykinase may be stimulated by increased level of oxaloacetate and/or GTP through the action of hormones to stimulate mitochondrial metabolism.

Evidence supporting the hypothesis that pyruvate kinase is a site of hormone action comes from studies which show a direct correlation between hormonal stimulation of gluconeogenesis and inhibition of pyruvate kinase activity (e.g. FELÍU et al. 1976; PILKIS et al. 1976b; CHAN and EXTON 1978). Evidence for an important role for pyruvate kinase was also obtained by GARRISON and BORLAND (1978) who found that catecholamines did not stimulate gluconeogenesis to the same extent as glucagon, but did stimulate mitochondrial $CO_2$ fixation to the same extent. The difference in the stimulation of gluconeogenesis by these hormones was ascribed to a smaller inhibition of pyruvate kinase activity by catecholamines. Glucagon inhibits flux through pyruvate kinase in intact cells (PILKIS et al. 1976b; ROGNSTAD and KATZ 1977; CLAUS et al. 1979) as well as inhibiting its activity in homogenates of hepatocytes treated with the hormones (BLAIR et al. 1976; FELÍU et al. 1976; PILKIS et al. 1976a, b; RIOU et al. 1976; VAN BERKEL et al. 1976, 1977a, b). The glucagon-induced changes in enzyme activity are identical to those induced by phosphorylation of the enzyme in vitro (RIOU et al. 1978; CLAUS et al. 1979). The addition of glucagon to perfused liver or hepatocytes also lowers the level of fructose-1,6-bisphosphate (BLAIR et al. 1973; PILKIS et al. 1976b; CLAUS et al. 1979). Although there is no direct evidence that lower levels of fructose-1,6-bisphosphate contribute to the hormone-induced inhibition of pyruvate kinase, it is known that this metabolite can control the activity of the enzyme in the cell under certain cir-

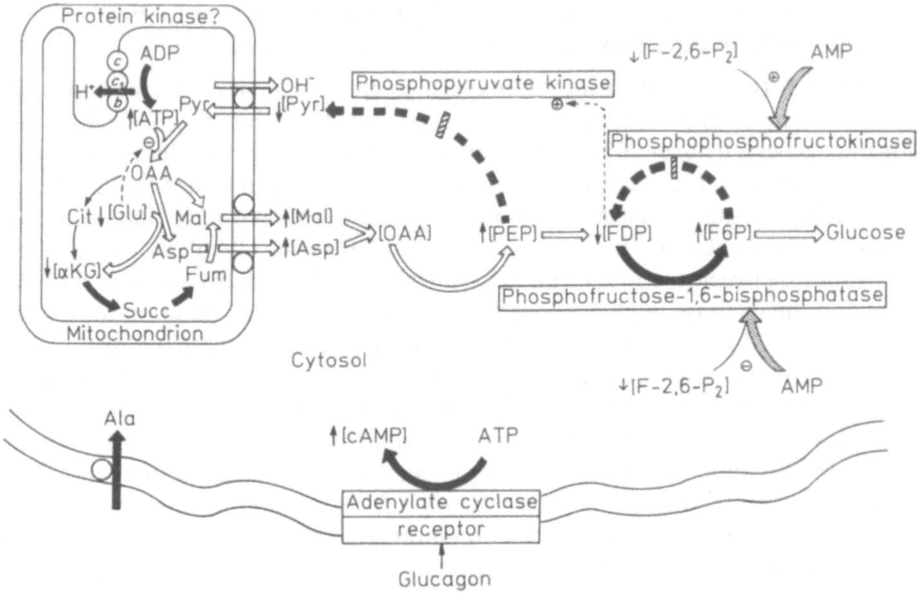


**Fig. 10.** Changes in the hepatic gluconeogenic pathway in response to glucagon. *Wavy lines* represent the plasma membrane. The *rectangle* represents the mitochondrion. *Thick open arrows* represent increased flow of substrates. *Thick full arrows* represent sites of glucagon action to accelerate a reaction, while *thick broken arrows* represent sites of glucagon action to inhibit a reaction. The *stippled arrows* represent the effect of effectors on enzyme activity. The symbols, *b*, *c*, and $c_1$ represent cytochromes; the other abbreviations are the same as in Fig. 1. For the sake of simplicity, the activation by cyclic AMP of the protein kinase reaction and the subsequent phosphorylation of pyruvate kinase, phosphofructokinase, and fructose-1,6-bisphosphase have been omitted

cumstances, and that it can affect the ability of glucagon to phosphorylate pyruvate kinase in intact hepatocytes (CLAUS et al. 1979).

There is now direct evidence that glucagon stimulates the phosphorylation of both phosphofructokinase (KAGIMOTO and UYEDA 1979; CLAUS et al. 1980) and fructose-1,6-bisphosphatase (CLAUS et al. 1981 b). However, the concentration of glucagon that was required for half-maximal stimulation of $^{32}P$ incorporation into both enzymes was greater than that required for half-maximal stimulation of gluconeogenesis. This observation made the phosphorylation of these enzymes as the mechanism for regulating their activities less certain. The discovery of fructose-2,6-bisphosphate as an activator of phosphofructokinase (VAN SCHAFTINGEN et al. 1980c; PILKIS et al. 1981 a) and an inhibitor of fructose-1,6-bisphosphatase (PILKIS et al. 1981 d) appears to provide a mechanism for the reciprocal regulation of these two enzymes by glucagon. Glucagon dramatically lowers the level of this effector (VAN SCHAFTINGEN et al. 1980 b; PILKIS et al. 1981 d, 1982; CLAUS et al. 1981 b), and this would lead to activation of fructose-1,6-bisphosphatase and inhibition of phosphofructokinase. The changes in activity of these two enzymes would result in the decrease in flux through phosphofructokinase and the increase in flux through fructose-1,6-bisphosphatase that is observed when glucagon is added to hepatocytes (ROGNSTAD and KATZ 1976; VAN SCHAFTINGEN et al. 1980b). These effects would also explain the glucagon-induced fall in fructose-1,6-bisphosphate levels. The role of phosphorylation of phosphofructokinase and fructose-1,6-bisphosphatase may be to offer a fine tuning of the enzymes by altering their affinity for fructose-2,6-bisphosphate.

As presented in Fig. 10, the regulation of hepatic gluconeogenesis by glucagon occurs at multiple sites and in several cellular compartments. These effects can be divided into three categories: (1) covalent changes in cytosolic enzymes; (2) changes in allosteric effectors; and (3) effects on mitochondrial energy production. The relative contribution of these categories to the acute effects of glucagon on gluconeogenesis will depend on the nutritional and hormonal status of the animal. For example, in the well-fed case, the levels of phosphofructokinase and pyruvate kinase and flux through these enzymes in intact cells are high. In this situation, pyruvate kinase is a major site of glucagon action since the hormone-induced decrease in flux through pyruvate kinase accounts for a large part of the increase in glucose synthesis (ROGNSTAD and KATZ 1977). The effect of glucagon to inactivate pyruvate kinase is potentiated by the ability of the hormone to lower fructose-1,6-bisphosphate levels by influencing the amount of cycling at the fructose-6-phosphate–fructose-1,6-bisphosphate level. Glucagon also acts to increase phosphoenolpyruvate production by influencing mitochondrial respiration. Thus, all the sites of action depicted in Fig. 10 are affected and glucagon has large effects on the rate of glucose synthesis (CLAUS et al. 1975; CLAUS and PILKIS 1976).

On the other hand, starvation results in an increase in the level of phosphoenolpyruvate carboxykinase and a decrease in pyruvate kinase and phosphofructokinase activity. Both the level of fructose-1,6-bisphosphate and that of fructose-2,6-bisphosphate are also depressed during starvation (Table 3; CLAUS et al. 1979; PILKIS 1981a). As a result of these changes, and perhaps others which are not defined yet, phosphoenolpyruvate production is enhanced, flux though pyruvate kinase and phosphofructokinase is very low, and the rate of glucose synthesis from physiologic substrates is elevated 2- to 3-fold (CLAUS and PILKIS 1976). In this situation, the stimulation of glucose synthesis by glucagon is much smaller than in the fed situation (CLAUS and PILKIS 1976) and the stimulation of mitochondrial metabolism probably makes a large contribution to it. However, even during starvation the inhibition of pyruvate kinase by glucagon contributes to the stimulation of glucose synthesis. Evidence for this comes from studies with hepatocytes from starved rats, where the effect of glucagon and catecholamines on mitochondrial pyruvate carboxylation is of the same magnitude, but the effect of glucagon on both gluconeogenesis and pyruvate kinase activity is substantially greater than that of catecholamines (GARRISON and BORLAND 1979). In the diabetic state, the rate of hepatic gluconeogenesis is enhanced primarily as a result of alterations in the level of various gluconeogenic and glycolytic enzymes (EXTON 1972; PILKIS et al. 1978b) and acute hormone effects on rates of gluconeogenesis in isolated liver systems are not observed. It should also be noted that the accelerated rates of gluconeogenesis seen in vivo during starvation and in diabetes are influenced by an increased substrate supply.

Although our knowledge of how, and at what sites, glucagon affects gluconeogenesis has greatly increased in the last few years, there are still many areas that merit further study. These include: elucidation of the mechanism of hormonal regulation at the level of the mitochondria; quantitation of substrate cycling at both the phosphoenolpyruvate–pyruvate and the fructose-6-phosphate–fructose-1,6-bisphosphate cycles; further characterization of the phosphorylation and hormonal control of phosphofructokinase and fructose-1,6-bisphosphatase, particularly with regard to the synthesis and degradation of fructose-2,6-bisphosphate; and identification, isolation, and characterization of the phosphoprotein phosphatases involved in dephosphorylation of pyruvate kinase, phosphofructokinase, and fructose-1,6-bisphosphatase. These studies should provide further insights into the mechanism of action of glucagon on hepatic gluconeogenesis.

*Acknowledgments.* The work of the authors' laboratories described in this chapter was supported by grant AM-18270 from the U.S. National Institutes of Health. We would like to thank Mrs. Patsy Barrett for her skillful typing of the manuscript.

## References

Adam PAJ, Haynes RC Jr (1969) Control of hepatic mitochondrial $CO_2$ fixation by glucagon, epinephrine, and cortisol. J Biol Chem 244:6444–6450

Babcock DF, Chen J-LJ, Yip BP, Lardy HA (1979) Evidence for mitochondrial localization of the hormone-responsive pool of $Ca^{2+}$ in isolated hepatocytes. J Biol Chem 254:8117–8120

Ballard FJ (1970) Kinetic studies with cytosol and mitochondrial phosphoenolpyruvate carboxykinases. Biochem J 120:809–814

Ballard FJ (1971) Electrophoretic and chromatographic separation of phosphoenolpyruvate carboxykinases. Biochim Biophys Acta 242:470–472

Ballard FJ, Hanson RW (1969) Purification of phosphoenolpyruvate carboxykinase from the cytosol fraction of rat liver and the immunochemical demonstration of differences between this enzyme and the mitochondrial phosphoenolpyruvate carboxykinase. J Biol Chem 244:5625–5630

Barritt GJ, Zander GL, Utter MF (1976) The regulation of pyruvate carboxylase activity in gluconeogenic tissues. In: Hanson RW, Mehlman MA (eds) Gluconeogenesis: its regulation in mammalian species. Wiley and Sons, New York London Sydney Toronto, pp 3–46

Bentle LA, Lardy HA (1976) Interaction of anions and divalent metal ions with phosphoenolpyruvate carboxykinase. J Biol Chem 251:2916–2921

Bentle LA, Lardy HA (1977) P-enolypruvate carboxykinase ferroactivator. Purification and some properties. J Biol Chem 252:1431–1440

Bentle LA, Snoke RE, Lardy HA (1976) A protein factor required for activation of phosphoenolpyruvate carboxykinase by ferrous ions. J Biol Chem 251:2922–2928

Berglund L, Ljungström O, Engström L (1977) Studies on the cyclic 3′: 5′-AMP-stimulated pig liver protein kinase reaction with pyruvate kinase as substrate. J Biol Chem 252:613–619

Bewsher PD, Ashmore J (1966) Ketogenic and lipolytic effects on glucagon on liver. Biochem Biophys Res Commun 24:431–436

Blackmore PF, Brumley FT, Marks JL, Exton JH (1978) Studies on α-adrenergic activation of hepatic glucose output. Relationship between α-adrenergic stimulation of calcium efflux and activation of phosphorylase in isolated rat liver parenchymal cells. J Biol Chem 253:4851–4858

Blackmore PF, Dehaye J-P, Strickland WG, Exton JH (1979) α-Adrenergic mobilization of hepatic mitochondrial calcium. FEBS Lett 100:117–120

Blair JB, Cook DE, Lardy HA (1973) Influence of glucagon on the metabolism of xylitol and dihydroxyacetone in the isolated perfused rat liver. J Biol Chem 248:3601–3607

Blair JB, Cimbala MA, Foster JL, Morgan RA (1976) Hepatic pyruvate kinase. Regulation by glucagon, cyclic adenosine 3′:5′-monophosphate, and insulin in the perfused rat liver. J Biol Chem 251:3756–3762

Bloxham DP, Lardy HA (1973) Phosphofructokinase. In: Boyer PD (ed) The enzymes, 3rd edn, vol VIII. Academic, New York London, pp 229–278

Brand IA, Söling H-D (1974) Rat liver phosphofructokinase. Purification and characterization of its reaction mechanism. J Biol Chem 249:7824–7831

Brock DJH (1969) Purification and properties of sheep liver phosphofructokinase. Biochem J 113:235–242

Bryla J, Harris EJ, Plumb JA (1977) The stimulatory effect of glucagon and dibutyryl cyclic AMP on ureogenesis and gluconeogenesis in relation to the mitochondrial ATP content. FEBS Lett 80:443–448

Bygrave FL, Tranter CJ (1978) The subcellular location, maturation and response to increased plasma glucagon of rutheniumred-insensitive calcium-ion transport in rat liver. Biochem J 174:1021–1030

Carlson CW, Baxter RC, Ulm EH, Pogell BM (1973) Role of oleate in the regulation of "neutral" rabbit liver fructose 1,6-diphosphatase activity. J Biol Chem 248:5555–5561

Castaño JG, Nieto A, Felíu JE (1979) Inactivation of phospofructokinase by glucagon in rat hepatocytes. J Biol Chem 254:5576–5579

Chan TM, Exton JH (1978) Studies on α-adrenergic activation of hepatic glucose output. Studies on α-adrenergic inhibition of hepatic pyruvate kinase and activation of gluconeogenesis. J Biol Chem 253:6396–6400

Chan TM, Bacon CB, Hill SA (1979) Glucagon stimulation of liver mitochondrial $CO_2$ fixation utilizing pyruvate generated inside the mitochondria. J Biol Chem 254:8730–8732

Chance B, Schoener B, Krejci K, Rüssmann W, Wesemann W, Schnitger H, Bücher TH (1965) Kinetics of fluorescence and metabolite changes in rat liver during a cycle of ischemia. Biochem Z 341:325–333

Chatterjee T, Datta AG (1978) Effect of glucagon administration on mice liver fructose-1,6-bisphosphatase. Biochem Biophys Res Commun 84:950–956

Chen J-LJ, Babcock DF, Lardy HA (1978) Norepinephrine, vasopressin, glucagon, and A 23187 induce efflux of calcium from an exchangeable pool in isolated rat hepatocytes. Proc Natl Acad Sci USA 75:2234–2238

Chiao Y-B (1976) The intracellular localization and kinetic properties of chicken liver phosphoenolpyruvate carboxykinase (Abstr 1522). Fed Proc 35:1655

Clark MG, Jarrett IG (1978) Responsiveness to glucagon by isolated rat hepatocytes controlled by the redox state of the cytosolic nicotinamide-adenine dinucleotide couple acting on adenosine 3′:5′-cyclic monophosphate phosphodiesterase. Biochem J 176:805–816

Clark MG, Bloxham DP, Holland PC, Lardy HA (1974a) Estimation of the fructose 1,6-diphosphatase-phosphofructokinase substrate cycle and its relationship to gluconeogenesis in rat liver *in vivo*. J Biol Chem 249:279–290

Clark MG, Kneer NM, Bosch AL, Lardy HA (1974b) The fructose 1,6-diphosphatase-phosphofructokinase substrate cycle. A site of regulation of hepatic gluconeogenesis by glucagon. J Biol Chem 249:5695–5703

Claus TH, Pilkis SJ (1976) Regulation by insulin of gluconeogenesis in isolated rat hepatocytes. Biochim Biophys Acta 421:246–262

Claus TH, Pilkis SJ (1977) Effect of dichloroacetate and glucagon on the incorporation of labeled substrates into glucose and on pyruvate dehydrogenase in hepatocytes from fed and starved rats. Arch Biochem Biophys 182:52–63

Claus TH, Pilkis SJ (1981) Hormonal control of hepatic gluconeogenesis. In: Litwack G (ed) Biochemical actions of hormones, vol VIII. Academic, New York, pp 209–271

Claus TH, Pilkis SJ, Park CR (1975) Characterization of gluconeogenesis in isolated rat hepatocytes. Biochim Biophys Acta 421:246–262

Claus TH, El-Maghrabi MR, Pilkis SJ (1979) Modulation of the phosphorylation state of rat hepatocyte pyruvate kinase by allosteric effectors and insulin. J Biol Chem 254:7855–7864

Claus TH, Schlumpf JR, El-Maghrabi MR, Pilkis J, Pilkis SJ (1980) Mechanism of action of glucagon on hepatocyte phosphofructokinase activity. Proc Natl Acad Sci USA 77:6501–6506
Claus TH, Schlumpf JR, El-Maghrabi MR, McGrane M, Pilkis SJ (1981 a) Glucagon stimulation of fructose 1,6-bisphosphatase phosphorylation in rat hepatocytes. Biochem Biophys Res Commun 101:716–723
Claus TH, Schlumpf JR, Pilkis J, Johnson RA, Pilkis SJ (1981 b) Evidence for a new activator of rat liver phosphofructokinase. Biochem Biophys Res Commun 98:359–366
Clifford AJ, Riumallo JA, Baliga BS, Munro HN, Brown PR (1972) Liver nucleotide metabolism in relation to amino acid supply. Biochim Biophys Acta 277:443–458
Cohen SM, Glynn P, Shulman RG (1981) $^{13}C$ NMR study of gluconeogenesis from labeled alanine in hepatocytes from euthyroid and hyperthyroid rats. Proc Natl Acad Sci USA 78:60–64
Crabb DW, Mapes JP, Boersma RW, Harris RA (1976) Effect of dichloroacetate on carbohydrate and lipid metabolism of isolated hepatocytes. Arch Biochem Biophys 173:658–665
Daile P, Carnegie PR (1974) Peptides from myelin basic protein as substrates for adenosine 3′,5′-cyclic monophosphate-dependent protein kinases. Biochem Biophys Res Commun 61:852–858
Daile P, Carnegie PR, Young JD (1975) Synthetic substrate for cyclic AMP-dependent protein kinase. Nature 257:416–418
Dambach G, Friedmann N (1974) The effects of varying ionic composition of the perfusate on liver membrane potential, gluconeogenesis and cyclic AMP responses. Biochim Biophys Acta 332:374–386
Datta AG, Abrams B, Sasaki T, van den Berg JWO, Pontremoli S, Horecker BL (1974) The activation of rabbit muscle, liver, and kidney fructose bisphosphatases by histidine and citrate. Arch Biochem Biophys 165:641–645
Debeer LJ, Thomas J, DeSchepper PJ, Mannaerts GP (1979) Lysosomal triacylglycerol lipase and lipolysis in isolated rat hepatocytes. J Biol Chem 254:8841–8846
DeRosa G, Swick RW (1975) Metabolic implications of the distribution of the alanine aminotransferase isozymes. J Biol Chem 250:7961–7967
DeWulf H, Keppens S (1976) Is calcium the second messenger in liver for cyclic AMP-independent glycogenolytic hormones? Arch Int Physiol 84:159–160
Diesterhaft M, Shrago E, Sallach HG (1971) Human liver phosphoenolpyruvate carboxykinase: evidence for a separate mitochondrial and cytosol enzyme. Biochem Med 5:297–303
Dieterle PM, Brawand F, Moser UK, Walter P (1978) Alanine metabolism in rat liver mitochondria. Eur J Biochem 88:467–473
Dunaway GA Jr, Weber G (1974) Rat liver phosphofructokinase isozymes. Arch Biochem Biophys 162:620–628
Edlund B, Andersson J, Titanji V, Dahlqvist U, Ekman P, Zetterqvist Ö, Engström L (1975) Amino acid sequence at the phosphorylated site of rat liver pyruvate kinase. Biochem Biophys Res Commun 67:1516–1521
Elliott KRF (1976) The effects of gelatin and bovine serum albumin on $Ca^{2+}$ stimulation of gluconeogenesis in isolated rat hepatocytes. FEBS Lett 64:62–64
El-Maghrabi MR, Haston WS, Flockhart DA, Claus TH, Pilkis SJ (1980) Studies on the phosphorylation and dephosphorylation of L-type pyruvate kinase by the catalytic subunit of cyclic AMP-dependent protein kinase. J Biol Chem 255:668–675
El-Maghrabi MR, Claus TH, Pilkis SJ (1981) Influence of phosphorylation on effector interactions with rat liver pyruvate kinase. J Biol Chem 257:233–240
Exton JH (1972) Gluconeogenesis. Metabolism 21:945–990
Exton JH (1979) Hormonal control of gluconeogenesis. In: Klachko DM, Anderson RR, Heimberg M (eds) Hormones and energy metabolism. Plenum, New York London, pp 125–167
Exton JH, Park CR (1966) D. The stimulation of gluconeogenesis from lactate by epinephrine, glucagon, and cyclic 3′,5′-adenylate in the perfused rat liver. Pharmacol Rev 18:181–188

Exton JH, Park CR (1967) Control of gluconeogenesis in liver. I. General features of gluconeogenesis in the perfused livers of rats. J Biol Chem 242:2622–2636
Exton JH, Park CR (1968) Control of gluconeogenesis in liver. II. Effects of glucagon, catecholamines, and adenosine 3′,5′-monophosphate on gluconeogenesis in the perfused rat liver. J Biol Chem 243:4189–4196
Exton JH, Park CR (1969) Control of gluconeogenesis in liver. III. Effects of lactate, pyruvate, fructose, glucagon, epinephrine, and adenosine 3′,5′-monophosphate on gluconeogenic intermediates in the perfused rat liver. J Biol Chem 244:1424–1433
Exton JH, Mallette LE, Jefferson LS, Wong EHA, Friedmann N, Miller TB Jr, Park CR (1970) The hormonal control of hepatic gluconeogenesis. Recent Prog Horm Res 26:411–461
Exton JH, Friedmann N, Wong EHA, Brineaux JP, Corbin JD, Park CR (1972) Interaction of glucocorticoids with glucagon and epinephrine in the control of gluconeogenesis and glycogenolysis in liver and lipolysis in adipose tissue. J Biol Chem 247:3579–3588
Felíu JE, Hue L, Hers H-G (1976) Hormonal control of pyruvate kinase activity and of gluconeogenesis in isolated hepatocytes. Proc Natl Acad Sci USA 73:2762–2766
Felíu JE, Hue L, Hers H-G (1977) Regulation in vitro and in vivo of adenosine 3′: 5′-monophosphate-dependent inactivation of rat liver pyruvate kinase type L. Eur J Biochem 81:609–617
Flory W, Peczon BD, Koeppe RE, Spivey HO (1974) Kinetic properties of rat liver pyruvate kinase at cellular concentrations of enzyme, substrates and modifiers. Biochem J 141:127–131
Foldes M, Barritt GJ (1977) Regulation by calcium ions of pyruvate carboxylation, pyruvate transport, and adenine nucleotide transport in isolated rat mitochondria. J Biol Chem 252:5372–5380
Foster DO, Lardy HA, Ray PD, Johnston JB (1967) Alteration of rat liver phosphoenolpyruvate carboxykinase activity by L-tryptophan in vivo and metals in vitro. Biochemistry 6:2120–2128
Foster JL, Blair JB (1978) Acute hormonal control of pyruvate kinase and lactate formation in the isolated rat hepatocyte. Arch Biochem Biophys 189:263–276
Friedman B, Goodman EH Jr, Saunders HL, Kostos V, Weinhouse S (1971) An estimation of pyruvate recycling during gluconeogenesis in the perfused rat liver. Arch Biochem Biophys 143:566–578
Friedmann N (1972) Effects of glucagon and cyclic AMP on ion fluxes in the perfused liver. Biochim Biophys Acta 274:214–225
Friedmann N, Dambach G (1973) Effects of glucagon, 3′,5′-AMP and 3′,5′-GMP on ion fluxes and transmembrane potential in perfused livers of normal and adrenalectomized rats. Biochim Biophys Acta 307:399–403
Friedmann N, Park CR (1968) Early effects of 3′,5′-adenosine monophosphate on the fluxes of calcium and potassium in the perfused liver of normal and adrenalectomized rats. Proc Natl Acad Sci USA 61:504–508
Friedmann N, Rasmussen H (1970) Calcium, manganese and hepatic gluconeogenesis. Biochim Biophys Acta 222:41–52
Friedmann N, Somylo AV, Somylo AP (1971) Cyclic adenosine and guanosine monophosphates and glucagon: effect on liver membrane potentials. Science 171:400–402
Friedrichs D (1976) Regulation of pyruvate kinase by glucagon and cyclic AMP during gluconeogenesis in isolated rat hepatocytes. In: Tager JM, Soling HD, Williamson JR (eds) Use of isolated liver cells and kidney tubules in metabolic studies. Elsevier/North Holland, Amsterdam Oxford New York, pp 444–447
Furuya E, Uyeda K (1980a) An activation factor of liver phosphofructokinase. Proc Natl Acad Sci USA 77:5861–5864
Furuya E, Uyeda K (1980b) Regulation of phosphofructokinase by a new mechanism. J Biol Chem 255:11656–11659
Garber AJ, Ballard FJ (1970) Regulation of phosphoenolpyruvate metabolism in mitochondria from guinea pig liver. J Biol Chem 245:2229–2240
Garcia A, Williamson JR, Cahill GF Jr (1966) Studies on the perfused rat liver. II. Effect of glucagon on gluconeogenesis. Diabetes 15:188–193

Garfinkel L, Kohn MC, Garfinkel D (1979) Computer simulation of the fructose bisphosphatase/phosphofructokinase couple in rat liver. Eur J Biochem 96:183–192

Garrison JC, Borland MK (1978) The effects of glucagon, catecholamines, and the calcium ionophore A 23187 on the phosphorylation of rat hepatocyte cytosolic proteins. J Biol Chem 253:7091–7100

Garrison JC, Borland MK (1979) Regulation of mitochondrial pyruvate carboxylation and gluconeogenesis in rat hepatocytes via an α-adrenergic, adenosine 3′: 5′-monophosphate-independent mechanism. J Biol Chem 254:1129–1133

Garrison JC, Haynes RC Jr (1975) The hormonal control of gluconeogenesis by regulation of mitochondrial pyruvate carboxylation in isolated rat liver cells. J Biol Chem 250:2769–2777

Garrison JC, Borland MK, Florio VA, Twible DA (1979) The role of calcium ions as a mediator of the effects of angiotensin II, catecholamines, and vasopressin on the phosphorylation and activity of enzymes in isolated hepatocytes. J Biol Chem 254:7147–7156

Gimpel JA, deHaan EJ, Tager JM (1973) Permeability of isolated mitochondria to oxaloacetate. Biochim Biophys Acta 292:582–591

Halestrap AP (1975) The mitochondrial pyruvate carrier. Kinetics and specificity for substrate and inhibitors. Biochem J 148:85–96

Halestrap AP (1978 a) Pyruvate and ketone-body transport across the mitochondrial membrane. Exchange properties, pH-dependence and mechanism of the carrier. Biochem J 172:377–387

Halestrap AP (1978 b) Stimulation of pyruvate transport in metabolizing mitochondria through changes in the transmembrane pH gradient induced by glucagon treatment of rats. Biochem J 172:389–398

Halestrap AP (1978 c) Stimulation of the respiratory chain of rat liver mitochondria between cytochrome $c_1$ and cytochrome c by glucagon treatment of rats. Biochem J 42:399–405

Halestrap AP (1978 d) Pyruvate transport across mitochondrial and plasma membranes. Fed Eur Biochem Soc Symp (Berl) 42:61–70

Halestrap AP, Scott RD, Thomas AP (1980) Mitochondrial pyruvate transport and its hormonal regulation. Int J Biochem 11:97–105

Harris RA (1975) Studies on the inhibition of hepatic lipogenesis by $N^6,O^2$-dibutyryl adenosine 3′,5′-monophosphate. Arch Biochem Biophys 169:168–180

Haynes RC Jr (1976) Mechanisms of hormonal control of gluconeogenesis. Metabolism [Suppl 1] 25:1361–1363

Hjelmqvist G, Andersson J, Edlund B, Engström L (1974) Amino acid sequence of a ($^{32}P$)phosphopeptide from pig liver pyruvate kinase phosphorylated by cyclic 3′,5′-AMP-stimulated protein kinase and γ-($^{32}P$)ATP. Biochem Biophys Res Commun 61:559–563

Holten DD, Nordlie RC (1965) Comparative studies of catalytic properties of guinea pig liver intra- and extramitochondrial phosphoenolpyruvate carboxykinase. Biochemistry 4:723–731

Horecker BL, Melloni E, Pontremoli S (1975) Fructose 1,6-bisphosphatase: properties of the neutral enzyme and its modification by proteolytic enzymes. Adv Enzymol 42:193–226

Hosey MM, Marcus F (1981) Fructose-bisphosphatases as a substrate of cyclic AMP-dependent protein kinase. Proc Natl Acad Sci USA 78:91–94

Hue L, Hers H-G (1974) On the use of [$^3H$, $^{14}C$] labelled glucose in the study of the so-called "futile cycles" in liver and muscle. Biochem Biophys Res Commun 58:532–539

Hue L, Van Schaftingen E, Blackmore PF (1981) Stimulation of glycolysis and accumulation of a stimulator of phosphofructokinase in hepatocytes incubated with vasopressin. Biochem J 194:1023–1026

Hughes BP, Barritt GJ (1978) Effects of glucagon and $N^6,O^2$-dibutyryladenosine 3′:5′-cyclic monophosphate on calcium transport in isolated rat liver mitochondria. Biochem J 176:295–304

Humble E, Berglund L, Titanji V, Ljungström O, Edlund B, Zetterqvist Ö, Engström L (1975) Non-dependence on native structure of pig liver pyruvate kinase when used as a substrate for cyclic 3′,5′-AMP-stimulated protein kinase. Biochem Biophys Res Commun 66:614–621

Humble E, Dahlqvist-Edberg U, Ekman P, Netzel R, Ragnarsson U, Engström L (1979) Amino acid sequence at the phosphorylated site of rat liver fructose 1,6-diphosphatase and phosphorylation of a corresponding synthetic peptide. Biochem Biophys Res Commun 90:1064–1072

Ibsen KH (1977) Interrelationships and functions of the pyruvate kinase isozymes and their variant forms: a review. Cancer Res 37:341–353

Imamura K, Taniuchi K, Tanaka T (1972) Multimolecular forms of pyruvate kinase. II. Purification of $M_2$-type pyruvate kinase from Yoshida ascites hepatoma 130 cells and comparative studies on the enzymological and immunological properties of the three types of pyruvate kinases, L, $M_1$, and $M_2$. J Biochem 72:1001–1015

Ishibashi H, Cottam GL (1978) Glucagon-stimulated phosphorylation of pyruvate kinase in hepatocytes. J Biol Chem 253:8767–8771

Ishiharai N, Kikuchi G (1968) Studies on the functional relationship between phosphopyruvate synthesis and the substrate level phosphorylation in guinea pig liver mitochondria. Biochim Biophys Acta 153:733–748

Jomain-Baum M, Schramm VL, Hanson RW (1976) Mechanism of 3-mercaptopicolinic acid inhibition of hepatic phospoenolpyruvate carboxykinase (GTP). J Biol Chem 251:37–44

Joseph SK, Bradford NM, McGivan JD (1978) Characteristics of the transport of alanine, serine and glutamine across the plasma membrane of isolated rat liver cells. Biochem J 176:827–836

Kagimoto T, Uyeda K (1979) Hormone-stimulated phosphorylation of liver phosphofructokinase in vivo. J Biol Chem 254:5584–5587

Kalant N (1954) Metabolic effects of the pancreatic hyperglycemic factor. Proc Soc Exp Biol 86:617–619

Katz J, Rognstad R (1976) Futile cycles in the metabolism of glucose. Curr Top Cell Regul 10:237–289

Katz J, Wals PA, Golden S, Rognstad R (1975) Recycling of glucose by rat hepatocytes. Eur J Biochem 60:91–101

Keech DB, Utter MF (1963) Pyruvate carboxylase. II. Properties. J Biol Chem 238:2609–2614

Kemp RG (1971) Rabbit liver phosphofructokinase. Comparison of some properties with those of muscle phosphofructokinase. J Biol Chem 246:245–252

Kemp RG (1975) Phosphofructokinase from rabbit liver. Methods Enzymol 42/C:67–71

Kido H, Vita A, Horecker BL (1980) A one-step procedure for the isolation of fructose 1,6-bisphosphatase and fructose 1,6-bisphosphate aldolase from rabbit liver. Anal Biochem 106:450–454

Kimmich GA, Rasmussen H (1969) Regulation of pyruvate carboxylase activity by calcium in intact rat liver mitochondria. J Biol Chem 244:190–199

Kono N, Uyeda K (1971) Cold labile phosphofructokinase. Biochem Biophys Res Commun 42:1095–1100

Kono N, Uyeda K (1973) Chicken liver phosphofructokinase. I. Crystallization and physiochemical properties. J Biol Chem 248:8592–8602

Kneer NM, Wagner MJ, Lardy HA (1979) Regulation by calcium of hormonal effects on gluconeogenesis. J Biol Chem 254:12160–12168

Leiter AB, Weinberg M, Isohashi F, Utter MF, Linn T (1978) Relationship between phosphorylation and activity of pyruvate dehydrogenase in rat liver mitochondria and the absence of such a relationship for pyruvate carboxylase. J Biol Chem 253:2716–2723

Linn TC, Pettit FH, Hucho F, Reed JL (1969a) α-Keto acid dehydrogenase complexes. XI. Comparative studies of regulatory properties of the pyruvate dehydrogenase complexes from kidney, heart, and liver mitochondria. Proc Natl Acad Sci USA 64:227–234

Linn TC, Pettit FH, Reed LJ (1969b) α-Keto acid dehydrogenase complexes. X. Regulation of the activity of the pyruvate dehydrogenase complex from beef kidney mitochondria by phosphorylation and dephosphorylation. Proc Natl Acad Sci USA 62:234–241

Ljungström O, Ekman P (1977) Glucagon-induced phosphorylation of pyruvate kinase (type L) in rat liver slices. Biochem Biophys Res Commun 78:1147–1155

Lungström O, Hjelmquist G, Engström L (1974) Phosphorylation of purified rat liver pyruvate kinase by cyclic 3′,5′-AMP-stimulated protein kinase. Biochim Biophys Acta 358:289–298

Ljungström O, Berglund L, Engström L (1976) Studies on the kinetic effects of adenosine-3′:5′-monophosphate dependent phosporylation of purified pig-liver pyruvate kinase type L. Eur J Biochem 68:497–506
MacDonald MJ, Bentle LA, Lardy HA (1978) P-enolpyruvate carboxykinase ferroactivator. Distribution, and the influence of diabetes and starvation. J Biol Chem 253:116–124
Mallette LE, Exton JH, Park CR (1969) Control of gluconeogenesis from amino acids in the perfused rat liver. J Biol Chem 244:5713–5723
Mansour TE (1965) Studies on heart phosphofructokinase. Active and inactive forms of the enzyme. J Biol Chem 240:2165–2172
Mansour TE (1972) Phosphofructokinase. Curr Top Cell Regul 5:1–46
Mapes JP, Harris RA (1976) Inhibition of gluconeogenesis and lactate formation from pyruvate by $N^6$,$O^{2'}$-dibutyryl adenosine 3′:5′-monophosphate. J Biol Chem 251:6189–6196
Marcus F, Hosey MM (1980) Purification and properties of liver fructose 1,6-bisphosphatase from C57BL/KsJ normal and diabetic mice. J Biol Chem 255:2481–2486
Massey TH, Deal WC Jr (1973) Unusual, metabolite-dependent solubility properties of phospofructokinase. The basis for a new and rapid purification from liver, kidney, and other tissues. J Biol Chem 248:56–62
Massey TH, Deal WC Jr (1975) Phosphofructokinase from porcine liver and kidney and from other mammalian tissues. Enzymol 42/C:99–110
McClure WR, Lardy HA (1971) Rat liver pyruvate carboxylase. IV. Factors affecting the regulation in vivo. J Biol Chem 246:3591–3596
Mendes-Mourão J, Halestrap AP, Crisp DM, Pogson CI (1975) The involvement of mitochondrial pyruvate transport in the pathways of gluconeogenesis from serine and alanine in isolated rat and mouse liver cells. FEBS Lett 53:29–32
Mendicino J, Vasarkely F (1963) Renal D-fructose 1,6-diphosphatase. J Biol Chem 238:3528–3534
Miller LL (1961) Some direct actions of insulin, glucagon, and hydrocortisone on the isolated perfused rat liver. Recent Prog Horm Res 17:539–563
Mörikofer-Zwez S, Kunin AS, Walter P (1973) Effects of calcium and sucrose on pyruvate carboxylase activity in intact rat liver mitochondria. J Biol Chem 248:7588–7594
Mowbray J (1975) A mitochondrial monocarboxylate transporter in rat liver and heart and its possible function in cell control. Biochem J 148:41–47
Müllhofer G, Loy E, Wollenberg P, Krämer R (1974) A possible role of the glycerol phosphate cycle in cyclic AMP-stimulate gluconeogenesis from lactate in perfused rat livers. Hoppe-Seyler's Z Physiol Chem 355:239–254
Nakashima K, Horecker BL, Traniello S, Pontremoli S (1970) Rabbit liver and rabbit kidney fructose diphosphatases: catalytic properties of enzymes activated by coenzyme A and acyl carrier protein. Arch Biochem Biophys 139:190–199
Newsholme EA, Gevers W (1967) Control of glycolysis and gluconeogenesis in liver and kidney cortex. Vitam Horm 25:1–87
Nieto A, Castaño JG (1980) Control in vivo of rat liver phosphofructokinase by glucagon and nutritional changes. Biochem J 186:953–957
Nimmo HG, Tipton KF (1975) The purification of fructose 1,6-diphosphatase from ox liver and its activation by ethylenediaminetetra-acetate. Biochem J 145:323–334
Paetkau V, Lardy HA (1967) Phosphofructokinase. Correlation of physical and enzymatic properties. J Biol Chem 242:2035–2042
Paetkau Y, Younathan ES, Lardy HA (1968) Phosphofructokinase: studies on the subunit structure. J Mol Biol 33:721–736
Papa S (1976) Proton translocation reactions in the respiratory chains. Biochim Biophys Acta 456:39–84
Papa S, Paradies G (1974) On the mechanism of translocation of pyruvate and other monocarboxylic acids in rat-liver mitochondria. Eur J Biochem 49:265–274
Papa S, Francavilla A, Paradies G, Meduri B (1971) The transport of pyruvate in rat liver mitochondria. FEBS Lett 12:285–288
Parrilla R, Jimenez M-I, Ayuso-Parrilla MS (1975) Glucagon and insulin control of gluconeogenesis in the perfused isolated rat liver. Effects on cellular metabolite distribution. Eur J Biochem 56:375–383

Parrilla R, Jimenez M-I, Ayuso-Parrilla MS (1976) Cellular redistribution of metabolites during glucagon and insulin control of gluconeogenesis in the isolated perfused rat liver. Arch Biochem Biophys 174:1–12
Passoneau JV, Lowry OH (1964) The role of phosphofructokinase in metabolite regulation. Adv Enzyme Regul 2:265–274
Pilkis SJ, Claus TH, Johnson RA, Park CR (1975) Hormonal control of cyclic 3′:5′-AMP levels and gluconeogenesis in isolated hepatocytes from fed rats. J Biol Chem 250:6328–6336
Pilkis SJ, Claus TH, Riou JP, Park CR (1976a) Possible role of pyruvate kinase in the hormone control of dihydroxyacetone gluconeogenesis in isolated hepatocytes. Metabolism [Suppl 1] 25:1355–1360
Pilkis SJ, Riou JP, Claus TH (1976b) Hormonal control of [$^{14}$C]glucose synthesis from [U-$^{14}$C]dihydroxyacetone and glycerol in isolated rat hepatocytes. J Biol Chem 251:7841–7852
Pilkis SJ, Claus TH, Riou JP, Cherrington AD, Chiasson JL, Liljenquist JE, Lacy WW, Park CR (1978a) Control of hepatic gluconeogenesis and glucose output by glucagon and insulin. Fed Eur Biochem Soc Symp (Berl) 42:13–29
Pilkis SJ, Park CR, Claus TH (1978b) Hormonal regulation of hepatic gluconeogenesis. Vitam Horm 36:383–460
Pilkis SJ, Pilkis J, Claus TH (1978c) Effect of fructose diphosphate and phosphoenolpyruvate on inactivation of pyruvate kinase in hepatocyte homogenates. Biochem Biophys Res Commun 81:139–147
Pilkis SJ, Schlumpf JR, Pilkis J, Claus TH (1979) Regulation of phosphofructokinase activity by glucagon in isolated rat hepatocytes. Biochem Biophys Res Commun 88:960–967
Pilkis SJ, El-Maghrabi MR, Claus TH, Tager HS, Steiner DE, Keim P, Henrikson R (1980a) Phosphorylation of rat hepatic fructose 1,6-bisphosphatase and pyruvate kinase. J Biol Chem 255:2770–2775
Pilkis SJ, Schlumpf J, Claus TH, El-Maghrabi R (1980b) Studies on the phosphorylation state of rat hepatic phosphofructokinase. (Abstr 1765) Fed Proc 39:1940
Pilkis SJ, El-Maghrabi MR, McGrane M, Pilkis J, Claus TH (1981b) The role of fructose 2,6-bisphosphate in regulation of fructose 1,6-bisphosphatase. J Biol Chem 256:11489–11495
Pilkis SJ, El-Maghrabi MR, Pilkis J, Claus TH (1981c) Inhibition of fructose 1,6-bisphosphatase by fructose 2,6-bisphosphate. J Biol Chem 256:3619–3622
Pilkis SJ, El-Maghrabi MR, Pilkis J, Claus TH, Cumming DA (1981d) Fructose 2,6-bisphosphate. A new activator of phosphofructokinase. J Biol Chem 256:3171–3174
Pilkis SJ, Schlumpf JR, El-Maghrabi MR, Pilkis J, Claus TH (1981e) Glucagon regulation of phosphofructokinase activity in isolated hepatocytes. Cold Spring Harbor Symp Quant Biol 8:547–560
Pilkis SJ, El-Maghrabi MR, McGrane M, Claus TH (1982) Regulation by glucagon of hepatic pyruvate kinase, phosphofructokinase, and fructose 1,6-bisphosphatase. Fed Proc 41:2623–2628
Pogell BM, Tanaka A, Siddons RC (1968) Natural activators of liver fructose 1,6-diphosphatase and the reversal of adenosine 5′-monophosphate inhibition by muscle phosphofructokinase. J Biol Chem 243:1356–1367
Pontremoli S, Grazi E, Accorsi A (1968) Fructose diphosphatase from rabbit liver. X. Isolation and kinetic properties of the enzyme-adenosine monophosphate complex. Biochemistry 7:3628–3633
Prpić V, Spencer TL, Bygrave FL (1978) Stable enhancement of calcium retention in mitochondria isolated from rat liver after the administration of glucagon to the intact animal. Biochem J 176:705–714
Racker E (1970) The two faces of the inner mitochondrial membrane. In: Campbell PN, Dickens F (eds) Essays in biochemistry, vol 6. Academic, London New York, pp 1–22
Ramaiah A (1974) Pasteur effect and phosphofructokinase. Curr Top Cell Regul 8:298–345
Ramaiah A, Tejwani GA (1970) Interconvertible forms of phosphofructokinase of rabbit liver. Biochem Biophys Res Commun 39:1149–1156
Rasmussen H (1970) Cell communication, calcium ion, and cyclic adenosine monophosphate. Science 170:404–412

Reinhart GD, Lardy HA (1980a) Rat liver phosphofructokinase: kinetic activity under near-physiological conditions. Biochemistry 19:1477–1484
Reinhart GD, Lardy HA (1980b) Rat liver phosphofructokinase: use of fluorescence polarization to study aggregation at low protein concentration. Biochemistry 19:1484–1490
Reinhart GD, Lardy HA (1980c) Rat liver phosphofructokinase: kinetic and physiological ramifications of the aggregation behavior. Biochemistry 19:1491–1495
Reinhart PH, Bygrave FL (1981) Glucagon stimulation of ruthenium red-insensitive calcium ion transport in developing rat liver. Biochem J 194:541–549
Reshef L, Hanson RW (1972) The interaction of catecholamines and adrenal corticosteroids in the induction of phosphopyruvate carboxylase in rat liver and adipose tissue. Biochem J 127:809–818
Richards CS, Uyeda K (1980) Changes in the concentration of activation factor for phosphofructokinase in hepatocytes in response to glucose and glucagon. Biochem Biophys Res Commun 97:1535–1540
Riou JP, Claus TH, Pilkis SJ (1976) Control of pyruvate kinase activity by glucagon in isolated hepatocytes. Biochem Biophys Res Commun 73:591–599
Riou JP, Claus TH, Flockhart DA, Corbin JD, Pilkis SJ (1977) In vivo and in vitro phosphorylation of rat liver fructose 1,6-bisphosphatase. Proc Natl Acad Sci USA 74:4615–4619
Riou JP, Claus TH, Pilkis SJ (1978) Stimulation by glucagon of in vivo phosphorylation of rat hepatic pyruvate kinase. J Biol Chem 253:656–659
Robison GA, Butcher RW, Sutherland EW (1971) Cyclic AMP. Academic, New York London
Rognstad R (1975) Cyclic AMP induced inhibition of pyruvate kinase flux in the intact liver cell. Biochem Biophys Res Commun 63:900–905
Rognstad R (1976) Control of pyruvate kinase flux during gluconeogenesis in isolated liver cells. Int J Biochem 7:403–408
Rognstad R (1979) Effects of medium pH on gluconeogenesis from dihydroxyacetone. Int J Biochem 10:619–621
Rognstad R, Katz J (1972) Gluconeogenesis in the kidney cortey. Quantitative estimation of carbon flow. J Biol Chem 247:6047–6054
Rognstad R, Katz J (1976) Effects of hormones and of ethanol on the fructose 6-P-fructose 1,6-$P_2$ futile cycle during gluconeogenesis in the liver. Arch Biochem Biophys 177:337–345
Rognstad R, Katz J (1977) Role of pyruvate kinase in the regulation of gluconeogenesis from L-lactate. J Biol Chem 252:1831–1833
Rose S, Nelson J (1956) Protein metabolism and hypoglycemia. Aus J Exp Biol Med Sci 34:33–42
Rosenberg JS, Tashima Y, Horecker BL, Pontremoli S (1973) Activation of rabbit kidney fructose diphosphatase by Mg-EDTA, Mn-EDTA and Co-EDTA complexes. Arch Biochem Biophys 154:283–291
Ross BD, Hems R, Krebs HA (1967) The rate of gluconeogenesis from various precursors in the perfused rat liver. Biochem J 102:942–951
Rozengurt E, Jimenez de Asua L, Carminatti H (1969) Some kinetic properties of liver pyruvate kinase (type L). II. Effect of pH on its allosteric behavior. J Biol Chem 244:3142–3147
Salter JM, Davidson IWF, Best CH (1957) The pathologic effects of large amounts of glucagon. Diabetes 6:248–252
Salzmann M, Carofoli E, Jakob A (1978) $Ca^{2+}$ redistribution and phenylephrine effects in perfused rat liver. Experientia 34:917–923
Schimassek H, Mitzkat HJ (1963) Über eine spezifische Wirkung des Glucagon auf die Embden-Myerhof-Kette in der Leber. Biochem Z 337:510–518
Scrutton MC, Utter MF (1968) The regulation of glycolysis and gluconeogenesis in animal tissues. Annu Rev Biochem 37:249–302
Scrutton MC, White MD (1974) Pyruvate carboxylase. Inhibition of the mammalian and avian liver enzymes by $\alpha$-ketoglutarate and L-glutamate. J Biol Chem 249:5405–5415
Seubert W, Schoner W (1971) The regulation of pyruvate kinase. Curr Top Cell Regul 3:237–267

Shrago E, Lardy HA, Nordlie RC, Foster DO (1963) Metabolic and hormonal control of phosphoenolpyruvate carboxykinase and malic enzyme in rat liver. J Biol Chem 238:3188–3192
Siess EA, Wieland OH (1978) Glucagon-induced stimulation of 2-oxoglutarate metabolism in mitochondria from rat liver. FEBS Lett 93:301–306
Siess EA, Wieland OH (1979) Isolated hepatocytes as a model for the study of stable glucagon effects on mitochondrial respiratory functions. FEBS Lett 101:277–281
Siess EA, Brocks DG, Wieland OH (1976) A sensitive and simple method for the study of oxaloacetate compartmentation in isolated hepatocytes. FEBS Lett 70:51–55
Siess EA, Brocks DG, Lattke HK, Wieland OH (1977) Effect of glucagon on metabolite compartmentation in isolated rat liver cells during gluconeogenesis from lactate. Biochem J 166:225–235
Siess EA, Brocks DG, Wieland OH (1978) Distinctive effects of glucagon on gluconeogenesis and ketogenesis in hepatocytes isolated from normal and biotin-deficient rats. Biochem J 172:517–521
Snoke RE, Johnston JB, Lardy HA (1971) Response of phosphopyruvate carboxylase to tryptophan metabolites and metal ions. Eur J Biochem 24:342–346
Söling H-D, Kleineke J (1976) Species dependent regulation of hepatic gluconeogenesis in higher animals. In: Hanson RW, Mehlman MA (eds) Gluconeogenesis: its regulation in mammalian species. Wiley, New York London, pp 369–462
Söling H-D, Willms B, Friedrichs D, Kleineke J (1968) Regulation of gluconeogenesis by fatty acid oxidation in isolated perfuses livers of non-starved rats. Eur J Biochem 4:364–372
Söling H-D, Bernhard G, Kuhn A, Lück H-J (1977) Inhibition of phosphofructokinase by fructose 1,6-diphosphatase in mammalian systems: protein-protein interaction or fructose 1,6-diphosphate trapping? Arch Biochem Biophys 182:563–572
Sols A, Marco R (1970) Concentrations of metabolites and binding sites. Implications in metabolic regulation. Curr Top Cell Regul 2:227–273
Somlyo AP, Somylo AV, Friedmann N (1971) Cyclic adenosine monophosphate, cyclic guanosine monophosphate, and glucagon: Effects on membrane potential and ion fluxes in the liver. Ann NY Acad Sci 185:108–114
Steiner KE, Chan TM, Claus TH, Exton JH, Pilkis SJ (1980) The role of phosphorylation in the α-adrenergic-mediated inhibition of rat hepatic pyruvate kinase. Biochem Biophys Acta 632:366–374
Struck E, Ashmore J, Wieland O (1966) Effects of glucagon and long chain fatty acids on glucose production by isolated, perfused rat liver. Adv Enzyme Regul 4:219–224
Sugano T, Shiota M, Khono H, Shimada M, Oshino N (1980a) Effects of calcium ions on the activation of gluconeogenesis by norepinephrine in perfused rat liver. J Biochem 87:465–472
Sugano T, Shiota M, Tanaka T, Miyamae Y, Shimada M, Oshino N (1980b) Intracellular redox state and stimulation of gluconeogenesis by glucagon and norepinephrine in the perfused rat liver. J Biochem 87:153–166
Swick RW, Barnstein PL, Stange JL (1965) The metabolism of mitochondrial proteins. I. Distribution and characterization of the isozymes of alanine amino transferase in rat liver. J Biol Chem 240:3334–3345
Taketa K, Pogell BM (1963) Reversible inactivation and inhibition of liver fructose-1,6-diphosphatase by adenosine nucleotides. Biochem Biophys Res Commun 12:229–235
Taketa K, Pogell BM (1965) Allosteric inhibition of rat liver fructose 1,6-diphosphatase by adenosine 5′-monophosphate. J Biol Chem 240:651–662
Taunton OD, Stifel FB, Greene HL, Herman RH (1972) Rapid reciprocal changes of rat hepatic glycolytic enzymes and fructose-1,6-diphosphatase following glucagon and insulin injection in vivo. Biochem Biophys Res Commun 48:1663–1669
Taunton OD, Stifel FB, Greene HL, Herman RH (1974) Rapid reciprocal changes in rat hepatic glycolytic enzymes and fructose diphosphatase activities following insulin and glucagon injection. J Biol Chem 249:7228–7239
Taylor WM, Bygrave FL, Blackmore PF, Exton JH (1979) Stable enhancement of ruthenium red-insensitive calcium transport in an endoplasmic reticulum-rich fraction following the exposure of isolated rat liver cells to glucagon. FEBS Lett 104:31–34

Taylor WM, Prpić V, Exton JH, Bygrave FL (1980) Stable changes to calcium fluxes in mitochondria isolated from rat livers perfused with α-adrenergic agonists and with glucagon. Biochem J 188:443–450

Tejwani GA, Pedrosa FO, Pontremoli S, Horecker BL (1976) The purification and properties of rat liver fructose 1,6-bisphosphatase. Arch Biochem Biophys 17:253–264

Tilghman SM, Hanson RW, Ballard FJ (1976) Hormonal regulation of phosphoenolpyruvate carboxykinase (GTP) in mammalian tissues. In: Hanson RW, Mehlman MA (eds) Gluconeogenesis: its regulation in mammalian species. Wiley, New York London, pp 47–91

Titanji VPK, Zetterqvist Ö, Engström L (1976) Regulation in vitro of rat liver pyruvate kinase by phosphorylation-dephosphorylation reactions, catalyzed by cyclic AMP-dependent protein kinases and a histone phosphatase. Biochim Biophys Acta 422:98–108

Titheradge MA, Coore HG (1975) Initial rates of pyruvate transport in mitochondria determined by an "inhibitor-stop" technique. Biochem J 150:553–556

Titheradge MA, Coore HG (1976a) The mitochondrial pyruvate carrier. Its exchange properties and its regulation by glucagon. FEBS Lett 63:45–50

Titheradge MA, Coore HG (1976b) Hormonal regulation of liver mitochondrial pyruvate carrier in relation to gluconeogenesis and lipogenesis. FEBS Lett 71:73–78

Titheradge MA, Haynes RC Jr (1980) The control of uncoupler-activated ATPase activity in rat liver mitochondria by adenine nucleotide transport. The effect of glucagon treatment. J Biol CHem 255:1471–1477

Titheradge MA, Binder SB, Yamazaki RK, Haynes RC Jr (1978) Glucagon treatment stimulates the metabolism of hepatic submitochondrial particles. J Biol Chem 253:3357–3360

Titheradge MA, Stringer JL, Haynes RC Jr (1979) The stimulation of the mitochondria uncoupler-dependent ATPase in isolated hepatocytes by catecholamines and glucagon and its relationship to gluconeogenesis. Eur J Biochem 102:117–124

Tolbert MEM, Fain JN (1974) Studies on the regulation of gluconeogenesis in isolated rat liver cells by epinephrine and glucagon. J Biol Chem 249:1162–1166

Traniello S, Pontremoli S, Tashima Y, Horecker BL (1971) Fructose 1,6-diphosphatase from liver: isolation of the native form with optimal activity at neutral pH. Arch Biochem Biophys 146:161–166

Triebwasser KC, Freedland RA (1977) The effect of glucagon on ureagenesis from ammonia by isolated rat hepatocytes. Biochem Biophys Res Commun 76:1159–1165

Trujillo JL, Deal WC Jr (1977) Pig liver phosphofructokinase: asymmetry properties, proof of rapid association-dissociation equilibria, and effect of temperature and protein concentration on the equilibria. Biochemistry 16:3098–3104

Tsai MY, Kemp RG (1973) Isozymes of rabbit phosphofructokinase. Electrophoretic and immunochemical studies. J Biol Chem 248:785–792

Tyberghein J (1953) Action du glucagon sur le métabolisme des protéines. Arch Int Physiol (Liège) 61:104–107

Ui M, Claus TH, Exton JH, Park CR (1973a) Studies on the mechanism of action of glucagon on gluconeogenesis. J Biol Chem 248:5344–5349

Ui M, Exton JH, Park CR (1973b) Effects of glucagon on glutamate metabolism in the perfused rat liver. J Biol Chem 248:5350–5359

Underwood AH, Newsholme EA (1965a) Some properties of fructose 1,6-diphosphatase of rat liver and their relation to the control of gluconeogenesis. Biochem J 95:767–774

Underwood AH, Newsholme EA (1965b) Properties of phosphofructokinase from rat liver and their relation to the control of glycolysis and gluconeogenesis. Biochem J 95:868–875

Utter MF, Kolenbrander HM (1972) Formation of oxalacetate by $CO_2$ fixation on phosphoenolpyruvate. In: Boyer PD (ed) The enzymes, 3rd edn, vol VI. Academic, New York London, pp 117–168

Utter MF, Kurahashi K (1953) Mechanism of the action of oxalacetic carboxylase from liver. J Am Chem Soc 75:758

Utter MF, Keech DB, Scrutton MC (1964) A possible role for acetyl CoA in the control of gluconeogenesis. Adv Enzyme Regul 2:44–68

Uyeda K, Luby LJ (1974) Studies on the effect of fructose 1,6-diphosphatase on phosphofructokinase. J Biol Chem 249:4562–4570
Van Berkel TJC, Koster JF, Kruyt JK, Hulsmann WC (1974) On the regulation and allosteric model of L-type pyruvate kinase from rat liver. Biochim Biophys Acta 370:450–458
Van Berkel TJC, Kruijt JK, Koster JF, Hülsmann WC (1976) Cyclic nucleotide-, pyruvate- and hormone-induced changes in pyruvate kinase activity in isolated rat hepatocytes. Biochem Biophys Res Commun 72:917–925
Van Berkel TJC, Kruijt JK, Koster JF (1977a) Hormone-induced changes in pyruvate kinase. Effects of glucagon and starvation. Eur J Biochem 81:423–432
Van Berkel TJC, Kruijt JK, Koster JF (1977b) Hormone-induced changes in pyruvate kinase activity in isolated hepatocytes. II. Relation to the hormonal regulation of gluconeogenesis. Biochim Biophys Acta 500:267–276
Van Schaftingen E, Hers H-G (1980) Synthesis of a stimulator of phosphofructokinase, most likely fructose 2,6-bisphosphate, from phosphoric acid and fructose 6-phosphoric acid. Biochem Biophys Res Commun 96:1524–1531
Van Schaftingen E, Hue L, Hers H-G (1980a) Study of the fructose 6-phosphate/fructose 1,6-bisphosphate cycle in the liver in vivo. Biochem J 192:263–271
Van Schaftingen E, Hue L, Hers H-G (1980b) Control of the fructose 6-phosphate/fructose 1,6-bisphosphate cycle in isolated hepatocytes by glucose and glucagon. Biochem J 192:887–895
Van Schaftingen E, Hue L, Hers H-G (1980c) Fructose 2,6-bisphosphate, the probable structure of the glucose- and glucagon-sensitive stimulator of phosphofructokinase. Biochem J 192:897–901
Vardanis A (1977) Protein kinase activity at the inner membrane of mammalian mitochondria. J Biol Chem 252:807–813
Veech RL, Raijman L, Krebs HA (1970) Equilibrium relations between the cytoplasmic adenine nucleotide system and nicotinamide-adenine nucleotide system in rat liver. Biochem J 117:499–503
Veech RL, Nielsen R, Harris RL (1975) Effects of pineal and other hormones on the free [$NADP^+$]/[NADPH]ratio in rat liver. In: Altschule MD (ed) Frontiers of pineal physiology. MIT Press, Cambridge London, pp 177–196
Walsh DA, Chen L-J (1971) A reinvestigation of the kinetic properties of phosphoenolpyruvate carboxykinase. Biochem Biophys Res Commun 45:669–675
Waltenbaugh A-MA, Friedmann N (1978) Hormone sensitive calcium uptake by liver microsomes. Biochem Biophys Res Commun 82:603–608
Whitehouse S, Randle PJ (1973) Activation of pyruvate dehydrogenase in perfused rat heart by dichloroacetate. Biochem J 134:651–653
Whitehouse S, Cooper RH, Randle PJ (1974) Mechanism of activation of pyruvate dehydrogenase by dichloroacetate and other halogenated carboxylic acids. Biochem J 141:761–774
Wicks WD, Lewis W, McKibbin JB (1972) Induction of phosphoenolpyruvate carboxykinase by $N^6,O^{2'}$-dibutyryl cyclic AMP in rat liver. Biochim Biophys Acta 264:177–185
Williams TF, Exton JH, Friedmann N, Park CR (1971) Effects of insulin and adenosine 3′,5′-monophosphate on $K^+$ flux and glucose output in perfused rat liver. Am J Physiol 221:1645–1651
Williamson JR, Herczeg B, Coles H, Danish R (1966a) Studies on the ketogenic effect of glucagon in intact rat liver. Biochem Biophys Res Commun 24:437–442
Williamson JR, Kreisberg RA, Felts PW (1966b) Mechanism for the stimulation of gluconeogenesis by fatty acids in perfused rat liver. Proc Natl Acad Sci USA 56:247–254
Williamson JR, Wright PH, Malaisse WJ, Ashmore J (1966c) Control of gluconeogenesis by acetyl CoA in rats treated with glucagon and anti-insulin serum. Biochem Biophys Res Commun 24:765–770
Williamson JR, Browning ET, Thurman RG, Scholz R (1969a) Inhibition of glucagon effects in perfused rat liver by (+)decanoylcarnitine. J Biol Chem 244:5055–5064
Williamson JR, Scholz R, Browning ET (1969b) Control mechanisms of gluconeogenesis and ketogenesis. II. Interactions between fatty acid oxidation and the citric acid in perfused rat liver. J Biol Chem 244:4617–4627

Wimhurst JM, Manchester KL (1970) Some aspects of the kinetics of rat liver pyruvate carboxylase. Biochem J 120:79–93

Yamazaki RK (1975) Glucagon stimulation of mitochondrial respiration. J Biol Chem 250:7924–7930

Yamazaki RK, Graetz GS (1977) Glucagon stimulation of citrulline formation in isolated hepatic mitochondria. Arch Biochem Biophys 178:19–25

Yamazaki RK, Haynes RC Jr (1975) Dissociation of pyruvate dehydrogenase from the glucagon stimulation of pyruvate carboxylation in rat liver mitochondria. Arch Biochem Biophys 166:575–583

Yamazaki RK, Sax RD, Hauser MA (1977) Glucagon stimulation of mitochondrial ATPase and potassium ion transport. FEBS Lett 75:295–299

Yeaman SJ, Cohen P, Walson DC, Dixon GH (1977) The substrate specificity of adenosine 3′:5′-cyclic monophosphate-dependent protein kinase of rabbit skeletal muscle. Biochem J 162:411–421

Zahlten RN, Hochberg AA, Stratman FW, Lardy HA (1972) Glucagon-stimulated phosphorylation of mitochondrial and lysosomal membranes of rat liver in vivo. Proc Natl Acad Sci USA 69:800–804

Zahlten RN, Stratman FW, Lardy HA (1973) Regulation of glucose synthesis in hormone-sensitive isolated rat hepatocytes. Proc Natl Acad Sci USA 70:3213–3218

Zahlten RN, Kneer NM, Stratman FW, Lardy HA (1974) The influence of ammonium and calcium ions on gluconeogenesis in isolated rat hepatocytes and their response to glucagon and epinephrine. Arch Biochem Biophys 161:528–535

Zetterqvist Ö, Ragnarsson U, Humble E, Berglund L, Engström L (1976) The minimum substrate of cyclic AMP-stimulated protein kinase, as studied by synthetic peptide representing the phosphorylatable site of pyruvate kinase (type L) of rat liver. Biochem Biophys Res Commun 70:696–703

CHAPTER 16

# Glucagon and Liver Glucose Output In Vivo

J. L. CHIASSON and A. D. CHERRINGTON

## A. Introduction

Human survival is dependent on maintaining a constant supply of fuel to various tissues of the body. During periods of feasting, ingested nutrients are stored in the liver, fat, and muscle. In between meals, and in periods of fasting, these substrates can be either used locally or distributed to different tissues for consumption. Of the various energy substrates available, glucose is perhaps the most important since it plays a crucial and essential role with regard to nervous tissue. Under most circumstances, the liver is the only organ which releases glucose into the circulation for distribution to other tissues. This glucose is derived from either glycogen stores within the liver (glycogenolysis) or from de novo synthesis of glucose (gluconeogenesis).

The flow of glucose from the liver to the various tissues of the body is primarily regulated by the two major pancreatic hormones: glucagon and insulin. It is the purpose of this chapter to describe pertinent observations characterizing the role of glucagon in regulating hepatic glucose output in vivo. We will therefore discuss sequentially the effect of glucagon on glycogenolysis and gluconeogenesis and its interaction with insulin in regulating these processes. We will also discuss briefly the role of glucagon in controlling glucose homeostasis in certain unique physiologic (feasting, fasting) and pathophysiologic (diabetes mellitus) conditions. Details of the mechanism of action on liver glycogenolysis and gluconeogenesis can be found in Chaps. 14 and 15, respectively.

## B. Effects of Glucagon on Hepatic Glucose Production

### I. Glycogenolysis

Though glucagon was known to raise the blood glucose as early as 1923 (MURLIN et al. 1923), it was not until 40 years later that this hormone was shown to have an effect on hepatic glucose production. DE BODO et al. (1963), using tracer techniques in the overnight fasted dog, were the first to show that glucagon could increase glucose production. Similar observations were made in overnight fasted humans by KIBLER et al. (1964) and later by CHIASSON et al. (1975) using the arterio-hepatic venous difference technique. In the mid-1970 s, glucagon was also shown to have a stimulatory effect on hepatic glucose output in overnight fasted diabetic humans (LILJENQUIST et al. 1974). Since in the postabsorptive state glycogenolysis accounts for 75%–85% of the glucose produced by the liver, these studies indicated that the hormone could affect glycogen breakdown in vivo. Although large doses of glucagon were employed, the data implied a physiologic role for the hormone in the regulation of glycogenolysis within the whole animal.

More recently, physiologic amounts of glucagon have been shown to elicit a glycogenolytic response in vivo. Using the arteriohepatic venous difference technique in humans, BOMBOY et al. (1977 a) and FELIG et al. (1976) demonstrated that a rise in plasma glucagon levels of 400 pg/ml, giving a level well within the range found in portal venous plasma as measured by BLACKARD et al. (1974), resulted in a greater than twofold increase in the rate of hepatic glucose production. These findings have been confirmed in overnight fasted dogs by means of tracer methods (CHERRINGTON et al. 1981; SACCA et al. 1979). It should be noted that in all of these studies glucagon was able to stimulate glucose production in spite of counterregulatory insulin release. As a result of the insulin mobilization, however, the true potential of glucagon to enhance glucose output was not established.

Assessment of the full physiologic potential of glucagon in regulating glycogenolysis in vivo has proved difficult because of the potent glucoregulatory feedback mechanisms which exist in the whole animal. Discovery of somatostatin (BRAZEAU et al. 1973), a polypeptide which inhibits the secretion of both glucagon and insulin when administered systemically (KOERKER et al. 1974), provided an invaluable tool with which the physiologic role of glucagon was ultimately discerned. Using this tool, CHERRINGTON et al. (1976) demonstrated for the first time that basal amounts of glucagon play an important role in regulating resting glucose production in the overnight fasted state. In these studies on conscious dogs, glucagon deficiency brought about in the presence of basal amounts of insulin was associated with a 35% drop in hepatic glucose production and a substantial fall (25 mg/dl) in the plasma glucose level (Fig. 1). Similar observations were soon made by ALTSZULER et al. (1976) and LICKLEY et al. (1979). In later studies, CHERRINGTON et al. (1979) showed that the drop (35%) in glucose production pursuant to the induction of selective glucagon deficiency was sustained for at least 4.5 h and that the magnitude of the decline was limited by the hypoglycemia which developed. If the latter was prevented by exogenous glucose infusion, hepatic glucose output declined by 75% rather than 35% in response to glucagon lack. LILJENQUIST et al. (1977) showed that the hormone was similarly important in overnight fasted humans. Although control studies validated the use of somatostatin in these experiments, it is worth noting that DERI et al. (1981) came to the same conclusion regarding the role of basal glucagon when a surgical rather than pharmacologic approach was used to control the endocrine pancreas. Taken together, these observations clearly showed an important role for glucagon in maintaining glucose production after an overnight fast and hence in regulating glycogenolysis in vivo.

Use of somatostatin has also allowed clarification of the physiologic potential of increments in plasma glucagon to regulate glycogenolysis in the whole animal. Normally, infusion of glucagon triggers counterregulatory insulin release and it is necessary to separate the effect of the infused glucagon from the opposing effects of the insulin if the true effect of glucagon is to be determined. In the study shown in Fig. 2, somatostatin was infused along with replacement amounts of both insulin and glucagon (intraportally) in order to break the glucoregulatory feedback loops involving the endocrine pancreas. After a control period in which both insulin and

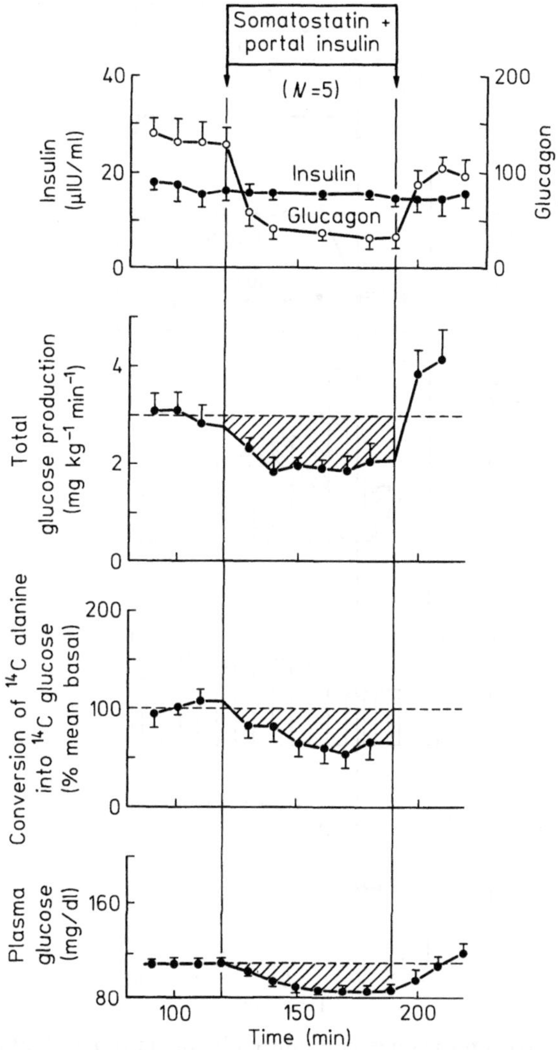


**Fig. 1.** Effect of selective glucagon deficiency brought about by infusion of somatostatin (0.8 μg $kg^{-1}$ $min^{-1}$) and intraportal insulin (300 μIU $kg^{-1}$ $min^{-1}$) on overall glucose production, the conversion of alanine $^{14}C$ to glucose and the plasma glucose level in five overnight fasted anesthetized dogs. The data are mean ± standard error. CHERRINGTON et al. (1978b)

glucagon were delivered in basal amounts, the glucagon infusion rate was increased fourfold and the insulin infusion rate was left unchanged. As a result of this selective 170 pg/ml rise in the plasma glucagon level, glucose production rose rapidly by 5–6 mg $kg^{-1}$ $min^{-1}$, a value 2–3 times that observed when the same glucagon infusion was given to normal animals. Similar results were again obtained when a surgical rather than a pharmacologic approach was used to control the endocrine pancreas (CHERRINGTON et al. 1981). While glucagon's effect on glucose production is potent and rapid, it is clear from Fig. 2 that, in the overnight fasted dog, the

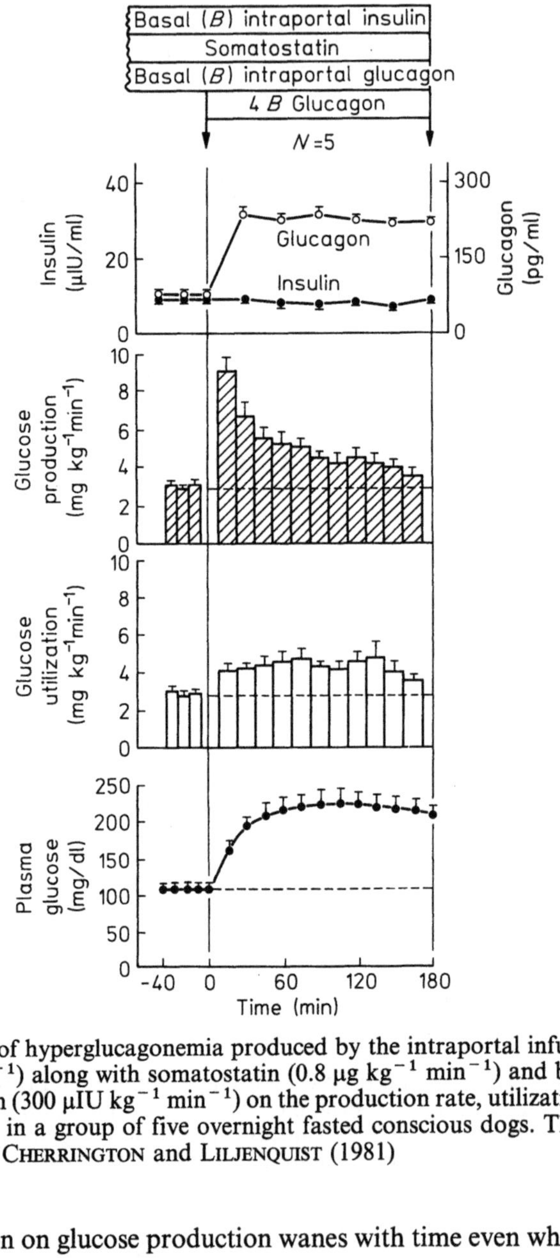


**Fig. 2.** The effect of hyperglucagonemia produced by the intraportal infusion of glucagon (2.6 ng $kg^{-1}$ $min^{-1}$) along with somatostatin (0.8 μg $kg^{-1}$ $min^{-1}$) and basal replacement infusions of insulin (300 μIU $kg^{-1}$ $min^{-1}$) on the production rate, utilization rate, and level of plasma glucose in a group of five overnight fasted conscious dogs. The data are mean ±standard error. CHERRINGTON and LILJENQUIST (1981)

effect of glucagon on glucose production wanes with time even when counterregulatory insulin secretion does not occur.

FELIG et al. (1976) observed a similar phenomenon in normal postabsorptive humans. In six subjects, glucagon infusion at 3 ng $kg^{-1}$ $min^{-1}$ caused a rapid, but transient, rise in splanchnic glucose output. Since the rise in glucose output waned prior to the rise in plasma insulin the authors suggested that the inhibition was not dependent on increased insulin secretion. This was confirmed by BOMBOY et al. (1977a) who showed, in insulin-dependent diabetic humans, that a similar temporary rise in glucose production occurred during glucagon infusion

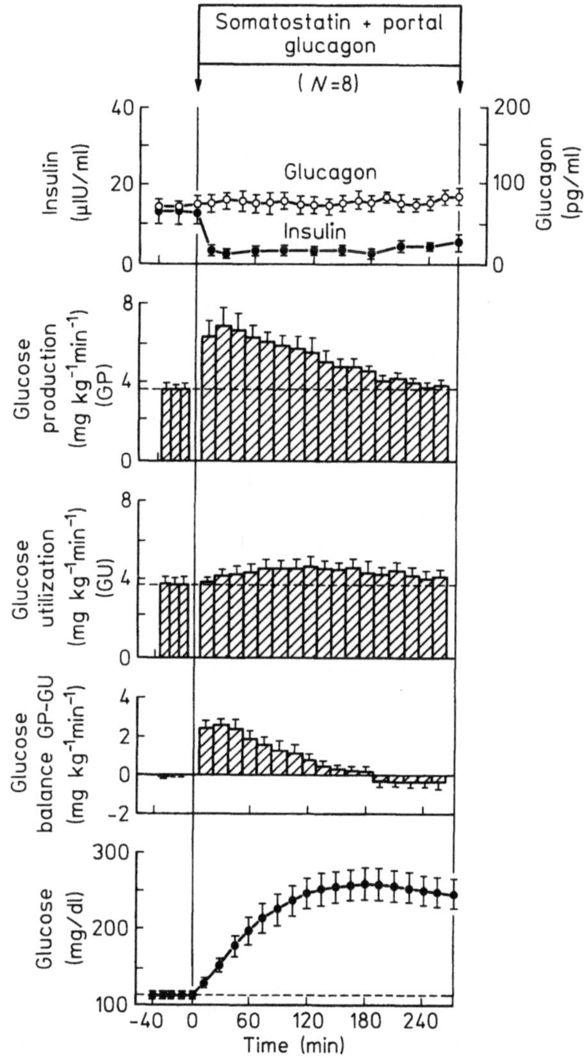


**Fig. 3.** Effects of insulin deficiency produced by combined somatostatin (0.8 μg $kg^{-1}$ $min^{-1}$) and intraportal glucagon infusion (0.65 ng $kg^{-1}$ $min^{-1}$) on arterial plasma insulin and glucagon levels and on the production, utilization, and concentration of plasma glucose in eight normal overnight fasted conscious dogs. The data are mean ± standard error. CHERRINGTON et al. (1978 a)

(5 ng $kg^{-1}$ $min^{-1}$), despite the fact that these subjects could not secrete insulin in response to the infused hormone. The transient nature of glucagon's effect on glucose production in overnight fasted dogs was also evident when the glucagon excess was relative rather than absolute (CHERRINGTON et al. 1978 a). As can be seen in Fig. 3, when basal glucagon was maintained during a period of insulin deficiency, hepatic glucose production increased rapidly, but fell back close to the baseline by 180 min. Since glucose production in overnight fasted humans or dogs is pri-

marily attributable to glycogenolysis, this time dependent fall in glucose production was felt to represent a decline in the rate of glycogen breakdown.

The waning of glucagon's effect on glucose production should not be interpreted to mean, however, that the increment in glucagon loses its physiologic importance. Restoration of the basal glucagon level after 3 h of selective hyperglucagonemia (i.e., insulin fixed at basal levels) was associated with a fall in glucose output to a rate significantly below the control rate (CHERRINGTON et al. 1981). In other words, the liver adapts to hyperglucagonemia and becomes dependent on it for the maintenance of basal glucose production. A similar "adaptation" phenomenon has been observed in humans (FRADKIN et al. 1980; RIZZA and GERICH 1979).

In summary, glucagon plays an important role in the regulation of glycogenolysis in vivo. Increases in the plasma glucagon concentration within the physiologic range can result in a prompt (several minutes), marked (up to fivefold), but short-lived increase in glycogenolysis. The exact nature of the "downregulatory" mechanism by which glycogenolysis is turned off remains unclear and has been discussed elsewhere (CHERRINGTON et al. 1981). It is now evident that the small amount of glucagon which is present in plasma during the postprandial period plays a vital role in stimulating glycogen breakdown and thus redistributing the glucose originally stored in the liver to the other tissues of the body. Glucagon is responsible for between two-thirds and three-quarters of total glucose output after an overnight fast and since the latter is primarily dependent on glycogen breakdown the hormone is responsible for as much as 90% of glycogenolysis at that time.

## II. Gluconeogenesis

While gluconeogenesis is of minor importance to overall hepatic glucose output after an overnight fast, it becomes important during prolonged periods of fasting, in exercise and in neonatal life, as well as under certain pathologic conditions such as diabetes mellitus (FELIG 1973). The first suggestion that glucagon could stimulate gluconeogenesis came from TYBERGHEIN (1953) who showed that the hormone could increase urea excretion in rabbits. SALTER et al. (1960) later made similar findings in humans. Despite these early observations indicating that glucagon could increase ureogenesis in vivo, and many in vitro observations suggesting that the hormone could indeed stimulate gluconeogenesis (EXTON 1972; EXTON et al. 1968; EXTON and PARK 1969; GARCIA et al. 1966; CLAUS and PILKIS 1981), studies directly establishing glucagon as a gluconeogenic hormone in vivo have only recently been performed.

In 1974, CHERRINGTON and VRANIC showed that, in overnight fasted dogs, large increments in plasma glucagon increased the rate at which glucose recycled, a finding consistent with an increase in gluconeogenesis. That same year, CHIASSON et al. (1974) published data demonstrating the ability of high levels of glucagon to stimulate the conversion of alanine $^{14}C$ to glucose $^{14}C$ in the intact dog. A year later, they confirmed their observation in overnight fasted humans (CHIASSON et al. 1975). In the latter study, glucagon (15–50 ng $kg^{-1}$ $min^{-1}$) caused a 93% increase in the rate of splanchnic glucose $^{14}C$ production by subjects being given a constant infusion of alanine $^{14}C$. This gluconeogenic effect of glucagon was exerted in the

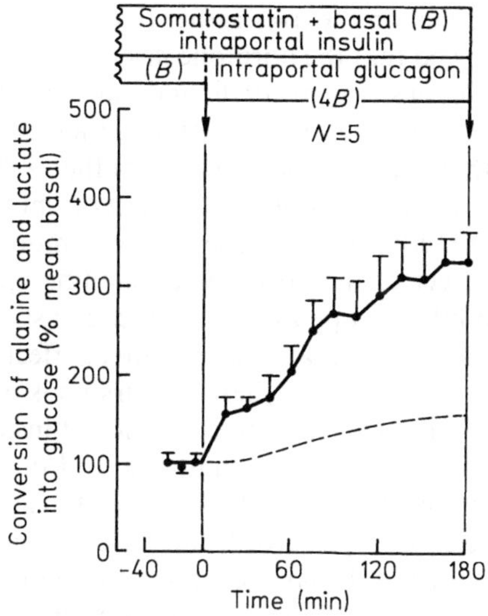


**Fig. 4.** The effect of hyperglucagonemia produced by the intraportal infusion of glucagon (2.6 ng $kg^{-1}$ $min^{-1}$) along with somatostatin (0.8 μg $kg^{-1}$ $min^{-1}$) and basal replacement infusions of insulin (300 μIU $kg^{-1}$ $min^{-1}$) on the conversion rate of circulating alanine and lactate to glucose in a group of five normal overnight fasted conscious dogs. The *broken line* represents the mean of five dogs, where glucagon was maintained at basal rate along with somatostatin and basal replacement infusion of insulin. The data are mean ± standard error. Adapted from CHERRINGTON et al. (1981)

absence of any change in net splanchnic alanine extraction, thus the hormone was shown to be capable of stimulating gluconeogenesis at an intrahepatic site.

All of these studies employed rather high doses of glucagon and thus the physiologic importance of the hormone as a gluconeogenic regulator remained unclear. Recently, however, more physiologic changes in the plasma level of the hormone have also been shown to stimulate gluconeogenesis in vivo (CHERRINGTON et al. 1981). The major reason for the difficulty in detecting the effect of small increments in glucagon on gluconeogenesis has been the potent inhibitory effect of the insulin normally mobilized by the hormone. It is now evident that the 8–10 μIU/ml rise in the peripheral plasma insulin level (25–30 μIU/ml in the portal vein) is enough to obliterate the effect of the hormone on gluconeogenesis completely. If the glucagon level is raised fourfold and insulin release is prevented, glucagon causes a threefold increase in the conversion rate of alanine $^{14}C$ and lactate $^{14}C$ to glucose $^{14}C$ (Fig. 4) within 3 h. Interestingly in this study, the prolonged but physiologic increment (200–250 pg/ml) in plasma glucagon caused a significant fall in the plasma alanine concentration (CHERRINGTON et al. 1981). This fall was probably attributable to a stimulatory effect of the hormone on hepatic alanine uptake rather than an inhibitory effect on peripheral alanine release since the concentration of radioactive alanine in plasma also fell, whereas the alanine $^{14}C$ specific activity remained unchanged. More recent studies have shown directly that a 200–250 pg/ml increment in plasma glucagon can cause a marked (twofold) increase in net hepatic alanine uptake (STEINER et al. 1981). The concept that glucagon alters amino acid metabolism through an effect on the liver rather than the periphery is further supported by data which showed that the hormone has no effect on net alanine release by the human forearm (POZEFSKY et al. 1976) and that it augments hepatic alanine uptake in vitro (MALLETTE et al. 1969; LECAM and FREYCHET 1976).

It should be noted that these studies do not conflict with the earlier study of CHIASSON et al. (1974) in which an action of glucagon on splanchnic alanine uptake was not detected. In that study, the response was only monitored for 1 h and it is now apparent that the effect of the hormone on alanine uptake requires at least 90 min to be manifest. While it is apparent that glucagon stimulates hepatic amino acid uptake in vivo, it is also clear that it has no direct effect on lactate or glycerol uptake by the liver since the uptake of these precursors is not regulated. In addition, it should be noted that glucagon in the presence of insulin has little if any effect on the release of lactate or glycerol from muscle and/or fat in dogs or humans. It should be mentioned, however, that in the absence of insulin, glucagon can stimulate lipolysis and the release of glycerol from fat (see Chap. 19).

It is also apparent from Fig. 4 that the stimulatory effect of glucagon on gluconeogenesis, unlike its effect on glycogenolysis, is not time dependent. This observation confirms that the "downregulation" observed with overall glucose production is indeed attributable to a fall in glycogenolysis. The explanation for the differential time course of the hormone's effect on the two processes is at present not clear (see CHERRINGTON et al. 1981). It is nevertheless evident that, by virtue of its ability to enhance glycogenolysis for only a short period while augmenting gluconeogenesis for a prolonged period, glucagon can rapidly change the liver from a glycogenolytic to a gluconeogenic organ.

Recently, the importance of basal amounts of glucagon in regulating gluconeogenesis in overnight fasted dogs has been examined (JENNINGS et al. 1977; CHERRINGTON et al. 1978 b). In both of these studies, isolated glucagon deficiency (basal insulin present) was associated with a modest decrease in the conversion rate of alanine to glucose as can be seen in Fig. 1. However, since gluconeogenesis is a minor contributor to overall glucose production after an overnight fast, even a 15%–40% decline in the process is of little consequence to overall glucose output.

Perhaps a better way to examine the gluconeogenic potential of basal amounts of glucagon is to examine the role the hormone plays in regulating the process during a more prolonged fast. WAHREN et al. (1977) infused somatostatin to inhibit both glucagon and insulin secretion in 60-h fasted humans and examined the consequences of this perturbation on splanchnic glucose balance. Despite the concomitant deficiency of insulin (40%), glucagon lack was associated with a 70% reduction in glucose output (Fig. 5). This drop in glucose production was associated with a moderate decline in splanchnic lactate (28%), pyruvate (42%), and alanine (15%) extraction, but an increase in the splanchnic uptake of glycerol (500%). The latter was attributable to an almost proportional rise in the plasma glycerol concentration, which itself undoubtedly reflected an increase in lipolysis subsequent to the induction of insulin deficiency. Since glucose output fell even though the total gluconeogenic precursor uptake increased modestly, these data suggest that the primary mechanism by which the inhibition of glucose production occurs in such fasted subjects is an intrahepatic inhibition of net flux up the gluconeogenic pathway. Similarly, in the 72-h fasted dog, selective glucagon deficiency was associ-

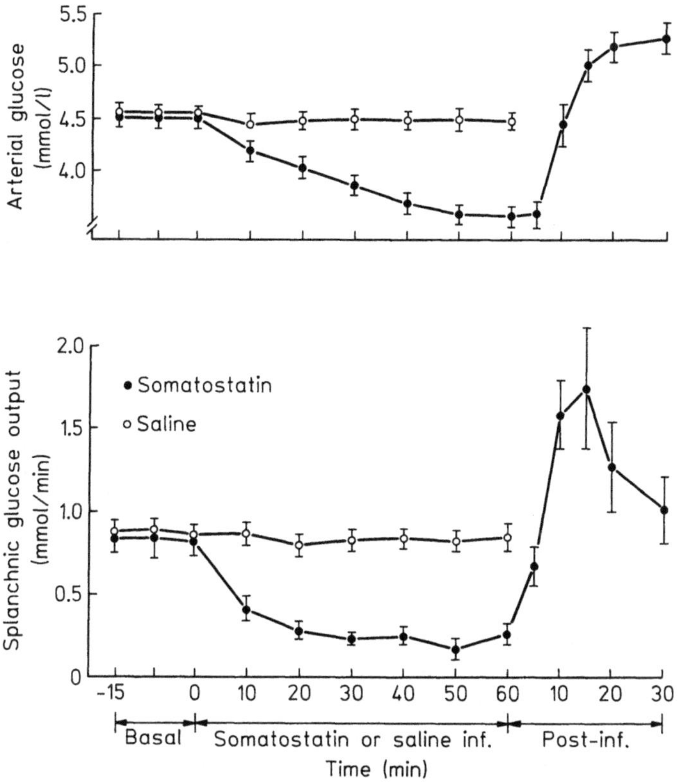


**Fig. 5.** Arterial glucose concentration and splanchnic glucose output in the basal state and during infusion of somatostatin at 10 μg/min or saline in 60-h fasted human subjects. The data are mean ±standard error. WAHREN et al. (1977)

ated with a decline in net hepatic glucose output (25%) as well as a drop in the plasma glucose level (KELLER et al. 1978).

In summary, the data currently available indicate that basal amounts of glucagon, as well as physiologic increments in the level of the hormone, play an important role in regulating gluconeogenesis in vivo. Glucagon appears to play a particularly important role in the enhancement of gluconeogenesis associated with prolonged fasting. After an overnight fast, the gluconeogenic rate is low as a result of a marked inhibition by insulin. At this time, glucagon has relatively little effect on the process. As the fast progresses, however, the plasma insulin level falls and in some species, most notably humans, the plasma glucagon level rises. As a result of these changes glucagon becomes an important gluconeogenic regulator. Glucagon exerts its stimulatory effect on gluconeogenesis at two points: on the enzymatic setting of the hepatocyte and on the amino acid transport process at the plasma membrane. It is now clear that, in contrast to its effect on glycogenolysis, the stimulatory action of an increment in the plasma level of the hormone on gluconeogenesis is slow in onset but sustained (both with regard to the enzymatic and transport changes). As a result, an increment of the hormone can switch the liver of an overnight fasted animal from a glycogenolytic to a gluconeogenic mode within 2–4 h.

## C. Glucagon – Insulin Interaction in the Regulation of Hepatic Glucose Production

In the preceeding sections, the effects of glucagon on glycogenolysis and gluconeogenesis have been described separately. In view of the importance of insulin and its opposing actions on the regulation of these two processes, we would be remiss not to consider the interaction of this hormone with glucagon in the regulation of glucose production in vivo. It is now evident that a dynamic interaction exists between the two hormones which is responsible for the fine regulation of both glycogenolysis and gluconeogenesis in the whole animal.

### I. Glycogenolysis

Several studies have examined the interaction of glucagon and insulin in regulating glycogenolysis in vitro (JEFFERSON et al. 1968; GLINSMANN and MORTIMORE 1968; WAGLE 1975). The interaction of the two hormones in regulating the process in vivo, however, has been more difficult to assess. It is only recently that BOMBOY et al. (1977b) presented human data demonstrating the potential ability of insulin to inhibit the glycogenolytic activity of glucagon. In this study, high levels of insulin (1,500 μIU/ml) were brought about by constant insulin infusion and normoglycemia was maintained by variable glucose infusion. Infusion of glucagon at 5 ng $kg^{-1}$ $min^{-1}$ had no effect on the suppressed splanchnic glucose output, but when glucagon was infused at 15 ng $kg^{-1}$ $min^{-1}$, splanchnic glucose production increased and insulin's inhibitory effect was overcome. Similar observations were made by FELIG et al. (1976) in normal humans in whom a small increment in the plasma glucagon level (272 pg/ml) was sufficient to partially prevent an insulin-induced suppression of net splanchnic glucose output. These observations were the first to suggest that small increments in circulating glucagon could reverse the inhibitory effect of insulin on glycogenolysis in vivo. More recently, two studies in the dog have examined the interaction of increments in insulin and glucagon in the regulation of glycogenolysis more closely. In one case, the effects of a given insulin: glucagon molar ratio brought about using different absolute levels of the two hormones was examined (STEINER et al. 1981) and in the other, the relative importance of the two phases of insulin release in countering the action of glucagon was assessed (STEINER et al. 1980).

In the first case, overnight fasted dogs were given somatostatin and basal replacement amounts of insulin and glucagon intraportally to break the glucoregulatory feedback loops involving the endocrine pancreas. After a control period in which the plasma levels of the two hormones were basal, they were increased fourfold (but the insulin: glucagon molar ratio was kept fixed) for 3 h. As a result of this change there was a marked suppression of glucose production. Thus, as the absolute levels of the two hormones increase insulin tends to dominate, at least with regard to glycogenolysis (STEINER et al. 1981). The concept of hepatic metabolism being proportional to the insulin: glucagon ratio, rather than to the absolute levels of plasma insulin and glucagon, is thus of limited value. Such has previously been suggested for other reasons by CHERRINGTON et al. (1974), EL-REFAI and BERGMAN (1979), and DERI et al. (1981).

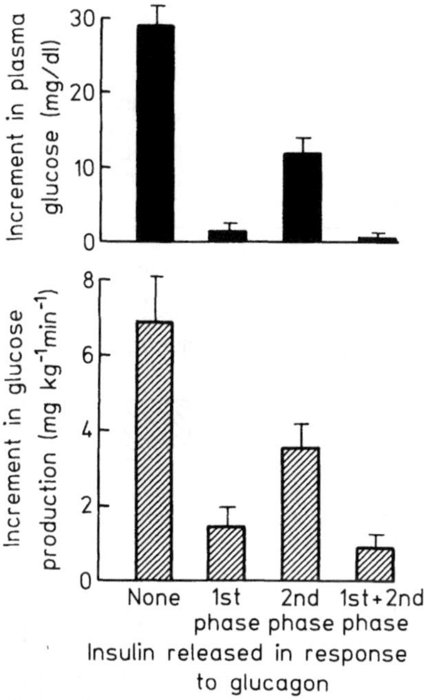


**Fig. 6.** Importance of first and second phase insulin secretion in countering the effects of glucagon on glucose production and the plasma glucose levels in overnight fasted conscious dogs. Animals were initially given somatostatin (0.8 μg $kg^{-1}$ $min^{-1}$) as well as insulin (250 μIU $kg^{-1}$ $min^{-1}$) and glucagon (0.65 ng $kg^{-1}$ $min^{-1}$) in basal amounts. After a 40-min control period, the glucagon infusion rate was increased to 2.6 ng $kg^{-1}$ $min^{-1}$ and insulin was given so as to simulate first (peak insulin level of 35 μIU/ml at 5 min), second (peak insulin level of 21 μIU/ml at 15 min), or combined first and second phase insulin release (peak at 5 min, 35 μIU/ml, and steady increment, 21 μIU/ml), respectively. The data are mean $\pm$ standard error from the initial point of measurement (7.5). Adapted from STEINER et al. (1982)

In the other study, the importance of the two phases of insulin release in countering glucagon-induced glycogenolysis were examined. In normal dogs, intraportal infusion of glucagon at a rate of 2.6 ng $kg^{-1}$ $min^{-1}$ causes a biphasic release of insulin with an initial peak ($35 \pm 5$ μIU/ml) at 7.5 min and the second peak ($21 \pm 3$ μIU/ml) at 30 min. In the study of STEINER et al. (1980), dogs were first given somatostatin and intraportal replacement infusions of insulin and glucagon in order to break the glucoregulatory feedback loops involving the endocrine pancreas. After a control period in which both hormones were given in basal amounts, glucagon was infused at a rate four times basal (2.6 ng $kg^{-1}$ $min^{-1}$) and insulin was given in such a way as to simulate the first, the second, or the first and second phases of insulin release, respectively. The initial increment in glucose production (glycogenolysis) caused by glucagon was reduced from ~6.9 to 1.5, 3.5, and 0.9 mg $kg^{-1}$ $min^{-1}$ by first phase, second phase, and combined first and second phase insulin release, respectively (Fig. 6). It is clear that only a small amount of insulin is required to inhibit the in vivo action of glucagon on glycogenolysis markedly. It is also clear from these experiments that the presence of the initial insulin spike facilitates glycemic control significantly by ensuring a quick and marked inhibition

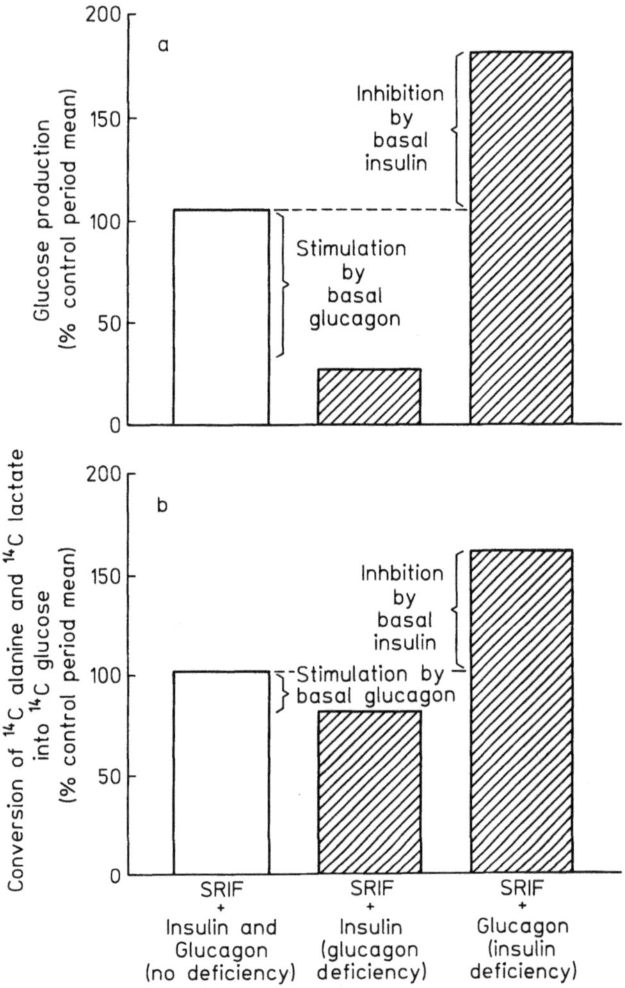


**Fig. 7 a, b.** Effects of insulin and glucagon deficiency on total glucose production (**a**) and the gluconeogenic conversion of alanine and lactate into glucose (**b**) in normal, overnight fasted conscious dogs. The data are mean ± standard error; SRIF = somatostatin. CHERRINGTON and LILJENQUIST (1981)

of glucagon action. The plasma glucose level was elevated by 61 mg/dl after 7.5 min when insulin mobilization was not possible, but rose by only 14, 33, and 2 mg/dl, respectively when first, second, and combined first and second phase insulin release were simulated (Fig. 6). In summary, although an increment in glucagon represents a potent glycogenolytic stimulus, its effects can be readily blocked, even by small amounts of insulin.

Figure 7 a demonstrates that interactions between insulin and glucagon occur even when the hormones are present in basal amounts. Acute removal of both insulin and glucagon from an overnight fasted dog results in a small initial drop in glucose production, followed by a return of the rate to the baseline after about 1 h. The explanation for the transient fall in glucose output is thought to be that the effects of glucagon lack are manifest more rapidly than are the consequences of insulin deficiency. Selective removal of glucagon, as shown in the figure, results in a marked and sustained fall in glucose production which is in fact a consequence of the inhibitory effect of the remaining insulin. Selective insulin deficiency, on the other hand, is associated with a marked rise in glucose production, followed by a decline to the baseline after 3 h (see Fig. 3). This rise is attributable to basal glucagon since, when insulin is made deficient in the absence of glucagon, there is no increase in glucose production. After an overnight fast, therefore, glucagon exerts a marked stimulatory effect on glycogenolysis while insulin restrains the process.

### II. Gluconeogenesis

Studies have also been carried out to address the question of glucagon–insulin interaction in the regulation of gluconeogenesis in vivo. Most of these studies, however, have examined control after an overnight fast when the gluconeogenic rate is low and the data should be interpreted with caution. Figure 7 b shows data which demonstrate the role of basal amounts of the two hormones in regulating the conversion of alanine and lactate to glucose in dogs fasted for 16 h. Glucagon only exerts a small stimulatory effect, presumably because of the potent gluconeogenic inhibition exerted by insulin. If insulin deficiency is produced in the presence of basal glucagon, gluconeogenesis rises markedly (Fig. 7) and in fact increases as much as fourfold within 4 h. It is primarily the unrestrained action of glucagon which is responsible for the rise in gluconeogenesis since, if glucagon deficiency is produced at the same time that insulin is withdrawn, gluconeogenesis rises by only 150% (CHERRINGTON et al. 1977). Selective insulin removal in 72-h fasted dogs results in a doubling of glucose production (KELLER et al. 1978) which undoubtedly reflects an increase in gluconeogenesis.

The interaction between elevated levels of insulin and glucagon in regulating gluconeogenesis in the whole animal has not been adequately studied. It is clear that the gluconeogenic effect of a fourfold rise in glucagon can be completely abolished by a small (15–30 μIU/ml) rise in the portal plasma insulin level (STEINER et al. 1980). It would thus seem that the gluconeogenic potential of an elevation in plasma glucagon is determined by the change in plasma insulin which accompanies it. If insulin drops (as with fasting), the gluconeogenic effect of glucagon is enhanced, but if insulin rises (as with protein feeding), the gluconeogenic effect of glucagon is inhibited and the hormone has a selective glycogenolytic effect. Definitive studies relating to the interplay of insulin–glucagon in the fasted state remain to be carried out.

## D. Role of Glucagon in the Regulation of Glucose Homeostasis

Glucose homeostasis refers to the maintenance of the plasma glucose concentration within a narrow range under a wide variety of conditions. To achieve this control, the body is equipped to store glucose when it is presented in excess and to release it or even synthesize it de novo when it is deficient. These processes are regulated, at least in part, by changes in the plasma glucagon level. We will now examine three different metabolic situations in which changes in the plasma glucagon concentration are thought to play a role in the regulation of the plasma glucose level. These conditions are:

a) Protein (increase in glucagon and insulin) or carbohydrate feasting (decrease in glucagon and increase in insulin)
b) Fasting (increase in glucagon and decrease in insulin)
c) Exercise (increase in glucagon and decrease in insulin)
d) Diabetes mellitus (increase in glucagon and deficiency of insulin).

## I. Feasting

UNGER et al. (1969) were the first to suggest that the stimulation of glucagon release, which occurs simultaneously with the stimulation of insulin secretion during protein ingestion, might be necessary to prevent hypoglycemia. This hypothesis was tested by CHERRINGTON and VRANIC (1973) who studied the effect of arginine on the release of the two major pancreatic hormones and their subsequent effects on glucose turnover. When arginine was infused intravenously into dogs, both glucagon and insulin release were stimulated, as were the rates of glucose production and utilization. Since the two rates increased in parallel, the plasma glucose level remained unchanged. When the study was repeated in depancreatized dogs in which glucagon and insulin were infused to achieve the same hormonal levels as were observed in normal dogs during arginine infusion, glucose production and utilization again increased in parallel and normoglycemia was again maintained (CHERRINGTON et al. 1974). However, inadequate replacement of glucagon resulted in a diminished increase in glucose production and consequent hypoglycemia. These studies thus supported the concept of UNGER and emphasized the importance of glucagon release in the maintenance of glucose homeostasis during ingestion of a protein meal. In normal humans, protein ingestion results in a modest increase in both glucagon and insulin with little change in glucose production, utilization, or the plasma glucose level (RASKIN et al. 1978; WAHREN et al. 1976). However when a protein meal is ingested by insulin-dependent diabetics, a rise in glucagon occurs in the absence of any increment in insulin and as a result glucose production increases, suggesting that glucagon may also play a role in stabilizing the blood glucose level during protein feeding in humans.

The ingestion of carbohydrates is usually associated with a decrease in the plasma glucagon concentration and an increase in the plasma insulin level. When glucagon was infused (3 ng $kg^{-1}$ $min^{-1}$) for 3 h prior to and during an oral glucose tolerance test, however, it did not induce an abnormality in glucose disposal (SHERWIN et al. 1977). This finding suggested that modification of the glucagon level is of little importance to glycemic control during carbohydrate feeding. Unfortunately, this study was not definitive since it did not exclude the possibility that an effect of glucagon was not apparent because the glucose load was given after the hormone's effect on glycogen metabolism had waned. This turned out not to be a problem, however, since in a study in which the plasma glucagon level was increased concomitantly with glucose, there was still no effect of glucagon on glucose tolerance (LILJENQUIST and RABIN 1979). Taken together, these studies indicate

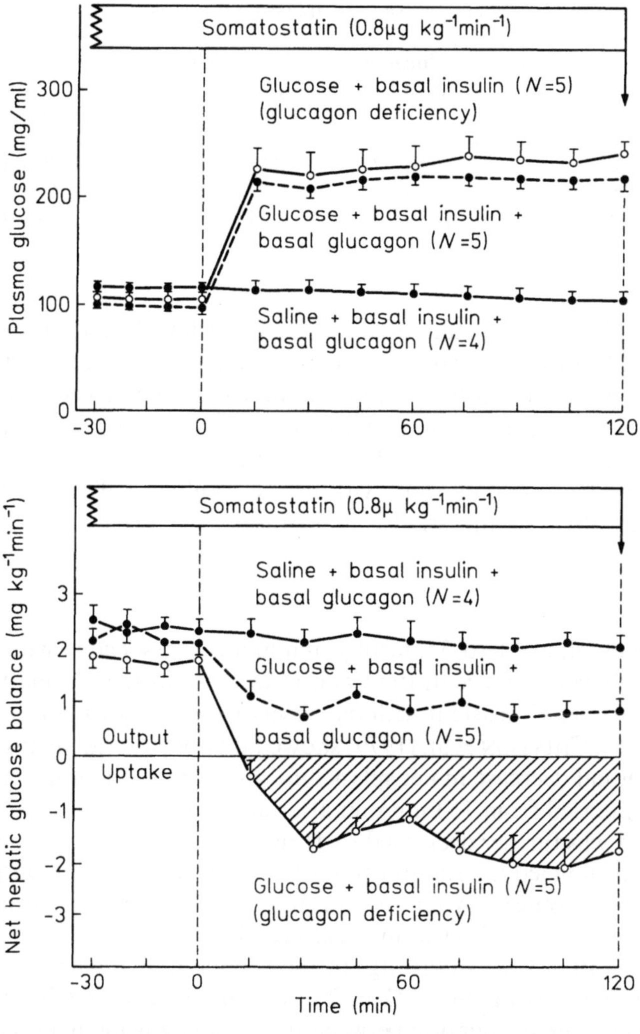


**Fig. 8.** The effect of basal glucagon on the ability of hyperglycemia to trigger net hepatic glucose uptake in the presence of basal amounts of insulin in overnight fasted conscious dogs. Animals were initially given somatostatin (0.8 µg $kg^{-1}$ $min^{-1}$) along with basal replacement amounts of insulin (300 µIU $kg^{-1}$ $min^{-1}$) and glucagon (1.0 ng $kg^{-1}$ $min^{-1}$). After a 30-min control period, three protocols were instituted; (1) the basal hormone infusions were continued and normoglycemia was maintained; (2) the basal hormone infusions were continued while hyperglycemia was established by glucose infusion; and (3) the insulin infusion was continued, but the glucagon infusion was discontinued while hyperglycemia was established as in protocol 2. SHULMAN et al. (1978)

that in a normal individual the disposal of a glucose load is not dependent on the fall in the plasma glucagon which usually accompanies carbohydrate ingestion.

Recently, however, SHULMAN et al. (1978) have shown that glucagon suppression can be important if glucose does not stimulate adequate insulin release. They raised the plasma glucose level to 220 mg/dl in the presence of basal insulin

**Table 1.** Glucagon and hepatic glucose output in fasting humans

| Length of fast (days) | Number of subjects | Glucagon (pg/ml) | Net splanchnic glucose production (mg/min) | Extracted alanine converted to glucose (%) |
|---|---|---|---|---|
| 0.5 | 8 | 102 ± 7 | 153 ± 10 | 39 ± 6 |
| 2 | 9 | 175 ± 7 | 72 ± 9 | 61 ± 4 |
| 18 | 3 | 157 ± 17 | 41 ± 13 | 54 ± 7 |

and either kept glucagon constant or allowed it to decline. Net hepatic glucose production fell by 56% when glucagon was kept constant, as can be seen in Fig. 8. On the other hand, when glucagon was reduced, the same hyperglycemic signal not only caused complete suppression of glucose production, but also caused net hepatic glucose uptake (2 mg $kg^{-1}$ $min^{-1}$). These experiments thus suggest that glucagon suppression may become important with regard to glucose disposal if the insulin response to carbohydrate ingestion is inadequate.

## II. Fasting

As reviewed in detail in Chap. 34, fasting in humans is associated with an increase in the plasma glucagon concentration and a decrease in the plasma insulin level (MARLISS et al. 1970). Consistent with the observations of other (FELIG et al. 1969; OWEN et al. 1969), CHIASSON et al. (1979) showed that prolonged fasting was associated with a gradual decrease in net splanchnic glucose production (Table 1). Glucose output decreased from 153 mg/min after an overnight fast, to 72 mg/min by 48 h fasting and was further decreased to 41 mg/min by 3 weeks fasting. However, as fasting was prolonged a larger portion of total hepatic glucose output was derived from gluconeogenesis. After an overnight fast, 39% of the extracted alanine was converted to glucose. This was increased to 61% after 48 h fasting and was still increased (54%) after 3 weeks fasting. In these same individuals, glucagon levels were 102, 175, and 157 pg/ml at 12 h, 2 days, and 3 weeks fasting respectively. There was thus a good correlation between the plasma glucagon level and the rate of gluconeogenesis as fasting progressed. The data suggest therefore, that during a fast glucagon increases to ensure maintenance of the plasma glucose level by means of its strong gluconeogenic effect. In agreement with this concept are the observations mentioned earlier which show clearly that: (1) an increase in glucagon markedly augments gluconeogenesis; (2) a fall in insulin substantially potentiates the effect of glucagon on gluconeogenesis; and (3) that the augmented gluconeogenesis apparent after a prolonged fast is dependent on the availability of glucagon.

## III. Exercise

VRANIC and his co-workers have attempted over the past few years to define the role of glucagon in regulating glucose production during exercise in the dog. Their studies have shown that strenuous exercise in normal dogs is associated with a 2-

to 3-fold rise in the circulating glucagon level, a 50% decline in the plasma insulin concentration, and a threefold increase in both glucose production and utilization. Similar changes have also been reported to occur in humans (FELIG et al. 1972; FELIG and WAHREN 1978; ZINNMAN et al. 1980). Since the latter two alterations occur in parallel, normoglycemia is maintained. When depancreatized dogs maintained on constant basal intraportal insulin infusions were exercised in a similar manner, they exhibited near normal increases in both glucose production and utilization. It should be noted that such animals have normal plasma glucagon levels, and although the hormone is derived from gastrointestinal A-cells, it is equipotent to 3,500 daltons glucagon from the pancreas (KEMMER and VRANIC 1981). Surprisingly the exercise-induced increase in glucose production occurred in the depancreatized animals, even though the plasma glucagon level did not change in response to exercise. Since glucose production rose normally, even in the absence of an increase in immunoreactive glucagon, these data suggest that the exercise-induced increment in plasma glucagon was not essential for the increase in glucose output. The data thus suggest that the increase in glucose production seen during exercise is mediated in large measure by factors other than the increment in circulating glucagon.

More recently ISSEKUTZ and VRANIC (1980) exercised dogs in which glucagon and insulin were markedly suppressed by somatostatin. Under these conditions, the maximal rate of glucose production observed during the exercise period was reduced by 2–3 mg $kg^{-1}$ $min^{-1}$. Since the acute removal of glucagon and insulin in a normal dog is associated with a drop in glucose production of 1 mg $kg^{-1}$ $min^{-1}$, these data suggest that, while the exercise-induced increment in glucose output occurs in part independently of glucagon, the hormone may be required in a permissive sense since it was required for full manifestation of the exercise-induced increase in glucose production. For a more complete review of this area see Chap. 38.

## IV. Diabetes Mellitus

Although it is not within the scope of this chapter to address the role of glucagon in the pathogenesis of diabetic hyperglycemia, a subject treated in Chaps. 37 and 44, some information regarding the regulation of hepatic glucose production by glucagon in diabetes seems warranted. Administration of somatostatin to insulin-dependent diabetics at the same time as their insulin is withdrawn results in an immediate suppression of plasma glucagon, an initial fall in the plasma glucose level, followed by a slow rise to levels slightly above basal (GERICH et al. 1974). Although glucose production was not measured in this study, the changes in plasma glucose suggested that glucose production had initially declined, but then returned to a basal rate, as it has been shown to do in normal humans after somatostatin-induced insulin and glucagon suppression (LILJENQUIST et al. 1977). Similar results have been reported in maturity-onset diabetic subjects in whom insulin and glucagon were also suppressed by somatostatin (TAMBORLANE et al. 1977). LUNDBAEK et al. (1976), on the other hand, reported that the inhibition of glucagon secretion in juvenile diabetics withdrawn from insulin for 60–72 h resulted in a progressive fall in plasma glucose such that it finally reached a plateau and did not rise again.

These data showed that suppression of glucagon secretion many hours after insulin withdrawal does in fact cause a sustained fall in glucose production.

Thus, the effect of glucagon suppression on glucose production in the diabetic appears to be influenced by the insulin status of the individual. In the totally insulin-deficient diabetic, glucose production is increased because of the increased glucagon level. Under these conditions, glucagon suppression will result in sustained decreased glucose production to the nonhormonal-stimulated rate. In the insulin-treated diabetic, glucose production is normal as well as the glucagon level. In this situation, glucagon suppression will result in an initial decrease in glucose production which will then rise back to the nonhormonal-stimulated rate. In this latter case, the rise in glucose production corresponds to a wearing off of the insulin effect. In both situations, the final glucose production rate is essentially the same.

The administration of glucagon in physiologic amounts to diabetic humans has been shown to increase splanchnic glucose production within 5 min (BOMBOY et al. 1977a). However, as in normal humans, glucose production rapidly declines, despite continuing hyperglucagonemia and the absence of endogenous insulin secretion, thus the diabetic has a noninsulin-mediated hepatic mechanism to limit glucagon-stimulated glucose production. As noted earlier, the details of this "down-regulatory" mechanism are not clear, but it may play a particularly important role in limiting glucagon-induced hyperglycemia in the diabetic who is unable to rely on insulin to counter the action of glucagon.

Even though the initial rise in hepatic glucose production attributable to glucagon-stimulated glycogenolysis is short-lived in the diabetic, the hyperglycemia that results may be sustained. As noted in Fig. 3, the induction of selective insulin deficiency (glucagon maintained at basal levels with insulin secretion markedly suppressed) resulted in only a short burst of glucose output. Since glucose production initially exceeded glucose utilization, the plasma glucose level rose markedly and in doing so enhanced glucose utilization. Since glucose output subsequently fell, the rates of glucose production and utilization soon became equal and the plasma glucose level plateaued. In the absence of an ability to release insulin in response to hyperglycemia and thus to increase glucose utilization, the hyperglycemia attributable to glucagon was of long duration. If the glucose level had continued to decline at the rate apparent during the fourth hour of selective insulin lack, it would have taken ~12 h to reach a level of 170 mg/dl, the maximal level achieved as a result of inducing insulin deficiency in the absence of glucagon. It is thus evident that when insulin is deficient, a brief burst of glucagon-stimulated hepatic glucose production (lasting only minutes) can result in prolonged hyperglycemia (lasting many hours).

In the clinical setting of diabetic humans, glucagon-stimulated glucose production may result in a worsening of the plasma glucose level in three situations. The first occurs when an insulin-dependent diabetic does not take insulin on time. The plasma insulin concentration will fall below the desired level and an increase in glucagon-induced glucose production will occur. Furthermore, since the A-cell is insulin sensitive (GERICH et al. 1976), insulin deficiency not only augments the effect of glucagon on the liver directly, but also causes a rise in the plasma glucagon level itself which further stimulates the liver. The second situation in which a deleterious effect of glucagon might occur is when a diabetic feeds at a time when insulin is inadequate. Consumption of a protein meal will result in enhanced glucagon release which will worsen glycemic control. In addition, failure of normal glucagon suppression during carbohydrate loading in an insulin-deficient situation, could impair disposition of the glucose load. The third situation in which glucagon may be particularly harmful is when a poorly controlled diabetic patient is subjected to stress. Catecholamines have been shown to cause marked glucagon release when insulin is deficient (PEREZ et al. 1981). Thus, diabetic patients subjected to stress can be exposed to the combined deleterious effects of catecholamines and glucagon.

It is evident that glucagon can be detrimental in the juvenile diabetic at various times, all of which are characterized by the presence of inadequate insulin. When glucagon is mobilized inappropriately in the diabetic, it causes a brief burst of glycogen breakdown which rapidly raises the patient's blood glucose. The hormone then pushes the patient into a state characterized by elevated gluconeogenesis. Insulin-dependent diabetics off insulin for 24 h exhibit a 40% increase in the fractional extraction of gluconeogenic amino acids by the liver (CHERRINGTON 1981). In addition it is now clear that intrahepatic gluconeogenesis is stimulated 3- to 4-fold in diabetic dogs off insulin for 24 h (CHERRINGTON et al. 1981). If the plasma glucagon level is decreased in such animals, gluconeogenesis falls markedly and overall glucose production is halved. Clearly therefore glucagon is a primary determinant of the accelerated gluconeogenesis evident in the uncontrolled diabetic. In view of the potentially deleterious effect of glucagon on the blood glucose level, somatostatin has been advocated for use in the diabetic as a means to suppress glucagon and thus enhance glucose control. While improvement in patients glycemic status has been reported with the agent (RIZZA and GERICH 1978), other problems associated with its use have prevented its widespread application. Furthermore, with glucagon eliminated, the body's first line of defense against hypoglycemia is absent and the patient may very well be at greater risk of hypoglycemia.

## References


Altszuler N, Gottlieb B, Hampshire J (1976) Interaction of somatostatin, glucagon and insulin on hepatic glucose output in the normal dog. Diabetes 25:116–121

Blackard WJ, Nelson NC, Andrews SS (1974) Portal and peripheral vein immunoreactive glucagon concentrations after arginine or glucose infusion. Diabetes 23:199–206

Bomboy JD, Lewis SB, Lacy WW, Sinclair-Smith BC, Liljenquist JE (1977a) Transient stimulatory effect of sustained hyperglucagonemia on splanchnic glucose production in normal and diabetic man. Diabetes 26:177–184

Bomboy JD, Lewis SB, Lacy WW, Sinclair-Smith BC, Liljenquist JE (1977b) Insulin-glucagon interaction in controlling splanchnic glucose production in normal man. J Clin Endocrinol Metab 44:474–480

Brazeau P, Vale W, Burgus R, Ling N, Butcher M, Rivier J, Guillemin R (1973) Hypothalamic peptide that inhibits the secretion of immunoreactive pituitary growth hormone. Science 179:77–79

Cherrington AD (1981) Glucgoneogenesis: its regulation by insulin and glucagon. In: Brownlee M (ed) Diabetes mellitus. Garland, New York, pp 49–117

Cherrington AD, Liljenquist JE (1981) Role of glucagon in regulating glucose production in vivo. In: Unger RH, Orci L (eds) Glucagon: physiology, pathophysiology and morphology of the pancreatic A cells. Current endocrinology. Elsevier, Amsterdam Oxford New York, pp 221–253

Cherrington AD, Vranic M (1973) Effect of arginine on glucose turnover and plasma free fatty acids in normal dogs. Diabetes 22:537–543
Cherrington AD, Vranic M (1974) Effect of interaction between insulin and glucagon on glucose turnover and FFA concentrations in normal and depancreatized dogs. Metabolism 23:729–744
Cherrington AD, Kawamori R, Pek S, Vranic M (1974) Arginine infusions in dogs: model for the role of insulin and glucagon in regulating glucose turnover and free fatty acid levels. Diabetes 23:805–815
Cherrington AD, Chiasson JL, Liljenquist JE, Hennings AS, Keller U, Lacy WW (1976) The role of insulin and glucagon in the regulation of basal glucose production in the postabsorptive dog. J Clin Invest 58:1407–1418
Cherrington AD, Williams PE, Liljenquist JE, Lacy WW (1977) The control of glycogenolysis and gluconeogenesis in vivo by insulin and glucagon. Int Congr Ser 459:179–191
Cherrington AD, Lacy WW, Chiasson JL (1978 a) Effect of glucagon on glucose production during insulin deficiency in the dog. J Clin Invest 62:664–677
Cherrington AD, Chiasson JL, Liljenquist JE, Lacy WW, Park CR (1978 b) Control of hepatic glucose output by glucagon and insulin in intact dog. Biochem Soc Symp 43:31–45
Cherrington AD, Liljenquist JR, Shulman GI, Williams PE, Lacy WW (1979) Importance of hypoglycemia-induced glucose production during isolated glucagon deficiency. Am J Physiol 236:E 263–E 271
Cherrington AD, Williams PE, Shulman GI, Lacy WW (1981) Differential time course of glucagon's effect on glycogenolysis and gluconeogenesis in the conscious dog. Diabetes 30:180–187
Chiasson JL, Cook J, Liljenquist JE, Lacy WW (1974) Glucagon stimulation of gluconeogenesis from alanine in the intact dog. Am J Physiol 227:19–23
Chiasson JL, Liljenquist JE, Sinclair-Smith BC, Lacy WW (1975) Gluconeogenesis from alanine in normal postabsorptive man. Intrahepatic stimulatory effect of glucagon. Diabetes 24:574–584
Chiasson JL, Atkinson RL, Cherrington AD, Keller U, Sinclair-Smith BC, Lacy WW, Liljenquist JE (1979) Effects of fasting on gluconeogenesis from alanine in nondiabetic man. Diabetes 28:56–60
Claus TH, Pilkis SJ (1981) Hormonal control of hepatic gluconeogenesis. In: Litwach GS (ed) Biochemical actions of hormones, vol 8. Academic, New York London, pp 209–271
De Bodo RC, Steele R, Altszuler N, Dunn A, Bishop JS (1963) On the hormonal regulation of carbohydrate metabolism: studies with $^{14}$C-glucose. Recent Prog Horm Res 19:445–488
Deri JJ, Williams PE, Steiner KE, Cherrington AD (1981) Altered ability of the liver to produce glucose following a period of glucagon deficiency. Diabetes 30:490–495
El-Refai M, Bergman RN (1979) Glucagon-stimulated glycogenolysis: time-dependent sensitivity to insulin. Am J Physiol 28:E 248–E 254
Exton JH (1972) Gluconeogenesis. Metabolism 21:945–990
Exton JH, Park CR (1969) Control of gluconeogenesis in liver. III. Effects of L-lactate, pyruvate, fructose, glucagon, epinephrine and adenosine 3′,5′-monophosphate on gluconeogenic intermediates in perfused rat liver. J Biol Chem 244:1424–1433
Exton JH, El-Refai MF, Park CR (1968) Control of gluconeogenesis in liver. II. Effects of glucagon, catecholamines and adenosine 3′,5′-monophosphate on gluconeogenesis in perfused rat liver. J Biol Chem 243:4189–4196
Felig P (1973) The glucose-alanine cycle. Metabolism 22:179–207
Felig P, Owen OE, Wahren J, Cahill GF (1969) Amino acid metabolism during prolonged starvation. J Clin Invest 48:584–594
Felig P, Wahren J, Hendler R, Ahlborg G (1972) Plasma glucagon levels in exercising man. N Engl J Med 286:184–185
Felig P, Wahren J, Hendler R (1976) Influence of physiologic hyperglucagonemia on basal and insulin-inhibited splanchnic glucose output in normal man. J Clin Invest 58:761–765
Fradkin J, Shamoon H, Felig P, Sherwin RS (1980) Evidence for an important role of changes rather than absolute concentrations of glucagon in the regulation of glucose production in humans. J Clin Endocrinol Metab 50:698–703

Garcia A, Williamson JR, Cahill GF (1966) Studies on the perfused rat liver. II. Effect of glucagon on gluconeogenesis. Diabetes 15:188–193

Gerich JE, Lorenzi M, Schneider V, Karam JH, Rivier J, Guillemin R, Forsham PH (1974) Effects of somatostatin on plasma glucose and glucagon levels in diabetes mellitus. N Engl J Med 291:544–547

Gerich JE, Langlois M, Noacco C, Lorenzi M, Karam JH, Forsham PH, with the technical assistance of Gustafson G (1976) Comparison of the suppressive effects of elevated plasma glucose and free fatty acid levels on glucagon secretion in normal and insulin-dependent diabetic subjects. Evidence for selective alpha cell insensitivity to glucose in diabetes mellitus. J Clin Invest 58:320–325

Glinsmann NH, Mortimore GE (1968) Influence of glucagon and 3′,5′-AMP on insulin responsiveness of the perfused rat liver. Am J Physiol 215:553–559

Issekutz B, Vranic M (1980) Role of glucagon in the regulation of glucose production in exercising dogs. Am J Physiol 238:E 13–E 20

Jefferson LS, Exton JH, Butcher RW, Sutherland EW, Park CR (1968) Role of adenosine 3′,5′-monophosphate in the effects of insulin and anti-insulin serum on liver metabolism. J Biol Chem 254:1031–1038

Jennings AS, Cherrington AD, Liljenquist JE, Keller U, Lacy WW, Chiasson JL (1977) The role of insulin and glucagon in the regulation of gluconeogenesis in the postabsorptive dog. Diabetes 26:847–856

Keller U, Chiasson JL, Liljenquist JE, Cherrington AD, Jennings AS, Crofford OB (1978) The role of insulin, glucagon and free fatty acids in the regulation of ketogenesis in dogs. Diabetes 26:1040–1051

Kemmer FW, Vranic M (1981) The role of glucagon and its relationship to other gluconeogenesis hormones in exercise. In: Unger RH, Orci L (eds) Glucagon: physiology, pathophysiology and morphology of the pancreatic A cells. Current endocrinology. Elsevier, Amsterdam Oxford New York, pp 297–331

Kibler RF, Taylor WJ, Mayers JD (1964) The effects of glucagon on net splanchnic balances of glucose, amino acid nitrogen, urea, ketones and oxygen in man. J Clin Invest 43:904–911

Koerker DJ, Ruch W, Chideckel E, Palmer J, Goodner CJ, Ensinck J, Gale CC (1974) Somatostatin: hypothalamic inhibitor of the endocrine pancreas. Science 184:482–484

LeCam A, Freychet P (1976) Glucagon stimulates the A-system for neutral amino acid transport in isolated hepatocytes of the adult rat. Biochem Biophys Res Commun 72:893–901

Lickley LLA, Ross GG, Vranic M (1979) Effects of selective insulin or glucagon deficiency on glucose turnover. Am J Physiol 236:E 255–E 262

Liljenquist JE, Rabin D (1979) Lack of a role for glucagon in the disposal of an oral glucose load in normal man. J Clin Endocrinol Metab 49:937–939

Liljenquist JE, Bomboy JD, Lewis SB, Sinclair-Smith BC, Felts PW, Lacy WW, Crofford OB, Liddle GW (1974) Effect of glucagon on net splanchnic cyclic AMP production in normal and diabetic man. J Clin Invest 53:198–204

Liljenquist JE, Mueller GL, Cherrington AD, Keller U, Chiasson JL, Perry JM, Lacy WW, Rabinowitz D (1977) Evidence for an important role for glucagon in the regulation of hepatic glucose production in normal man. J Clin Invest 59:369–372

Lundbaek K, Hansen AP, Orskow H, Christiansen S, Tuerson J, Seyer-Hausen K, Alberti KGMM, Whitefoot R (1976) Failure of somatostatin to correct manifest diabetic ketoacidosis. Lancet 1:215–218

Mallette LE, Exton JH, Park CR (1969) Effects of glucagon on amino acid transport and utilization in the perfused rat liver. J Biol Chem 244:5724–5728

Marliss EB, Aoki TT, Unger RH, Soeldner JS (1970) Glucagon levels and metabolic effects in fasting man. J Clin Invest 49:2256–2270

Murlin JR, Clough HD, Gibbs CBF, Stokes AM (1923) Aquous extracts of pancreas. I. Influence on carbohydrate metabolism of depancreatized animals. J Biol Chem 56:253–260

Owen OE, Felig P, Morgan AP, Wahren J, Cahill GR (1969) Liver and kidney metabolism during prolonged starvation. J Clin Invest 48:574–583

Perez G, Kemmer FW, Lickley HLA, Vranic M (1981) Importance of glucagon in mediating epinephrine-induced hyperglycemic in alloxan-diabetic dogs. Am Physiol 4:E 328–335
Pozefsky T, Tancredi RG, Moxley RT, Dupre J, Tobin JD (1976) Metabolism in forearm tissues in man. Studies with glucagon. Diabetes 25:125–128
Raskin P, Aydin I, Yamamoto T, Unger RH (1978) Abnormal alpha-cell function in human diabetes. The response to oral protein. Am J Med 64:988–997
Rizza RA, Gerich JE (1978) Somatostatin and diabetes. Med Clin North Am 62:735–746
Rizza RA, Gerich JE (1979) Persistent effect of sustained hyperglucagonemia on glucose production in man. J Clin Endocrinol Metab 48:352–355
Sacca L, Sherwin RS, Felig P (1979) Influence of somatostatin on glucagon and epinephrine stimulated hepatic glucose output in the dog. Am J Physiol 236:E 113–117
Salter JM, Ezrin C, Laidwal JC (1960) Metabolic effects on glucagon in human subjects. Metabolism 9:753–757
Sherwin RS, Hendler R, Defronzo R, Wahren J, Felig P (1977) Glucose homeostasis during prolonged suppression of glucagon and insulin secretion by somatostatin. Proc Natl Acad Sci USA 74:348–358
Shulman GI, Liljenquist JE, Williams PE, Lacy WW, Cherrington AD (1978) Glucose disposal during insulinopenia in somatostatin treated dogs: the roles of glucose and glucagon. J Clin Invest 60:487–491
Steiner KE, Mouton S, Williams PE, Cherrington AD (1980) Relative importance of the two phases of insulin release in glycemic control. Diabetes [Suppl 2] 29:17 A
Steiner KE, Williams PE, Lacy WW, Cherrington AD (1981) Effects of the insulin/glucagon molar ratio on glucose production in the dog. Fed Proc 40:3481
Steiner KE, Mouton SM, Bowles CR, Williams PE, Cherrington AD (1982) in: The relative importance of first- and second phase insulin secretion in countering the action of glucagon on glucose turnover in the conscious dog. Diabetes 31:964–972
Tamborlane W, Sherwin R, Hendler R, Felig P (1977) Metabolic effects of somatostatin in maturity-onset diabetes. N Engl J Med 297:181–183
Tyberghein J (1953) Action du glucagon sur le métabolisme des proteines. Arch Int Physiol 61:104–110
Unger RH, Ohneda A, Aguilar-Parada E, Eisentraut AM (1969) The role of aminogenic glucagon secretion in blood glucose homeostasis. J Clin Invest 48:510–822
Wagle SR (1975) Interrelationship of insulin and glucagon ratios on carbohydrate metabolism in isolated hepatocytes containing high glucagon. Biochem Biophys Res Commun 67:1019–1027
Wahren J, Felig P, Hagenfeldt L (1976) Effect of protein ingestion on splanchnic and leg metabolism in normal man and in patients with diabetes mellitus. J Clin Invest 57:987–999
Wahren J, Efendic S, Luft R, Hagenfeldt L, Björkman O, Felig P (1977) Influence of somatostatin on splanchnic glucose metabolism in post absorptive and 60 h fasted humans. J Clin Invest 59:299–307
Zinman B, Murry FF, Vranic M, Albisser M, Leibel B, McClean PA, Marliss EB (1979) Glucoregulation during moderate exercise. Diabetes [Suppl 1] 28:76–81

CHAPTER 17

# Glucagon and Ketogenesis

J. D. McGarry and D. W. Foster

## A. Introduction

Although suspected to exist as a pancreatic hyperglycemic factor as early as 1923, isolated and crystallized by 1953, and measurable by radioimmunoassay since 1959, only in the last decade has glucagon become firmly established as a critical hormone in the minute-to-minute regulation of the blood glucose concentration. This aspect of glucagon physiology has been discussed in detail elsewhere in this volume. The purpose of the present chapter is to review a second area of fuel homeostasis in which the A-cell hormone has again emerged as a central regulatory factor, namely, the control of hepatic fatty acid metabolism and ketone body production. From an historical standpoint, the function of glucagon in this area gained recognition even more recently than its role in the regulation of glucose metabolism, most of the significant advances having been made only in the last 15 years or so. Not surprisingly, a number of interesting parallelisms between the two systems have begun to evolve. Thus, it now appears that the same bihormonal mechanism (reciprocal changes in the levels of insulin and glucagon), instrumental in permitting the liver to respond smoothly to periods of food deprivation with accelerated release of glucose, also induces production of ketone bodies to support the energy needs of the brain and peripheral tissues. By the same token, severe imbalance between the two hormones, such as occurs in uncontrolled diabetes, is now recognized as a primary factor in the etiology of the two major metabolic derangements associated with this disorder, namely, hyperglycemia and ketoacidosis.

In the sections that follow we shall first review recent developments that provide insight into the biochemical mechanisms operative within the liver cell for the regulation of the ketogenic process itself. Second, we shall summarize the evolution of current concepts regarding the overall hormonal control of ketone body production in the intact organism, with emphasis on the interplay between insulin and glucagon. Finally, ongoing studies into how the A- and B-cell hormones interact at the level of the liver to coordinate the interlocking pathways of carbohydrate and lipid metabolism will be discussed.

## B. Intrahepatic Factors in the Regulation of Ketogenesis

It was early recognized that the ketone bodies, acetoacetic and β-hydroxybutyric acids, arise almost exclusively from the oxidation of fatty acids in the liver. The fact that ketotic states are characterized by accelerated rates of lipolysis in adipose tissue, with attendant elevation of plasma free fatty acid (FFA) levels, understand-

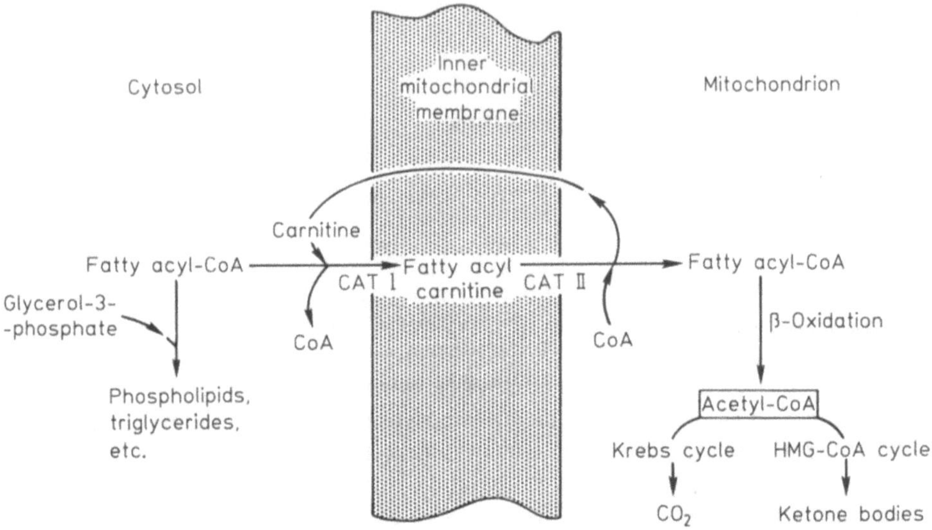


**Fig. 1.** Major pathways of fatty acid metabolism in liver. CAT I and CAT II refer to carnitine acyltransferases I and II located on the outer and inner aspects of the inner mitochondrial membrane, respectively. McGarry (1978)

ably led to the view that the output of ketone bodies by the liver was a passive consequence of the input of fatty acid precursors (Fritz 1961; Bewsher et al. 1966). While such a formulation still remains true in a broad sense, numerous studies, beginning with the classic experiments of Mayes and Felts in 1967, have shown that the control of ketogenesis is somewhat more complicated. Thus, by the early 1970 s it had become clear that the disposition of a fatty acid load within the liver is greatly influenced by nutritional and hormonal status. In the fed state, the bulk of the fatty acids taken up are reesterified with glycerol-3-phosphate to form triglycerides, whereas in ketotic states (e.g., starvation and uncontrolled diabetes) a much greater fraction is oxidized to ketone bodies with concomitantly less esterification (McGarry and Foster 1972). Such findings focused attention on the branch point between the pathways of esterification and oxidation as an important control site in the overall regulation of hepatic fatty acid metabolism (Fig. 1).

On which arm of this metabolic crossroads is control exerted? Originally, emphasis was placed on the esterification pathway which was thought to be less efficient in ketotic states so that incoming fatty acids were preferentially shunted into oxidative metabolism (Mayes and Felts 1967; Wieland and Matschinsky 1962; Tzur et al. 1964). However, for reasons detailed in previous reviews (McGarry and Foster 1977, 1980), this notion did not withstand rigorous experimental testing and, by 1973, a considerable body of evidence pointed to the oxidative pathway as the primary site of control. More specifically, the carnitine acyltransferase (CAT) step, which catalyzes the transfer of long chain fatty acids from the cytosol into the mitochondrial matrix (the site of $\beta$-oxidation and ketogenesis) was considered a logical candidate. It was postulated that this transport mechanism was somehow turned on and off in ketotic and nonketotic states, respectively (McGarry et al. 1973).

If the CAT system was the major control site for hepatic fatty acid oxidation how might its activity be regulated? Two important facts had already been established. First, the total quantity of the enzyme was essentially unchanged in livers from normal and ketotic animals (DIMARCO and HOPPEL 1975). Second, the $\beta$-oxidation capacity of liver mitochondria appeared to be fixed and large (MCGARRY and FOSTER 1971). These observations suggested that in ketotic states either an activator for CAT is generated or an inhibitor of the enzyme is lost; a combination of both events also had to be considered. In support of the first of these possibilities it was found that the liver content of carnitine, a substrate for CAT I, fluctuated in proportion to the ketogenic capacity of the tissue in fed, fasted, and diabetic rats (MCGARRY et al. 1975)[1]. Moreover, addition of carnitine to the perfusion fluid stimulated the oxidation of oleate in perfused livers from fed animals (MCGARRY et al. 1975). However, subsequent studies with lactating and neonatal rats (ROBLES-VALDES et al. 1976), which will not be elaborated upon here, revealed that an elevated carnitine level could not be the only factor responsible for the accelerated hepatic fatty acid oxidation in ketotic states. The same experiments suggested that full activation of ketogenesis required, in addition to carnitine enrichment, glycogen depletion of the liver. We suspected that a metabolite of glucose, whose concentration paralleled that of glycogen, acted as an inhibitor of the CAT step. Loss of this compound in ketotic states, coupled with a rise in the tissue carnitine content, were seen as the two major prerequisites for "turning on" the process of hepatic fatty acid oxidation and ketogenesis (MCGARRY and FOSTER 1976).

The putative glucose-derived suppressor of the CAT reaction turned out to be a simple molecule, malonyl-CoA, the first committed intermediate in the conversion of glucose into fat (MCGARRY et al. 1977). It proved to be a potent, specific, and reversible inhibitor of the CAT I reaction (the initial step in the pathway of mitochondrial fatty acid oxidation), acting competitively towards long chain acyl-CoA (MCGARRY et al. 1978a). That this finding had physiologic relevance for the control of ketogenesis in vivo was supported by two already established facts. First, it had been recognized for many years that in ketotic states (where the capacity for hepatic fatty acid oxidation is high) the ability of the liver to synthesize fatty acids is low (MASORO 1965). Second, GUYNN et al. (1972) had shown that during periods of feeding and fasting the concentration of malonyl-CoA in rat liver fluctuated in line with the rate of lipogenesis. Taken together, these observations provided a plausible biochemical explanation for how, at least in the rat, the activities of two complex, opposing pathways of liver metabolism (fatty acid synthesis and oxidation) could be reciprocally controlled in simple fashion.

The postulated interactions are shown schematically in Fig. 2 and may be summarized as follows. With carbohydrate feeding a major function of the liver is to convert excess carbon derived from dietary glucose into glycogen and fat. Under these conditions the tissue malonyl-CoA level rises and, by blocking the reoxidation of newly formed fatty acids, ensures their conversion into triglycerides which are then transported in the form of very low density lipoproteins to adipose tissue for storage. Conversely, in starvation or uncontrolled diabetes, hepatic fatty acid synthesis comes to a halt because the malonyl-CoA level falls, and the tissue con-

1 The reason for the elevation of liver carnitine levels in ketotic rats is still unclear

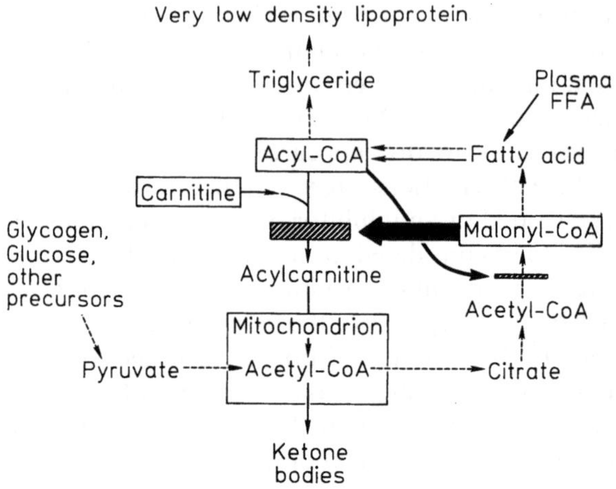


**Fig. 2.** Interrelations between the pathways of fatty acid synthesis *(broken lines)* and oxidation *(solid lines)* in liver

tent of carnitine rises; these changes prime the liver for the efficient oxidation of fatty acids (now being delivered from the breakdown of adipose tissue triglycerides) and accelerated production of the ketone bodies. Thus, within the liver cell the rate of fatty acid oxidation is governed primarily by the relative concentrations of three molecular species: (1) [long chain acyl-CoA], substrate for CAT I which is set by the extracellular concentration of fatty acid; (2) [carnitine], cosubstrate for CAT I, and governed by factors yet to be delineated; and (3) [malonyl-CoA], competitive inhibitor of CAT I. The fall in [malonyl-CoA] in ketotic states results from reduced carbon flow through the acetyl-CoA carboxylase step in response to hormonal influences on the liver (see Sect. D) and elevation in [long chain acyl-CoA], itself an inhibitor of the carboxylase reaction. It seems likely that for any given level of carnitine the rate of $\beta$-oxidation, and therefore the rate of ketogenesis [2], will be dictated by the relative concentrations of long chain acyl-CoA and malonyl-CoA. For a more complete discussion of the interactions between these metabolic parameters the reader is referred to an earlier review (McGarry and Foster 1980).

To summarize, it is now clear that two requirements must be met if accelerated ketone body synthesis is to occur. First, there must be an adequate supply of fatty acid substrate to the liver. Second, there must be a change in the metabolic profile of the liver such that incoming fatty acids are directed into the $\beta$-oxidation sequence rather than into the pathway of reesterification. The important new observation is that this switch in the intrahepatic disposal of fatty acids hinges mainly on shutdown of the de novo pathway for fatty acid biosynthesis; more specifically, it requires the cellular depletion of malonyl-CoA.

2 Although primary control of ketogenesis is considered to be exerted at the CAT I step, secondary modulation is imparted at the branch point of acetyl-CoA metabolism through the hydroxymethylglutaryl (HMG)-CoA and tricarboxylic acid cycles. In general, however, the rate of ketone body production will reflect the flux of fatty acid carbon through the $\beta$-oxidation pathway (McGarry and Foster 1980)

## C. Emergence of a Ketogenic Role for Glucagon

Running parallel with the biochemical advances already outlined, interest in the hormonal signals responsible for the "on–off" control of the ketogenic process was growing rapidly. Of particular concern was whether insulin deficiency, long recognized as the hallmark of all ketotic states (whether physiologic or pathologic), could still be considered the sole hormonal disturbance underlying their development. Certainly, there was indirect evidence to support this notion. For example, one of the most potent in vivo effects of insulin is its ability to suppress lipolysis in adipose tissue. Moreover, both prerequisites for accelerated ketogenesis, namely, activation of lipolysis and the switch of liver metabolism from a nonketogenic to a ketogenic mode (see Sect. B), appeared to occur simultaneously with reduction in the circulating level of insulin (McGarry et al. 1973). On the other hand, the fact that such conditions were usually found to be accompanied by elevated levels of glucagon together with the demonstration by several groups that glucagon stimulated ketogenesis in isolated liver preparations (Bewsher and Ashmore 1966; Heimberg et al. 1969; Menahan and Wieland 1969), raised the possibility that the A-cell hormone might also be important in regulating ketone body production in vivo. Compelling evidence to this effect has since been obtained from studies both in experimental animals and in humans.

### I. Studies in Animals

The first clear-cut demonstration that activation of the liver's ketogenic machinery could be accomplished in vivo in the absence of accelerated adipose tissue lipolysis came from studies in which fed rats were infused either with anti-insulin serum (AIS) or glucagon for short periods of time (1–3 h) (McGarry et al. 1975). Treatment with AIS produced the expected results: depletion of liver glycogen, profound hyperglycemia, marked elevation of plasma free fatty acids and ketones, and a striking enhancement of the liver's ketogenic capacity, as determined by subsequent perfusion with oleic acid. Glucagon-treated animals displayed a very different picture in vivo: transient hyperglycemia, low plasma FFA levels, and complete absence of ketonemia. Yet the livers were found to be glycogen depleted and to have the same high ketogenic capacity when studied in vitro as those from AIS-treated animals. That the failure of the blood ketone body level to rise must have been due to lack of ketogenic substrate was confirmed when saline- and glucagon-treated rats were infused with a triglyceride emulsion plus heparin to cause artificial elevation of the plasma FFA concentration. In contrast to the control animals, those treated with glucagon exhibited a rapid elevation of plasma acetoacetate and $\beta$-hydroxybutyrate concentrations.

The data were interpreted as follows. Glucagon infusion triggered the dissipation of liver glycogen stores. The ensuing hyperglycemia was rapidly corrected because of increased insulin secretion from the pancreas; this also served to keep plasma FFA concentrations suppressed. Despite normal or even elevated insulin levels, the liver had been switched into a ketogenic pattern because of the simultaneous presence of high glucagon concentrations. The possibility was therefore raised that activation of hepatic ketogenesis in the AIS-treated animals resulted not simply from insulin deficiency, but rather from an action of glucagon exerted on the liver in the absence of restraint by insulin. We further concluded that mobilization of FFA from fat depots was triggered primarily by an absolute fall in circulating insulin (since even pharmacologic amounts of glucagon are not lipolytic in the presence of physiologic quantities of insulin, see Chap. 19), and that activation of hepatic ketogenesis responded to elevation of the [glucagon]: [insulin] ratio whether or not the absolute level of the B-cell hormone changed. Stated in another way, insulin deficiency appeared to be essential for ketosis to develop, but was not necessary for the switch of liver metabolism from a nonketogenic to a ketogenic profile; the latter required a relative excess of glucagon over insulin.

The interactions of insulin and glucagon in the regulation of hepatic ketone body production were also examined by Keller et al. (1977) in conscious, anesthetized dogs. The animals were infused with somatostatin to suppress endogenous hormone secretion and with triglyceride plus heparin to raise plasma FFA levels. There was little increase in the plasma ketone concentration when the animals were rendered deficient in both insulin and glucagon. In contrast, insulin deficiency coupled with replacement levels of glucagon produced a marked increase in ketone production that was independent of changes in plasma FFA levels. Thus, the potent ketogenic effect on the liver of a normal concentration of glucagon was revealed when the opposing influence of insulin was removed.

## II. Studies in Humans

Strong evidence that glucagon also plays a key role in the development of ketoacidosis in humans is now available. For example, Gerich et al. (1975) and Alberti et al. (1975) showed that insulin withdrawal from insulin-dependent diabetics resulted in parallel increases in plasma glucagon, FFA, and ketone body levels. Prevention of the rise in glucagon by somatostatin infusion greatly obtunded the development of ketonemia. Although this was also associated with a less striking rise in plasma FFA concentrations (possibly reflecting a lipolytic action of glucagon in the absence of insulin, see Chap. 19), the important point is that for any given concentration of FFA the concentration of ketone bodies was lower when glucagon was suppressed. Taking a different approach, Schade and Eaton (1975) and also Harano et al. (1977) observed that administration of glucagon enhanced the conversion of fatty acids into ketone bodies in diabetic humans, exactly as in the experiments with rats and dogs described earlier. Also of interest in this context is the report by Boden et al. (1977) who studied a patient with an islet cell carcinoma in whom preoperative plasma levels of glucagon were 1,000–5,000 pg/ml (normal values are 50–200 pg/ml). At a time when plasma FFA levels were no higher than in a normal individual after an overnight fast, plasma ketone body levels were 400% of normal. A marked reduction occurred after tumor resection.

Taken together, these in vivo studies emphasized the importance of two organ systems, adipose tissue and liver, and two hormones, insulin and glucagon, in the acute "on–off" regulation of ketone body production in the whole organism.[3]

3 Other hormones, particularly those associated with stress, can also modulate fatty acid metabolism at the level of adipose tissue and liver. However, available evidence indicates that the primary signals responsible for acute "on–off" regulation of the ketogenic process are insulin and glucagon

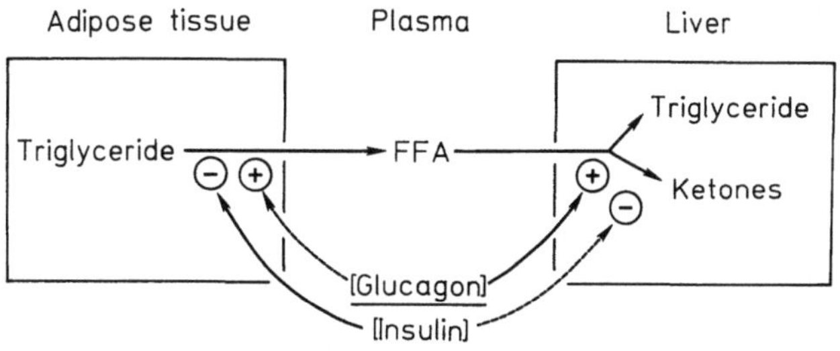


**Fig. 3.** Postulated roles of insulin and glucagon in the control of ketone body production. Primary and secondary actions are represented by the *solid* and *broken lines*, respectively

With reference to the model depicted in Fig. 3, key interactions between these components can be summarized as follows. Adipose tissue lipolysis is potently suppressed by insulin such that the first requirement for ketogenesis, namely, an increased supply of FFA to the liver, requires an absolute reduction in the concentration of this hormone. The second requirement (activation of the liver's capacity for fatty acid oxidation) can be accomplished without a rise in glucagon levels, as long as there is a fall in insulin secretion. Alternatively, it can be brought about experimentally with normal levels of insulin if a sufficient excess of glucagon is provided. It follows that ketosis will not develop under normoinsulinemic conditions, even in the face of extremely high glucagon concentrations, since adipose tissue lipolysis would be suppressed. Further, once the liver's ketogenic potential has been enhanced, the production of ketone bodies will be governed by the supply of fatty acids which, in turn, will be dictated by the degree of insulin deficiency. Herein lies the basis for the pathologic increase in ketone body synthesis associated with severe insulinopenia compared with the controlled production rate in the physiologic ketosis of starvation (McGarry and Foster 1977).

## D. Interactions of Insulin and Glucagon on Hepatic Metabolism

The models proposed in Figs. 2 and 3 implied that during the transition from the nonketotic (e.g., fed) to the ketotic (e.g., fasted) state, activation of hepatic ketogenesis is triggered by elevation of the circulating [glucagon]: [insulin] ratio and is secondary to suppression of the lipogenic process in liver. By the same token, refeeding after a fast (which rapidly reduces the liver's fatty acid oxidation capacity) would be expected to be mediated by lowering of the [glucagon]: [insulin] ratio with concomitant reinduction of hepatic lipogenesis. Both formulations have been tested experimentally using hepatocytes isolated from fed and fasted rats. For convenience these studies will be addressed in two parts. First, we shall examine the role of glucagon in effecting the "fed to fasted" transition of hepatic fatty acid metabolism in vitro. Subsequently, studies designed to elucidate the relative importance of insulin, glucagon, and various substrates in the reversal of liver metabolism from a "fasted" to a "fed" profile will be discussed.

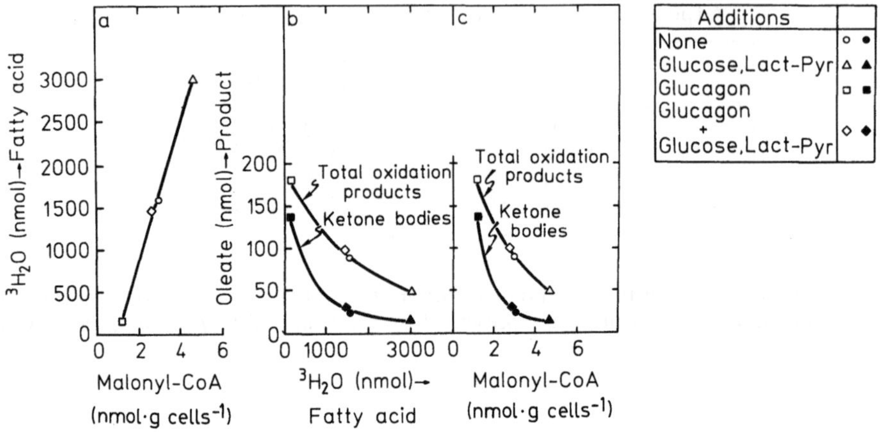


**Fig. 4a–c.** Relationship between fatty acid synthesis, fatty acid oxidation, and malonyl-CoA in hepatocytes from fed rats. Hepatocytes (200 mg wet wt) were incubated with 0.4 m*M* oleic acid (labeled or unlabeled) bound to albumin in the presence or absence of the agents indicated. Glucose, lactate, pyruvate, and glucagon were used in concentrations of 10 m*M*, 10 m*M*, 1 m*M*, and 4 µg/ml, respectively. Rates of fatty acid synthesis ($^3H_2O$→-fatty acids), fatty acid oxidation (oleate $^{14}C$-1→acid-soluble products + $CO_2$), and ketogenesis (oleate $^{14}C$-1→ketone bodies) were measured over the 15- to 60-min interval. Malonyl-CoA values are averages of those found at 15, 30, 45, and 60 min. Adapted from McGarry and Foster (1979)

## I. The "Fed to Fasted" Transition

Before the malonyl-CoA link between hepatic fatty acid synthesis and oxidation had been established it was known from isolated reports that glucagon had the capacity both to suppress lipogenesis (Harris et al. 1979) and to stimulate ketogenesis (Bewsher and Ashmore 1966; Heimberg et al. 1969; Menahan and Wieland 1969; Woodside 1979) in liver preparations in vitro. However, demonstration that the two events were causally related and mediated through changes in the cellular content of malonyl-CoA required that all of these metabolic parameters be shown to change in appropriate fashion when measured in parallel in the same experiments. To this end, hepatocytes from fed rats were treated in such a way that rates of fatty acid synthesis could be manipulated over a wide range, and simultaneous measurements made of the tissue level of malonyl-CoA, together with the rates of fatty acid oxidation and ketogenesis (McGarry et al. 1978b; McGarry and Foster 1979). Typical results are shown in Fig. 4.

It is seen that maximal rates of fatty acid synthesis and minimal rates of fatty acid oxidation were achieved in the presence of glucose, lactate, and pyruvate; the situation was reversed when the cells were exposed to glucagon (Fig. 4a). Addition of all components restored both processes to control rates, but not to the extremes seen when the substrates were added in the absence of the hormone. In keeping with the findings of Guynn et al. (1972), the rate of lipogenesis was directly related to the tissue malonyl-CoA content. As expected, the amount of oleate converted into total oxidation products (i.e., the rate of $\beta$-oxidation) and the extent of ketogenesis were both inversely related to the rate of fatty acid synthesis and to the tissue malonyl-CoA content (Figs. 4b and c). Importantly, the range through which the tissue malonyl-CoA level fluctuated (about 0.5–6 nmol/g), coupled with the previously determined inhibitory constant for malonyl-CoA on the CAT I activity of isolated mitochondria (1–2 $\mu M$) (McGarry et al. 1978a), was entirely consistent with the postulated physiologic role of this molecule in the regulation of hepatic ketogenesis (Figs. 2).[4] It should also be noted that in these same experiments the cellular content of long chain acylcarnitine (the product of the CAT I reaction) was related directly to the rate of oleate oxidation and inversely to the malonyl-CoA level (McGarry and Foster 1979), further supporting the model shown in Fig. 2. Also as predicted, maximal ketogenesis was achieved by incubating the cells with glucagon together with exogenous carnitine. Under these conditions, rates approached those seen in hepatocytes from fasted rats which were already depleted of malonyl-CoA and enriched with carnitine (McGarry and Foster 1979).

Based on this type of study, the question of how glucagon acutely activates hepatic fatty acid oxidation can now be asked in another way: how does the hormone suppress the lipogenic process in liver? Although firm details are lacking, a considerably body of evidence, accumulated only in the last 3–4 years, suggests that this action of glucagon stems from its inhibitory effects on two components of the lipogenic pathway: the glycolytic process and the acetyl-CoA carboxylase reaction. Support for this notion comes from the following observations. Coincident with its ability to suppress lipogenesis in hepatocytes from fed animals, glucagon also activated glycogenolysis, but completely prevented lactate production (McGarry et al. 1978b; McGarry and Foster 1979), suggesting that net carbon flow from glucose-6-phosphate to pyruvate had stopped. That a step in the lipogenic pathway distal to pyruvate had also been suppressed follows from the inability of pyruvate (or its precursors) to restore maximal rates of lipogenesis when glucagon was also present (Fig. 4; McGarry et al. 1978b; McGarry and Foster 1979). Almost certainly, the additional block occurred at the level of acetyl-CoA carboxylase, the reaction long considered to be rate limiting for fatty acid synthesis. Since all of these glucagon effects could be reproduced by dibutyryl cyclic AMP, it is probable that in all cases the intracellular mediator of glucagon action was the cyclic nucleotide itself.

It now appears that glucagon suppresses glycolysis in liver through cyclic AMP-mediated inhibition of two key enzymes in the pathway, phosphofructokinase (PFK) and pyruvate kinase (PK). The work of Uyeda and colleagues revealed that the former enzyme is subject to a novel form of control in that it requires an activation factor which binds tightly to the dephosphorylated form of the protein, but not to the phosphorylated species (Richards and Uyeda 1980; Furuya and Uyeda 1980). The factor has recently been identified as fructose-2,6-diphosphate (Van Schaftingen et al. 1980; Pilkis et al. 1981). Evidently, glucagon inhibits the activity of PFK by a dual mechanism: it promotes the cyclic AMP-dependent phosphorylation of the enzyme, thus reducing its affinity for fructose-2,6-diphosphate and simultaneously causes the rapid disappearance of the phosphate

4 The much higher concentrations of malonyl-CoA required by other groups (Cook et al. 1980; Ontko and Johns 1980) for the inhibition of mitochondrial fatty acid oxidation are probably artifacts due to inadequacies in experimental design (McGarry and Foster 1981)

ester (Richards and Uyeda 1980; Furuya and Uyeda 1980; Van Schaftingen et al. 1980; Pilkis et al. 1981). These interesting new observations promise to shed fresh light on many nagging questions surrounding the long suspected, but poorly understood role of phosphofructokinase in the regulation of hepatic glycolysis.

The L form of pyruvate kinase (the predominant isozyme present in rat liver) is also subject to inactivation through cyclic AMP-dependent phosphorylation (Feliu et al. 1976). However, regulation of this enzyme appears to be dependent upon the activity of PFK since the product of the latter reaction, fructose-1,6-diphosphate, allosterically activates PK and prevents its inactivation by phosphorylation (see Geelen et al. 1980 for review). Thus, the primary control point at which glucagon exerts its negative effect on glycolysis is probably the PFK step. It should also be noted that both PFK and PK catalyze reactions involved in potentially futile cycles of the glycolytic pathway. Inhibition of both steps by glucagon not only suppresses carbon flow from glucose-6-phosphate to pyruvate, but also facilitates the opposing pathway of gluconeogenesis, as detailed in Chap. 15.

Acetyl-CoA carboxylase is subject to many different forms of control, including allosteric modification, protomer–polymer transformation, and covalent modulation by phosphorylation. As discussed by Geelen et al. (1980), the relative importance of phosphorylation–dephosphorylation versus fluctuation in the concentrations of positive and negative effectors (such as citrate and long chain acyl-CoA, respectively) in the overall control of its activity is far from clear. While all of these factors might be contributory there is good reason to believe that cyclic AMP-dependent phosphorylation of the polymeric enzyme followed by its rapid dissociation into inactive protomers represents at least one component of the mechanism through which glucagon acutely inhibits the reaction in rat liver (Lent et al. 1978).

Regarding the inhibitory action of glucagon on the lipogenic process several additional points should be emphasized. First, in most of the studies cited pharmacologic levels of glucagon were employed, simply to obtain maximal effects and to circumvent the loss of hormone due to degradation during the hepatocyte incubations. However, when more physiologic concentrations of the hormone were used, qualitatively similar results were obtained (Geelen and Gibson 1975; Beynen et al. 1979). Second, although the inhibition of lipogenesis (and concomitant activation of ketogenesis) induced by high concentrations of glucagon cannot be reversed by insulin, antagonism between the two hormones has been demonstrated when glucagon was present at physiologic levels (Beynen et al. 1979). Although the mechanism through which insulin offsets the catabolic action of glucagon has yet to be established it probably stems, at least in part, from the ability of the B-cell hormone to prevent the rise in tissue cyclic AMP levels caused by glucagon (Exton and Park 1972). Finally, the observed effects of glucagon on hepatic fatty acid synthesis are rapid (occurring in minutes) and constitute a short-term control of this pathway that does not involve changes in the quantity of regulatory enzymes. Prolonged exposure of the liver to an environment of low insulin and high glucagon levels (e.g., lengthy starvation or uncontrolled diabetes) results in loss of enzyme protein that may take hours or days to correct. Such events fall into the category of long-term control.

To summarize, glucagon acting on the liver in the absence of the restraining influence of insulin rapidly activates the ketogenic machinery. The sequence of events appears to be as follows:

$$\uparrow \frac{[\text{Glucagon}]}{[\text{Insulin}]} \rightarrow \uparrow [\text{Cyclic AMP}] - \left\langle \begin{array}{l} \downarrow \text{Glycolysis} \\ \\ \downarrow \text{Acetyl-CoA} \\ \quad \text{carboxylase} \end{array} \right\rangle \rightarrow$$

$$\rightarrow \downarrow [\text{Malonyl-CoA}] \rightarrow \uparrow \text{CAT I} \rightarrow \uparrow \beta\text{-Oxidation} \rightarrow$$
$$\rightarrow \uparrow \text{Ketogenic capacity.}$$

When such conditions are combined with elevated plasma FFA levels, as occurs with absolute insulin deficiency, marked acceleration of ketone body production will ensue.

## II. The "Fasted to Fed" Transition

In the rat, the switch of liver metabolism from an anabolic (high glycogen levels, active lipogenesis, suppressed fatty acid oxidation) to a catabolic set (glycogen depletion, minimal lipogenesis, active fatty acid oxidation) begins approximately 6 h after food withdrawal and is complete some 4–6 h later (McGarry et al. 1973). (As already noted, these events can be greatly accelerated when isolated hepatocytes are exposed to glucagon.) When fasted rats are refed, the time required for reversal of liver metabolism to an anabolic profile is also short (about 3–4 h), but surprisingly little is known about the substrates and hormones involved. Although historically glucose and insulin have been assigned primary roles, one is struck by the following observations. First, glucose, even in high concentrations, has always proved to be a poor substrate, either for glycogen synthesis or for lipogenesis in isolated liver preparations (Hems et al. 1972; Seglen 1974; Katz et al. 1976, 1979; Clark et al. 1974; Hems 1977). Second, although insulin has frequently been reported as stimulatory for both processes (Beynen et al. 1980; Exton et al. 1970; Miller and Larner 1973; Davidson and Berliner 1974), the magnitude of this effect has generally not been impressive. Indeed, to our knowledge insulin has never been shown to cause a significant *net* increase in liver glycogen deposition in vitro.

We have recently examined these issues using isolated hepatocytes from 18-h fasted rats (Boyd et al. 1981). In agreement with others (Hems et al. 1972; Seglen 1974; Katz et al. 1976, 1979) we too observed little glycogen formation when the cells were incubated for 2 h in the presence of 20 m$M$ glucose, the value being about 2–3 mg glycogen/g tissue. This low rate was not improved by addition of insulin and is to be contrasted with a value of some 25 mg/g liver when the animals were allowed to eat ad libitum for the same period of time. The difference between the in vitro and in vivo rates becomes even more striking in light of the fact that during the feeding period the portal plasma glucose concentration seldom exceeded 10 m$M$. Good rates of glycogen synthesis could be obtained in hepatocytes, how-

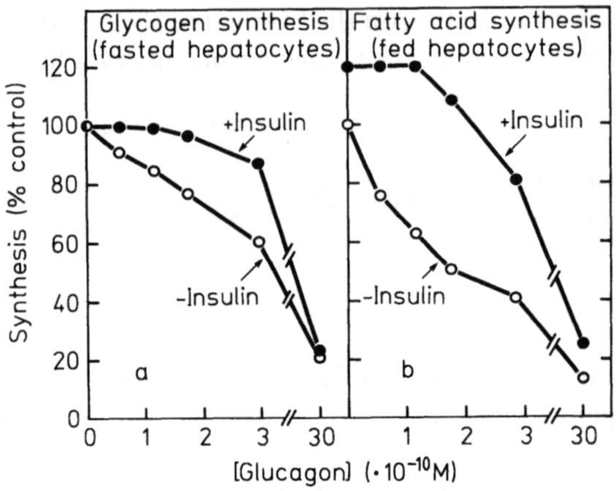


**Fig. 5.** Antagonistic effects of insulin and glucagon on glycogen and fatty acid synthesis in hepatocytes. Glycogen synthesis was measured in hepatocytes from fasted rats incubated with 20 m*M* glucose, 5 m*M* fructose, and 10 m*M* glutamine for a period of 2 h. Fatty acid synthesis was measured over the same time period in hepatocytes from fed animals incubated in the absence of substrates. Insulin (2,000 μIU/ml) and the indicated concentration of glucagon were added every 30 min. Adapted from Boyd et al. (1981)

ever, provided that the flasks contained, in addition to glucose, a gluconeogenic substrate, such as fructose or lactate, and an amino acid, such as glutamine or alanine. In this case the glycogen was derived mainly from the glucose precursors. Again, insulin was neither necessary nor stimulatory. These findings were also reminiscent of earlier reports (Katz et al. 1976, 1979; Hems 1977; Geelen 1977).

A similar picture emerged with regard to the reversal of fatty acid metabolism in these hepatocytes. Lipogenesis responded poorly to glucose as the sole substrate, but was greatly enhanced when lactate and glutamine were also provided. Under such conditions the tissue malonyl-CoA level rose and its capacity for oleate oxidation and ketogenesis was greatly reduced, in accord with the model of Fig. 2. Once again, however, insulin was not needed for the reinduction of fatty acid synthesis and, when added, produced only marginal enhancement of the process.

In contrast to the poor effectiveness of insulin when present as the sole hormone, both glycogen synthesis and lipogenesis in these hepatocyte preparations were extremely sensitive to the inhibitory action of glucagon. Moreover, when the A-cell hormone was present in physiologic concentrations, a striking protective influence of insulin could now be observed (Fig. 5). Such findings were entirely consistent with the early studies of Exton and Park (1972) in which clear-cut effects of insulin on various aspects of hepatic metabolism could only be seen against a background of perturbation by glucagon.

At the present time it is difficult to reconcile these in vitro observations with the long-standing concept that glucose and insulin, acting *in and of themselves*, are the two key factors responsible for the reversal of the fasting state of liver metabolism. On the question of substrates, the data suggest that, in the rat at least, glucose is not the immediate precursor of liver glycogen, but must first be converted into lactate (either by the intestine or peripheral tissues) which subsequently traverses the gluconeogenic pathway in the liver. The glucose-6-phosphate thus formed might then be "directed" into glycogen formation if the glucose-induced activation of hepatic glycogen synthase as proposed by HERS (1976) is operative physiologically. Should the diet contain sucrose, fructose would also act as a glycogen precursor by this mechanism. Amino acids might also be contributory. The same non-glucose substrates could readily serve as precursors for fat synthesis in the liver. On the other hand, it is possible that an as yet unidentified factor (perhaps of gastric origin) operates in vivo to facilitate the direct hepatic uptake and utilization of glucose. If such a factor exists, as has been suggested from studies in humans (DEFRONZO et al. 1978), its absence from in vitro incubations might explain the poor utility of glucose as a precursor, either for glycogen synthesis or lipogenesis in isolated liver preparations.

As to the hormonal signals at work, the essentiality of insulin for the in vivo reversal of liver metabolism from a "fasted" to a "fed" setting is clearly established since this change will not occur in the absence of the B-cell hormone. What is not clear is whether insulin acts as a direct driving force, as is commonly believed, or, rather, in indirect fashion through its ability both to suppress the pancreatic secretion of glucagon and to offset the catabolic action of the A-cell hormone at the level of the liver. We tend to favor the latter formulation.

## E. Overview

There can no longer be any doubt that the pancreatic A-cell hormone, glucagon, must be assigned equal status with its B-cell partner, insulin, for the overall control of glucose metabolism in the mammalian organism. The studies discussed in this chapter indicate, that this same bihormonal system has even broader regulatory significance for fuel homeostasis in that its sphere of influence also encompasses the overall control of fatty acid and ketone body metabolism. In retrospect, this is hardly surprising since the pathways of carbohydrate and lipid metabolism have long been suspected to be intimately linked. What has only recently begun to emerge is the biochemical basis for this linkage and a coherent picture of how the hormonal perturbation of one system automatically leads to appropriate modulation of the other. Thus, when carbohydrate is in short supply, the pancreas curtails its secretion of insulin and increases its output of glucagon. The fall in insulin signals muscle to release gluconeogenic precursors in the form of amino acids, and fat depots to mobilize stored energy in the form of fatty acids. The increase in the [glucagon]: [insulin] ratio causes elevation of hepatic cyclic AMP levels with the result that glycogen breakdown is accelerated, glycolysis is blocked, the gluconeogenic pathway is stimulated, and acetyl-CoA carboxylase is suppressed. The attendant fall in the tissue content of malonyl-CoA brings fatty acid synthesis to a halt and causes derepression of the CAT reaction such that incoming fatty acids are efficiently converted into acetoacetate and $\beta$-hydroxybutyrate. Through this remarkable series of coordinated metabolic adaptations, peripheral tissues are provided with adequate fuel in the form of fatty acids and ketone bodies, thus sparing newly synthesized glucose for consumption by the central nervous system.

Under normal circumstances all of these events are reversed with reingestion of carbohydrate and the attendant readjustment of circulating insulin and glucagon levels to postprandial values (although questions remain regarding the precise substrate mixture utilized by the liver for anabolic purposes). On the other hand, they are grossly exaggerated, with pathologic consequences, when insulin production falls to zero as in uncontrolled type 1 diabetes.

Finally, implicit in this formulation is that insulin controls peripheral events (e.g., amino acid balance in muscle and the esterification – lipolysis system in adipose tissue) by actions on these target tissues that are independent of glucagon. At the level of the liver, however, evidence is mounting that the acute effects of insulin on carbohydrate and lipid metabolism are secondary to its ability to offset the catabolic action of the A-cell hormone.

*Acknowledgments.* The author's studies discussed in this chapter were supported by grants AM-18573 and AM-07307 from the U.S. Public Health Service.

## References

Alberti KGMM, Christensen NJ, Iversen J, Orskov H (1975) Role of glucagon and other hormones in development of diabetic ketoacidosis. Lancet 1:1307–1311
Bewsher PD, Ashmore J (1966) Ketogenic and lipolytic effects of glucagon on liver. Biochem Biophys Res Commun 24:431–436
Bewsher PD, Tarrant ME, Ashmore J (1966) Effect of fat mobilization on liver acetate metabolism. Diabetes 15:346–350
Beynen AC, Vaartjes WJ, Geelen MJH (1979) Opposite effects of insulin and glucagon in acute hormonal control of hepatic lipogenesis. Diabetes 28:828–835
Beynen AC, Geelen MJH, Van den Bergh SG (1980) Short-term control of hepatic lipogenesis by insulin. Trends Biochem Sci 5:288–290
Boden G, Owen OE, Rezvani I, Elfenbein BI, Quickel KE (1977) An islet cell carcinoma containing glucagon and insulin. Diabetes 26:128–137
Boyd ME, Albright EB, Foster DW, McGarry JD (1981) In vitro reversal of the fasting state of liver metabolism in the rat. Reevaluation of the roles of insulin and glucose. J Clin Invest 68:142–152
Clark DG, Rognstad R, Katz J (1974) Lipogenesis in rat hepatocytes. J Biol Chem 249:2028–2036
Cook GA, Otto DA, Cornell NW (1980) Differential inhibition of ketogenesis by malonyl-CoA in mitochondria from fed and starved rats. Biochem J 192:955–958
Davidson MB, Berliner JH (1974) Acute effects of insulin on carbohydrate metabolism in rat liver slices: independence from glucagon. Am J Physiol 227:79–87
DeFronzo RA, Ferrannini E, Hendler R, Wahren J, Felig P (1978) Influence of hyperinsulinemia, hyperglycemia, and the route of glucose administration on splanchnic glucose exchange. Proc Natl Acad Sci USA 75:5173–5177
DiMarco JP, Hoppel C (1975) Hepatic mitochondrial function in ketogenic states. Diabetes, starvation and after growth hormone administration. J Clin Invest 55:1237–1244
Exton JH, Park CR (1972) Interaction of insulin and glucagon in the control of liver metabolism. In: Geiger SR (ed) Handbook of physiology, vol 1. American Physiological Society, Washington DC, pp 437–455
Exton JH, Mallette LE, Jefferson LS, Wong EHA, Friedmann N, Miller TB Jr, Park CR (1970) The hormonal control of hepatic gluconeogenesis. Recent Prog Horm Res 26:411–461
Feliu JE, Hue L, Hers HG (1976) Hormonal control of pyruvate kinase activity and of gluconeogenesis in isolated hepatocytes. Proc Natl Acad Sci USA 73:2762–2766
Fritz IB (1961) Factors influencing the rates of long-chain fatty acid oxidation and synthesis in mammalian systems. Physiol Rev 41:52–129
Furuya E, Uyeda K (1980) An activation factor of liver phosphofructokinase. Proc Natl Acad Sci USA 77:5861–5864
Geelen MJH (1977) Restoration of glycogenesis in hepatocytes from starved rats. Life Sci 20:1027–1034

Geelen MJH, Gibson DM (1975) Lipogenesis in maintenance cultures of rat hepatocytes. FEBS Lett 58:334–339

Geelen MJH, Harris RA, Beynen AC, McCune SA (1980) Short-term hormonal control of hepatic lipogenesis. Diabetes 29:1006–1022

Gerich JE, Lorenzi M, Bier DM, Schneider V, Tsalikian E, Karam JH, Forsham PH (1975) Prevention of human diabetic ketoacidosis by somatostatin – evidence for an essential role of glucagon. N Engl J Med 292:985–989

Guynn RW, Veloso D, Veech RL (1972) The concentration of malonyl-Coenzyme A and the control of fatty acid synthesis in vivo. J Biol Chem 247:7325–7331

Harano Y, Ohgaku S, Shimizu Y, Izumi K, Takahashi J, Shichiri M, Fukuchi M, Shigeta Y, Abe H (1977) Ketogenic action of glucagon in insulin-dependent diabetic subjects. Endocrinol Jpn 24:121–128

Harris RA, Mapes JP, Ochs RS, Crabb DW, Stropes L (1979) Hormonal control of hepatic lipogenesis. Adv Exp Med Biol 111:17

Heimberg M, Weinstein I, Kohout M (1969) The effects of glucagon, dibutyryl cyclic adenosine 3′,5′-monophosphate, and concentration of free fatty acid on hepatic lipid metabolism. J Biol Chem 244:5131–5139

Hems DA (1977) Short-term hormonal control of hepatic carbohydrate and lipid catabolism. FEBS Lett 80:237–245

Hems DA, Whitton PD, Taylor EA (1972) Glycogen synthesis in the perfused liver of the starved rat. Biochem J 129:529–538

Hers HG (1976) The control of glycogen metabolism in the liver. Annu Rev Biochem 45:167

Katz J, Golden S, Wals PA (1976) Stimulation of hepatic glycogen synthesis by amino acids. Proc Natl Acad Sci USA 73:3433–3437

Katz J, Golden S, Wals PA (1979) Glycogen synthesis by rat hepatocytes. Biochem J 180:389–402

Keller U, Chiasson JL, Liljenquist JE, Cherrington AD, Jennings AS, Crofford OB (1977) The roles of insulin, glucagon, and free fatty acids in the regulation of ketogenesis in dogs. Diabetes 26:1040–1051

Lent BA, Lee K-H, Kim R-H (1978) Regulation of rat liver acetyl-CoA carboxylase: stimulation of phosphorylation and subsequent inactivation of liver acetyl-CoA carboxylase by cyclic 3′,5′-monophosphate and effect on the structure of the enzyme. J Biol Chem 253:8149–8156

Masoro EJ (1965) Mechanisms related to the homeostatic regulation of lipogenesis. Ann NY Acad Sci 131:199–206

Mayes PA, Felts JM (1967) Regulation of fat metabolism in the liver. Nature 215:716–718

McGarry JD (1978) New perspectives in the regulation of ketogenesis. Lilly Lecture 1978. Diabetes 28:517–523

McGarry JD, Foster DW (1971) Regulation of ketogenesis from octanoic acid. The role of the tricarboxylic acid cycle and fatty acid synthesis. J Biol Chem 246:1149–1159

McGarry JD, Foster DW (1972) Regulation of ketogenesis and clinical aspects of the ketotic state. Metabolism 21:471–489

McGarry JD, Foster DW (1976) Ketogenesis and its regulation. Am J Med 61:9–13

McGarry JD, Foster DW (1977) Hormonal control of ketogenesis – biochemical considerations. Arch Intern Med 137:495–501

McGarry JD, Foster DW (1979) In support of the roles of malonyl-CoA and carnitine acyltransferase I in the regulation of hepatic fatty acid oxidation and ketogenesis. J Biol Chem 254:8163–8168

McGarry JD, Foster DW (1980) Regulation of hepatic fatty acid oxidation and ketone body production. Annu Rev Biochem 49:395–420

McGarry JD, Foster DW (1981) Reaffirmation of the extreme sensitivity of mitochondrial fatty acid oxidation to inhibition by malonyl-CoA. Potential pitfalls in demonstrating the effect. Biochem J 200:217–223

McGarry JD, Meier JM, Foster DW (1973) The effects of starvation and refeeding on carbohydrate and lipid metabolism *in vivo* and in the perfused rat liver. The relationship between fatty acid oxidation and esterification in the regulation of ketogenesis. J Biol Chem 248:270–278

McGarry JD, Robles-Valdes C, Foster DW (1975) The role of carnitine in hepatic ketogenesis. Proc Natl Acad Sci USA 72:4385–4388
McGarry JD, Wright PH, Foster DW (1975) Hormonal control of ketogenesis. Rapid activation of hepatic ketogenic capacity in fed rats by antiinsulin serum and glucagon. J Clin Invest 55:1202–1209
McGarry JD, Mannaerts GP, Foster DW (1977) A possible role for malonyl-CoA in the regulation of hepatic fatty acid oxidation and ketogenesis. J Clin Invest 60:265–270
McGarry JD, Leatherman GF, Foster DW (1978a) Carnitine palmitoyltransferase I: the site of inhibition of hepatic fatty acid oxidation by malonyl-CoA. J Biol Chem 253:4128–4136
McGarry JD, Takabayashi Y, Foster DW (1978b) The role of malonyl-CoA in the coordination of fatty acid synthesis and oxidation in isolated rat hepatocytes. J Biol Chem 253:8294–8300
Menahan LA, Wieland O (1969) The role of endogenous lipid in gluconeogenesis and ketogenesis of perfused rat liver. Eur J Biochem 9:182–188
Miller TB Jr, Larner J (1973) Mechanism of control of hepatic glycogenesis by insulin. J Biol Chem 248:3483–3488
Ontko JA, Johns ML (1980) Evaluation of malonyl-CoA in the regulation of long-chain fatty acid oxidation in the liver. Biochem J 192:959–962
Pilkis SJ, El-Maghrabi MR, Pilkis J, Claus TH, Cumming DA (1981) Fructose 2,6-bisphosphate, a new activator of phosphofructokinase. J Biol Chem 256:3171–3174
Richards CS, Uyeda K (1980) Changes in the concentration of activation factor for phosphofructokinase in hepatocytes in response to glucose and glucagon. Biochem Biophys Res Commun 97:1535–1540
Robles-Valdes C, McGarry JD, Foster DW (1976) Maternal-fetal carnitine relationships and neonatal ketosis in the rat. J Biol Chem 251:6007–6012
Schade DS, Eaton RP (1975) Glucagon regulation of plasma ketone body concentration in human diabetes. J Clin Invest 56:1340–1344
Seglen PO (1974) Autoregulation of glycolysis, respiration, gluconeogenesis and glycogen synthesis in isolated parenchymal rat liver cells and aerobic and anaerobic conditions. Biochim Biophys Acta 338:317–336
Tzur R, Tal E, Shapiro B (1964) Alpha-glycerophosphate as regulatory factor in fatty acid esterification. Biochim Biophys Acta 84:18–23
Van Schaftingen E, Hue L, Hers H-G (1980) Fructose 2,6-bisphosphate, the probable structure of the glucose- and glucagon-sensitive stimulator of phosphofructokinase. Biochem J 192:897–901
Wieland O, Matschinsky F (1962) Zur Natur der antiketogenen Wirkung von Glycerin und Fructose. Life Sci 2:49–54
Woodside WF (1979) Influence of insulin and glucagon on ketogenesis by isolated rat hepatocytes. Adv Exp Med Biol 111:97

CHAPTER 18

# Glucagon and Amino Acid Metabolism

G. F. CAHILL, JR., T. T. AOKI, and R. J. SMITH

## A. Introduction

Glucagon is thought to play two major metabolic roles (UNGER and ORCI 1981). One, as shown over 30 years ago by SUTHERLAND and associates (SUTHERLAND 1950; SUTHERLAND et al. 1968), is to increase blood glucose levels by initiating hepatic glycogenolysis. The other is to promote gluconeogenesis, thus serving to maintain glucose production, even in the face of increased insulin levels, such as after a high protein meal when insulin and glucagon are both increased. Therefore, glucagon has a very significant metabolic role, especially in carbohydrate deficiency, be it in the fed or in the fasted state. Glucagon has also been found to decrease amino acid levels after its infusion in humans (Fig. 1) and experimental animals (HELMER et al. 1957; WEINGES 1959; LANDAU and LUGIBIHL 1960; BROMER and CHANCE 1969; MARLISS et al. 1970; AOKI et al. 1974). This effect is apparent in subjects with glucagon-producing tumors, who have been found to have strikingly low concentrations of all or almost all amino acids (MALLINSON et al. 1974; HOLST 1978, 1979; KESSINGER et al. 1977; BODEN et al. 1978; RIDDLE et al. 1978; HOLST et al. 1979; STACPOOLE et al. 1981; BHATHENA et al. 1981). This response could result both from increased hepatic amino acid trapping and, possibly, from a global effect of glucagon in regulating the gradients of amino acids across all cells, or even in regulating protein synthesis and breakdown.

Although insulin certainly is preeminent, regulation of fuel metabolism in normal humans is simplified conceptually by the postulation of a bihormonal mechanism (UNGER 1976; UNGER et al. 1978; UNGER and ORCI 1981). This hypothesis takes into account the significant changes in concentrations of both insulin and glucagon in normal humans (fed and fasted states), in diabetes, and in trauma and other stressful states. Depending on the hormonal environment, glucagon appears to have both anabolic and catabolic functions. Specifically, glucagon promotes inward transport of amino acids (FREYCHET and LECAM 1977; FEHLMANN et al. 1979 a, b; KELLEY et al. 1980 a, b) and stimulation of protein synthesis in the presence of insulin (KILBERG and NEUHAUS 1977) in hepatocytes, and increased hepatic proteolysis at elevated concentrations in the absence of insulin (MALLETTE et al. 1969 a, b). Tissue responsiveness to both hormones is further dictated by the nutritional and activity state of the tissue in question. All of these factors must be considered.

As noted elsewhere in this volume (particularly in Chaps. 13 and 27), following an acute rise in glucagon levels, glucagon binds to the hepatocyte and initiates adenylate cyclase activity leading to increased levels of cyclic AMP. Protein kinase

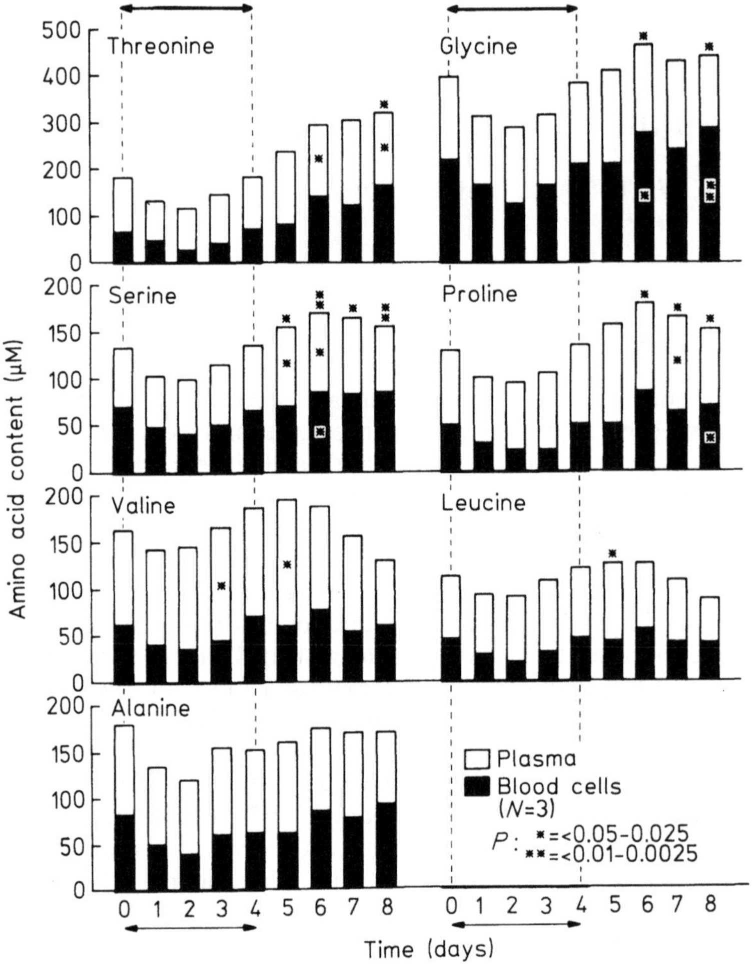


**Fig. 1.** Changes in content of plasma and blood cell amino acid compartments induced by the infusion of 0.1 mg glucagon every 24 h for 4 days (*double-headed arrows*) in 4–6 week fasted subjects. AOKI et al. (1974)

activation then accelerates the phoshorylase cascade of glycogenolysis, ultimately leading to increased concentrations of glucose-1-phosphate and glucose. It is also clear that glucagon plays another important role, although it may be primarily permissive, at other enzymatic steps in liver, to promote gluconeogenesis (EXTON 1972; DERI et al. 1981). These two widely recognized actions of glucagon that are important in glucose homeostasis, especially in fasting and emergency situations, were the first to be described. The glycogenolytic action of glucagon has suggested a "catabolic" or "crisis" role analogous to the glucocorticoids, but with a much more immediate time constant. Its increased levels following burns and infection, and its role in promoting gluconeogenesis during fasting and in diabetes reinforce this "emergency" concept, as does the marked influence of the sympathetic nervous system on glucagon secretion (see Chaps. 30 and 38).

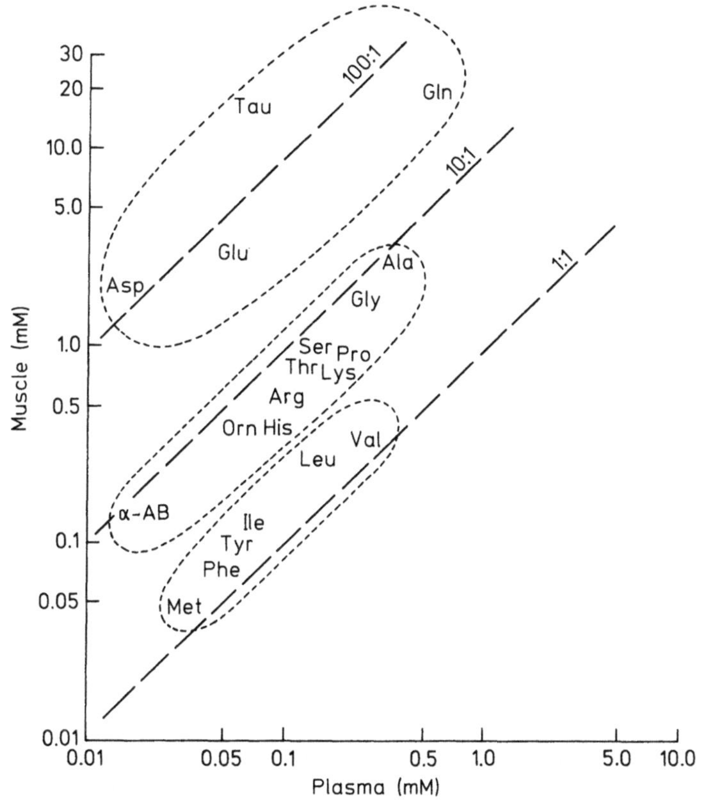


**Fig. 2.** Ratios of plasma: muscle amino acid content on logarithmic scales, showing the clustering into three broad categories; those with little increase in concentration falling on the 1:1 line; those with a first-order increase in muscle; and those with a second-order increase, namely aspartic and glutamic acids and glutamine and taurine. Asparagine, not shown here, also falls into this group. *α-AB*, α-aminobutyrate

## B. Effects of Glucagon on Amino Acid Levels

A major problem in studying glucagon effects on amino acid metabolism is that many parameters must be considered simultaneously, unlike the study of single substrates like glucose, pyruvate, lactate, or fatty acids. Another complicating factor is the intracellular–extracellular amino acid gradient, which varies with the different amino acids. For a given amino acid, the gradient may vary from one metabolic state to another and from one organ or tissue to another. The problem can be somewhat simplified, at least for skeletal muscle, by grouping the amino acids into three broad categories: (1) those amino acids which are only minimally accumulated in muscle compared with extracellular fluid; (2) those amino acids which are concentrated by approximately one order of magnitude; and (3) those amino acids which are concentrated by two orders of magnitude. Generally, the essential amino acids have intracellular: extracellular ratios approximating unity and the nonessential amino acids have high gradients (Fig. 2) with arginine, threonine, and lysine being the exceptions (Bergström et al. 1974; Askanazi et al. 1980b).

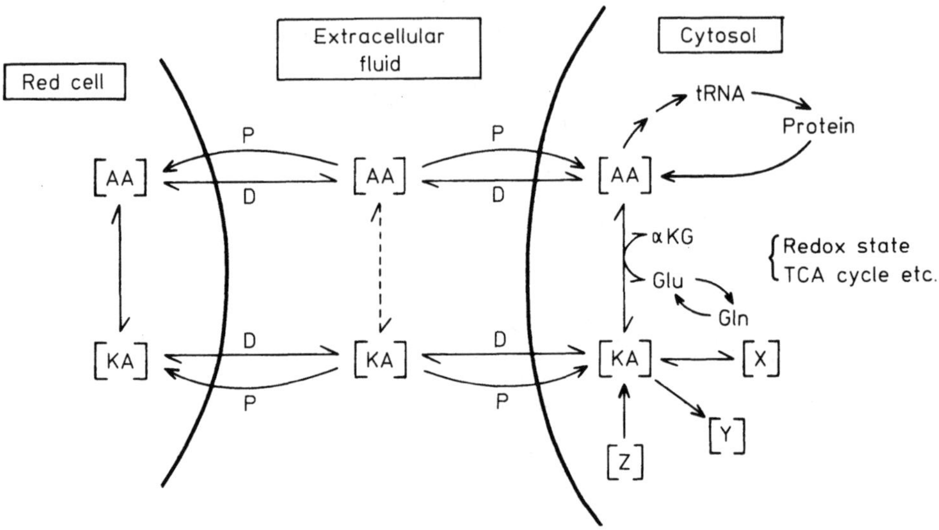


**Fig. 3.** Processes regulating levels of amino acids (AA) and their keto analogs (KA) between red cells, plasma, and cytosol, showing that the two dominant events are the gradients of the amino acids as regulated by pumps (P) and diffusion (D) and the formation and removal of the amino acids in cells by transamination, a function of the glutamate: α-ketoglutarate (Glu: α-KG) ratio and the absolute concentration of the keto analog for the specific amino acid. CAHILL et al. (1981)

Two separate processes appear to control the relative concentration of intracellular and extracellular amino acids. Since most amino acids undergo transamination with the α-ketoglutarate/glutamate couple and since transaminase activities are generally in vast excess, the reactions are usually close to equilibrium (WILLIAMSON 1967; KREBS 1969). Thus, the intracellular level of alanine, for example (Fig. 3), is dependent upon the intracellular level of pyruvate and the ratio of α-ketoglutarate to glutamate. Therefore, one of the major controls of the level of an amino acid inside the cell is the concentration of its keto analog and the "transamination potential," which is a function of the ratio of α-ketoglutarate to glutamate. Since α-ketoglutarate is an intermediate in the tricarboxylic acid cycle, it is governed in turn by the other intermediates of the cycle, such as citrate, and by the activities of the various enzymes involved, particularly α-ketoglutarate dehydrogenase.

Glutamate is also involved in many other reactions. Critical among these is the glutamate dehydrogenase reaction, which usually favors glutamate formation and ammonia removal. Of particular significance in the dehydrogenase reaction is the participation of the pyridine coenzyme couple, $NAD^+$/NADH, which links the α-ketoglutarate/glutamate couple to the redox potential, similar to the case of pyruvate and lactate. Thus, the level of a single amino acid such as alanine or glutamate is the result of a complex interaction between interlocking equilibria, influenced by both the redox potential in the cytosol and the transamination potential. To complicate matters further, there is compartmentation into intramitochondrial and extramitochondrial spaces, a topic that is beyond the scope of this chapter (DEROSA and SWICK 1975; DIETERLE et al. 1978).

**Table 1.** Plasma amino acids in humans

| | Amino acids (μ*M*) | | |
|---|---|---|---|
| | Arterial (*A*) | Forearm venous (*V*) | *A* − *V* |
| Glutamine | 550 | 650 | −100 |
| Glutamate | 55 | 20 | 35 |
| Alanine | 270 | 350 | − 80 |
| Glycine | 220 | 240 | − 20 |
| Threonine | 150 | 175 | − 25 |
| Serine | 140 | 145 | − 5 |
| Proline | 195 | 225 | − 30 |
| Valine | 245 | 250 | − 5 |
| Isoleucine | 63 | 70 | − 7 |
| Leucine | 130 | 140 | − 10 |
| Methionine | 30 | 36 | − 6 |
| Tyrosine | 55 | 60 | − 5 |
| Phenylalanine | 50 | 60 | − 10 |
| Lysine | 190 | 215 | − 25 |
| Histidine | 90 | 100 | − 10 |
| Citrulline | 35 | 35 | 0 |
| Ornithine | 60 | 60 | 0 |
| Arginine | 80 | 100 | − 20 |

A second factor of major importance in the control of intracellular amino acid concentrations is the plasma membrane transport system (Oxender and Christensen 1963; Christensen 1973, 1979, 1981; Kilberg et al. 1981). Christensen and colleagues have described a number of different carrier systems, such as the A system for alanine and certain other amino acids and the L system for leucine and other branched chain amino acids. These transport processes have been characterized mainly on the initial velocities of transport of radiolabeled amino acids and not on the steady state intracellular–extracellular gradients. Thus, what is being measured is the rate of exchange of amino acid from the extracellular to the intracellular space which, in a steady state, should be matched by the velocity in the opposite direction. It is difficult to correlate this process with net flux since the latter may occur in either direction without a significant change in the transfer velocity, but simply as a result of an alteration in the concentration in one or the other space secondary to production or removal of the amino acid. Not until recently have physiologic studies been done to measure net movement of amino acids by means of arteriovenous differences in addition to determination of amino acid profiles in both extracellular and intracellular water.

Of interest, glucagon has been reported to increase inward transport of amino acids into the hepatocyte (Freychet and LeCam 1977; Fehlmann et al. 1979a, b; Kelley et al. 1980a, b), perhaps by mobilization of intracellular calcium (Kelley 1980a) and, interestingly, by a system different from insulin (Kilberg and Neuhaus 1977). The importance of this latter observation with respect to protein meals is considerable for it suggests that glucagon may play an important role in the re-

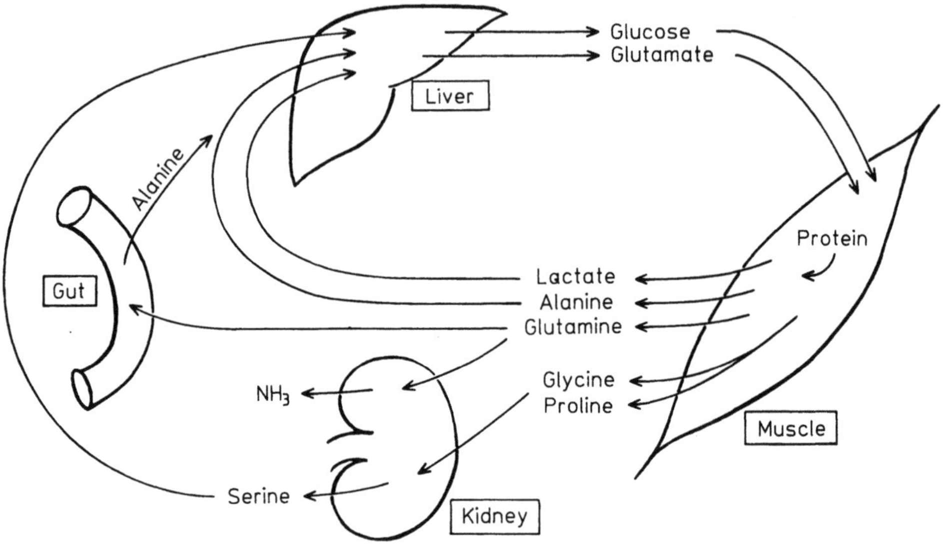


**Fig. 4.** Net amino acid flow between organs as determined by arteriovenous differences across organs in postabsorptive humans. Data are derived from the many references in the text. Glutamate is also released with alanine from the gut. CAHILL et al. (1981)

tention of large quantities of amino acids, presumably in the liver, following the ingestion of a protein meal (DEUEL et al. 1926).

There is augmented intracellular accumulation of nonmetabolized amino acids such as α-aminoisobutyric acid (CHAMBERS et al. 1968) in response to glucagon administration in vivo or addition to liver slices in vitro (TEWS et al. 1970). Parenthetically, this effect is inhibited by amiloride (FEHLMANN et al. 1981), colchicine (PRENTKI et al. 1981), dexamethasone (KELLEY et al. 1980a, b), and cycloleucine. More importantly, cyclic AMP and epinephrine have been reported to increase intracellular levels of alanine and lysine under conditions in which the steady state flux of these compounds into the cell is raised. From the foregoing, it is clear that stimulation of amino acid transport into cells occurs in response to glucagon, but the mechanism by which it is accomplished has not yet been elucidated. The potential role of cyclic AMP in mediating this effect of glucagon remains to be clarified.

Plasma amino acid levels in arterial blood and forearm arteriovenous differences taken from normal humans after an overnight fast (the postabsorptive state) are shown in Table 1. The amino acids are clustered into physiologically related groups. Glutamine and alanine are the dominant amino acids and play the major role in transport of amino acid carbon and nitrogen from the periphery to liver and kidney (Fig. 4). Glutamate, although much lower in concentration, is included because of its relationship to glutamine. Like alanine, which is related to pyruvate, glutamate concentration is proportional to the concentration of α-ketoglutarate. Thus, in the postabsorptive state, arterial glutamate and α-ketoglutarate concentrations are 50 and 10 $\mu M$, respectively while those of alanine and pyruvate are 270 and 50 $\mu M$. The levels of all of these correlate with the rate of glucose metabolism, being increased with high glycolytic rates and falling with fasting or carbohydrate deprivation (YOUNG and SCRIMSHAW 1968; ADIBI 1968; CAHILL and OWEN 1968; FELIG et al. 1969; SMITH et al. 1974; FELIG 1975; AOKI et al. 1975; CAHILL et al. 1981). It should also be added that in pathologic states such as lactic acidosis, which is usually associated with high pyruvate levels, alanine levels are similarly markedly increased (MARLISS et al. 1972).

## C. Glucagon and Liver Amino Acid Metabolism

The liver has been considered to be the primary determinant of circulating amino acid levels since the studies of SCHIMASSEK and GEROK (1965) who demonstrated that the isolated perfused rat liver contributed amino acids to the perfusate in proportion to their in vivo plasma concentrations. Unlike insulin, which has easily demonstrable rapid effects on muscle, glucagon primarily affects the liver. With the attachment of a glucagon molecule to a specific glucagon receptor on the hepatocyte, there is: (1) increased transport of amino acids into the hepatocyte (LECAM and FREYCHET 1976; FREYCHET and LECAM 1977; FEHLMANN et al. 1979 a, b; KELLEY et al. 1980 a, b); (2) inhibition of net protein synthesis (PRYOR and BERTHET 1960) and increased protein catabolism (MALLETTE et al. 1969 b); (3) augmented ureogenesis (SNODGRASS et al. 1978), and (4) marked stimulation of gluconeogenesis, as reviewed in a number of chapters in this volume. It is difficult to separate these metabolic effects of glucagon into primary and secondary components since, for example, stimulation of hepatic gluconeogenesis may have profound effects on protein synthesis, protein catabolism, amino acid transport, intra-cellular amino acid concentrations, and related metabolic pathways. Nevertheless, in view of its known actions, the stage is set for glucagon, acting with other hormones, to direct circulating amino acid concentrations.

As previously discussed, glucagon promotes de novo gluconeogenesis and net hepatic glucose output. EXTON, PARK and co-workers, and others (EXTON and PARK 1967–1969, ROSS et al. 1967; MALLETTE et al. 1969 a, b; TOLBERT and FAIN 1974) identified a number of steps accelerated by glucagon in the gluconeogenic pathway that mainly involved the formation of phosphoenolpyruvate. In more detailed studies, substrate crossovers were found at this locus with physiologic levels of alanine (Fig. 5 a). With saturating alanine concentrations (Fig. 5 b), both transamination and phosphoenolpyruvate formation were found to be rate limiting (UI et al. 1973 a, b). Levels of glutamate in liver were decreased strikingly (UI et al. 1973 b) as were levels of α-ketoglutarate (SIESS et al. 1977; SIESS and WIELAND 1978). Other studies (TEWS et al. 1970; AIKAWA et al. 1972; PARILLA et al. 1975, 1976; MACDONALD et al. 1976) have supported the role of glucagon in initiating amino acid uptake into liver and thus stimulating the formation of phosphoenolpyruvate.

Using isolated hepatocytes, FEHLMANN et al. (1979 b) showed direct stimulation by glucagon of more than one transport system for alanine. Using the nonmetabolized amino acid, α-aminoisobutyric acid, they found an increase in activity of high affinity systems. Thus, glucagon promotes amino acid trapping by the liver, at least as expressed by α-aminoisobutyric acid uptake or alanine uptake. To date there are no data demonstrating an altered total amino acid pattern in liver after glucagon stimulation. The dramatically decreased levels of all amino acids in plasma of

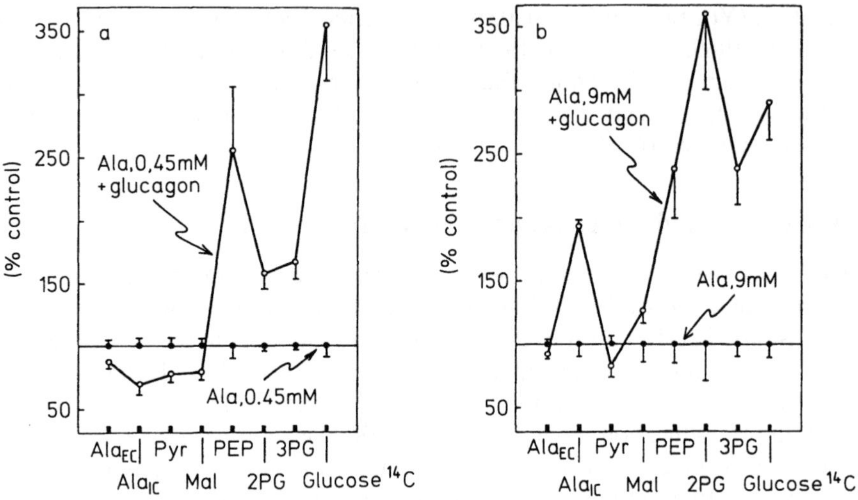


**Figs. 5 a, b.** Crossover plots of metabolic intermediates in the perfused liver as altered by glucagon with both physiologic (0.45 m*M*, **a**) and saturating (9 m*M*, **b**) levels of alanine (Ala) as substrate. With the former, there is a single crossover at the formation of phosphoenolpyruvate (PEP). With the saturating concentrations of alanine, (**b**), are demonstrated a rate-limiting step at transamination of intracellular alanine ($Ala_{IC}$) to pyruvate (Pyr) as well as a crossover between pyruvate and phosphoenolpyruvate. $Ala_{EC}$, extracellular alanine; *Mal*, malate; *2PG*, 2-phosphoglycerate; *3PG*, 3-phosphoglycerate. UI et al. (1973a)

patients with glucagonoma (as previously discussed), and a striking decrease in amino acids following glucagon administration to animals or humans (see Fig. 1) is consistent with the hepatic trapping effect. A possible explanation for the decrease in branched chain amino acids is the concomitant stimulation of insulin release (see Chap. 22), which has been shown to result in lowered levels of branched chain amino acids, probably by promoting their uptake and incorporation into protein in muscle (MUNRO and THOMSON 1953; CROFFORD et al. 1964; CARLSTEN et al. 1966; POZEFSKY et al. 1968; MANCHESTER 1970; CAHILL 1971; CAHILL et al. 1972).

MALLETTE et al. (1969 a), using perfused rat livers with medium containing normal plasma amino acid concentrations, were able to document half-maximal gluconeogenesis. Glucose production approached saturation when amino acids were increased to three times normal concentration, and urea production increased concomitantly. The administration of glucagon at $4 \times 10^{-9}$ *M* resulted in the stimulation of both glucose and urea production at all levels of administered amino acids, increasing therefore overall the $V_{max}$, but not the apparent affinity $K_m$ of the system. Specifically, glucagon resulted in a threefold increase in the conversion of alanine to glucose. Measurement of steady state levels of gluconeogenic intermediates suggested that the conversion of pyruvate to phosphoenolpyruvate was stimulated. In the presence of elevated concentrations of alanine, there was also marked conversion of alanine to glucose. These observations suggest that there was a net increase in uptake of alanine and this leads to increased gluconeogenesis. Since a similar increase in glucose production from alanine could be seen in the presence of 1 m*M* cyclic AMP, it appeared that the effect of glucagon might be entirely mediated by the activation of the adenylate cyclase system.

MALLETTE et al. (1969 b) have also examined the influence of glucagon on the fate of both natural and nonmetabolized amino acids. Using media containing amino acids in concentrations found in normal plasma, they were able to show that glucagon induced: (1) intracellular utilization of glycine, alanine, glutamate, and phenylalanine; (2) intracellular production of leucine, isoleucine, and valine; and (3) inward transfer of lysine, α-aminoisobutyric acid, and cycloleucine. Since valine is neither synthesized nor catabolized by the liver to a significant degree, its intracellular production and subsequent release from this tissue may be considered to be an index of hepatic proteolysis. This observation may explain the increase in branched chain amino acid levels seen in briefly fasted subjects (FELIG et al. 1969).

Glucagon has been reported to stimulate not only enzymes associated with gluconeogenesis and urea synthesis (SNODGRASS et al. 1978), but also enzymes associated with the catabolism of amino acids. In rats, glucagon administration results in significant increases in the activity of serine dehydratase, homoserine dehydratase, tyrosine transaminase, ornithine transaminase, and cytoplasmic aspartate and alanine transaminases. With the simultaneous administration of glucose and glucagon, the stimulatory effect of glucagon on these enzyme activities is diminished. It is interesting to speculate that excess synthesis of such enzymes may generate a labile pool of protein that is readily broken down later. Parenthetically, it should be noted that insulin appears to oppose the stimulatory effect of glucagon on enzymatic activities, providing some support for the bihormonal hypothesis originally proposed by UNGER (1976).

While evidence has been presented for glucagon-stimulated protein synthesis, a great deal of information is now available for documenting the apparent catabolic effect of this particular hormone. Reference has previously been made to increased branched chain amino acid output from livers perfused with glucagon, suggesting that hepatic proteolysis has been stimulated. In addition, the work of PRYOR and BERTHET (1960) showed an inhibition of the incorporation of amino acids into liver protein in the presence of glucagon. The prospect of hormonal control over the regulation of rates of liver proteolysis is especially intriguing. The work of SCHIMKE (1970) as well as other investigators suggests that there is regulation of the breakdown of a number of enzymes of variable half-lives that are responsive to specific substrate and hormonal cues. In addition, the work of WOODSIDE and MORTIMORE (1970) showed that glucagon-induced valine release, presumably secondary to hepatic proteolysis, could be inhibited by the presence in the perfusing media of elevated concentrations of valine alone or, interestingly, an amino acid mixture that did not contain any of the branched chain amino acids.

Glucagon is also widely recognized as a potent stimulator of lysosomal activity (ASHFORD and PORTER 1962). It is used for the induction of autophagocytosis (SCHWORER and MORTIMORE 1979), i.e., formation of phagolytic vacuoles containing cytoplasmic components such as mitochondria or endoplasmic reticulum. The appearance of such vacuoles correlates with a demonstrable increase in activity of enzymes of cellular catabolism. The induced lytic enzymes include acid phosphatase, lysosomal lipase, and *N*-acetyl-β-D-glucosaminidase.

Of interest, with respect to the mechanism of these changes is the observation of ROSA (1971), who produced the same change with cyclic AMP and epinephrine by means of perfused isolated livers. Lysosomal glucosaminidase activity does not increase (GUDER et al. 1970) in fed or adrenalectomized animals, a condition in which glucagon is not particularly effective in increasing cyclic AMP levels.

## D. Glucagon and Muscle Amino Acid Metabolism

The de novo synthesis of glucose by the liver in response to glucagon and other hormones requires a continuous supply of substrate molecules, predominantly in the form of amino acids. The most important source of these gluconeogenic precursors is skeletal muscle, which constitutes more than 40% of body weight in normal humans (MUNRO 1964) and thus represents a substantial portion of total body protein. Since major repositories of glucogenic amino acids analogous to triglyceride in adipose tissue do not exist, these amino acids must be derived through catabolism of functional proteins when exogenous sources are unavailable. In the extreme, as exemplified by patients with glucagon-producing tumors, the protein-catabolic effects of glucagon can lead to marked muscle wasting (see Sect. E and Chap. 43).

In spite of early reports to the contrary (PETERSON et al. 1963; BEATTY et al. 1963), it is now generally accepted that physiologic doses of glucagon have no direct effects on the metabolism of skeletal muscle protein and amino acids in normal postabsorptive humans. The most convincing evidence comes from the work of POZEFSKY et al. (1976), who infused glucagon into the brachial artery of normal subjects to produce increments within the forearm of 1,216 and 784 pg/ml over basal levels in overnight and 60-h fasted subjects, respectively. During a 2-h infusion, glucagon had no effect on net amino acid release from the forearm. Consistent with these results is the apparent absence of a glucagon-responsive adenylate cyclase in skeletal muscle (SUTHERLAND and ROBISON 1969; ROBISON et al. 1971). Thus, the catabolic effect of glucagon on skeletal muscle protein occurs as a secondary phenomenon that most probably is a response to the lowered free amino acid levels.

In contrast to skeletal muscle, the myocardium has been shown to respond to glucagon with elevated cyclic AMP levels (SUTHERLAND and RALL 1960; LEVEY and EPSTEIN 1969), increased contractility (FARAH and TUTTLE 1960), increased mitochondrial respiration (KRUTY et al. 1978), accelerated branched chain amino acid oxidation (BUSE et al. 1973), and decreased amino acid incorporation into protein (HAIT et al. 1972). All these aspects are reviewed in detail in Chap. 53.

## E. Physiologic and Pathologic States in Which Glucagon Affects Amino Acid Metabolism

### I. Glucagon Deficiency

Early studies in experimental animals first demonstrated the consequences of isolated glucagon deficiency. In rats fasted for 48 h to deplete hepatic glycogen stores, the injection of glucagon antibodies caused hypoglycemia, presumably as a result of decreased gluconeogenesis (GREY et al. 1970). Destruction of the separate A-cell-containing islets of birds has even more dramatic results, causing death from hypoglycemia within a few hours (MIALHE 1958; MIKAMI and ONO 1962). The frequency and consequences of glucagon deficiency in humans have not yet been defined. In spite of the general availability of radioimmunoassays for glucagon, very few cases of spontaneous glucagon deficiency have been described (see review in Chap. 42).

There have been a few reports of isolated glucagon deficiency presenting in the neonatal period (WAGNER et al. 1969; GOTLIN and SILVER 1970; ZUPPINGER 1975; VIDNES 1976; KOLLEE et al. 1978). In two cases, severe hypoglycemia resolved during treatment with zinc protamine glucagon (VIDNES 1976; KOLLEE et al. 1978). Although there have been sporadic cases of hypoglycemia and glucagon deficiency in adults, in most instances, hypoglucagonemia has not been clearly documented (LEVY et al. 1969), or other causes of hypoglycemia have not been ruled out (LORETI et al. 1974). In a recent study, FOÀ et al. (1980) demonstrated abnormally low glucagon levels in seven patients with reactive hypoglycemia. Establishment of the causal relationship between the decreased glucagon levels and hypoglycemia, as well as the frequency of this disorder, will require further study. Unfortunately, detailed amino acid levels have not been reported in any cases of spontaneous glucagon deficiency.

There is an increase in the concentration of several amino acids, especially the glucogenic amino acids, in patients who are partially glucagon deficient as a result of total pancreatectomy (MÜLLER et al. 1979; BODEN et al. 1980). The amino acid levels are returned toward normal following the infusion of physiologic amounts of glucagon (BODEN et al. 1980), suggesting a role for basal glucagon in setting the concentrations of circulating amino acids. This hypothesis has been supported by studies with somatostatin, which inhibits both glucagon and insulin secretion (KOERKER et al. 1974; MORTIMER et al. 1974). When somatostatin was infused in normal humans for 6 h, plasma alanine concentration increased by 14% (SHERWIN et al. 1977). Simultaneous replacement of glucagon at slightly above basal levels produced a decline in alanine levels to 14% below control. Thus, normal basal glucagon appears to have an important influence on circulating concentrations of amino acids, probably via its effects on hepatic uptake.

## II. Glucagon Excess

### 1. Diabetes Mellitus

The most common cause of glucagon hypersecretion is diabetes mellitus, especially during diabetic ketoacidosis, when glucagon levels are inappropriately elevated in the presence of systemic hyperglycemia (UNGER et al. 1970). In addition to the effects of insulin deficiency, excessive glucagon probably contributes to the hypoaminoacidemia of uncontrolled diabetes (FELIG et al. 1970) by stimulating hepatic gluconeogenesis. This concept is supported by studies of GERICH et al. (1975) in diabetic humans. When insulin therapy was discontinued, blood glucose levels rose at a much slower rate if glucagon secretion was suppressed by the infusion of somatostatin. It should be pointed out, however, that somatostatin also decreases splanchnic blood flow, which could contribute to the decreased rate of hepatic glucose production (WAHREN and FELIG 1976).

### 2. Glucagonoma

As reviewed in Chap. 43, the most extreme elevations of glucagon occur in patients with glucagon-secreting tumors of the pancreatic A-cells (MCGAVRAN et al. 1966). The full glucagonoma syndrome was first described by MALLINSON et al. (1974), who recognized among other things the association of high glucagon levels, low plasma amino acid levels, and a characteristic skin rash. Although augmented hepatic gluconeogenesis has not been documented, it is generally believed to be the cause of the hypoaminoacidemia (BODEN et al. 1977).

In most cases, the decline in amino acid levels is generalized, with concentrations of alanine and several other amino acids being reduced as much as fivefold (MALLINSON et al. 1974; BODEN et al. 1977; HOLST et al. 1979; NORTON et al. 1979). The low levels of branched chain amino acids observed in some patients can be attributed to the effects of excess insulin, since many patients develop secondary diabetes. There is good evidence that the characteristic skin rash is caused by the hypoaminoacidemia, since it resolves following: (1) removal of the tumor (BHATHENA et al. 1981); (2) suppression of glucagon levels with somatostatin (BODEN and OWEN 1977); or (3) partial restoration of amino acids with parenteral amino acid solutions (STACPOOLE et al. 1981) or total parenteral nutrition (NORTON et al. 1979). With the last two regimens, the cutaneous lesions resolve in spite of persistently elevated glucagon, thus ruling out a direct effect of glucagon on the skin.

Another characteristic finding in patients with glucagonomas is marked weight loss and muscle wasting (BHATHENA et al. 1981). Successful extirpation of the tumors is followed by a gradual return to normal weight over a period of several weeks. Since glucagon has not been shown to have direct effects on skeletal muscle (see Sect. D), one can speculate that the muscle wasting develops from the influence of amino acid concentrations on protein synthesis and degradation. It would be interesting to determine the concentrations of amino acids in the muscle of patients with glucagonomas.

### 3. Trauma and Sepsis

Following severe trauma, sepsis, burns, or other catastrophic illnesses, there is a catabolic stress response characterized by markedly negative nitrogen balance (CUTHBERTSON 1932; LAWSON 1965; KINNEY et al. 1968). The loss of nitrogen in these states is greater than can be accounted for by simple nutritional deprivation (RICHARDS 1980), the bulk of nitrogen being excreted as urinary urea and thus reflecting accelerated gluconeogenesis (CAHILL et al. 1970). This is consistent with studies that have shown accelerated glucose turnover (LONG et al. 1971) and a failure of infused glucose to suppress hepatic gluconeogenesis (HINTON et al. 1971) in injured patients. It is thought that the catabolic stress response is hormonally mediated. Since glucagon secretion increases after surgery, burns, and infections in proportion to the severity of the illness (KINNEY and FELIG 1979), it was at one time suggested that glucagon may be a primary mediator of the stress response

(RUSSEL et al. 1975). A major role for glucagon now appears to be unlikely, however, since the time course of changes in glucagon levels does not correlate well with the progression of the stress response (KINNEY and FELIG 1979).

Recent studies of muscle and plasma amino acid levels following severe illnesses should rekindle interest in at least a contributory role for glucagon. BERGSTRÖM and KINNEY, and their collaborators have used a needle biopsy technique to study muscle amino acids in patients after surgery, trauma, burns, and sepsis (VINNARS et al. 1975; ASKANAZI et al. 1978, 1980a, b; FÜRST et al. 1979). They have demonstrated a marked decline in the concentrations of most amino acids in muscle that is associated with a decrease in plasma amino acid concentrations. Considering the documented inverse relationship between glucagon and amino acid levels in other physiologic states, it is important to reconsider the role of glucagon in influencing amino acid metabolism following stress and trauma.

## F. Summary

In summary, the metabolic role played by glucagon in amino acid metabolism in humans remains unclear. What is known, however, is that it can profoundly affect both extracellular and intracellular concentrations of amino acids, some of which have marked influences on both protein synthesis and degradation. In this latter regard, the elevated level of this hormone in patients with glucagonomas is accompanied by a marked decrease in circulating concentrations of amino acids, a finding shared with those individuals suffering from severe trauma (with the exception that in the latter instance branched chain amino acid levels are also frequently elevated). Finally, aside from its potent effects on glycogenolysis and gluconeogenesis, glucagon appears to have a primary role in the processing and storage of dietary protein. As noted elsewhere, investigators have long known (DEUEL et al. 1926) that the ingestion of large quantities of protein results in the acute retention of nitrogen within the body, presumably in the form of proteins, since circulating concentrations of free amino acids are not increased in proportion to their content in the meal. Since glucagon is the hormone most markedly elevated in normal individuals following the ingestion of a protein meal, it appears to be largely responsible, in conjunction with insulin, for the retention of these amino acids in the form of proteins.

*Acknowledgments.* Support for the studies described in this manuscript derive in part from U.S. National Institutes of Health grants AM 15191 and BRSG S07 RR-05673, and the Howard Hughes Medical Institute.

## References

Adibi SA (1968) Influence of dietary deprivation on plasma concentration of free amino acids of man. J Appl Physiol 25:52–57

Aikawa T, Matsutaka H, Takezawa K, Ishikawa E (1972) Gluconeogenesis and amino acid metabolism. I. Comparison of various precursors for hepatic gluconeogenesis in vivo. Biochim Biophys Acta 279:234–244

Aoki TT, Müller WA, Brennan MF, Cahill GF Jr (1974) Effect of glucagon on amino acid and nitrogen metabolism in fasting man. Metabolism 23:805–814

Aoki TT, Toews CJ, Rossini AA, Ruderman NB, Cahill GF Jr (1975) Gluconeogenic substrate levels in fasting man. Adv Enzyme Regul 13:329–336

Ashford TP, Porter KR (1962) Cytoplasmic components in hepatic cell lysosomes. J Cell Biol 12:198–202
Askanazi J, Elwyn DH, Kinney JM, Gump FE, Michelsen CB, Stinchfield FE (1978) Muscle and plasma amino acids after injury. The role of inactivity. Ann Surg 188:797–803
Askanazi J, Fürst P, Michelson CB, Elwyn DH, Vinnars E, Gump FE, Stinchfield FE, Kinney JM (1980 a) Muscle and plasma amino acids after injury. Hypocaloric glucose vs. amino acid infusion. Ann Surg 191:465–472
Askanazi J, Carpentier YA, Michelsen CB, Elwyn DH, Fürst P, Kantrowitz LR, Gump FE, Kinney JM (1980 b) Muscle and plasma amino acids following injury. Influence of intercurrent infection. Ann Surg 192:78–85
Beatty CH, Peterson RD, Bocek RM, Craig NC, Weleber R (1963) Effect of glucagon on incorporation of glycine-$C^{14}$ into protein of voluntary skeletal muscle. Endocrinology 73:721–726
Bergström J, Fürst P, Norée L-O, Vinnars E (1974) The intracellular free amino acid concentration in human muscle tissue. J Appl Physiol 36:693–697
Bhathena SJ, Higgins GA, Recant L (1981) Glucagonoma and glucagonoma syndrome. In: Unger RH, Orci L (eds) Glucagon. Elsevier/North Holland, Amsterdam Oxford New York, pp 413–438
Boden G, Owen OE (1977) Familial hyperglucagonemia – an autosomal dominant disorder. N Engl J Med 296:534–538
Boden G, Owen OE, Rezvani I, Elfenbein EI, Quickel KE (1977) An islet cell carcinoma containing glucagon and insulin. Chronic glucagon excess and glucose homeostasis. Diabetes 26:128–137
Boden G, Wilson RM, Owen OE (1978) Effects of chronic glucagon excess on hepatic metabolism. Diabetes 27:643–648
Boden G, Master RW, Rezvani I, Palmer JP, Lobe TE, Owen OE (1980) Glucagon deficiency and hyperaminocidemia after total pancreatectomy. J Clin Invest 65:706–716
Bromer WW, Chance RE (1969) Zinc glucagon depression of blood amino acids in rabbits. Diabetes 18:748–754
Buse MG, Biggers JF, Drier C, Buse JF (1973) The effect of epinephrine, glucagon, and the nutritional state on the oxidation of branched chain amino acids and pyruvate by isolated hearts and diaphragms of the rat. J Biol Chem 248:697–706
Cahill GF Jr (1971) Physiology of insulin in man. Diabetes 20:785–799
Cahill GF Jr, Owen OE (1968) Some observations on carbohydrate metabolism in man. In: Dickens F, Randle PJ, Whelan WJ (eds) Carbohydrate metabolism and its disorders, chap 16. Academic Press, New York London, pp 497–556
Cahill GF Jr, Felig P, Marliss EB (1970) Some physiological principles of parenteral nutrition. In: Fox CL Jr, Nahas GG (eds) Fluid replacement in the surgical patient: a symposium, 1969. Grune and Stratton, New York, pp 286–295
Cahill GF Jr, Marliss EB, Aoki TT (1972) Insulin and muscle protein. In: Steiner DF, Freinkel N (eds) Endocrinology. American Physiology Society, Washington, DC (Handbook of physiology, sect 7, vol 1, pp 563–577)
Cahill GF Jr, Aoki TT, Smith RJ (1981) Amino acid cycles in man. Curr Top Cell Regul 18:389–400
Carlsten A, Hallgren B, Jagenburg R, Svanborg A, Werkö L (1966) Amino acids in plasma in diabetes. I. The effect of insulin on arterial levels. Acta Med Scand 179:361–370
Chambers JW, Georg RH, Bass AD (1968) Effects of catecholamines and glucagon on amino acid transport in the liver. Endocrinology 83:1185–1192
Christensen HN (1973) On the development of amino acid transport systems. Fed Proc 2:19–28
Christensen HN (1979) Exploiting amino acid structure to learn about membrane transport. Adv Enzymol 49:41–101
Christensen HN (1981) Membrane domination of biological energy exchanges: a message of the 1978 Nobel award in chemistry. Perspect Biol Med 24:358–373
Crofford DB, Felts PW, Lacy WW (1964) Effect of glucose infusion on the individual free amino acids in man. Proc Soc Exp Biol Med 117:11–14

Cuthbertson DP (1932) Observations on the disturbance of metabolism produced by injury to the limbs. Biochem J 25:233–246
Deri JJ, Williams PE, Steiner KE, Cherrington AD (1981) Altered ability of the liver to produce glucose following a period of glucagon deficiency. Diabetes 30:490–495
DeRosa G, Swick RW (1975) Metabolic implications of the distribution of the alanine aminotransferase isozymes. J Biol Chem 250:7961–7967
Deuel HJ Jr, Sandiford K, Sandiford I, Boothby WM (1926) Deposit protein: the effect of thyroxin on the deposit protein after reduction of the nitrogen excretion to a minimal level by a prolonged protein-free diet. J Biol Chem 67:XXIII–XXIV
Dieterle PM, Brawand F, Moser UK, Walter P (1978) Alanine metabolism in rat liver mitochondria. Eur J Biochem 88:467–473
Exton JH (1972) Gluconeogenesis. Metabolism 21:945–990
Exton JH, Park CR (1967) Control of gluconeogenesis in liver. I. General features of gluconeogenesis in the perfused livers of rats. J Biol Chem 242:2622–2636
Exton JH, Park CR (1968) Control of gluconeogenesis in liver. II. Effects of glucagon, catecholamines, and adenosine 3′,5′-monophosphate on gluconeogenesis in the perfused rat liver. J Biol Chem 243:4189–4196
Exton JH, Park CR (1969) Control of gluconeogenesis in liver. III. Effects of lactate, pyruvate, fructose, glucagon, epinephrine and adenosine 3′,5′-monophosphate on gluconeogenic intermediates in the perfused rat liver. J Biol Chem 244:1424–1433
Farah A, Tuttle R (1960) Studies on the pharmacology of glucagon. J Pharmacol Exp Ther 129:49–55
Fehlmann M, LeCam A, Kitabgi P, Rey J-F, Freychet P (1979 a) Regulation of amino acid transport in the liver. Emergence of a high affinity transport system in isolated hepatocytes from fasting rats. J Biol Chem 254:401–407
Fehlmann M, LeCam A, Freychet P (1979 b) Insulin and glucagon stimulation of amino acid transport in isolated rat hepatocytes. Synthesis of a high affinity component of transport. J Biol Chem 254:10431–10437
Fehlmann M, Samson M, Koch KS, Leffert HL, Freychet P (1981) The effect of amiloride on hormonal regulation of amino acid transport in isolated and cultured adult rat hepatocytes. Biochim Biophys Acta 642:88–95
Felig P (1975) Amino acid metabolism in man. Annu Rev Biochem 44:933–955
Felig P, Owen OE, Wahren J, Cahill GF Jr (1969) Amino acid metabolism during prolonged starvation. J Clin Invest 48:584–594
Felig P, Marliss E, Ohman J, Cahill GF Jr (1970) Plasma amino acid levels in diabetic ketoacidosis. Diabetes 19:727–729
Foà PP, Dunbar JC, Klein SP, Levy SH, Malik MA, Campbell BB, Foà NL (1980) Reactive hypoglycemia and A-cell ("pancreatic") glucagon deficiency in the adult. JAMA 244:2281–2285
Freychet P, LeCam A (1977) Amino acid transport in isolated hepatocytes: effect of glucagon. Ciba Found Symp 55:247–262
Fürst P, Bergström J, Chao L, Larsson J, Liljedahl SO, Neuhauser M, Schildt B, Vinnars E (1979) Influence of amino acid supply on nitrogen and amino acid metabolism in severe trauma. Acta Chir Scand [Suppl] 494:136–141
Gerich JE, Lorenzi M, Bier DM, Schneider V, Tsalikian E, Karam JH, Forsham PH (1975) Prevention of human diabetic keotacidosis by somatostatin. Evidence for an essential role of glucagon. N Engl J Med 292:985–989
Gotlin RW, Silver HK (1970) Neonatal hypoglycemia, hyperinsulinism and absence of pancreatic alpha cells. Lancet 1:1346
Grey N, McGuigan JE, Kipnis DM (1970) Neutralization of endogenous glucagon by high titer glucagon antiserum. Endocrinology 86:1383–1388
Guder W, Hepp KD, Wieland O (1970) The catabolic action of glucagon in rat liver. Biochim Biophys Acta 222:593
Hait G, Kypson J, Massik R (1972) Amino acid incorporation into myocardium: effect of insulin, glucagon, and dibutyryl 3′,5′-AMP. Am J Physiol 222:404–408
Helmer OM, Kirtley WR, Ridolfo AS (1957) Clinical and metabolic changes induced by glucagon in patients with rheumatoid arthritis. J Lab Clin Med 50:824

Hinton P, Allison SP, Littlejohn S, Lloyd J (1971) Insulin and glucose to reduce catabolic response to injury in burned patients. Lancet 1:767–769
Holst JJ (1978) Glucagonomas. In: Bloom SR (ed) Gut hormones. Churchill Livingstone, Edinburgh London, pp 599–694
Holst JJ (1979) Gut endocrine tumour syndromes. Clin Endocrinol Metab 8:413–432
Holst JJ, Helland S, Ingemannson S, Pedersen NB, Von Schenk H (1979) Functional studies in patients with the glucagonoma syndrome. Diabetologia 17:151–156
Kelley DS, Evanson T, Potter VR (1980a) Calcium-dependent hormonal regulation of amino acid transport and cyclic AMP accumulation in rat hepatocyte monolayer cultures. Proc Natl Acad Sci USA 77:5953–5957
Kelley DS, Shull JD, Potter VR (1980b) Hormonal regulation of amino acid transport and cAMP production in monolayer cultures of rat hepatocytes. J Cell Physiol 103:159–168
Kessinger A, Lemon HM, Foley JF (1977) The glucagonoma syndrome and its management. J Surg Oncol 9:419–424
Kilberg MS, Neuhaus OW (1977) Hormonal regulation of hepatic amino acid transport. J Supramol Struct 6:191–204
Kilberg MS, Handlogten ME, Christensen HN (1981) Characteristics of system ASC for transport of neutral amino acids in the isolated rat hepatocyte. J Biol Chem 256:3304–3312
Kinney JM, Felig P (1979) The metabolic response to injury and infection. In: DeGroot LJ et al. (eds) Endocrinology. Grune and Stratton, New York, pp 1963–1986
Kinney JM, Long CL, Gump FE, Duke JH Jr (1968) Tissue composition of surgical patients. I. Elective operation. Ann Surg 168:459–474
Koerker DJ, Ruch W, Chideckel E, Palmer J, Goodner CJ, Ensinck J, Gale CC (1974) Somatostatin: hypothalamic inhibitor of the endocrine pancreas. Science 184:482–484
Kollee LA, Monnens LA, Cejka V, Wilms RH (1978) Persistent neonatal hypoglycemia due to glucagon deficiency. Arch Dis Child 53:422–424
Krebs HA (1969) The role of equilibria in the regulation of metabolism. Curr Top Cell Regul 6:45–55
Kruty F, Gvozdjak A, Bada V, Niederland TR, Gvozdjak J, Kaplan M (1978) The effect of glucagon on the heart muscle: relation between metabolic processes and contractility. Biochem Pharmacol 27:2153–2155
Landau RL, Lugibihl K (1960) Effect of glucagon on concentration of several free amino acids in plasma. Metabolism 18:265–276
Lawson LJ (1965) Parenteral nutrition in surgery. Br J Surg 52:795–799
LeCam A, Freychet P (1976) Glucagon stimulates the A system for neutral amino acid transport in isolated hepatocytes of the adult rat. Biochem Biophys Res Commun 72:893–901
Levey GS, Epstein SE (1969) Activation of adenyl cyclase by glucagon in cat and human heart. Circ Res 24:151–156
Levy LJ, Zarowitz H, Bleicher SJ, Spergel G (1969) Possible combined alpha and beta cell deficiency. Clin Res 17:590
Long CL, Spencer JL, Kinney JM, Geiger JW (1971) Carbohydrate metabolism in man: effects of elective operations and major injury. J Appl Physiol 31:110–116
Loreti L, Klein SP, Foà PP (1974) Hypoglycemia associated with glucagon insufficiency. Clin Res 22:475
MacDonald M, Neufeldt N, Park BN, Berger M, Ruderman NB (1976) Alanine metabolism and gluconeogenesis in the rat. Am J Physiol 231:619–625
Mallette LE, Exton JH, Park CR (1969a) Control of gluconeogenesis from amino acids in the perfused rat liver. J Biol Chem 244:5713–5723
Mallette LE, Exton JH, Park CR (1969b) Effects of glucagon on amino acid transport and utilization in the perfused rat liver. J Biol Chem 244:5724–5728
Mallinson CH, Bloom SR, Warin AP, Salmon PR, Cox B (1974) A glucagonoma syndrome. Lancet 2:1–5
Manchester KL (1970) Control by insulin of amino acid accumulation in muscle. Biochem J 117:457–465

Marliss EB, Aoki TT, Unger RH, Soeldner JS, Cahill GF Jr (1970) Glucagon levels and metabolic effects in fasting man. J Clin Invest 49:2256–2270
Marliss EB, Aoki TT, Toews CJ, Felig P, Connon JJ, Kyner J, Huckabee WE, Cahill GF Jr (1972) Amino acid metabolism in lactic acidosis. Am J Med 52:474–481
McGavran MH, Unger RH, Recant L, Polk HC, Kilo C, Levin ME (1966) A glucagon-secreting alpha-cell carcinoma of the pancreas. N Engl J Med 274:1408–1413
Mialhe P (1958) Glucagon, insuline, et regulation endocrine de la glycémie chez le canard. Acta Endocrinol (Copenh) 28:(36)139
Mikami SI, Ono K (1962) Glucagon deficiency induced by extirpation of alpha islets of the fowl pancreas. Endocrinology 71:464–475
Mortimer CH, Carr D, Lind T, Bloom SR, Mallinson CN, Schally AV, Tunbridge WMG, Yeomans L, Coy DH, Kastin A, Besser GM, Hale R (1974) Effects of growth-hormone release-inhibiting hormone on circulating glucagon, insulin, and growth hormone in normal, diabetic, acromegalic, and hypopituitary patients. Lancet 1:697–701
Müller WA, Berger M, Suter P, Cuppers HJ, Reiter J, Wyss T, Berchtold P, Schmidt F, Assal J-P, Renold AE (1979) Glucagon immunoreactivities and amino acid profile in plasma of duodenopancreatectomized patients. J Clin Invest 53:820–827
Munro HN (1964) General aspects of the regulation of protein metabolism by diet and by hormones. In: Munro HN, Allison JB (eds) Mammalian protein metabolism, vol 1. Academic, New York London, p 406
Munro HN, Thomson WST (1953) Influence of glucose on amino acid metabolism. Metabolism 2:354–361
Norton JA, Kahn CR, Schiebinger R, Gorschboth C, Brennan MF (1979) Amino acid deficiency and the skin rash associated with glucagonoma. Ann Intern Med 91:213–215
Oxender DL, Christensen HN (1963) Distinct mediating systems for the transport of neutral amino acids by the Ehrlich cell. J Biol Chem 238:3686–3699
Parilla R, Jimenez M-K, Ayuso-Parrilla MS (1975) Glucagon and insulin control of gluconeogenesis in the perfused isolated rat liver. Effects on cellular metabolite distribution. Eur J Biochem 56:375–383
Parilla R, Jimenez M-K, Ayuso-Parrila MS (1976) Cellular redistribution of metabolites during glucagon and insulin control of gluconeogenesis in the isolated perfused rat liver. Arch Biochem Biophys 174:1–12
Peterson RD, Beatty CH, Bocek RM (1963) Effects of insulin and glucagon on carbohydrate and protein metabolism of adductor muscle and diaphragm. Endocrinology 72:71–77
Pozefsky T, Felig P, Soeldner JS, Cahill GF Jr (1968) Insulin blockade of amino acid release by human forearm tissues. Trans Assoc Am Physicians 82:258–264
Pozefsky T, Tancredi RG, Moxley RT, Dupre J, Tobin JD (1976) Metabolism in forearm tissues in man. Studies with glucagon. Diabetes 25:125–128
Prentki M, Crettaz M, Jeanrenaud B (1981) Role of microtubules in insulin and glucagon stimulation of amino acid transport in isolated rat hepatocytes. J Biol Chem 256:4336–4340
Pryor J, Berthet J (1960) The action of adenosine 3′,5′-monophosphate on the incorporation of leucine into liver proteins. Biochim Biophys Acta 43:556–557
Richards JR (1980) Current concepts in the metabolic responses in injury, infection and starvation. Proc Nutr Soc 39:113–123
Riddle MC, Golper TA, Fletcher WS, Ensinck JW, Smith PH (1978) Glucagonoma syndrome in a 19 year old woman. West J Med 129:68–72
Robison GA, Butcher RW, Sutherland EW (1971) Cyclic AMP. Academic, New York London, pp 232–284
Rosa F (1971) Ultrastructural changes produced by glucagon, cyclic 3′,5′-AMP and epinephrine on perfused rat livers. Ultrastruc Res 34:205
Ross BD, Hems R, Krebs HA (1967) The rate of gluconeogenesis from various precursors in the perfused rat liver. Biochem J 102:942–951
Russell RCG, Walker CJ, Bloom SR (1975) Hyperglucagonemia in the surgical patient. Br Med J 1:10–12

Schimassek H, Gerok W (1965) Control of the levels of free amino acids in plasma by the liver. Biochem Z 342:407–415
Schimke RT (1970) Regulation of protein degradation in mammalian tissues. In: Munro HN, Allison JB (eds) Mammalian protein metabolism. Academic, New York London, pp 177–228
Schworer CM, Mortimore GE (1979) Glucagon-induced autophagy and proteolysis in rat liver: mediation by selective deprivation of intracellular amino acids. Proc Natl Acad Sci USA 76:3169–3173
Sherwin RS, Tamborlane W, Hendler R, Sacca L, DeFronzo RA, Felig P (1977) Influence of glucagon replacement on the hyperglycemic and hyperketonemic response to prolonged somatostatin infusion in normal man. J Clin Endocrinol Metab 45:1104–1107
Siess EA, Wieland OH (1978) Glucagon-induced stimulation of 2-oxoglutarate metabolism in mitochondria from rat liver. FEBS Lett 93:301–306
Siess EA, Brocks DG, Lattke HK, Wieland OH (1977) Effect of glucagon on metabolite compartmentation in isolated rat liver cells during gluconeogenesis from lactate. Biochem J 166:225–235
Smith SR, Pozefsky T, Chhetri MK (1974) Nitrogen and amino acid metabolism in adults with protein-calorie malnutrition. Metabolism 23:603–618
Snodgrass PJ, Lin RC, Müller WA, Aoki TT (1978) Induction of urea cycle enzymes of rat liver by glucagon. J Biol Chem 253:2748–2753
Stacpoole PW, Jaspan J, Kasselberg AG, Halter SA, Polonsky K, Gluck FW, Liljenquist JE, Rabin D (1981) A familial glucagonoma syndrome. Genetic, clinical and biochemical features. Am J Med 70:1017–1026
Sutherland EW (1970) The effect of the hyperglycemic factor of the pancreas and of epinephrine on glycogenolysis. Recent Prog Horm Res 5:441–463
Sutherland EW, Rall TW (1960) The relation of adenosine-3′,5′-monophosphate and phosphorylase to the actions of catecholamines and other hormones. Pharmacol Rev 12:265–299
Sutherland EW, Robison GA (1969) The role of cyclic AMP in the control of carbohydrate metabolism. Diabetes 18:797–819
Sutherland EW, Robison GA, Butcher RW (1968) Some aspects of the biological role of adenosine 3′,5′-monophosphate (cyclic AMP). Circulation 37:279–306
Tews JK, Woodcock NH, Harper AE (1970) Stimulation of amino acid transport in rat liver slices by epinephrine, glucagon and adenosine-3′,5′-monophosphate. J Biol Chem 245:3026–3032
Tolbert MEM, Fain JN (1974) Studies on the regulation of gluconeogenesis in isolated rat liver cells by epinephrine and glucagon. J Biol Chem 249:1162–1166
Ui M, Claus TH, Exton JH, Park CR (1973a) Studies on the mechanism of action of glucagon on gluconeogenesis. J Biol Chem 248:5344–5349
Ui M, Exton JH, Park CR (1973b) Effects of glucagon on glutamate metabolism in the perfused rat liver. J Biol Chem 248:5350–5359
Unger RH (1976) Diabetes and the alpha cell. The Banting Memorial Lecture 1975. Diabetes 25:136–151
Unger RH, Orci L (1981) Glucagon and the A cell. N Engl J Med 304:1518–1524, 1575–1580
Unger RH, Aguilar-Parada E, Müller WA, Eisentraut AM (1970) Studies of pancreatic alpha-cell function in normal and diabetic subjects. J Clin Invest 49:837–848
Unger RH, Dobbs RE, Orci L (1978) Insulin, glucagon, and somatostatin secretion in the regulation of metabolism. Annu Rev Physiol 40:307–343
Vidnes J (1976) Persistent hereditary neonatal hypoglycemia caused by glucagon deficiency. Pediatr Res 10:881
Vinnars E, Bergström J, Fürst P (1975) Influence of postoperative state on the intracellular free amino acids in human muscle tissue. Ann Surg 182:665–671
Wagner T, Spranger J, Brunck HJ (1969) Kongenitaler Alpha-Zellmangel als Ursache einer chronischen infantilen Hypoglykämie. Monatsschr Kinderheilk 117:236–238
Wahren J, Felig P (1976) Influence of somatostatin on carbohydrate disposal and absorption in diabetes mellitus. Lancet 2:1213–1216

Weinges KF (1959) Action of glucagon on the total amino acids in the serum. Arch Exp Pathol Pharmacol 237:17–21
Williamson JR (1967) Effects of fatty acids, glucagon and anti-insulin serum on the control of gluconeogenesis and ketogenesis in rat liver. Adv Enzyme Regul 5:229–255
Woodside KH, Mortimore GE (1970) Control of proteolysis in the perfused rat liver: influence of amino acids, insulin and glucagon (Abstr). Fed Proc 29:379
Young VR, Scrimshaw NS (1968) Endogenous nitrogen metabolism and plasma free amino acids in young adults given a protein-free diet. Br J Nutr 22:9–20
Zuppinger KA (1975) Hypoglycemia in childhood. Evaluation of diagnostic procedures. Monogr Paediatr 4. Karger, Basel

CHAPTER 19

# Glucagon and Adipose Tissue Lipolysis

P. J. LEFEBVRE

## A. Introduction

Although discovered much later than its effects on carbohydrate metabolism, the action of glucagon on adipose tissue lipolysis seems particularly relevant and its importance in physiology has been stressed in various reviews and monographs (LEFEBVRE 1966a, 1967, 1972, 1975; WEINGES 1968; FOÀ 1968; FELIG 1970; LEFEBVRE and LUYCKX 1971; SCHADE et al. 1979; UNGER 1981). In the present chapter, we will discuss the effects of glucagon on adipose tissue and on circulating free fatty acids (FFA) and attempt to delineate the physiologic importance of these phenomena.

## B. Glucagon and Adipose Tissue

Glucagon belongs to the group of lipolytic hormones, i.e., it stimulates directly glycerol and FFA release from the adipocyte.

### I. Species Variations in Glucagon-Induced Lipolysis

The lipolytic effect of glucagon was first described and has mainly been studied on white adipose tissue from rats and other rodents (STEINBERG et al. 1959; HAGEN 1961; WEINGES 1961; VAUGHAN and STEINBERG 1963; ROBBELL and JONES 1966; LEFEBVRE 1966a, 1972, 1975; LEFEBVRE and LUYCKX 1969a). As reviewed by RUDMAN and DI GIROLAMO (1967), there exist major species differences in adipose tissue responsiveness to glucagon. Dog adipose tissue responds poorly to glucagon (LEFEBVRE and LUYCKX 1969a; PRIGGE and GRANDE 1971) while rabbit adipose tissue is clearly responsive (LEWIS and MATTHEWS 1968; PRIGGE and GRANDE 1971). Avian adipose tissue (pigeon, chicken, duck, goose, owl, etc.) is particularly sensitive to glucagon (GOODRIDGE and BALL 1965; LANGSLOW and HALES 1969, 1970; DESBALS et al. 1970; LANGSLOW 1972; BOYD et al. 1975; LANGSLOW et al. 1979, STROSSER et al. 1981). In view of their extreme sensitivity to glucagon and insensitivity to other lipolytic hormones and to insulin, chicken adipose cells have been proposed as a bioassay system for glucagon by LANGSLOW and HALES (1970). Glucagon is also lipolytic when tested on the interscapular brown adipose tissue of the rat (JOEL 1966; KUROSHIMA et al. 1977; FRIEDLI et al. 1978; KUROSHIMA and YAHATA 1979). On human adipose tissue, conflicting results have been obtained, both a lack of effect (MOSINGER et al. 1965; LEFEBVRE and LUYCKX 1969a; GHIDONI et al. 1970; POZZA et al. 1971a; NYBERG et al. 1977; VIZEK et al. 1979) and a stimulation of li-

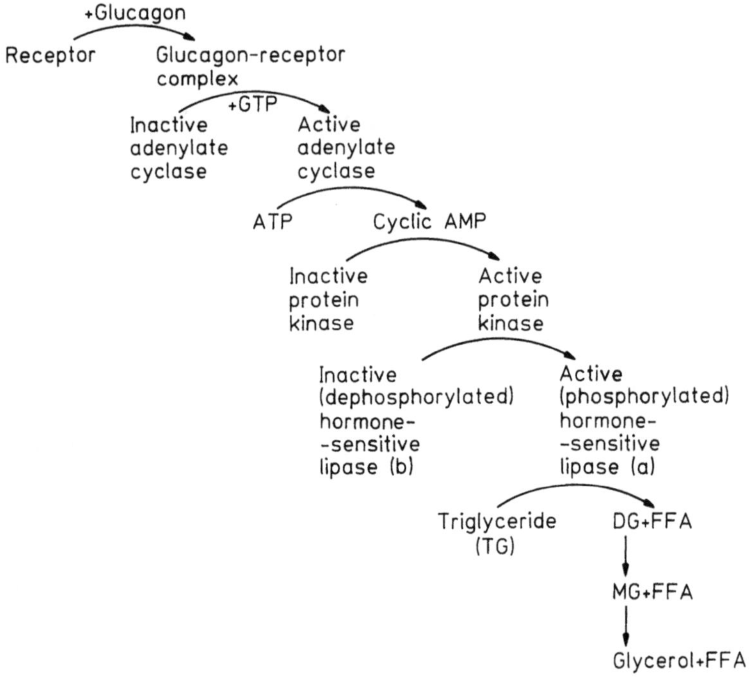


**Fig. 1.** The glucagon-induced lipolytic cascade. *GTP*, guanosine triphosphate; *ATP*, adenosine triphosphate; *DG*, diglyceride, *MG*, monoglyceride; *FFA*, free fatty acids. See details in Chap. 13. Adapted from STEINBERG and HUTTUNEN (1972)

polysis (GRIES 1968; BJÖRNTORP et al. 1969) have been reported (see also LEFEBVRE 1975). The minimal effective concentration of glucagon on glycerol release is 0.002 μg/ml on pieces of rat adipose tissue, 0.001 μg/ml on rat adipose cells, 0.001 μg/ml on pieces of chicken adipose tissue, and 0.0001 μg/ml on chicken adipose cells (LEFEBVRE 1972).

In rat perifused fat cells, maximum lipolysis was observed at $10^{-7}$ *M* concentrations in the perifusion medium (HECKEMEYER et al. 1981 a). Recent data of HECKEMEYER et al. (1981 b) have indicated that glucagon was rapidly degraded by rat adipose cells. Using a perifusion system, a continuous stimulus of intact glucagon can be ensured; under these conditions, $5 \times 10^{-11}$ *M* concentrations of glucagon increase glycerol production 300% above baseline levels.

## II. Mechanisms Involved in the Lipolytic Action of Glucagon

The first step in the action of glucagon on the adipocyte is the binding of the hormone to a specific receptor located on the outer surface of the adipose cell plasma membrane (BIRNBAUMER and RODBELL 1969; BÄR and HECHTER 1969; RODBELL et al. 1970). The intimate mechanisms in the interaction of glucagon with its receptor and the immediate subsequent events have been reviewed in detail in Chap. 13.

As a consequence of the glucagon-receptor interaction, a cascade of subsequent events, illustrated in Fig. 1, will ultimately result in the stimulation of lipolysis. It is beyond the scope of the present chapter to describe in detail the succession of events occurring in lipolysis, the so-called lipolytic cascade. These events involve successively adenylate cyclase activation, cyclic AMP synthesis from ATP, protein kinase activation, triacylglycerol lipase (hormone sensitive lipase) activation (by phosphorylation), and finally triglyceride breakdown. Hydrolysis of diglyceride and monoglyceride involve diacylglycerol and monoacylglycerol lipases. Details of the various enzymatic steps involved can be found in the recent extensive review of BELFIORE (1980). BUTCHER et al. (1965) have demonstrated that the rate of glucagon-induced lipolysis and intracellular cyclic AMP levels correlate positively with each other: temporally, qualitatively, and quantitatively while ZUMSTEIN et al. (1974) have shown that glucagon-induced lipolysis was accompanied by a significant release of cyclic AMP in the incubation medium. Theophylline, a classical phosphodiesterase inhibitor, enhances the lipolytic effect of glucagon (RODBELL and JONES 1966; LEFEBVRE and LUYCKX 1969a) while imidazole, a phosphodiesterase activator, reduces it (CAMU 1969a).

As recently reviewed (BELFIORE 1980; GAION 1981; GAION and KRISHNA 1981), studies on the mechanisms involved in the control of fat cell metabolism have shown that the interrelationships between the system of cellular second messengers is much more complex than had been envisioned previously. Various products of fat cell metabolism: fatty acids, adenosine, prostaglandins (see Sect. B.IV), an unidentified "feedback regulator" (FR), as well as calcium ions and 3′,5′-guanosine monophosphate (cyclic GMP) are involved in the modulation of adipocyte lipolysis (see also Sect. B.IV).

### III. Glucagon-Induced Lipolysis and Glucose Metabolism in the Adipocyte

Glucagon-induced lipolysis is accompanied by a marked stimulation of adipose cell glucose uptake (VAUGHAN 1960; LEE et al. 1960; WÖRNER and WEINGES 1961). Using C-1- and C-6-labeled glucose, it has been shown that the oxidation of C-6 is stimulated proportionally more than C-1 oxidation (FROESCH et al. 1960; LEFEBVRE and LUYCKX 1969a). This suggests that glucagon stimulates the glycolytic pathway more markedly than the shunt pathway in the adipocyte. Part of the glucose that is utilized by the adipose cell exposed to glucagon provides the α-glycerophosphate necessary for the esterification (or reesterification) process (Fig. 2). It is generally accepted that white adipose cells are devoid of significant amounts of active glycerokinase (WIELAND and SUYTER 1957; MARGOLIS and VAUGHAN 1962) and are therefore unable to phosphorylate glycerol to produce the α-glycerophosphate which is a prerequisite for fatty acid esterification or reesterification (for review, see BELFIORE 1980). The rate of esterification or reesterification can be calculated from simultaneous measurements of the net changes in glycerol and FFA accumulation in adipose tissue and release into the incubation medium (VAUGHAN and STEINBERG 1963). A similar estimate can be obtained using a radioactive procedure (LEFEBVRE 1966a). The two methods gave very similar results (see LEFEBVRE 1972). The "futile cycle" lipolysis-reesterification may be involved in local heat production (see Sect. B.VI). However, experiments based on the use of various inhibitors of lipolysis, have clearly shown that glucagon can also stimulate the entry

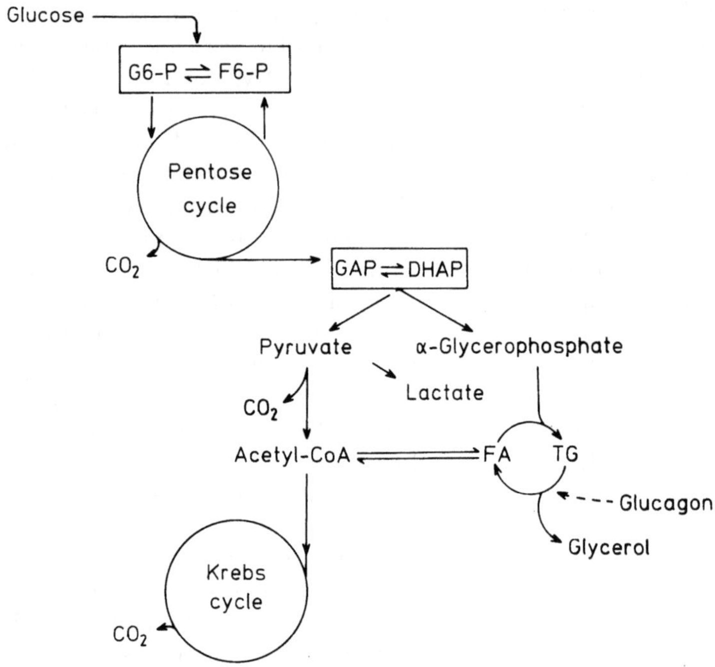


**Fig. 2.** Schematic metabolic pathways in the fat cell. *G6-P*, glucose-6-phosphate; *F6-P*, fructose-6-phosphate; *GAP*, glyceraldehydephosphate; *DHAP*, dihydroxyacetone phosphate; *TG*, triglyceride; *FA*, fatty acid

and overall oxidation of glucose by a mechanism which is distinct from its lipolysis-stimulating mechanism (BLECHER et al. 1969). In that respect, a direct activation of adipose tissue phosphofructokinase has been recently reported (BELFIORE et al. 1981), an important finding which needs confirmation.

## IV. Factors Affecting Glucagon-Induced Adipose Tissue Lipolysis

### 1. Size of the Adipocytes, Age and Nutritional Status of the Animals

Large rat adipocytes exhibit a marked resistance to the lipolytic action of glucagon when compared with small cells (MANGANIELLO and VAUGHAN 1972). Since in these experiments, small fat cells were prepared from young animals and large cells from adult animals, it was difficult to decide whether the resistance to glucagon was linked to the size of the cell or to the age of the animal. Investigations of HOLM et al. (1975) showed that, in rats, for a given adipocyte size, the lipolytic effect of glucagon was considerably increased in the young animals. Furthermore, irrespective of fat cell size, the lipolytic action of glucagon was reduced in old animals. Studies of LIVINGSTON et al. (1974) indicated clearly that large cells bind less $^{125}$I-labeled glucagon than small cells. The diminished binding is not a consequence of increased glucagon degradation since similar amounts of the tracer were degraded by both cell types. Surprisingly, the decrease in $^{125}$I-labeled glucagon binding did

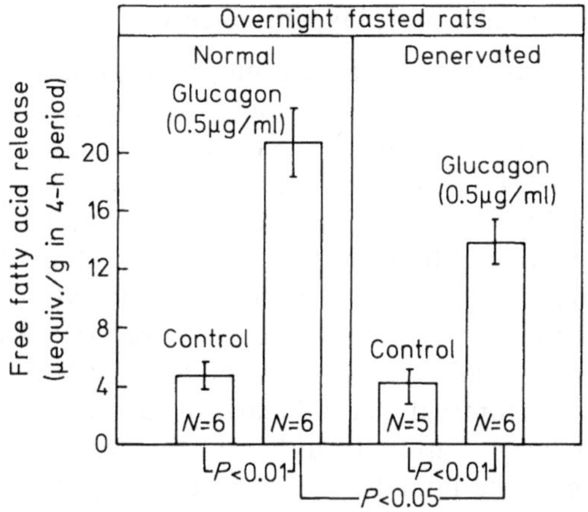


**Fig. 3.** Significant reduction of free fatty acid release by denervated lumbar fat of the rat exposed to glucagon in vitro. LEFEBVRE et al. (1973)

not parallel the decrease in the lipolytic response to glucagon. Therefore, diminished binding explains only in part the marked resistance to glucagon of large rat fat cells. A possible explanation may be found in the fact that phosphodiesterase activity, which degrades cyclic AMP, is greater in large cells than in small cells (DE SANTIS et al. 1974). COOPER et al. (1977) reported that, in adipocyte ghosts prepared from rats fed ad libitum, the responsiveness of adenylate cyclase to glucagon markedly declined from 1 to 4 months of age. Restriction of food starting at 1 month of age markedly promoted the glucagon responsiveness of the adenylate cyclase activity of these ghosts, but did not prevent a decline in the activity from occurring between 2 and 4 months of age. In contrast, BERTRAND et al. (1980) recently reported that, in rats, food restriction starting at 6 weeks of age markedly delays the age-related loss in lipolytic responsiveness to glucagon since the adipocytes from 6- and 12-month-old rats had the same lipolytic response to glucagon as those of 6-week-old rats fed ad libitum. COHAN and SAGGERSON (1982) recently reported that, in rats, activation of adipocyte adenylate cyclase by physiological concentrations of glucagon was substantially enhanced in the fasted state.

## 2. Innervation of the Adipose Tissue

Unilateral denervation of the lumbar fat body in the rat results in a significant weight increase as compared with the contralateral, normally innervated fat deposit (LEFEBVRE et al. 1973). In vitro glucagon-induced glycerol release is not affected by fat denervation; FFA release, in contrast, is significantly reduced (Fig. 3). Glucose uptake by denervated adipose tissue exposed to glucagon is significantly increased. These results were interpreted as the consequence of a relative increase in reesterification of FFA by denervated adipose tissue. Such a mechanism was suggested to be responsible, at least in part, for the relative weight gain (or the reduced weight loss) of denervated adipose tissue (LEFEBVRE et al. 1973).

### 3. Hypophysectomy, Adrenalectomy, Hypothyroidism

The response to glucagon is reduced, or even completely abolished, by previous hypophysectomy or adrenalectomy and in hypothyroid animals. The mechanism by which hypophysectomy produces this effect is not completely understood; data of GORIN and GOODMAN (1974) suggest that decreased protein kinase activity may be, at least partly, responsible for the decreased cyclic AMP-induced activation of lipase found in adipose tissue of hypophysectomized rats. Regarding the adrenals, it is generally accepted that the "permissive effect" of glucocorticoid is required for a normal response to lipolytic hormones (SHAFRIR and STEINBERG 1960; SHAFRIR et al. 1960; MAICKEL et al. 1967). FERNANDEZ and SAGGERSON (1978) showed that adipocytes from adrenalectomized rats had markedly decreased lipolytic responses to concentrations of glucagon up to 0.5 μg/ml. In the presence of adenosine deaminase, the difference in the lipolytic response to glucagon of adipocytes from adrenalectomized and sham-operated rats was abolished. These findings suggested that changes in adipocyte glucagon responsiveness after adrenalectomy may result from changes in adenosine metabolism or release[1]. The disability of hypothyroid animals with regard to glucagon-induced fat mobilization from adipose tissue is apparently not linked to a defect in the adenylate cyclase system; it has been attributed to the particularly high activity in hypothyroid animals of a membrane-associated, high affinity, cyclic AMP phosphodiesterase (ARMSTRONG et al. 1974).

### 4. Insulin

Early studies on the effects of insulin on glucagon-induced lipolysis have produced conflicting results. FAIN et al. (1966) first reported that insulin does not reduce the lipolytic action of glucagon on rat adipose tissue, but the glucagon concentrations used by these authors were particularly high. RODBELL and JONES (1966), using submaximal, but still high, concentrations of glucagon on isolated rat adipose cells, have found an inhibitory action of small concentrations of insulin. LEFEBVRE and LUYCKX (1969b) using still lower concentrations of glucagon (2–10 ng/ml) and insulin (25–100 μIU/ml) have demonstrated a marked reduction or even a complete abolition of the lipolytic effect of glucagon by insulin on pieces of rat epidydimal fat pad. At a glucose concentration of 5.5 *M*, insulin completely inhibits the lipolytic action of glucagon providing the molar fraction of insulin/glucagon in the incubation medium is higher than 0.12 (Table 1). These data indicate that, with respect to their common activities upon the adipocyte, insulin is 5–6 times more potent than glucagon on a molar basis. This is in contrast to estimates of their relative potencies on the liver (see Chap. 16) in which glucagon is regarded as more potent than insulin. At higher glucose concentrations (16.5 *M*), the situation is complicated by the stimulatory effect which glucose itself exerts on lipolysis (CHLOUVERAKIS 1967; LEFEBVRE and LUYCKX 1969b, CAMU 1969b). The antilipolytic effect

1 In fact, HONNOR and SAGGERSON (1980) have demonstrated that low doses of adenosine deaminase, which alone slightly increased basal lipolysis, rendered the rat adipocytes more sensitive to glucagon an effect recently confirmed by SHECHTER (1982). This effect was particularly apparent in cells from starved animals. This finding suggests that the in vitro accumulation of adenosine released by the adipocytes may artifactually reduce the sensitivity of these cells to glucagon

**Table 1.** Effect of insulin on the lipolytic action of glucagon on rat epidydimal fat pads in the presence of 100 mg/dl glucose (Lefebvre and Luyckx 1969a)

| Insulin (μIU/ml) | Glucagon (μg/ml) | Glycerol release[a] μM/g in 4-h period | *P* | Molar fraction Insulin/glucagon |
|---|---|---|---|---|
| 0 | 0 | 5.19±0.39 (22) | | |
| 0 | 0.002 | 11.01±1.00 (12) | <0.01 | 0 |
| 0 | 0.005 | 22.84±3.35 (6) | <0.01 | 0 |
| 0 | 0.010 | 20.25±2.37 (6) | <0.01 | 0 |
| 25 | 0 | 3.87±0.87 (6) | | |
| 25 | 0.002 | 3.86±0.44 (6) | N.S. | 0.29 |
| 25 | 0.005 | 6.26±0.46 (6) | <0.01 | 0.12 |
| 25 | 0.010 | 5.83±0.47 (6) | <0.01 | 0.06 |
| 100 | 0 | 4.86±0.28 (6) | | |
| 100 | 0.002 | 4.73±0.21 (6) | N.S. | 1.16 |
| 100 | 0.005 | 4.54±0.18 (6) | N.S. | 0.48 |
| 100 | 0.010 | 4.90±0.14 (6) | N.S. | 0.23 |

[a] Mean ± standard error of the mean, the number of determinations is indicated in parentheses
N.S. = not significant

of insulin on glucagon-induced lipolysis was confirmed in a perifused rat fat cell system by Allen and Gardner (1979). In birds, there is general agreement that insulin does not inhibit the lipolytic effect of glucagon which, as previously seen, is particularly potent in avian species (Langslow and Hales 1969; Prigge and Grande 1971; Desbals et al. 1970; Boyd et al. 1975; Langslow et al. 1979; Strosser et al. 1981).

## 5. Prostaglandins and Prostaglandin Synthesis Inhibitors

Prostaglandins are potent inhibitors of hormone-stimulated lipolysis in vitro in the adipose tissue of the rat (Fig. 4; Steinberg et al. 1963; Mülbachova et al. 1967; Lefebvre and Luyckx 1971, 1974), rabbit (Micheli 1970; Boberg et al. 1970), bird (Gascon 1971; Langslow 1971; Grande and Prigge 1972), and human (Bergström and Carlson 1965; Micheli et al. 1969). Since it has been demonstrated that lipolysis, caused by hormones and by nerve stimulation, is accompanied by the release of significant quantities of prostaglandins in the incubation medium (Ramwell et al. 1966; Shaw and Ramwell 1968; Christ and Nugteren 1970) it has been suggested that the local release of prostaglandins during lipolysis may modulate the action of the original stimulus by a negative feedback mechanism (Horton 1969; Ramwell and Shaw 1970):

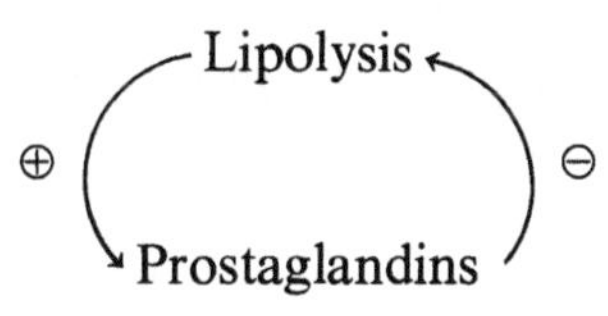

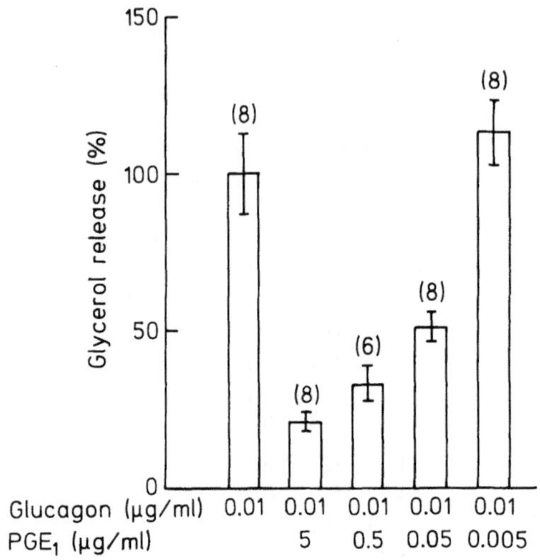


**Fig. 4.** Inhibition by prostaglandin $E_1$ ($PGE_1$) of glucagon-induced glycerol release from pieces of rat epididymal fat pad. LEFEBVRE and LUYCKX (1974)

**Table 2.** Effect of long-term administration of L8027 (an inhibitor of prostaglandin synthesis) on glucagon-induced lipolysis *in vitro* (LEFEBVRE and LUYCKX 1974)

| | Glucagon effect[a] 0.5 μg/ml | |
|---|---|---|
| | Control rats | L 8027-treated rats |
| Glycerol release (μmol/g in 2-h period) | +2.83±1.19 (13) ($P<0.05$) | +8.03±1.73 (14) ($P<0.01$) |
| | $P<0.05$ | |
| FFA release (μequiv./g in 2-h period) | +3.85±0.99 (15) ($P<0.01$) | +6.16±2.15 (14) ($P<0.01$) |
| | N.S. | |

[a] Results (mean±standard error) are expressed as "glucagon effect", *i.e.* change in glycerol or FFA release induced by glucagon on epididymal adipose tissue originating from the same animal (paired comparison). Treated rats received oral L8027 (50 mg/kg for 15 days); control rats received solvent alone. The number of animals is indicated in parentheses. Significance was determined by Student's *t* test. N.S. = not significant

Such a concept was supported by the findings of STONE et al. (1969), who showed that sodium salicylate inhibited glucagon-induced lipolysis, and by those of ILIANO and CUATRECASAS (1971) who demonstrated that blocking endogenous prostaglandin biosynthesis by indomethacin or competitively antagonizing the action of prostaglandins by specific inhibitors such as 7-oxa-13-prostynoic acid or SC-19220, a dibenzoazepine hydrazide, resulted in enhanced lipolysis induced by epinephrine or adrenocorticotropic hormone. Similarly, LEFEBVRE and LUYCKX (1974) investigated the effect of L8027, a pyridyindolyl ketone which is a potent inhibitor of prostaglandin biosynthesis, on basal and glucagon-stimulated adipose tissue li-

**Table 3.** Effect[a] of long-term administration of L 8027 (an inhibitor of prostaglandin synthesis) on glycerol release, FFA release, and glucose uptake by adipose tissue of rats submitted to a 60-min forced swim (LEFEBVRE and LUYCKX 1974)

| | Control rats | L 8027-treated rats | Statistical comparison |
|---|---|---|---|
| Glycerol release (μmol/g in 2-h period) | 4.87 ± 0.59 | 8.53 ± 0.68 | $P<0.01$ |
| FFA release (μequiv./g in 2-h period | 2.09 ± 0.46 | 5.59 ± 0.88 | $P<0.01$ |
| Glucose uptake (mg/g in 2-h period) | 2.22 ± 0.15 | 1.59 ± 0.17 | $P<0.01$ |

[a] Results are expressed as mean ± standard error of the mean. Two pieces of epididymal adipose tissue were taken from each rat. Eight animals were thus studied in each series, $n=16$. Treated rats received oral L 8027 (50 mg/kg for 15 days). Control rats received solvent alone. All animals were submitted to a 60-min forced swim. Significance was determined by Student's $t$ test

polysis in rats. Long-term administration of this compound to rats resulted in a significant increase in lipolysis when adipose tissue of treated rats was subsequently incubated in the presence of glucagon (Table 2). Long-term treatment with L8027 also enhanced exercise-induced lipolysis, a situation in which glucagon may be involved (Table 3). These data thus support the concept that endogenous prostaglandins may be concerned with the regulation of glucagon-induced lipolysis. The theory that endogenous prostaglandins in general exert a local negative feedback on adipose tissue lipolysis has not been supported by several other authors (FAIN et al. 1973; DALTON and HOPE 1973; FREDHOLM and HEDQVIST 1975; LIPINSKI and MATHIAS 1978).

## 6. Other Factors

At high concentrations, $\beta$-adrenoreceptor blocking agents like propranolol or 1-(3-methyl phenoxy)-3-isopropylaminopropanol-(2) hydrochloride inhibit glucagon-induced lipolyis in isolated rat fat cells (BLECHER et al. 1969). In the rat, the lipolytic effect of glucagon is also reduced or abolished by several nucleotides (KAPPELER and BALLY 1965, cited by CARLSON and BALLY 1965), by potassium cyanide (CAMU 1969a), and by nicotinic acid (CARLSON and BALLY 1965) although nicotinic acid does not inhibit glucagon-induced lipolysis in several other species like rabbit, goose, duck, and owl (PRIGGE and GRANDE 1971). On chicken adipose cells, the lipolytic effect of glucagon is inhibited, in a dose-dependent manner, by somatostatin while glucagon binding to the plasma membrane receptor was unaffected (STROSSER et al. 1981). Somatostatin markedly decreases the glucagon-induced cyclic AMP accumulation within the adipocyte (STROSSER et al. 1981). The effects of modifications of the primary structure of glucagon on its binding to the adipocyte receptor, adenylate cyclase activation, and subsequent lipolysis have been studied in detail by FRANDSEN et al. (1981) and are summarized in Table 4. Further data on the glucagon structure-function relationships can be found in Chaps. 1 and 13.

**Table 4.** Effects of modifications of the primary structure of glucagon (after FRANDSEN et al. 1981)

| Preparation | Receptor binding (%) | Adenylate cyclase activation (%) | Lipolysis (%) |
|---|---|---|---|
| Native porcine glucagon | 100 | 100 | 100 |
| Synthetic glucagon | 108 | | |
| des-(1–4)-glucagon | 5.7 | < 0.001 | < 0.01 |
| des-(5–9)-glucagon | 0.47 | < 0.001 | < 0.01 |
| des-(10–15)-glucagon | 0.0028 | < 0.001 | < 0.01 |
| des-(16–21)-glucagon | 0.0017 | < 0.001 | < 0.01 |
| des-(22–26)-glucagon | 0.0006 | < 0.001 | < 0.01 |
| des-(27–29)-glucagon | 0.92 | 0.16 | 0.20 |

## V. Glucagon-Like Peptides and Adipose Tissue Lipolysis

LEFEBVRE et al. (1969) reported that material identified as "peak I" and "peak II" in partially purified extracts of dog jejunum "glucagon-like immunoreactive material" (VALVERDE et al. 1968) does posses lipolytic acitivity. Similarly, GUTMAN et al. (1973) isolated, by chromatography on Sephadex G-50 columns, two peaks of glucagon-like immunoreactivity (GLI). Both GLI peaks increased FFA release from rat epididymal fat pad and cyclic AMP levels in adipose tissue. Only peak II GLI activity was suppressed by anti-glucagon antibody. In these experiments, the possibility of in vitro conversion of GLI molecules to glucagon should always be considered. In contrast, KRUG and MIAHLE (1977) have suggested that gut GLI may play the role of an inhibitor of lipolysis. Additional information on this subject can be found in Chaps. 7 and 12.

## VI. Brown Adipose Tissue and Nonshivering Thermogenesis

Early experiments of JOEL (1966) have shown that glucagon stimulates lipolysis from the interscapular brown adipose tissue of the rat in vitro. KUROSHIMA et al. (1975) have demonstrated an enhanced sensitivity to the lipolytic action of glucagon in cold-acclimatized rats. In these conditions, there is evidence of enhanced glucose consumption by the brown adipose depots (KUROSHIMA et al. 1977) suggesting an active "futile cycle" consisting of enhanced lipolysis and subsequent active reesterification of FFA (see Sect. B.III). Glucagon is a potent agent which stimulates heat production of brown adipose tissue both in vivo (HEIM and HULL 1966; COCKBURN et al. 1968) and in vitro (JOEL 1966; FRIEDLI et al. 1978). It is interesting to note that glucagon exerts on rat brown adipose tissue a larger maximum response than norepinephrine (KUROSHIMA and YAHATA 1979), suggesting that this hormone plays a crucial role in regulating nonshivering thermogenesis in that species. KUROSHIMA et al. (1979) have also shown that the maximal calorigenic response of rat adipocytes to glucagon was increased in cold acclimatization and reduced in heat acclimatization.

### VII. Adipose Tissue Lipoprotein Lipase Activity

It has been reported that glucagon enhances lipoprotein lipase activity in rat in vivo (BORENSZTAJN et al. 1973; RAULT et al. 1974). In contrast, in vitro glucagon markedly reduces lipoprotein lipase activity of adipose tissue (NESTEL and AUSTIN 1969), an effect which has been confirmed at low concentrations (500 pg/ml) of glucagon (MURASE et al. 1981). It is likely that the increase in lipoprotein lipase activity observed after in vivo injection of glucagon represents an indirect effect mediated through glucagon-induced insulin release (MURASE et al. 1981).

## C. Glucagon and Circulating Free Fatty Acids

Judging from its lipolytic action on adipose tissue (see Sect. B), one would expect a clear-cut rise in plasma FFA after in vivo administration of glucagon. Numerous factors, however, complicate the situation. Some of these factors tend to lower plasma FFA while others tend to raise them. Among the factors tending to lower plasma glucagon, we have to consider: (1) the possible contamination of glucagon preparation by insulin, mainly if high, pharmacologic doses are used; (2) the hyperglycemic response to glucagon; (3) the potent stimulatory effect that glucagon exerts on insulin secretion (see Chaps. 22 and 55). The FFA-increasing factors include: (1) the nonspecific FFA rise after any injection in some subjects (so-called needle reactors); (2) the stimulation by glucagon of various stress hormones which may in turn stimulate lipolysis, these hormones include epinephrine and norepinephrine (see Chap. 51), cortisol and growth hormone (see Chap. 52). In addition, the artifactual character of all experiments in which glucagon is injected outside its physiologic site of secretion, i.e., the portal area, must always be considered (LEFEBVRE 1972; SCHADE et al. 1979).

### I. Intravenous, Subcutaneous, or Intramuscular Injection of High Doses of Glucagon in Mammals

In mammals, including humans, the intravenous or intramuscular injection of relatively high doses of glucagon (0.1–1 mg) induces a triphasic pattern of plasma FFA: (1) an early rise lasting 10–20 min; (2) a secondary drop lasting 90–120 min; and (3) a subsequent rise lasting 3–5 h.

#### 1. The Early Rise in Plasma FFA

The early and transient increase in plasma FFA after intravenous injection of glucagon has been initially described by LEFEBVRE and LUYCKX (1969a) and WHITTY et al. (1969). It has been studied in detail by SCHADE and EATON (1975a, b). These authors convincingly demonstrated that the early rise in plasma FFA after intravenous glucagon administration was modulated by the amplitude of the insulin response (Fig. 5). In normal subjects who responded with little insulin secretion and in diabetic subjects who did not have any insulin response, glucagon injection induced a brisk plasma FFA rise. In contrast, obese subjects had an excessive insulin response to glucagon and no early rise in plasma FFA. Similar findings were also

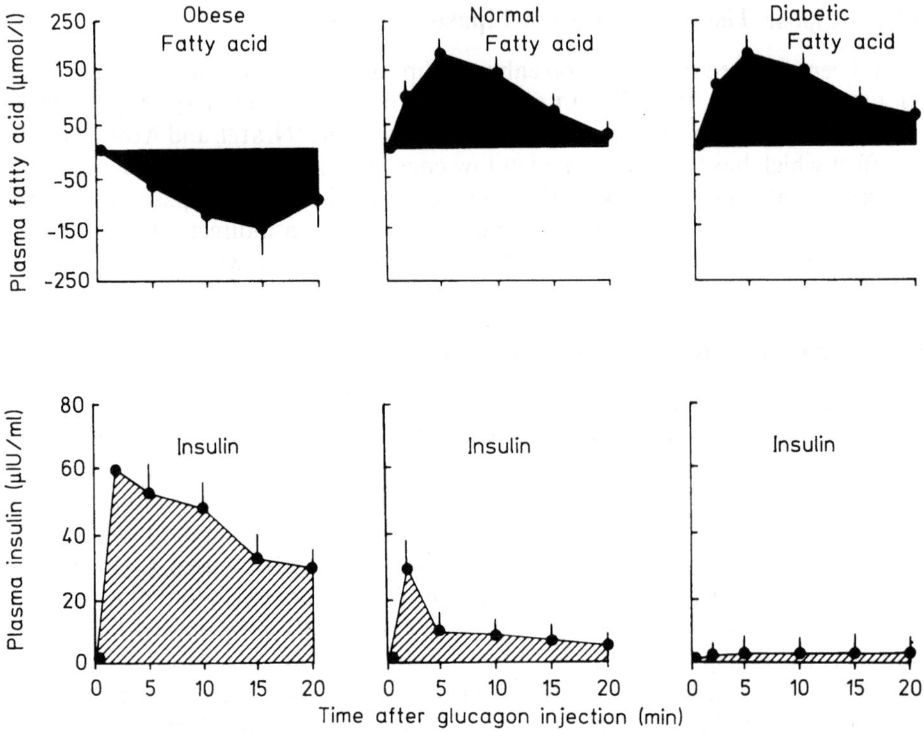


**Fig. 5.** Early changes in plasma FFA after intravenous injection of 1.0 µg/kg glucagon. In the normal subjects, with low insulin response, and in the insulin-dependent diabetics, with no insulin response, the injection of glucagon induces an early rise in plasma FFA. In contrast, in the obese subjects, with massive insulin release, no rise in plasma FFA and, on the contrary, an immediate drop in the FFA circulating levels are observed. SCHADE and EATON (1975a)

reported by LILJENQUIST et al. (1974). It should be emphasized, however, that at these doses and under these conditions, glucagon induces a rapid and transient rise in circulating catecholamines that may certainly contribute to the FFA response (LEFEBVRE 1972). The early rise in plasma FFA after intraperitoneal injection of glucagon is enhanced in cold-acclimatized rats (KUROSHIMA et al. 1975).

## 2. The Secondary Drop in Plasma FFA

The secondary drop in plasma FFA has long been known (DREILING and BIERMAN 1957; BIERMAN et al. 1957; DREILING et al. 1962; LEFEBVRE 1965, 1966a, 1972; DE PLAEN and GALANSINO 1966; WHITTY et al. 1967, 1969; LEFEBVRE and LUYCKX 1969a; SCHADE and EATON 1975a, b). It is observed using highly purified (cystein-treated) glucagon (LEFEBVRE 1966a) and is undoubtedly due to the combination of hyperglycemia and hyperinsulinemia (LEFEBVRE 1966a; LEFEBVRE and LUYCKX 1969a; SCHADE and EATON 1975a, b). As emphasized by SCHADE and EATON (1975a, b), the magnitude of the insulin response is a major determinant of the degree of plasma FFA drop after intravenous administration of glucagon.

### 3. The Late Increase in Plasma FFA

The secondary drop in plasma FFA observed after intravenous or intramuscular injection of glucagon is followed by a late increase lasting 3–5 h (DREILING and BIERMAN 1957; BIERMAN et al. 1957; LEFEBVRE 1965; LEFEBVRE and LUYCKX 1969 a). It is delayed in obese nondiabetic subjects and is present, although somewhat reduced, in patients with panhypopituitarism (LEFEBVRE and LUYCKX 1969 a). It is much more pronounced than the increase which follows hyperglycemia of comparable magnitude and duration (LEFEBVRE 1965, 1967). At this time, it is not still clear if it really represents a late lipolytic effect of glucagon.

## II. Intraarterial, Intravenous, or Intraportal Infusion of Low Doses of Glucagon in Mammals

POZZA et al. (1971 a, b) were the first to demonstrate that, in humans, infusion of glucagon into the brachial artery produces a transient and immediate rise in plasma FFA levels in the corresponding (but not the contralateral) vein. Using a similar technique, POZEFSKY et al. (1976) were unable to demonstrate lipolytic effects of glucagon in either postabsorptive or 50-h fasted individuals. The most convincing studies with this model have been reported by SCHNEIDER et al. (1981). In seven normal postabsorptive males, the intraarterial infusion of glucagon (mean glucagon increment $691 \pm 50$ pg/ml) resulted in a clear-cut increased mobilization of FFA from both the deep and the superficial compartments of the forearm and of glycerol from the deep compartment; these effects were markedly reversed by the intraarterial infusion of insulin (mean insulin increment of $105 \pm 13$ μIU/ml). This latter study clearly indicates the sensitivity of human peripheral adipose tissue to glucagon in vivo, in contrast with the conflicting results obtained in vitro (see Sect. B.II).

In humans also, after 3 days total starvation when the insulinotropic effect of glucagon is at its lowest, intravenous glucagon infusion at the rate of 0.1 μg/min induced an immediate and sustained rise in plasma FFA levels (SAMOLS et al. 1969). However, such a rise in response to glucagon infusion (0.1 mg in 24 h) has not been found after a more prolonged fast. In insulin-dependent diabetics, LILJENQUIST et al. (1974) have demonstrated that glucagon, at relatively high concentrations ($1.3 \times 10^{-9}$ *M*), is capable of stimulating lipolysis. However, a rise in plasma FFA has not been observed in insulin-dependent diabetics infused intravenously with glucagon in order to reach peripheral plasma levels between 200 and 1600 pg/ml; in these experiments, complete insulin deprivation has not been unequivocally established (LUYCKX et al. 1980). In normal humans, physiologic quantities of glucagon infused intravenously, along with somatostatin in order to inhibit simultaneously endogenous insulin and glucagon secretion, resulted in rises in plasma FFA and glycerol levels which were 2–3-fold greater than those observed during infusion of somatostatin alone. Thus, under appropriate conditions, physiological levels of glucagon can stimulate lipolysis in humans, an effect which is strongly antagonized by low levels of insulin (GERICH et al. 1976).

The intraportal infusion of 2 ng $kg^{-1}min^{-1}$ of cystein-treated glucagon – a dose that can be considered physiologic (LEFEBVRE and LUYCKX 1969 a) – raises plasma FFA levels markedly in the peripheral blood of normal dogs fasted 18 h and anesthetized with pentobarbital (LEFEBVRE 1966 b). In this type of protocol, the duration of fast and the route of administration are probably critical since these data were not confirmed in overnight fasted dogs by MÜLLER et al. (1977) who infused 3.5 ng $kg^{-1}min^{-1}$ glucagon by a catheter in the right auricle. Infusion of 150 µg/h glucagon in a mesenteric vein in sheep increases arterial glycerol levels (BROCKMAN et al. 1975).

### III. Intravenous Injection or Infusion of Glucagon in Birds

In birds, the intravenous injection or infusion of glucagon causes a dramatic rise in plasma FFA levels (HEALD et al. 1965; HOAK et al. 1968; GRANDE and PRIGGE 1970) which is related to the extreme sensitivity of avian adipose tissue to glucagon (see Sect. B.I). In birds, the striking rise in plasma FFA causes a marked accumulation of liver triglycerides and a profound secondary rise in serum lipids and lipoprotein triglycerides (HOAK et al. 1968; GRANDE and PRIGGE 1970). Functional hepatectomy completely suppresses the secondary rise in plasma triglycerides observed after glucagon administration in normal geese (DE OYA et al. 1971).

## D. Physiologic Importance of Glucagon-Induced Lipolysis

As already emphasized, glucagon has been demonstrated to be a potent lipolytic hormone when studied in vitro using adipose tissue or cells from several animal species or in vivo under various experimental conditions. Strong arguments in favor of the physiological relevance of glucagon-induced lipolysis have been brought by numerous experiments in vitro and in vivo demonstrating that FFA, the final products of lipolysis, were indeed able to modulate glucagon secretion in a manner similar to the effect of glucose levels on glucagon release (see Chap. 23 and 25).

Thus, the original concept of glucagon as a "hormone of glucose need" as proposed by UNGER and EISENTRAUT (1964) has been extended to that of glucagon "hormone of fuel need" (LEFEBVRE and LUYCKX 1971) based on these observations which show that glucagon is able to promote not only the mobilization of glucose from liver stores (see Chaps. 14, 15, and 16), but also that of FFA from adipose tissue stores and that both glucose and FFA have been demonstrated to modulate glucagon secretion via a negative feedback system (see Chaps. 23 and 25). Among the various physiologic conditions, where there undoubtedly exists a "fuel need" and where glucagon is likely to be involved as an important lipid-mobilizing hormone, we may briefly consider fasting, muscular exercise, and adaption of mammals to extrauterine life.

## I. Fasting

Fasting is par excellence a condition of fuel need. It is associated with an unequivocal rise in both plasma FFA and plasma glucagon (see Chap. 34). Investigations of KOERKER et al. (1974) in fasting baboons have indicated that FFA and glycerol levels cycle in phase with glucagon levels with a significant correlation, thus suggesting that the two events, glucagon secretion and lipolysis, are either causally related, or else share a common (perhaps neural) regulatory mechanism. Since fasting is accompanied by an inhibition of insulin secretion, the insulin/glucagon molar fraction in peripheral blood is substantially decreased. In the experiments of LUYCKX (1974) in rats, it falls from 6–8 in the postabsorptive state to values of 1.5–2 after 2 or 3 days starvation. This ratio would seem to favor adipose tissue lipolysis, but we have to recall here that in vitro, when rat adipose tissue is incubated with various concentrations of glucagon and insulin, lipolysis increases above basal values, only when the molar fraction is below 0.2 (LEFEBVRE and LUYCKX 1969 b). A value as low as this is rarely observed under physiologic conditions. Nevertheless, if one recalls that insulin is the major, if not the only circulating antilipolytic hormone, glucagon is far from being the only circulating lipolytic hormone. Therefore we suggest considering the following equation as the physiologically important one as far as adipose tissue lipolysis is concerned:

$$\frac{\text{Antilipolytic hormones}}{\text{Lipolytic hormones}} = \frac{\text{Insulin (+ other antilipolytic hormones?)}}{\text{Glucagon + other lipolytic hormones}}.$$

On this basis, we consider that glucagon can reasonably be assumed to be a contributory factor to lipolysis under fasting conditions in a given species when: (1) adipose cells of the species considered are sensitive to minimal levels of glucagon; and (2) fasting induces an increase in peripheral plasma glucagon levels in that species. While these phenomena occur in rats and in birds, there is no definite proof that they occur in other species, including humans.

## II. Muscular Exercise

Prolonged muscular exercise is accompanied by a striking rise in plasma glucagon (see Chap. 38). There is evidence that the exercise-induced rise in plasma glucagon may be due to at least two basic mechanisms: (1) a stimulation of the adrenergic system which in turn stimulates the $\beta$-adrenoceptors of the A-cell, resulting in glucagon release (so-called neural stimulation); and (2) a decrease in circulating glucose (so-called metabolic stimulation). Extensive studies from our laboratory (LEFEBVRE et al. 1972; LUYCKX and LEFEBVRE 1973, 1974; LUYCKX et al. 1975) using rats forced to swim as a model of physical exercise, have demonstrated that, under these conditions and in that species, "neural stimulation" is the predominant factor (LUYCKX and LEFEBVRE 1974; LUYCKX et al. 1975). With more strenuous exercise, "metabolic stimulation" may also become important. In a series of studies, we investigated the role of glucagon in exercise, studying various factors which control glucagon release, adipose tissue FFA mobilization, or both (LUYCKX and LEFEBVRE 1974; LUYCKX et al. 1975). Some of our findings are summarized in Table 5 which clearly demonstrates that the exercise-induced glucagon rise may be dissociated

**Table 5.** Comparison of the changes in plasma glucagon and plasma FFA as observed after muscular exercise (forced swim) in rats

| | References |
|---|---|
| I. Simultaneous rise in plasma glucagon and plasma FFA | |
| Normal, untrained rats | [1–4] |
| Normal, trained rats | [4] |
| Medulloadrenalectomy | [4] |
| (+)-Propranolol (5 mg/kg) | [4] |
| II. Normal rise in plasma glucagon, but no rise in plasma FFA | |
| Adrenalectomy | [3] |
| Hypophysectomy | [3] |
| Phentolamine (5 mg/kg) | [3] |
| III. No rise in plasma glucagon, but normal rise in plasma FFA | |
| Immunosympathectomy | [4] |
| (+, −)-Propranolol (5 mg/kg) | [3] |
| Pindolol (0.1 mg/kg) | [4] |

[1] LEFEBVRE et al. (1972)
[2] LUYCKX and LEFEBVRE (1973)
[3] LUYCKX and LEFEBVRE (1974)
[4] LUYCKX et al. (1975)

from the exercise-induced FFA rise. The fact that FFA levels may be increased with exercise while glucagon levels remain at baseline values suggest that glucagon is not "essential" for adipose tissue FFA release during exercise, but does not exclude the possiblity that glucagon, during the period when its levels are raised, could contribute, along with other factors, to the rise in FFA observed as a consequence of exercise.

Exercise-induced FFA rise is a complex phenomenon which is the resultant of various changes induced by exercise (LEFEBVRE et al. 1972): (1) a stimulation of lipolysis induced by circulating lipolytic hormones whose secretion is stimulated (besides glucagon, circulating catecholamines, growth hormone, and others) or by the local release of norepinephrine from nerve endings in adipose tissue; (2) a decrease in the "antilipolytic tone", essentially due to the exercise-induced decrease in circulating insulin; and (3) a decrease in esterification or reesterification of FFA within adipose tissue (LEFEBVRE and FEDERSPIL 1971), a factor to which the decreased circulating levels of insulin, the lesser availability of glucose for adipose tissue, the increased circulating levels of corticosterone, and perhaps the release of catecholamines from nerve endings within adipose tissue contribute. The multiplicity of factors involved in metabolic regulation during exercise and the interplay between them satisfactorily explains the observation that when only one of these factors has been neutralized, the final result (i.e., FFA mobilization) may be unaffected. On the contrary, when the FFA mobilization itself is experimentally blocked (unresponsiveness of adipose tissue in adrenalectomized or hypophysectomized animals for instance), the circulating hormones (glucagon for instance) may be increased without showing their effects.

## III. Adaptation to Extrauterine Life

The adaptation of newborn mammals to extrauterine life is another situation in which an acute fuel need is present. Evidence is now available showing that glucagon is already secreted at birth and that its level increases during the first hours of extrauterine life. This has been demonstrated in various species (see Chap. 36). Several lines of evidence indicate that glucagon is involved in the release of glucose from the liver at this crucial period of life. In order to prove that fatty acid mobilization from adipose tissue, a process which is extremely active in the newborn, results at least partially from increased glucagon, it must be demonstrated that adipocytes in the neonate are sensitive to minimal concentrations of glucagon. As noted in Sect. B.I, conflicting results have been reported in this matter. Moreover, the simple *presence* of white adipose tissue must also be considered; the newborn rat, for instance, is known to be very poor in white adipose cells.

## References


Allen DO, Gardner EA (1979) Antilipolytic action of insulin in the perifused fat cell system. Biochem Pharmacol 617–622

Amstrong KJ, Stouffer JE, Van Iwegen RG, Thompson WJ, Robison GA (1974) Effect of thyroid hormone deficiency on cyclic adenosine 3′:5′-monophosphate and control of lipolysis in fat cells. J Biol Chem 249:4226–4231

Bär HP, Hechter O (1969) Adenylcyclase and hormone action. I. Effects of adrenocorticotropic hormone, glucagon and epinephrine on the plasma membrane of the rat fat cell. Proc Natl Acad Sci USA 63:350–356

Belfiore F (1980) Enzyme regulation and metabolic diseases. Karger, Basel, New York

Belfiore F, Rabuazzo AM, Borzi V, Iannello S (1981) Regulation of two key-enzymes involved in triglyceride synthesis in human adipose tissue: phosphofructokinase and phosphatidate phosphatase (Abstr). Third European Symposium on Metabolism, Padua May 28–30, p 17

Bergström S, Carlson LA (1965) Inhibitory action of prostaglandin $E_1$ on the mobilization of free fatty acids and glycerol from human adipose tissue in vitro. Prostaglandin and related factors. Acta Physiol Scand 63:195–196

Bertrand HA, Masoro EJ, Byung PYU (1980) Maintenance of glucagon-promoted lipolysis in adipocytes by food restriction. Endocrinology 107:591–595

Bierman E, Dole VP, Roberts TN (1957) An abnormality of non esterified fatty acid metabolism in diabetes mellitus. Diabetes 6:475–479

Birnbaumer L, Rodbell M (1969) Adenylcyclase in fat cell. II: Hormone receptor. J Biol Chem 244:3477–3482

Björntorp P, Karlsson M, Hovden A (1969) Quantitative aspects of lipolysis and reesterifaction in human adipose tissue in vitro. Acta Med Scand 185:89–97

Blecher M, Merlino NS, Ro'Ane JT, Flynn PD (1969) Independance of the effects of epinephrine, glucagon and adrenocorticotropin on glucose utilization from those of lipolysis in isolated rat adipose cells. J Biol Chem 244:3423–3429

Boberg J, Micheli H, Rammer L (1970) Effect of nicotinic acid on ACTH and noradrenaline stimulated lipolysis in the rabbit. II. In vitro studies including comparison with Prostaglandin $E_1$. Acta Physiol Scand 79:299–304

Borensztajn J, Keig P, Rubenstein AH (1973) The role of glucagon in the regulation of myocardial lipoprotein lipase activity. Biochem Biophys Res Commun 53:603–608

Boyd TA, Wieser PB, Fain JN (1975) Lipolysis and cyclic AMP accumulation in isolated fat cells from chicks. Gen Comp Endocrinol 26:243–247

Brockman RP, Bergman EN, Joo PK, Manns JG (1975) Effects of glucagon and insulin on net hepatic metabolism of glucose precursors in sheep. Am J Physiol 229:1344–1350

Butcher RW, Ho RJ, Meng HC, Sutherland EW (1965) Adenosine 3′, 5′-monophosphate in biological materials. II. The measurement of adenosine 3′, 5′-monophosphate in tissues and the role of the cyclic nucleotide in the lipolytic response of fat to epinephrine. J Biol Chem 240:4515–4523

Camu F (1969 a) Effects of imidazole and cyanide upon rat adipose tissue lipolysis. Arch Int Physiol Biochim 77:663–669

Camu F (1969 b) Interactions du glucagon, de l'insuline et du glucose sur la lipolyse de tissus adipeux de rats in vitro. Arch Int Pharmacodyn Ther 178:370–381
Carlson LA, Bally PR (1965) Inhibition of lipid mobilization. In: Renold AE, Cahill JF Jr (eds) Adipose tissue. American Physiological Society, Washington, DC (Handbook of physiology, sect 5, pp 557–574)
Chlouverakis C (1967) The action of glucose on lipolysis. Metabolism 16:469–472
Chohan P, Saggerson D (1982) Increased sensitivity of adipocyte adenylate cyclase to glucagon in the fasted state. FEBS Lett 146:357–360
Christ EJ, Nugteren DH (1970) The biosynthesis and possible function of prostaglandins in adipose tissue. Biochim Biophys Acta 218:296–307
Cockburn F, Hull D, Walton I (1968) The effect of lipolytic hormones and theophylline on heart production in brown adipose tissue in vitro. Br J Pharmacol 31:568–577
Cooper B, Weinblatt F, Gregerman RI (1977) Enhanced activity of hormone-sensitive adenylate cyclase during dietary restriction in the rat: dependance on age and relation to cell size. J Clin Invest 59:467–474
Dalton C, Hope HH (1973) Inability of prostaglandin synthesis inhibition to affect adipose tissue lipolysis. Prostaglandins 4:461–651
de Oya M, Prigge WF, Grande F (1971) Suppression by hepatectomy of glucagon-induced hypertriglyceridemia in geese. Proc Soc Exp Biol Med 136:107–110
De Plaen J, Galansino G (1966) Effect of glucagon and epinephrine in fasted dogs. Proc Soc Exp Biol Med 121:501–504
De Santis RA, Gorenstein T, Livingston JN, Lockwood DH (1974) Role of phosphodiesterase in glucagon resistance of large adipocytes. J Lipid Res 15:33–38
Desbals P, Miahle P, Desbals B (1970) Sensitivity of avian adipose tissue to lipolytic hormones (Abstr). Diabetologia 6:625
Dreiling DA, Bierman EL (1957) Correlation between plasma amylase activity and concentration of non esterified fatty acids (NEFA). Proc Soc Exp Biol Med 95:496–497
Dreiling DA, Bierman EL, Debons AF, Elsbach P, Schwartz IL (1962) Effect of ACTH, hydrocortisone and glucagon on plasma non esterified fatty acid concentration (NEFA) in normal subjects and in patients with liver disease. Metabolism 11:572–578
Fain JN, Kovacev VP, Scow RO (1966) Antilipolytic effect of insulin on isolated fat cells of the rat. Endocrinology 78:773–778
Fain JN, Psychoyos S, Czernik AJ, Frost S, Cash WD (1973) Indomethacin, lipolysis and cyclic AMP accumulation in white fat cells. Endocrinology 93:632–639
Felig P (1970) Glucagon: physiologic and diabetogenic role. N Engl J Med 283:149–150
Fernandez BM, Saggerson ED (1978) Alteration in response of rat white adipocytes to insulin, noradrenalin, corticotropin and glucagon after adrenalectomy. Correction of these changes by adenosine deaminase. Biochem J 174:111–118
Foà PP (1968) Glucagon. Ergeb Physiol 60:141–219
Frandsen EK, Grønvald FC, Heding LG, Johansen NL, Lundt BF, Moody AJ, Markussen J, Vølund AA (1981) Glucagon: structure-function relationships investigated by sequence deletions. Hoppe-Seylers Z Physiol Chem 362:665–667
Fredholm BB, Hedqvist P (1975) Indomethacin and the role of prostaglandins in adipose tissue. Biochem Pharmacol 24:61–66
Friedli C, Chinet A, Girardier L (1978) Comparative measurements of in vitro thermogenesis of brown adipose tissue from control and cold-adapted rats. Experientia [suppl] 32:259–266
Froesch E, Bally P, Guhl U, Ramseier E, Labhart A (1960) Die Wirkung des Glukagons auf die Fettgewebe. Schweiz Med Wochenschr 90:1329–1332
Gaion RM (1981) Calcium and cyclic GMP as an alternative control system in fat cell metabolism. In: Enzi G, Crepaldi G, Pozza G, Renold AE (eds) Obesity: pathogenesis and treatment. Academic, New York London, pp 149–160
Gaion RM, Krishna G (1981) The role of cyclic nucleotides in the regulation of lipolysis. In: Enzi G, Crepaldi G, Pozza G, Renold AE (eds) Obesity: pathogenesis and treatment. Academic, New York London. pp 135–148
Gascon S (1971) Influencia de la prostaglandine $E_1$ en el control hormonal de la lipolisis en aves. Rev Esp Fisiol 27:69–72

Gerich JE, Lorenzi M, Bier D, Taslikian E, Schneider V, Karam JH, Forsham PH (1976) Effects of physiologic levels of glucagon and growth hormone on human carbohydrate and lipid metabolism. Studies involving administration of exogenous hormone during suppression of endogenous hormone secretion with somatostatin. J Clin Invest 57:875–884

Ghidoni A, Pappalettera A, Sanesi E, Tognetti A, Pozza G (1970) Metabolismo di cellule adipose umane isolate in vitro. II. Regolazione della lipolisi. Acta Diabetol Lat 7:271–278

Goodrige AG, Ball EG (1965) Studies on the metabolism of adipose tissue: XVIII: In vitro effects of insulin, epinephrine and glucagon on lipolysis and glycolysis in pigeon's adipose tissue. Comp Biochem Physiol 16:367–381

Gorin E, Goodman HM (1974) Protein kinase in adipose tissue: effect of hypophysectomy. Horm Metab Res 6:146–150

Grande F, Prigge WF (1970) Glucagon infusion, plasma FFA and triglycerides, blood sugar and liver lipids in birds. Am J Physiol 218:1406–1411

Grande F, Prigge WF (1972) Influence of prostaglandin $E_1$ on the adipokinetic effect of glucagon in birds. Proc Soc Exp Biol Med 140:999–1004

Gries FA (1968) Wirkungen von Glukagon auf den Stoffwechsel des Fettgewebes. In: 14. Symposion der Deutschen Gesellschaft für Endokrinologie. Springer, Berlin Heidelberg New York, pp 235–241

Gutman RA, Fink G, Voyles N, Selawry H, Penhos JC, Lepp A, Recant L (1973) Specific biologic effects of intestinal glucagon-like materials. J Clin Invest 52:1165–1175

Hagen JH (1961) Effects of glucagon on the metabolism of adipose tissue. J Biol Chem 236:1023–1027

Heald PJ, McLachlan PM, Rookledge KA (1965) The effect of insulin, glucagon and adrenocorticotrophic hormone on the plasma glucose and free fatty acids of the domestic fowl. J Endocrinol 33:83–95

Heckemeyer CM, Barker JA, Duckworth WC, Solomon SS (1981 a) Lipolytic effect of glucagon on perifused isolated fat cells. Clin Res 29:505A

Heckemeyer CM, Barker JA, Duckworth WC, Solomon SS (1981 b) Glucagon degradation by the perifused isolated fat cell. Diabetes [Suppl] 30:105A

Heim T, Hull D (1966) The effects of propranolol on the calorigenic response in brown adipose tissue of newborn rabbits to catecholamines, glucagon, corticotrophin and cold exposure. J Physiol (Lond) 187:271–283

Hoak JC, Connor WE, Warner ED (1968) Toxic effects of glucagon-induced acute lipid mobilization in geese. J Clin Invest 47:2701–2710

Holm G, Jacobsson B, Björntorp P, Smith U (1975) Effect of age and cell size on rat adipose tissue metabolism. J Lipid Res 16:461–464

Honnor RC, Saggerson ED (1980) Altered lipolytic response to glucagon and adenosine deaminase in adipocytes from starved rats. Biochem J 188:757–761

Horton EW (1969) Hypotheses on physiological roles of prostaglandins. Physiol Rev 49:122–161

Iliano G, Cuatrecasas P (1971) Endogenous prostaglandins modulate lipolytic processes in adipose tissue. Nature New Biol 234:72–74

Joel CD (1966) Stimulation of metabolism of rat brown adipose tissue by addition of lipolytic hormones in vitro. J Biol Chem 241:814–821

Koerker DJ, Goodner CJ, Koivola PTK, Gale CC, Ensinck JW (1974) Adaptation to fasting in baboon. I. Influence of feeding schedule. Am J Physiol 227:520–530

Krug E, Miahle P (1979) A possible role for gut GLI: an inhibitor of lipolysis. Horm Metab Res 6:465–469

Kuroshima A, Yahata T (1979) Thermogenic responses of brown adipocytes to noradrenaline and glucagon in heat-acclimated and cold-acclimated rats. Jpn J Physiol 29:683–690

Kuroshima A, Doi K, Kurahashi M, Ohno T (1975) In vivo lipolytic effect of glucagon in warm-adapted and cold-adapted rats. Jpn J Physiol 25:275–285

Kuroshima A, Ohno T, Doi K (1977) In vivo lipolytic action of glucagon in brown adipose tissue of warm-acclimatized and cold-acclimatized rats. Experientia 33:240–241

Langslow DR (1971) The antilipolytic action of prostaglandin $E_1$ on isolated fat cells. Biochim Biophys Acta 239:33–37

Langslow DR (1972) The development of lipolytic sensitivity in the isolated fat cells of Gallus domesticus during the foetal and neonatal period. Comp Biochem Physiol [B] 43:689–701

Langslow DR, Hales CN (1969) Lipolysis in chicken adipose tissue in vitro. J Endocrinol 43:285–294

Langslow DR, Hales CN (1970) Bioassay of glucagon in human serum. Lancet 1:1151–1152

Langslow DR, Gramb G, Siddle K (1979) Possible mechanisms for the increased sensitivity to glucagon and catecholamines of chicken adipose tissue during hatching. Gen Com Endocrinol 39:527–533

Lee HM, Ellis RM, Bromer WW (1960) Insulin-like effect of crystalline glucagon as measured with epididymal fat pad preparation. Proc Soc Exp Biol Med 104:4–6

Lefèbvre P (1965) Glucagon et taux sanguin des acides gras non estérifiés chez l'Homme. Ann Endocrinol (Paris) 26:602–608

Lefèbvre P (1966a) Contribution à l'étude du rôle physiologique du glucagon. Arscia, Brussels

Lefèbvre P (1966b) The physiological effect of glucagon on fat mobilization. Diabetologia 2:130–132

Lefèbvre P (1967) Le glucagon, seconde hormone pancréatique. Maloine, Paris, Arscia, Brussels

Lefèbvre P (1972) Glucagon and lipid metabolism. In: Lefèbvre PH, Unger RH (eds) Glucagon: molecular physiology, clinical and therapeutic implications. Pergamon, Oxford New York, pp 109–121

Lefèbvre PJ (1975) Commentary: glucagon and adipose tissue. Biochem Pharmacol 24:1261–1266

Lefèbvre PJ, Federspil G (1971) Effets d'un exercise musculaire continu (nage forcée) sur le métabolisme du tissu adipeux du rat. Arch Int Physiol Biochim 79:565–572

Lefèbvre P, Luyckx A (1969a) Lipolytic action of glucagon in vitro and in vivo. In: Vague J (ed) Physiopathology of adipose tissue. Excerpta Medica, Amsterdam London New York, pp 257–267

Lefèbvre P, Luyckx A (1969b) Effect of insulin on glucagon-enhanced lipolysis in vitro. Diabetologia 5:195–197

Lefèbvre P, Luyckx A (1971) The role of glucagon in clinical medicine. In: Foà PP (ed) The action of hormones: genes to population. Thomas, Springfield, pp 315–351

Lefèbvre PJ, Luyckx AS (1974) Effect of L8027, a new potent inhibitor of prostaglandin biosynthesis, on the metabolism and response to glucagon of rat adipose tissue. Biochem Pharmacol 23:2119–2125

Lefèbvre PJ, Unger RH, Valverde I, Rigopoulou D, Luyckx AS, Eisentraut A (1969) Effect of dog jejenum "glucagon-like immunoreactive material" on adipose tissue metabolism. Horm Metab Res 1:143–144

Lefèbvre PJ, Luyckx AS, Federspil G (1972) Muscular exercise and pancreatic function in rats. Isr J Med Sci 8:390–398

Lefèbvre PJ, Luyckx AS, Bacq ZM (1973) Effects of denervation on the metabolism and the response to glucagon of white adipose tissue of rats. Horm Metab Res 5:245–250

Lewis GP, Matthews J (1968) The mobilization of free fatty acids from rabbit adipose tissue in situ. Br J Pharmacol 34:564–578

Liljenquist JE, Bomboy JD, Lewis SB, Sinclair-Smith BC, Felts PW, Lacy WW, Crofford OB, Liddle GW (1974) Effects of glucagon on lipolysis and ketogenesis in normal and diabetic men. J Clin Invest 53:190–197

Lipinski BA, Mathias MM (1978) Prostaglandin production and lipolysis in isolated rat adipocytes as affected by dietary fat. Prostaglandins 16:957–963

Livingston JN, Cuatrecasas P, Lockwood DH (1974) Studies of glucagon resistance in large adipocytes: $^{125}$I-labeled glucagon binding and lipolytic activity. J Lipid Res 15:26–32

Luyckx A (1974) Etude de la sécrétion de l'insuline et du glucagon. Thesis, University of Liège

Luyckx AS, Lefèbvre PJ (1973) Exercise-induced glucagon secretion. Postgrad Med J 49:620–623

Luyckx AS, Lefèbvre PJ (1974) Mechanisms involved in the exercise-induced increase in glucagon secretion in rats. Diabetes 23:81–91
Luyckx AS, Dresse A, Cession-Fossion A, Lefèbvre PJ (1975) Catecholamines and exercise-induced glucagon and fatty acid mobilization in the rat. Am J Physiol 229:376–383
Luyckx AS, Giugliano D, Lefèbvre PJ (1980) Influence of glucagon on the metabolic abnormalities of insulin-dependent diabetics. In: Waldhausl WK (ed) Diabetes 1979. Excerpta Medica, Amsterdam London New York, pp 409–414
Maickel RP, Stern DN, Takabatake E, Brodie BB (1967) The sympathetic nervous system as a homeostatic mechanism. II. Effect of adrenocortical hormones on body temperature maintencance of cold-exposed adrenalectomized rats. J Pharmacol 157:111–116
Manganiello V, Vaughan M (1973) Selective loss of adipose cell responsiveness to glucagon with growth in the rat. J Lipid Res 13:12–16
Margolis S, Vaughan M (1962) α-glycerophosphate synthesis and breakdown in homogenates of adipose tissue. J Biol Chem 237:44–48
Micheli H (1970) Some characteristics of lipolysis in rabbit adipose tissue. Effects of noradrenaline, ACTH, theophylline and prostaglandin $E_1$. Acta Physiol Scand 79:289–298
Micheli H, Carlson LA, Hallberg D (1969) Comparison of lipolysis in human subcutaneous and omental adipose tissue with regard to effects of noradrenaline, theophylline, prostaglandin $E_1$ and AGE. Acta Chir Scand 135:663–670
Mosinger B, Kuhn E, Kujalova V (1965) Actions of adipokinetic hormones on human adipose tissue in vitro. J Lab Clin Med 66:380–389
Mülbachova E, Solyom AL, Puglisi L (1967) Investigations on the mechanism of the prostaglandin $E_1$ antagonism to norepinephrine and theophylline-induced lipolysis. Eur J Pharmacol 1:321–325
Müller WA, Aoki TT, Egdahl RH, Cahill GF Jr (1977) Effects of exogenous glucagon and epinephrine in physiological amounts on the blood levels of free fatty acids and glycerol in dogs. Diabetologia 13:55–58
Murase T, Tanaka K, Iwamoto Y, Akanuma Y, Kosaka K (1981) Reciproqual changes caused by insulin and glucagon of adipose tissue lipoprotein lipase in rats in vitro. Horm Metab Res 13:212–213
Nestel PJ, Austin W (1969) Relationship between adipose tissue lipase activity and compounds which affect intracellular lipolysis. Life Sci 8:157–164
Nyberg G, Häger A, Smith U (1977) Effect of age on human adipose tissue metabolism and hormonal responsiveness. Acta Paediatr Scand 66:495–500
Pozefsky T, Tancredi RG, Moxley RT, Dupre J, Tobin JD (1976) Metabolism of forearm tissues in man. Studies with glucagon. Diabetes 25:128–135
Pozza G, Pappalettera A, Ghidoni A, Sanesi E, Tognetti A, Melogli O (1971 a) La regolazione dell lipolisi nel tessuto adiposo humano, con particolare riguardo allazione del glucagone. In: Austoni M, Scandellari C, Federspil G, Trisotto A (eds) Current topics on glucagon. Cedam, Padova, pp 93–105
Pozza G, Pappalettera A, Melogli O, Viberti G, Ghidoni A (1971 b) Lipolytic effect of intraarterial injection of glucogon in man. Horm Metab Res 3:291–292
Prigge WF, Grande F (1971) Effects of glucagon, epinephrine and insulin on in vitro lipolysis of adipose tissue from mammals and birds. Comp Biochem Physiol 39B:69–82
Ramwell PW, Shaw JE (1970) Biological significance of the prostaglandins. Recent Prog Horm Res 26:139–187
Ramwell PW, Shaw JE, Douglas WW, Poisner AM (1966) Efflux of prostaglandin from adrenal glands stimulated with acetlycholine. Nature 210:273–274
Rault C, Fruchart JC, Dewailly P, Jaillard J, Sezille G (1974) Experimental studies of myocardial and adipose tissue lipoprotein lipase activities in rat. Biochem Biophys Res Comm 59:160–166
Rodbell M, Jones AB (1966) Metabolism of isolated fat cells. III: The similar inhibitory action of phospholipase C (*Clostridium perfringens α toxin*) and of insulin on lipolysis stimulated by lipolytic hormones and theophylline. J Biol Chem 241:140–142
Rodbell M, Birnbaumer L, Pohl SL (1970) Adenylcyclase in fat cells. Stimulation by secretin and the effects of trypsin on the receptors for lipolytic hormones. J Biol Chem 245:718–718
Rudman D, Di Girolamo M (1967) Comparative studies on the physiology of adipose tissue. Adv Lipid Res 5:35–117

Samols E, Tyler JM, Marks V, Miahle P (1969) The physiological role of glucagon in different species. In: Gual C (eds) Progress in endocrinology. Excerpta Medica, Amsterdam London New York, pp 206–219
Schade DS, Eaton RP (1975g a) The contribution of endogenous insulin secretion to the ketogenic response to glucagon in man. Diabetologia 11:555–559
Schade DS, Eaton RP (1975 b) Modulation of fatty acid metabolism by glucagon in man. I. Effects in normal subjects. Diabetes 24:502–509
Schade DS, Woodside W, Eaton RH (1979) The role of glucagon in the regulation of plasma lipids. Metabolism 28:874–886
Schneider SH, Fineberg SE, Blackburn GL (1981) The acute metabolic effects of glucagon and its interactions with insulin in forearm tissue. Diabetologia 20:616–624
Shafrir E, Steinberg D (1960) The essential role of the adrenal cortex in the response of plasma free fatty acids, cholesterol and phospholipids to epinephrine injection. J Clin Invest 39:210–219
Shafrir E, Sussman EK, Steinberg D (1960) Role of the pituitary and the adrenals in the mobilization of free fatty acids and lipoproteins. J Lipid Res 1:459–465
Shaw JE, Ramwell PW (1968) Release of prostaglandin from rat epididymal fat pad on nervous and hormonal stimulation. J Biol Chem 243:1498–1503
Shechter Y (1982) Evaluation of adenosine or related nucleosides as physiological regulators of lipolysis in adipose tissue. Endocrinology 110:1579–1583
Steinberg D, Huttunen JK (1972) The role of cyclic AMP in activation of hormone-sensitive lipase of adipose tissue. Adv Cyclic Nucleotide Res 1:47–62
Steinberg DM, Shafrir E, Vaughan M (1969) Direct effect of glucagon on release of unesterified fatty acids (UFA) from adipose tissue. Clin Res 7:250
Steinberg D, Vaughan M, Nestel PJ, Bergström S (1963) Effects of prostaglandin E opposing those of catecholamine on blood pressure and on triglyceride breakdown in adipose tissue. Biochem Pharmacol 12:764–766
Stone DB, Brown JD, Steele AA (1969) Effect of sodium salicylate on induced lipolysis in isolated fat cells of the rat. Metabolism 18:620–624
Strosser MT, Cohen L, Koch B, Miahle P (1981) La somatostatine: une hormone pancréatique qui pourrait être importante. Journ Annu Diabetol Hôtel Dieu 197–210
Unger RH (1981) The milieu interieur and the islets of Langerhans (Claude Bernard Memorial Lecture). Diabetologia 20:1–12
Unger RH, Eisentraut AM (1964) Studies of the physiological role of glucagon. Diabetes 13:563–568
Valverde I, Rigopoulou D, Exton J, Ohneda A, Eisentraut AM, Unger RH (1968) Demonstration and characterization of a second fraction of glucagon-like immunoreactivity in jejunal extracts. Am J Med Sci 255:415–420
Vaughan M (1960) Effect of hormones on phosphorylase activity in adipose tissue. J Biol Chem 235:3049–3053
Vaughan M, Steinberg D (1963) Effect of hormones on lipolysis and esterification of free fatty acids during incubation of adipose tissue in vitro. J Lipid Res 4:193–199
Vizek K, Razova M, Melichar V (1979) Lipolytic effect of TSH, glucagon and hydrocortisone on the adipose tissue of new borns and adults in vitro. Physiol Bohemoslov 28:325–331
Weinges KF (1961) The effect of glucagon and insulin on the metabolism of non-esterified fatty acids in isolated fatty tissue of the rat in vitro. Klin Wochenschr 39:293–298
Weinges KF (1968) Glucagon. Thieme, Stuttgart
Whitty AJ, Shima K, Trubow M, Foà PP (1967) Effect of glucagon on serum FFA in normal and depancreatized dogs. Fed Proc 26:15
Whitty AJ, Shima K, Trubow M, Foà PP (1969) Effect of glucagon and of insulin of serum free fatty acids in normal and depancreatized dogs. Proc Soc Exp Biol Med 130:55–61
Wieland O, Suyter M (1957) Glycerokinase: its isolation and properties. Biochem Z 329:320–331
Wörner H, Weinges KF (1961) Über den Einfluß von Insulin und Glukagon auf den Glukosestoffwechsel des epididymalen Fettanhanges der Ratte in vitro. Der Einfluß von Glukagon im Vergleich mit dem Insulineffekt. Klin Wochenschr 39:243–246
Zumstein P, Zapf J, Froesch ER (1974) Effects of hormones on cyclic AMP release from rat adipose tissue in vitro. FEBS Lett 41:65–69

CHAPTER 20

# Glucagon and Lipoprotein Metabolism

A. TIENGO and R. NOSADINI

## A. Introduction

The hepatic very low density lipoprotein (VLDL) synthesis seems to depend not only on the concentration of substrates affluent to the liver, but on the nutritional and hormonal milieu as well. It has long been known that changes in the circulating levels of insulin and glucagon can shift glucose metabolism in the liver from storage to release. It has been recognized only recently that this same hormone couple also plays a pivotal role in hepatic fat metabolism (EATON 1977; MCGARRY 1979). Glucagon and insulin exert opposite effects upon lipid metabolism both in peripheral tissues and in liver (Fig. 1). Insulin and glucagon have to be secreted, and transported to target tissues, in which they must interact with specific receptors to initiate a series of postreceptor events. Defects in any of these processes can impair lipoprotein metabolism.

Thus, it has been observed that insulin-resistant states, associated with hyperinsulinemia, are characterized by an enhanced lipogenic insulin action on the liver

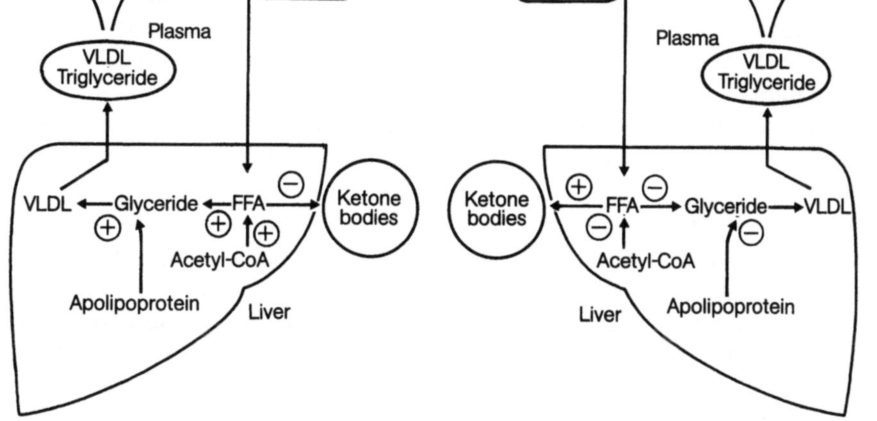


**Fig. 1.** Insulin and glucagon action on lipid metabolism

and by increased triglyceride–very low density lipoprotein (TG-VLDL) liver production (Olefsky et al. 1974). On the other hand, a decreased glucagon activity could also contribute to "gear up" hepatic lipoprotein synthesis (Eaton and Schade 1973). A more physiologic role is played by insulin and glucagon throughout the day in the regulation of lipoprotein metabolism during postprandial and postabsorptive states.

The triglyceride-rich lipoproteins can enter the plasma either as chylomicrons, derived from dietary fat originating in the gut, or as endogenous VLDL synthesized mainly in liver and gut from glucose and circulating free fatty acids (FFA) (Windmueller et al. 1973). During and after meals, FFA enter the liver and small intestine where they may be esterified with glycerol to form triglycerides. Lipogenesis also occurs in the liver, where fatty acids are synthesized from glucose or acetate (Stein and Stein 1967). Both these processes appear to be regulated by changes in insulin and glucagon levels associated with feeding. Fatty acids in the cytosol of the hepatocyte can either enter mitochondria, where oxidation occurs, or be esterified in the cytosol to triglycerides. The fate of fatty acids appears to be under hormonal control inside hepatic cells as well: glucagon enhances and insulin prevents mitochondrial fatty acid uptake and subsequent oxidation (McGarry and Foster 1976; see also Chap. 17).

The plasma concentration and the pattern of lipoproteins result not only from lipoprotein production, but also from their transport, removal, and catabolism. The triglycerides that enter the plasma as chylomicrons and endogenously synthesized VLDL are primarily transported to adipose tissue and muscle for storage and utilization. Calorie intake and the resultant insulin secretion increases lipoprotein lipase activity, while the activity of the hormone-sensitive lipase decreases. Between meals, when insulin levels are low and plasma glucagon rises, lipoprotein lipase activity decreases, while hormone-sensitive lipase activity increases. In a number of clinical situations, abnormalities in lipoprotein lipase may be manifest as hypertriglyceridemia due generally to a lack of insulin availability. Insulin deficiency appears to be associated with a decrease in lipoprotein lipase release from fat cells and from muscle that accounts for most of the elevation of triglyceride-rich lipoprotein seen in untreated diabetes (Brunzell et al. 1975). Conflicting results on the other hand have been reported for the effect of glucagon on the peripheral clearance of lipoproteins (Schade and Eaton 1977b). Glucagon could enhance lipid removal from blood either by enhancing triglyceride adhesion to platelet membranes (Caren and Corbo 1970) or by potentiating the action of lipoprotein lipase in muscle. Borensztajn et al. (1973) demonstrated in vitro that glucagon activates myocardial lipoprotein lipase. An opposite effect has been observed by Hallberg (1970), who demonstrated a slight decrease of the maximal removal rate of exogenous triglyceride after glucagon administration (see also Chapter 46).

Glucagon, by means of its catabolic effect on protein metabolism, could also modulate the synthesis of the VLDL protein carriers, that is, the apoproteins which limit hepatic lipoprotein production. Eaton (1973) reported that glucagon causes a depression in hepatic apoprotein synthesis and a decrease in circulating VLDL apoprotein consistent with this postulated mechanism of the hypolipemic action of glucagon at the level of protein synthesis. This hypothesis needs further investigations in vivo as well as in vitro.

On the other hand it has to be underlined that the main effect of glucagon on lipid metabolism seems to be on the liver. Glucagon shifts the liver, through a short-term control mechanism opposed by insulin, from an organ primarily active in glycogenesis, glycolysis, fatty acid synthesis, fatty acid esterification, and cholesterogenesis into one primarily active in glycogenolysis, gluconeogenesis, fatty acid oxidation, and ketogenesis (GEELEN et al. 1980).

## B. Glucagon Effects on Liver Lipid Metabolism

Glucagon is undoubtedly involved in the regulation of hepatic lipid metabolism in different nutritional states. The effects of glucagon administration to the animal in vivo are modified by other endogenously released hormones such as insulin. The changes in liver metabolism induced by glucagon cannot be explained solely by an increased FFA afflux to the liver due to enhanced lipolysis of adipose tissue triacylglycerols. Both adaptive changes in enzyme activities and short-term hormone effects seem to be involved. Insulin stimulates lipogenesis whereas glucagon inhibits this biosynthetic process. The inhibiting effect of glucagon is antagonized by insulin. The relative concentrations of these two hormones in the portal blood control the disposition of key nutrients and endogenous substrates in a manner appropriate to the prevailing exogenous fuel supply and energy requirement (UNGER and ORCI 1976).

Glucagon, as well as dibutyryl cyclic AMP and cyclic AMP, acutely inhibits fatty acid synthesis in liver slices (HARRIS and YOUNT 1975), isolated hepatocytes (MULLER et al. 1976; GEELEN et al. 1978; BEYNEN et al. 1979; WITTERS et al. 1979), perfused liver (HEIMBERG et al. 1969; MA et al. 1978), and the intact animal (KLAIN 1977; COOK et al. 1977). The action of glucagon is presumably mediated through the elevation of cyclic AMP.

Two hypotheses have been advanced to explain the inhibition of fatty acid synthesis by glucagon (Fig. 2). The first states that the hormone exerts its effect at a point in the glycolytic sequence prior to the formation of pyruvate (WATKINS et al. 1977). Inactivation of pyruvate kinase and phosphofructokinase appears to explain in large part the inhibition of aerobic glycolysis by glucagon and dibutyryl cyclic AMP (LJÜNGSTRÖM et al. 1974; OCHS and HARRIS 1978), resulting in a depressed availability of pyruvate and lactate for fatty acid synthesis. This is reflected by a decreased level of citrate, a precursor of fatty acid synthesis. The second theory postulates that the hormone acts at a point distal to pyruvate and prior to malonyl-CoA formation, most likely at the acetyl-CoA carboxylase step (WATKINS et al. 1977).

Acetyl-CoA carboxylase may be considered a target in the acute control of fatty acid biosynthesis by glucagon and insulin (BEYNEN et al. 1979). The observation that hepatic concentrations of malonyl-CoA are rapidly depressed after glucagon treatment in vivo or in isolated liver cells (COOK et al. 1977, 1978) implicated acetyl-CoA carboxylase as the site at which hormonal inhibition is exerted (WITTERS et al. 1979). The acetyl-CoA carboxylase activity correlates with the rate of fatty acid synthesis which in turn correlates with the cellular level of malonyl-CoA. The glucagon-induced decrease in acetyl-CoA carboxylase activity with subsequent reduction in malonyl-CoA concentrations would lead to an activation of carnitine pal-

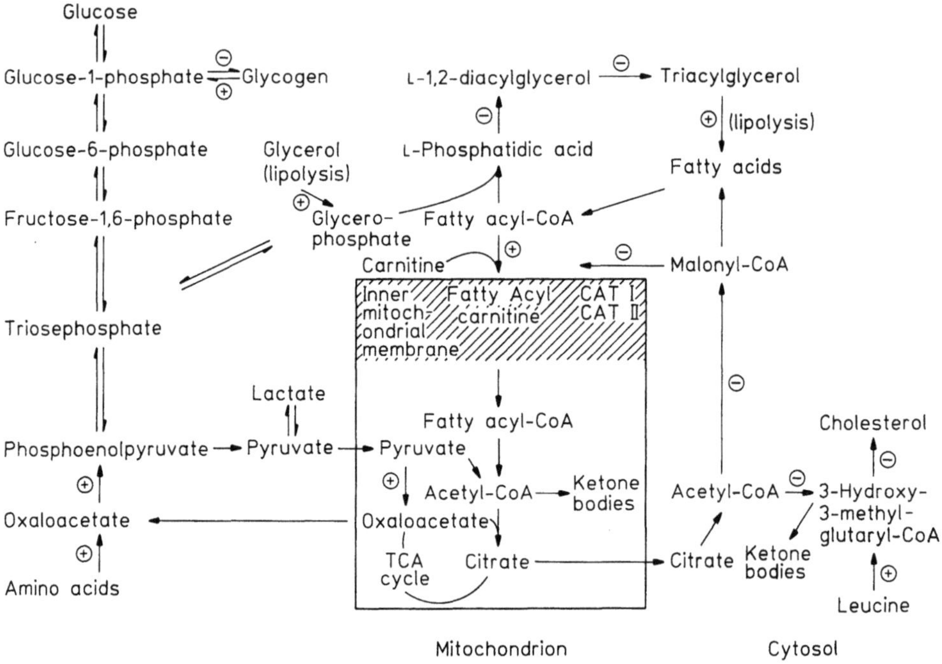


**Fig. 2.** Interrelations between the pathways of fatty acid synthesis and oxidation, lipolysis, glycolysis, and gluconeogenesis in the liver. *Plus* and *minus* signs indicate glucagon effects. (CAT, carnitine acyltransferase)

mitoyltransferase and an enhanced mitochondrial $\beta$-oxidation of fatty acids. According to McGARRY et al. (1978) the stimulatory effect of glucagon on hepatic fatty acid oxidation is secondary to the suppressive action of the hormone on the pathway of fatty acid synthesis. The rate of fatty acid synthesis appears directly related to the concentration of malonyl-CoA which depends on the activity of acetyl-CoA carboxylase, which in turn may be determined by substrate supply. The ketogenic action of glucagon on liver appears to reside in the ability to suppress the synthesis of malonyl-CoA.

With carbohydrate feeding (low glucagon: insulin ratio) hepatic malonyl-CoA levels are high, fatty acid synthesis is geared up, and the pathway of fatty acid oxidation is shut down. Conversely, in fasting and diabetes (high glucagon: insulin ratio) hepatic malonyl-CoA levels and rates of fatty acid synthesis are diminished, with the result that the opposite pathway of fatty acid oxidation and ketone body production is activated (McGARRY et al. 1978). On the other hand, two different research groups suggested recently that malonyl-CoA does not play the same role in starved animals that it does in fed animals (COOK et al. 1980; ONTKO and JOHNS 1980) and they stated that its significance in ketotic states remains to be determined. Although carnitine palmitoyltransferase is present in excess in mitochondria from both fed and starved rats, malonyl-CoA can make this enzyme rate limiting in the fed state, but in the starved state the rate of ketogenesis must be related elsewhere, either inside or outside the mitochondria (COOK et al. 1980). Conversely, it has been reported that fatty acid synthesis could be regulated by the intracellular

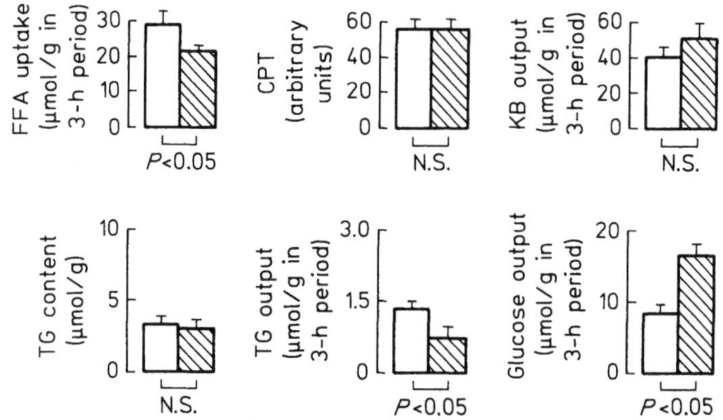


**Fig. 3.** FFA uptake, mitochondrial carnitine palmitoyltransferase (CPT) activity, triglyceride content, ketone body (KB) – triglyceride (TG), and glucose production (mean ± standard error) by the isolated liver following a 3-h perfusion with oleate alone (1.67 μequiv./min, *open columns*) and oleate + glucagon (150 μg over 3 h, *hatched columns*) in ten Sprague-Dawley rats. (*N.S.*, not significant)

concentration of long chain acyl-CoA esters, as suggested by the finding that triglyceride-rich lipoproteins inhibit lipogenesis from isolated hepatocytes by an increase of long chain acyl-CoA esters (MAYES and TOPPING 1974).

Partitioning of incoming fatty acids between esterification and oxidation may be regulated by coordinated changes in the activity of carnitine acyltransferase and glycerophosphate acyltransferase. In the transition from starvation to refeeding, the activity of carnitine palmitoyltransferase decreases whereas the activity of glycerophosphate acyltransferase changes in the opposite direction (AAS and DAAE 1971; NORUM 1965). Hormonal factors also influence the distribution of fatty acids between oxidation and esterification. HEIMBERG et al. (1969) reported an increase in the oxidation and a decrease in the esterification of palmitic acid in the perfused liver due to the addition of glucagon. The same effect on esterification has been observed with isolated liver cells. In our studies with isolated rat liver perfused with oleate (NOSADINI et al. 1979, 1980b; TIENGO et al. 1980) a marked inhibitory effect on triglyceride output by glucagon was demonstrated without any significant ketogenic effect and activation of carnitine palmitoyltransferase, in agreement with MCGARRY et al. (1975) (Fig. 3). Our data demonstrated that glucagon moreover exerts an inhibitory action on liver FFA uptake whereas it does not affect liver triglyceride content. MCGARRY et al. (1975) postulated the existence of an extra factor which is involved in hormone action upon fatty acid oxidation and which is missing from the medium used in perfusion studies. The determination of the components of the lipoproteins produced by the isolated perfused liver showed a significant specific inhibitory effect of glucagon on VLDL triglyceride and on VLDL cholesterol (Fig. 4). Glucagon could affect FFA metabolism and triglyceride synthesis by modulating the hepatic content of both glycerophosphate and glycerophosphate acyltransferase activity rather than increasing free fatty acid oxidation (CHRISTIANSEN 1977). Our observations (TESSARI et al. 1981) demonstrated that pharmacologic doses of glucagon inhibit the glycerophosphate acyltransferase activity in perfused rat liver. Moreover, SOLER-ARGILAGA et al. (1978) reported, in

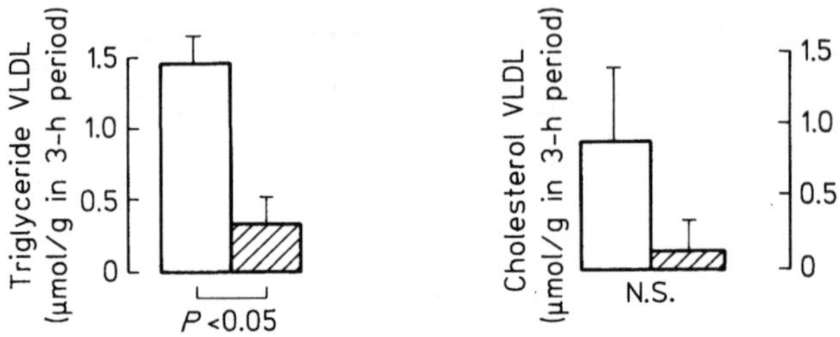


**Fig. 4.** Glucagon effect (150 µg over 3 h) on triglyceride VLDL and cholesterol VLDL production by the rat liver perfused with oleate (1.67 µEquiv./min for 3 h). *N.S.*, not significant

studies with the perfused liver, that dibutyryl cyclic AMP decreases the activity of microsomal glycerophosphate acyltransferase. On the contrary, insulin specifically increases the activity of the mitochondrial glycerophosphate acyltransferase in the perfused liver system.

Although glucagon gives a marked decrease in the rate of triacylglycerol synthesis, it has little or no effect on the total synthesis of phospholipids. However, it has been reported that glucagon stimulates the entry of labeled glycerol in phospholipids. Thus, one can suggest that under conditions of increased glucagon concentrations, diacylglycerols may be utilized preferentially for phospholipid synthesis (GEELEN et al. 1978) at the expense of triacylglycerol formation.

Glucagon, like norepinephrine and dibutyryl cyclic AMP, inhibits the efflux of cholesterol from isolated rat hepatocytes (EDWARDS et al. 1979). Reduction in the generation of cytosol acetyl-CoA, coupled with inhibition of acetyl-CoA carboxylase by glucagon, would not be expected to alter greatly the flow of acetyl-CoA to cholesterol. Glucagon blocked acetate $^{14}$C incorporation into fatty acids by 98% whereas acetate $^{14}$C incorporation into cholesterol was decreased by only 30%. In rat hepatocytes, glucagon and dibutyryl cyclic AMP inhibited the induction of 3-hydroxy-3-methylglutaryl-CoA (HMG-CoA) reductase, the rate-controlling enzyme for cholesterol biosynthesis under most physiologic conditions (EDWARDS et al. 1979; INGEBRITSEN et al. 1979). Since both reductase and reductase kinase are cyclic AMP-independent enzymes, a possible route for glucagon regulation is through phosphoprotein phosphatase, which catalyzes dephosphorylation of both reductase and reductase kinase (GIBSON and INGEBRITSEN 1978; INGEBRITSEN et al. 1979). However other reports are not consistent with this hypothesized mechanism of action of glucagon on cholesterol synthesis. On the other hand, RASKIN et al. (1974) contribute to the clarification of this problem by demonstrating that glucagon had no significant effect on reductase activity during the first 60 min of hepatocyte incubation, but the inhibitory effect was observed after 3 h incubation under conditions that promoted HMG-CoA reductase induction.

## C. Metabolic Effects of Glucagon In Vivo

As reviewed in Chap. 17 and 19, glucagon has been shown to stimulate lipolysis and ketogenesis in many animal species in vivo (WILLIAMSON 1967; GRANDE 1969; LEFEBVRE 1966) and to decrease triglyceride and cholesterol circulating levels

(AMATUZIO et al. 1962; PALOYAN et al. 1962). However it is difficult to assess the role played by glucagon itself, irrespective of other hormones, since glucagon injection may promote the release of insulin (SAMOLS et al. 1966), growth hormone (MITCHELL et al. 1971), and catecholamine (SARCIONE et al. 1963). In order to investigate the effect of glucagon on lipolysis and ketogenesis in the presence and absence of glucagon-induced insulin release, LILJENQUIST et al. (1974) performed a continuous glucagon infusion in normal and insulin-dependent diabetic humans. Net splanchnic glucose production rose markedly during glucagon stimulation both in normal subjects and diabetics, despite the marked rise in insulin in the normal subjects, whereas glycerol and ketone body blood concentrations were enhanced only in the diabetic group. The lipolytic and ketogenic effect of glucagon was evident only when insulin secretion was absent, but glucagon (on a molar basis) had greater stimulatory activity on hepatic glucose release. SCHADE and EATON (1975 a) sought to clarify the problem using scalar rates of glucagon infusion ranging from pharmacologic to near physiologic doses in normal humans. Glucagon was able to decrease triglyceride circulating concentrations only after injection of doses higher than 0.5 μg/kg, resulting in circulating levels of glucagon higher than 1,000 pg/ml. A ketogenic and hyperglycemic effect of glucagon was on the contrary evident with lower levels of glucagon, even when the hypotriglyceridemic effect was not present.

In obese patients, glucagon was able to decrease blood circulating levels of triglyceride and to enhance glucose, but not to influence $\beta$-hydroxybutyrate and free fatty acids after bolus intravenous injection of 1 μg/kg glucagon (SCHADE and EATON 1977a). This behavior contrasts with the normal lipolytic and ketogenic response observed in control subjects (SCHADE and EATON 1977b). When a higher dose of glucagon was administered these abnormal responses of obese subjects were corrected. These results suggest a relative rather than absolute "resistance" to glucagon in obesity, most likely related to the insulin hypersecretion that occurs after glucagon challenge. On the contrary, obese subjects demonstrated a significantly greater decline in plasma triglyceride than control subjects at both high and low doses of glucagon, suggesting that the hypolipemic action is not impaired in obesity and that this metabolic effect can be independent of a concomitant effect on circulating levels of free fatty acids and ketone bodies.

In normal and diabetic hypertriglyceridemic subjects, we observed that pharmacologic doses of glucagon induce an enhanced hyperglycemic effect, a lack of lipolytic effect, and a preserved ketogenic and hypotriglyceridemic effect in comparison with controls (NOSADINI et al. 1978). A significant, inverse correlation was present between fasting insulin and free fatty acid increment following glucagon injection, suggesting that the lack of lipolysis in hyperlipemics is the expression of the prevalence of the insulin-induced antilipolytic action over the lipolytic effect. It is interesting to point out that the hypotriglyceridemic effect was not related to free fatty acid mobilization as in obese subjects. Moreover the persistance of the hypolipidemic effect of glucagon invalidates the hypothesis of a resistance to exogenous glucagon and of a reduced hormonal activity in hyperlipemia. A similar lipolytic, ketogenic, and hypotriglyceridemic effect was observed following glucagon administration in insulin-dependent diabetic subjects (SCHADE and EATON 1975 b).

Despite the fact that peripheral concentrations of glucagon higher than 1,000 pg/ml are rarely observed in normal humans, except in severe shock and infections, it has to be pointed out that intraportal concentrations of glucagon have been reported to be significantly higher than peripheral ones (BLACKARD et al. 1974). It is possible to suggest therefore that glucagon plays a physiologic role in humans both on lipid and carbohydrate homeostasis although previous studies could not rule out that the metabolic effects observed following glucagon injection may reflect stimulation of growth hormone (MITCHELL et al. 1971) and catecholamine (SARCIONE et al. 1963).

A new approach has been proposed by GERICH et al. (1975) who infused glucagon in insulin-dependent diabetic subjects along with somatostatin to inhibit growth hormone stimulation. The circulating levels of glucagon reached during exogenous glucagon infusion were about 250 pg/ml, hormonal pattern close to the physiologic concentration observed in diabetic subjects and often in normal subjects as well. Glucagon was able to promote peripheral release of free fatty acids and glycerol, with a concomitant increase in ketone body circulating levels. It was not possible in this study to assess whether the ketogenic effect was due either to a shift of free fatty acids from esterification to oxidation at the hepatic level or to the enhanced delivery of lipolytic metabolites from peripheral tissue.

GEY et al. (1977) confirmed in the rat in vivo that the hypolipemic effect of glucagon does not appear to depend upon the presence of insulin, as indicated by experiments in streptozotocin-induced diabetic rats. This lipid-lowering effect of glucagon is present after long-term treatment as well, becomes maximal after 2–3 days of repeated injections of pancreatic glucagon twice daily, and lasts for at least 4 weeks. Plasma lipid depression occurs at glucagon doses which have no effect on glucose, free fatty acids, and urea. These authors showed that glucagon on a molar basis is more potent than other lipid-lowering hormones like thyroxine and estradiol.

More recently it has been suggested that glucagon can promote other metabolic effects on lipid metabolism in vivo. KLAIN (1977) reported that glucagon reduced hepatic fatty acid synthesis and acetyl-CoA carboxylase activity in both fed and fasted rats, and that exogenous insulin antagonized the glucagon effects. This effect was evident in the liver, but not in muscle. Further investigations are needed in humans to clarify the role played by glucagon in fatty acid synthesis. The role that endogenous glucagon may play in the pathogenesis of hyperlipidemic states is discussed in detail in Chap. 46.

## References

Aas M, Daae LNM (1971) Fatty acid activation and acyl transfer in organs from rats in different nutritional states. Biochim Biophys Acta 239:208–216

Amatuzio DS, Grande F, Wada S (1962) Effect of glucagon on the serum lipids in essential hyperlipaemia and in hyperglycaemia. Metabolism 11:1240–1249

Beynen AC, Vaartjes WJ, Geelen MJH (1979) Opposite effects of insulin and glucagon in acute hormonal control of hepatic lipogenesis. Diabetes 28:828–835

Blackard WG, Nelson NC, Andrews SS (1974). Portal and peripheral vein immunoreactive glucagon after arginine or glucose infusions. Diabetes 23:199–202

Borensztajn JP, Keig P, Rubenstein AH (1973) The role of glucagon in the regulation of myocardial lipoprotein lipase activity. Biochem Biophys Res Commun 53:603–608

Brunzell JD, Porte D Jr, Bierman EL (1975) Reversible abnormalities in postheparin lipolytic activity during the late phase of release in diabetes mellitus. Metabolism 24:1123–1137

Caren R, Corbo L (1970) Glucagon and plasma lipoprotein lipase. Proc Soc Exp Biol Med 146:1106–1110

Christiansen RZ (1977) Regulation of palmitate metabolism by carnitine and glucagon in hepatocytes isolated from fasted and carbohydrate refed rats. Biochim Biophys Acta 488:249–262

Cook GA, Nielsen RC, Hawkins RA, Mehlman MA, Lakshmanan MR, Veech RF (1977) Effect of glucagon on hepatic malonyl-coenzyme A concentration and on lipid synthesis. J Biol Chem 252:4421–4424

Cook GA, King MT, Veech RL (1978) Ketogenesis and malonyl-coenzyme A content of isolated rat hepatocytes. J Biochem 253:2529–2531

Cook GA, Otto DA, Cornell NW (1980) Differential inhibition of ketogenesis by malonyl-CoA in mitochondria from fed and starved rats. Biochem J 192:955–958

Eaton RP (1973) Hypolipemic action of glucagon in experimental endogenous lipemia in the rat. J Lipid Res 14:312–318

Eaton RP (1977). Glucagon and lipoprotein regulation in man. In: Foà PP, Bajaj JS, Foà NL (eds) Glucagon: its role in physiology and clinical medicine. Springer, Berlin Heidelberg New York, pp 533

Eaton RP, Schade DS (1973) Glucagon resistance: a hormonal basis for endogenous hyperlipemia. Lancet 1:973–974

Edwards PA, Lemongello D, Fogelman AM (1979) The effect of glucagon, norepinephrine, and dibutyryl cyclic AMP on cholesterol efflux and on the activity of 3-hydroxy-3-methylglutaryl CoA reductase in rat hepatocytes. J Lipid Res 20:2–7

Geelen MJH, Beynen AC, Christiansen RS, Lepreau-Jose MJ, Gibson DM (1978) Short-term effects of insulin and glucagon on lipid synthesis in isolated rat hepatocytes. Covariance of acetyl-CoA carboxylase activity and the rate of $^3H_2O$ incorporation into fatty acids. FEBS Lett 95:326–330

Geelen MJH, Harris RA, Beynen AC, McCune S (1980) Short-term hormonal control of hepatic lipogenesis. Diabetes 29:1006–1022

Gerich JE, Lorenzi M, Bier DM, Tsalikian E, Schneider V, Karam JH, Forsham PH (1975) Effects of physiologic levels of glucagon and growth hormone on human carbohydrate and lipid metabolism. J Clin Invest 57:875–884

Gey F, Georgi H, Buhler E (1977) Lowering of plasma lipids, the major effect of repeated glucagon administration in rats. In: Foà PP, Bajaj JJ, Foà NL (eds) Glucagon: its role in physiology and clinical medicine. Springer, Berlin Heidelberg New York, pp 517–532

Gibson DM, Ingebritsen TS (1978) Minireview: reversible modulation of liver hydroxymethylglutaryl CoA reductase. Life Sci 23:2649–2664

Grande F (1969) Lack of insulin effect on free fatty acid mobilization produced by glucagon in birds. Proc Soc Exp Biol Med 130:711

Hallberg D (1970) Insulin and glucagon in the regulation of removal rate of exogenous lipids from the blood in dogs. Acta Chir Scand 136:291–297

Harris RA, Yount RA (1975) Inhibition of hepatic lipogenesis by adenine nucleotides. Lipids 10:673–680

Heimberg M, Weinstein I, Kohout M (1969) The effects of glucagon, dibutyril cyclic adenosine 3′,5′-monophosphate and the concentration of free fatty acid on hepatic lipid metabolism. J Biol Chem 244:5131–5139

Ingebritsen TS, Geelen MJH, Parker RA, Evenson KJ, Gibson DM (1979) Modulation of hydroxymethylglutaryl CoA reductase activity, reductase kinase activity and cholesterol synthesis in rat hepatocytes in response to insulin and glucagon. J Biol Chem 254:9986–9989

Klain GJ (1977) In vivo effects of glucagon on fatty acids synthesis in fasted and refed rats. J Nutr 107:942–948

Lefèbvre P (1966) The physiological effect of glucagon on fat mobilization. Diabetologia 2:130–132

Liljenquist JE, Bomboy JD, Lewis SB, Sinclair-Smith BC, Felts PW, Lacy WW, Crofford OB, Liddle GW (1974) Effects of glucagon on lipolysis and ketogenesis in normal and diabetic men. J Clin Invest 53:190–197

Ljüngström O, Hjelmquist G, Engström L (1974) Phosphorylation of purified rat liver pyruvate kinase by cyclic 3′-5′-AMP-stimulated protein kinase. Biochim Biophys Acta 358:289–298

Ma GY, Gove CD, Hems DA (1978) Effects of glucagon and insulin on fatty acid synthesis and glycogen degradation in the perfused liver of normal and genetically obese (ob/ob) mice. Biochem J 174:761–768

Mayes PA, Topping DL (1974) Regulation of hepatic lipogenesis by plasma free fatty acids: simultaneous studies on lipoprotein secretion, cholesterol synthesis, ketogenesis and gluconeogenesis. Biochem J. 140:111–114

McGarry JD (1979) New perspectives in the regulation of ketogenesis. Diabetes 28:517–523

McGarry JD, Foster DW (1976) Ketogenesis and its regulation. Am J Med 61:9–13

McGarry JD, Wright PH, Foster DW (1975) Hormonal control of ketogenesis. Rapid activation of hepatic ketogenic capacity in fed rats by antiinsulin serum and glucagon. J Clin Invest 55:1202–1209

McGarry JD, Takabayashi Y, Foster DW (1978) The role of malonyl-CoA in the coordination of fatty acid synthesis and oxidation in isolated rat hepatocytes. J Biol Chem 253:8294–8300

Mitchell ML, Suvunrungsi P, Sawin CT (1971) Effect of propranol on the response of serum growth hormone to glucagon. J Clin Endocrinol Metab 32:470–475

Muller P, Singh A, Orci L, Jeanrenaud B (1976) Secretory process, carbohydrate and lipid metabolism in isolated mouse hepatocytes. Aspects of regulation by glucagon and insulin. Biochim Biophys Acta 428:480–494

Norum KR (1965) Activation of palmityl-CoA: carnitine palmityltransferase in livers from fasted fat-fed or diabetic rats. Biochim Biophys Acta 98:652–654

Nosadini R, Soldà G, De Biasi F, Tiengo A (1978) Metabolic effect of glucagon in endogenous hypertriglyceridemia. Acta Diabetol Lat 15:251–258

Nosadini R, Ursini F, Tessari P, Tiengo A, Gregolin CM (1979) Perfused liver carnitine palmitoyltransferase activity and ketogenesis in streptozotocin-treated and genetic hyperinsulinemic rats. Effect of glucagon. Horm Metab Res 11:661–664

Nosadini R, Meneghel A, Del Prato S, Fedele D, Tiengo A (1980) Hypolipemic glucagon activity in isolated liver of genetic or acquired hypertriglyceridemic rats. Diabete Metab 6:129–133

Ochs RS, Harris RA (1978) Glucagon and $N^6$,$O^2$-dibutyryl adenosine 3′:5′-monophosphate inhibition of lipogenesis and phosphofructokinase activity of hepatocytes from meal-fed rats. Lipids 15:504–510

Olefsky JM, Farquhar JW, Reaven GM (1974) Reappraisal of the role of insulin in hypertriglyceridemia. Am J Med 57:551–560

Ontko JA, Johns ML (1980) Evaluation of malonyl-CoA in the regulation of long-chain fatty acid oxidation in the liver. Biochem J 192:959–962

Paloyan E, Dumbrys N, Gallagher TF Jr, Rodgers RE, Harper PV (1962) The effect of glucagon on hyperlipaemic states. Fed Proc 21:200

Raskin P, McGarry JD, Foster DW (1974) Independence of cholesterol and fatty acid biosynthesis from cyclic AMP concentration in the perfused rat liver. J Biol Chem 249:6029–6032

Samols E, Marri G, Marks V (1966) Interrelationship of glucagon, insulin and glucose. The insulinogenic effect of glucagon. Diabetes 15:855–866

Sarcione EJ, Back N, Sokal JE, Mehlman B, Knoblock E (1963) Elevation of plasma epinephrine levels produced by glucagon in vivo. Endocrinology 72:523–526

Schade DS, Eaton RP (1975a) Modulation of fatty acid metabolism by glucagon in man. I. Effects in normal subjects. Diabetes 24:502–509

Schade DS, Eaton RP (1975b) Modulation of fatty acid metabolism by glucagon in man. II. Effects in insulin-deficient diabetics. Diabetes 24:510–515

Schade DS, Eaton PR (1977a) Modulation of the catabolic activity of glucagon by endogenous insulin secretion in obese man. Acta Diabetol Lat 14:62–72

Schade DS, Eaton RP (1977b) The effects of short term physiological elevation of plasma glucagon concentration on plasma triglyceride concentration in normal and diabetic man. Horm Metab Res 9:253–257

Soler-Argilaga C, Russell R, Heimberg M (1978) Enzymatic aspects of the reduction of microsomal glycerolipid biosynthesis after perfusion of the liver with dibutyril adenosine-3′,5′-monophosphate. Arch Biochem Biophys 190:367–372

Stein O, Stein Y (1967) Lipid synthesis, intracellular transport and secretion. J Cell Biol 33:319–339

Tessari P, Meneghel A, Del Prato S, Avogaro P, Tuzzato MR, Tiengo A (1981) Glycerophosphate acyltransferase activity in perfused livers of normal and hyperlipemic rats: glucagon effect. Acta Diabet Lat 18:357–363

Tiengo A, Nosadini R, Meneghel, Tessari P, Del Prato S, Fedele D, Crepaldi G (1980) In: Andreani D, Lefèbvre P, Marks V (eds) Current views on hypoglycemia and glucagon. Academic Press, London, pp 151–161

Unger RH, Orci L (1976) Physiology and pathophysiology of glucagon. Physiol Rev 56:778–826

Watkins PA, Tarlow DM, Lane MD (1977) Mechanism for acute control of fatty acid synthesis by glucagon and 3′:5′-cyclic AMP in the liver cell. Proc Natl Acad Sci USA 74:1497–1501

Williamson JR (1967) Effects of fatty acids, glucagon and anti-insulin serum on the control of gluconeogenesis and ketogenesis in rat liver. Adv Enzyme Regul 5:229–255

Windmueller HG, Herbert PN, Levy RI (1973) Biosynthesis of lymph and plasma lipoprotein aproproteins by isolated perfused rat liver and intestine. J Lipid Res 14:215–223

Witters LA, Moriarity D, Martin DB (1979) Regulation of hepatic acetyl-CoA carboxylase by insulin and glucagon. J Biol Chem 254:6644–6649

CHAPTER 21

# Glucagon and Liver Regeneration

H. L. LEFFERT, K. S. KOCH, P. J. LAD, B. DE HEMPTINNE, and H. SKELLY

## A. Nature of Liver Regeneration

### I. Phenomenology

#### 1. Key Questions

This chapter focuses on three central questions:
1) What is the evidence that glucagon controls liver regeneration?
2) What growth control parameters does glucagon affect?
3) How does glucagon work?

Because glucagon acts together with other endogenous hepatoproliferogens, and because the proliferative response is complex, a brief review is given to put these questions into their proper perspective.

#### 2. Background

Liver regeneration is a compensatory response to liver ablation or injury (SLATER 1978). It consists of an explosive but limited proliferation of hepatocytes and non-parenchymal cells. For example, using adult rats [1] as experimental models (BUCHER and MALT 1971), the near-maximal stimulus of 70% hepatectomy converts the $G_0$ remnant, wherein $\simeq 5/10{,}000$ and $5/100{,}000$ hepatocytes synthesize DNA semiconservatively ($S$-phase) and divide ($M$-phase), respectively, – into a tissue whose labeled nuclear and labeled mitotic indices (L.I. and L.M.I.) are 600-fold higher within 24 h (Table 1 defines these and other "growth control" terms). In other words, regeneration is so fast that a normal intact liver would take almost 600 days to produce the numbers of new hepatocytes formed by regenerating liver within 24–48 h.

Liver cells near portal tracts proliferate sooner than midzonal cells. The latter, in turn, proliferate before centrilobular hepatocytes (GRISHAM 1962). These observations, together with the results of cross-circulation (SAKAI 1970), heterotopic autograft (LEONG et al. 1964), and portal blood flow reversal studies (SIGEL et al. 1968) were for many years the basis of the controversial notion that blood-borne factors, especially those of portal origin, control liver regeneration (FISHER et al. 1971). During the last decade, unequivocal evidence has accumulated to support this concept (for review, see LEFFERT et al. 1979). One of the controlling factors is glucagon (molecular weight 3,600 daltons), which originates from and is secreted by pancreatic islet A-cells. Glucagon's known and postulated role as a hepatoproliferogen is the subject of this chapter.

1 Unless noted otherwise, all experimental work discussed in this chapter refers to studies with rats and rat hepatocytes

**Table 1.** Growth control terminology

| Term | Definition [a] | Method of measurement [b] |
|---|---|---|
| $G$ | A temporal gap in the "cell cycle" | – |
| $G_0$ | Quiescent time interval, when cellular progression toward $S$ is at a low or minimal value | Radioactive dT labeling [c]; kinetic studies [c] |
| $G_1$ | Time interval between $M$ (or $G_0$) and $S$; its duration varies, depending on cell type and environmental conditions | Radioactive dT labeling; kinetic studies; DNA dye binding |
| $S$ | Phase of semiconservative nuclear DNA replication, when genetic material is duplicated, that follows $G_1$ and precedes $M$ | Radioactive dT labeling; kinetic studies; DNA dye binding |
| $G_2$ | Time interval between $S$ and $M$ | Radioactive dT labeling; kinetic studies; DNA dye binding |
| $M$ | Phase when chromosomes form and segregate into daughter cells (mitosis); it follows $G_2$ and precedes $G_0$ or $G_1$ | Cytochemical staining |
| L.I. | Nuclear labeling index: the fraction of (mononucleated) cells whose nuclei contain radioactive thymidine ([$^3$H]dT) because they are in $S$ | Radioactive dT pulse labeling; radioautography |
| M.I. | Mitotic index: the fraction of cells in $M$ | Cytochemical staining |
| L.M.I. | Labeled mitotic index: the fraction of cells in $M$ that previously passed through $S$ | Radioactive dT pulse labeling; radioautography; cytochemical staining |
| $S_t$ | Duration of time between the application of a known proliferogenic stimulus and a detectable L.I. change [d] | Radioactive dT pulse labeling; radioautography; kinetic studies |
| $S_\Delta$ | Fraction of cells entering $S$-phase per unit time [d] | Radioactive dT pulse labeling; radioautography; kinetic studies |
| $M_t$ | Duration of time between the application of a known proliferogenic stimulus and a detectable L.M.I. change [d] | Radioactive dT pulse labeling; radioautography; cytochemical staining; kinetic studies |
| $M_\Delta$ | Fraction of cells entering $M$-phase per unit time [d] | Radioactive dT pulse labeling; radioautography; cytochemical staining; kinetic studies |

[a] See LEFFERT et al. (1982)
[b] See KOCH and LEFFERT (1979, 1980)
[c] Both methods are indirect (no specific $G_0$ "markers" are known currently)
[d] See Fig. 1 and KOCH and LEFFERT (1980) for more details

## 3. Kinetic Aspects

To understand the relation between glucagon and hepatic proliferation, some fundamental kinetic properties of regenerative responses following conventional stimuli like 70% hepatectomy will be outlined. The interested reader should consult the excellent reviews of this subject by BUCHER and MALT (1971), and BECKER (1973).

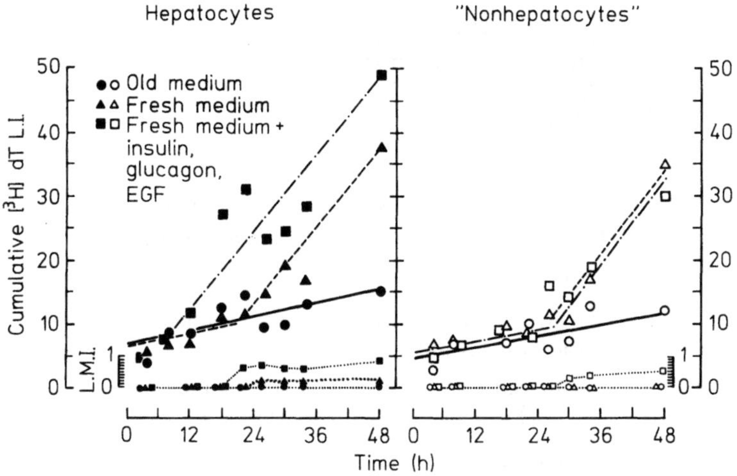


Growth transition constants

| | Hepatocytes | | "Nonhepatocytes" | |
|---|---|---|---|---|
| $S_t$ (h) | ●—● | N.D. | ○—○ | N.D. |
| | ▲—▲ | 21 | △—△ | 27 |
| | ■—■ | 11 | □—□ | 27 |
| $S_\Delta$ (h$^{-1}$) | ●—● | 0.2 | ○—○ | 0.2 |
| | ▲—▲ | 1.0 | △—△ | 1.1 |
| | ■—■ | 1.0 | □—□ | 1.1 |
| $M_t$ (h) | ●—● | N.D. | ○—○ | N.D. |
| | ▲—▲ | 26 | △—△ | N.D. |
| | ■—■ | 22 | □—□ | 30 |
| $M_\Delta$ (h$^{-1}$) | ●—● | N.D. | ○—○ | N.D. |
| | ▲—▲ | 0.25 | △—△ | N.D. |
| | ■—■ | 0.7 | □—□ | 0.4 |

**Fig. 1.** Kinetics of liver cell proliferation in primary culture under chemically defined conditions; 12-day-old stationary phase cultures were used. Exact details are given elsewhere (KOCH and LEFFERT 1980). Peptides were added *(squares)* at 0 h at final concentrations of 50 ng/ml. Terminology is defined further in Table 1 and in the text. *N.D.*, not detectable

Although the proliferative response is not liver cell specific, it is time ordered with respect to cell type. However, our discussion will emphasize, events mainly concerning hepatocytes.

Following partial hepatectomy, a time delay occurs before nuclear DNA synthesis rates change. The duration of this delay ($S_t$; often called the prereplicative phase) is about 12–18 h. Hepatocytes then enter $S$-phase with rates ($S_\Delta$) that, unlike $S_t$, are proportional to stimulus dose (% hepatectomy). $S$-phase is followed by another temporal gap ($G_2$) before mitosis begins. For a single cell, the durations of $S$, $G_2$, and $M$ are fairly constant, about 8, 5, and 0.5 h, respectively (GRISHAM 1962; FABRIKANT 1968; RABES 1978). Thus, $S_\Delta$ reflects d[L.I.]/d$t$ (in other words, the frequency of $S$-phase entry) whereas $S$ reflects the temporal duration of nuclear DNA synthesis per cell (the average DNA replication rate).

If a newly formed hepatocyte proliferates again following its first poststimulus mitosis, it is said to enter $G_1$. A small fraction of regenerating hepatocytes undergoes second and third rounds of growth during normal regeneration (FABRIKANT

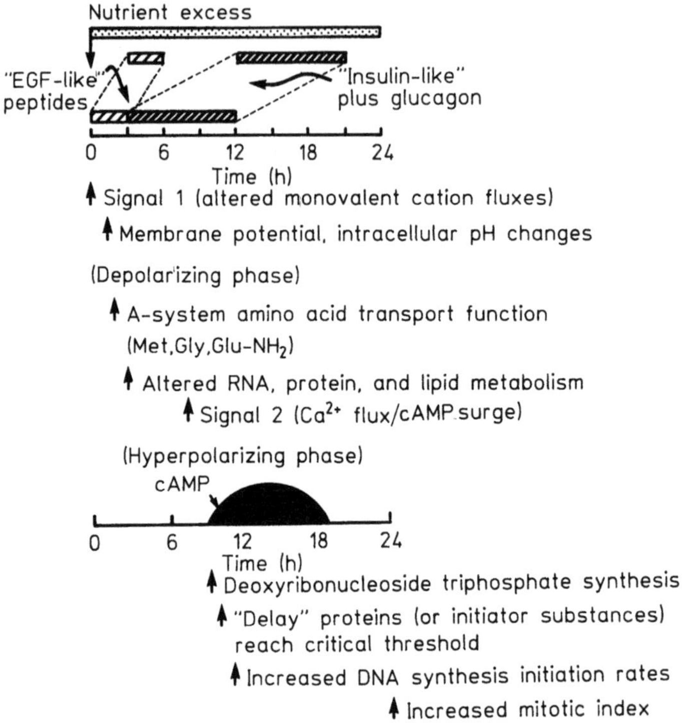


**Fig. 2.** A schematic regeneration control model. *Vertical arrows* indicate approximate event times relative to mitogen exposure. Further details are given in the text and elsewhere (KOCH and LEFFERT 1980; LEFFERT and KOCH 1980, 1981). The exact durations of the depolarizing and hyperpolarizing phases are unknown; they might span the ranges 0–3 h and 3–24 h, respectively

1968; RABES 1978). The $G_1$ duration is less than $S_t$, presumably because growing cells are less limited by one or more intracellular factors that exist in subthreshold amounts in $G_0$ cells (see Sect. A.II.3).

Proliferative landmarks are measured by pulse labeling with standard DNA precursors (e.g., tritiated thymidine, [$^3$H]dT) or by DNA-specific dye binding (e.g., mithramycin, CRISSMAN 1975), and cytochemical techniques (e.g., Feulgen and hematoxylin–eosin staining, DE ROBERTIS et al. 1963). Using in vitro primary "monolayer" liver cell culture systems, in vivo kinetic changes can be simulated (KOCH and LEFFERT 1979, 1980) and direct measurements of operational cell "cycle" parameters made easily (e.g., $M_t$ and $M_\Delta$; see Fig. 1).

## II. The Endocrine Hypothesis

### 1. Concerted Control by Peptide Hormones

Figure 2 shows a schematic regeneration control model developed to account for the results of our experiments and those of other investigators (LEFFERT et al. 1979; KOCH and LEFFERT 1980; LEFFERT and KOCH 1980). The model's principal elements are that a repertoire of peptide hormones acts concertedly in a time-ordered fashion to stimulate $G_{0,1} \rightarrow S \rightarrow M$ transitions (LEFFERT 1974). Accordingly, "EGF-like" peptides (particularly epidermal growth factor, EGF; CARPENTER and COHEN 1979; ZAPF et al. 1978) plus authentic insulin, are necessary to permit significant glucagon-potentiated hepatoproliferogenesis. Various kinetic and molecular predictions of this model have been reviewed recently (LEFFERT and KOCH 1981) and are discussed in Sect. A.II.3 and Sect. B–D.

In vitro systems provide ways in which direct evidence can be obtained to support or refute this model. For example, under chemically defined conditions, hepatocyte $G_{0,1} \rightarrow S \rightarrow M$ transitions are stimulated only by fresh culture media ("nutrient excess"; see Figs. 1 and 2) and three peptides: EGF, insulin, and glucagon, acting at nanomolar concentrations (KOCH and LEFFERT 1979). Using these growth "reinitiation" assays, it can be shown that differentiated hepatocytes respond preferentially to the presentation of EGF (first) followed by the presentation of glucagon *plus* insulin (second) (Fig. 2; see KOCH and LEFFERT 1980; LEFFERT and KOCH 1980, 1981, for experimental details). Glucagon's "delayed" and synergistic effects (with insulin) on hepatocyte $G_{0,1} \rightarrow S$ transitions in vitro are also observed under special conditions in vivo and will be reviewed in Sect. B.II, D.I, and D.II. The model predicts that many physiologic and xenobiotic substances, known and unknown, may exert proliferogenic effects by mimicking the actions of these three peptides (see Sect. C.III.1).

### 2. Interactions with Nutrients

Liver cell proliferation is stimulated by high protein diets (LEDUC 1949) or by amino acids (SHORT et al. 1972, 1973). Bioassays of $G_{0,1} \rightarrow S$ transitions with primary hepatocyte cultures show that amino acids (and still additional nutrients) exert proliferogenic effects directly, even in the absence of added hormones (see, e.g., KOCH and LEFFERT 1974, 1979; LEFFERT 1974; Fig. 1). However, maximal kinetic and quantitative responses require hormonal *plus* nutritional interactions. The "removal" of one or more proliferation inhibitors (LEFFERT and WEINSTEIN 1976; NADAL 1979) from the extracellular milieu may constitute an additional variable in such interactions, but this is not known yet with certainty in vivo or in vitro (for reviews, see LEFFERT et al. 1979; LEFFERT and KOCH 1978). Mechanisms by which peptide–amino acid interactions stimulate $G_{0,1} \rightarrow S$ transitions are described in Sect. D.III.1.

### 3. Two Sequential Rate-Limiting Events

#### a) Signal 1

Pharmacologic and $^{22}Na^{+}$ and $^{86}Rb^{+}$ flux studies in vitro and in vivo (KOCH and LEFFERT 1979, 1980; LEFFERT and KOCH 1980; HASEGAWA et al. 1980; HASEGAWA and KOGA 1979; FEHLMANN et al. 1981a), plus peptide/nutrient repletion studies in vitro (KOCH and LEFFERT 1979, 1980; LEFFERT and KOCH 1980), are consistent with the hypothesis that an early (perhaps, the initial) rate-limiting proliferogenic transition (e.g., $G_0 \rightarrow G_1$) requires a transient bidirectional stimulation of monovalent ion fluxes across the plasma membrane (KOCH and LEFFERT 1979). We call this postulated *necessary* proliferogenic event "Signal 1" (see Fig. 2) and have discussed its implications and the evidence for it elsewhere (KOCH and LEFFERT 1980; LEFFERT and KOCH 1980, 1981).

EGF, but not glucagon, seems to be involved in this activation process since EGF, but not glucagon, stimulates $^{22}Na^+$ influx in adult hepatocyte cultures (LEFFERT and KOCH 1980; FEHLMANN et al. 1981 a) with a median effective dose $ED_{50} \simeq 1$ n$M$ that is close to the $K_{D[apparent]}$ of $^{125}$I-labeled EGF binding to its cell surface membrane receptors (FEHLMANN et al. 1981 a). Current evidence suggests that many early $G_1$ changes "secondary" to increased $Na^+$ influx occur inside hepatocytes which, if sustained appropriately, lead within 12–18 h to suprathreshold concentrations of putative DNA synthesis "initiator" substances ["delay proteins", LEFFERT and KOCH 1980); see Fig. 2]. Though the conclusions are tentative, direct pharmacologic blockade studies, using the $Na^+$ diuretic, amiloride (GLITZER and STEELMAN 1966), the $Na^+/H^+$ and $K^+/H^+$ exchange ionophores, monensin and nigericin (PRESSMAN 1976), the $Na^+$, $K^+$-ATPase inhibitor, ouabain (POST et al. 1960), and actinomycin D and cycloheximide (KOCH and LEFFERT 1979), suggest collectively that such early secondary events include plasma membrane depolarization (KOCH and LEFFERT 1980), intracellular increases in pH (KOCH and LEFFERT 1980) and in rates of $Na^+$-dependent amino acid cotransport (KOCH and LEFFERT 1979, FEHLMANN et al. 1981 b), and altered RNA, protein, and lipid metabolism (KOCH and LEFFERT 1979, LEFFERT and WEINSTEIN 1976). Promotion of these early events (the sudden acceleration of what may be a set of spontaneously occurring oscillating reactions (LEFFERT and KOCH 1981) by exposure to increased levels of EGF and nutrients is insufficient, however, to produce "normal" proliferogenic kinetic transitions with respect to maximal and minimal changes in values of $S_\Delta$ and $S_t$, respectively (e.g., see Fig. 1). Normal transitions require additional activation of at least another rate-limiting process.

b) Signal 2

This second process or "Signal 2" occurs late in the prereplicative or $G_1$ phase. It may be similar to what others have termed the "restriction point" (PARDEE 1974) or A-state (BROOKS et al. 1980). From the available data, it seems that this second process involves a transient $Ca^{2+}$-dependent rise in the levels of hepatocellular 3′,5′-cyclic adenosine monophosphate (cAMP). This biochemical change is needed for the $G_1 \rightarrow S$ transition. The model in Fig. 2 postulates that glucagon plus insulin (or, their mimetics) mediate Signal 2 (see Sects. D.I, II, and III). Furthermore, both pancreatic hormones may potentiate Signal 1, by accelerating a family of spontaneously occurring and Signal 1-dependent oscillating reactions. Thus, hepatoproliferogenesis is "on schedule" and is maximized when glucagon and insulin are added simultaneously with or after (but not before) EGF, together with amino acids and other nutrients (KOCH and LEFFERT 1979; LEFFERT and KOCH 1980).

# B. Regulatory Evidence

## I. Direct

Table 2 summarizes different lines of evidence for glucagon's involvement in controlling liver regeneration. More details are presented in Sect. C.I. and II. Strictly

**Table 2.** Summary of evidence that glucagon is hepatoproliferogenic

| |
|---|
| Direct glucagon effects on hepatocyte proliferation |
| Endocrine organ ablation plus glucagon repletion |
| Glucagon infusion studies |
| Stimulus dose-dependent changes in arterial and portal venous blood glucagon levels |
| Stimulus dose-dependent changes in hepatocyte glucagon receptor or receptor-associated components |
| Stimulus dose-dependent changes in hepatic glucagon uptake and/or turnover |

**Table 3.** Correlations between glucagonemia and hepatocyte proliferation

| Physiologic state | Plasma glucagon levels[a] | Hepatocyte thymidine $^3$H-labeling index |
|---|---|---|
| Development | | |
| Postnatal | ↑↑ to ↑ | ↑↑ to ↑ |
| Adult | ↓ | ↓ |
| Old adult | ? | ↓ |
| Regeneration | | |
| Slow | ↑ | ↑ |
| Rapid | ↑↑↑ | ↑↑↑ |
| Diabetes | ↑ | ↑ |
| Nutritional deficiency | | |
| Lipotrope | ↑↑ to ↑ | ↑↑ to ↑ |
| Protein (acute) | ↑ | ↑ |
| Chronic portacaval shunt | ↑ | ↑ |
| Genetic obesity | 0 or ↓ | ↓ |
| Chemical hepatocarcinogenesis | ↑[b] | ↑[c] |
| Carbon tetrachloride toxicity | ↑↑ | ↑↑ |

[a] See LEFFERT and KOCH (1978)
[b] LEFFERT et al. (1982a)
[c] Refers to hyperplastic nodule and/or "oval" cells (SELL et al. 1981)

speaking, only one of these experimental approaches, i.e., in vitro culture systems with functional hepatocytes (LEFFERT and PAUL 1972; BISSELL et al. 1973; LEFFERT et al. 1977), provides direct evidence that glucagon acts alone and in combination with other factors, in a dose-dependent manner, directly on hepatocytes to regulate their proliferative transitions (Fig. 1). Depending on extracellular conditions, hepatocyte age, and peptide dose, glucagon's effects in such systems (and in vivo) are both stimulatory and inhibitory (LEFFERT 1974; PAUL and WALTER 1975; RICHMAN et al. 1976; LEFFERT et al. 1979; KOCH and LEFFERT 1979; PRICE et al. 1972; PRICE 1976; BUCHER and SWAFFIELD 1975; STRECKER et al. 1979). This chapter will concentrate on glucagon's stimulatory role because its inhibitory effects are currently less well analyzed in mechanistic terms (Sect. D.IV).

## II. Indirect

Six kinds of observations in vivo provide indirect evidence that glucagon controls hepatocyte regeneration (see Tables 2 and 3). First, glucagon (plus insulin) re-

**Table 4.** Rapid changes in plasma glucagon after 70% hepatectomy (adapted from LEFFERT et al. 1975)

| Plasma glucagon (pg/ml) by radioimmunoassay | Time after operation (min) | | | | |
|---|---|---|---|---|---|
| | 0 | 1 | 10 | 20 | 60 |
| After laparotomy | 191 | 257 | 205 | 253 | 281 |
| After 70% hepatectomy | 186 | 162 | 150 | 680 | 575 |

moval from circulating blood, following surgical evisceration of the pancreas and gastrointestinal tract, severely blunts the induction and maintenance of regeneration after 70% hepatectomy. However, glucagon (plus insulin) repletion early or late in the prereplicative phase in such animals restores control hepatocyte DNA synthesis rates, measured by [$^3$H]dT uptake into liver cell DNA (BUCHER and SWAFFIELD 1975). Second, intravenous glucagon infusions, combined with amino acids (SHORT et al. 1972; LEFFERT and WEINSTEIN 1976) or with EGF plus insulin (BUCHER et al. 1978), stimulate some hepatocyte DNA synthesis in a dose-dependent manner. If neutralizing monoclonal glucagon antibodies become available soon, they will be useful to substantiate further the obligatory requirement of endogenous glucagon. Third, fourth, and fifth, stimulus dose-dependent changes in plasma blood glucagon levels, hepatocyte glucagon receptor or receptor-associated components, and hepatic glucagon uptake and/or "turnover" occur, respectively.

The sixth set of observations is outlined in Table 3. It consists mainly of findings that different physiologic states reveal a glucagonemic "pattern" correlated with hepatic mitotic activity as measured by L.I. values (LEFFERT 1977; LEFFERT and KOCH 1978). Thus, if plasma glucagon is elevated or decreased (relative to control values), hepatocyte DNA synthesis is elevated or decreased coordinately. These correlations are compelling because they are independent of different "stimulus" conditions (Table 3). For example, rapid surgically induced regeneration is preceded by rapid increases in plasma glucagon (Table 4). By contrast, hepatic proliferation rates decline during development, after plasma glucagon levels begin to fall (see curves for "normal diet" in Fig. 3; SELL et al. 1974).

Five other interesting examples underscore the "endocrine pattern" concept with respect to glucagon (LEFFERT and KOCH 1978): (1) liver proliferative maturation is "reversed" by placing weaned rats onto lipotrope-deficient diets (LEFFERT 1977; LEFFERT et al. 1978), for hepatocyte L.I. and plasma glucagon values rise abnormally during the next 6 weeks instead of falling normally (see "lipotrope deficiency" curves in Fig. 3); (2) if young adult rats are shifted onto lipotrope-deficient diets of varying severity, the liver becomes more "sensitive" to the stimulus of 70% hepatectomy (as seen in Fig. 4b by a leftward shift of the hepatectomy dose–response curve) and this increased sensitivity is correlated with elevated preoperative plasma glucagon levels (LEFFERT 1977; Fig. 4a); (3) basal hepatocyte L.I. and plasma glucagon levels both rise in animals subjected to chronic portacaval shunt conditions (STARZL et al. 1976); (4) both plasma glucagon and hepatic L.I. values rise during hepatocarcinogenic ethionine (CRAVEN and DE ROBERTIS 1977) or *N*-acetyl-2-aminofluorene feeding regimens (LEFFERT et al. 1982a); and (5) hypergluca-

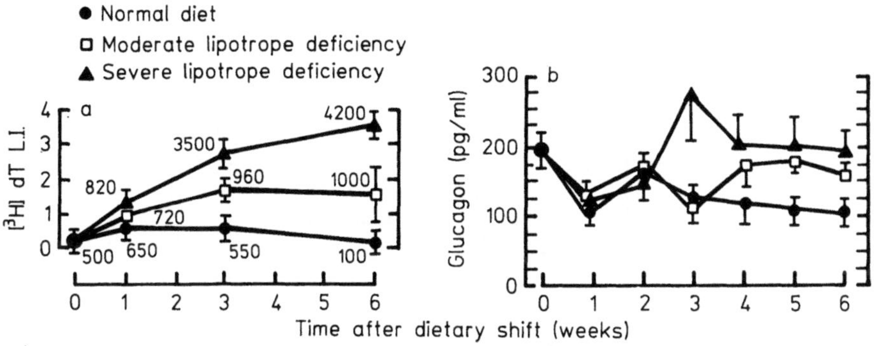


**Fig. 3a, b.** Modification of liver proliferative development and glucagonemia by lipotrope-deficient diets. Weaned immature rats were shifted onto the feeding regimens indicated. Hepatocyte L.I. and plasma immunoreactive glucagon values **b** were determined over the next 6 weeks of life until the animals reached adult body weights (≧130 g). Numbers **a** next to points on the curves refer to cpm [$^3$H]dT uptake/mg DNA in that particular tissue from which the L.I. value was determined. Further details are given in the text and elsewhere (LEFFERT 1977, LEFFERT et al. 1978)

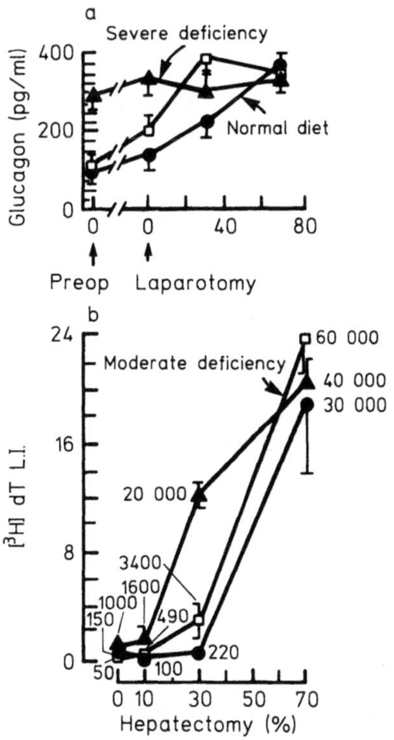


**Fig. 4a, b.** Modification of liver regenerative responses to partial hepatectomy by lipotrope-deficient diets. Young adult rats were shifted onto diets for 6 weeks (as in Fig. 3). Hepatocyte L.I. and plasma glucagon values were determined (Preop) and the rats were divided into four groups: laparotomy (0% hepatectomy), and 10, 30, and 70% hepatectomy. About 24 h postsurgery, values of hepatocyte L.I., plasma glucagon **a** and cpm [$^3$H]dT **b** uptake/mg DNA (numbers next to points in **b**) were determined again. Further details are given in the text and elsewhere (LEFFERT 1977)

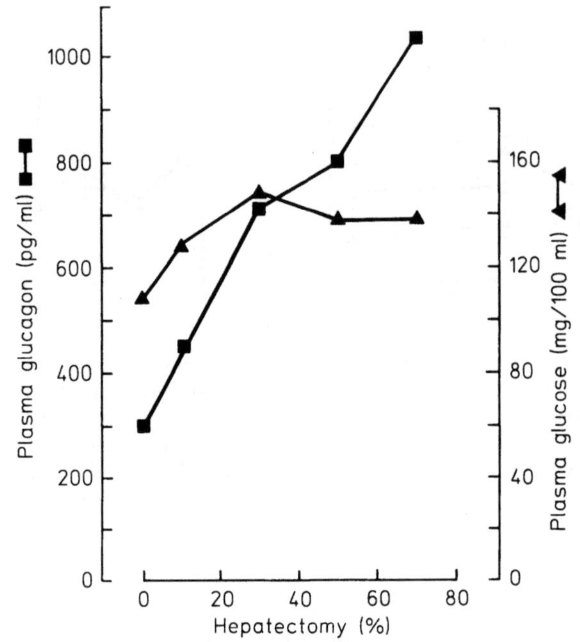


**Fig. 5.** Effects of partial hepatectomy on plasma glucagon and glucose levels. See the text for further details. Adapted from LEFFERT et al. (1975)

gonemia precedes rapid hepatic regeneration induced by carbon tetrachloride poisoning (MOURELLE and RUBALCAVA 1979, 1981) and, in certain instances, cirrhosis (SHERWIN et al. 1978).

In summary, correlations between hyperglucagonemia and increased hepatic proliferation are rather consistent (Table 3). Notably, positive correlations like these do not alone satisfy hepatoproliferogenic criteria for a given factor (LEFFERT et al. 1981) since, in other cases, peptide plasma levels of obligatory hepatoproliferogens, for example, insulin, are correlated inversely with regenerative stimuli (for a quantitative discussion of this point see LEFFERT and KOCH 1978).

# C. Physiology

## I. Extrahepatic

### 1. Bloodstream

Table 4 shows that 70% hepatectomy induces a rapid rise in plasma glucagon (LEFFERT et al. 1975). This sustained response is not seen after laparotomy, partial splenectomy, or unilateral adrenonephrectomy. These observations were confirmed by different investigators subsequently (MORLEY et al. 1975; BUCHER and WEIR 1976). Plasma glucagon levels rise to peak values 3–4 h postoperatively (LEFFERT et al. 1975), are proportional to stimulus dose (see Fig. 5), persist for at least 24 h after 70% hepatectomy, and fall gradually within 3–4 days to preoperative levels.

### 2. Responsive Tissues

Hyperglucagonemia induced by 70% hepatectomy occurs both in portal and in systemic blood. This suggests increased pancreatic islet A-cell secretion. Catecholamines, gastrointestinal peptides, or neurohumoral mechanisms might underly A-cell secretory responses to liver loss (LEFFERT et al. 1979). Hypoglycemia alone does not appear to account for hyperglucagonemia because above 30% hepatectomy (a nonstimulating "dose" if [$^3$H]dT uptake is measured at 24 h; Fig. 4), plasma glucose levels are neither decreased nor elevated significantly in comparison with controls (Fig. 5; LEFFERT et al. 1975). Thus far, no reports have determined what or how many secretagogues are involved. Additional evidence that increased pancreatic glucagon secretion facilitates liver regeneration comes from findings that chronic somatostatin infusions into 70% hepatectomized rats reduce blood glucagon levels and hepatic DNA synthesis (GOLDMAN 1978). Although physiologic levels of somatostatin do not block hepatocyte DNA synthesis directly, on the basis of validated adult hepatocyte growth reinitiation conditions in vitro (KOCH and LEFFERT 1979), in vivo results should be interpreted cautiously until studies are performed with glucagon-specific inhibitory somatostatin analogs (LEFFERT et al. 1979; BEX et al. 1981). Many endocrine functions change after 70% hepatectomy. These changes, especially those involving iodothronine receptor physiology which may be influenced by glucagon (DILLMAN et al. 1978; DILLMAN and OPPENHEIMER 1979), and their possible relevance to control of liver regeneration are discussed elsewhere (LEFFERT et al. 1979).

## II. Hepatic

### 1. Hepatocytes

a) Receptors

Liver glucagon receptors are found almost exclusively in hepatocytes, not in nonparenchymal cells (CHRISTOFFERSEN and BERG 1974; BARRAZZONE et al. 1980). After 70% hepatectomy, a gradual decline in $^{125}$I-labeled glucagon binding to liver plasma membrane receptors occurs (Fig. 6). By 24 h, decreases of 40%–50% are seen. This trend is reversible and normal binding is restored within 10 days. Scatchard analyses at 25 h poststimulus show that decreased glucagon binding results from reductions in $B_{max}$ (Fig. 7; see also LEFFERT and KOCH 1978), with minimal changes, if any, in $K_{D[apparent]}$ (Table 5 lists these parameters for experimental and control conditions). Equilibrium binding data agree well with $K_{D[apparent]}$ values determined kinetically from association and dissociation rate constants. Decreased glucagon binding is due neither to contamination of membranes with endogenous glucagon nor to proteolytic activities that degrade ligand or receptor; nor is it observed after 24–48 h starvation of unoperated rats (B. RUBALCAVA and H. L. LEFFERT 1976, unpublished work). Decreased glucagon $B_{max}$ levels also are seen following hyperglucagonemia caused by acute intoxication with carbon tetrachloride (MOURELLE and RUBALCAVA 1979, 1981), a chemical inducer of pseudoparasynchronous liver regeneration. Decreased glucagon $B_{max}$ levels are also seen in neonatal life, during active chronic hepatic proliferation prior to adult liver maturation and proliferative quiescence (BLASQUEZ et al. 1976). The functional conse-

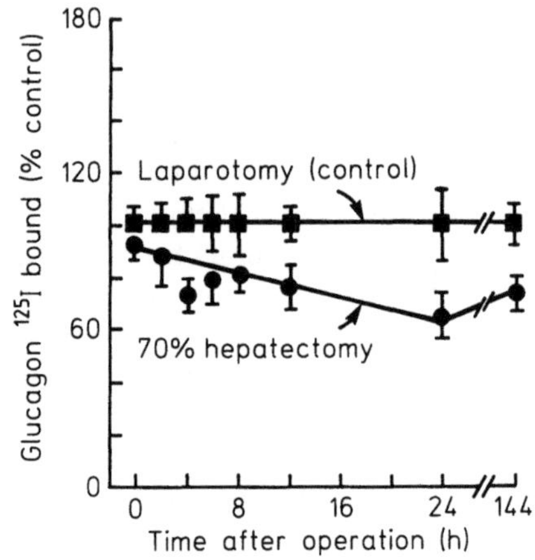


**Fig. 6.** Effects of 70% hepatectomy on the binding of an $^{125}$I-labeled glucagon tracer to liver plasma membranes. See the text for further details. Adapted from LEFFERT et al. (1979)

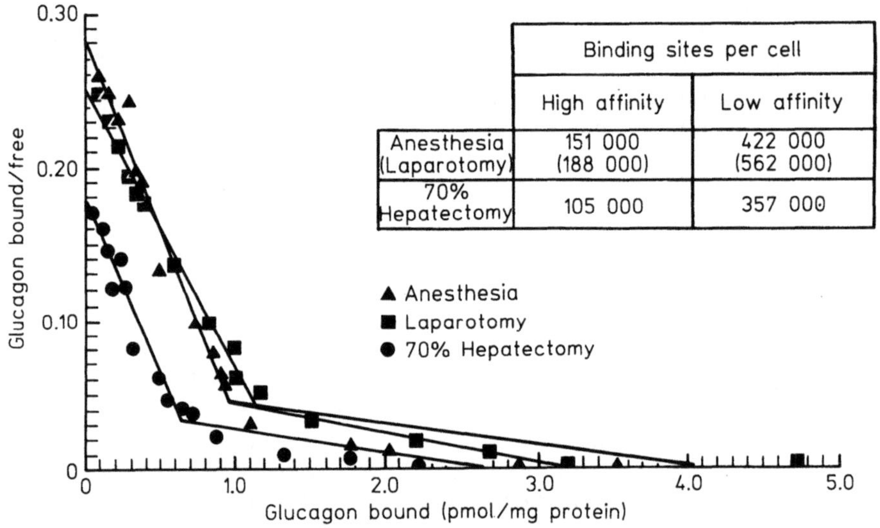


**Fig. 7.** Scatchard analyses of $^{125}$I-labeled glucagon binding to liver plasma membranes 24 h after 70% hepatectomy. See the text and Table 5. Adapted from LEFFERT and KOCH (1978) and LEFFERT et al. (1979)

quences of these changes to acute hepatocyte $G_1 \rightarrow S$ transitions, for example, decreased adenylate cyclase activation by glucagon, remains unclear (LEFFERT and KOCH 1978). Further work is needed, especially with intact hepatocytes, to establish what significance, if any, glucagon receptor "downregulation" phenomena have upon the regeneration process.

b) Uptake

Table 6 shows experimental results that suggest decreased "disappearance" of $^{125}$I-labeled glucagon from remnant liver soon after 67% hepatectomy (LEFFERT et al.

**Table 5.** Parameters of glucagon binding to liver plasma membrane fractions at 30 °C[a] (adapted from LEFFERT et al. 1979)

| Manipulation | Parameter | Class of sites | |
|---|---|---|---|
| | | High affinity – low capacity | Low affinity – high capacity |
| Anesthesia | $K_{a[apparent]}$ | $4.8 \times 10^{8}\ M^{-1}$ | $3.8 \times 10^{7}\ M^{-1}$ |
| | $K_{D[apparent]}$ | $2.1 \times 10^{-9}\ M$ | $2.6 \times 10^{-8}\ M$ |
| | $B_{max}$ | $1.1 \times 10^{-12}$ mol/mg | $2.0 \times 10^{-12}$ mol/mg |
| Laparotomy | $K_a$ | $3.9 \times 10^{8}\ M^{-1}$ | $2.7 \times 10^{7}\ M^{-1}$ |
| | $K_D$ | $2.5 \times 10^{-9}\ M$ | $3.7 \times 10^{-8}\ M$ |
| | $B_{max}$ | $1.4 \times 10^{-12}$ mol/mg | $2.7 \times 10^{-12}$ mol/mg |
| 70% Hepatectomy | $K_a$ | $4.3 \times 10^{8}\ M^{-1}$ | $3.1 \times 10^{7}\ M^{-1}$ |
| | $K_D$ | $2.3 \times 10^{-9}\ M$ | $3.2 \times 10^{-8}\ M$ |
| | $B_{max}$ | $0.7 \times 10^{-12}$ mol/mg | $1.8 \times 10^{-12}$ mol/mg |

[a] Data were calculated from Scatchard plots shown in Fig. 7

**Table 6.** Hepatic disappearance of $^{125}$I-labeled glucagon after 67% hepatectomy

| $t_{1/2}$ (min)[a] | |
|---|---|
| Laparotomy | 67% Hepatectomy |
| $3.7 \pm 0.7$ | $6.4 \pm 0.9$ |

[a] Data were calculated from semilogarithmic linear curves, fit ($r > 0.95$) to at least three time points, according to the relation $t_{1/2} = \ln 2/k$, where $k$ is assumed to be a first-order rate constant determined graphically. Posthepatectomy time intervals for measurements were 20–40 min. Data points (mean ± standard error) were obtained from measurements in six rats. See LEFFERT et al. (1979) for more details

1976, 1979). Since detectable increases in apparent $K_D$ or $B_{max}$ do not occur at these very early times poststimulus, the results must be due to: (1) increased blood flow; (2) altered rates of glucagon exchange from fluid boundary layers between the Disse space and sinusoidal blood; and/or (3) decreased hepatic processing of the peptide (e.g., decreased glucagon or glucagon/receptor turnover). The first possibility is supported indirectly by blood flow studies (BLUMGART 1978; CARUANA and GAGE 1980). However, direct measurements of intact cellular glucagon are needed in these experiments. The second possibility is intriguing (e.g., if endothelial fenestrations are enlarged), but difficult to test. The third possibility is reasonable, especially in view of decreased protein turnover rates that accompany liver regeneration (SCORNIK 1972) and hepatocyte proliferation in vitro (LEFFERT et al. 1982b). Our results support the latter possibility thus far since we have been unable to obtain evidence (by radioligand, immunochemical, and biochemical studies) for net increases in intrahepatic glucagon levels in vivo (LEFFERT et al. 1976, 1979). These findings contrast with results of similar experiments using $^{125}$I-labeled EGF, which indicate net EGF uptake shortly after 70% hepatectomy (POTTER et al. 1982).

### 2. Nonparenchymal Cells

Sinusoidal and vascular endothelial cells, Kupffer's cells, and biliary and cholangiolar epithelia lack glucagon receptors and proliferate after hepatocytes. Three indirect mechanisms for an effect of glucagon on nonparenchymal cell regeneration might exist. One is that glucagon acts on extrahepatic cells and their functional responses promote nonparenchymal cell growth. Another is that glucagon, once degraded inside hepatocytes, generates proliferogenic fragments that pass into nonparenchymal cells through extracellular routes or intercellular channels, e.g., at the hepatocholangiolar junctions (STEINER and CARRUTHERS 1961). Third, positive or negative regulatory substances produced by glucagon-treated hepatocytes might control nonparenchymal cell growth (see LEFFERT et al. 1979, 1981).

## III. Specificity

### 1. Mimetics

Many studies indicate that glucagon's effects are replaced fully by cAMP or its analogs (see Sect. D.II.1). However, no peptide is known yet that replaces glucagon's proliferogenic actions on hepatocytes. In this respect, since insulin and EGF are both likely to be mitogenic for a variety of nonhepatocyte systems in vivo (CARPENTER and COHEN 1979; SCHEVING et al. 1980), glucagon can be considered to be "hepatospecific". Whether this property partly explains the proliferation of heterotopic liver autografts in 70% hepatectomized rats and whether it is critical to regeneration kinetics has not been established, for still other liver cAMP-elevating factors like norepinephrine and neuroadrenergic agents are implicated in regenerative control (see Sect. D.II.1.a; LEFFERT et al. 1979).

### 2. Nonhepatocyte Targets

Glucagon receptors are found in plasma membranes of adipocytes (LIVINGSTON et al. 1974), renal epithelium (DUCKWORTH 1978), and cardiac muscle cells (KRIESBERG and WILLIAMSON 1964). Whether glucagon's interactions with these extrahepatic receptors after 70% hepatectomy alters the regenerative response, for example, by promoting lipolysis, decreasing renal glucagon clearance, or increasing cardiac inotropy, respectively, is unknown. A report suggesting that glucagon stimulates rebound parathyroid hormone (PTH) secretion has appeared (AVIOLI et al. 1969), but further work with specific PTH immunoassays is needed to substantiate this interesting observation. Such findings might help to explain putative indirect effects of glucagon in the maintenance of plasma $Ca^{2+}$ levels, for the latter are necessary for regeneration (see Sects. D.II.1.a and D.II.2). Along more speculative lines, glucagon has been reported to stimulate liver hematopoiesis (YEOH and OLIVER 1971) and roles for leukocyte factors have been proposed in immunologic stimulus liver regeneration models (FISHER et al. 1979; MAKOWKA et al. 1980). The relevance of these intriguing relations to regeneration remains obscure.

# D. Mechanisms of Action

## I. Potentiation of Signal 2

### 1. Kinetic Evidence

From the considerations already discussed, it seems reasonable to conclude that elevated intrahepatic glucagon, acting largely if not exclusively through hepatocyte cell surface glucagon receptors, potentiates $G_1 \rightarrow S$ transitions during liver regeneration. Two specific kinds of experiments indicate that: (1) glucagon's actions are not required until late in the prereplicative phase (or $G_1$); and (2) continued exposure to glucagon of a last 9 h, together with insulin is needed. From in vitro studies using chemically defined conditions, the time interval and duration of peptide exposure to hepatocytes can be examined directly by appropriate "addition washout" studies (KOCH and LEFFERT 1979, LEFFERT and KOCH 1980). The hatched bars in Fig. 2 indicate that this 9-h exposure interval occurs in vitro between 3 and 21 h after application of the initial stimulus (fresh medium plus EGF). Under these conditions, the quantity of [$^3$H]dT incorporated into hepatocyte DNA between 12 and 24 h poststimulus (a reflection of the hepatocyte L.I.) is similar to that of cultures exposed to all three peptides from zero time (see Fig. 2; KOCH and LEFFERT 1980; LEFFERT and KOCH 1980, 1981). In the second experimental study, solutions of glucagon plus insulin can be infused into eviscerated, 70% hepatectomized rats as long as 6–8 h poststimulus without significant decreases in control liver nuclear DNA synthesis rates measured about 24 h poststimulus (BUCHER and SWAFFIELD 1975).

The in vitro studies reveal another interesting phenomenon. Hepatocyte DNA synthesis reinitiation responses are higher if the peptides are presented sequentially: first, EGF (≧0–3 h); then, glucagon plus insulin (≧3–12 h), relative to the initial zero time stimulus of fresh medium change (KOCH and LEFFERT 1979; LEFFERT and KOCH 1980). In other words, glucagon stimulates DNA synthesis better in the presence of insulin if the hepatocytes are first activated by EGF.

Thus, four broad physiochemical questions arise: (1) what are the reasons for the delayed requirements of glucagon plus insulin? (2) why must both peptides be present continuously for at least 9 h? (3) what accounts for cellular preference for the order of factor presentation? and (4) what explains glucagon's synergistic effects with insulin? At this time, none of these questions has been answered definitively. The "delay" phenomenon is subject to several interpretations. For example, proliferogenic reactions stimulated by glucagon plus insulin (G+I) could be limited by the accumulation of a Signal 1-produced substance and thus not required until some hours after Signal 1. Or, internal cellular "clocks" entrained by Signal 1, that permit late $G_1$ proliferogenic events, may not become competent to respond to the actions of (G+I) until some amount of time has elapsed.

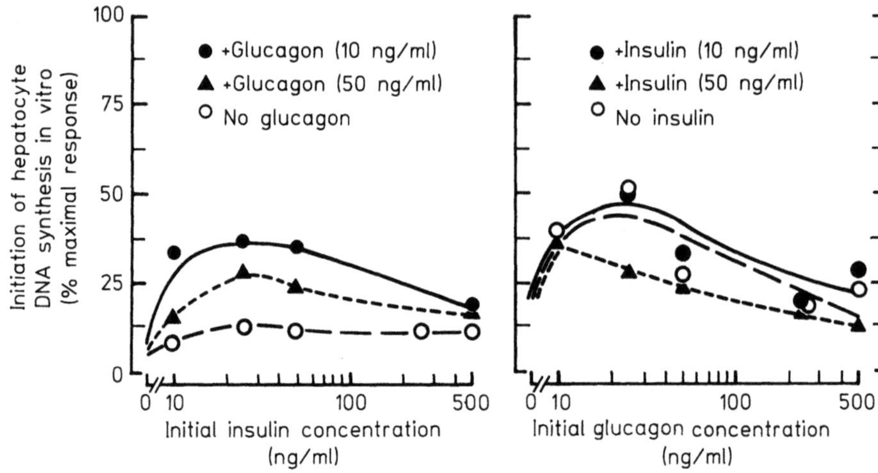


**Fig. 8.** Effects of glucagon and insulin on hepatocyte growth reinitiation in primary culture under chemically defined conditions; 11-day-old stationary phase cultures were used and percentage maximal responses calculated exactly as described elsewhere (KOCH and LEFFERT 1979). [$^3$H]dT ($3 \times 10^{-6}$ *M*; 1.25 μCi/ml [specific activity 20–40 Ci/mmol]) pulse times were 20–24 h later. Control values ± standard deviations ($N = 3$–6) at that time were: no medium change (2,890 ± 550 cpm/$10^6$ cells); fresh medium only (9,700 ± 670 cpm/$10^6$ cells); fresh medium plus 50 ng/ml each of glucagon, insulin, and EGF (38,500 ± 200 cpm/$10^6$ cells). Errors of measurement ranged ± 7% (with respect to cpm/$10^6$ cells). Radioautographic results (not shown) indicate that DNA synthesis measured under these conditions reflects mainly hepatocyte *S*-phase entry (see also Fig. 1)

Continuous (G+I) exposure probably is needed to sustain one or more rate-limiting processes that decay rapidly (see Sect. D.II and D.III). That combined (G+I) addition can be postponed long after the initial stimulus (see Fig. 2; KOCH and LEFFERT 1980; LEFFERT and KOCH 1980, 1981) is not inconsistent with this interpretation if it is assumed that: (1) (G+I) actions are rapid in onset (short delay times); and (2) that intracellular regulators or regulatory changes produced by Signal 1 decay slower and become rate limiting (e.g., fall below required threshold concentrations) ≧20–22 h poststimulus.

The order of factor presentation, EGF→(G+I), a phenomenon observed earlier by us (KOCH and LEFFERT 1979, 1980; LEFFERT and KOCH 1980) and recently by other laboratories investigating growth regulation in other cell culture systems (PHILLIPS et al. 1980; H. ARMELIN 1980, personal communication; ANDREIS and ARMATO 1981) suggests that a "priming" substance is needed to make hepatocytes more responsive to (G+I). Alternatively, this phenomenon might reflect a (G+I)-induced degradation or decreased production of a labile $G_1 \rightarrow S$ transition inhibitor. Much work is needed to test these and other plausible mechanisms that account for such observations. As to the physicochemical nature of the (G+I) synergism, little is known with respect to proliferogenesis.

### 2. Synergisms with Other Peptide Hormones

Positive or negative synergistic effects of (G+I) on $G_{0,1} \rightarrow S$ transitions occur in adult hepatocyte cultures under chemically defined conditions. For example, re-

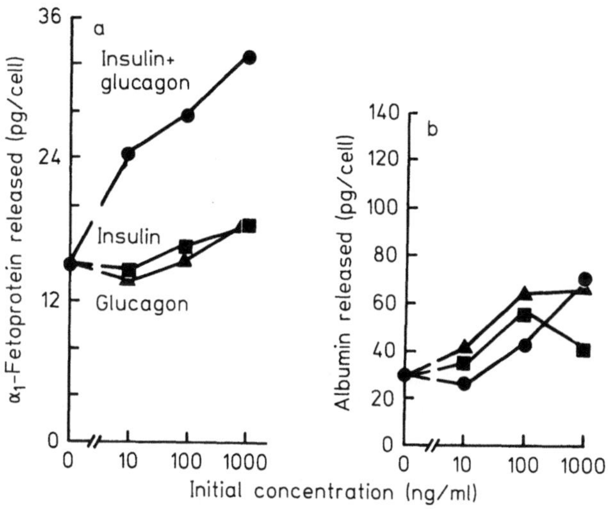


**Fig. 9 a, b.** Effects of glucagon and insulin on $\alpha_1$-fetoprotein (**a**) and albumin (**b**) production in primary fetal hepatocyte cultures. Adapted from LEFFERT et al. (1978)

sults shown in Fig. 8 indicate stimulatory and inhibitory interactions. Notably, stimulated adult hepatocyte DNA synthesis, like the fetal hepatocyte response (LEFFERT 1974; PAUL and WALTER 1975) is reduced by high glucagon concentrations ($>1\times10^{-7}$ *M*). These observations suggest that the DNA synthesis response is quite sensitive to the absolute and relative proportions of both peptides, as suggested earlier (LEFFERT 1974). However, the interpretation is complicated by the possibility that hepatocyte (and mesenchymal cell) growth state and cell density also determine the nature of peptide interactions shown in Fig. 8 (H. L. LEFFERT, T. MORAN, P. J. LAD 1978, unpublished work). Similar conclusions apply to glucagon's synergistic effects with EGF. Thus, hepatocellular proliferative L.I. and L.M.I. frequencies seem to be determined by exquisitely evolved chemical equilibria between these protein ligands, and their cellular receptors and degradation sites, but more work is needed to understand the bases of these equilibria (see Sects. D.II.1. and D.IV).

Hepatocyte DNA synthesis is not the only function that shows a positive synergistic response to (G+I), or a blunted response to excessive glucagon levels. For example, as shown in Fig. 9a, physiologic concentrations of (G+I) stimulate synergistically $\alpha_1$-fetoprotein production by cultured proliferating fetal rat hepatocytes. This response is specific since albumin production by the same hepatocytes is affected differently under identical experimental conditions (Fig. 9b). $\alpha_1$-Fetoprotein is a marker for normal and abnormal proliferating hepatocytes. Its blood levels are associated strongly with a variety of hepatic growth stimulus conditions. By contrast, albumin production is diminished in proliferating hepatocytes (SCHREIBER et al. 1971; LEFFERT et al. 1982b). It is likely, therefore, that analysis of the positive (G+I) synergy on $\alpha_1$-fetoprotein production will illuminate mechanisms by which these peptides act concertedly to promote hepatocyte proliferation (see Sects. D.II and D.III).

Recently, mitogenic conditions similar to those shown in Figs. 2 and 8 have revealed that while lactate gluconeogenesis is stimulated in culture by $10^{-9}$ *M* glucagon, as expected from in vivo studies after 70% hepatectomy (KATZ 1979), glucagon levels $\geqq 2 \times 10^{-7}$ *M* inhibit the gluconeogenic response (LAD et al. 1981). Similar glucagon receptor "desensitization" phenomena, detected by cAMP formation, are observed in short-term nonproliferating hepatocyte cultures (GURR and RUH 1980). We return to the "inhibitory" effects of glucagon on hepatocyte proliferation in Sect. D.IV.

#### 3. Nonrequirement of Glucocorticoids

Many hepatocyte functions regulated by glucagon are enhanced by appropriate concentrations of glucocorticoids, for example, glucose synthesis (PILKIS et al. 1978) or neutral amino acid transport (FREYCHET and LECAM 1978). Glucocorticoid enhancement of insulin- and purine-stimulated fetal hepatocyte DNA synthesis has also been observed directly (KOCH et al. 1976), but under different assay conditions fetal hepatocyte proliferation is inhibited (LEFFERT et al. 1978) in a fashion similar to that observed for the inhibition of neonatal liver proliferation by dexamethasone administered exogenously (CASTELLANO et al. 1978). Apparent complications of interpretation were added when it was found that cultured adult hepatocytes required hydrocortisone initially at plating for growth cycle proliferation (LEFFERT et al. 1977; see also RICHMAN et al. 1976), but not for reinitiation of proliferation after the cultures had reached stationary phase (see Fig. 1, KOCH and LEFFERT 1979). Since adult rats regenerate liver normally after gonadectomy and adrenalectomy (TERPSTRA et al. 1979), it would appear that glucocorticoids are not required for peptide-induced hepatocyte proliferation. This condition seems to apply, however, only to situations where conditions do not limit hepatocellular shape and topological relationships to nonparenchymal cells. Glucocorticoids may, therefore, stimulate hepatocyte or nonparenchymal cell production of extracellular basement-membrane-like proteins that facilitate normal cell–cell morphogenetic relationships in vitro (see Sect. E.I.2). Consequently, we conclude that there is no evidence, thus far, to indicate that Signal 2 potentiation by glucagon requires glucocorticoids. The mechanism or mechanisms by which high levels of glucocorticoids directly inhibit immature hepatocyte proliferation are as yet unknown.

## II. Molecular

### 1. cAMP-Dependent

#### a) The Late-$G_1$ cAMP "Surge"

Two sequential cAMP elevations occur during the prereplicative period after 70% hepatectomy (MACMANUS et al. 1973). However, only the latter one, observed 8–18 h poststimulus that peak at 12–14 h, is necessary for changes in $S_\Delta$. This second transient cAMP elevation is called the late $G_1$ cAMP "surge", to indicate that the rise and fall of intrahepatic cAMP levels is needed to stimulate hepatocyte DNA synthesis (WHITFIELD et al. 1980). Pharmacologic evidence for this conclusion has been presented and reviewed in detail elsewhere (WHITFIELD et al. 1980; KOCH and

LEFFERT 1980; LEFFERT and KOCH 1979). The duration of the surge presumably reflects the asynchrony of individual cAMP responses in single cells.

It seems that α- or β-adrenergic receptors (putatively hepatocytic) regulate this "surge", but the actual endogenous ligands (blood-borne and/or neurohumoral) have yet to be identified definitively. Glucagon is one of the reasonable extracellular candidates; together with insulin (KOCH and LEFFERT 1979, 1980; LEFFERT and KOCH 1980), and $Ca^{2+}$ (WHITFIELD et al. 1980), it might account for such biphasic cyclic nucleotide alterations. That this surge in vivo somewhat follows the time interval of (G+I) response seen in vitro (3–12 h) is not inconsistent with such results, if it is assumed that (G+I) potentiates an endogenous, spontaneously occurring set of events (Sect. A.II.3.b).

Attempts to explain the biochemical and biophysical basis of this "surge" are additional important avenues to understanding how (G+I) synergisms potentiate Signal 2 to "start" hepatocyte DNA replication. One might imagine interactions among at least nine cellular proteins, including the glucagon or appropriate adrenergic receptor (MACMANUS et al. 1973; WHITFIELD et al. 1980); the GTP-coupling protein (RODBELL 1980); adenylate cyclase; and particulate and soluble cAMP-dependent phosphodiesterases, protein kinases (see Sect. D.II.1.c), and phosphatases (ROSEN and KREBS 1981).

### b) Membrane Potential and $K^+$ Flux

Glucagon or cAMP cause hyperpolarizing changes across the plasma membrane of rat hepatocytes (PETERSEN 1974; SOMLYO et al. 1971). Resting membrane potential ($\Delta\Psi$) values in $G_0$ tissue of $-30$ to $-50$ mV have been reported by various investigators using electrophysiologic methods (CLARET et al. 1973; LAMBOTTE 1977). Stationary phase hepatocyte cultures show values of $-65$ mV (KOCH and LEFFERT 1980) using chemical methods of lipophilic cation uptake, e.g., $[^3H]TPP^+$ (LICHTSTEIN et al. 1979). The reasons for these discrepant values are not known presently. In culture, mitogenic conditions in the absence of peptides cause rapid depolarization within 10 min ($-65 \rightarrow -7$ mV). This depolarization is blocked significantly by amiloride, a DNA synthesis initiation and $^{22}Na^+$ influx blocker as well (KOCH and LEFFERT 1979, 1980; see also HASEGAWA et al. 1980). However, the addition of peptides (see Fig. 1) causes less depolarization ($-65 \rightarrow -32$ mV; KOCH and LEFFERT 1980) owing to the presence of glucagon. Late in the hepatocyte prereplicative phase in vivo, hyperpolarizing changes are observed, relative to sham-hepatectomized controls, by electrophysiologic methods (WONDERGEM and HARDER 1980). Thus, phases of depolarization and hyperpolarization seem to be occurring early (Signal 1) and late (Signal 2) in the prereplicative phase (see Fig. 2; KIEFER et al. 1980).

The putative glucagon-induced hyperpolarizing changes are consistent with the sustained hyperglucagonemic changes already shown (Table 4 and Fig. 5). One mechanism by which glucagon (presumably through its second messenger, cAMP) hyperpolarizes hepatocytes may be mediated by a sustained stimulation of the membrane $Na^+$, $K^+$-ATPase. Earlier (BERG and IVERSEN 1976) and more recent studies with glucagon and ouabain-treated hepatocytes in vitro (IHLENFELDT 1981) are consistent with this hypothesis, but do not prove it. The functional importance of such changes in stimulating $G_1 \rightarrow S$ transitions may involve enhanced $K^+$ influxes, as discussed in Sect. D.III.1.

c) Phosphorylation

Recent results suggest that a "surge" of cAMP-dependent autophosphorylation of liver cell membrane proteins (molecular weights 35,000, 48,000, and 100,000 daltons) occurs after 70% hepatectomy (MACMANUS and WHITFIELD 1981). Given glucagon's well-known role in the phosphorylation (and altered functioning) of many hepatocyte proteins, it is reasonable to think that glucagon plays a role in these autophosphorylations too. However, no direct evidence indicates yet that glucagon (plus insulin) mediates these events, nor that (G+I)-integrated actions of the protein repertoire previously mentioned are involved (see Sect. D.II.1.a; LEFFERT 1980). It is important to determine if glucagon mediation is involved here, because the phosphorylation "surge" parallels the onset of both the DNA synthesis initiation-dependent cAMP surge and the hyperpolarizing phase induced by 70% hepatectomy.

#### 2. cAMP-Independent

Thus far, most work linking glucagon to control of regeneration indicates that the hormone's effects result mainly from cAMP production; specifically, an insulin and $Ca^{2+}$-dependent late $G_1$ production and degradation of the cyclic nucleotide. But, glucagon might act proliferogenically via non-cAMP-mediated mechanisms, for example, by directly modulating $Ca^{2+}$ flux (KELLEY et al. 1980) or through another second messenger somewhat analogous to that proposed for insulin recently (SEALS and CZECH 1980). However, no direct evidence is available to support either of these possibilities thus far.

### III. Functional Linkages

#### 1. $Na^+$ Gradient-Dependent Amino Acid Cotransport

Many observations indicate that hepatocyte DNA synthesis stimulated in vivo by 70% hepatectomy or in vitro by glucagon, insulin, and EGF, is preceded by activation of $Na^+$ gradient-dependent amino acid cotransport. Recent in vitro studies suggest that these transport processes (e.g., methionine and glutamine) are partly necessary for the stimulation of $G_{0,1} \rightarrow S$ transitions (KOCH and LEFFERT 1979). So-called A-system-mediated transports, usually studied with the nonmetabolizable analog, α-aminoisobutyric acid (AIB), are known to fluctuate with growth state in most animal cell systems: the rates are low in the $G_0$ and high in the prereplicative and $G_1$ phase (reviewed in LEFFERT and KOCH 1977). Glucagon alone stimulates A-system function when studied both in isolated hepatocyte suspensions (LECAM and FREYCHET 1976) and under chemically defined conditions using growth reinitiation assays (Fig. 1; see FEHLMANN et al. 1981 b). A $Na^+$ gradient is necessary for this stimulation (LECAM and FREYCHET 1977). Both insulin and EGF potentiate glucagon's effects in monolayer cultures under growth reinitiation conditions (KOCH and LEFFERT 1979; FEHLMANN et al. 1981 b).

The mechanism by which glucagon stimulates [$^{14}$C] AIB transport during the prereplicative phase seems complex and is poorly understood. Some possible generalizations can be made, using present knowledge gained from other model system studies (LEVER and KABACK 1979). Glucagon stimulates the formation of a high "affinity" AIB transport system, determined by kinetic studies (FEHLMANN et al. 1981 b; LECAM et al. 1979). The formation of the amino acid transport component (or components) requires protein synthesis and is time and hormone dose dependent (FEHLMANN et al. 1981 b; LECAM et al. 1979). Removal of glucagon from contact with hepatocytes causes the system to decay (PARIZA et al. 1976). Whether an increase in A-system transport components ($V_{\max\,[\text{apparent}]}$) also occurs late in the prereplicative phase is not yet known. Since A-system transports are driven by energy derived from both the $Na^+$ gradient across the cell surface ($[Na^+]_o > [Na^+]_i$; in turn, one consequence of operating modes of the $Na^+$, $K^+$-ATPase) plus the magnitude of the $\Delta\Psi$ (interior negative), a component of which may be determined electrogenically as well, it is reasonable to suspect that a late $G_1$ hyperpolarizing phase contributes to energizing this transport further. Direct evidence is needed to substantiate this point.

An additional point is worth noting. By speeding ouabain-sensitive $^{86}Rb^+$ (a $K^+$ marker) influx because of faster "pumping" (assuming no change in the exchange ratio of $Na^+$:$K^+$ = 3:2), the hyperpolarized state may also be accompanied by increased intracellular $K^+$ concentrations (compartmented or free). Such measurements have not yet been made accurately in proliferation-competent hepatocyte systems in vitro, but could be quite revealing, for recent work in other model systems suggests that $[K^+]_i$ levels might regulate protein chain elongation rates (CAHN and LUBIN 1978).

In summary, some of glucagon's late $G_1$ cAMP-mediated effects may stimulate protein synthesis by a variety of cell surface and intracellular mechanisms (also dependent on insulin) that are testable experimentally with available technology. The apparent causal relationships between the cAMP surge and late $G_1$ $K^+$ influxes, increased $\Delta\Psi$, $Na^+$-dependent amino acid transport, and membrane protein autophosphorylation need to be explored rigorously. If causal relationships are found, they must then be related to changes in $S_\Delta$ and/or $S_t$ at the biochemical level. There is no a priori reason to believe that these functions are uniquely critical to proliferogenesis. In fact, it is likely that additional changes are necessary to stimulate regeneration. Two of these are discussed in Sects. D.III.2 and D.III.3.

## 2. Altered Hepatocyte Lipid Metabolism

Though glucagon is known to affect many enzymes that control hepatic lipid metabolism (GEELEN et al. 1980) and, therefore, possibly membrane lipid composition and fluidity, there is no rigorous evidence (other than correlative) that identifies any particular change as necessary to stimulate hepatocyte $G_{0,1} \rightarrow S$ transitions. One of these correlations involves the decreased production (or secretion) of hepatic very low density lipoprotein (VLDL) after 70% hepatectomy. VLDL and lipids extracted from it (gangliosides and acidic phospholipids) inhibit proliferogenesis by acting during the prereplicative phase in a number of cell systems (LEFFERT and WEINSTEIN 1976; CHISARI 1977; ZUCKER et al. 1980). The specific component or components involved are still unknown. Glucagon infusions reduce plasma VLDL levels rapidly, as do a variety of different hepatoproliferogenic stimuli. This subject has been reviewed in detail elsewhere (LEFFERT et al. 1976, 1979; LEFFERT and KOCH 1978) and requires further work to clarify its link to the proliferogenic role of glucagon. In this regard, one of glucagon's paradoxical stimulatory effects (LEFFERT et al. 1976) might be to reduce the formation of an endogenous hepatocyte lipid growth inhibitor (LEFFERT et al. 1979; KOCH and LEFFERT 1978).

#### 3. Deoxyribonucleoside Triphosphate Production

Glucagon (plus insulin) effects late in $G_1$ may also be needed to stimulate the $Ca^{2+}$-dependent formation of intracellular deoxyribonucleosides, the precursors needed for DNA replication (see Fig. 2). This possibility is based mainly upon evidence reviewed in detail elsewhere (WHITFIELD et al. 1980). Most of this work represents studies after 70% hepatectomy. Some is direct, however, having been performed with a liver-derived epithelioid T51B cell line (BOYNTON and WHITFIELD 1979) that responds to glucagon (A. L. BOYNTON 1980, personal communication).

### IV. Inhibitory Aspects

#### 1. Hepatocyte Loci

##### a) Developmental Variables

Embryonic liver originates from the same germinal layer that forms pancreatic islet tissue, and at about the same time (LANGMAN 1963). It is not known yet, however, if endogenous glucagon exerts a proliferogenic role on hepatocyte proliferation in the fetus. Cell culture experiments indicate possible differences is responsiveness. For example, insulin-stimulated fetal hepatocyte DNA synthesis is inhibited by glucagon (LEFFERT 1974; PAUL and WALTER 1975) at levels that stimulate adult cells (KOCH and LEFFERT 1979; RICHMAN et al. 1976). Cycling neonatal cells, however, respond to very low levels of glucagon with increased rates of DNA synthesis (ANDREIS and ARMATO 1981). Further work is needed to understand these differences, artifactual or otherwise (see Sect. D.I.2).

##### b) Shifts in $ED_{50}$

Any of these differences might result from different glucagon $ED_{50}$ values. Some evidence suggests that this occurs (LAD et al. 1981). Detailed comparative glucagon dose–response studies are needed using identical stationary cell culture conditions for fetal, neonatal, and adult hepatocytes in attempts to answer this question. Such studies alone will not provide unequivocal answers, but if they are combined with additional measurements (see Sect. D.IV.1.c) useful information should emerge.

##### c) Receptor "Desensitization"

Some of the additional measurements needed involve glucagon's interactions with its receptors, preferably under in vitro conditions in which hepatocyte proliferation is observed. Work along these lines has just begun. In nonproliferating, short-term systems, preexposure to either cAMP or epinephrine reduces subsequent glucagon-inducible cAMP production. This response is time and dose dependent, and reversible (GURR and RUH 1980).

In proliferation-competent long-term cultures, glucagon's proliferogenic effects are detected only after the cultures "mature" and reach stationary phase (KOCH and LEFFERT 1979; LAD et al. 1981). In contrast to DNA synthesis, but with respect to lactate gluconeogenesis, this type of culture system responds to glucagon both in "young" and "old" lag and stationary phases, respectively. In addition, bell-shaped glucagon dose–response curves are observed only in lag phase cultures. And, after growth reinitiation in stationary phase, such cultures show a gradual leftward shift in the bell-shaped dose–response curve for stimulation of gluconeogenesis (LAD et al. 1981). Such findings suggest that hepatocyte "developmental" transitions (at least in culture) are accompanied by what appear to be both "desensitization" and "resensitization" of the glucagon response. Although glucagon receptors or receptor-associated components are implicated in these changes, direct evidence is needed to confirm this view.

#### 2. Desynchronization of Diurnal Rhythms

[$^3$H]dT incorporation into hepatocyte DNA fluctuates diurnally. This is best revealed under rigidly controlled light/dark and feeding exposure schedules (HOPKINS et al. 1973). Variations in observed DNA "synthesis rates" may reflect changes in blood flow, [$^3$H]dT transport rates, phosphorylation to dTMP, increases in specific activity of dTTP pools, or even decreased rates of tracer degradation. Alternatively, these oscillations actually may reflect the low basal rate of normal hepatocyte turnover (see Sect. A.I.2). If the latter is true, it is of interest to learn more about extracellular factors that regulate proliferative diurnal rhythms, for in view of recent findings of differential incorporation of [$^3$H]dT into a variety of gastrointestinal tissues after exogenously administered EGF, it would appear that cellular responses to mitogenic peptides depend upon circadian rhythms (SCHEVING et al. 1980; GOODNER et al. 1977). Thus, it would not be surprising to find that glucagon enhances or diminishes liver regeneration, depending on the relative time of its elevated exposure to hepatocytes. Furthermore, circadian mitotic rhythms might be controlled by endogenous interactions among fluctuating levels of peptide proliferogens.

## E. Future Goals

### I. Role of Other Factors

#### 1. High Molecular Weight Glucagon

Various immunoreactive forms of glucagon exist with molecular weights greater than 3,500 daltons (see Chaps. 7 and 11). They are situated at and released into the bloodstream from various gastrointestinal sites. Whether macromolecular glucagon is hepatoproliferogenic is unknown. Cell culture studies with these purified molecules and their antibodies, when available, should answer this question.

### 2. Intrahepatic Loci

a) Cytoskeletal and Extracellular Matrix Components

Studies of these molecules in vivo and in vitro are accumulating rapidly (for review, see KLEINMAN et al. 1981). A number of detailed reports of cytoskeletal, e.g., actin (POLLARD and WEIHING 1974), α-actinin (LAZARIDES and BURRIDGE 1975), filamin (WANG et al. 1975), myosin (SOBIESZK and BREMEL 1975), and vimentin (BENNETT et al. 1978), and extracellular basement-membrane-associated components, e.g., fibronectin (YAMADA and OLDEN 1978), laminin (TIMPL et al. 1979), collagen type IV (BORNSTEIN and TRAUB 1980), and vinculin (BURRIDGE and FERAMISCO 1980) have appeared. Morphological arrangements of fibronectin and actin filaments appear to be specific for hepatocytes and epithelial cells (MARCEAU et al. 1980). The latter two proteins, plus collagen type IV and (possibly) laminin, are produced by liver cells and may undergo topological shifts during the course of liver regeneration (CARLSSON et al. 1981). Hormonal control of the production of these proteins, and, perhaps, their deposition, is suggested, but it is unclear yet how and if glucagon is involved.

b) Cell–Cell Communication

Extracellular (diffusional) and intercellular (gap and/or tight junction) routes of liver cell communication exist (LOEWENSTEIN 1979). Both electrical coupling and the passage of molecules (molecular weight $\leq$ 1,000 daltons) between hepatocytes occurs, but whether similar hepatocyte–nonparenchymal coupling occurs is unclear. Since hepatocytes extrude cAMP in response to glucagon or adrenergic agents (BROADUS et al. 1970) it is conceivable that cAMP levels are abnormally and chronically high in the Disse space during early phases of liver regeneration. Under these conditions, nonparenchymal cells exposed to this cAMP might actually be delayed in their $G_{0,1} \rightarrow S$ transitions ($S_t$ and/or $S_\Delta$). This testable model to explain the different entry times of different liver cell populations has not yet been explored rigorously (LEFFERT et al. 1979). It would be interesting also to look more carefully into the possibility that glucagon promotes the production of other short-lived arachidonic acid-derived lipids (BOYNTON and WHITFIELD 1980), e.g., thromboxanes and/or prostacyclines (MCGIFF 1981) or, prostaglandins of the E series (LEFFERT et al. 1976), that exert local effects within or between hepatocytes and nonparenchymal cells to control proliferogenesis, positively or negatively.

## II. Additional Problems

### 1. Complexity of Animal Cell Regulatory Processes

Since many variables control hepatocyte regeneration (both outside and inside the cell), it is reasonable to continue to approach this fundamental biologic problem through as many experimental routes as are available. Working models like those already discussed (see Fig. 2) may ultimately prove too complex, too simple, or they may need revision following subsequent work. They are useful now because they focus experimental work upon what is currently the available evidence to link glucagon to the control of regeneration in a causal role. Obviously, glucagon exerts many more changes in hepatocyte function and structure than have been discussed here, e.g., histone phosphorylation. But in cases like these, no evidence has yet been obtained that histone phosphorylation, a marvel of complexity by itself (ROSEN and KREBS 1981; LANGAN 1976), is in fact necessary for the initiation of hepatocyte DNA synthesis.

### 2. Limitations of Current Experimental Models

Regarding these statements, investigators of regulatory problems in biology know how difficult it is to establish cause and effect. Besides correlations, there exist only two main tools to do this: the use of "specific" drugs or mutational analysis. While pharmacologic experiments can be quite reveasling, it is an old adage that as one learns more about the mechanism or mechanisms of drug action the less "specific" the drug becomes (though this need not be a general rule!). Mutational analysis, the approach established by phage and prokaryotic biologists (DELBRÜCK and BAILEY 1946), is somewhat limited presently in animal cell systems, especially primary epithelial cell culture, except for those models involving certain transforming viruses inserted into eukaryotic genomes. This latter approach may prove helpful regarding recent success in the transformation by temperature-sensitive SV40 viral mutants of normal fetal hepatocytes. These cells retain some adult and fetal hepatocyte-specific differentiated function (CHOU and SCHLEGEL-HAUTER 1981). Rodent animal mutants also exist, but few workers have yet managed to combine in vivo and in vitro approaches with these models, promising as they are (LEFFERT and WEINSTEIN 1976).

## III. Reasons for Optimism

Much has been learned in the last decade about the requirement and the role of glucagon in controlling liver regeneration. Despite complications of interpretation of results and the wide variety of seemingly unconnected observations, we are still at an early stage in our attempts to understand regulation of such complicated processes. Powerful investigational tools are being developed, including monoclonal antibodies, cell culture models and molecular biologic technology, and more workers are entering this field rapidly. Even if now the "light at the end of the tunnel seems dim" (M. DELBRÜCK 1980, personal communication), the next decade should brighten it considerably, considering the progress made during the last ten years.

# F. Summary

The native low molecular weight form of glucagon participates in the control of liver regeneration. Glucagon does not act alone. It requires additional collaboration with insulin and possibly EGF, or factors like EGF. Hyperglucagonemia appears to potentiate cAMP-mediated, sodium ion-dependent amino acid transport, specific protein synthesis, and the $Ca^{2+}$-dependent formation of deoxyribonucleoside triphosphates: three known events which are necessary to stimulate hepatocyte proliferation. Causal relationships between other late $G_1$ glucagon-mediated changes (inside and outside hepatocytes) are suggested, for example, membrane $Na^+$, $K^+$-ATPase-stimulated $K^+$ influx and plasma membrane hyperpolarization, but not proven definitively. Combinations of new technologies, in cell culture, immunology, and molecular biology, and greater use of animal mutants should provide many possibilities for further defined experimental analysis of causal relationships between glucagon's actions and liver regeneration.

## References

Andreis PG, Armato V (1981) Effects of epidermal growth factor/urogastrone and associated pancreatic hormones on mitotic cycle phases and proliferation kinetics of neonatal rat hepatocytes in primary culture. Endocrinology 108:1954–1964

Avioli LV, Birge SJ, Scott S, Shieber W (1969) Role of the thyroid gland during glucagon-induced hypocalcemia in the dog. Am J Physiol 216:939–945

Barrazzone P, Gorden P, Carpentier JL, Orci L, Freychet P, Canivet B (1980) Binding, internalization and lysozomal association of $^{125}$I-glucagon in isolated rat hepatocytes. J Clin Invest 66:1081–1093

Becker FF (1973) Humoral aspects of liver regeneration. In: LoBue J, Gordon AS (eds) Humoral control of growth and differentiation, vol I. Academic, New York London, pp 249–256

Bennett GS, Fellini SA, Croop JM, Otto JJ, Bryan J, Holtzer H (1978) Differences among 100-Å filament subunits from different cell types. Proc Natl Acad Sci USA 25:4364–4368

Berg T, Iversen JG (1976) $K^+$ transport in isolated rat liver cells stimulated by glucagon and insulin in vitro. Acta Physiol Scand 97:202–208

Bex FJ, Corbin A, Sarantakis D, Lien E (1981) Absence of luteinizing hormone-releasing and anti-fertility properties in a glucagon-selective somatostatin analogue. Nature 291:672–673

Bissell DM, Hammaker LE, Meyer VA (1973) Parenchymal cells from adult rat liver in non-proliferating monolayer culture. I. Functional studies. J Cell Biol 59:722–734

Blasquez E, Rubalcava B, Montesano R, Orci L, Unger RH (1976) Development of insulin and glucagon binding and the adenylate cyclase response in liver membranes of the prenatal, postnatal, and adult rat: evidence of glucagon "resistance". Endocrinology 98:1014–1023

Blumgart LH (1978) Liver atrophy, hypertrophy and regenerative hyperplasia in the rat: the relevance of blood flow. CIBA Found Symp 55:181–215

Bornstein P, Traub W (1980) Structurally distinct collagen types. Annu Rev Biochem 49:957–1003

Boynton AL, Whitfield JF (1979) The cyclic AMP-dependent initiation of DNA synthesis by T51B rat liver epithelioid cells. J Cell Physiol 101:139–148

Boynton AL, Whitfield JF (1980) Possible involvement of arachidonic acid in the initiation of DNA synthesis by rat liver cells. Exp Cell Res 129:474–478

Broadus AE, Kaminsky NI, Northcutt RC, Hardman JG, Sutherland EW, Liddle GW (1970) Effects of glucagon on adenosine 3′,5′-monophosphate and guanosine 3′,5′-monophosphate in human plasma and urine. J Clin Invest 49:2237–2245

Brooks RF, Bennett DC, Smith JA (1980) Mammalian cell cycles need two random transitions. Cell 19:493–504

Bucher NLR, Malt RA (1971) Regeneration of liver and kidney. N Engl J Med, Med Progr Ser. Little, Brown, Boston, pp 1–176

Bucher NLR, Swaffield MN (1975) Regulation of hepatic regeneration by synergistic action of insulin and glucagon. Proc Natl Acad Sci USA 72:1157–1160

Bucher NLR, Weir GC (1976) Insulin, glucagon, liver regeneration and DNA synthesis, Metabolism 25:1423–1425

Bucher NLR, Patel U, Cohen S (1978) Hormonal factors concerned with liver regeneration. CIBA Found Symp 55:95–107

Burridge K, Feramisco JR (1980) Microinjection and localization of a 130 K protein in living fibroblasts: a relationship to actin and fibronectin. Cell 19:587–595
Cahn F, Lubin M (1978) Inhibition of elongation steps of protein synthesis at reduced potassium concentrations in reticulocytes and reticulocyte lysate. J Biol Chem 253:7798–7803
Carlsson R, Engvall E, Freeman A, Ruoslahti E (1981) Laminin and fibronectin in cell adhesion: enhanced adhesion of cells from regenerating liver to laminin. Proc Natl Acad Sci USA 78:2403–2406
Carpenter G, Cohen S (1979) Epidermal growth factor. Annu Rev Biochem 48:193–216
Caruana JA, Gage AA (1980) Increased uptake of insulin and glucagon by the liver as a signal for liver regeneration. Surg Gynecol Obstet 150:390–394
Castellano TJ, Schiffman RL, Jacob MC (1978) Suppression of liver cell proliferation by glucocorticoid hormones: a comparison of normally growing and regenerating tissue in the immature rat. Endocrinology 102:1107–1112
Chisari FV (1977) Immunoregulatory properties of human plasma in very low density lipoproteins. J Immunol 119:2129–2136
Chou JY, Schlegel-Hauter SE (1981) Study of liver differentiation in vitro. J Cell Biol 89:216–222
Christoffersen T, Berg T (1974) Glucagon control of cyclic AMP accumulation in isolated intact rat liver parenchymal cells in vitro. Biochim Biophys Acta 338:408–417
Claret B, Claret M, Mazet JL (1973) Ionic transport and membrane potential of rat liver cells in normal and low chloride solutions. J Physiol (Lond) 230:87–101
Craven PA, De Robertis FR (1977) Hyperglucagonemia and altered responsiveness of hepatic adenylate cyclase-adenosine 3′,5′-monophosphate system to hormonal stimulation during chronic ingestion of DL-ethionine. Biochim Biophys Acta 497:415–427
Crissman H (1975) Cell preparation and staining for flow systems. In: Richmond CR et al. (eds) Mammalian cells: probes and problems. First Los Alamos Sci Symp, Technical Information Center, OPA. US Energy Research and Development Administration, Washington, DC, pp 94–106
Delbrück M, Bailey WT (1946) Induced mutations in bacterial viruses. Cold Spring Harbor Symp Quant Biol 11:33–37
De Robertis EDP, Nowinski WW, Saez FA (eds) (1973) General cytology. Saunders, Philadelphia London
Dillman WH, Oppenheimer JH (1979) Glucagon influences the expression of thyroid hormone action: discrepancy between nuclear triodothyronine receptor number and enzyme responses. Endocrinology 105:74–79
Dillman WH, Bonner RA, Oppenheimer JH (1978) Glucagon administration decreases hepatic nuclear triodothyronine binding capacity. Endocrinology 102:1633–1636
Duckworth WC (1978) Insulin and glucagon binding and degradation by kidney cell membranes. Endocrinology 102:1766–1774
Fabrikant J (1968) The kinetics of cellular proliferation in regenerating liver. J Cell Biol 36:551–565
Fehlmann M, Canivet B, Freychet P (1981 a) Epidermal growth factor stimulates monovalent cation transport in isolated rat hepatocytes. Biochem Biophys Res Commun 100:254–260
Fehlmann M, Samson M, Koch KS, Leffert HL, Freychet P (1981 b) Hormonal regulation of amino acid transport in isolated and cultured adult rat hepatocytes: effect of amiloride. Biochim Biophys Acta 642:88–95
Fisher B, Szuch P, Fisher ER (1971) Evaluation of a humoral factor in liver regeneration utilizing liver transplants. Cancer Res 31:322–331
Fisher B, Gebhardt MC, Saffer EA, Fisher ER (1979) Effect of *Corynebacterium parvum* in liver proliferation and regeneration. Cancer Res 39:1361–1368
Freychet P, LeCam A (1978) Amino acid transport in isolated hepatocytes: effect of glucagon. Ciba Found Symp 55:247–260
Geelen MJH, Harris RA, Beynen AC, McCune SA (1980) Short-term hormonal control of hepatic lipogenesis. Diabetes 29:1006–1022

Glitzer MS, Steelman SL (1966) N-amidino-3,5-diamino-6′-chloropyrazinecarboxamide: an active diuretic in the carboxamide series. Nature 212:191–193
Goldman AL (1978) Influence of somatostatin on hepatic regeneration in the adult rat. Surg Forum 29:402–404
Goodner CJ, Walike BC, Koerker DJ, Ensinck JW, Brown AC, Chideckel EW, Palmer J, Kalnasy L (1977) Insulin, glucagon and glucose exhibit synchronous, sustained oscillations in fasting monkeys. Science 195:177–179
Grisham JW (1962) A morphologic study of deoxyribonucleic acid synthesis and cell proliferation in regenerating rat liver; autoradiography with thymidine-$H^3$. Cancer Res 22:842–849
Gurr JA, Ruh TA (1980) Desensitization of primary cultures of adult rat liver parenchymal cells to stimulation of adenosine 3′,5′-monophosphate production by glucagon and epinephrine. Endocrinology 107:1309–1319
Hasegawa K, Koga M (1979) Relationship between the induction of DNA synthesis and changes in the concentration of intracellular $Na^+$ in the liver of intact rats. J Physiol Soc Jpn 41:282
Hasegawa K, Namai K, Koga M (1980) Induction of DNA synthesis in adult rat hepatocytes cultured in a serum-free medium. Biochem Biophys Res Commun 95:243–249
Hopkins HA, Bonney RJ, Walker PR, Yager JD Jr, Potter VR (1973) Food and light as separate entrainment signals for rat liver enzymes. Adv Enzyme Regul 11:169–191
Ihlenfeldt MJA (1981) Stimulation of $Rb^+$ transport by glucagon in isolated rat hepatocytes. J Biol Chem 256:2213–2218
Katz N (1979) Correlation between rates and enzyme levels of increased gluconeogenesis in rat liver and kidney after partial hepatectomy. Eur J Biochem 98:535–542
Kelley DS, Evanson T, Potter VR (1980) Calcium-dependent hormonal regulation of amino acid transport and cyclic AMP accumulation in rat hepatocyte monolayer cultures. Proc Natl Acad Sci USA 77:5953–5957
Kiefer H, Blume AJ, Kaback HR (1980) Membrane potential changes during mitogenic stimulation of mouse spleen lymphocytes. Proc Natl Acad Sci USA 77:2200–2204
Kleinman HK, Klebe RJ Martin GR (1981) Role of collagenous matrices in the adhesion and growth of cells, J Cell Biol 88:473–485
Koch K, Leffert HL (1974) Growth control of differentiated fetal rat hepatocytes in primary monolayer culture. VI. Studies with conditioned medium and its functional interactions with serum factors. J Cell Biol 62:780–791
Koch KS, Leffert HL (1979) Increased sodium ion influx is necessary to initiate rat hepatocyte proliferation. Cell 18:153–163
Koch KS, Leffert HL (1980) Growth regulation of adult rat hepatocytes in primary culture. Ann NY Acad Sci 349:111–127
Koch KS, Leffert HL, Moran T (1976) Hepatic proliferation control by purines, hormones and nutrients. In: Fishman W, Sell S (eds) Oncodevelopment gene expression, 3rd meeting, International Study Group for Carcinoembryonic Proteins. Academic, New York London, pp 21–33
Kreisberg RA, Williamson JR (1964) Metabolic effects of glucagon in the perfused rat heart. Am J Physiol 207:721–727
Lad PJ, Brown JW, Blasquez E, Lin M, Koch KS, Skelly H, deHemptinne B, Leffert HL (1981) Adult rat hepatocytes in primary culture. VI. Lactate gluconeogenesis: developmental changes and glucagon-responsiveness during the growth cycle. Submitted
Lambotte L (1977) Effect of anoxia and ATP depletion on the membrane potential and permeability of dog liver. J Physiol (Lond) 269:53–70
Langan TA (1976) Methods for the assessment of site-specific histone phosphorylation. Methods Cell Biol 19:127–142
Langman J (1963) Medical embryology. Williams & Wilkins, Baltimore
Lazarides E, Burridge K (1975) α-Actinin: immunofluorescent localization of a muscle structural protein in non-muscle cells. Cell 6:289–298
LeCam A, Freychet P (1976) Glucagon stimulates the A system for neutral amino acid transport in isolated hepatocytes of adult rat. Biochem Biophys Res Commun 72:893–901

LeCam A, Freychet P (1977) Neutral amino acid transport. J Biol Chem 252:148–156
LeCam A, Rey JF, Fehlmann M, Kitabgi M, Freychet P (1979) Amino acid transport in isolated hepatocytes after partial hepatectomy in the rat. Am J Physiol 236:E594–E602
Leduc EH (1949) Mitotic activity in the liver of the mouse during inanition followed by refeeding with different levels of protein. Am J Anat 84:397–430
Leffert HL (1974) Growth control of differentiated fetal rat hepatocytes in primary monolayer culture. VII. Hormonal control of DNA synthesis and its possible significance to the problem of liver regeneration. J Cell Biol 62:792–801
Leffert HL (1977) Glucagon, insulin, and their hepatic "receptors": an endocrine pattern characterizing hepatoproliferative transitions in the rat. In: Foá PP, Bajaj JS, Foá NL (eds) Glucagon, its role in physiology and clinical medicine. Springer, Berlin Heidelberg New York, pp 305–319
Leffert HL (ed) (1980) Growth regulation by ion fluxes. Ann NY Acad Sci 339:261
Leffert HL, Koch KS (1977) Control of animal cell proliferation. In: Rothblat GH, Cristofalo VJ (eds) Growth, nutrition and metabolism of cells in culture, vol 3. Academic, New York London San Francisco, pp 225–294
Leffert HL, Koch KS (1978) Proliferation of hepatocytes. CIBA Found Symp 55:61–94
Leffert HL, Koch KS (1980) Ionic events at the membrane initiate rat liver regeneration. Ann NY Acad Sci 339:201–215
Leffert HL, Koch KS (1981) Two ionic signals as prominent regulators of liver regeneration. In: Berk PD (ed) Frontiers of science and the liver. Thieme-Straton, New York
Leffert HL, Paul D (1972) Studies on primary cultures of differentiated fetal liver cells. J Cell Biol 52:559–568
Leffert HL, Weinstein DB (1976) Growth control of differentiated fetal rat hepatocytes in primary monolayer culture. IX. Specific inhibition of DNA synthesis initiation by very low density lipoprotein and possible significance to the problem of liver regeneration. J Cell Biol 70:20–32
Leffert H, Alexander MN, Faloona G, Rubalcava B (1975) Specific endocrine and hormonal receptor changes associated with liver regeneration in adult rats. Proc Natl Acad Sci USA 72:4033–4036
Leffert HL, Koch KS, Rubalcava B (1976) Present paradoxes in the environmental control of hepatic proliferation. Cancer Res 36:4250–4255
Leffert HL, Moran T, Boorstein R, Koch KS (1977) Procarcinogen activation and hormonal control of cell proliferation in differentiated primary adult rat liver cell cultures. Nature 267:58–61
Leffert HL, Koch KS, Rubalcava B, Sell S, Moran T, Boorstein R (1978) Hepatocyte growth control: in vitro approach to problems of liver regeneration and function. Natl Cancer Inst Monogr 48:87–101
Leffert HL, Koch KS, Moran T, Rubalcava B (1979) Hormonal control of rat liver regeneration. Gastroenterology 76:1470–1482
Leffert HL, Koch KS, Lad PJ, Skelly H, de Hemptinne B (1982) Hepatocyte growth factors. In: Zakim D, Boyer TD (eds) Hepatology, chap 3, pp 64–75
Leffert HL, Sell S, Skelly H, Moran T (1982a) Adult rat hepatocytes in primary culture. XIII. N-acetyl-2-aminofluorene (AAF): multicycle characterization of short-term proliferative and functional properties of cells obtained from rats fed a carcinogenic diet. In preparation
Leffert HL, Brown JW, Sela-Trepat J, Lad PJ, Sell S, Skelly H, Moran T (1982b) Adult rat hepatocytes in primary culture. IX. A developmental program for the synthesis, secretion and turnover of albumin. In preparation
Leong GF, Grisham JW, Hole BV, Albright ML (1964) Effect of partial hepatectomy on DNA synthesis and mitosis in heterotopic partial autografts of rat liver. Cancer Res 24:1495–1501
Lever JE, Kaback HR (1979) The use of membrane vesicles in transport studies. CRC Rev Biol 7:187–246
Lichtstein D, Kaback HR, Blume AJ (1979) Use of a lipophilic cation for determination of membrane potential in neuroblastoma-glioma hybrid cell suspensions. Proc Natl Acad Sci USA 76:650–654

Livingston JN, Cuatrecasas P, Lockwood DH (1974) Studies of glucagon resistance in large rat adipocytes: $^{125}$I-labeled glucagon binding and lipolytic capacity. J Lipid Res 15:26–32
Loewenstein WR (1979) Junctional intercellular communication and the control of growth. Biochim Biophys Acta 560:1–65
MacManus JP, Whitfield JF (1981) Stimulation of autophosphorylation of liver cell membrane proteins by calcium and partial hepatectomy. J Cell Physiol 106:33–40
MacManus JP, Braceland BM, Youdale T, Whitfield JF (1973) Adrenergic antagonists and a possible link between the increase in cyclic adenosine 3′:5′-monophosphate and DNA synthesis during liver regeneration. J Cell Physiol 82:157–164
Makowka L, Falk RE, Rotstein LE, Falk JA, Nossal N, Langer B, Blendis LM, Phillips MJ (1980) Cellular transplantation in the treatment of experimental hepatic failure. Science 210:901–903
Marceau N, Goyette R, Dêschenes J, Valet JP (1980) Morphological differences between epithelial and fibroblast cells in rat liver cultures and the roles of cell surface fibronectin and cytoskeletal element organization in cell shape. Ann NY Acad Sci 349:138–152
McGiff JC (1981) Prostaglandins, prostacyclin, and thromboxanes. Annu Rev Pharmacol Toxicol 21:479–509
Morley LGD, Kuku S, Rubenstein AH (1975) Serum hormone levels following partial hepatectomy in the rat. Biochem Biophys Res Commun 67:653–661
Mourelle M, Rubalcava B (1979) Changes in the insulin and glucagon receptors in the regenerating liver following intoxication with carbon tetrachloride. Biochem Biophys Res Commun 88:189–198
Mourelle M, Rubalcava B (1981) Regeneration of the liver after carbon tetrachloride. J Biol Chem 256:1656–1660
Nadal C (1979) Control of liver growth by growth inhibitors (chalones). Arch Toxicol [Suppl] 2:131–142
Pardee AB (1974) A restriction point for control of normal animal cell proliferation. Proc Natl Acad Sci USA 71:1286–1290
Pariza MW, Butcher FR, Kletzien RF, Becker JE, Potter VR (1976) Induction and decay of glucagon-induced amino acid transport in primary cultures of adult rat liver cells. Proc Natl Acad Sci USA 73:4511–4515
Paul D, Walter S (1975) Growth control in primary fetal rat liver cells in culture. J Cell Physiol 85:113–123
Petersen OH (1974) The effect of glucagon on the liver cell membrane potential. J Physiol (Lond) 239:647–656
Phillips PD, Gorman S, Cristofalo VJ (1980) Regulation of WI-38 cell growth in a serum-free medium by EGF. J Cell Biol 87: Abstr 9a
Pilkis SJ, Park CR, Claus TH (1978) Hormonal control of hepatic gluconeogenesis. Vitam Horm 36:383–460
Pollard TD, Weihing RR (1974) Actin and myosin and cell movement. CRC Crit Rev Biochem 2:1–65
Post RL, Merritt CR, Kinsolving CR, Albright CD (1960) Membrane adenosine triphosphatase as a participant in the active transport of sodium and potassium in the human erythrocyte. J Biol Chem 235:1796–1802
Potter S, Moran T, Leffert HL (1982) Selective hepatic uptake of $^{125}$I-labelled epidermal growth factor following partial hepatectomy in the adult rat. In preparation
Pressman BC (1976) Biological applications of ionophores, Annu Rev Biochem 45:501–530
Price JB (1976) Insulin and glucagon as modifiers of DNA synthesis in regenerating rat liver. Metabolism 25:1427–1428
Price JB, Takeshige K, Max MH, Voorhees AB (1972) Glucagon as the portal factor modifying hepatic regeneration. Surgery 72:74–82
Rabes HM (1978) Kinetics of hepatocellular proliferation as a function of the microvascular structure and functional state of the liver. CIBA Found Symp 55:31–53
Richman RA, Claus TH, Pilkis SJ, Friedman DL (1976) Hormonal stimulation of DNA synthesis in primary cultures of adult rat hepatocytes. Proc Natl Acad Sci USA 73:3589–3593

Rodbell M (1980) The role of hormone receptors and GTP-regulatory proteins in membrane transduction. Nature 284:17–22
Rosen OM, Krebs EG (eds) (1981) Protein phosphorylation. Cold Spring Harbor Conf Cell Proliferation 8
Sakai A (1970) Humoral factor triggering DNA synthesis after partial hepatectomy in the rat. Nature 228:1186
Scheving LA, Yeh YC, Tsai TH, Scheving LE (1980) Circadian phase-dependent stimulatory effects of epidermal growth factor on deoxyribonucleic acid synthesis in the duodenum, jejunum, ileum, cecum, colon, and rectum of the adult male mouse. Endocrinology 106:1498–1503
Schreiber G, Urban J, Zähringer J, Reutter W, Frosh U (1971) The secretion of serum protein and the synthesis of albumin and total protein in regenerating liver. J Biol Chem 246:4531–4538
Scornik OA (1972) Decreased in vivo disappearance of labelled liver protein after partial hepatectomy. Biochem Biophys Res Commun 47:1063–1066
Seals JR, Czech MP (1980) Evidence that insulin activates an intrinsic plasma membrane protease in generating a secondary chemical mediator. J Biol Chem 255:6529–6531
Sell S, Nichols M, Becker FF, Leffert HL (1974) Hepatocyte proliferation and $\alpha_1$-fetoprotein in pregnant, neonatal, and partially hepatectomized rats. Cancer Res 34:865–871
Sell S, Osborn K, Leffert HL (1981) Autoradiography of oval cells appearing rapidly in the livers of rats fed N-2-fluorenylactamide in a choline-devoid diet. Carcinogenesis 2:7–14
Sherwin RS, Fisher M, Bessof J, Snyder N, Hendler R, Conn HO, Felig P (1978) Hyperglucagonemia in cirrhosis: altered secretion and sensitivity to glucagon. Gastroenterology 74:1224–1228
Short J, Brown RF, Husakova A, Gilbertson JR, Zemel R, Lieberman I (1972) Induction of DNA synthesis in the liver of the intact animal. J Biol Chem 247:1757–1766
Short J, Armstrong NB, Zemel R, Lieberman I (1973) A role for amino acids in the induction of deoxyribonucleic acid synthesis in liver. Biochem Biophys Res Commun 50:430–437
Sigel B, Baldia LB Brightman SA, Dunn MR, Price RIM (1968) Effect of blood flow reversal in liver autotransplants upon the site of hepatocyte regeneration. J Clin Invest 46:1231–1237
Slater TF (ed) (1978) Biochemical mechanisms of liver injury. Academic, London New York
Sobieszk A, Bremel RD (1975) Preparation and properties of vertebrate smooth-muscle myofibrils and actomyosin. Eur J Biochem 55:49–60
Somlyo AP, Somlyo AV, Friedmann N (1971) Cyclic adenosine monophosphate, cyclic guanosine monophosphate and glucagon: effects on membrane potential and ion fluxes in the liver. Ann NY Acad Sci 185:108–114
Starzl TE, Porter KA, Watanabe K, Putnam CW (1976) Effects of insulin, glucagon, and insulin/glucagon infusions on liver morphology and cell division after complete portacaval shunt in dogs. Lancet 1:821–825
Steiner JW, Carruthers JS (1961) Studies of the fine structure of the terminal branches of the biliary tree. I. The morphology of normal bile canaliculi, bile preductules (Ducts of Hering) and bile ductules, Am J Pathol 6:639–661
Strecker W, Goldberg M, Feeny DA, Ruhenstroth-Bauer G (1976) The influence of extended glucagon infusion on liver cell regeneration after partial hepatectomy in the rat. Acta Hepatogastroenterol (Stuttg) 26:439–441
Terpstra OT, Malt RA, Bucher NLR (1979) Negligible role of adrenal hormones in regulation of DNA synthesis in liver of partially hepatectomized rats. Proc Soc Exp Biol Med 161:326–331
Timpl R, Rohde H, Robey PG, Renard SI, Foidart JM, Martin GR (1979) Laminin – a glycoprotein from basement membranes. J Biol Chem 254:9933–9937
Wang K, Ash JF, Singer SJ (1975) Filamin, a new high molecular weight protein found in smooth muscle and non-muscle cells. Proc Natl Acad Sci USA 72:4483–3386
Whitfield JF, Boynton AL, MacManus JP, Rixon RH, Sikorska M, Tsang B, Walker RR (1980) The roles of calcium and cyclic AMP in cell proliferation. Ann NY Acad Sci 339:216–240

Wondergem R, Harder DR (1980) Membrane potential measurements during rat liver regeneration. J Cell Physiol 102:193–197
Yamada KM, Olden K (1978) Fibronectins-adhesive glycoproteins of cell surface and blood. Nature 275:179–184
Yeoh G, Oliver IT (1971) A stimulatory effect of glucagon on DNA synthesis in neonatal rat liver. Comp Biochem Physiol [A] 39:723–733
Zapf J, Rinderknecht E, Humbel RE (1978) Nonsuppressible insulin-like activity (NSILA) from human serum: recent accomplishments and their physiologic implication. Metabolism 27:1803–1827
Zucker S, Lysik RM, Chikkappa G, Glucksman MJ, Gomez-Reino J, Distefano JF (1980) Very low density lipoprotein hematopoiesis inhibitor from rat plasma. J Cell Sci 8:895–905

CHAPTER 22

# Glucagon and Insulin Secretion

E. SAMOLS

## A. Introduction and Historical Overview

There have been suggestions, in view of the striking anatomic juxtaposition of the A- and B-cells (HAIST 1965; EPPLE 1968; ORCI et al. 1975) that these cells might influence the secretion of one another. The studies of SAMOLS, MARKS and colleagues, based on the demonstrations that glucagon stimulated insulin secretion (SAMOLS et al. 1965a, 1966a) and that insulin suppressed glucagon secretion (SAMOLS et al. 1970, 1972; SAMOLS and HARRISON 1976) and of the conditions favoring or inhibiting these actions (SAMOLS et al. 1965b, 1966b, 1969a, b; SAMOLS and MARKS 1967; PORTE et al. 1966) led to the construction of their hypothesis of positive-negative insulin-glucagon feedback within the islets (SAMOLS et al. 1970, 1972). Not surprisingly, in view of its multiple separations from orthodox concepts, and because of its implications, this hypothesis has been controversial. On the one hand, this hypothesis and its subsequent development continues to be contentious (more so in terms of the insulinotropic effect of glucagon than of the glucagonosuppressive effect of insulin). On the other hand, the discovery of pancreatic somatostatin renewed interest in the concept of paracrine insular control mechanisms (UNGER and ORCI 1977). Consideration of the possible paracrine interrelationships of somatostatin, insulin, and glucagon, is presented in Chap. 31. The present chapter is primarily concerned with the interrelationships between glucagon and insulin, in particular with regard to the hypothesis of SAMOLS and MARKS, and to its further development in terms of more recent information and concepts.

Is there a physiologic relationship between glucagon and insulin, as is implied by the title of the chapter? One view is that there is none. According to orthodox concepts, the blood glucose concentration controls insulin secretion (upper limit setting), and blood glucose separately controls glucagon secretion (lower limit setting). Glucagon and insulin are considered to be antipodal antagonistic hormones, independent of one another. If glucagon is regarded as a hormone of glucose need, to provide glucose for the noninsulin-dependent brain, it seemed appropiate that hypoglycemia stimulated, while hyperglycemia per se inhibited glucagon secretion (UNGER and EISENTRAUT 1965). A primary (insulin-independent) defect of the A-cells, i.e., resistance to the suppressive action of hyperglycemia, was proposed in genetically diabetic humans (UNGER 1972) and supported by many studies (FRANKEL et al. 1974; BRAATEN et al. 1974; BENNETT et al. 1976; GERICH et al. 1976; UNGER and ORCI 1976; OHNEDA et al. 1978; SEINO et al. 1978). The possibility that glucagon plays an important and essential role in the normal modulation of insulin secretion has not been entertained in several major reviews (CAHILL 1976; LAMBERT 1976; MALAISSE et al. 1979).

In contrast to these views, an alternative hypothesis was advanced, proposing a fundamental physiologic interrelationship between glucagon and insulin. It was suggested that an intraislet positive-negative insulin-glucagon feedback governs the relative concentrations of glucagon and insulin perfusing the portal vein, and hence acting on the economy of all organs, according to the following concept (SAMOLS and MARKS 1972):

> Reciprocal correspondence in the concentrations of glucagon and insulin has frequently been observed. One permutation – insulin levels increased, glucagon levels decreased – of such changes is identical with that occurring during infusions of tolbutamide and exogenous insulin and could, therefore, be interpreted as the manifestation of the negative feedback loop. For example, according to this concept it is likely that it is the insulin secretion provoked by glucose infusion that suppresses pancreatic glucagon secretion rather than, as has been suggested, the hyperglycemia itself... Our hypothesis [is] that the secretion of glucagon and insulin by their respective cells of origin is interdependent... Glucagon acts as a signal sensor and amplifier of certain stimuli to insulin secretion, thereby enormously increasing the sensitivity of the insulin-secreting system to insulin requirements. A major role of glucagon in man is to regulate insulin secretion in response to the ingestion of food without the necessity for comparatively large changes in substrate concentration in the blood. This would not only be wasteful, due to loss of amino acids and carbohydrates in the urine, but possibly harmful as a result of large changes in the composition of the milieu interieur. While direct stimulation of the B cells by glucagon provides a method for amplifying the signal to insulin release, the inhibition of glucagon secretion by insulin provides all the advantages of a negative feedback control, namely a delicate mechanism for achieving and maintaining precise homeostasis. Like all servo-systems so far studied it can be overridden by stress... Under these circumstances there is increased sympathonervous activity which, by enhancing glucagon secretion (SAMOLS et al. 1970) at the same time as it blocks its insulinotropic effect (PORTE et al. 1966), increasing adipose tissue lipolysis (LEFEBVRE 1966, 1968; SAMOLS et al. 1969a) and decreasing glucose uptake by non-glucose dependent organs.

There was another concept integral to the new hypothesis, which emphasized that mutual feedback probably occurred within the islets, because the anatomic evidence then available showed that each type of islet cell shared a common extracellular space (HAIST 1965; ORCI et al. 1967; EPPLE 1968). There were three predictions inherent in this "short-loop" feedback:

(1) Amplification of insulin secretion by glucagon is restricted by the ambient arterial glucose concentration. The stimulation of insulin secretion directly, as well as indirectly through its hyperglycemic action, allows glucagon to amplify insulin secretion or increase hepatic glucose production preemptively. In turn, insulin may inhibit glucagon preemptively.

(2) Primary insulin deficiency (absolute or relative) causes secondary glucagon excess (absolute or relative).

(3) Primary hyperglucagonism causes secondary hyperinsulinism.

During the past decade, there have been some developments consistent with, or compatible with certain aspects of the new hypothesis. There have also been a number of studies which have been interpreted as evidence against this hypothesis. Although we cannot comment in detail on our reevaluation of publications which disagree with our interpretation, we do recognize that the proposed profound physiologic interrelationship between glucagon and insulin remains a hypothesis. We shall treat each limb of the feedback almost as a separate entity: (1) for the sake of simplicity of discussion; (2) because the nature of the pro or con arguments may be different; and (3) because some authorities consider the anatomic, physiologic, and pathologic evidence in favor of the negative limb more satisfactory than that for the positive limb. However this form of presentation should not obscure the possibility of considering both limbs as a dynamic feedback, making the hypothesis as a whole rather difficult to prove, or disprove, directly.

There is not much dispute about two basic observations relevant to glucagon and insulin in vivo: (1) exogenous glucagon will promote insulin secretion; and (2) exogenous and endogenous insulin will inhibit glucagon secretion. The problems are whether either or both of these observations are relevant to normal or pathologic physiology, or whether (1) is a "physiologic anachronism" and (2) a "permissive effect". It is hoped that research in the next decade will answer these authentic questions.

## B. Experimental Stimulation of Insulin Secretion by Glucagon In Vivo and In Vitro

Glucagon is a direct potent stimulus to insulin secretion (SAMOLS et al. 1965a). Although the insulinotropic effect was demonstrated to occur independently of the induced hepatogenous hyperglycemia, it was apparent from the outset that the prevailing blood glucose level modulated the insulinotropic effect (SAMOLS et al. 1965a, b, 1966a), so that in general terms (to be qualified in the following discussion) hyperglycemia enhances and hypoglycemia suppresses the insulin response to the administration of glucagon (SAMOLS et al. 1965b, 1966a, 1969b; SAMOLS and MARKS 1967; GOLDFINE et al. 1972a; OAKLEY et al. 1972). Similarly, in vitro, the insulinotropic effect, which is dose responsive, is best demonstrated at normal and supranormal levels of glucose in the incubation medium (TURNER and MCINTYRE 1966; MALAISSE et al. 1967; GRODSKY et al. 1967; CURRY 1970).

In the early studies, the insulinotropic effect of glucagon appeared to be much more readily demonstrated, and at much lower glucagon concentrations, in vivo than in vitro (SAMOLS et al. 1972). However, presumably because of improved viability of certain in vitro preparations, an insulinotropic effect has recently been demonstrated at a glucagon concentration of 90 pg/ml (KAWAI and ROUILLER 1981) in the isolated dog pancreas. The "physiologic" concentration of glucagon at the B-cell surface remains hypothetical, but in vivo studies in humans and animals, using progressively smaller doses of glucagon, whether given by a bolus (KETTERER et al. 1967) or by continuous infusion (SAMOLS et al. 1965b, 1966a, 1969a; LEFEBVRE and LUYCKX 1969) demonstrate an insulinotropic effect at concentrations calculated to be raised by 300–1000 pg/ml. It has been difficult to find data establishing the lowest exogenous glucagon concentration capable of exerting an insulinotropic effect in humans. During combined infusion of somatostatin and glucagon (SHERWIN et al. 1977), an increase in peripheral glucagon levels of approximately 70 pg/ml above baseline has been interpreted as being insulinotropic (UNGER 1978), although the insulinotropic effect may have been enhanced by hyperglycemia.

## I. Effect of Food and Fasting

The insulinotropic effect of glucagon is potentiated by hyperglycemia or by recent feeding (SAMOLS et al. 1965b, 1966a, 1969b; SAMOLS and MARKS 1967; RYAN et al. 1967; OAKLEY et al. 1972; GARCIA et al. 1973). It is reduced by hypoglycemia (GOLDFINE et al. 1972a) or fasting sufficiently long to cause depletion of glycogen stores. In normal adults, 2–4 days of starvation is necessary before the insulinotropic effect of glucagon is severely reduced (MARKS and SAMOLS 1968a; SAMOLS et al. 1969a). This time period coincides with the length of time for the glycogenolytic-hyperglycemic effect of glucagon to disappear, and its lipolytic activity to be "unmasked" by the absence of insulinotropy. With more prolonged fasting there is: (1) a reestablishment of intracellular glycogen stores and restoration of the hyperglycemic (HARO et al. 1965) and insulinotropic properties of exogenous glucagon (SAMOLS and MARKS 1967; METZ et al. 1968; CORREDOR et al. 1969); (2) a return of circulating glucagon levels to normal. In infants and children, as well as in adults, during "accelerated fasting" (ARKY and FREINKEL 1966) produced either by alcohol or thyrotoxicosis, the period of fasting (MARKS and SAMOLS 1968a) required to deplete intracellular glycogen stores is much shorter, sometimes hours. This factor needs to be considered when the insulinotropic effect of glucagon is being used diagnostically in children. By glucose feeding animals before harvesting their islets, it is possible to stimulate insulin release in the absence of extracellular glucose, plausibly by glucagon-provoked glycogenolysis and glycolysis from enriched endogenous stores of B-cell glycogen (MALAISSE et al. 1967).

## II. Adrenergic Effects

α-Adrenergic stimulation, whether induced by epinephrine or norepinephrine infusions (PORTE et al. 1966) or by increased sympathetic nervous activity (PORTE 1969; MAJID et al. 1970), inhibits the insulinotropy of glucagon. The islets of Langerhans are richly innervated by sympathetic nerve terminals (ESTERHUIZEN et al. 1968). Although typical synaptic clefts have not been described in mammals, SAMOLS and STAGNER (1981) have recently shown that both pancreatic arterioles and insulin secretion are governed by presynaptic-postsynaptic modulation, confirming that there is almost certainly a local release of norepinephrine capable of reaching very much higher concentrations than can be achieved by circulating catecholamines, as has been shown to occur in other organs with presynaptic adrenergic neuromodulation. Thus, sympathoadrenergic stimulation provides a vital escape mechanism, allowing for restraint of the insulinotropic effects of glucagon during stress. The original demonstrations that catecholamines stimulate glucagon secretion (SAMOLS and MARKS 1967; SAMOLS et al. 1970) have been confirmed (IVERSEN 1973) and both β-adrenergic and α-adrenergic agonism are stimulatory to glucagon (SAMOLS and WEIR 1979).

Although the typical stresses [fight, flight, knock (trauma), shock] induce a "catabolic" insulin:glucagon ratio in the portal vein, it should be emphasized that the effects of α-adrenergic agonism versus insulinotropy appear to be dose responsive on either side of the equation. OAKLEY et al. (1972) demonstrated an insulinotropic effect of 1 mg glucagon (i.v.) in humans during α-adrenergic agonism (induced by epinephrine plus propranolol infusion) of sufficient degree to inhibit glucose-induced release. It is therefore tempting to speculate that the insulinotropy of stress-induced hyperglucagonemia maintains the essential minimal insulin levels so vital during acute starvation or other stresses. The studies of SORENSON et al. (1979), using perifused isolated rat islets, support this suggestion. Under both high and low glucose conditions, glucagon release was maximally stimulated by 10 $\mu M$ norepinephrine, whereas A-cell stimulation by much lower concentrations of norepinephrine is dampened by a high concentration of glucose in the perifusing medium (IVERSEN 1973). Therefore, agonism by the very high local norepinephrine concentration likely to influence the A-cell during sympathetic stress should be relatively independent of the ambient glucose level. SORENSON et al. (1979) noted that, during the initial maximal stimulation of glucagon by norepinephrine, the expected inhibition of insulin was prevented, possibly by a paracrine effect of the high level of glucagon released.

It is usually assumed that glucopenia is directly recognized by the A-cell, which immediately secretes glucagon. Perhaps underestimated is the contribution by local intrapancreatic autonomic neuroregulators which identify glucopenia with resultant local outpouring of norepinephrine. The release of large amounts of norepinephrine from the isolated perfused canine pancreas during glucose deprivation has been convincingly demonstrated (CHRISTENSEN and IVERSEN 1973). Glucopenia, therefore, itself inhibits insulin secretion by the B-cell (MALAISSE et al. 1979) and reinforces the inhibition by local and central (WOODS and PORTE 1974) autonomic regulation. Yet insulin secretion in vivo is rarely, if ever, completely abolished during glucopenia (SAMOLS 1965). It is not known which mechanisms promote insulin secretion under such unfavorable conditions, but glucagon is presumably one of the candidates. Supporting this concept is the effect of anti-glucagon serum on C-peptide levels during insulin-induced hypoglycemia in rats (NISHINO et al. 1980). Although the fall and return of blood glucose levels after exogenous insulin was comparable to controls, the C-peptide levels were reduced when endogenous glucagon was neutralized by antiserum. This suggests that: (1) endogenous glucagon does have an insulinotropic effect during hypoglycemia; and (2) the combined lowering of glucagon and insulin achieves glucose concentrations which are similar to controls.

## III. Insulinotropic Effect of Glucagon in Disease

In liver diseases of various etiologies with limited capacity for hepatic glycogenolysis, the insulinotropic effect of glucagon may still be readily demonstrable (SAMOLS and HOLDSWORTH 1968). In obesity, the insulinotropic effect of glucagon is exaggerated (MARKS and SAMOLS 1968a), as it is in other conditions (acromegaly, cirrhosis, Cushing's syndrome) in which the plasma insulin response to glucose or arginine is magnified (SAMOLS and HOLDSWORTH 1968; SEINO et al. 1975). Chronic growth hormone treatment (PIERLUISSI and CAMPBELL 1980) augments, but acute growth hormone administration suppresses (ADAMSON and CERASI 1975) glucagon-induced insulin release.

As analysed in detail in Chapter 55, the insulinotropic effect of glucagon has been found to be useful as a diagnostic adjunct in insulinomas (MARKS and SAMOLS

1968b; KUMAR et al. 1974). After administration of 1 mg glucagon i.v. and sampling at 2, 5, 10, 15, 20, 30, 60 min, a peak in insulin ≧ 130 μIU/ml or a rise of >100 μIU/ml above baseline is very suggestive of insulinoma. In a review of 32 published cases, KUMAR et al. concluded that over 80% of tested insulinoma cases had positive tests. Two-thirds of all reported patients showed maximum insulin values during the first 15 min, but the other third had delayed peaks at 20–30 min, thus blood samples should be drawn every 5 min for the first 30 min. Infrequent sampling may have accounted for inconclusive results in six patients. Fasting hypoglycemia may mildly inhibit, but does not usually abolish, the diagnostically excessive insulin response to 1 mg glucagon (NELSON et al. 1980). The glucagon test is the safest of the stimulative tests, as it does not induce early hypoglycemia, but the test should be terminated by 60 min lest hypoglycemia develop. Patients with other causes of fasting hypoglycemia do not show an exaggerated and/or delayed insulin response to glucagon.

The insulinotropic effect of 1 mg glucagon i.v. has been used to quantitate insulin secreting capacity insulin-dependent diabetes mellitus (IDDM) by measurement of C-peptide response, with results that agree well with the use of a mixed meal (HENDRIKSEN et al. 1977). In the case of pronounced B-cell failure, glucagon seemed to exert a greater stimulus to the B-cells than the test meal. On follow-up testing with mixed meals, it was shown that an initial negative glucagon test (i.e., absent C-peptide response) invariably predicted complete failure of endogenous insulin secretion. A positive initial glucagon test was followed by demonstrable endogenous insulin during meals in 84% of cases.

The insulinotropy of various doses of glucagon has been studied in patients with noninsulin-dependent diabetes mellitus (NIDDM) (CROCKFORD et al. 1969). In a study which compared groups of NIDDM, using appropriate controls by weight, the insulinotropy of glucagon was, like that of other insulin secretagogues, reduced. However a dissociation between the insulinotropy of glucagon and glucose in individual diabetic patients, i.e., negligible insulin response to glucose, "good" or normal insulin response to glucagon, is not uncommon (SAMOLS et al. 1966a; SEINO et al. 1975), leading HALTER et al. (1979) to conclude that basal levels of such patients and their insulin responses to nonglucose stimuli, may be indistinguishable from those of normal subjects of comparable age and body weight. The pathophysiologic significance of this finding is discussed in Sect. D.

## IV. Insulinotropic Effect of Glucagon in the Child, Neonate, and Fetus

Most of the features of the insulinotropic effect of glucagon in adults, e.g., the excessive insulin response in obesity have been confirmed in children (CUTILLO et al. 1968). In addition, various methods of glucagon infusion, in particular intramuscular glucagon (0.1 ng/kg body weight) have been used to evaluate insulin and growth hormone reserve in Belgian (VANDERSCHEUREN-LODEWEYCKX et al. 1974) and Thai (PREEYASOMBAT et al. 1975) children. Insulin responses to intravenous glucagon were impaired in growth hormone-deficient children (AUGUST and HUNG 1976), while a normal insulinotropic effect of glucagon, but low insulin responses to glucose and arginine, were reported in children with familial constitutional short stature (KARP et al. 1975). BLUM et al. (1978) stressed that glucose preloading, but not arginine preloading, increased the insulin response to glucagon by a mechanism which is growth hormone dependent. Impaired insulin secretion and elevated mean free fatty acid levels during intramuscular (i.m.) glucagon and i.v. tolbutamide stimulation tests has been presented as evidence of insulin deficiency in children with untreated XO-Turner's syndrome, whose blood glucose and serum growth hormone responses did not differ from controls (AVRUSKIN et al. 1979).

The normal fetal and the neonatal B-cell, unlike that of the child or adult, responds poorly to glucose. This insulin response to glucose is considerably enhanced by glucagon and/or theophylline in vivo (GRASSO et al. 1970) or in vitro (LAMBERT 1976; ESPINOSA DE LOS MONTEROS et al. 1970). Neither the reason for this ontogenic variation in sensitivity of B-cells, nor the significance of its correction by glucagon, is presently understood.

## V. Mechanism of Insulinotropic Effect of Glucagon

There is good evidence to suggest that the primary physiologic function of glucagon (if any) as an insulinotrope, is not to act independently, but to augment or "promote" glucose-induced and amino acid-induced insulin secretion, as prolonged periods of aglycemia are unusual in vivo. As reviewed in Chaps. 13 and 14, glucagon stimulates glycogenolysis in the liver by promoting (see last two paragraphs of this subsection) the activation of membrane adenylate cyclase, which converts ATP to cyclic AMP (cAMP), which in turn activates protein kinases (SUTHERLAND et al. 1965). With the recognition of the insulinotropic effect of glucagon, it was inferred that a similar activation was occurring on the B-cell membrane (SAMOLS et al. 1966a), an effect which appeared to be confirmed when TURTLE and KIPNIS (1967) reported that glucagon and theophylline increased the concentration of cAMP in islet tissue. In addition to its relevance to the insulinotropic effect of glucagon, it was speculated (SAMOLS et al. 1966a) whether adenylate cyclase might be pertinent to the glucose-induced metabolic signal initiating or perpetuating insulin secretion. Assuming the substrate hypothesis to be valid, the suggested pathway of glucose metabolism was via glycogen synthesis and breakdown, with a proposed compartmentation of glucose-6-phosphate pools.

It has been shown that glucose stimulates adenylate cyclase activity in homogenates of mouse pancreatic islets (CAPITO and HEDESKOV 1977) and it is clear that the level of cAMP within the islets of Langerhans is increased following acute stimulation with glucose (CHARLES et al. 1973; GRILL and CERASI 1974). Proponents of the substrate-site model (i.e., that a glucose metabolite is the intracellular trigger of insulin release) have noted that, of glucose metabolites so far studied, only glucose-6-phosphate and phosphoenolpyruvate are capable of the requisite interaction, namely modifying: (1) insulin release from an artifically reconstructed in vitro release system; and (2) the activity of islet adenylate cyclase (SUGDEN and ASHCROFT 1977). On the other hand, MALAISSE et al. (1979), while acknowledging that changes in the concentration of glucose metabolic intermediates and cyclic nucleotides, or both, could play a modulatory role upon stimulated insulin release, maintain that the initiation of insulin release depends on the generation of two essential factors: $H^+$ and reduced pyridine nucleotides.

Although the precise link between the primary signal of glucagon (known) and the primary signal or signals of glucose (uncertain) remains controversial, there are a number of experimental observations defining glucagon's potentiation of glucose-induced insulin secretion. Prolonged continuous insulin secretion in vitro occurs only in the presence of metabolizable substrates known to have the capacity to act as a fuel in the islet cells. Glucagon or adenylate cyclase activitation stimulates insulin secretion from isolated islets in vitro only in the presence of metabolizable substrate, or when pancreas slices or islets are obtained from animals or islets preperfused with hyperglucosia, thereby augmenting glycogen within the B-cells (MALAISSE et al. 1967, 1977). Pyruvate and lactate can substitute for glucose as the metabolizable substrate but, among the amino acids, L-leucine alone has this capacity (MILNER 1969). When normal rats are fasted for 48–72 h, or their islets are incubated in a low glucose concentration, there is a decrease in insulin secretion and in islet adenylate cyclase activity and cAMP content (LACY 1977), perhaps because enhanced somatostatin secretion restrains glucagon secretion (ITOH 1981).

Even though cAMP appears not to "initiate" insulin release (MALAISSE et al. 1979), there is excellent evidence that cAMP can profoundly amplify insulin secretion. During circumstances in which the effects of glucose itself are relatively weak as an insulin stimulant, in vivo in diabetes or premature infants (GRASSO et al. 1970) or in vitro in fetal pancreas (LAMBERT et al. 1967; ESPINOSA DE LOS MONTEROS et al. 1970) the administration of agents (glucagon, phosphodiesterase inhibitors, cAMP) known to increase intracellular cAMP will greatly potentiate insulin secretion. If local intraislet extracellular glucagon is available to the B-cell, this paracrine glucagon would be the logical major regulator of B-cell cAMP, and thus of amplification of insulin secretion.

Because glucagon promotes glucose-induced insulin secretion when glucose is perfusing the B-cell, it is not surprising that the insulinotropic effect of glucagon is diminished or abolished by competitive inhibitors of glucose metabolism such as mannoheptulose or 2-deoxyglucose, by $\alpha$-adrenergic agonists, and by diazoxide – all recognized inhibitors of glucose-induced insulin secretion in vitro and in vivo (MARKS and SAMOLS 1968c).

Although there has been some question as to whether glucagon stimulates insulin synthesis (CURRY 1970; SCHATZ et al. 1973) it is logical, if glucagon promotes glucose-induced insulin secretion, that it should promote glucose-induced synthesis. It has recently been reported that glucagon will increase glucose-stimulated leucine $^{3}$H incorporation into proinsulin (TRIMBLE et al. 1981).

The specificity of glucagon action in the B-cell adenylate cyclase system resides in the properties of the glucagon receptor linked to the system. GOLDFINE et al. (1972b) studied the relationship between glucagon binding and glucagon activation of adenylate cyclase in tissue from the insulin-secreting tissue of the Syrian (golden) hamster. In the B-cell, half-maximal glucagon concentrations for displacement of glucagon $^{125}$I binding and glucagon activation of adenylate cyclase were close to those seen for both functions in the liver. In B-cell or liver systems, glucagon 1–21, glucagon 20–29, and secretin had little or no activity, whereas glucagon 2–29 bound to receptors, but did not activate adenylate cyclase. More recent studies on the glucagon "message" have shown that glucagon 1–21 and glucagon 1–6 bind to the glucagon receptor and activate adenylate cyclase in liver membranes, although with affinities 3–5 orders of magnitude less than the native hormone (WRIGHT and RODBELL 1979). Specificity of glucagon binding to the receptor also seems to reside in the first six residues. [desHis[1]] glucagon has been used for partial inhibition of the insulinotropic effect of glucagon in the isolated perfused rat pancreas (FUSSGÄNGER et al. 1976). In contrast to the liver, where rapid binding and dissociation requires the addition of ATP or GTP, binding and dissociation in the B-cell occur very rapidly without addition of nucleotides (GOLDFINE et al. 1972b).

Originally, it was thought that glucagon action was defined by a relatively simple model: the glucagon-receptor complex diffused within the cell membrane to activate the catalytic subunit, i.e., to convert ATP to cAMP. However, during the past decade (RODBELL 1981) it has been recognized that, in the liver cell at least, the glucagon-receptor complex primarily stimulates GTP binding to an intermediate regulatory protein which, in turn, activates adenylate cyclase. Thus, adenylate cyclase activity is "turned on" by glucagon-stimulated GTP binding to a regulatory protein and is "turned off" by GTP hydrolysis at the regulatory site. The "turn off" reaction is inhibited by nonphysiologic agents such as Gpp (NH)p (an analog of GTP) and cholera toxin (CASSEL et al. 1977; MOSS and VAUGHAN 1979).

As reviewed in Chapter 13, guanosyl nucleotides also influence the binding kinetics and affinity of the glucagon-receptor interaction in the liver cell. GTP reduces the "tightness" of glucagon binding, from a high to a low affinity state. The low affinity receptor state favors activation of adenylate cyclase, whereas the high affinity form of the receptor is apparently associated with inactive adenylate cyclase. Other factors which regulate adenylate cyclase activity include: (1) divalent cations, e.g., calcium ion in submicromolar concentrations activates adenylate cyclase, and at 10 $\mu M$ concentrations (PERKINS 1973) inhibits adenylate cyclase; (2) adenosine which exerts at least three different effects on the activity of membrane-bound adenylate cyclase systems (RODBELL 1981). Two of these effects, stimulation and inhibition, are exerted through outer surface receptors for adenosine. The third type of effect is exerted at a recently discovered site, the so-called P-site, which seems to be related to the metal ion binding sites. Glucagon potentiates adenosine inhibition at the P-site, suggesting that there may be a negative feedback in glucagon action, i.e., glucagon inhibits its own action.

## C. Consideration of the Evidence for the Endogenous Insulinotropic Effect of Glucagon

### I. Inherent Difficulties in Testing the Hypothesis

It may reasonably be asked why it has been so difficult to either "prove" or "disprove" the physiopathologic existence of the positive links of the proposed feedback, given that the potent promotion of insulin secretion by glucagon in vivo and in vitro is incontrovertible. The answer is that the hypothesis had first to accommodate known changes in insulin and glucagon function in the periphery, as well as allow for sophisticated reconceptions. By and large, insulin and glucagon are physiologically simultaneously raised for relatively prolonged periods only during a protein meal. Our hypothesis was that "glucagon serves as a signal sensor and amplifier of certain stimuli to insulin secretion". This is different from a simple feedback, such as the glucose-insulin feedback, in that we were proposing as additional features: (1) an amplification phenomenon governed by the ambient glucose and amino acid levels, associated with (2) a positive force (insulin secretion) for negative feedback. This concept is also different from a single amplification system, such as the gut hormone amplification, in that additional consideration needs to be given to both: (1) the paracrine potential in terms of concentrations of hormone, and (2) the negative link of the glucagon-insulin feedback. Moreover, this paracrine concept allows for: (1) the possibility that prevailing local concentrations of glucagon (not necessarily reflected by peripheral levels) may influence the response of the B-cell to a particular stimulus, and (2) that rapid local dynamic changes (first positive, then the negative link of the feedback) may occur and be difficult to detect in the systemic bloodstream.

The difficulties in performing an analysis based on relative glucagon and insulin concentrations in the blood are compounded by the potential for an adrenergic stress escape mechanism. Even if the latter is operative, how can it be established that a particular low level of insulin, associated with a high glucagon level during adrenergic stress, would not have been even lower without the glucagon? In vivo testing therefore requires that the glucose level be kept constant (e.g., by glucose clamp) and that an additional independent influence be exerted on either the A- or the B-cell.

Yet another problem to be confronted in the paracrine concept is that of hormone concentration. For example, it has been questioned whether glucagon, which may potentiate insulin secretion at a concentration of 90 pg/ml, is too potent to be allowed paracrine access to the B-cell. A decade ago this dilemma was perceived only for glucagon, but as similar concentration problems are now recognized to be valid for somatostatin and insulin as paracrine hormones, solutions (see Sect. G–I) are being proposed. The evidence to be discussed in Sects. C.II–C.VIII represents, roughly in order, evidence favoring, evidence consistent with, evidence compatible with, and evidence against an insulinotropic effect of endogenous glucagon.

## II. Anatomy of the Islets of Langerhans

Influenced by the studies of HAIST (1965), ORCI et al. (1967), and EPPLE (1968), and aware of the effects of exogenous glucagon or insulin, we were unable to grasp how the A- and B-cells could not be influenced by one another, as they appeared to be contiguous, and to share a common extracellular space. With recent advances in anatomy, the question of paracrine access of A-cell secretion to the B-cell membrane needs to be raised again, and is discussed fully in Chap. 31. The current interpretation of islet microanatomy permitting paracrine interactions can be summarized in terms of probabilities, ignoring the possibility of any paracrine interaction between the *inner* core of B-cells, and the mantle of A- and D-cells. The highest probability is of a reciprocal paracrine interaction between A- and D-cells in the mantle. The lowest probability is that glucagon or somatostatin, from the peripheral A- or D-cells, cross the interstitium to reach the juxtaposed outer core of B-cells. The reverse passage, that of insulin crossing the same interstitium from the outer core of B-cells to reach the juxtaposed A-cells, is assigned an intermediate probability.

The distinguished contributions of ORCI's laboratory (reviewed by ORCI and PERRELET 1981; BONNER-WEIR and WEIR 1979) have shown regional ontogenetically derived differences in islet composition. In rats, and humans, there are two types of islets: "dorsal" (in the tail, body, and superior part of the head) and "ventral" (lower part of the head). The dorsal and ventral islets demonstrate similarities in that B-cells constitute the bulk of the islet and are situated in the islet centers, while D-cells are at the periphery. However, the dorsal islets are rich in peripheral A-cells and poor in pancreatic polypeptide cells (PP-cells), while the converse obtains in the ventral islets. TRIMBLE et al. (1981) used this observation as a natural experimental model to test whether functional differences occurred between rat islets of ventral and dorsal pancreatic origin and, if so, whether these reflected endogenous glucagon contact. Glucose-induced insulin release from dorsal islets was 50% greater than that from ventral islets, despite similar insulin content and basal insulin release. Glucose-stimulated leucine $^3$H incorporation into proinsulin was 50% greater in dorsal than in ventral islets, as were glucose utilization rates and ATP levels. The dorsal islets had a much greater content of glucagon, which could be responsible for these effects because: (1) both basal and glucose-stimulated cyclic AMP levels were significantly higher in dorsal than in ventral islets; (2) addition of exogenous glucagon eliminated the differences in insulin release and proinsulin biosynthesis. These findings strongly suggest that changes in the relative proportions of the different islet endocrine cells exert profound effects on endocrine function. In the light of the known effects of glucagon and pancreactic polypeptide on insulin secretion, these findings also suggest that local glucagon modulates not only glucose-induced insulin secretion, but also glucose-induced insulin synthesis. Assuming that glucagon is indeed the potentiation factor in the dorsal islets, it is remarkable that its augmentation of insulin secretion occurs under conditions which are least favorable, i.e., suppression of glucagon release by the higher insulin/higher glucose ambience. Perhaps the latter – the negative limb in the feedback – prevents an undesirable "explosive" augmentation of insulin secretion.

If as seems likely, glucagon is the amplifier of glucose-stimulated insulin release by dorsal islets in vitro, then either the current interpretation of islet microanatomy is defective, or the in vitro system is producing a deceptive artifact. It could be argued that TRIMBLE's findings may possibly have been caused by "leaching" of glucagon, i.e., from the mantle of dorsal islets into the perfusion medium, thence to the B-cells. Confirmation of the dorsal islet amplification in the vascularly perfused pancreas may help to counter the pro-artifact argument. Indeed, there is very recent supportive evidence for the biologic validity of the work of TRIMBLE et al. (1981) from studies using an isolated perfused canine ventral lobe compared with an almost equal-sized isolated perfused dorsal lobe segment (STAGNER and SAMOLS 1982). When the perfusate glucose was increased from 88 to 200 mg/dl, both the absolute and relative increase in insulin secretion from the dorsal lobe segment was significantly greater than the increase observed from the ventral lobe. The morphological differences between the dorsal and ventral pancreatic islets at last present a new opportunity for studying the functional significance of the positive limb of the feedback.

## III. Dynamic and Magnitudinal Changes in Glucagon and Insulin After Secretagogue Administration

Many studies have attempted analysis of the sequences and magnitudes and patterns of acute-phase hormone (glucagon and insulin release) after the administration of secretagogues which stimulate both hormones. In trying to establish whether one hormone precedes the other, most studies are of little value, because they have been performed in vivo, when damping and flow problems may readily conceal the sequence of changes. In the large canine pancreas in vitro, there is blurring of the very finely defined first and second phases of acute-phase secretion seen in the rat pancreas in vitro, presumably because of the time required for flow from different portions of the dog pancreas. Therefore, the study by PEK et al. (1976), by far the most rigorous, is far superior to all others in that the perfused rat pancreas was used, with a shorter sampling time interval than any other study.

PEK's study was remarkable in that it showed that glucagon was invariably released before insulin after infusion of diverse dual secretagogues, including leucine, arginine, prostaglandins $F_{2\alpha}$ and $E_2$, bovine growth hormone, and isoproterenol. However, because the timing and the magnitude of glucagon and insulin release did not correlate statistically, PEK et al. concluded: (1) that antecedent release of glucagon is not the principal mediator of insulin release in response to stimuli common to both hormones; and (2) that endogenous glucagon may at best modify the release of insulin evoked by certain secretagogues. Unfortunately, PEK et al. did not consider the possible effects of the negative limb feedback, i.e., inhibition of glucagon by insulin, in their analysis. As there may have been differing degrees of paracrine suppression of glucagon secretion by insulin, recognition of the feedback servo-system may require more complicated methodology than that used by PEK et al. (1976). Nevertheless, they did note that the first significant increase of insulin occurred synchronously with, or 12 s subsequent to, the mean *maximal* levels of glucagon, and suggested that the attainment of a certain "permissive" concentration of glucagon in the extracellular fluid plays a role in the initiation of secretion of insulin. These conclusions may be especially significant when one recalls the suggestion that glucose-induced insulin release may be augmented by paracrine glucagon (TRIMBLE et al. 1981) even though glucagon secretion, as estimated by efflux measurement, is presumably promptly inhibited.

The fact that glucagon is secreted before insulin when dual secretagogues are tested certainly does not prove that glucagon either provokes or amplifies the insulinotropic effect of the secretagogue, but this notable precedence is compatible with such an action. Subsequent magnitudinal analysis becomes complex. How should "dissociation" of the glucagon- and insulin-stimulatory properties of an agent (secretagogue) be interpreted? If the agent shows, under different conditions, a decrease in its glucagonotropic and an increase in insulinotropic effect, this change could reflect dominance of the negative limb of the feedback. Conversely, an increase in the glucagonotropic, with a decrease in insulinotropic potency, could reflect a decline in the strength of the negative limb of the feedback, as proposed (see Sect. E) by ASPLIN et al. (1981) for arginine.

"Parallel dissociation" has been cited (SAMOLS et al. 1972) as compatible with the feedback hypothesis. Whereas the glucagonotropic effect of arginine is enhanced in human subjects who have fasted for 3 days, its insulinotropic effect is, like that of exogenous glucagon, completely abolished (AGUILAR-PARADAR et al. 1969), even though exogenous glucose can still stimulate insulin secretion in these individuals. The loss of the insulinotropic effect of glucagon could be explained by temporary exhaustion of glycogen within the B-cells (see Sect. B) or downregulation of B-cell receptors, while the augmentation of the glucagon response to arginine would naturally result from the removal of the negative (insulin) feedback.

## IV. Augmentation of Insulin Secretion by Protein and/or Fat Meals

After a "pure" protein meal, the degree of augmentation of insulin secretion, in terms of the relatively normal levels of glucose, is greater at these glucose levels than after a pure carbohydrate meal, or after oral glucose administration. This augmentation may be attributed, at least in part, to the observed increase in glucagon secretion induced by hyperaminoacidemia and/or gut hormones and/or parasympathetic agonism. Questioning the physiologic relevance of pure protein (FUJITA et al. 1975) mixed with pure glucose meals, DAY et al. (1978) tested "normal, attractive" mixed meals and still found a higher increase in insulin per unit of glucose in those meals containing 12 g or more of protein, concluding that this was unlikely to be due to the direct effect of amino acid stimulation of insulin release, and likely to be due to a gut hormone. Their observation of very early stimulation of glucagon release prior to insulin release, regardless of mixed meal content, is consistent with, but does not prove, interpretations presented in detail (SAMOLS and MARKS 1972; MARKS 1972; FUJIMOTO et al. 1978) that the insulinotropic effect of gut hormones, liberated by protein, fat, or carbohydrate ingestion, could be mediated by glucagon.

## V. Active and Passive Immunization

EPAND and DOUGLAS (1973) reported that rabbits actively immunized against glucagon have normal fasting levels of blood glucose, but a lowered level of insulin. They interpreted these observations as suggesting that the decreased insulin levels were the result of the decreased insulonotropic effect of the neutralized glucagon, resulting in a normal insulin:glucagon ratio, which maintained a normal blood glucose level.

That rabbit antiserum contains large amounts of endogenous glucagon, originating in the rabbit in which the antiserum is raised, was confirmed by HOLST et al. (1978). Passive immunization of rats with rabbit anti-glucagon serum, even after the rabbit glucagon was stripped from the antiserum, had no effect on the blood glucose in fasting rats. As the glycemic and insulinemic response to exogenous glucagon was abolished by antiserum, HOLST et al. concluded that either circulating glucagon is of limited significance for the maintenance of normoglycemia in the rat or, conversely, that the effect of glucagon may readily be compensated for by the diminished insulin secretion, because of the decrease in available glucagon. The latter concept is supported by the studies of NISHINO et al. (1980) who also recorded essentially unchanged blood glucose levels, but found reduced C-peptide levels, after neutralization of endogenous glucagon by anti-glucagon (COH-terminal-specific) serum during fasting and after exogenous insulin injection to normal rats.

Studies of "glucagon deficiency" induced by somatostatin infusions have suggested that basal glucagon secretion is important in the maintenance of normoglycemia (ALFORD et al. 1974; GERICH et al. 1974b). On the other hand, studies using anti-glucagon serum (HOLST et al. 1978; GREY et al. 1970; BARLING and BELOFF-CHAIN 1973) have not affected blood glucose concentrations in normal animals. Thus, there may be problems in the interpretation of either the somatostatin and/or the antiserum experiments. Moreover, the effects of circulating antibody are unlikely to reach the paracrine compartment in rats or rabbits, so that it could be argued: (1) that these antiserum studies are compatible with an insulinotropic effect of glucagon which occurs by a long loop, not a short loop, pathway; or conversely (2) that the remarkable compensation between glucose, insulin, and rapidly neutralized glucagon could only occur by a complex paracrine mechanism.

## VI. Diabetes Mellitus

In diabetes mellitus, there is almost always relative or absolute hyperglucagonism (UNGER and ORCI 1981a, b) which is often reversible by normalization of insulin action (see review in Chap. 44). While this does not prove that the observed hyperglucagonism is insulinotropic in fact or intent, it is compatible with the hypothesis. In NIDDM, despite the absence of B-cell response to i.v. glucose, the insulin response to either exogenous glucagon or a mixed meal is often partially or completely retained. CHERRINGTON and LILJENQUIST (1981) aver that glucagon undoubtedly plays a role in driving hepatic glucose production in diabetes. If glucagon is still capable of stimulating insulin secretion in the diabetic there is therefore no cogent reason why hyperglucagonism is not providing an insulinotropic function. The postulated insulinotropic effect of glucagon could presumably also occur within a paracrine setting in diabetes. There is no reason to suspect that glucagon plays any less important a role than other nonglucose stimuli (e.g., arginine, gut hormones) in promoting insulin secretion, particularly in diabetes mellitus. When "glucagon-selective" somatostatin analogs have been tested in rats (SARANTAKIS et al. 1978), the insulin response to arginine has not been reduced as much as the glucagon response, but it has been subnormal, either because of the loss of the glucagon-releasing activity of arginine, or because of remaining nonselective suppression of the B-Cell.

## VII. Glucagonoma-Induced Hyperglucagonemia and Hyperinsulinemia

In the majority of patients with the "glucagonoma syndrome", diabetes is mild, and untreated other than by diet (BHATHENA et al. 1981; HIGGINS et al. 1979). Varying degrees of hyperinsulinemia are found (MCGAVRAN et al. 1966; RIDDLE et al. 1978; BINNICK et al. 1977; BLOOM and POLAK 1978). The latter may be attributable to the insulinotropic effect of glucagon, because the diabetes may become far more severe when insulin secretion is compromised by genetic diabetes or by tumor replacing the pancreas (BHATHENA et al. 1981). Although hyperinsulinism may also be caused by multihormone-secreting tumors (TIENGO et al. 1976) in some cases, in others, insulin (and somatostatin) may be absent from the tumor, therefore heterogeneity of tumor cell types may, in selected cases, not be responsible for the hyperinsulinism (RIDDLE et al. 1978).

Hyperglucagonism from glucagonoma may be only remotely relevant to normal physiology. In the case of large, malignant, undifferentiated tumors causing the glucagonoma syndrome, the tumor glucagon presumably reaches the B-cells by the long-loop route. It has been suggested that abnormal rises in plasma immunoreactive glucagon (IRG) after various stimuli, or the failure of insulin hypoglycemia to elicit an increase in IRG in glucagonoma patients, may reflect the tumor's lack of customary islet paracrine influence on glucagon by insulin and/or somatostatin (RIDDLE et al. 1978; ASPLIN et al. 1981).

### VIII. Evidence Against an Insulinotropic Effect of Endogenous Glucagon

A reasonable objection to an insulinotropic effect for glucagon is that "it goes the wrong way" in the common experimental situation, i.e., administration of exogenous glucose. If enhanced glucose-induced insulin secretion from isolated islets from the "dorsal pancreas" (TRIMBLE et al. 1981) is indeed caused by glucagon, and if this is also true under in vivo conditions, this objection is overruled. Another reasonable objection is that endogenous glucagon may increase, while endogenous insulin levels remain low. This combination is almost invariably related to adrenergic stress (Sect. B.II), including hypoglycemic states. However, it is also possible that the stress hyperglucagonism is insulinotropic in the sense of helping to maintain a minimum insulin secretion.

It has taken 50 years to produce evidence that endogenous glucagon normally promotes hepatic glucose production. The probability that glucagon normally promotes insulin secretion is not, in terms of objective evidence, as high, nor is the probability that glucagon is insulinotropic via a paracrine pathway. Yet the circumstantial evidence favoring an intraislet insulinotropic effect by glucagon is compelling enough to merit further research.

## D. Potential Physiologic and Pathologic Significance of the Insulinotropic Effect of Glucagon

Teleologically, the amplification of insulin secretion is an ideal role for glucagon, as the amplification stimulus would be accompanied by an increase in hepatic glucose production, further enhancing amplification, until the increase in insulin and glucose causes feedback inhibition. The relatively brief glycogenolytic effect of glucagon (BOMBOY et al. 1977) would be sufficient for immediate elevation of glucose levels, but would be limited and therefore help (together with the insulin feedback) to prevent an accelerated escape effect. Spurts of secretion of glucagon would most effectively correct glucopenia and/or insulinopenia. There are indeed primary pancreatic oscillations in secretion of glucagon, and of insulin, but these are slightly out of phase (STAGNER et al. 1980), with mean periods of cycles of 10 min for insulin and 8.6 min for glucagon. This suggests that the oscillations are not driven by the glucagon-insulin feedback mechanism, but by an independent pacemaker and coordinator. Teleologically, this is particularly useful for periodic amplification of either the positive loop or of the negative loop of the feedback, when the appropriate phases of the cycles coincide with high or low glucose, preventing spontaneous induction of a vicious circle, and allowing insulin and glucagon to preemptively control the glucose concentration (see discussion later in this section).

The physiologic concomitants of a mixed normal meal: parasympathetic agonism, increased gut hormones, hyperaminoacidemia, all enhance glucagon and insulin secretion. Amplified insulinotropy is biologically necessary for fuel storage and homeostatically desirable to avoid hyperglycemia. Glucagonotropy is equally homeostatically desirable, to avoid hypoglycemia, and to aid in transport, i.e., transferring carbohydrate from the liver to the periphery. Preemptive amplification of insulin dosage (an early large peak immediately preceding the meal) has been shown to be most effective in controlling blood glucose concentrations in open-loop delivery systems for the treatment of IDDM (RASKIN et al. 1979) and for the normal suppression of glucagon by exogenous insulin in IDDM (AYDIN et al. 1977). While parasympathetic modulation and gut peptides would usefully assist amplification of insulin secretion, the accomplishment of preemptive amplification would best be via paracrine glucagon. The very early stimulation of glucagon secretion during a mixed meal, even when its carbohydrate content is relatively high (DAY et al. 1978) would theoretically be ideal for preemptive, immediate amplification of insulin secretion. Additional advantages of paracrine feedback regulation during a mixed meal include: (1) inhibiting "runaway" excessive insulin secretion, by its own paracrine suppression of glucagon (this minimizes the danger of excessive hepatic fuel storage, as occurs in diabetics treated with excessive insulin); (2) minimizing the danger of hypoglycemia by amino acid-associated secretion of glucagon when blood glucose levels are normal, with preemptive paracrine amplification of glucagon secretion when insulin secretion is decreased (ASPLIN et al. 1981) by a falling blood glucose.

It has been proposed that, in NIDDM (HALTER and PORTE 1981), the basal glucose level rises "high enough to provide sufficient potentiation of the nonglucose signals so that basal insulin secretion is compensated" to explain the relatively stable hyperglycemic level in each patient. We would suggest that there is a step prior to this equilibrium, because, of the "nonglucose signals", glucagon could be the most important. Glucagon would simultaneously increase the hyperglycemia and amplify insulin secretion. The degree of hyperglucagonism (dictated by the need to correct hypoinsulinism) would establish the level of hyperglycemia. Therefore, instead of the glucose level "rising" until basal insulin is adequate, we propose that hyperglucagonism is the force pushing to correct the relative hypoinsulinism, so that the equilibrium (diabetic) basal glucose level could be regarded more as the result of, than the cause of, a paracrine feedback adjustment to achieve normoinsulinism.

## E. Effect of Insulin on Glucagon Secretion

At the time when glucagon was considered to be primarily a hormone of glucose need, the possibility that insulin might mediate glucose-induced glucagon suppression was not recognized (UNGER and EISENTRAUT 1965). Because of the insulinotropic effects of glucagon, it was reasonable to examine the reverse situation, leading to the demonstration of the importance of insulin in the suppression of glucagon (SAMOLS et al. 1969a, b, 1970, 1972). These observations, accompanied by evidence that the hyperglucagonemia of diabetic ketoacidosis (ASSAN et al. 1969; MÜLLER et al. 1973) was not suppressed by hyperglycemia, but that insulin therapy was essential, led to a general recognition that the A-cell is "insulin sensitive". However, there were still two divergent hypotheses whose evolution is briefly described:

(1) In the original bihormonal hypothesis, UNGER proposed a causal primary A-cell defect in diabetes mellitus because it was thought that "physiologic" exogenous insulin failed to correct: (a) impaired glucagon suppression by glucose in IDDM (UNGER 1972); and (b) abnormal glucagon secretion after administration of arginine, protein, or glucose in NIDDM (UNGER et al. 1972; RASKIN et al. 1975, 1976).

(2) The intraislet glucagon-insulin feedback hypothesis implied that absolute or relative glucagon excess was a result of local absolute or relative insulin deficiency. Requirements for "supraphysiologic insulin" concentrations could be related to in vitro studies (WEIR et al. 1976; SAMOLS and HARRISON 1976a, 1977), suggesting a relative insulin resistance of intraislet A-cells normally exposed to very high concentrations of insulin.

With improvement in techniques of exogenous insulin administration, the concept of a primary A-cell defect in IDDM has been revised (UNGER and ORCI 1981a, b), as suppression of glucagon by hyperglycemia can be normalized by use of the "open-loop" system of insulin administration (RASKIN et al. 1979). However, it is still proposed that a primary A-cell defect exists in obese NIDDM with failure of exogenous insulin to "normalize" the diabetic glucagon hyperresponse to arginine or protein (RASKIN et al. 1978; UNGER and ORCI 1981b).

These phenomena are better explained by intraislet negative feedback than by a primary A-cell defect (also see Chap. 31). Briefly, in all forms of diabetes, other than obese NIDDM, the pancreatic A-cells may be visualized as relatively free of local insulin feedback because of severe B-cell deficiency. Exogenous insulin suppresses A-cells if it is administered in a way that mimics physiologic systemic levels, only because paracrine B-cell feedback is absent. In contrast, in a *normal* person, administration of exogenous insulin will not "normalize" the glucagon response to arginine (or protein meals) but will, by paracrine feedback (ASPLIN et al. 1981) *enhance* the glucagon response to arginine. In obese NIDDM, intraislet secretion may be low enough to allow hyperresponsiveness of the A-cell to arginine, but high enough so that the effects of exogenous insulin on the glucagon response to arginine balances out between: (1) that seen in IDDM (termed "normalization" but really inhibition); and (2) that in normal subjects (potentiation, which is the real normal response).

## I. Effect of Exogenous Insulin on Glucagon Release In Vivo

The negative feedback of insulin on glucagon secretion was demonstrated directly by infusing exogenous porcine insulin into the pancreatic artery of the duck and dog (SAMOLS et al. 1972). The exogenous insulin infusion caused an immediate suppression of pancreaticoduodenal venous blood glucagon levels, when glucose levels were unchanged or mildly decreasing. Similar studies in geese (SITBON and MIALHE 1978) were confirmatory. In dogs with phlorhizin-induced hypoglycemia, glucagon levels in excess of 500 pg/ml are common, even though the hypoglycemia is much less severe than in insulin-induced hypoglycemia. An infusion of insulin to a phlorhizin-induced-hypoglycemic dog decreases the hyperglucagonemia, even though it worsens the hypoglycemia. Thus, during hypoglycemia caused by high insulin levels, the hyperglucagonemia is less than in hypoglycemia caused by low insulin levels (DOBBS 1981).

Insulin-induced hypoglycemia in humans produces an increase in glucagon, despite the direct glucagon-suppressing action of insulin. Ultimately, however, glucagon levels decline even though hypoglycemia is maintained (GERICH et al. 1974b), presumably a consequence of the high insulin levels. By raising exogenous insulin levels by 200 μU/ml, SERVICE et al. (1978) showed that, if the level of glucose was maintained in the normal fasting range by normoglycemic clamp, there was a direct suppression of glucagon and C-peptide. The same dose of insulin, infused without a normoglycemic clamp, caused hypoglycemia, and a larger suppression of C-peptide, but the inhibition of glucagon was "overcome by hypoglycemia". We prefer to interpret the breakthrough of glucagon during insulin-induced suppression of glucagon as being a consequence of adrenergic stress, having emphasized that sympathoadrenergic stimulation may override either limb of the insulin-glucagon feedback system (SAMOLS et al. 1970, 1972). However, the degree of dependence (see Sect. B.II) or independence (BENSON et al. 1977) of glucopenia-induced glucagon secretion on catecholamines (local and circulating) remains controversial.

ASPLIN et al. (1981) have used the normoglycemic clamp to illustrate a major advance in the search for evidence of paracrine function. First, they infused exogenous insulin, raising the plasma insulin level by 30–100 μU/ml in normal humans. Both circulating glucagon and C-peptide (i.e., endogenous insulin) decreased. Next, by infusing arginine, which normally stimulates glucagon and insulin secretion, they found that the glucagon response to arginine was potentiated in proportion to the fall in C-peptide. To show that this augmentation of the glucagon response to arginine was not due to a metabolic effect of the infused insulin, similar studies were performed in IDDM patients who were C-peptide negative and thus had no paracrine insulin secretion. In these patients, the glucagon response to arginine was progressively inhibited by increasing concentrations of exogenous insulin. Their elegant conclusion is: hyperinsulinemia within the physiologic range achieved by the infusion of insulin inhibits endogenous insulin secretion which, via a paracrine mechanism, potentiates glucagon secretion. The problem of interpretation of the many experiments using exogenous insulin without: (1) C-peptide measurements; or (2) considering other A-cell stimuli, is evident. Whether the fall in basal (i.e., before the arginine stimulus) IRG is pancreatic or nonpancreatic (ASPLIN et al. 1981) in origin is still moot, but in humans we think the quantitative drop ($\sim$30 pg/ml) suggests a pancreatic origin.

In the presence of insulin deficiency diabetes, whether IDDM in humans, or alloxan-induced (BRAATEN et al. 1974) or streptozotocin-induced diabetes in dogs or rats, the hyperglucagonemia is readily reduced to normal by low doses of exogenous insulin (RASKIN et al. 1979; PFEIFER et al. 1979). In IDDM, the exquisite insulin sensitivity of the pancreatic A-cell may be similar to that of the canine gastric A-cell (LEFEBVRE and LUYCKX 1978) because of the absence of paracrine insulin which would normally downregulate insulin receptors on the A-cell. After total pancreatectomy in dogs, gastric A-cell glucagon secretion is readily suppressed by exogenous insulin (BLAZQUEZ et al. 1976).

## II. Effect of Endogenous Insulin on Glucagon Release In Vivo

Several combination patterns of endogenous glucose and insulin responses suggest that glucose does not suppress glucagon in the absence of an appropriate endogenous insulin response. Although increased glucose and insulin levels during carbohydrate meals are associated with suppression of endogenous glucagon, the combination of increased glucose but inhibited insulin (adrenergic stress) is associated with stimulated endogenous glucagon. In diabetes mellitus, hyperglycemia is associated with absolute or relative hyperglucagonism and with absolute or relative hypoinsulinism.

Several attempts have been made (NELSON et al. 1980; CLOUTIER et al. 1979; HAYASHI et al. 1977) to study interrelationships between endogenous insulin and glucagon using as a "model" patients with insulinoma. The human studies' conclusions were that no direct suppressive effect of insulin on glucagon was demonstrable. Glucagon levels tended to be normal (CLOUTIER et al. 1979) or elevated (NELSON et al. 1980) in most patients, but: (1) control values were based on normoglycemic subjects; (2) glucagon secretion by the tumors could not be excluded; and (3) statistical correlations showed that hyperinsulinism was not the cause of hypoglycemia in patients with insulinoma, indicating the fallibility of this type of statistical correlation. In contrast, in a rat insulinoma model, YOSHINA et al. (1979) concluded that nontumorous A-cell secretion was restrained by tumorous B-cell secretion. However, the relevance of these "models" to insulin-glucagon relations is inherently limited. The insulinoma probably does not mimic paracrine suppression, and these studies do not consider the effects of adrenergic stress during hypoglycemia. As the insulinoma "model" should not involve paracrine suppression, many of the findings described in individual patients were in accord with (a) those described for infusions of exogenous insulin, and (b) the reiterated proviso that sympathicoadrenergic activity will tend to override insulin-induced suppression of glucagon in normal islets.

## III. Effect of Exogenous Insulin on Glucagon Secretion In Vitro

In general, it is difficult to suppress glucagon secretion using exogenous insulin in a variety of in vitro systems, and relatively easy to do so in vivo in health and disease. The "customary concentration" theory, that the pancreatic A-cell is normally exposed to very high concentrations of endogenous insulin, is frequently but not invariably tenable. WEIR et al. (1976) estimated that the insulin concentration surrounding the A-cells might at times exceed 100 mIU/ml, and SAMOLS and HARRISON (1976a) inferred, from in vitro studies during infusions, that an intraislet increase of at least 25 mIU/ml insulin was critical to suppress the A-cell during glucopenia. Most in vitro studies testing A-cell suppression have used a perfusate insulin concentration of approximately 25 mIU/ml.

Suppression of glucagon secretion by exogenous insulin has been reported in a number of studies, under conditions favoring a low basal endogenous insulin secretion, using pancreas slices (BUCHANAN and MAWHINNEY 1973a), perfused rat pancreas – epinephrine or streptozotocin-treated by WEIR et al. (1976), alloxan-treated by PAGLIARA et al. (1975), normal rats (FUSSGÄNGER et al. 1976), perfused dog pancreas (SAMOLS and HARRISON 1976a), and perfused chicken pancreas (HONEY and WEIR 1979). Perhaps the low basal insulin secretion increases the number of insulin receptors usually down-regulated on the A-cell (WEIR et al. 1976). Although SAMOLS and HARRISON concluded that the negative feedback "operates best" during glucopenia and insulinopenia, mild glucagon suppression could also be demonstrated in the absence of glucose, as shown also by PAGLIARA et al. (1975).

However, there is a lack of consistency of glucagon suppression by exogenous insulin in vitro, whether for isolated normal dog pancreas (WEIR et al. 1979), streptozotocin-treated dog pancreas (HERMANSEN et al. 1979), or streptozotocin-treated rat pancreas (MATSCHINSKY et al. 1976). Differences in procedure, e.g., times of starvation of animals and of chronicity of streptozotocin may be pertinent. Moreover, reinterpretation of the study of HERMANSEN et al. (1979) suggests that exogenous insulin did slowly and steadily suppress glucagon secretion, especially compared with the stimulation of glucagon secretion by glucose alone. The slow A-cell response may in part be related to streptozotocin toxicity to D-cells, as the normal increase in somatostatin response to glucose was absent. As discussed in Chap. 31, paracrine somatostatin may contribute to the rapidity of suppression of the A-cell by glucose.

Why does there appear to be a difference in the sensitivity of the A-cell to exogenous insulin in vivo (maintaining glucose constant with a normoglycemic clamp) and in vitro? There are several possibilities: (1) the canine gastric A-cell and the A-cells of the IDDM patient are not juxtaposed to B-cells – suppression of immunoreactive glucagon by low concentrations of exogenous insulin in NIDDM subjects may reflect suppression of a (still questionable) nonpancreatic source (ASPLIN et al. 1981) of glucagon; (2) in vivo the exogenous insulin may change concentrations of nonglucose fuels (free fatty acids, ketone bodies, amino acids) in blood or modulate extrapancreatic hormones which could suppress immunoreactive glucagon; (3) the difference is neurally modulated – using an in situ isolated canine pancreas (i.e., innervated, but with a separate independent perfusion) we have obtained significant mild (−25%) suppression of glucagon release by an exogenous insulin perfusion which raised efflux insulin levels by approximately 100 μIU/ml (E. SAMOLS and J. STAGNER 1981, unpublished work); (4) the paracrine balance is different in vitro. During infusion of exogenous insulin, the net output of glucagon from the pancreas may depend on the balance of the degree of A-cell inhibition (by exogenous insulin) versus the degree of A-cell stimulation (caused by the decrease in paracrine endogenous insulin). The "force" stimulating the A-cell may be particularly augmented in vitro by glucopenic or amino acid-rich perfusates, unmodulated in vitro neuroendocrine agonism, nonrecirculation of normal pancreatic peptides, e.g., absence of normal autosuppression by circulating glucagon. On the other hand, the striking sensitivity of B- and D-cells to exogenous glucagon, and of A- and B-cells to exogenous somatostatin, is retained in vitro.

Studies on insulin suppression of canine gastric A-cell in vitro favor the neural (3) and/or paracrine (4) explanations. Using the isolated perfused canine stomach, LEFEBVRE and LUYCKX (1978) showed the gastric A-cell to be exquisitely sensitive to suppression by insulin. The gastric A-cell differs from the pancreatic A-cell in (a) being free of the proposed effects of paracrine insulin and (b) having different neural control (LEFEBVRE and LUYCKX 1980; LEFEBVRE et al. 1978).

In vitro studies have been able to separate the effects on the A-cell of glucose alone from those of glucose plus insulin. The canine gastric A-cell is not suppressed by glucose in the absence of insulin (LEFEBVRE and LUYCKX 1978, 1981). Using isolated rat islets, BUCHANAN and MAWHINNEY (1973a, b) found that glucagon release was increased from islets which were severely insulin deficient, but not from islets showing only moderate insulin insufficiency after streptozotocin treatment. The normal suppressive effect of glucose on glucagon release was also abolished in the severely insulin-deficient islets. Incubation of diabetic islets with insulin, 1 mIU/ml, suppressed glucagon release at high glucose concentrations. Addition of insulin to control islets had no effect, nor had the administration of insulin to diabetic animals before decapitation. Using guinea pig isolated islets, ÖSTENSON (1979) showed that in A-cell-rich islets (from streptozotocin-treated animals) 30 mIU/ml insulin exerted a powerful depressant action on glucagon release in the presence of glucose, whereas no significant effect of glucose alone was observed. In normal islets, glucagon release was not affected by exogenous insulin, but was suppressed by glucose. The lack of insulin effects on normal islets were presumably due to accumulation in these preparations of endogenous insulin sufficient to saturate insulin receptors on the A-cells.

## IV. Mechanism of the Glucagonosuppressive Effect of Insulin

At one end of the spectrum, insulin is not considered to have an important role in the suppression of glucagon by glucose (LECLERQ-MEYER et al. 1979; MATSCHINSKY et al. 1980). At the other, we have proposed (a) that the inhibitory effect of glucose on the A-cell is directly mediated by insulin (SAMOLS et al. 1972) and (b) that insulin may also inhibit glucagon in the absence of glucose (PAGLIARA et al. 1975; SAMOLS and HARRISON 1976a). Both (a) and (b) are "direct" suppression, but different mechanisms may be involved. It is possible that (a) occurs by increasing glucose entry into the cell (LEFEBVRE and LUYCKX 1979), or by influencing the rate or direction of intracellular substrate metabolism, or by furnishing energy. It is speculative whether (b) includes insulin-stimulated somatostatin release, cAMP decrease, miscellaneous receptor-directed events, etc. The evidence in support of our hypothesis is: (1) it is the rule, not the exception, for glucose not to suppress glucagon in the absence of insulin; (2) at a constant glucose level, an increase in insulin suppresses glucagon; (3) it is possible to suppress glucagon by increasing insulin in a medium devoid of glucose; (4) during hyperosmolar diabetic "comas", glucose entry and metabolism are accelerated by hyperosmolarity (KUZUYA and SAMOLS 1965), yet glucagon secretion is maximal.

Conceptually, use of the term "direct" suppression of glucagon by insulin emphasizes the relevance of the paracrine feedback concept in the pathogenesis of diabetes. It suggests that the aim in therapy should be restoration of normal insulin action of the A-cells. However, this should not detract from the potential role of intraislet somatostatin in suppression of the A-cell by glucose (Chap. 31).

It seems unlikely that A-cell suppression is induced by a simple increase in glucose entry into the cell, or increased glucose utilization, or ATP generation. Using A-cell-rich islets, ÖSTENSON (1979) showed that either increasing the glucose concentration or exogenous insulin resulted in an increase in glucose utilization and ATP concentration, yet only the insulin effectively suppressed A-cell secretion. Somatostatin suppressed glucagon secretion without changes in glucose utilization or ATP.

KUZUYA and SAMOLS (1965) have shown that hyperosmolarity per se increased glucose entry and its metabolism in adipose tissue and muscle. Comparison of the effects of hyperosmolarity, of insulin, and of increasing glucose concentrations without insulin in vitro revealed striking similarities in effects, but distinct dissimilarities were also noted. In adipose tissue, oxidation of glucose C-6 was increased by hyperosmolarity or increasing glucose concentrations, but not by insulin, while glycogen synthesis was much greater during insulin stimulation. In muscle, $CO_2$ production was stimulated by hyperosmolarity and increasing glucose concentrations, but not by insulin, whereas only insulin caused a marked stimulation of glycogen synthesis. Stimulation of substrate metabolism without glycogen synthesis within the A-cell may explain the paradoxical stimulatory effect of hyperglycemia on the A-cell in insulin-deprived, alloxan-induced diabetic dogs (BRAATEN et al. 1974), as the energy-providing metabolic intermediates, fumarate, glutamate and pyruvate (which are not efficient at glycogen synthesis) stimulate glucagon secretion.

The evidence presented here, the arguments against key relevance of a glucoreceptor (MALAISSE et al. 1979), and the fact that normally in vivo glucose does not suppress glucagon unless insulin is stimulated by it, lead us to propose the following relationship between glucose, insulin, and glucagon suppression. Insulin suppresses glucagon "directly", in the presence of glucose, by two mechanisms: (a) after accelerating glucose entry (not in itself of significance), by influencing the rate and *direction* of metabolism; and (b) influencing surface peptide receptors and paracrine events, e.g., somatostatin release (HONEY and WEIR 1979) for which glucose itself is augmentory. According to current peptide receptor theory (KING and CUATRECASAS 1981), the effect of (b) would likely be more prompt and of (a) would likely be delayed (10–30 min) because of the need for aggregation and internalization of the receptor to influence metabolic events directly. These two types of "direct" effects appear to be dissociable. Only (a) may occur in subjects with insulin deficiency diabetes, who, provided they have been treated with "background" insulin, may show after intravenous glucose administration, a very much delayed (after 30 min) suppression of glucagon (LARKINS et al. 1978).

## F. Evidence for and Against an Intraislet Negative Insulin-Glucagon Feedback

In contrast to healthy skepticism about the positive loop of the proposed feedback there has been broader acceptance of the negative loop perhaps because: (1) increased insulin and decreased glucagon levels are familiar relatives in common experimental situations after glucose infusions; (2) insulin is necessary for suppression of glucagon by hyperglycemia in diabetes; and (3) current anatomy is interpreted as favoring an effect of B-cell secretion on the A-cells (Chap. 31).

The argument that the negative loop is short, i.e., "paracrine", is supported by the fact that in vitro suppression of pancreatic A-cells requires high concentrations of exogenous insulin, and conditions favoring low basal endogenous insulin secretion. The failure to find suppression by exogenous insulin when endogenous insulin secretion is high suggests that maximal suppression of A-cells by endogenous insulin has already occurred (WEIR et al. 1976). Moreover, exogenous insulin decreases endogenous insulin which would augment forces stimulating glucagon secretion, unless enormous concentrations of offsetting exogenous insulin were given.

The effect of exogenous insulin on the glucagon response to arginine in humans is currently best explained by a local intraislet mechanism. In the presence of normal paracrine B-cells, exogenous insulin potentiates the glucagon response to arginine, probably via suppression of paracrine B-cells (ASPLIN et al. 1981). In the absence of paracrine B-cells, in C-peptide-negative IDDM patients exogenous insulin inhibits (or "normalizes", a misleading term) the glucagon hyperresponse to arginine by simple long-loop suppression. In the presence of modestly defective paracrine B-cells, in obese NIDDM patients exogenous insulin has no effect on the glucagon hyperresponse to arginine. This hyperresponse is presumably caused by a relative insulin deficiency, and is neither potentiated (a normal paracrine response) nor inhibited (an absent paracrine response) by exogenous insulin (see Sect. E).

Strongly suggesting intraislet negative feedback in the isolated pancreas is the sequence of changes during tolbutamide infusion. Stimulation of the A-cells by tolbutamide, followed by their presumed paracrine suppression by B-cell (SAMOLS and HARRISON 1976a) and/or D-cell stimulation (SAMOLS et al. 1978) is described in Chap. 31. Also favoring intraislet negative feedback is the fact that in the isolated perfused canine pancreas, perfused with a glucose-Ringer-bicarbonate medium for 5–6 h, elevation of the glucose concentration causes increased insulin (and somatostatin) plus suppression of glucagon. If at least "some insulin is necessary for glucose to suppress the A-cell" (LEFEBVRE and LUYCKX 1979), the only source of insulin is the endogenous islets.

Studies using anti-insulin serum, superficially a potential tool for the study of intraislet negative feedback, have not been helpful in either "proving" or "disproving" the thesis. Passive immunization of rats with anti-insulin serum did increase endogenous glucagon secretion (UNGER and LEFEBVRE 1972). Even if (for argument, as the phenomenon may be complex) the A-cell stimulation does reflect the simple consequence of acute insulin deficiency, this phenomenon presumably would favor a long-loop negative insulin-glucagon feedback. We have (E. SAMOLS, 1975, unpublished work) been unable to detect an effect of anti-insulin serum on glucagon secretion by the isolated perfused canine pancreas, presumably because globulins remain in the vascular compartment. Although we have not been able to find published data on the effect of anti-insulin serum on A-cell secretion by isolated islets, it seems that, regardless of the result, interpretation will be difficult in terms of normal intraislet glucagon-insulin interrelationships. The potential technical problems would be similar to those bedeviling the interpretation of the studies of TRIMBLE et al. (1981). LECLERCQ-MEYER et al. (1973) failed to find any significant enhancement of glucagon release after the addition of anti-insulin serum to the incubation medium surrounding isolated pieces of pancreas, a finding which could be interpreted as being for, against, or of no relevance to the negative feedback in question.

In essence, the "pro-paracrine" argument relies heavily on the following points: (1) glucose alone will not suppress the A-cell; (2) glucose suppresses the A-cell in the isolated perfused pancreas; and (3) recognized A-cell suppressors, insulin and somatostatin, are present in the islets and their release is stimulated by glucose. As discussed in Chap. 31, both B- and D-cells may well complement or augment one another to facilitate suppression.

In a nutshell, the "anti-paracrine" argument, for insulin at least, suggests that the need of the A-cell for insulin is "permissive", i.e., the A-cell requires a minimal insulin level for suppression (LARKINS et al. 1980) or "optimal" (MATSCHINSKY et al. 1980) insulin level for maximal glucose sensitivity. These arguments depend upon the demonstration of suppression of glucagon by glucose in the presence of constant very low levels of insulin in the circulation.

At present, both pro and anti-hypotheses remain viable. It is not inconceivable that both are correct, e.g., if there is an optimal concentration of background insulin, an acute rise in local somatostatin, released by glucose, could acutely suppress the A-cell. Moreover, an acute rise in local insulin, if it occurred, might also acutely suppress the A-cell.

## G. New Hypotheses and Concepts in Local Intraislet Regulation of A- and B-cells

### I. Compartmentalization: Systemic Versus Local Intraislet Effects

There is a dilemma regarding the question of physiologic "paracrine" levels: on the one horn, SAMOLS and HARRISON (1976b) emphasized the remarkable potency of influx somatostatin on the B-cell, while on the other horn, SAMOLS and HARRISON (1976a) argued that very high levels of influx insulin were necessary to mimic paracrine physiology! KAWAI and ROUILLER (1981) logically argue that if the islet interstitium were a single compartment through which islet hormones could reach all islet cells, islet cell function should not be affected by changes in the arterial concentration of an islet hormone that is but a small fraction of the intraislet level. They propose a two-compartment model with functional separation of receptors in the "arterial space" from hormones secreted into the "venous space". Possible anatomic or physiologic methods of compartmentalization, at this stage both intriguing and speculative, are discussed in Chap. 31 including homocellular core versus heterocellular mantle, flexible tight junctions, polarity or specificity of receptors, unique islet vasculature, and coordination through gap junctions. It is usually assumed that the paracrine environs may contain very high concentrations of insulin and/or glucagon in the form in which they normally circulate in blood. However there may also be: (1) steric differences of the hormone (e.g., dimers, hexamers of insulin) in the interstitium; and (2) prohormones and their remnants, which by modulating specific receptors, would create a "functional" intraislet compartment.

### II. The Principle of Uncertainty in Studies of Local Islet Interactions

It is possible that investigational procedures which disrupt the extraordinary complex islet organ may inherently mask normal physiologic augmentation or suppressive mechanisms. Cellular toxins are not specific for either the B- or A-cell, and introduction of specific antiserum to the local intraislet compartments is problematic. Isolated islet or cellular preparations disrupt normal anatomic barriers and could be misleading in terms of normal function. Independent modulation by glucose, of either B- or A-cells, free of surrounding glucagon, somatostatin, or insulin, has not been demonstrated to date.

### III. Possible Paracrine Effects on Growth and Differentiation

If local intraislet interactions influence secretion responses, one would expect that "paracrine" effects could also modulate cell growth and differentiation. As glucagon appears to be the first islet hormone to appear in fetal rat pancreas (LIKE and ORCI 1972; RALL et al. 1973; DOI et al. 1978) and as the glucagon cell population throughout life is relatively stable compared with the other cell types (MALAISSE-LAGAE et al. 1979; ORCI and PERRELET 1981), it is possible that the A-cell supplies an elemental signal for differentiation and growth. In this respect, see Chap. 21 on the effects of glucagon on liver regeneration.

## H. Synopsis and Conclusions

There are two incontrovertible facts regarding the interrelationship of glucagon and insulin. One is that glucagon will promote the secretion of insulin and the other is that insulin, at least in vivo, will suppress secretion of glucagon. The insulinotropic action of glucagon has been utilized as a test of secretion in a wide variety of disease states, notably as an adjunct in the diagnosis of insulinoma, or of functional B-cell reserve in IDDM. The glucagonosuppressive effect of insulin is thought to be an additional reason for the efficacy of insulin in the treatment of diabetes mellitus.

How has the hypothesis of SAMOLS and MARKS, of an intraislet positive-negative glucagon-insulin feedback (see Sect. A) fared in the past decade? On the one hand, reinforced by the attractiveness of a paracrine role for intraislet D-cells, the concept of intraislet interactions of A- and B-cells has received attention and recognition and there has been some further support for the hypothesis. On the other, there is no direct proof of, and there is also evidence against, the hypothesis. However, we conclude that the hypothesis is still viable; that currently the best evidence is that for an intraislet negative feedback, i.e., insulin inhibiting glucagon secretion. Except for the studies of TRIMBLE et al. (1981), the evidence that glucagon stimulates insulin secretion is either explicable by long-loop stimulation (active and passive immunization) or is disconcertingly circumstantial. Given the difficulty of either proving or disproving the hypothesis, we should perhaps apologize for proceeding into an explanation of the potential relevance of the hypothesis to physiology and pathology.

In humans, the secretion of insulin into the portal blood is not directly proportional to the blood glucose level, but will be influenced by the degree of inhibition (adrenergic stress) or the degree of amplification (feeding) of glucose-stimulated insulin secretion. Assuming that the insulinotropic effect of glucagon is expressed via an intraislet pathway, this would provide an extraordinarily effective preemptive (i.e., anticipatory) amplification of insulin secretion so that there are only minor increases in blood glucose concentration during a mixed meal. Under conditions (e.g., hypoglycemia) which activate powerful mechanisms for the inhibition of insulin secretion, the insulinotropic effect of glucagon, although diminished by the low glucose and α-adrenergic activation, may be useful to ensure that insulin secretion does not completely cease. In the interprandial period, when glucose, insulin, and glucagon levels appear to oscillate, glucagon would be the ideal amplifier of insulin secretion, because it enhances its own insulinotropic effect by temporarily raising blood glucose.

Glucagon concentrations in the portal blood are frequently not inversely proportional to the glucose concentration during protein meals and stress situations. Glucose itself does not properly suppress glucagon secretion in diabetes. Insulin is important (regardless of whether its action is "direct" or "permissive" or whether there is a contribution by somatostatin) in determining the degree and rapidity of suppression of the A-cell by glucose. An intraislet negative insulin-glucagon feedback would provide an exquisitely sensitive mechanism for precise determination of the concentration of portal insulin and glucagon, a ratio which seems to be of primary importance in glucose homeostasis. This feedback should allow amplification of the positive or negative loop, depending on the blood glucose level, and spontaneous corrective autoregulation. Alternating changes in glucagon-insulin feedback would permit glucose stored in the liver to be transferred to the periphery with minimal fluctuations in blood glucose, since the glucose released by brief periods of glycogenolysis (higher glucose: insulin ratio) would be efficiently assimilated (high insulin). Similarly, the very early release of glucagon during ingestion of a mixed meal could (higher glucose: insulin ratio) transport glycogen from the liver as well as encourage its deposition in the periphery (high insulin) with minimal perturbation of circulating blood glucose.

If there were no negative paracrine insulin feedback, then the secretory responses of the A-cell, as in IDDM, could be feeble, late, or excessive/paradoxical; i.e., in the absence of the normal inhibitory force, the stimulatory forces become prominent, as in the glucagon response to arginine (excessive)/oral glucose (paradoxical). The responses of the A-cell to exogenous insulin therapy in IDDM, NIDDM, and normal subjects is currently best explained by the degree of destruction or retention of normal intraislet B-cell negative feedback.

*Acknowledgment*. The author thanks Dr. G. C. Weir for many hours of discussion and criticism, appreciates the superb secretarial assistance of C. S. Meyers, and is grateful to G. H. Samols for inestimable research assistance.

## References

Adamson V, Cerasi E (1975) Acute suppressive effect of human growth hormone on insulin release induced by glucagon and tolbutamide in man. Diab Metab 1:51–56
Aguilar-Paradar E, Eisentraut AM, Unger RH (1969) Effects of starvation on plasma pancreatic glucagon in normal man. Diabetes 18:717–723
Alford FP, Bloom SR, Nabarro JDN, Hall R, Besser GM, Coy DH, Kasting AJ, Schally AV (1974) Glucagon control of fasting glucose in man. Lancet 2:974–977
Arky RA, Freinkel N (1966) Alcohol hypoglycemia. V. Alcohol infusion to test gluconeogenesis in starvation with special reference to obesity. N Engl J Med 274:426–433
Asplin CM, Paquette TL, Palmer JP (1981) In vivo inhibition of glucagon secretion by paracrine beta cell activity in man. J Clin Invest 68:314–318
Assan R, Hautecouverture G, Guillemant S, Dauchy F, Protin P, Derot M (1969) Evolution of hormonal parameters (glucagon, cortisol, growth hormone) and energetic parameters (glucose, fatty acids, free glycerol) in 10 severe cases of diabetic acido-ketosis under treatment. Pathol Biol (Paris) 17:1095–1105
August GP, Hung W (1976) Impaired glucose, insulin and adenosine 3′,5′-monophosphate responses to glucagon in growth hormone deficient children. J Clin Endocrinol Metab 43:1029–1035
AvRuskin TW, Crigler JR Jr, Soeldner JS (1979) Turner's syndrome and carbohydrate metabolism. I. Impaired insulin secretion after tolbutamide and glucagon stimulation tests: evidence of insulin deficiency. Am J Med Sci 277:145–162
Aydin I, Raskin P, Unger RH (1977) The effect of short-term intravenous insulin administration on the glucagon responses to a carbohydrate meal in adult-onset and juvenile type diabetes. Diabetologia 13:629–636
Barling P, Beloff-Chain A (1973) Studies on the administration of glucagon and insulin antibodies to rats. Horm Metab Res 5:154–159
Bennett PH, Aronoff SL, Unger RH (1976) Evidence for an insulin-independent alpha-cell abnormality in human diabetes. Metabolism [Suppl 1] 25:1527–1529
Benson JW Jr, Johnson DG, Palmer JP, Werner PL, Ensinck JW (1977) Glucagon and catecholamine secretion during hypoglycemia in normal and diabetic man. J Clin Endocrinol Metab 44:459–464
Bhathena SJ, Higgins GA, Recant L (1981) Glucagonoma and glucagonoma syndrome. In: Unger RH, Orci L (eds) Glucagon. Physiology, pathophysiology and morphology of the pancreatic A-cells. Elsevier/North Holland, Amsterdam Oxford New York, pp 413–438
Binnick AN, Spencer SK, Dennison WL, Horton ES (1977) Glucagonoma syndrome. Report of two cases and literature review. Arch Dermatol 113:749–754
Blazquez E, Muñoz-Barragan L, Patton GS, Orci L, Dobbs RE, Unger RH (1976) Gastric A-cell function in insulin-deprived depancreatized dogs. Endocrinology 99:1182–1188
Bloom SR, Polak JM (1978) The glucagonoma syndrome. Adv Exp Med Biol 106:183–194
Blum I, Doron M, Laron Z (1978) Lack of influence of arginine preloading on the insulin response to i.v. glugacon in children and adolescents. Acta Diabetol Lat 15:68–71
Bomboy JD Jr, Lewis SB, Lacy WW, Liljenquist JE (1977) Transient stimulatory effect of sustained hyperglucagonemia on splanchnic glucose production in normal and diabetic man. Diabetes 26:177–184
Bonner-Weir S, Weir GC (1979) The organization of the endocrine pancreas: a hypothetical unifying view of the phylogenetic differences. Gen Comp Endocrinol 38:28–37
Braaten JT, Faloona GR, Unger RH (1974) The effect of insulin on the alpha-cell response to hyperglycemia in long-standing alloxan diabetes. J Clin Invest 53:1017–1021
Buchanan KD, Mawhinney WAA (1973a) Insulin control of glucagon release from insulin-deficient rat islets. Diabetes 22:801–803
Buchanan KD, Mawhinney WAA (1973b) Glucagon release from isolated pancreas in streptozotocin-treated rats. Diabetes 22:797–800
Cahill GF Jr (1976) Insulin and glucagon. In: Parsons JA (ed) Peptide hormones. University Park Press, Baltimore, pp 85–103
Capito K, Hedeskov CJ (1977) The effect of glucose, glucose metabolites and calcium ion on adenylate cyclase activity in homogenates of mouse pancreatic islets. Biochem J 162:569–573

Cassel D, Levkovitz H, Selinger Z (1977) The regulatory GTPase cycle of turkey erythrocyte adenylate cyclase. J Cyclic Nucleotide Res 3:393–406

Charles MA, Fanska R, Schmid FG, Forsham PH, Grodsky GM (1973) Adenosine-3′: 5′-monophosphate in pancreatic islets: glucose-induced insulin release. Science 179:569–571

Cherrington AD, Liljenquist JE (1981) Role of glucagon in regulating glucose production in vivo. In: Unger RH, Orci L (eds) Glucagon. Physiology, pathophysiology and morphology of the pancreatic A-cells. Elsevier/North Holland, Amsterdam Oxford New York, pp 221–253

Christensen NJ, Iversen J (1973) Release of large amounts of noradrenaline from the isolated perfused canine pancreas during glucose deprivation. Diabetologia 9:396–399

Cloutier MG, Pek S, Crowther RL, Floyd JC, Fajans SS (1979) Glucagon-insulin interactions in patients with insulin-producing pancreatic islet lesions. J Clin Endrinol Metab 48:201–206

Corredor DG, Sabeh G, Mendelsohn LV, Wasserman RE, Sunder JH, Danowski TS (1969) Enhanced postglucose hypophosphatemia during starvation therapy of obesity. Metabolism 18:754–763

Crockford PM, Hazzard WR, Williams RH (1969) Insulin response to glucagon. The opposing effects of diabetes and obesity. Diabetes 18:216–224

Curry DL (1970) Glucagon potentiation of insulin secretion by the perfused rat pancreas. Diabetes 19:420–428

Cutillo S, Ansanelli V, Stoppoloni G, Pacelli V, D'Onofrio F (1968) Insulin response to glucagon in obese children. Lancet 1:1188

Day JL, Johansen K, Ganda OP, Soeldner JS, Gleason RE, Midgley W (1978) Factors governing insulin and glucagon responses during normal meals. Clin Endocrinol 9:443–454

Dobbs RE (1981) Control of glucagon secretion: nutrients, gastroenteropancreatic hormones, calcium, and prostaglandins. In: Unger RH, Orci L (eds) Glucagon. Physiology, pathophysiology and morphology of the pancreatic A-cells. Elsevier/North Holland, Amsterdam London New York, pp 115–133

Doi K, Yoshida M, Utsumi M, Kawara A, Fujii S, Sakoda M, Baba S (1978) Developmental patterns of insulin, glucagon and somatostatin in fetal newborn and adult rat pancreas. In: Baba S, Kaneko T, Yanaihara N (eds) Proinsulin, insulin and C-peptide. Excerpta Medica, Amsterdam London New York, pp 432–441

Epand RM, Douglas RJ (1973) The effect of glucagon antibodies on plasma glucose and insulin levels. Biochim Biophys Acta 320:741–744

Epple A (1968) Comparative studies on the pancreatic islets. Endocrinol Jpn 15:107–122

Espinosa De Los Monteros A, Driscoll SG, Steinke J (1970) Insulin release from isolated human fetal pancreatic islets. Science 168:1111–1112

Esterhuizen AC, Spriggs TL, Lever JD (1968) Nature of islet-cell innervation in the cat pancreas. Diabetes 17:33–36

Frankel BJ, Gerich JE, Hagura R, Fanska RE, Gerritsen GC, Grodsky GM (1974) Abnormal secretion of insulin and glucagon by the in vitro perfused pancreas of the genetically diabetic Chinese hamster. J Clin Invest 53:1637–1646

Fujimoto WY, Ensinck JW, Merchant FW, Williams RH, Smith PH, Johnson DG (1978) Stimulation by gastric inhibitory polypeptide of insulin and glucagon secretion by rat islet cultures. Proc Soc Exp Biol Med 157:89–93

Fujita Y, Gotto A, Unger RH (1975) Relationships of insulin and glucagon to triglyceride levels during a high and low carbohydrate intake. Diabetes 24:552–558

Fussgänger RD, Süssmann H, Hager G, Heinze E, Schleyer M, Pfeiffer EF (1976) Glucagon-insulin interrelationships in the regulation of islet cell functions. In: Proceedings of the Vth Int Congr of Endocrinol, Hamburg, July 18–24, 1976, vol 2. Excerpta Medica, Amsterdam London New York, pp 558–567

Garcia MJ, Czerwinski C, De Santis R, Lan VV, Ramey E, Penhos JC (1973) Hyperglycemic and insulinogenic effects of intravenous glucagon at different blood glucose levels. Proc Soc Exp Biol Med 143:707–710

Gerich JE, Lorenzi M, Schneider V, Forsham PH (1974a) Effect of somatostatin on plasma glucose and insulin responses to glucagon and tolbutamide in man. J Clin Endocrinol Metab 39:1057–1060

Gerich JE, Schneider V, Dippe SE, Langlois M, Noacco C, Karam JH, Forsham PH (1974b) Characterization of the glucagon response to hypoglycemia in man. J Clin Endocrinol Metab 38:77–82
Gerich JE, Langlois M, Naocco C, Lorenzi M, Karam JH, Forsham PH (1976) Comparison of the suppressive effects of elevated plasma glucose and free fatty acid levels on glucagon secretion in normal and insulin-dependent diabetic subjects. Evidence for selective α-cell insensitivity to glucose in diabetes mellitus. J Clin Invest 58:320–325
Goldfine ID, Cerasi E, Luft R (1972a) Glucagon stimulation of insulin release in man: inhibition during hypoglycemia. J Clin Endocrinol Metab 35:312–315
Goldfine ID, Roth J, Birnbaumer L (1972b) Glucagon receptors in β-cells. J Biol Chem 247:1211–1218
Grasso S, Messina A, Saporito N, Reitano G (1970) Effect of theophylline, glucagon and theophylline plus glucagon on insulin secretion in the premature infant. Diabetes 19:837–841
Grey N, McGuigan JE, Kipnis DM (1970) Neutralization of endogenous glucagon by high titer glucagon antiserum. Endocrinology 86:1383–1388
Grill V, Cerasi E (1974) Stimulation by D-glucose of cyclic adenosine 3′:5′-monophosphate accumulation and insulin release in isolated pancreatic islets of the rat. J Biol Chem 249:4196–4201
Grodsky GM, Bennett LL, Smith DF, Schmid FG (1967) Effect of pulse administration of glucose or glucagon on insulin secretion in vitro. Metabolism 16:222–223
Haist RE (1965) Effects of changes in stimulation on the structure and function of islet cells. In: Leibel BS, Wrenshall GA (eds) On the nature and treatment of diabetes. Excerpta Medica, Amsterdam London New York, pp 12–30
Halter JB, Porte D Jr (1981) Current concepts of insulin secretion in diabetes mellitus. In: Rifkin H, Raskin P (eds) Diabetes mellitus, vol V. Brady, Bowie, Maryland, pp 33–42
Halter JB, Graf RJ, Porte D Jr (1979) Potentiation of insulin secretory responses by plasma glucose levels in man: evidence that hyperglycemia in diabetes compensates for impaired glucose potentiation. J Clin Endocrinol Metab 48:946–954
Haro EN, Blum SF, Falcon WW (1965) The glucagon response of fasting obese subjects. Metabolism 14:976–984
Hayashi M, Floyd JC Jr, Pek S, Fajans SS (1977) Insulin, proinsulin, glucagon and gastrin in pancreatic tumors and in plasma of patients with organic hyperinsulinism. J Clin Endocrinol Metab 44:681–694
Hendriksen C, Faber OK, Drejer J, Binder C (1977) Prevalence of residual B-cell function in insulin-treated diabetics evaluated by the plasma C-peptide response to intravenous glucagon. Diabetologica 13:615–619
Hermansen K, Ørskov H, Christensen SE (1979) Streptozotocin diabetes: a glucoreceptor dysfunction affecting D-cells as well as B- and A-cells. Diabetologia 17:385–389
Higgins GA, Recant L, Fischman AB (1979) The glucagonoma syndrome: surgically curable diabetes. Am J Surg 137:142–148
Holst JJ, Galbo H, Richter EA (1978) Neutralization of glucagon by antiserum as a tool in glucagon physiology. Lack of depression of basal blood glucose after antiserum treatment in rats. J Clin Invest 62:182–190
Honey RN, Weir GC (1979) Insulin stimulates somatostatin and inhibits glucagon secretion from the perfused chicken pancreas-duodenum. Life Sci 24:1747–1750
Itoh M (1981) Role of pancreatic somatostatin in starvation induced changes in insulin and glucagon secretion. Diabetes [Suppl 1] 30:117A
Iversen J (1973) Adrenergic receptors and the secretion of glucagon and insulin from the isolated perfused canine pancreas. J Clin Invest 52:2102–2116
Karp M, Laron Z, Doron M (1975) Insulin response to intravenous glucagon in children with familial constitutional short stature. Arch Dis Child 50:805–808
Kawai K, Rouiller D (1981) Evidence that the islet interstitium contains functionally separate "arterial" and "venous" compartments. Diabetes [Suppl 1] 30:14A
Ketterer H, Eisentraut AM, Unger RH (1967) Effect upon insulin secretion of physiologic doses of glucagon administered via the portal vein. Diabetes 16:283–288
King AC, Cuatrecasas P (1981) Peptide hormone-induced receptor mobility, aggregation, and internalization. N Engl J Med 305:77–88

Kumar D, Mehtalia SD, Miller LV (1974) Diagnostic use of glucagon-induced insulin response. Studies in patients with insulinoma or other hypoglycemic conditions. Ann Intern Med 80:697–701
Kuzuya T, Samols E (1965) Stimulation by hyperosmolarity of glucose metabolism in rat adipose tissue and diaphragm in vitro. J Biol Chem 240: 2277–2283
Lacy PE (1977) The physiology of insulin release. In: Volk BW, Wellman KF (eds) The diabetic pancreas. Plenum, New York, pp 211–230
Lambert AE (1976) The regulation of insulin secretion. Rev Physiol Biochem Pharmacol 75:98–159
Larkins RG, Martin FIR, Alford FP, Chisholm DJ (1978) Relationship between α and β cell function before and after metabolic control in ketotic diabetic subjects. J Clin Endocrinol Metab 46:131–139
Leclerq-Meyer V, Marchand J, Malaisse WJ (1973) The effect of calcium and magnesium on glucagon secretion. Endocrinology 93:1360–1370
Leclerq-Meyer V, Marchand J, Leclerq R, Malaisse WJ (1979) Interactions of α-ketoisocaproate, glucose and arginine in the secretion of glucagon and insulin from the perfused rat pancreas. Diabetologia 17:121–126
Lefèbvre P (1966) The physiological effect of glucagon on fat-mobilisation. Diabetologia 2:130–132
Lefèbvre P (1968) Current theories concerning the physiological role of glucagon. Acta Diabetol Lat 5:143–159
Lefèbvre P, Luyckx A (1969) Lipolytic action of glucagon in vitro and in vivo. In: Vague J (ed) Physiopathology of adipose tissue. Excerpta Medica, Amsterdam London New York, pp 257–267
Lefèbvre PJ, Luyckx AS (1978) Glucose and insulin in the regulation of glucagon release from the isolated perfused dog stomach. Endocrinology 103:1579–1582
Lefèbvre PJ, Luyckx AS (1979) Glucagon and diabetes: a reappraisal. Diabetologia 16:347–354
Lefèbvre PJ, Luyckx AS (1980) Neurotransmitters and glucagon release from the isolated, perfused canine stomach. Diabetes 29:697–701
Lefèbvre PJ, Luyckx AS (1981) The physiology of extrapancreatic glucagon. In: Unger RH, Orci L (eds) Glucagon. Physiology, pathophysiology and morphology of the pancreatic A-cells. Elsevier/North Holland, Amsterdam Oxford New York, pp 335–348
Lefèbvre PJ, Luyckx AS, Brassinne AH (1978) Vagal stimulation and its role in eliciting gastrin but not glucagon release from the isolated perfused dog stomach. Gut 19:185–188
Like AA, Orci L (1972) Embryogenesis of the human pancreatic islets: a light and electron microscopic study. Diabetes [Suppl 2] 21:511–534
Majid PA, Saxton C, Dykes JRW, Galvin MC, Taylor SH (1970) Autonomic control of insulin secretion and the treatment of heart failure. Br Med J 4:328–334
Malaisse WJ, Malaisse-Lagae F, Mayhew D (1967) A possible role for the adenyl-cyclase system in insulin secretion. J Clin Invest 46:1724–1734
Malaisse WJ, Sener A, Koser M, Ravazzola M, Malaisse-Lagae F (1977) The stimulus-secretion coupling of glucose-induced insulin release. Insulin release due to glycogenolysis in glucose-deprived islets. Biochem J 164:447–454
Malaisse WJ, Sener A, Herchuelz A, Hutton JC (1979) Insulin release: the fuel hypothesis. Metabolism 28:373–386
Malaisse-Lagae F, Stefan Y, Cox J, Perrelet A, Orci L (1979) Identification of a lobe in the adult human pancreas rich in pancreatic polypeptide. Diabetologia 17:361–365
Marks V (1972) Glucagon. Clin Endocrinol Metab 1(3):829–845
Marks V, Samols E (1968 a) The insulinogenic effect of glucagon and plasma immunoreactive glucagon in obesity. Rev Med 21:1373–1382
Marks V, Samols E (1968 b) The glucagon test for insulinoma. J Clin Pathol 21:346–352
Marks V, Samols E (1968 c) Diazoxide in the management of intractable hypoglycemia. Ann NY Acad Sci 150:442–454
Matschinsky FM, Pagliara AS, Hover BA, Pace CS, Ferrendelli JA, Williams A (1976) Hormone secretion and glucose metabolism in islets of Langerhans of the isolated perfused pancreas from normal and streptozotocin diabetic rats. J Biol Chem 254:6053–6064

Matschinsky FM, Rujanavech C, Pagliara AS, Norfleet WT (1980) Adaptations of $\alpha_2$- and $\beta$-cells of rat and mouse pancreatic islets to starvation, to refeeding after starvation, and to obesity. J Clin Invest 65:207–218
McGavran MH, Unger RH, Recant L, Polk HC, Kilo C, Levin ME (1966) A glucagon-secreting alpha-cell carcinoma of the pancreas. N Engl J Med 274:1408–1413
Metz R, Nice M, Berger S, Mako M (1968) Preservation of insulin and liver glycogen stores during very long fasts. J Lab Clin Med 71:573–581
Milner RD (1969) Stimulation of insulin secretion in vitro by essential aminoacids. Lancet 1:1075–1076
Moss J, Vaughan M (1979) Activation of adenylate cyclase by choleragen. Annu Rev Biochem 48:581–600
Müller WA, Faloona GR, Unger RH (1973) Hyperglucagonemia in diabetic ketoacidosis. Am J Med 54:52–57
Nelson RL, Service FJ, Go VLW (1980) Interrelationships among insulin, glucagon, and gastric inhibitory polypeptide in insulinoma. Mayo Clin Proc 55:138–145
Nishino T, Shima K, Shimomura Y, Tanaka R, Kodaira T, Imagawa K, Kumahara Y (1980) A study of the physiological role of glucagon. Passive immunization with antiserum specific for pancreatic glucagon in rats. Endorcrinol Jpn 1:109–113
Oakley NW, Harrigan P, Kissbah AH, Kissin EA, Adams PW (1972) Factors affecting insulin response to glucagon in man. Metabolism 21:1001–1007
Ohneda A, Watanabe K, Horigome K, Sakai T, Kai Y, Oikawa S-I (1978) Abnormal response of pancreatic glucagon to glycemic changes in diabetes mellitus. J Clin Endocrinol Metab 46:504–510
Orci L, Perrelet A (1981) The morphology of the A-cell. In: Unger RH, Orci L (eds) Glucagon. Physiology, pathophysiology, and morphology of the Pancreatic A-cells. Elsevier/North Holland, Amsterdam Oxford New York, pp 3–36
Orci L, Pictet R, Rouiller C (1967) Demonstration of the extracellular space in the islets of Langerhans. Its suggested importance in normal and pathological situations. Excerpta Int Congr Ser 140:97
Orci L, Malaisse-Lagae F, Ravazzola M, Rouiller D, Renold AE, Perrelet A, Unger RH (1975) A morphological basis for the intercellular communication between $\alpha$- and $\beta$-cells in the endocrine pancreas. J Clin Invest 56:1066–1070
Östenson C-G (1979) Regulation of glucagon release: effects of insulin in the pancreatic $A_2$-cell of the guinea pig. Diabetologia 17:325–330
Pagliara AS, Stillings SN, Haymond MW, Hover BA, Matschinsky FM (1975) Insulin and glucose as modulators of the amino acid-induced glucagon release in the isolated pancreas of alloxan and streptozotocin diabetic rats. J Clin Invest 55:244–255
Pek S, Tai T-Y, Crowther R, Fajans SS (1976) Glucagon release precedes insulin release in response to common secretagogues. Diabetes 25:764–770
Perkins JP (1973) Adenyl cyclase. Adv Cyclic Nucleotide Res 3:1–64
Pfeifer MA, Samols E, Wolter CF, Winkler CF (1979) Low dose versus high dose insulin therapy for diabetic ketoacidosis. South Med J 72:149–154
Pierluissi J, Campbell J (1980) Metasomatrophic diabetes and its induction: basal insulin secretion and insulin release responses to glucose, glucagon, arginine and meals. Diabetologia 18:223–228
Porte D (1969) Regulation of insulin secretion in vivo by glucose. In: Gual C, Ebling FJG (eds) Progress in endocrinology. Excerpta Medica, Amsterdam London New York, pp 192–201
Porte D Jr, Graber AL, Kuzuya T, Williams RH (1966) The effect of epinephrine on immunoreactive insulin levels in man. J Clin Invest 45:228–236
Preeyasombat C, Pitchayayothin N, Veravekin A (1975) The glucagon stimulation test in normal Thai children: effect on blood glucose, growth hormone and insulin. J Med Assoc Thai 58:459–464
Rall LB, Pictet RL, Williams RH, Rutter WJ (1973) Early differentiation of glucagon-producing cells in embryonic pancreas: a possible developmental role for glucagon. Proc Natl Acad Sci USA 70:3478–3482
Raskin P, Fujita Y, Unger RH (1975) Effect of insulin-glucose infusions on plasma glucagon levls in fasting diabetics and nondiabetics. J Clin Invest 56:1132–1138

Raskin P, Aydin I, Unger RH (1976) Effect of insulin on the exaggerated glucagon response to arginine stimulation in diabetes mellitus. Diabetes 25:227–229
Raskin P, Aydin I, Yamamoto T, Unger RH (1978) Abnormal alpha cell function in human diabetes: the response to oral protein. Am J Med 64:988–997
Raskin P, Pietri A, Unger RH (1979) Changes in glucagon levels after 4–5 weeks of glucoregulation by portable insulin infusion pumps. Diabetes 28:1033–1035
Riddle MC, Golper TA, Fletcher WS, Ensinck JW, Smith PH (1978) Glucagonoma syndrome in a 19-year-old woman. West J Med 129:68–72
Rodbell M (1981) The actions of glucagon on the adenylate cyclase system. In: Unger RH, Orci L (eds) Glucagon. Physiology, pathophysiology, and morphology of the pancreatic A-cells. Elsevier, Amsterdam Oxford New York, pp 177–193
Ryan WG, Nibbe AF, Schwartz TB (1976) Beta-cytotrophic effects of glucose, glucagon, and tolbutamide in man. Lancet 1:1255–1256
Samols E (1965) Immunochemical aspects of insulin. In: Leibel BS, Wrenshall GA (eds) On the nature and treatment of diabetes. Excerpta Medica, Amsterdam London New York, pp 227–249
Samols E, Harrison J (1976a) Intraislet negative insulin-glucagon feedback. Metabolism 25:1443–1447
Samols E, Harrison J (1976b) Remarkable potency of somatostatin as a glucagon suppressant. Metabolism 25(11):1495–1497
Samols E, Harrison J (1977) Tolbutamide: stimulator and suppressor of glucagon secretion. In: Foà PP, Bajaj JS, Foà NL (eds) Glucagon. Its role in physiology and clinical medicine. Springer, Berlin Heidelberg New York, pp 699–710
Samols E, Holdsworth D (1968) Disturbances in carbohydrate metabolism: liver disease. In: Dickens F, Randle PJ, Whelan WJ (eds) Carbohydrate metabolism and its disorders, vol 2. Academic, New York London, pp 289–336
Samols E, Marks V (1967) New concepts of the function of pancreatic and nonpancreatic glucagon. J Annu Diabetol Hôtel Dieu 7:43–66
Samols E, Stagner J (1981) Modulation of insulin: local presynaptic $\alpha_2$ but postsynaptic $\alpha_2$ adrenoceptors. Diabetes [Suppl 1] 30:44A
Samols E, Weir GC (1979) Adrenergic modulation of pancreatic A-, B-, and D-cells. α-adrenergic suppression and β-adrenergic stimulation of somatostatin secretion, α-adrenergic stimulation of glucagon secretion in the perfused dog pancreas. J Clin Invest 63:230–238
Samols E, Marri G, Marks V (1965a) Promotion of insulin secretion by glucagon. Lancet 2:415–416
Samols E, Tyler J, Marri G, Marks V (1965b) Stimulation of glucagon secretion by oral glucose. Lancet 2:1257–1259
Samols E, Marri G, Marks V (1966a) The interrelationship of glucagon, insulin and glucose. Diabetes 15:855–866
Samols E, Tyler J, Megyesi C, Marks V (1966b) Immunochemical glucagon in human pancreas, gut and plasma. Lancet 2:727–729
Samols E, Tyler J, Marks V, Mialhe P (1969a) The physiologic role of glucagon in different species. In: Gual C, Ebling FJG (eds) Progress in endocrinology. Excerpta Medica, Amsterdam London New York, pp 206–219
Samols E, Tyler JM, Rege VP, Marks V (1969b) The possible role of glucagon (pancreatic and non-pancreatic) in insulin secretion. In: Östman J (ed) Diabetes. Proc 6th Int Diabetes Fed Congr. Excerpta Medica, Amsterdam London New York, pp 446–454
Samols E, Tyler J, Kajinuma H (1970) Influence of the sulfonamides on pancreatic humoral secretion and evidence of an insulin-glucagon feedback system. Excerpta Medica Int Congr Ser 231:636–655
Samols E, Tyler J, Marks V (1972) Glucagon-insulin interrelationships. In: Lefèbvre PJ, Unger RH (eds) Glucagon. Molecular physiology, clinical and therapeutic implications. Pergamon, Oxford New York, pp 151–173
Samols E, Weir GC, Ramseur R, Day JA, Patel YC (1978) Modulation of pancreatic somatostatin by adrenergic and cholinergic agonism and by hyper- and hypoglycemic sulfonamides. Metabolism [Suppl 1] 27:1219–1221

Sarantakis D, Teichman J, Fenichel R, Lien E (1978) (des-ALA[1], GLY[2])-HIS[4,5] D-TRP[8]-somatostatin. A glucagon-specific and long-acting somatostatin analog. FEBS Lett 92:153–155

Schatz H, Maier V, Hinz M, Nierle C, Pfeiffer EF (1973) Stimulation of H-3-leucine incorporation into the proinsulin and insulin fraction of isolated pancreatic mouse islets in the presence of glucagon, theophylline and cyclic AMP. Diabetes 22:433–441

Seino Y, Kurahachi H, Goto Y, Taminato T, Ikeda M, Imura H (1975) Comparative insulinogenic effects of glucose, arginine and glucagon in patients with diabetes mellitus, endocrine disorders and liver disease. Acta Diabetol Lat 12:89–99

Seino Y, Ikeda M, Kurahachi H, Taminato T, Sakurai H, Goto Y, Inoue Y, Kadowaki S, Mori K, Imura H (1978) Failure to suppress plasma glucagon concentrations by orally administered glucose in diabetic patients after treatment. Diabetes 27:1145–1150

Service FJ, Nelson RL, Rubenstein AH, Go VLW (1978) Direct effect of insulin on secretion of insulin, glucagon, gastric inhibitory polypeptide, and gastrin during maintenance of normoglycemia. J Clin Endocrinol Metab 47:488–493

Sherwin RS, Tamborlane W, Hendler R, Sacca L, DeFronzo RA, Felig P (1977) Influence of glucagon replacement on the hyperglycemic and hyperketonemic response to prolonged somatostatin infusion in normal man. J Clin Endocrinol Metab 45:1104–1107

Sitbon G, Mialhe P (1978) Pancreatic hormones and plasma glucose: regulation mechanisms in the goose under physiological conditions. III. Inhibitory effect of insulin on glucagon secretion. Horm Metab Res 10:473–477

Sorenson RL, Elde RP, Seybold V (1979) Effect of norepinephrine on insulin, glucagon, and somatostatin secretion in isolated perifused rat islets. Diabetes 28:899–904

Stagner J, Samols E, Weir G (1980) Sustained oscillations of insulin, glucagon and somatostatin from the isolated canine pancreas during exposure to a constant glucose concentration. J Clin Invest 65:939–942

Stagner J, Samols E (1982) Differential glucagon and insulin release from the isolated lobes of the in vitro canine pancreas. Diabetes 31 (Suppl 2):39A

Sugden MC, Ashcroft SJH (1977) Phosphoenolpyruvate in rat pancreatic islets: a possible intracellular trigger of insulin release. Diabetologia 13:481–486

Sutherland EW, Oye I, Butcher RW (1965) The action of epinephrine and the role of the adenyl cyclase system in hormone action. Recent Prog Horm Res 21:623–646

Tiengo A, Fedele D, Marchiori E, Nosadini R, Muggeo M (1976) Suppression and stimulation mechanisms controlling glucagon secretion in a case of islet-cell tumor producing glucagon, insulin, and gastrin. Diabetes 25:408–412

Trimble ER, Halban PA, Wollheim CB, Renold AE (1981) Functional differences between rat islets of ventral and dorsal pancreatic origin: a reflection of endogenous glucagon content? Diabetes [Suppl 1] 30:14A

Turner DS, McIntyre N (1966) Stimulation by glucagon of insulin release from rabbit pacreas in vitro. Lancet 1:351–352

Turtle JR, Kipnis DM (1967) An adrenergic receptor mechanism for the control of cyclic 3′,5′ adenosine monophosphate synthesis in tissues. Biochem Biophys Res Commun 28:797–802

Unger RH (1972) Pancreatic alpha-cell function in diabetes mellitus. In: Lefèbvre PJ, Unger RH (eds) Glucagon. Molecular physiology, clinical and therapeutic implications. Pergamon, Oxford New York, pp 245–257

Unger RH (1978) Role of glucagon in the pathogenesis of diabetes: the status of the controversy. Metabolism 27:1691–1709

Unger RH, Eisentraut AM (1965) Studies on the physiologic role of glucagon. In: Leibel BS, Wrenshall GA (eds) On the nature and treatment of diabetes. Excerpta Medica, Amsterdam London New York, pp 274–282

Unger RH, Lefèbvre PJ (1972) Glucagon physiology. In: Lefèbvre PJ, Unger RH (eds) Glucagon. Molecular physiology, clinical and therapeutic implications. Pergamon, Oxford New York, pp 213–244

Unger RH, Orci L (1976) Physiology and pathophysiology of glucagon. Physiol Rev 56:778–826

Unger RH, Orci L (1977) Hypothesis: the possible role of the pancreatic D-cell in the normal and diabetic states. Diabetes 26:241–244
Unger RH, Orci L (1981 a) Glucagon and the A cell. Physiology and pathophysiology. Part I. N Engl J Med 304(25):1518–1524
Unger RH, Orci L (1981 b) Glucagon and the A cell. Physiology and pathophysiology. Part II. N Engl J Med 304(26):1575–1580
Unger RH, Ohneda A, Valverde I, Eisentraut AM (1967) Mechanisms of response of plasma "glucagons" to ingested nutrients: evidence for enteric control of islet hormone secretion. J Clin Invest 46:1125–1126
Unger R, Madison L, Müller W (1972) Abnormal alpha cell function in diabetes. Response to insulin. Diabetes 21:301–307
Vanderscheuren-Lodeweyckx M, Wolter R, Malvaux P, Eggermont E, Eeckels R (1974) The glucagon stimulation test: effect of plasma growth hormone and on immunoreactive insulin, cortisol and glucose in children. J Pediatr 85:182–187
Weir GC, Knowlton SD, Atkins RF, McKennan KX, Martin DB (1976) Glucagon secretion from the perfused pancreas of streptozotocin-treated rats. Diabetes 25:275–282
Weir GC, Samols E, Loo S, Patel YC, Gabbay KH (1979) Somatostatin and pancreatic polypeptide secretion. Effects of glucagon, insulin, and arginine. Diabetes 28:35–40
Woods SC, Porte D Jr (1974) Neural control of the endocrine pancreas. Physiol Rev 54:596–619
Wright DE, Rodbell M (1979) Glucagon 1–6 binds to the glucagon receptor and activates hepatic adenylate cyclase. J Biol Chem 254:268–269
Yoshino G, Kazumi T, Morita A, Kobayashi N, Terashi K, Baba S (1979) Glucagon secretion during the development of insulin-secreting tumors induced by streptozotocin and nicotinamide. Endocrinol Jpn 26:655–660

# Subject Index

# Handbook of Experimental Pharmacology

Continuation of "Handbuch der experimentellen Pharmakologie"





Springer-Verlag
Berlin
Heidelberg
New York

# Handbook of Experimental Pharmacology

Continuation of "Handbuch der experimentellen Pharmakologie"





Springer-Verlag
Berlin
Heidelberg
New York

MIX
Papier aus verantwortungsvollen Quellen
Paper from responsible sources
FSC® C105338

If you have any concerns about our products,
you can contact us on
**ProductSafety@springernature.com**

In case Publisher is established outside the EU,
the EU authorized representative is:
**Springer Nature Customer Service Center GmbH**
**Europaplatz 3, 69115 Heidelberg, Germany**

Printed by Libri Plureos GmbH
in Hamburg, Germany